Scientific Basis for Nuclear Waste Management

Volume 2

Scientific Basis for Nuclear Waste Management

Volume 2

Edited by

Clyde J.M. Northrup, Jr.
Sandia National Laboratories
Albuquerque, New Mexico

Associate Editors

John G. Moore
Oak Ridge National Laboratory
Oak Ridge, Tennessee

Harry C. Burkholder
Battelle Memorial Institute
Columbus, Ohio

Stephen V. Topp
Savannah River Laboratory
Aiken, South Carolina

Gregory J. McCarthy
North Dakota State University
Fargo, North Dakota

Ernest A. Bryant
Los Alamos Scientific Laboratory
Los Alamos, New Mexico

Wallace W. Schulz
Rockwell Hanford Operations
Richland, Washington

Richard L. Schwoebel
Sandia National Laboratories
Albuquerque, New Mexico

Editorial Assistants

Mary M. Wood
Sandia National Laboratories
Albuquerque, New Mexico

Regina G. Reilly
Letter Perfect
Albuquerque, New Mexico

PLENUM PRESS · NEW YORK AND LONDON

Library of Congress Cataloging in Publication Data

Symposium on Science Underlying Radioactive Waste Management, Boston, 1978.
Scientific basis for nuclear waste management.

Proceedings of the symposium sponsored by the Materials Research Society.
Vol. 2- contains proceedings of the symposium under its later name: International Symposium on the Scientific Basis for Nuclear Waste Management, held in Boston, 1979-
Vol. 2- edited by C. J. M. Northrup.
Includes indexes.
1. Radioactive waste disposal—Congresses. I. McCarthy, Gregory J. II. Northrup, Clyde J. M. III. Materials Research Society. IV. International Symposium on the Scientific Basis for Nuclear Waste Management. V. Title.
TD898.S896 621.48'38 79-12440
ISBN 0-306-40550-4 (v. 2)

Proceedings of the International Symposium on the Scientific Basis for Nuclear Waste Management, sponsored by the Materials Research Society, and held in Boston, Massachusetts, November 27—30, 1979.

A Division of Plenum Publishing Corporation
227 West 17th Street, New York, N.Y. 10011

Printed in the United States of America

PREFACE

The Symposium on the Scientific Basis for Nuclear Waste Management was held in the fall of 1979 in Boston, Massachusetts and was one of a number of symposia included in the Annual Meeting of the Materials Research Society. The thrust of this annual Symposium is unique in the area of waste management. Recognizing that this is an area of great complexity which requires contributions from scientists with many different backgrounds some of which are not normally associated with nuclear energy, the Materials Research Society provides a forum for discussions of a wide range of materials behavior and transport phenomena. As can be seen from the list of references in each paper, the authors draw heavily on contributions associated with professional societies in addition to the Materials Research Society, and this annual meeting encourages the cross-fertilization between disciplines that are essential to an adequate treatment of the problems associated with nuclear waste management. The proceedings of the first Symposium that was held in 1978 was designated as Volume 1 in this series. The third Symposium is scheduled for 1980.

The scope of the 1979 Symposium was guided by the Steering Committee:

R. L. Schwoebel, Sandia Laboratories, USA (Chairman)
W. Carbiener, Battelle Memorial Institute, Columbus, USA
D. Ferguson, Oak Ridge National Laboratory, USA
W. Heimerl, DWK, Mol, Belgium
W. Lutze, Hahn Meitner Institut, Berlin, W. Germany
J. D. Mather, Institute of Geological Sciences, Harwell, UK
G. Oertel, Department of Energy, USA
R. Roy, Pennsylvania State University, USA
C. Sombret, Centre d'Etudes Nucleaires, Marcoule, France
V. I. Spitsyn, Academy of Sciences, Moscow, USSR
D. B. Stewart, U. S. Geological Survey, USA
W. D. Weart, Sandia Laboratories, USA
T. Westermark, Royal Institute of Technology, Stockholm, Sweden

The program was organized by the Program Committee:

C. J. Northrup, Sandia Laboratories, USA (Chairman)
E. A. Bryant, Los Alamos Scientific Laboratory, USA
H. C. Burkholder, Battelle Memorial Institute, Columbus, USA
G. J. McCarthy, Pennsylvania State University, USA
J. G. Moore, Oak Ridge National Laboratory, USA
R. W. Potter, U. S. Geological Survey USA
W. Shulz, Rockwell International, Hanford, USA
S. V. Topp, Savannah River Laboratory, USA

The session co-chairman played a vital part in the smooth functioning of the Symposium. In addition to the members of the Program and Steering Committees, the following served in this capacity:

John Mendel, Battelle Pacific Northwest Laboratory
Robert G. Dosch, Sandia National Laboratories
Douglas G. Brookins, University of New Mexico
Clyde Clairborne, Oak Ridge National Laboratory
Della Roy, Pennsylvania State University

There are 110 papers in these Proceedings and they have been divided into 11 sections. These span many of the areas of nuclear waste management, from the processes that produce the wastes to the production and characterization of various potential waste forms, the materials behavior of various engineered barriers, the nature of the repository and its response to the nuclear wastes, and, finally, to our predictive capability for assessing the consequences of emplacing wastes in repositories. These detailed technical papers are prefaced by a short series of overview papers which discuss current policies, activities and criteria for nuclear waste management. All of the contributions have been refereed and the authors and editorial staff have attempted to make the documents error-free, and therefore this collection of papers is both an excellent source of current information in the field as well as a statement of our understanding of the problems and some of the proposed solutions. Authors of technical papers have been encouraged to liberally reference their work to provide a sound background for their observations as well as to provide the interested reader with an entry into the key literature. In addition a number of definitive papers in nuclear management are presented thus making this volume an excellent reference. Preparation of the manuscripts was aided greatly by financial support from the U.S. Department of Energy.

Clyde J.M. Northrup
Chemical Technology Division
Sandia National Laboratories
Albuquerque, NM 87185, U.S.A.
April 1979

UNPUBLISHED PAPERS

For a variety of reasons, the following papers presented at the Symposium were not published in this volume:

"Distribution of Rad Waste Elements in SYNROC," W. Durham, C. Hoenig, G. Smith and J. Tewhey, Lawrence Livermore Laboratory, Livermore, CA.

"Novel Metal and Metal Matrix Waste Forms," N. Macmillan and R. Roy, Materials Research Laboratory, Penn State University, PA, P. T. B. Shaffer, Carborundum Co. Electro-Minerals Division, Niagra Falls, NY.

"A Slagging Pyrolysis Incinerator for Processing Low Level Transuranic Waste," M. D. McCormack, E.G.&G. Idaho, Inc.

"The Study of Radionuclide Transport, Why and How," E. A. Bryant, Los Alamos Scientific Laboratory, Los Alamos, NM.

"Computer Modeling of Nuclear Waste Storage Canister Corrosion," P. T. Cottrell, FAMU, Box 658, Dept. of Chemistry, Tallahassee, FL, W. D. Ludemann & R. D. McCright, Lawrence Livermore Laboratory, CA.

CONTENTS

OVERVIEWS OF NUCLEAR WASTE MANAGEMENT

WASTE FORMS

Vitreous Materials

Ceramic Materials

Transport of Radionuclides in the Geomedia

MODELING AND SAFETY ASSESSMENT

PROCESSING OF NUCLEAR WASTES

REMARKS ABOUT DEPARTMENT OF ENERGY POLICY ON HIGH-LEVEL NUCLEAR WASTE MANAGEMENT

G. K. Oertel

United States Department of Energy
Washington, DC 20545

INTRODUCTION

At the request of Dr. C. Northrup, it is a pleasure to make a few impromptu remarks about the Department of Energy's (DOE's) policy on nuclear waste management and about this meeting.

First, let me say that Steve Goldberg regrets that he will be unable to be here tomorrow to give his paper. Steve sends his regards. He is on assignment in the United States Congress and, unfortunately, has a schedule conflict.

As most of you know, the "Report to the President by the Interagency Review Group on Nuclear Waste Management" was published in March 1979 (1). Its findings, conclusions and recommendations have been summarized comprehensively elsewhere, and I will not repeat them here. The Interagency Review Group (IRG) offered options on two issues, the strategy for the development of National high level waste repositories and the scope and timing of the Waste Isolation Pilot Plant (WIPP). A Presidential position on both of these issues is expected to part of a Presidential policy statement on nuclear waste management that will be issued in the near future. In the interim, we are proceeding to implement the policy recommendations of the IRG in other areas.

SYSTEMS APPROACH

Let me mention a few highlights that may be of interest to you in light of the preceding two days of discussions.

As the IRG has recommended, we will employ a systems approach with respect to high level waste in a repository. We will also give appropriate attention to the system that involves the steps from removal of waste from interim storage tanks through processing, immobilization, packaging, and transportation to the repository: the waste form, the logistics system, and the repository must be compatible with one another. Repository and high level waste technology will be developed aggressively to assure that a broad spectrum of programmatic options will be available. Special emphasis will be placed on geologic media, waste forms, and additional engineered barriers.

Our repository program is being managed by the Richland/Columbus, Ohio office and the high level waste technology program, including immobilization, by the Savannah River Operations Office in South Carolina. The support contractors are Battelle/Columbus and E. I. Dupont de Nemours' Savannah River Laboratory, respectively. There is broad participation by Federal agencies, States, laboratories, universities, and the industry in accord with their interests and capabilities.

Our strategy is to:

- Aggressively develop alternatives to glass and salt.
- Make technically conservative choices consistent with laws and regulations and on a sound scientific basis.
- Follow the National Environmental Policy Act (NEPA) process conscientiously to assure a broad participation in the decision-making process,
- Set up procedures and tests to assure that waste forms and repositories will perform as designed, will be compatible with each other, and will satisfy regulatory requirements; and
- Proceed with the design of at least one major immobilization facility without delay while keeping waste form options open until, a) regulatory requirements are better defined, and b) the development of alternatives to monolithic, borosilicate glass has identified a few such alternatives that are ready to be selected. The choice will, of course, be made in full compliance with the National Enviornmental Policy Act (NEPA).

Our program to develop alternatives to borosilicate glass has been described for you earlier by Jack Crandall of E. I. Dupont de Nemours. This high priority program is now supporting exploratory research on a wide variety of waste forms. Within the next two years it will narrow down to a few prime candidates that will undergo intensive laboratory and, eventually, engineering development. The goal is to know enough about at least a few waste forms to be able to select from among them and borosilicate glass.

At the first immobilization facility, the Defense Waste Processing Facility planned for Savannah River, South Carolina, the major portion of the plant will serve to remove the waste from tanks, wash it, concentrate the activity while cleaning up the large quantity of bulk salt and to calcine the waste. The equipment for these steps will occupy about 90% of the space. The remaining two out of 20 shielded cells will be devoted to immobilizing the calcine and are being designed flexibly to permit the substitution of any immobilization technique that can handle calcine.

Since regulatory constraints are still developing while waste form and repository technology are being further improved, and since a geologic repository may not be in operation until the 1990's, it is tempting to defer the immobilization of existing high level wastes once again. This argument could probably be used each year for some time to come.

CONCLUSIONS

The IRG has noted that "since final processing of defense waste has been deferred for three decades, the IRG also recommends that remedial action, including immobilization of the waste, should begin as soon as practicable." We believe that our approach to immobilization permits technical conservatism by maintaining options on waste forms and geologic media without further deferral of the design and construction of a first immobilization plant that could begin to operate no sooner than the late 1980's.

Unlike some national and international organizations represented at this meeting, we are not committed either to salt as the repository medium or to glass as the waste form. On the other hand, there is no reason to rule out either salt or glass for all types of radioactive waste. Instead, the systems approach we expect to apply will consider waste properties, waste form, packaging, repository engineering, geologic medium, and other relevant factors in tailoring waste types and disposal sites to each other.

The papers given at this meeting show that a strong scientific basis for nuclear waste management is developing. The quantity and quality of papers at this meeting are encouraging.

Low level waste has become more important and visible since the planning for this meeting began. I have noticed only one paper on that subject in contrast to the many papers on high level waste immobilization and repositories. We would like to see a strong scientific basis for low level waste management as well. I imagine that there will be considerably more discussion of that subject at next year's meeting.

Thank you for the opportunity to make these remarks, to hear many good papers, and to meet many of you personally.

REFERENCES

1. _Report to the President by the Interagency Review Group on Nuclear Waste Management_, TID-29442 (1979).

REGULATION OF GEOLOGIC REPOSITORIES FOR THE DISPOSAL OF HIGH-LEVEL RADIOACTIVE WASTES

Lawrence A. White, Michael J. Bell and David M. Rohrer

High Level Waste Technical Development Branch
Division of Waste Management
U. S. Nuclear Regulatory Commission
Washington, D.C. 20555

INTRODUCTION

The U. S. Nuclear Regulatory Commission (NRC) is vested with licensing and regulatory authority over certain U. S. Department of Energy (DOE) facilities by Sections 202 (3) and (4) of the Energy Reorganization Act of 1974. These sections refer to: 1) facilities used primarily for the receipt and storage (including disposal) of high level radioactive wastes (HLW) and 2) retrievable surface storage and other facilities authorized for the express purpose of long term storage of HLW. Geologic repositories would not be licensed as "production" or "utilization" facilities. Rather, they would be licensed under those provisions of the Atomic Energy Act dealing with receipt and possession of "byproduct" and "special nuclear" materials. The NRC has the responsibility to evaluate for all aspects of repository performance which could impact the public health and safety. As a first step in carrying out its responsibility, the NRC staff is developing the regulations under which a geologic repository will be licensed. This regulation will be known as 10 CFR Part 60 - Disposal of High Level Radioactive Wastes in Geologic Repositories.

PURPOSES OF THIS PAPER

The primary purposes of this paper are to:

1. Discuss the NRC staff's proposed strategy for regulating the disposal of radioactive wastes in geologic repositories;
2. Discuss the basic technical requirements being considered in developing the regulation;

3. Discuss the rationale for the major technical requirements with regard to engineered systems; and

4. Solicit comments and recommendations from interested parties.

PROPOSED NRC STAFF STRATEGY FOR REGULATING GEOLOGIC DISPOSAL OF RADIOACTIVE WASTES

The overall performance objective for the repository (i.e., the allowable release of radionuclides to the biosphere) will be established by the U.S. Environmental Protection Agency (EPA) in their Environmental Radiation Protection Standards. The NRC staff will implement these requirements in developing its regulations. The NRC staff has considered a number of alternative approaches by which geologic disposal could be licensed. For example, the staff could merely state that the standards established by the EPA must be satisfied. In that case, the NRC would regulate on an ad hoc basis. Another alternative is for the staff to develop a general regulation without specific numerical performance objectives or criteria. Both of these alternatives offer substantial flexibility; however, they provide neither the DOE nor the public any substantive guidance as to how the NRC (and its licensing boards) will make their findings. A third alternative which the NRC staff has chosen is to develop a more detailed regulation including specific numerical performance requirements. This approach may constrain DOE's options; however, it does provide guidance as to what the NRC will find acceptable. Before I discuss our proposed regulatory approach, I would like to emphasize that our regulation at this point is tentative. The technical requirements I am going to discuss are still evolving and are subject to change. Our technical support and research programs will be aimed at refining these technical requirements.

Multibarrier Approach

Until very recently, the designs of deep geologic repositories have placed almost total reliance for containment of the radionuclides on the site characteristics and surrounding geology. As a result, insufficient attention has been given to waste forms, containers and other engineered barriers to significantly contribute to the containment and isolation of the radionuclides for extensive periods of time. For example, credit was not normally given for the waste form and its associated packaging being able to contain the radionuclides for more than a few tens of years. One does not have to search very far in the scientific literature regarding the geologic aspects of disposing of HLW before discovering that there is still a great deal of scientific uncertainty surrounding many of the assumptions that need to be made to predict long term geologic performance. First, we cannot define geologic conditions exactly and therefore must make certain assumptions which cannot be completely verified. Second, the earth scientist is called upon to describe how the geologic environment will change in the future. Such predictions introduce another source of uncertainty. The Interagency Review Group confirmed this assessment of the uncertainties in making geologic assessments in their report to the President in March 1979. In order to compensate for the uncertainty in predicting the behavior of geologic systems over long periods of time, the NRC staff has adopted a conservative multibarrier approach. In this approach, the staff views the repository to be composed of three major barriers: 1) the waste package, 2) the engineered repository structure and 3) the site and its environs. The NRC will establish minimum performance objectives for each of these major barriers. While we may foreclose some options, we do not intend to abandon the "systems approach" by setting performance objectives for these major parts of the system. DOE would have flexibility in the details of design of the waste package and repository structure so long as the performance objectives are met and the waste package, repository design and site are compatible with one another. The "Multiple Barrier Approach" was adopted for a number of reasons. It provides "targets" for scientists and engineers working on specific parts of the overall problem of engineered and geologic isolation. The engineers and material scientists know what is required of the engineered system, and the earth scientists know what is required of the site. Also, from the NRC's point of view, the approach permits inclusion of safety margins at convenient points in the system to compensate for uncertainty in assessing performance of the overall system.

The three major barriers are defined as follows: The "Waste Package" is the first major barrier and consists of the physical waste form and its container, as well as any ancillary enclosures, shielding or overpacks such as an outside container. The "Engineered Repository Structure" is the sum total of the many design features and engineered barriers of the repository which act to protect the waste package from the geologic environment and to retard the movement of radionuclides into the surrounding geologic environment. Engineered features could include such items as the design and layout of underground openings, treatment of the rock, structural support systems and materials emplaced to retard ground water movement and to prevent radionuclide migration. Borehole and shaft seals are also part the engineered repository; however, a separate objective will be established for their performance since they are primary potential pathways for water or nuclide movement into or away from the repository structure. The "Site and Its Environs" is the sum total of the many natural geological, hydrological and geochemical barriers which provide long travel times for ground water movement and act to retard the movement of radionuclides along those pathways from the engineered repository to the accessible environment. These barriers could include such items as low hydraulic gradients, low permeability rocks, high geochemical retardation capability, the ability to dissipate heat and a geologic setting which is simple to evaluate, highly stable and has the necessary characteristics needed to permit sealing of boreholes, shafts and underground openings.

PERFORMANCE OBJECTIVES

In general, our planned approach is to set performance objectives, first, for the period when the hazard is dominated by the fission products and, second, for the long term performance when the hazard is relatively constant and dominated by the actinides. The staff is putting emphasis on: 1) engineered <u>containment</u> of radionuclides during the fission product pulse when the hazard is the greatest and 2) assurance of a <u>controlled release</u> thereafter. This simplifies analysis and reduces uncertainties introduced into the analysis of the total system. During the period of engineered containment of the waste, the site geology should provide sufficient backup to account for those scenarios which may result in loss of engineered containment. Thereafter, the site geology should also have the capacity to retard the movement of the long lived radionuclides to the accessible environment so that the EPA standard is not exceeded.

The staff views the waste form and container as a major component of the engineered containment system. A major advantage of putting emphasis on the waste package is that it can be manufctured under closely controlled conditions and can be physically tested to verify its performance, thus reducing uncertainty in the performance of the disposed system. Other engineered barriers can be designed to have uniform properties and can be placed under controlled conditions; however, complex systems models may be needed to project their performance. The site geology is the most difficult component to deal with in that we cannot accurately define the geologic environment nor test the gross characteristics, given a limited number of exploratory boreholes over the several thousands of square miles that will be involved in the transport analyses. In summary, the waste package and repository design are inherently knowable and under man's control, whereas the site performance will involve much greater uncertainty and certainly is not under man's control. We are therefore encouraging DOE to: 1) develop and experiment with alternative waste forms and containers to see exactly how much reliance can be put on this component and 2) put more emphasis on other engineered barriers as part of the repository design to protect the waste package over the short term and help control releases over the long term. However, we are not neglecting the site. We anticipate that DOE will choose very stable sites which inherently have multiple natural barriers to radionuclide migration.

With regard to the long term performance objectives for the engineered systems, the staff is considering several alternatives generally along the approach of specifying a maximum near-field release rate, applied either to the whole engineered system or to the waste package. Placing a release rate restraint on the waste package has the advantage that the release rate can be physically measured under aggravated or accelerated conditions. On the other hand, applying a release restraint on the total engineered system permits flexibility in giving credit for other engineered barriers in retarding radionuclide migration.

With regard to the site, the performance objective will be largely controlled by the EPA standard. The site should provide sufficient protection to back up the engineered system considering all credible scenarios and the conditions of the engineered system for those scenarios. In addition to specifying performance objectives for the site, the NRC staff is incorporating preferred site characteristics and exclusion requirements to help guide DOE to selection of sites that meet the performance objectives.

The performance objective for the period during which the fission products dominate the hazard is one of containment within the engineered barriers. The combination of the waste package and repository structure is intended to provide assurance that radionuclides will not be released into the near-field geologic environment for at least the first 1000 years. For credible scenarios which breach the engineered system, the site should provide sufficient backup and retention of radionuclides so that releases to the accessible environment are within the EPA standard.

In order to ensure engineered containment of the radionuclides, we plan to specify a minimum performance objective for the waste package. Each waste package should be designed to provide reasonable assurance of containment of radionuclides for at least 1000 years and for as long as reasonably achievable beyond that, assuming that the repository becomes saturated shortly after closure. Other assumptions concerning ground water chemistry, flow rates and in situ stresses, which are necessary to define the environment in which the waste package must meet its performance objective, should be based on site conditions and the state of the repository structure under appropriately chosen scenarios. The intent here is to envelop potential perturbation to the site and design so as to establish design basis conditions.

The repository structure should be designed to protect the waste package and to provide reasonable assurance that radionuclides will not enter the geosphere for at least 1000 years in the event any of the waste packages should fail to meet their performance objective. The idea is to design the repository so as to compensate for any credible scenario which results in degradation of the waste package by itself. Under conservative assumptions on the conditions of the waste package for such scenarios, there should be sufficient barriers in the repository structure to mitigate the consequences so that there is high assurance that radionuclides will not be released into the geologic environment for the first 1000 years. This can be accomplished by: providing barriers to retard the movement of ground water to and from the waste, zoning backfill materials so as to prevent water from coming in contact with the waste or by providing zones of high ion exchange materials to capture radionuclides. Also, designing a waste package that would be extremely leach resistant, even though it may structurally degrade or become chemically altered, could mitigate the consequence of package failure. Borehole and shaft seals are also a part of the repository design which will affect the rate at which ground water from overlying or underlying aquifers may reach the repository structure or the waste package as well as affect the rate of movement of radionuclides to the biosphere once they are released to the geologic environment. The performance objective for borehole and shaft seals, therefore, is to provide a barrier to fluid migration equivalent to that of the undisturbed section of rock through which the borehole or shaft passes.

The performance objective for the engineered system after the first 1000 years is to limit the release of radionuclides to the geologic environment. After the first 1000 years, the waste package and other engineered barriers together should retard the movement of the radioactivity and limit its release from the confines of the engineered system to as low as reasonably achievable but to, at most, a yearly release rate of one part in 10^5 to 10^7 of the package inventory. It is recognized that it may ultimately be difficult to demonstrate that this performance objective is met because of difficulty in analyzing the performance of engineered barriers over the long term. One way of reducing the uncertainty is to limit the release rate of the waste form or package since this can be subjected to accelerated testing. Therefore, we are encouraging DOE to examine the release resistance of alternative waste forms and to see how much emphasis may be placed on the waste form to limit the release of radionuclides over the long term.

The influence of the release rate on containment of wastes can be evaluated as follows. Denote the Release Rate by RR and assume for the sake of simplicity a constant release rate over time; i.e., for each gram of waste <u>initially</u> present, RR grams of waste will be released each year. The total amount of waste will then be released in a time period called the Release Time (RT) where RT = 1/RR. Combining this relationship with the law of radioactive decay, one can calculate the fraction of a specific radionuclide inventory which will be released before the radionuclide decays. This relationship is given by the equation:

$$FR = \frac{1.443}{RT/t_{1/2}} \left[1-e^{-0.693(RT/t_{1/2})}\right]$$

where FR is the fraction of the initial nuclide inventory released before decay, $t_{1/2}$ is the half-life of the nuclide and RT is the release time. The relationship between FR and $RT/t_{1/2}$ indicates that in order to provide any significant containment of a nuclide, the release time must substantially exceed the half-life of the nuclide. If, for example, the release time of the waste is 15 times the half-life of a nuclide, approximately 90% of that nuclide will decay before being released to the geologic environment.

Table 1 evaluates the influence of the release time on the releases of several long lived nuclides present in high level wastes. The table shows that a relase time of 10^6 years would provide virtually total containment of the plutonium nuclides and about 70 percent containment of ^{99}Tc. A release time of 10^7 years would be necessary to achieve nearly complete containment of ^{99}Tc, and a release time exceeding 10^8 years would be necessary to provide substantial containment of ^{129}I. This information can be used to evaluate the effectiveness of proposed release rate requirements. If, for example, it is considered desirable to achieve virtually total containment of the ^{99}Tc and ^{129}I nuclides within the waste form, it would be necessary to limit the release rate to a value of about 10^{-7} and 10^{-8} per year, respectively. Alternatively, Table 1 indicates that release rates greater than 10^{-5} per year would provide little containment of these long lived nuclides and that nearly all of the initial radioactive inventory of these nuclides would ultimately be released from the waste form. This derivation is based on a simple model in which a constant number of grams of waste is released per year per gram of waste initially emplaced in a repository. While a more realistic model would need to include such factors as temperature and the release rate of the waste form, this simple model seems adequate as a rough estimate of the effectiveness of a release rate restriction.

Lowering the release limit allows more of the initial radioactivity to decay and thereby reduces the amount of radioactivity available for transport to the environment. As indicated above, release limits greater than 10^{-5} per year result in little reduction in the fraction of long lived nuclides which are ultimately released. In order to achieve virtually total containment of the plutonium nuclides, it would be necessary to limit the release to a value of about 10^{-6} per year. Similarly, substantial containment of ^{99}Tc would require a release limit of 10^{-7} per year or lower. Because the plutonium nuclides and ^{99}Tc could be the major long lived contributors to the population dose from waste disposal, the NRC staff considers a release limit on the order of 10^{-6} or 10^{-7} per year to be appropriate in order to minimize the impact of these nuclides.

The selection of good sites is a prerequisite to meeting the design performance objective. Repository sites should be chosen which 1) provide a high level of stability so as to enhance the engineered containment of radionuclides, 2) permit design of a stable re-

Table 1. Relationship Between Release Time and Total Releases From the Engineered Repository

Release Time = 10^5 years

Nuclide	Half-Life	Fraction Released
^{14}C	$5.7x10^3y$	0.08
^{240}Pu	$6.6x10^3y$	0.10
^{239}Pu	$2.4x10^4y$	0.27
^{99}Tc	$2.1x10^5y$	0.88
^{129}I	$1.7x10^7y$	1.0

Release Time = 10^6 years

Nuclide	Half-Life	Fraction Released
^{14}C	$5.7x10^3y$	0.008
^{240}Pu	$6.6x10^3y$	0.010
^{239}Pu	$2.4x10^4y$	0.035
^{99}Tc	$2.1x10^5y$	0.29
^{129}I	$1.7x10^7y$	0.98

Release Time = 10^7 years

Nuclide	Half-Life	Fraction Released
^{14}C	$5.7x10^3y$	0.0008
^{240}Pu	$6.6x10^3y$	0.0010
^{239}Pu	$2.4x10^4y$	0.0035
^{99}Tc	$2.1x10^5y$	0.030
^{129}I	$1.7x10^7y$	0.82

pository and 3) provide capability to contain the waste in the event of breach of the engineered system for the first 1000 years and to

isolate the waste so as to meet the EPA standard thereafter. In addition to the above performance objectives, there is another performance requirement for the design of the repository. The repository should be designed to preserve the option to retrieve the waste throughout operation of the repository and for 50 years thereafter. The length of time, 50 years, is tentative. One effect of this requirement is to ensure an extremely stable repository design that will preserve the option to monitor the performance of the repository after it is filled with waste, will allow corrective measures if necessary and, as a last resort, will allow all or part of the waste to be removed.

RATIONALE FOR MAJOR TECHNICAL REQUIREMENTS

The 1000 year performance requirement on the waste package was established because the waste form and packaging are engineered items which can be physically tested, their performance and reliability evaluated and the system optimized for the specific emplacement environment. Also, the manufacture of these components can be strictly controlled. Such performance of the waste form and packaging during the period of time when the wastes would be subjected to the greatest potential for detrimental alteration (i.e., the period of high thermal flux and of potential chemical interaction) would compensate for the uncertainties in predicting the complex waste/rock interactions and their consequences during this time period.

The staff also feels that requiring the waste package to have a reasonably extended expected life is one of the easiest ways to compensate for the inevitable uncertainties in predicting the behavior of the geologic systems over long periods of time. More specifically, containment of the radionuclides by the package for the first 1000 years is desirable from the standpoint that:

1. The thermal output of the waste packages will have dropped by over two orders of magnitude. (The leach rates of waste forms are highly temperature dependent.)

2. The temperature of the emplacement environment near the waste package will have reached a maximum and have begun to drop. However, under most conditions, temperatures will not have dropped to near ambient temperatures. For this reason, we are asking DOE to examine the possibility of developing a waste package which would last much longer than 1000 years. For spent fuel, it will take on the order of 10,000 years for the rock to return to near ambient temperatures.

3. The total radioactivity (curies) in the packages will have decreased by two orders of magnitude and the radiotoxicity of the wastes will have decreased by over three orders of magnitude.

The NRC staff does not believe that the changes in the waste package needed to meet the thousand year objective will be unduly expensive. In any case, DOE's present estimate of the costs associated with a repository shows the waste encapsulation (including the waste container) accounts for only about nine percent of the total cost of disposal. Thus, the financial impact of the 1000 years expected life for the package on total waste isolation costs is expected to be a small fraction of the cost and therefore a small price to pay for the protection. Recently the NRC staff has acted, through a contract at the Brookhaven National Laboratory, to see whether the expected life of the package might reasonably be increased to 10,000 years or more. In this effort, the NRC staff will continue to ask what is necessary to achieve waste isolation without limiting its requirements to those which can be met using only the presently demonstrated technology. It should be noted that on September 26, 1979, at the OECD Nuclear Energy Agency (NEA) meeting in Paris, the British announced that they are also placing major emphasis on the waste package. Their intent, which was developed quite independently of the NRC, is that the waste package provide redundant protection for at least 1000 years. Further, they are also actively working to see if that number can be pushed out to 10,000 years or more.

The rationale for emphasizing the need for a well designed engineered repository is rather simple. We are applying practical engineering principles in a conservative manner to gain the confidence needed that the repository will work the way it is intended. We view a repository as civil construction as opposed to a conventional mine. As a civil structure, the repository performs the functions of 1) protecting the waste package during the 1000 year containment, 2) mitigating the effects of any geologic events or processes which may affect the waste package during this period and 3) over the long term acts with the waste package to limit the release of radionuclides to the geologic environment. It also provides for the retrievability of waste until such time as a decision can be made that we have done everything right and the repository will work in containing the waste as planned. It is our intent to encourage the construction of a structure which 1) is stable, 2) provides a hydraulic barrier to ground water movement and 3) captures radionuclides or provides a long tortuous path for nuclides to reach the geologic environment. In view of the uncertainty involved in systems modeling, a well designed engineered structure will provide the needed confidence that releases are, in fact, within the EPA standard. The NRC staff believes that current geotechnial engineering principles can be used to design and construct a repository to meet its performance objectives. For example, rock fracturing can be reduced to an acceptable level by control of the construction process and heat load. We believe there is much room for technology transfer from experiences gained in underground civil construction. Research and development should be aimed at testing these design principles as applied to a repository.

The purpose of the long term performance objective of limiting the release of radionuclides to the geologic environment is similar to the containment objective for the first 1000 years; that is, to gain confidence that the EPA standard will be met. The radiotoxicity of the waste decreases with time, which is fortunate because so does our ability to predict repository performance. Our goal is to provide a level of safety commensurate with the hazard over the long term. We also want to compensate for uncertainty in systems analyses to be consistent with the degree of protection that is needed. The toxicity of the waste after 1000 years has been reduced to a level near to that of a natural ore body. However, the waste is much more concentrated and the inventory of nuclides is different. In fact, some nuclides are known to be more mobile than those of a natural ore body. The intent of limiting the release at this point is to 1) build in engineered barriers to reduce the mobility of the radionuclides as much as practical and 2) compensate for the uncertainty in predicting how well the geology will contain the radionuclides. The characteristics of the geologic and geochemical environment around an ore body is the reason the ore body was formed in the first place. Therefore, there are natural site characteristics which contain the waste. Repository sites, however, will be selected by man, and may not have comparable retention characteristics.

In addition to the performance objectives described in the paper, the NRC staff is developing guidance for the technical community regarding how to meet these objectives. The criteria and requirements that follow contain preliminary supporting guidance on waste forms and packaging for your purusal and comment.

Waste Form Acceptance Criteria

Solidification. Pursuant to 10 CFR 50, Appendix F, and 10 CFR 71.42(a), all high level liquid radioactive wastes shall be converted to a dry solid and placed in a sealed container prior to transfer to a federal repository. This criterion shall extend to all radioactive wastes (including TRU wastes and other non-high level wastes) emplaced in a repository.

Stabilization. Finely divided waste forms shall be stabilized by incorporation into a containing marix so as to minimize the production and availability of respirable fines during any accident condition. Particles of sizes smaller than about 10 microns present a potential for serious health hazards should the container be breached during a tranportation, handling or emplacement accident. Such fines can become airborne, enter the lungs through inhalation and reside in the pulmonary region of the lung for very long periods of time, resulting in doses to the lung in excess of those allowed under 10 CFR 20. Further, the aggregate surface area of the finely divided particles is large relative to the mass of the particles, and if a process which is surface area dependent (such as dissolution or leaching processes might be), the rate of reaction can be accelerated by the wastes being in the form of finely divided particles.

Free Liquid. Radioactive wastes containing free liquid shall not be accepted at the repository. The removal of these free unbounded liquids is required to 1) reduce the potential of radionuclide release during any accident which may breach the waste container, 2) lessen the potential for or extent of pressurization of the waste container due to hydrolysis and radiolysis, 3) reduce the potential of any criticality event taking place in the waste, 4) decrease the potential for internal corrosion of the waste container and 5) reduce the potential for dissolution or leaching of the wastes.

Combustibles. All solid or solidified radioactive waste classified as combustible shall be incinerated or otherwise reduced to a non-combustible "ash" which shall be fixed in a stabilizing matrix, or the original combustible wastes and their associated packaging shall be such that:

1. The incidence of a fire and/or explosion involving the wastes shall not unduly affect the health and safety of the repository operating personnel, and

2. A fire and/or explosion involving a single container cannot migrate to involve other containers, and

3. The organic content of the material shall be shown to be incapable of interacting with the available radionuclides in such a manner as to form organic complexes which may make those radionuclides chemically mobile.

If combustible wastes are to be allowed in the repository at all, the consequences of any potential fire and/or explosion event upon the health and safety of the operating personnel is of paramount importance and must be within acceptable limits (i.e., 10 CFR 20, etc.). Combustible wastes also present the potential for two other distinctly different detrimental effects which must be avoided. First, problems should be avoided which could arise from a fire or explosion taking place within the waste which could cause an uncontrolled release of radionuclides, volatilize parts of the waste or spread the fire to involve other nearby wastes. The second major effect is the potential for organic compounds within the waste to be altered by the chemical, thermal and radiation environments and subsequently form compounds which could react with the wastes or the radionuclides and make them more mobile.

Explosive, Pyrophoric and Toxic Materials. There shall be no known explosive or pyrophoric materials or conditions existing in the radioactive waste, nor shall there by any gross quantities of highly toxic chemical wastes. The exclusion of explosive and phyrophoric materials is a straightforward requirement for safety reasons. The exclusion of toxic materials is a more difficult problem. The staff's intention is that the wastes not contain gross quantities of highly toxic chemicals such as cyanides and other similar material which could become an occupational safety problem if the container were to be breached by an accident during the operational phase of the repository.

Chemical, Thermal and Radiological Stability. The waste form and its associated packaging shall be chemically non-reactive to the maximum extent practicable (including leachability and solubility) when exposed to the emplacement environment. The stability of the waste package shall be analyzed under expected in situ conditions considering the characteristics and properties of the waste form itself, the primary containment and subsequent packaging, the emplacement medium, the surrounding ground water and the radiation and thermal fields. Phase changes, loss of mechanical integrity, enhanced chemical activity and offgassing are only a few examples of conditions which could adversely affect the ability of the waste form and package to isolate the waste and must be avoided. The waste form and its associated packaging must be mutually compatible and, to the extent possible, compatible with both the emplacement environment and the repository design in order to provide substantial radionuclide isolation.

Container Design Requirement

Physical Dimensions and Weight. The physical dimensions and weight of the container and its contents shall be such that handling of the material at the repository can be performed safely with a minimum potential for handling accidents. Additional consideration should be given to the techniques and equipment required to retrieve the waste container should that become necessary for safety reasons. The maximum allowable weight is controlled by the hoisting capacity of the lifts used to transport the waste container from the repository surface facilities to the emplacement horizon and by the carrying capacity of the emplacement equipment itself. The container size and shape are limited by the physical layout and dimensions of the emplacement rooms and corridors.

Mechanical Strength, Heat Resistance and Fabrication. The container and packaging shall be designed and fabricated to the specification of acceptable codes and standards (ASME Boiler and Pressure Vessel Code, ANSI, ASTM, DOT, etc.) where they are applicable to existing containers of a similar type and function. The mechanical ability of the waste container to survive an accidental drop as well as routine emplacement operations and the stresses of retrieval operations without loss of its design performance capability is important to the protection of the public and operational personnel alike. Further, since corrosion rates are strongly temperature dependent, the temperature at the container surface is an important design parameter. The heat fluxes and temperatures can be controlled by limiting the amount, geometry and age of nuclear waste in the container. The staff feels that existing design and fabrication codes and standards developed by the ASTM, the ASME and other similar standards organizations for other purposes may well contain areas which could be appropriate for these waste containers.

Materials of Construction and Corrosion Control. The materials used to fabricate the container and packaging shall meet the specifications of acceptable codes (10 CFR 71.31, ANSI, ASTM, ASME, etc.) where they are applicable to existing containers of a similar type and function.

The primary failure mode of the emplaced container is expected to be corrosion. The requirement for retrievabilty severely constrains the selection of container materials. Appropriate packaging materials shall be selected to assure compatibility with the container material and the waste form, thereby reducing the potential for corrosion. Some designs may rely on corrosion to form insoluble, impermeable barriers. However, care must be taken to avoid situations where corrosion products could adversely affect the geochemical retardation of the radionuclide transport in the medium, increase the corrosion rate of the packaging material or increase the dissolution rate of the waste form in the ground water. Any of these situations could significantly degrade the waste isolation performance of the repository.

Mechanical Handling. The waste container shall include features and devices which enhance the capabilities for safely lifting and moving the container and its contents. These features and devices should be in compliance with 10 CFR 71, Subpart C, and 10 CFR 71.31, such that they do not provide a means for easily damaging the container should a handling accident occur. The ease of handling of the waste container during transport and emplacement is important to the safety of the operations personnel should an accident take place resulting in the breach of the container and release of the radionuclides. The use of appropriate lifting and handling devices and features could reduce the potential for handling accidents. However, these devices and features must be such that they themselves do not increase the potential for damage to the container during such a handling accident.

Criticality Control. The maximum allowable quantity of fissile material in individual waste container shall meet the requirements of criticality safety established in 10 CFR 71, 49 CFR 173 and other applicable standards. Calculations and measurements made by the shipper to assure criticality safety shall be available to the repository operator upon request. The possiblity of a criticality accident in the repository operations area must be avoided. This may be done by limiting the total inventory of fissile materials in each container or in the room itself, or by using neutron poisons, criticality safe package geometries and emplacement arrangements.

Surface Contamination. There shall be no significant removable radioactive surface contamination on the exterior of the container. Removable (non-fixed) radioactive contamination is considered significant if the level of contamination exceeds the requirements of 49 CFR 173.397.

Penetrating Radiation. The penetrating radiation level from an unshielded container shall be controlled to provide for the health and safety of the general public and the operational personnel at the repository. The provisions and requirements of 10 CFR 20, 10 CFR 71, applicable DOT standards and other applicable codes and standards shall be followed.

Unique Identification. A label or other means of identification shall be installed on each container. The label shall not impair the integrity of the container or make the surface irregular and shall be attached in such a manner that labels and descriptions thereon will be legible at least to the end of the retrievable storage period. Each label shall contain specific information considered necessary for identification and traceability of the packages in addition to a unique serial number for reference to a set of permanent records which contain detailed information regarding the specific packages.

CONCLUSION

It must be emphasized that the performance objectives presented in this paper are preliminary. They are subject to change in light of information we receive from our technical contractors and from comments and suggestions we receive from both the technical community at large as well as the interested public. We urge your review and critique of these positions and look forward to receiving your comments.

ACKNOWLEDGEMENT

The authors wish to express their appreciation to Dr. Daniel J. Fehringer of the NRC staff for his input on Appendix A of this paper and to Mr. John B. Martin, Director, Division of Waste Management, for his encouragement.

SURVEY OF HIGH-LEVEL RADIOACTIVE WASTE FORMS AND THEIR ROLE IN WASTE MANAGEMENT*

H. W. Levi

Hahn-Meitner-Institut für Kernforschung Berlin GmbH
Glienicker Str. 100, 1000 Berlin 39, Fed. Rep. of Germany

INTRODUCTION

Long lived fission products and actinide elements are generated in nuclear fuel while it supports operation of a reactor. Whatever concept of the fuel cycle is chosen, these radionuclides or a substantial fraction thereof will eventually form some kind of high level radioactive waste to be isolated from man for an extended period of time. This paper will present a European view mainly of one aspect of this isolation, namely the waste form.

Many European countries will choose deep underground geologic formations isolated from circulating groundwater to host the waste repository. These formations will serve as a principal barrier, especially after the repository has been finally sealed. For example, in the Federal Republic of Germany rock salt domes have been chosen as host formations. A second natural barrier system in the long term waste isolation concept is the depth of the repository and the physical and chemical properties of the surrounding and overlying material.

The third barrier system is the waste package which is defined to consist of the solid and stable radioactive material, the waste form, the canister, and an overpack when in the repository. This is an engineered barrier system which is important not only in the repository but also for the whole sequence of waste management operations before disposal. In accordance with the fuel cycle policy of

*Invited paper presented at the symposium on "Scientific Basis of Radioactive Waste Management", Nov. 24-27, 1979, Cambridge, MA.

most European countries, spent fuel will not be considered a waste form in this paper. It is rather assumed that high level waste forms are based on solidification products from high level liquid waste generated in a reprocessing plant for commercial nuclear fuel.

Research on HLW solidification has a long tradition and a great variety of solidification products have been developed on a laboratory scale. For very few, however, the general technology of preparation has been demonstrated on a technical or even on a pilot-plant scale. The canister and even more so the overpack have found serious interest only recently.

The general requirement of the waste package is stability against destructive influences which a highly radioactive material may be exposed to, i.e.:

- radiation stability
- thermal stability
- mechanical stability
- chemical stability

According to present policy, the waste form is supposed to be the principal carrier of these properties.

The purpose of this paper is to identify the functions of the waste form and their relative importance, to give a brief review of basic waste forms and their characteristics, and to assess their suitability for use in a waste management scheme.

FUNCTIONS OF THE WASTE FORM

The general purpose of the waste form is retention of radioactivity in an immobile state through all phases of a waste management scheme. In this sense, it has three functions.

<u>The first function</u> of the waste form is related to safe handling and in particular, to safe transportation of HLW from the reprocessing plant to the disposal site. This is a short term function in which the integrity of the canister plays the primary role. The solidification product generally has a secondary role. However, the mechanical stability of the solidification product is most important because it guarantees retrievability of the radioactive material in case of an accident.

<u>The second function</u> of the waste form is a medium term function. The waste form serves as a principal barrier while the waste is temporarily stored in an engineered surface facility. Temporary surface storage may range from a few years to a few decades depending on the

philosophy of the waste management system considered. It also has an important barrier function in the geologic repository during the operational phase which may last on the order of 100 years. Characteristically, during the operational phase of the repository, shafts, galleries and disposal holes are open. As neither the geologic formation nor the overlying rock which is penetrated by open shafts are fully effective as barriers, the barrier function of the waste form mainly in terms of leach resistance, may have a heavy weight during this period.

It has to be noted that it is this 100-year period when the waste form is exposed to the most unfavorable conditions in terms of heat and radiation. Heat and radiation are the effects most likely to cause alterations of the solidification product properties or to accelerate those alterations. Deterioration of the waste form will become less likely with further age.

The third function of the waste form is a long term function. In the sealed repository, when the waste is irretrievably stored, the waste form serves as a still important but clearly redundant barrier. The principal barriers in the final phase of the repository are the natural ones, the geologic formation preventing water from serving as a vehicle for radioactivity and the migration path strongly delaying radionuclide transport towards the biosphere.

For purposes of illustration, consider the situation in which the values for the fluid velocity, migration path length, sorption distribution coefficient, decay constant and other system parameters are such that the assumption of local equilibrium between bulk liquid and solid phases is a reasonable approximation (1), and for a constant source strength, the radionuclide concentration along the migration path approaches a steady-state profile which is independent of time. In such a situation, the effectiveness of the transportation delay due to sorption in comparison to that of mobilization delay due to leach resistance may be demonstrated by a relatively simple mathematical treatment of radionuclide transport in the geosphere. The following equation has been obtained by integration of a three-dimensional set of differential equations including terms for diffusion, convection, sorption and radioactive decay (2). To solve the differential equation the simplifying assumptions have been made that the source strength remains constant and that the concentration in a distance x from the source, C(x), has reached a stationary value. The latter means that the concentration in the solid phase is in equilibrium with that in the liquid phase. Before the stationary state is reached, apparently the concentration will rather be smaller.

Equation 1) describes the concentration C(x) along a migration path:

$$C(x) = \frac{B \cdot A \cdot r}{x} \; e^{-\left(\frac{R_d \cdot x}{v \cdot T_{1/2}}\right)} \qquad 1)$$

$C(x)$ = concentration of a radionuclide at distance x from a source (Ci/m^3)
A = radioactivity of the source (Ci)
r = leach rate (kg/m^2 yr)

$B = \frac{S}{4\pi\phi \; v\alpha}$ S = specific surface (m^2/kg)
ϕ = porosity
v = water velocity (m/yr)
α = dispersivity (m)

$T_{1/2}$ = half life of the radionuclide (yr)

$R_d = 1 + \frac{\rho \cdot K_d(1-\phi)}{\phi}$ ρ = density of sorbing material (kg/m^3)
K_d = distribution coefficient (m^3/kg)

The equation shows that the radionuclide concentration C(x) depends linearly on the leach rate but exponentially on the ratio $R_d/T_{1/2}$. That means that only for very large values of $T_{1/2}$ (half life) and/or very small values of K_d (sorption coefficient) the leach rate influences significantly the radionuclide concentrations in a distance x from the repository. More detailed evaluations for individual radionuclides are in progress.

REVIEW OF WASTE FORMS

The solidification products seriously considered for use in a waste management scheme are glasses, crystalline products or composites. In Table 1 the more important product alternatives are classified according to two basic process lines - calcine (no melting involved) and glass (melting involved).

Calcines are products obtained by removing the volatile components from the waste, i.e. water and nitrate, at temperatures between 400 and 900°C. The result is a mixture of oxides of fission products, actinides and corrosion products in particulate form. The plain calcine is not very stable chemically because of its large surface areas and the chemical properties of some of the oxides, and it is highly friable. To improve the properties of calcines advanced forms are developed. One such product is the so-called multi-barrier waste form, a composite consisting of calcine particles with inert coatings such as pyrocarbon, silicon carbide or aluminum, embedded in a metal matrix.

Table 1. Solidification product alternatives

Alternative	Product Line	
	Calcine	Glass
Basic	Fluidized bed (particles) Pot (cake)	Borosilicate (cylinder) Nepheline-syenite (cylinder) Phosphate (cylinder)
Composite	Multibarrier waste form (coated particles in metal matrix)	Vitromet (glass particles in metal matrix)
Ceramic	Supercalcine (chemical additives, high temperature ceramic product) SYNROCK (chemical additives, isostatically pressed ceramic product)	Borosilicate glass ceramic (cylinder)

Supercalcine and SYNROC are ceramics derived from calcines. Supercalcine is obtained by adding appropriate chemicals to the HLW to to form refractory compounds of fission products and actinides when fired at 1000°C after calcination. Supercalcine requires consolidation by embedding in a matrix but is supposed to have inherent chemical stability. SYNROC is obtained as a ceramic pellet by wet milling of radioactive calcine and additives, cold isostatically pressing and finally hot pressing at 1300°C. Supercalcine has a much higher fission product content than borosilicate glass, whereas the fission product content of SYNROC is relatively small.

Glasses are products obtained by melting waste oxides together with additives like SiO_2, B_2O_3, Al_2O_3, P_2O_5, Na_2O and CaO. Upon solidification, the melt forms a glass. Borosilicate glass is the type of solidification product most thoroughly studied all over the world. Phosphate glass has been abandoned in most countries. An advanced development along the glass line is vitromet, i.e. glass beads embedded in a metal matrix, with extremely high thermal conductivity and mechanical strength. A typical fission product content envisaged for borosilicate glass is 20%.

Glasses due to their lack of thermodynamic stability may devitrify forming crystalline phases of higher thermodynamic stability. A ceramic product derived from glass to avoid this drawback is glass ceramic. Glass ceramics are obtained by holding glasses of appropriate composition somewhere above the transformation point for hours causing controlled crystallization.

It is the common principle of supercalcine, SYNROC and glass ceramic to be composed of stable crystalline phases hosting fission product and actinide elements as well as probably of varying fractions of amorphous phases also containing radionuclides.

Experimental data related to the four stability categories defined before are available for various solidification products in the open literature and are not summarized here again. Borosilicate glass (and phosphate glass) has been more thoroughly investigated than any of the other products and a greater supply of data is available.

In terms of irradiation stability borosilicate glass does not show significant effects and particularly no significant deterioration of chemical or mechanical stability after being exposed to the simulated 1000-year α-dose. No doubt that further investigations of the irradiation behavior particularly in combination with other effects are still desirable because this is the most unconventional influence a waste form is exposed to. However, similar and sufficiently conclusive experimental data on ceramic materials are not yet available, and it has still to be proved that the crystalline host structures will survive irradiation. It has also still to be proved that transmutation of fission products will not adversely affect their crystalline host phase. Most of the irradiation stability evidence for the ceramic phases is derived from the stability of natural analogies.

As far as thermal stability is concerned, there is experimental evidence that borosilicate glasses (and even more so phosphate glasses) may devitrify within hours or days at temperatures above 500°C depending on the composition. In general, this leads to some increase of leach rates, though for borosilicate glasses this effect is not dramatic. Slow devitrification may of course occur also at lower temperatures, though this cannot be supported by experimental results. Apparently such processes are not to be expected with ceramic material.

Thermal properties such as thermal conductivity do not seem to differ significantly between glasses and ceramics. Metal matrix composites like vitromet, however, show much higher thermal conductivities. The thermal conductivity is relevant for the maximum temperature in the waste form.

Mechanically vitromet and probably to some extent also ceramics are superior to borosilicate glass as far as the consequences of a mechanical impact are concerned. Another mechanical effect closely related to thermal properties is cracking due to thermal stress. In a full-scale glass cylinder some cracking can hardly be avoided causing an increase of the potentially leachable surface and a decrease of the effective thermal conductivity. A glass ceramic, for instance, is expected to be somewhat better in this respect because of a smaller expansion coefficient. Apparently in a metal matrix product this effect will not occur.

Chemical stability, i.e. leach resistance against various liquors possibly occurring in a waste repository has not yet been fully explored, particularly not under conditions possibly prevailing in a rock salt formation. This is true for any of the solidification products. There is experimental evidence that glass will probably undergo some sort of decomposion when treated with groundwater or brine at 300°C or more under the corresponding pressure. The maximum temperature of the waste form under repository conditions, however, is a design parameter like others and may be chosen within reasonable limits. Leaching experiments with borosilicate glass in brines at 150-200°C in various laboratories seem to indicate that after a few days of considerable mass loss (up to about 2%) the leaching process slows down so much that no further effect can be observed. This result requires further confirmation. It is certainly desirable to gain a deeper understanding of waste glass leaching which would give more confidence in long term extrapolation of leach experiments. However, the prospects of those studies seem to be limited.

Ceramics are predicted to be more stable than glass under extreme leaching conditions, though this has not yet been proved experimentally.

As to the preparation technique, it has to be noted that glasses due to their disordered state may be less selective than ceramics towards incorporation of the great variety of waste oxides. Also, the processes to obtain ceramic products and composites are more complex, especially when mechanical operations such as pressing are involved.

It has been said before that little effort has been spent to develop containers with high chemical stability for the solidification product. The materials considered for HLW disposal canisters have been essentially stainless steel and cast iron. During this symposium some attention will be paid to container development.

As to the overpack, (and/or backfill) even the idea is quite new and the only material now under consideration is bentonite because of its swelling behavior and high sorption capacity for cations. There might be a lot of benefit especially from a highly adsorbing backfill. Again, during this conference considerable attention will be paid to overpacks and backfills.

SUITABILITY ASSESSMENT OF WASTE FORMS

Criteria to assess the suitability of a waste form to maintain the immobility of radioactivity throughout the waste management scheme are:

- chemical and mechanical stability which are directly relevant for potential remobilization of radioactivity.
- stability against heat and radiation which may deteriorate the chemical and mechanical stability.
- technical and economical feasibility.

If we start discussing the list of suitability criteria from the bottom, borosilicate glass is presently clearly superior to all other solidification products in terms of technical feasibility. Production of borosilicate glass with full radioactivity has been demonstrated on a pilot plant scale at several places and is now routinely performed on a technical scale with modestly radioactive, but true HLW in Marcoule. No other solidification product is even near to this stage of development.

Regarding the stability against heat and radiation we have to note the lack of thermodynamical stability of the glassy state which may cause thermal transformations. Regarding the chemical and mechanical stability, the possibility of hydrothermal decomposition seems to be the only serious drawback of borosilicate glasses. The likelihood of cracking due to thermal stress may be also mentioned.

Composites and ceramics have probably the potential to be better solidification products in some respects than borosilicate glass. Composites may be tailored so that a certain property of the plain material will be significantly improved. Ceramic products may be expected according to first principles to be superior to glasses, in terms of thermal, mechanical and, maybe, chemical stability. As yet, however, there is little experimental evidence of this superiority, at least at lower temperatures. Moreover, there is no evidence yet that any of these advanced products will become technically feasible in an appropriate time.

It is certainly desirable to make further efforts to develop advanced solidification products to a point where they are true alternatives to borosilicate glass. At present, however, borosilicate glass is the only available solidification product and a borosilicate glass block with up to 20% waste oxides in a stainless steel container is a suitable waste form for handling, transportation, interim storage and final disposal in a rock salt repository.

CONCLUSIONS

In conclusion, a few points should be made to keep in mind when the potential benefits of advanced waste forms in comparison to borosilicate glass are discussed.

(i) Demonstrated feasibility is an advantage of a solidification product often underestimated by the scientific community.

(ii) If under a given set of conditions the stability of a waste form is found to be insufficient, then there are basically two ways to improve the situation. Further efforts may either be spent to develop a better waste form or to change the conditions so that the available waste form performs adequately. The major deficiencies of borosilicate glass are in one way or the other related to temperature. There is no technical reason not to reduce this temperature considerably preferentially by aging the solidified waste in a surface facility prior to disposal. This will be beneficial anyway for the stability of the geologic host formation.

(iii) No doubt that there remains some uncertainty how a borosilicate glass will behave over a very long period of time. However, it is important to realize that this uncertainty can only be gradually reduced with any advanced waste form, because long term behavior can never be verified experimentally.

(iv) If it is felt that a more leach resistant waste form is required, it must be seriously considered whether the effort should be concentrated on the improvement of the solidification product or on that of the container. The material of the container does not contain any radioactivity which will highly facilitate production relative to that of the radioactive waste form. Moreover, a highly durable container may even corrode to some extent and will not release any radioactivity. A solidification product, however stable it may be, will release some radioactivity from the very beginning of a leaching process.

(v) As long term leach resistance is the prime incentive for the development of advanced waste forms it should be noted that leach resistance may be less effective in preventing radioactivity from reaching the biosphere than sorption of migrating radionuclides. Therefore, a well-designed backfill may be of more benefit than a highly sophisticated waste form.

(vi) The time scale we mean when we are talking about long term stability of waste packages is on the order of 10^3 to 10^4 years rather than of millions of years.

(vii) If one worries about the long term stability of HLW solidification products it has to be realized that the long term hazard associated with some of the α-bearing wastes is on the same order of magnitude. Consequently, solidification products for these types of waste would then deserve more attention than they presently get.

(viii) The goal of research and development in the field of waste forms is to obtain an "appropriate" waste form rather than the "best" one.

There are observations which deserve attention when the policy of further research and development on nuclear waste forms is to be defined.

REFERENCES

1. K. L. Erickson, "Preliminary Rate of Expressions for Analysis of Radionuclide Migration Resulting from Fluid Flow Through Jointed Media," Scientific Basis for Nuclear Waste Management (G. J. McCarthy, ed; Plenum Press, NY, Vol. 2, 1980).
2. G. Memmert, personal communication.

RADIOACTIVE WASTE DISPOSAL IN GEOLOGICAL FORMATIONS: RESEARCH ACTIVITIES OF THE COMMISSION OF THE EUROPEAN COMMUNITIES

M. Bresesti, F. Girardi, S. Orlowski and P. Venet

Commission of European Communities
Bruxelles

INTRODUCTION

The Commission of European Communities (CEC) conducts radioactive waste management R and D in two programs:

1. The Joint Research Center, which is directly operated by the Commission, has carried out a radioactive waste management program since 1973 under a four-year program plan sequence. The program is essentially directed to study the long term risks of geologic disposal of radioactive waste and to explore the feasibility and interest of an alternative waste management strategy (nuclear transmutation of actinides).

2. The indirect action program is a coordinated set of contractual activities carried out since 1975 in the European national laboratories under partial CEC financing (about 50%) under a five-year plan sequence.

An advisory committee for program management assures coordination between the two programs. The generalities of the CEC program have already been outlined in several publications and conference proceedings and annual status reports (1, 2, 3, 4). In this paper, those aspects of the program are discussed which aim at providing the necessary scientific understanding for a rational and safe management of the radioactive waste generated by the commercial fuel cycle. Recent results obtained and perspectives for the next few years are also briefly outlined.

WASTE CONDITIONING TECHNOLOGY AND PRODUCT CHARACTERIZATION

The technology for conditioning high level waste as borosilicate glass is well developed, and pre-industrial installations have been satisfactorily operated in France. For this reason, CEC is not financing R and D efforts to develop alternative technologies.

On the contrary, a strong community effort is underway to supplement the knowledge of the long term stability, both chemical and physical, of the glass blocks for the conditions that are expected in the long term in geological repositories. Physical stability and low leach rates are, in fact, considered a strong barrier against radionuclides release from geological repositories; and the knowledge of the barrier's value in the long term (up to 100,000 years) is important for long term risk assessment, which is required in a very early stage of the complex approval process for a repository.

The leaching behavior is being studied by various CEC laboratories both in simulated conditions and in fully active large blocks of conditioned waste corresponding to several glass compositions prepared by the CEA Marcoule Establishment. The results obtained indicate that in steady state conditions a homogeneous dissolution of glass has to be taken into consideration and that weight loss rate measurements seem to be most indicative of product degradation. The importance of the physical stability of the superficial gel layer has been verified.

The effect of α-radiation on the physical and chemical properties is studied by a coordinated effort of several laboratories which independently developed simulation methods based on α-spiking, accelerated ion bombardment and fission fragments. When measured effects are correlated to the number of displaced atoms (DPA), the three methods give equivalent results. Radiation damage equivalent to 500,000 y of decay can be reached in reasonable times by the fission fragment method. A study on possible dose-rate effects and identification of structural defects in glasses is reported separately. The results obtained up to now on radiation damage of vitrified waste show that:

1. The leaching coefficient is not affected by radiation damage.
2. Only minor density variations have been found.
3. Energy storage of 20-30 cal/g have been measured for high exposures. These values are far below dangerous levels.

While conditioning techniques for HLW are well established, other wastes of the fuel cycle are presently conditioned by methods which are not well adapted to a large scale nuclear industry. R and D efforts are therefore devoted to develop better technologies, aiming at:

1. Minimizing waste generation.
2. Minimizing the content of long-lived hazardous radionuclides.
3. Ameliorating the operational safety.
4. Improving the quality of the conditioned products, particularly the leaching resistance.

Table 1 shows a list of studies underway on various types of wastes and involving several laboratories of the E. C. member states. Among them, solid combustible α-bearing waste is receiving special attention due to the particularly nasty character of this type of waste. It is, in fact, produced in large quantities, the Pu loss can be rather high and conditioning techniques are not yet optimized in view of a satisfactory final disposal. For these reasons, several alternatives are being studied, aiming at either producing a highly leach resistant material or optimizing Pu recovery. The importance of this type of waste on the overall risk of geologic disposal should not be minimized. Indeed, preliminary assessment done by JRC indicates that their importance may be comparable to or even greater than that of the HLW. Conclusive statements cannot be made, however, since uncertainties on the conditioning processes are too large.

The problem of routine monitoring of α-emitters in waste is also receiving special attention. Technical guides for design and analysis of monitoring systems (5) have been prepared by the Joint Research Center (JRC). The JRC has also participated in or organized several monitoring campaigns in fuel process facilities. An integral experiment at the Dounreay Nuclear Power Development Establishment is presently being carried out under a JRC-DNPDE agreement, and such types of exercises are considered especially important in optimizing the entire monitoring system.

WASTE DISPOSAL IN GEOLOGIC FORMATION

The largest share of the available resources of the CEC indirect action program is being spent in studies related to geologic disposal, and this reflects the importance that CEC attributes to an early establishment of acceptable disposal strategies.

Three types of formations (salt, clay and hard rock) are being studied by concerted efforts of groups of countries, each one studying one type of formation.

A catalog of the geological formations within the European Community having favorable characteristics for the disposal of highly active solid radioactive wastes has been prepared by the national geological surveys under the CEC program. This compilation has shown the relative abundance, both in number and extent, of the three rock type categories taken into consideration, judged on the geological

Table 1. Development of Conditioning Technologies for Wastes Other Than HLW (CEC Indirect Action Program)

Low level - solid waste:	- imbedding in plastic resins
Cladding hulls:	- press compaction and encapsulation by low-melting alloy - rolling and embedding in concrete - embedding in borosilicate glass - eutectic fusion and decontamination by molten glass
Pu-contaminated combustible waste:	- high temperature incineration to produce a highly leach resistant product - incineration to produce a readily soluble product (Pu recovery) - incineration in molten salts (Pu recovery) - acid digestion (Pu recovery) - radiation-assisted incineration
Gaseous waste:	- Kr incorporation in a metallic matrix - Kr storage in pressurized cylinders - Kr absorption on charcoal in pressurized cylinders - encapsulation in zeolites - I incorporation in low-melting glass - I incorporation in epoxy resins

criteria adopted for selection. These formations are all well represented within the Community, although to differing degrees.

Table 2 shows the partition of research efforts between the various countries and formations. It should be noted that, in general, the selection of a certain type of formation by a country does not reflect any commitment of the country for the final disposal solution but simply an optimized share of the available resources, taking into account the competences existing in the various national laboratories.

The JRC is concentrating on risk analysis and related experimental studies for improving the data base and the knowledge of reaction mechanisms, which is essential for long range extrapolation of short term experiments. A generalized risk analysis model has been developed (6) which includes both a probabilistic treatment of failure of geological containment and a deterministic consequence analysis

Table 2. Research Underway on Waste Disposal in Various Types of Geological Formations (CEC Indirect Action Program)

Clay Formations

Belgium: complete geological-geochemical-hydrological prospection of a site, shaft drilling and preparation of an underground cavern for in-situ testing, laboratory studies of mechanical-physico-chemical properties of clays and model studies of radionuclide diffusion.

Italy: geological prospection at Trisaia site, exploratory borehole drilling, in-situ tests of heat dissipation and laboratory studies.

Denmark: clay properties.

Salt Formations

Federal Republic of Germany: in-situ testing (Asse mine), temperature distribution, convergence, stress analysis, brine migration, emplacement techniques for HLW canisters and material corrosion studies.

Netherlands: model studies, exploratory borehole drillings (planned) and risk assessment.

Hard Rocks

France: geological prospection of three sites, modeling of radionuclide migration, artificial barriers and in-situ tests (hydrology and radionuclide transport).

United Kingdom: hydrology of fissured rocks, water composition and dating, repository-formation interaction modeling, in-situ tests of heat effects, corrosion tests and risk assessment.

---

model. An application of the model is presently being made on the Boom clay formation in Belgium in cooperation with Belgian scientists, and it is reported in a separate paper at this symposium.

A working group on risk analysis of geologic disposal was formed in 1977, and European experts periodically meet to exchange information and discuss problems connected with risk evaluation. An international workshop on the subject was organized by the JRC in July 1977. The working group has repeatedly stressed the importance of a good scientific understanding of all phenomena related to geological containment and release consequence.

The long term interaction between leached radionuclides and geological media is receiving special attention. Scientific information on the interaction mechanisms is not sufficient to allow a realistic long term extrapolation of the potential value of geological media surrounding the repository as a barrier against radionuclide migration. Laboratory experiments and observation of natural phenomena indicate that the value of the barrier can be extremely high and hence the importance of the relevant studies. The subject is being tackled both by the JRC and by national laboratories in a concerted way. Preliminary results of JRC activities on the subject are reported in a separate paper.

FUTURE ACTIVITIES

Both the JRC and the indirect action program plans terminate at the end of 1979, and proposals for the next period (1980-1984) have already been presented to the Council of Ministers. In the JRC proposal, the allocated staff and budget have been kept at the present level of about 50 technical staff and 20 million European Units of Account (MEUA) (approximately $28 million, U.S.). The termination of the project on nuclear transmutation as an alternative waste management technology in 1980 will allow a considerable reinforcement of both theoretical and experimental studies related to risk assessment.

The proposal of the indirect action program foresees a strong increase of budget, from 19.6 MEUA to 53.5 MEUA (approximately $27.4 to $74.9 million, U.S.).

The main reorientation will be:

1. The reinforcement of actions on conditioning of low and medium level wastes as well as product qualification.
2. The conservation of the effort on geological disposal at the present level but with more emphasis on in-situ experiments and addition of limited studies on the into-seabed disposal concept.

Decisions on both program plans are expected by the end of 1979.

REFERENCES

1. S. Orlowski, F. Girardi and M. Bresesti, "CEC Program on Radioactive Waste Management: Disposal and Risk Assessment," Proceedings of Symposium Waste Management 1976, University of Arizona, Tucson (1976).
2. S. Orlowski, "Le Programme de la Commission des Communautés Européennes en Matière de Déchets Radioactifs," Rev. Générale Nucléaire 6, 520 (1976).

3. P. Venet, E. Della Loggia, W. Falke, B. Haiytink and P. Masure, "The R and D Program of the European Communities on Disposal of Radioactive Waste in Geological Formations," International Symposium on Underground Disposal of Radioactive Waste, IAEA/NEA, Otaniemi, in press (1979).
4. The Community's R and D Program on Radioactive Waste Management and Storage, Annual Reports EUR 5749 (1977), EUR 6128 (1978), in press (1979).
5. A Technical Guide for Design and Analysis of Monitoring Systems, I-V, Technical Reports EUR 5635 (1977), EUR 5636 (1977), EUR 5637 (1977), EUR 6027, V in press (1978).
6. F. Girardi, G. Bertozzi and M. D'Alessandro, Long Term Risk Assessment of Radioactive Waste Disposal in Geological Formations, Technical Report EUR 5902 (1977).
7. "Risk Analysis and Geologic Modeling in Relation to the Disposal of Radioactive Wastes into Geological Formations," Workshop Proceedings, NEA, Paris (1977).

DEVELOPMENT OF SOLID RADIONUCLIDE WASTE FORMS IN THE UNITED STATES*

J. L. Crandall

E. I. du Pont de Nemours & Co., Inc.
Savannah River Laboratory
Aiken, SC 29801

INTRODUCTION

Until fairly recently, there were really only two nuclear waste forms:

- Solid waste, which was by definition what went into a burial ground even if the waste happened to be mostly liquids, and

- Liquid waste, which was by definition what went into a waste tank even if the waste happened to be mostly solids.

In detail, of course, the solid wastes subdivided into beta-gamma wastes, transuranic wastes, medical wastes, reactor wastes, R and D wastes and what have you; while the liquid wastes subdivided into high level wastes, intermediate level wastes, acid wastes, alkaline wastes, etc. However, the overall division remained clear.

Obviously, this situation no longer holds. Gaseous wastes have been added to the list, and a new nuclear waste category (spent fuel) has been invented which does not fit into any of the previous categories. More important, the implied end points for waste disposal have changed. Except in circumscribed cases involving highly localized, very low level or very short lived activity, simple burial or simple tank containment of the wastes is no longer considered adequate. Instead, the end points now being looked at include geologic,

*The information contained in this article was developed during the course of work under Contract No. DE-AC09-76SR00001 with the U.S. Department of Energy.

ocean bed, and even ice sheet or space repositories (1). Transmutation of the waste radionuclides is another further-out possibility (2). All of these new end points require extensive reworking of the wastes to reduce their volume, to prepare them for shipment from the generating point to the disposal point, and to provide additional barriers against radionuclide dispersion. This extensive reworking of the wastes requires a new classification in terms of the final waste forms. The candidate forms are surveyed in this paper.

I will first address the types of waste forms, then who is working on them, next the waste form requirements, and finally waste form selection.

WASTE FORM DESCRIPTIONS

Encapsulated Waste Forms

For many of the solid wastes, the radionuclide levels are so low or they are so tightly fixed to the waste that these wastes can be directly packaged with, at most, a volume reduction and surface decontamination (3). The same approach may also apply to spent reactor fuels which were, of course, originally fabricated in very high integrity forms to withstand the extremely rigorous reactor conditions (4). The next stage beyond direct packaging is encapsulation of the waste in a suitable binder material. The binder serves both to impede leaching and to prevent mechanical dispersion of small waste particles. It may also provide a contamination-free surface. Technology for such encapsulation is well advanced in the low level and transuranic waste programs. It is in general limited to these programs. The two binders in common use are concrete and organic polymers. Their advantages and disadvantages are discussed below.

Concrete Binders (5). The concretes are readily formed with existing technology and present few processing problems except encrustation of the mixing equipment. For the most part, concrete binders offer excellent fire resistance, good adhesion and shielding, and intermediate leach rates and radiation resistance. Their main disadvantages are weight and the potential for high gas release from both irradiation and hot fires.

Organic Binders. The organic binders include the natural bitumens (6 and 8), urea formaldehydes (7 and 8), thermosetting polyester-styrene resins and vinylesters (8). They are easy to use, are light-weight and have excellent leach resistance. However, their radiation resistance is limited to 10^{10} rad or less, and they may be flammable or chemically unstable, particularly if oxidizing agents are present. There are also questions on the very long term stability of any organics.

In-Place Solidification Waste Forms

Although current Department of Energy (DOE) planning is aimed at placing essentially all high level and transuranic wastes in geologic repositories, it may turn out that for some of the older defense wastes, the risks of removal and treatment are greater than those of immobilizing the wastes in place. The same considerations apply even more strongly for low level and intermediate level wastes. Several options including concrete, clay additives and calcines have been considered for in-place solidification of liquid wastes under these conditions. These options are described below.

Concrete. The Oak Ridge National Laboratory has for a number of years been conducting a successful program to dispose of its intermediate level wastes by grouting them into special concrete formulations and hydrofracturing the grout into shale formations below the laboratory (9). Although limited to particular geologies, this is an inexpensive and straightforward disposal method that has given uniformly good results to date. Concretes have also been given limited consideration for in-place solidification of the high level wastes in the Hanford tanks as part of a tank "hardening" program.

Clay Solidification. Clay additives have also received limited attention as a solidifying agent for the Hanford high level wastes, offering the possibility of lower waste volumes and reduced reaction temperatures than with concrete solidification (10).

Calcines. Calcines are formed by firing the wastes to a mixture of (mainly) metal oxides at 300° to 700°C (11). Calcines can be either an intermediate in the preparation of more advanced forms or a waste form in their own right. They have the disadvantage of water solubility and of small particle size, which can lead to high dispersability. However, calculations at the Idaho Chemical Processing Plant, which has pursued a long term program of calcining its wastes, indicate that the risks of storing the calcine under their geologic conditions are small and can be further reduced by pelletizing and chemically treating the calcine. A limited program is investigating these possibilities.

Glass and Ceramic Waste Forms

Except for the very limited effort on in-place solidification of older defnese wastes described in the last section, essentially all of the effort on high level waste forms both in the U.S. and abroad is aimed at chemically binding the waste atoms into a high integrity lattice which will strongly resist mechanical dispersion, volatilization or leaching. The same approach is also expected to be applicable to other "chemical" waste such as incinerator ash from burning low level and transuranic wastes, decontaminant solutions and treated cladding hulls. The simplest and most studied high integrity waste lattices are the ceramics and glasses. Borosilicate glass, in particular, has received extensive development, is in commercial use for waste management by the French, and is the reference material in most other national programs for high level waste forms. The various glasses are described below, and the ceramics currently being investigated for the waste forms are discussed immediately after.

Phosphate Glass (12). Phosphate glass has received considerable attention as a waste form. It is produced by mixing the waste with phosphorus pentoxide or phosphoric acid and melting the mixture at about 1100°C. Phosphate glasses will incorporate about the same waste loadings as borosilicate glasses and are better at accommodating molybdates and sulfates. Waste leachability and volatility from the phosphate glasses are also similar to those for borosilicate glasses. However, two major deficiencies have largely ended work with the phosphate glasses. First, they are extremely corrosive, so that almost the only materials that can be in contact with the melt are platinum and some refractories. Second, the phosphate glasses devitrify at about 400° to 500°C to produce a 1000-fold increase in leachability.

Borosilicate Glass (13). Borosilicate glass is currently the furthest developed waste form for high integrity radionuclide containment. Typically a borosilicate waste glass might consist of about 20 to 30% waste oxides, 30 to 40% silica, 5 to 10% boron oxide and 10 to 15% alkali oxides, plus other additives. Formation temperatures are in the range 1100° to 1200°C for these compositions, and two main fabrication methods are in use: external melters from which the waste glass is poured into its final form and in-can melters in which the waste glass is formed in place in the final canisters. Large glass monoliths up to three feet in diameter and twelve feet high can be formed with either external or in-can melters. The external melters can also produce marbles for use in matrix forms. The advantages of borosilicate glass waste forms are low leachability, excellent radiation stability, low volatility, relative insensitivity to composition and, in the canisters, good mechanical strength. The primary concerns about borosilicate glass are devitrification, cracking, and leaching or chemical attack by high temperature, high pressure water or brine. There are also some secondary phases which usually form within the glass. These concerns are quite serious for high heat, high level waste loadings for which surface temperature in the repositories may reach 250° to 350°C. Such temperatures will definitely lead to short lifetimes for the glass in contact with pressurized repository water. However, the low heat, high level wastes from the U.S. defense program can easily be stored at surface temperatures of no more than 100°C. At these temperatures, borosilicate glass has demonstrated excellent characteristics in all tests to date. Similarly, as is being done in the French program (14) for incorporating their reprocessing wastes in borosilicate glass, it is possible to decay high heat wastes for 40 to 50 years before final emplacement and again to achieve high predicted repository lifetimes with 100°C surface temperature.

High Silica Glass (15). High silica natural glasses (obsidians and tektites) are known to have persisted for long periods in both terrestrial and lunar environments and to have better leaching characteristics than the borosilicate glasses. However, the high silica glasses are formed at about 1600°C, a temperature high enough to drive off most of the ruthenium and cesium radionuclides from the wastes. Thus, the key to producing these glasses as waste forms is to find low temperature processes for producing them. Several proprietary processes are being investigated for this purpose, but the products have not yet been thoroughly characterized.

Glass Ceramics (16). One form of glass ceramic can be made by sintering or hot-pressing the mixture of waste and glass frit at comparatively low temperatures rather than melting it as in normal glass making practice. The advantages are reduced radionuclide volatilization and chemical corrosion, although at some loss in form integrity. The more common forms of glass ceramics are formed by adding nucleating agents to the glass melt and then cooling the glass in a controlled manner so as to produce an assembly of small crystals held together by a glassy phase. The chief advantages of the glass ceramics are good thermal shock resistance, low leach rates and high temperature stability. Their disadvantages are complex processing and relatively small sizes.

Hot-Pressed (17) and Special Concretes (18). Studies at the Savannah River Laboratory showed that normal concretes give an inferior product at higher cost than borosilicate glass for immobilizing high level wastes. Accordingly, all current studies of concrete high level waste forms are aimed at advanced forms. Two possibilities are being explored. The one is to hot-press the concrete so as to eliminate voids and non-chemically combined water. The aims are to decrease the waste form volume, decrease the effective leaching areas and reduce steam volatilization in hot fires. The other approach is to preform the waste into particularly leach resistant particles (such as "supercalcine") before incorporating it into concrete. The aim is for these particles to retain enough chemical integrity to improve the concrete leach rates.

Clay Ceramics (19). Adding aluminum silicate clays such as kaolin or bentonite to the waste produces an insoluble cancrinite-type material. This material can be fired to a nephaline-type material. Typically, the waste is mixed with the clay and water, extruded into strings, chopped into pellets, dried and fired at about 1200°C. The product is markedly inferior to glass or the artificial minerals as regards leaching, dusting, surface area, etc. Accordingly, little attention is being given these forms.

Mineral Waste Forms

Although limited archaeological data are available on some glasses and ceramics for periods of up to several thousand years, most of the materials data for long time periods are in the geologic record. Accordingly, there are several efforts underway to model waste forms after geologic minerals, particularly after those minerals which have experienced natural radioactive decay without extensive lattice damage. It is hoped that this approach can lead to products with appreciably higher stability and durability than is exhibited by the current waste forms such as borosilicate glass. The two major artificial mineral systems which have been examined in this regard are the silicates as exemplified by the supercalcines and the titanates as exemplified by SYNROC.

Supercalcine (20). Pennsylvania State University, working in cooperation with Pacific Northwest Laboratories, has developed a series of waste forms they call supercalcines. The supercalcines are formed by calcining the waste with suitable additives such as Ca, Al, Sr and Si oxides and then heating the calcines at 900° to 1250°C to produce a polycrystalline form. Typical crystal phases include pollucites, monazites, apatites, scheelites, sodalites, fluorites, spinels and corundums. The supercalcines offer the advantages of very high waste loadings, up to 60%, good leach resistance, high thermal stability and, if properly formed, high thermodynamic stability. They can be directly produced as small particles suitable for the matrix forms discussed in the next section, or they can be pelletized or hot-pressed to larger forms. Three main questions arise for the supercalcine waste forms. The first concerns the multiplicity of crystalline species and the difficulty of assuring that an equilibrium assemblage of the desired species has been formed. The second question concerns the stability of the various crystals under heat, radiation, transmutation and leaching stress. Finally, the third question concerns the stability of pollucite crystals used to immobilize ^{137}Cs. Pollucite has been shown to have high leach rates in hot brines and in water at high temperatures and pressures.

SYNROC (21). Prof. Ringwood at the Australian National University has developed several assemblages of synthetic titanate minerals which he calls SYNROC. The current reference SYNROC composition is a mixture of fairly large (several millimeter) crystals of hollandite, perovskite and zirconolite. Typically, the SYNROCS are formed at very low waste loadings, 5 to 10%, but they may be able to accommodate quite high loadings of the alkaline defense high level wastes by using the inert alumina and ion phases in the waste as carriers for the SYNROC particles. SYNROC can be formed by melting (1600°C), hot isostatic pressing (1100° to 1350°C) and possibly by hot sintering at about 1000° to 1200°C. SYNROC has an outstanding resistance to high temperature leaching, showing no changes at 800°C and 1000 bar for 24 hours. It also has a high density and can be expected to show high thermal and radiation stability. Relatively little work has yet been done on SYNROC, and a large number of questions remain about its ability to handle a large variety of waste species, about its long term thermal and radiation stability, and about its fabrication techniques.

Titanates, Niobates, Zirconates (22). Sandia Laboratories have developed a series of hydrous oxide inorganic ion exchangers with the proposal that they be used to concentrate waste radionuclides and then be sintered into low leachability ceramic waste forms. Mixing the radionuclide loaded resins with about 30% glass and sintering at 900° to 1100°C at pressures of 7 to 14 MPa give low leachability products. Optimum formulations are still being developed. Also, the ^{137}Cs is in a pollucite phase which may be leachable under hydrothermal repository conditions.

Matrix Waste Forms

Many of the waste forms may be difficult to fabricate in large monoliths, and even for those where monoliths are possible, quality assurance testing of the product will in general be easier for one- or two-inch pieces than for very large pieces. It may therefore be necessary to go to matrix waste forms simply for mechanical reasons. Other more positive reasons are that the matrix material can offer another layer of protection and, if it is a metal, can offer better heat transfer. Multibarrier materials in which the waste particle is coated with additional protective layers before incorporation in the matrix are still further elaborations. For all the matrix forms, the trade-off is between the advantages of the matrix configuration against the greater waste surface areas and processing complexities involved in going to the matrix configuration. The advantages and disadvantages of the matrix forms are discussed below.

Metal Matrices (23). Most of the waste forms discussed to date, such as calcines, concretes, glasses, ceramics and artificial minerals, can be dispersed in a metal matrix for better heat transfer, reduced frangibility, easier sampling and additional leaching barriers. Low melting alloys such as Pb-Sb/Sn or Al-Si can be cast around the waste particles, while higher melting metals can be sintered around the particles at temperatures of about two-thirds the metal melting temperature. Aside from the process complexity, the main questions for these forms are securing void-free matrices, loss of low melting matrices in hot fires, radionuclide volatility with high melting matrices and the corrosion lifetimes of the matrix metals. The matrix forms will also by definition have lower waste loadings than the direct waste form.

Multibarrier Forms (23). Perhaps the most complex waste forms are made by coating the waste particles with impervious materials such as carbon, alumina or silicon carbide before placing the waste particles in a matrix. Such coatings provide additional barriers against waste leaching and also allow the use of high temperature matrix formation processing by reducing radionuclide volatilization. Concrete, sintered ceramics, glasses and other materials in addition to metals can be used as the matrices when coated particles are used. The main questions about the multibarrier forms are their high complexity, both as regards costs and as regards producing an acceptable product. The product quality questions are particularly acute because most of the existing particle coating technology involves coating very small particles having very large surface areas. A coating failure of such a product could greatly increase radionuclide release over that for a monolith.

Cermets (24). Cermet high level waste forms are a particular metal matrix form in which very fine waste particles are dispersed in the metal either by in situ reduction of the metal from the waste or by in situ precipitation of the waste from the metal. Ideally, these forms should be easier to make than the other matrix forms and should have very good bonding between the metal and waste. The main questions on this form versus the other metal matrices are the small size with resultant large surface area of the waste particles and the fact that their composition may not be as closely tailored as if they had been formed in a separate step.

Cladding Waste Matrices. Most of the current proposals for reprocessing fuels from commercial light water power reactors involve shearing the fuel assemblies and leaching out the uranium oxide, leaving the zircaloy cladding hulls and other hardware as waste. The hulls can be treated to recover the radionuclides for incorporation in the waste forms discussed to date, or the hulls themselves can be used as the waste particles to be incorporated in a suitable matrix. Relatively little work has been done on the cladding waste matrices; but some of the matrix materials being considered include lead alloys, sand, concrete, bitumen and glass (possibly even the waste glass material) (8, 25). Stability and containment problems still need to be examined for all these materials.

Gaseous Waste Forms

A number of waste radionuclides including ^{129}I, tritium and ^{14}C (and even ^{106}Ru and ^{137}Cs in high temperature processes) potentially can exist in gaseous form. However, it is believed that most of these materials can either be fixed in solid form for incorporation in the waste forms already discussed or, in the case of the shorter lived radionuclides such as tritium and ^{106}Ru, simply be held for decay. The nobles gas ^{85}Kr, which is presently being released to the atmosphere without exceeding discharge limits, may, however, have to be specially stored for a hundred years or so if a full scale nuclear fuel reprocessing industry develops. Possible waste forms for ^{85}Kr, still being developed, include pressurized gas cylinders, glass, concrete, metal or plastic "foams," molecular sieves and clathrates.

Canisters and Engineered Barriers

The waste forms are normally housed in a canister which provides a contamination-free handling surface, adds mechanical strength and supplies extra containment during shipping, handling and storage. This canister may in turn be surrounded by additional isolation materials, secondary canisters and other engineered barriers. In some cases, the canisters and engineered barriers can be designed to provide radionuclide containment equivalent to that of the waste form itself. They are also, in general, much less expensive to prepare than the waste forms because most of the work, except final installation, can be done in a nonradioactive environment. Hence, it is important to consider these barriers as an integral part of the waste form development.

The engineered barriers, some of which may be installed as part of the repository, also form the buffer between the waste form and the repository. Here they act to restrict and condition any flow of water from the repository to the waste form and similarly to restrict and condition any flow of leached radionuclides from the waste form to the repository.

WASTE FORM DEVELOPERS IN THE UNITED STATES

Currently, the largest expenditures for waste form development both in the United States and in the rest of the world are being made to engineer borosilicate glass as a possible waste form for high level waste and to engineer slagging pyrolysis incinerators for transuranic waste. However, the U.S. is also spending about $15 million a year to develop and characterize other waste forms which may offer improved radionuclide immobilization. The immediate target for this effort is to select, by 1984, the waste form for the initial DOE high level waste immobilization plant, the Defense Waste Processing Facility (DWPF), at the Savannah River plantsite. Studies are also expected to continue over a much longer period to take advantage of developing technology for second and later generation waste forms.

The U.S. programs for waste form development are widely disposed through the waste processing sites, the DOE national laboratories, industrial laboratories and universities in order to secure the widest possible input. Table 1 lists the major participants in the U.S. high level waste program (26).

Table 1. U.S. High Level Waste Form Developers

Waste Form	Developer
Concrete (Hot Pressed)	ORNL, PSU
Sintered and Pelletized Calcine	ICPP
Borosilicate Glass	PNL, SRL
Supercalcine	Rockwell, PSU
SYNROC	LLL, ANL
Titanate Ion Exchangers	Sandia
Supercalcine Coating and Matrices	PNL
Class-Marble/Metal Matrices	ANL, PNL
Cermets	ORNL

These U.S. development programs are the responsibility of the Division of Waste Products in the DOE Office of Nuclear Waste Management. The Division of Waste Products has in turn established lead offices for this work at the Savannah River Plant for high level waste, at Rocky Flats for transuranic waste, and at EG&G and ORNL for low level waste. Close coordination is also required with the transportation lead office at Sandia and the Office of Nuclear Waste Isolation at Battelle-Columbus.

WASTE FORM REQUIREMENTS

The waste forms must be considered as one part of a total nuclear waste disposal system. Their primary role within this system is to provide the initial barrier against radionuclide release. Here the overwhelming emphasis has been on reducing the availability of the waste radionuclides to water leaching when the wastes are disposed of in a repository. Current multibarrier philosophy is, in fact, that the waste form should provide a completely redundant backup to the repository barriers until the waste hazard has been greatly reduced by radioactive decay. The Nuclear Regulatory Commission has informally proposed a goal repository integrity period of 1000 years for the waste form plus its engineered barriers (26); 1000 years is roughly equivalent (for reprocessed waste) to decay to the radioactivity level of the ore from which the reprocessed fuel was originally prepared. Other possible goals are 300 to 500 years to allow for ^{137}Cs and ^{90}Sr decay or 10,000 years or more, as suggested by EPA, to allow for partial decay of the very long lived nuclides.

Overall requirements on the waste form include low dispersability and good impact resistance in transport, storage and sabotage situations; low radionuclide volatility in hot fires; long term stability; low sensitivity to waste variations; good waste density loadings and technology availability. Process factors are equally important and include processing safety both to the operator and the public, the ability to produce a fully licensable product at good yield, acceptable costs and minimum secondary waste generation.

WASTE FORM SELECTION

Plans for waste form selection in the U.S. are discussed below.

Waste Form Characterization

At the present stage of the U.S. waste forms programs, the main emphasis is on defining and characterizing the forms. This effort will enable reliable comparisons among the different forms and against the requirements being established by the process, transportation and repository groups on the one hand and by the regulatory agencies on the other. It is expected that most of the characterization will be performed by the forms developers themselves; however, in order to systemize the process, the Battelle Pacific Northwest Laboratories have been designated to prepare standard procedures, perform quality assurance testing and publish certified waste form data for the U.S. Also, a Materials Review Board has been set up to referee the data.

Waste Form Production

The worst danger of a paper such as this one is to lose sight of the fact that waste form development is not an end in itself. Its purpose is to dispose of the nuclear wastes at the earliest reasonable time schedule. Therefore, once the characterization work has supplied enough data (the goal date in the U.S. is 1982), the next phase is to enter into detailed engineering development of a few selected waste forms. After about two more years, the data should be available to select (by a suitable technical, regulatory, legislative and public process) the final waste forms for the actual waste disposal plants. As mentioned earlier, the target data for final selection of the waste form for the initial U.S. high level waste form plant at the Savannah River plantsite is in 1984. France has already made its decision for borosilicate glass, and a number of other countries seem close to a decision. Hopefully, then, we can see the waste forms work culminate within about 10 years.

REFERENCES

1. Draft Environmental Impact Statement, Management of Commercially Generated Radioactive Waste, Report DOE/EIS-0046D, U.S. Department of Energy (1979).

2. H. C. Clairborne, Effect of Actinide Removal on the Long Term Hazards of High Level Waste, USERDA Report ORNL-TM-4727, Oak Ridge National Laboratory, Oak Ridge, TN (1975).
3. Barnwell Nuclear Fuels Plant Applicability Study, Report DOE-ET0040, U.S. Department of Energy (1978).
4. Alternatives for Managing Wastes from Reactor and Post Fission Operations in the LWR Fuel Cycle, Report EDTA 76-43, U.S. Energy Research & Development Administration (1976).
5. J. A. Stone, Evaluation of Concrete as a Matrix for Solidification of Savannah River Plant Waste, USDOE Report DP-1448, E. I. du Pont de Nemours, Savannah River Laboratory, Aiken, SC.
6. E. D. Peltonen, "Incorporation of Radioactive Wastes from Nuclear Power Plants into Concrete and Bitumen," Management of Radioactive Wastes from the Nuclear Fuel Cycle II, IAEA, Vienna (1976).
7. P. Colombo and R. M. Neilson, Critical Review of the Properties of Solidified Radioactive Waste Packages Generated at Nuclear Power Reactors, USDOE Report BNL-NUREG 50591 (1976).
8. K. R. Yates, Characterization of Class A Waste From the Commercial Fuel Cycle, DOE Report ONWI-6 (2), Battelle-Columbus Laboratories (1979).
9. H. O. Weeren, Shale Fracturing Injections at Oak Ridge National Laboratory, 1972, USAEC Report ORNL-TM-4467, Oak Ridge National Laboratory, Oak Ridge, TN (1974).
10. Alternatives for Long Term Management of Defense High Level Radioactive Waste at the Hanford Reservation, USERDA Report ERDA 77-44, Washington, D.C. (1977).
11. Program Plan: Calcined Solids Retrieval and Handling, Allied Chemical Corp. Report ACI-224, Rev. 1 (1978).
12. I. Kostantinovich, "Features of a Process for Vitrifying Radioactive Waste Without Precalcination and Radionuclide Behavior in the Process," Management of Radioactive Waste From the Nuclear Fuel Cycle, IAEA, Vienna (1976).
13. J. E. Mendel, "High Level Waste Glass," Nucl. Tech. 32, 72 (1977).
14. C. Sombret, "Large Scale Waste Glass Production," Conference on High Level Radioactive Solid Waste Forms, USNRC Report NUREG/CP-0005, Denver, CO (1978).
15. J. H. Simmons, "Fixation of Radioactive Waste in High Silica Glasses," Nature 278, 729 (1979).
16. D. Gombert, Vitrification of High Level ICPP Calcined Wastes, USERDA Report ICP-1177, INEL, Allied Chemical Corp., Idaho Falls, ID.
17. J. G. Moore, Radioactive Waste Fixation in FUETAP (Formed Under Elevated Temperature and Pressures) Concretes -- Experimental Program and Initial Results, USDOE Report ORNL-TM-6573, Oak Ridge National Laboratory, Oak Ridge, TN.
18. D. M. Roy and G. R. Gouda, "High Level Radioactive Waste Incorporation Into Special Cements," Nucl. Tech. 40, 214 (1978).

19. D. M. Strachan, "Crystalline Materials for the Long Term Storage of Hanford Nuclear Defense Waste," Proceedings of National Meeting of American Ceramic Society, Detroit, MI, Report RHO-SA-13, May (1978).
20. G. J. McCarthy, "High Level Waste Ceramics: Materials Considerations, Process Simulations and Product Characterization," Nucl. Tech. 32, 92 (1977).
21. A. E. Ringwood, "Immobilization of High Level Nuclear Reactor Wastes in SYNROC," Nature 278, 219 (1979).
22. R. L. Schwoebel and J. K. Johnstone, "The Sandia Titanate: A Brief Overview," Ceramic and Glass Radioactive Waste Forms, ERDA Report CONF-77-102, 101 (1977).
23. J. M. Rusin, Multibarrier Waste Forms: Part II. Characterization and Evaluation, USDOE Report PNL-2668-2, Pacific Northwest Laboratory, Richland, WA.
24. W. S. Aaron, "Development of Cermets for High Level Radioactive Waste Fixation," Proceedings of Int. Symposium on Ceramics in Nuclear Waste Management, Cincinnati, OH, April 30-May 2 (1979).
25. J. van Geel, "Incorporation of Solid High Level Wastes Into Metal and Non-Metal Matrices," Conference on High Level Radioactive Solid Waste Forms, Denver, CO, USNRC Report NUREG/CP-0005 (1978).
26. Long Term High Level Waste Technology Program Strategy Document, USDOE Report DOE/SR-WM-79-3, E. I. du Pont de Nemours & Co., Savannah River Laboratory, Aiken, SC (1979).

THE ENVIRONMENTAL EFFECTS ASSOCIATED WITH THE TRANSPORTATION OF RADIOACTIVE MATERIAL*

James D. McClure, Ronald B. Pope and H. Richard Yoshimura

Nuclear Materials Transportation Technology Department
Transportation Technology Center
Sandia Laboratories
Albuquerque, NM 87185

INTRODUCTION

As a result of the wide geographical separation of nuclear fuel cycle facilities, transportation of nuclear materials is an essential link between each of the steps of the nuclear fuel cycle. Historically, the shipments of radioactive materials have moved with relative ease within the United States using shipping containers which are approved by the Nuclear Regulatory Commission or the Department of Transportation and are transported by common carrier or special transport vehicles.

In spite of this history, growing concern is being expressed regarding transportation of nuclear materials. These concerns are related to the ability of both industry and government to meet the requirements of shipping spent reactor fuel, transuranic materials and radioactive wastes. A recent U.S. Department of Energy task force (1) has indicated that special attention should be given to transportation as a potential problem area in the waste management system, since there exists a significant and growing backlog of radioactive materials from the fuel cycle, including spent fuel, which must eventually be shipped. In addition, according to current plans, transuranic waste from government programs and high level DOE waste will be shipped to repositories. The transportation of these commercial and defense waste materials could become the limiting factor in current operations and in maintaining the option of nuclear energy.

*This work supported by the U.S. Department of Energy.

The information presented in this paper provides a quantitative statement of the risks to the public associated with the transportation of radioactive materials and compares these risks to others more commonly accepted by the general public.

RADIOACTIVE MATERIALS TRANSPORTATION SYSTEMS AND REGULATIONS

The facilities which are currently operating and those which are planned for future operations in the commercial nuclear fuel cycle are located in widely separated sites in the United States. Such transportation involves a wide variety of chemical and physical forms. These materials include uranium ore concentrate, uranium hexafluoride, fresh fuel, spent fuel and many forms of radioactive waste. The principal transport modes for materials in the nuclear fuel cycle are truck and train, but there have been some special cases where shipments have been made by barge. Fuel cycle materials are not routinely shipped by air within the U.S. The majority of radioactive material shipments are made by truck, which provides a flexible and economical transport mode. An excellent review of the transportation requirements for the nuclear fuel cycle is presented in Reference 2., and a very detailed study of the quantities of radioactive material transported in the U.S. is given in Reference 3.

Regulations have been developed over the years which govern the transportation of radioactive materials. Within the United States, examples of these regulations are 10 CFR 71 and 49 CFR 173, while internationally the regulations are given in IAEA Safety Series No. 6. One principle involved in the regulations is that the amount of protection provided is proportional to the risk posed by the quantity of radioactive material involved. That is, if a package contains enough radioactivity to present a significant risk of injury or large property loss <u>if released</u>, then the package must be designed to retain its contents and provide adequate shielding during and following exposure to severe transportation accident environments (4). Such packages are known as "Type B" packages (i.e., they contain "Type B" quantities). Examples of "Type B" packages are spent fuel shipping casks and transuranic waste shipping systems.

Quantities of radioactive materials which could be less dangerous, if released, are known as "Type A" quantities, and less stringent packaging requirements are imposed (4). This is not meant to imply that Type A packages are not of high quality but rather are designed to withstand only the normal transport environment. In addition, both Type A and Type B packages are required to completely retain their contents during all normal conditions of transport.

ACCIDENT RATE DATA

There are approximately 500 billion packages of all commodities shipped each year in the United States (4). About 100 million of these shipments involve hazardous materials which include flammables, explosives, poisons, corrosives and radioactive materials. In 1975 there were over two million packages of radioactive materials transported in the U.S. Thus, about two percent of all hazardous materials shipments involve radioactive materials. Most of these shipments involve radiopharmaceuticals. In 1971 the Department of Transportation (DOT), acting under the requirements of federal regulations (49 CFR, Parts 100-199), instituted a uniform, multimodal hazardous materials incident (HMI) reporting system. Under this system, carriers of hazardous materials are required to provide reports to the DOT of transport incidents involving hazardous materials which result in death, personal injury or property damage. In the case of radioactive materials (RAM), even <u>suspected</u> radioactive contamination must be reported (see 49 CFR 174.750). This wording undoubtedly leads to the reporting of many "incidents" which in reality have not produced any radioactive contamination or hazard.

During the period 1971-1978, more than 77,000 HMI reports were filed with the DOT (5, 6) for all types of materials. During this same period, there were 389 HMI reports filed for radioactive materials, approximately 0.50 percent of the total sample. The hazardous material classifications and a comparison of the radioactive material incident experience with that of other hazardous materials is shown in Table 1.

Table 1. Eight-Year Total HMI Reports by Classification of Hazardous Material

Classification	No. of Reports	Percent of Total
Explosives	177	0.2
Flammable Compressed Gas	1 425	1.8
Non-Flammable Compressed Gas	1 030	1.3
Flammable Liquid	40 859	52.7
Flammable Solid	407	0.5
Oxidizing Material	1 344	1.7
Poisons Class A, B	3 776	4.9
Corrosive Materials	25 342	32.6
Radioactive Materials	389	0.5
Other	2 952	3.8
	77 701	100.0

It is noteworthy that, although radioactive materials constituted about two percent of the hazardous materials packages transported, these packages were involved in only 0.5 percent of the incidents reported. This lower incident rate may be attributed, in part, to the greater care exercised in the handling and transport of radioactive materials.

SPENT FUEL SHIPPING CASK TESTING PROGRAM (7, 8)

In order to evaluate transport system response to severe accident environments, Sandia Laboratories conducted, for the Department of Energy, a series of full scale crash tests and a fire test involving lead shielded, spent nuclear fuel shipping systems. These tests simulated extremely severe accident conditions which might involve both rail and highway carriers. The test environments exceeded the severity of the current licensing requirements for radioactive material shipping packages with respect to both the impact and fire environment. One primary purpose of the tests was to evaluate the ability of designers to predict package response in extreme accident environments using both analytical modeling methods and scale model testing.

Detailed scale model tests and mathematical analyses were first performed followed by full scale tests. The full scale test program consisted of three vehicle crash scenarios and one fire test. Out-of-service truck and rail spent fuel shipping systems dating back to the 1960's were used for economic reasons. Structurally these systems are similar to present-day units.

1. A tractor-trailer crash into a solid concrete barrier while carrying a spent fuel shipping cask (two separate tests of one cask). Impact velocities: 61 mph (98 kph) and 84 mph (135 kph), respectively.

2. A grade crossing impact between a high speed locomotive and a tractor-trailer carrying a large spent fuel shipping cask. Impact velocity: 81 mph (131 kph).

3. A high speed collision followed by a fire involving a special railcar carrying a spent fuel shipping cask. Impact velocity: 81 mph (131 kph). Fire duration: 120 minutes.

The following sections will describe some of the scale model testing, analysis and full scale testing for the truck-trailer system.

Risk Analysis

Risks associated with the transportation of radioactive materials have been assessed and compared with other risks to which society is subjected. These quantitative risk statements are the results of a systematic analysis of nuclear fuel cycle transportation operations. Much of the information necessary for risk analyses are obtained from the knowledge of fuel cycle operations, an examination of transport accident rates and package accident response described earlier. The methods of risk analysis as employed here are discussed in References 4. and 9. A working definition of risk analysis might be, "a systems approach to the examination of risk."

Model Tests (7)

A series of rocket sled track tests using 1/8 scale models were conducted in order to model the dynamics and the cask damage to be encountered in the full scale tests. The models running on a single rail were accelerated by a pusher sled to a velocity slightly greater than the intended impact velocity and allowed to coast into the target and impact at the desired test velocity.

Analysis (7)

The mathematical analyses of the crash tests were accomplished using two approaches. The first of these was the mathematical representation of the overall system response using a lumped parameter model. The second approach was to perform a more detailed dynamic analysis of the shipping cask using a finite element representation. The details of these analyses are found in Reference 7.

Full Scale Crash Tests (7, 8)

The full scale tests were conducted at the Sandia Sled Track facility where a road was constructed to the level of the top of the rails. In the truck tests, the vehicle wheels bridged a trench and a section of track leading to the target. The system was guided by the rails but rolled on its tires. Rocket motors mounted on a sled attached to the truck were fired in stages to accelerate the system up to a speed slightly greater than the impact velocity. The acceleration was controlled to about 0.5 g. Different impact velocities were achieved by varying the number of rockets and the accelerating distance. The rocket sled, which was connected to the rails with track shoes, pushed against the rear of the trailer. The sled was disengaged from the trailer before impact by a sand brake built into the trench between the rails.

To make the cask more representative of modern equipment, the cask was equipped with balsa wood impact limiters on each end. The 45,000 lb. (20,500 kg) cask was attached to the 11,000 lb. (5,000 kg) trailer using the tiedowns designed for normal service. The tractor was a used commercial unit weighing 13,000 lbs. (5,900 kg). In order to accelerate the system in a straight line into the target, the tractor was guided at its front and rear axles by the rails through a system of track shoes. Extensive high speed photo coverage and instrumentation were provided for the tests.

Quantification of Risk

Risk can be thought of as the product of the probability of the occurrence of an event and the consequence of the event occurrence. Some fraction of the population can be considered at risk if an unlikely event, such as a transportation accident involving radioactive material, were to occur. The units of risk can be expected number of early fatalities or latent cancer fatalities. A model was used to assign failure of package and quantity of radioactive materials discharged as a function of accident environment severity. Some results of a detailed risk assessment (4) are illustrated in Table 2 and show the individual risk of early fatalities from various causes. In addition, it was calculated that .0054 (1975 estimates of shipping activity) and .017 (1985 estimates of shipping activity) latent cancer fatalities, as a result of accidents, could be expected from a single year of shipment activity at the given level. These data demonstrate that the predicted fatalities, both early and latent, are extremely small compared to other risks commonly experienced by society.

The model used in Reference 4. is extremely conservative. For example, it was assumed that 100 percent of the releasable material for truck casks would be discharged to the environment if the impact velocity exceeded 40 mph (64 km/hr). Yet the results of the full scale truck crash tests, conducted after Reference 4. was complete, showed only small quantities of coolant were discharged when the system impact velocities were as high as 84 mph (135 km/hr). With the data from the transportation accident statistics, the probabilities of occurrence for the various accident scenarios considered in the full scale spent fuel cask testing program can be calculated. These probabilities are presented in tabular form in Table 3.

The purpose of presenting this type of risk information is to make such information available to government decision making bodies to allow regulatory and program decisions to be made. The decision making process must necessarily include trade-offs -- balancing of risks, costs and benefits. The more quantitative the risk statement, the more readily the trade-off process can be accomplished.

Table 2. Individual Risk of Early Fatality by Various Causes (4)

Accident Type	Number/Year	Individual Risk/Year
Motor Vehicle	55 000.	1 in 4,000
Falls	18 000.	1 in 10,000
Fires	7 500.	1 in 25,000
Drowning	6 200.	1 in 30,000
Air Travel	1 800.	1 in 100,000
Falling Objects	1 300.	1 in 160,000
Electrocution	1 100.	1 in 160,000
Lightning	160.	1 in 2,000,000
Tornadoes	91.	1 in 2,500,000
Hurricanes	93.	1 in 2,500,000
100 Nuclear Reactors	.003*	1 in 5,000,000,000
Transportation of RAM (from radioactive causes)	.00035**	1 in 200,000,000,000***

*Statistical estimate.
**Statistical estimate for 1975.
***Using a population at risk of 75 million people.

Table 3. Accident Probabilities

Accident Scenarios	Approximate Interval* (average number of years between accidents)
98 km/h Truck Impact	70
135 km/h Truck Impact	1 000
131 km/h Grade Crossing	4 500
Combined 131 km/h Special Railcar Impact and 120 minute Fire	1 000 000

*Assuming 11 x 10^6 km transport distance per year.

CONCLUDING REMARKS

The primary aim of this paper has been to describe some of the background information concerning nuclear materials transportation systems, accident statistics, accident severities and test information -- all of which when combined yield an environmental statement of the risks associated with the transportation of radioactive materials. The results of the ultimate risk analysis are expressed in terms of numbers of fatalities and, in that sense at least, tend to be an absolute measure of risk. When these risks are compared with other accepted societal risks, the relative risks associated with radioactive material transportation can be established. This information can be used to make decisions at the governmental level and to inform an interested public about these risks. It can be concluded from the work described in this paper that the risks associated with the transportation of radioactive material are low relative to the other risks that society has already accepted.

REFERENCES

1. Report of Task Force for Review of Nuclear Waste Management, DOE/ER-0004/D, U.S. Department of Energy, Directorate of Energy Research, Draft, Feb. (1978).
2. R. E. Rhoads, An Overview of Transportation in the Nuclear Fuel Cycle, BNWL-2066, Battelle Memorial Institute, Pacific Northwest Laboratories, Richland, WA, May (1977).
3. Transport of Radioactive Material in the U.S., A Detailed Summary of Radioactive Material Shipments in the U.S., BNWL-1972, NUREG-0073, U.S. Nuclear Regulatory Commission, Office of Standards Development, May (1976).
4. Final Environmental Statement on the Transportation of Radioactive Material by Air and Other Modes, 1, NUREG-0170, U.S. Nuclear Regulatory Commission, Office of Standards Development, Dec. (1977).
5. A. W. Grella, "A Review of Five Years' Accident Experience in the U.S.A. Involving Nuclear Transportation (1971-1975)," Office of Hazardous Materials Operations, U.S. Department of Transportation, Washington, D.C., a paper presented at an IAEA Seminar, Vienna, Austria, IAEA-SR-10/5, Aug. (1976).
6. W. C. Morgan, Compilation of Data on RAM HMI Reports, U.S. Department of Transportation, Safety Data Management Branch, April (1979).
7. Michael Huerta, Analysis, Scale Modeling and Full Scale Tests of a Truck Spent Fuel/Nuclear Fuel Shipping System in High Velocity Impacts Against a Rigid Barrier, SAND77-0170, Sandia Laboratories, Albuquerque, NM, April (1978).
8. H. R. Yoshimura, "Full Scale Simulations of Accidents on Spent Nuclear Fuel Shipping Systems," Proceedings of the Fifth International Symposium on Packaging and Transportation of Radioactive Materials, Las Vegas, NV, May (1978).
9. J. H. N. Selman, "Decision Risk Analysis," Proceedings of the Conference on Hazard Evaluation and Risk Analysis, National Research Council, National Academy of Sciences, Washington, D.C., Aug. (1976).

THE PERSPECTIVE OF WASTE ISOLATION RESEARCH ISSUES AND ASSESSMENT OF CONSEQUENCES FOR RADIONUCLIDE RELEASE*

T. O. Hunter

Sandia Laboratories
Albuquerque, New Mexico 87185

INTRODUCTION

Nuclear waste management has recently experienced a proliferation of technical and social issues which have seemingly confounded the stabilization and implementation of effective policy. Advocates and critics alike have shown an inexhaustable capacity to arouse public and scientific interest in these matters. This paper is restricted to the examination of some current technical issues in the perspective of their role in determining consequences of radionuclide release to the biosphere. The consequence assessments done for the Waste Isolation Pilot Plant (WIPP) site are used as a reference and compared with some current data on various technical issues. The comparisons and conclusions drawn apply specifically to WIPP configurations at the WIPP site. In addition, these initial assessments provide the basis for a method to identify key issues and eliminate inconsequential ones, thereby allowing commensurate allocation of research and development resources.

Determination of the potential hazard from disposal of nuclear waste is addressed by considering various scenarios in which the primary isolation mechanism--the media immediately surrounding the repository--is assumed to be breached. A series of these scenarios have been identified for WIPP (1). Consequences of the radionuclide release in the scenarios have been determined and published in the WIPP Draft Environmental Impact Statement (DEIS) (2). This document presents the results of five different scenarios.

*This work supported by the U. S. Department of Energy.

1. An open borehole which penetrated the repository allowing fluid transmission between lower and upper aquifers.
2. Two interconnected penetrations of overlying formations that allow fluids to enter the repository at one location and escape into the upper aquifer at another.
3. Diffusion of radionuclides from a flooded repository into overlying aquifers.
4. A bounding case in which the total flow in the overlying aquifer is assumed to enter the repository, mobilize radionuclides, and reenter the overlying formations.
5. A direct access scenario in which a drilling operation intercepts a high concentration of radionuclides, thereby exposing the drill crew.

The scenario chosen for comparison is depicted below in Fig. 1. In this scenario, a 9-inch uncased borehole allows up to 600 ft^3/day unsaturated brine (230,000 ppm) to enter the repository from below, dissolve the salt in the repository, and enter the overlying aquifer at 410,000 ppm. The assumptions within this scenario will now be compared with the related issues of waste form stability, brine migration, radionuclide migration, and borehole plugging.

WASTE FORM STABILITY

A solid waste form that chemically or mechanically binds radionuclides has been proposed as one of the "multiple barriers" in a

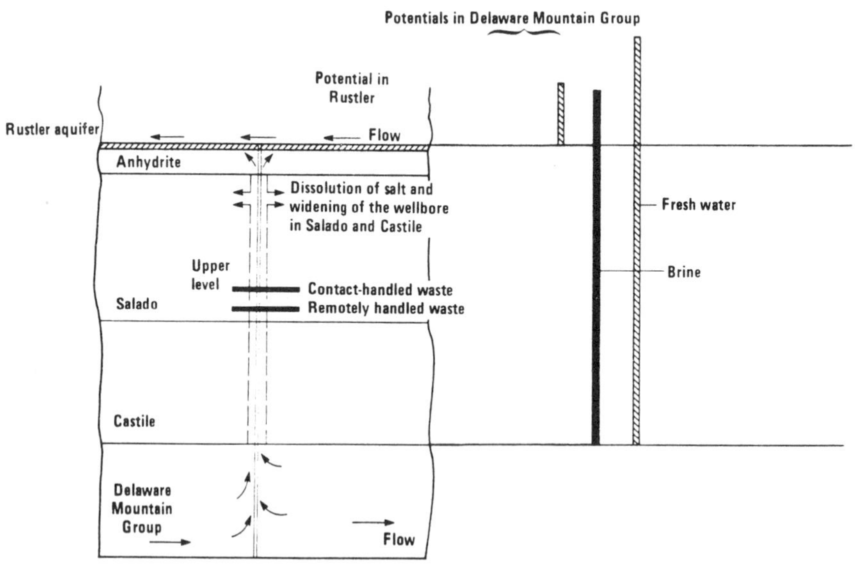

Fig. 1. Schematic Representation of Scenario 1

repository. While stable waste forms, under appropriately specified conditions, can no doubt serve that function, the degree of reliance on that barrier has not yet been assessed. In the WIPP consequence assessments, the existing defense transuranic (TRU) wastes were not of sufficient uniformity to specify any resistance to entrainment or solution by breaching fluids; consequently, the waste was assumed to go into solution at the same rate as the salt. For scenario 1, the assumptions of constant flow rate imposed by overlying aquifers and constant salt injection rates result in a gradual increase in effective borehole diameter and a commensurate reduction in surface dissolutioning rate as the surface area increases. The effective dissolution rate and borehole radius from these simple assumptions are shown in Fig. 2.

Also shown in Fig. 2. are some representative data on glass from Westsik and Turcotte (3) at various temperatures. Admittedly, leach rate under controlled laboratory conditions may not be the most appropriate indicator of transport of ions from solids into solution. The effects of conditions imposed by waste degradation and interaction with the host rock, which include Eh, pH, extent of devitrification, and radionuclide speciation must be considered. However, any emphasis of detailed interaction chemistry must be done only after

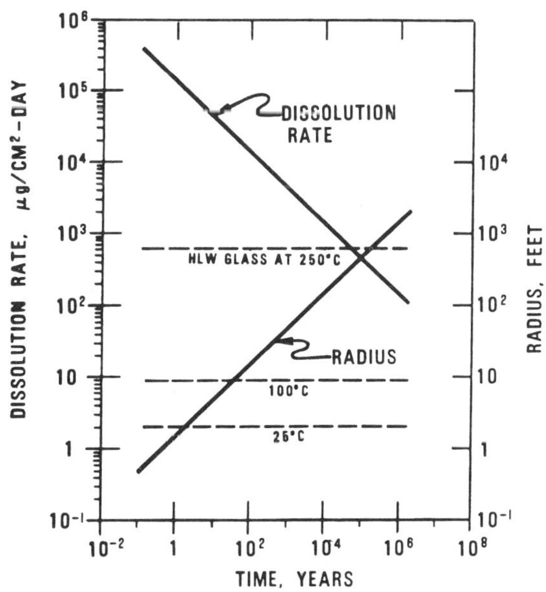

Fig. 2. Waste Form Stability (Scenario 1)

appropriate consideration of the importance of specific parameters in providing overall radionuclide isolation.

In the WIPP case, the scenarios analyzed have very small predicted doses (less than 0.10 millirem per year for a maximally exposed individual in scenario 1), yet the stability of the waste form played no role. Moreover, the assumed dissolution rates were several orders of magnitude higher than could be anticipated with waste forms such as glass. An analogous problem is the role of protective canisters surrounding the waste form. WIPP scenarios to date have not relied on any protection from canisters. Nevertheless, recent experimental tests on various alloys (4) have identified several candidates which exhibit corrosion lifetimes of several hundred years. These tests were performed under conditions which simulated brine inundation and varying degrees of oxidation potential including the effects of radiolysis. While it is unlikely that an objective assessment of consequences from a repository would rely entirely on an impervious canister, the technology for enhanced life canisters under repository conditions can be developed and relied upon for increased confidence in the isolation system. Canister material development and selection can, however, play an important role in future repository applications by assuring compliance with institutional constraints such as retrievability.

BRINE MIGRATION

Repositories in salt have been the subject of intense controversy over the potential migration and accumulation of brine around heat-producing wastes. At present, the mechanisms which determine the initiation or propagation of fluid motion are being studied experimentally (5) but have not yet been sufficiently investigated to provide minute resolution of rates or amounts of influx. The salient aspect of brine migration is that for long term isolation, it plays an insignificant role in releasing radionuclides when compared with the fluids which are assumed to intrude in consequence assessment scenarios. Most assessments using analytical models for brine influx to HLW canisters have predicted integrated flows less than 10-50 liters. The scenarios for WIPP assumed that up to 6×10^6 liters/year of fluids were allowed to enter the repository via a natural or man-made breach of the salt.

Recent experiments with a 1.5 kW heater in a 1-meter salt block (6) at Sandia Laboratories used higher temperatures and gradients than anticipated for HLW canisters. Fluid migration rates of approximately 1 gm/day (decreasing to 1/2 gm/day at the end of testing) were observed over a period of 100 days. More significant, however, is that these tests revealed the possibility of mechanisms such as vapor phase transport through the salt. These correlations indicate that moisture collected during the experiment may be induced by the maintenance of very low vapor pressure near the collecting surfaces and that fluid release is greatly enhanced by changing the stress field surrounding the heated source, especially upon power shutdown. Both of these conditions, which promote experimental observation of moisture collection, will not occur with actual waste canisters.

The most significant impact of brine migration would be the potential complication of retrieval operations. However, this impact has been considered by studies on waste form degradation or canister corrosion that have assumed brine is present. The results of such investigation indicate that both materials and waste forms have sufficient durability to easily endure the retrieval period under worst-case conditions. If further confidence is needed, engineering barriers or reduction of power densities can be used to eliminate concern about brine migration.

RADIONUCLIDE MIGRATION

Consequence assessments for WIPP have assumed that the fluids which violate the repository and entrain radionuclides are introduced into overlying aquifers. This hydrologic flow system is characterized by descriptions of aquifer properties and by the sorptive capacity of the rocks. Laboratory studies of distribution coefficients (7) (K_d) have been used to establish reference values for the Rustler aquifer which overlies the WIPP site. The sensitivity of the consequences to the values assumed has recently been addressed by Fortney (8). This study investigated the concern over potential reduction in sorption by complex formation or radiolytic induction of redox reactions. The results serve to place in perspective the importance of precise resolution of sorptive properties.

In this analysis, the sorptive properties of the Rustler aquifer were systematically reduced, and the effect on consequences predicted by scenarios for a TRU repository was examined. Cases considered are shown in Table 1.

The results indicated in Fig. 3. reveal the impact on individual 50-year dose commitments and emphasize the sensitivity to changes in plutonium retardation. Maximum whole body doses in Case 0 in which all radionuclides moved at the same velocity as the fluids did not exceed 3 mrem/yr. A K_d of 10 for Pu (Case 3) was sufficient to reduce the dose to 0.002 mrem/yr, essentially the same level as with a value of 2100. Hence, while a reduction of K_d to 0 and 1 result in a thousandfold increase in dose, the dose was still inconsequential, and reasonably small K_d's are sufficient to dramatically reduce predicted doses.

Table 1. Sorption Properties Used for the Rustler Formation

	Element	K_d(ml/g)	Element	K_d(ml/g)
DEIS VALUES	Np	700	I	0
	U	1	Tc	0
	Th	2200	Ra	25
	Pu	2100	Sr	0
	Cs	15		
Case 3	Pu*	10		
Case 2	Pu*	1		
Case 1	Pu*	0		
Case 0	All	0		

*All others at DEIS values

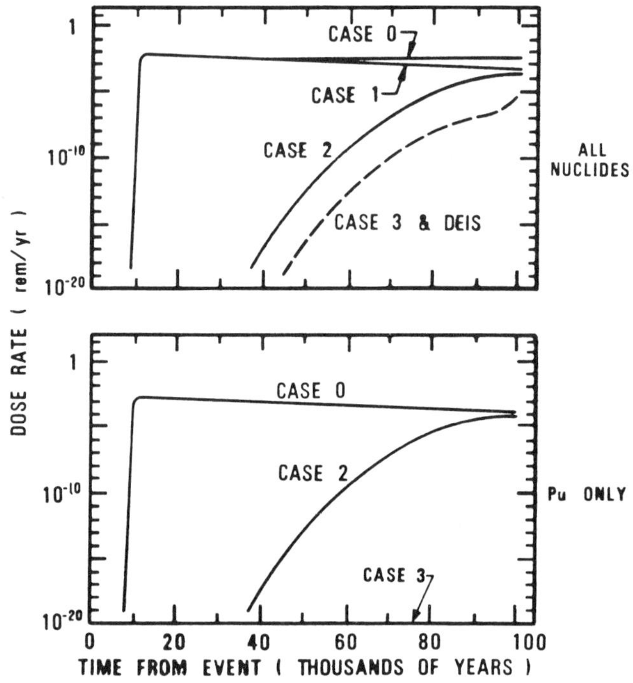

Fig. 3. Whole Body Dose, Scenario 1

Sensitivity studies of this type are just beginning, but these initial studies point out that certain ranges of parameters are not important to overall system consequences.

BOREHOLE PLUGGING

A long standing issue in waste management is the requirement of sealing boreholes and shafts. Experimental programs (9) have been initiated to investigate the feasibility of emplacing and monitoring performance of plugs. These programs, when coupled with supporting programs to assess long term stability of plugging materials, will provide the basis for sealing repositories upon decommissioning. While adequate plugs appear technically feasible, it should be stated that in all consequence assessments performed for WIPP, totally unsealed penetrations were assumed. Hence, the existence of a competent seal was not required to limit consequences to a small fraction of natural background. Borehole plugs can therefore be relied upon as an additional measure of protection, yielding further confidence beyond that provided by bounding case studies. They should not be viewed as essential components in assuming isolation.

SUMMARY

The finiteness of research resources to address technical issues for nuclear waste management requires a realistic assessment of their importance. The most appropriate method for making this assessment is the combination of system studies, which include consideration of the role of all protective barriers, and consequence assessment models that can estimate the potential hazard imposed upon man for assumed violations of these barriers. Sensitivity analyses on certain issues can then be performed to determine the impact of various descriptive parameters.

Work commensurate with this strategy has been initiated at Sandia. Some conclusions drawn from these studies specifically applicable to WIPP for the scenarios considered are that:

1. Waste form and protective canisters are not essential as barriers. Assumed radionuclide dissolutioning is far in excess of that possible with developmental waste forms.

2. Brine migration plays no significant role in long term isolation; complications to retrievability can be avoided with numerous technical alternatives.

3. Radionuclide sorption in geomedia near repositories plays an important role in reducing consequences, but sites with favorable geologic and hydrologic settings such as WIPP can tolerate essen-

tially no reliance on sorption and still achieve inconsequential predicted doses. Plutonium sorption with K_d greater than 10 are sufficient to eliminate direct doses from plutonium.

4. Borehole plugs are not required for primary protection.

Continued scientific resolution of both real and perceived issues can, however, provide additional assurance and public confidence that consequences can be reduced beyond those predicted in bounding scenarios. In addition, such investigation will assure that appropriate limits are instituted for defining parameters and that all assumptions in predictive models are technically sound. Nevertheless, all who participate in a socially significant technical problem such as nuclear waste management should endeavor to clearly distinguish between critical issues and those which merely require additional refinements.

REFERENCES

1. F. W. Bingham and G. W. Barr, Scenarios for Long Term Release of Radionuclides from a Nuclear Waste Repository in the Los Medanos Region of New Mexico, SAND78-1730, Sandia Laboratories, Albuquerque, NM (1979).
2. DOE/EIS-0026-D, Draft Environmental Impact Statement, Waste Isolation Pilot Plant, 1-2, U.S. Department of Energy, Washington, DC (1979).
3. J. H. Westsik, Jr., and R. P. Turcotte, Hydrothermal Reactions of Nuclear Waste Solids--A Preliminary Study, PNL-2759, Pacific Northwest Laboratory, Richland, WA (1978).
4. J. W. Braithwaite and M. A. Molecke, Nuclear Waste Canister Corrosion Studies Pertinent to Geologic Isolation, SAND79-1935J, Sandia Laboratories, Albuquerque, NM (1979).
5. S. J. Lambert and H. C. Shefelbine, A Strategy for Investigation of Fluid Migration in Evaporites (Waste Isolation Pilot Plant--WIPP), SAND79-1889, Sandia Laboratories, Albuquerque, NM (under review).
6. J. J. Hohlfelder, Salt Block II, Preliminary Results, Sandia Laboratories, Albuquerque, NM (in preparation).
7. R. G. Dosch, Interaction of Radionuclides with Geomedia Associated with the Waste Isolation Pilot Plant (WIPP) Site in NM, SAND78-0297, Sandia Laboratories, Albuquerque, NM (1978).
8. Summary of Research and Development Activities in Support of Waste Acceptance Criteria for WIPP, ch.--Immobilization, SAND79-1305 (1979).
9. C. L. Christensen and T. O. Hunter, Waste Isolation Pilot Plant (WIPP), Borehole Plugging Program Description, SAND79-0640, Sandia Laboratories, Albuquerque, NM (1979).

STATUS REPORT ON STUDIES TO ASSESS THE FEASIBILITY OF STORING NUCLEAR WASTE IN COLUMBIA PLATEAU BASALTS

Raul A. Deju and George C. Evans

Basalt Waste Isolation Project
Rockwell Hanford Operations
Richland, WA 99352

INTRODUCTION

The Basalt Waste Isolation Project, operated by Rockwell International for the United States Department of Energy, is part of the National Waste Terminal Storage Program. The program, as presently structured, is aimed at assessing the feasibility and providing the technology needed to design and construct a repository for storage of commercial radioactive waste in the extensive basalts beneath the Hanford Site of the United States Department of Energy.

The Hanford Site of the United States Department of Energy has been the location of nuclear energy work for several decades, and is centrally located within the Pasco Basin in the midst of the large basaltic flows (Fig. 1.). The previous and present commitments of this site to the nuclear field make it a logical place to be examined for its potential use as a repository site.

The program is presently in the research and development phase to assess feasibility. Geologic site selection studies are to be completed in September 1981, to allow a feasibility decision at that time, whereupon if feasibility is proven and the United States Department of Energy goes forward with the project, we would move into the licensing phase of the project. Ultimately, it would be up to the United States Nuclear Regulatory Commission to grant a construction license and, later on, an operating license.

During the research and development phase, the program includes seven areas of study; namely, geosciences, hydrology, multiple engineered barriers, test facility design and construction, engineering testing, systems integration, and preliminary repository engineering

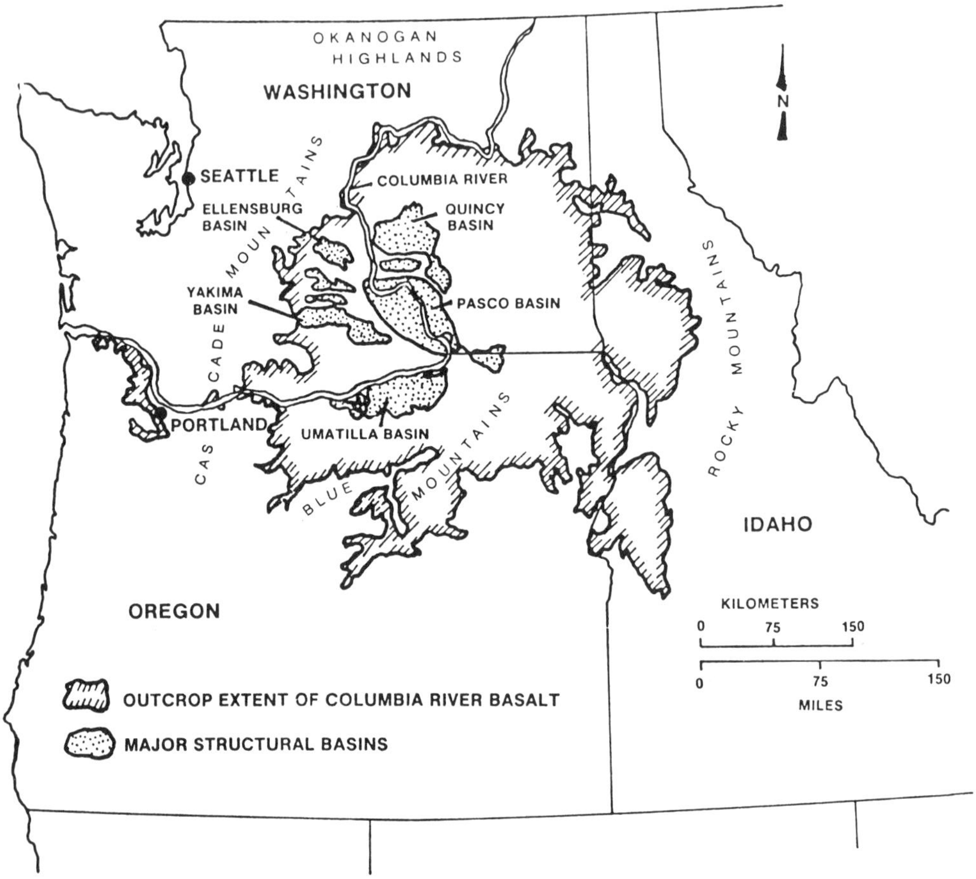

Fig. 1. The Columbia Plateau

design. In the next few paragraphs, I will attempt to detail the progress made to date in each of these areas.

GEOSCIENCES

The Geosciences Program is aimed at gathering the data required for selection and evaluation of potential repository sites in basalt. These studies lead to identifying repository target areas and then when a site is selected they will thoroughly characterize the area to determine in great detail the extensiveness of individual basalt flows, the stability of the region, and the presence or absence of potentially hazardous geologic structures.

Studies to date have included reconnaissance regional studies and local studies of a more intensive nature within the Pasco Basin where the Hanford Site is located. As part of the regional studies, a survey of published and unpublished documents concerning the geology of the Columbia Plateau was completed. In addition, mapping of the basalt within the Columbia Plateau and the overlying late Cenozoic sediments has been conducted and is essentially completed. The stratigraphy of the Columbia River Basalt Group has been revised and its understanding considerably expanded.

The Pasco Basin studies have included a definition of the local stratigraphy of the basalts and an assessment of the viability of using the chemical and magnetic properties of the basalts for stratigraphic definition. Extensive mapping has been conducted and continues to be conducted in structurally significant areas of the Pasco Basin. Drilled holes within the Hanford Site and vicinity are being used to understand the structural relations of individual basalt flows. The geologic mapping in the field is also aided by geophysical studies, including seismic, aeromagnetic, magnetotelluric and gravity prospecting.

Prior to site selection, all of the geological information will be incorporated into an extensive report, which will be subjected to extensive peer review.

HYDROLOGY

The Hydrology Program parallels the geologic studies. The program has emphasized the gathering of data to characterize the groundwater systems underlying the Pasco Basin and the modeling of such data so as to evaluate radiocontaminant migration to the biosphere. Studies to date indicate that the basalts in the Pasco Basin are extremely tight with the fractures in the deeper basalts being generally clay-filled.

The Hydrology Program has heavily depended on the existence of test holes at Hanford, and in its vicinity for obtaining all needed property measurements. Holes tested range in depth from 3,000 to over 10,000 feet (Fig. 2.). Additional holes are being planned for the very near future.

These holes form the basis of information being used to model the hydrology of the Pasco Basin. Rockwell Hanford Operations interpreted drill stem tests from Well RSH-1 southwest of Hanford. Our analysis shows hydraulic conductivity values between 10^{-7} to 10^{-9} cm/sec within the dense basalt flows themselves. Science Applications, Inc. conducted more sensitive tests and found hydraulic conductivities ranging from 10^{-7} to 10^{-13} cm/sec in the zones tested. At present, they are conducting additional tests. In parallel with

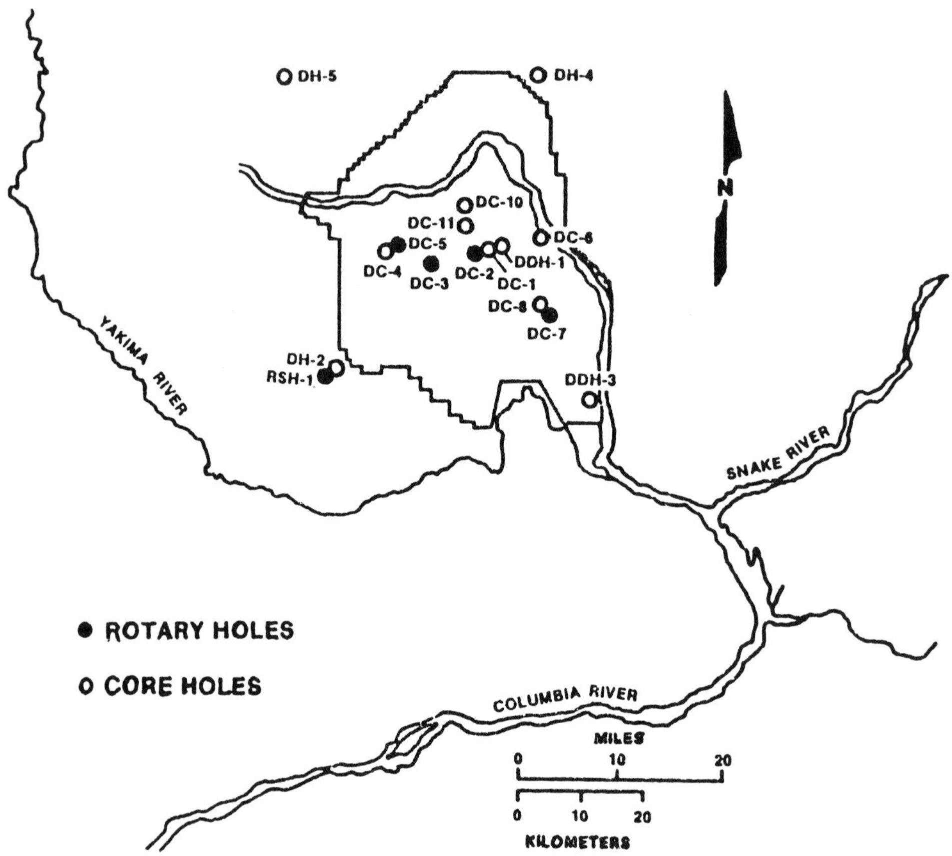

Fig. 2. Drilling Locations

the field program, we are conducting model adaptation and sensitivity analysis of our computer model for the hydrology of the basin.

MULTIPLE ENGINEERED. BARRIERS

The emplacement of nuclear waste in a geologic repository may cause physiochemical perturbations to the surrounding environment. The Multiple Engineered Barriers Program attempts to identify, from a physiochemical standpoint, the features of various barriers to transport of radioactive contaminants. The program is looking at four potential barriers: the waste form, the container, the overpack, and the rock. In addition, a borehole plugging system is being developed as a final barrier once the repository is sealed and abandoned.

The Multiple Barriers Program includes (Fig. 3.) a material characterization task where all the components of the multiple barrier system are analyzed, and a reaction study where radionuclide sorption tests and hydrothermal tests are being conducted to provide the source terms required for radionuclide transport calculations. Decision analysis techniques are being used to aid in optimizing the selection of a multiple barriers system.

TEST FACILITY DESIGN AND CONSTRUCTION

During the early phase of the Basalt Waste Isolation Project, the need for in situ thermal and mechanical testing of basalt was identified. This need for engineering data, to qualify basalt as a repository medium and to provide the basis for repository design, could be met by construction of a near-surface, multipurpose facility for in situ testing of basalt. Further, the facility would serve for the demonstration of placement, storage, and retrieval of nuclear waste canisters in an underground basalt environment and the demonstration of the effectiveness of waste monitoring systems. Detailed planning on this facility was started in October 1977. Construction of the facility began in June 1978, and the first test is scheduled for a mid-1980 startup. Work is presently on schedule.

The Near-Surface Test Facility is located approximately 340 feet below the land surface, and approximately 500 feet into the Pomona

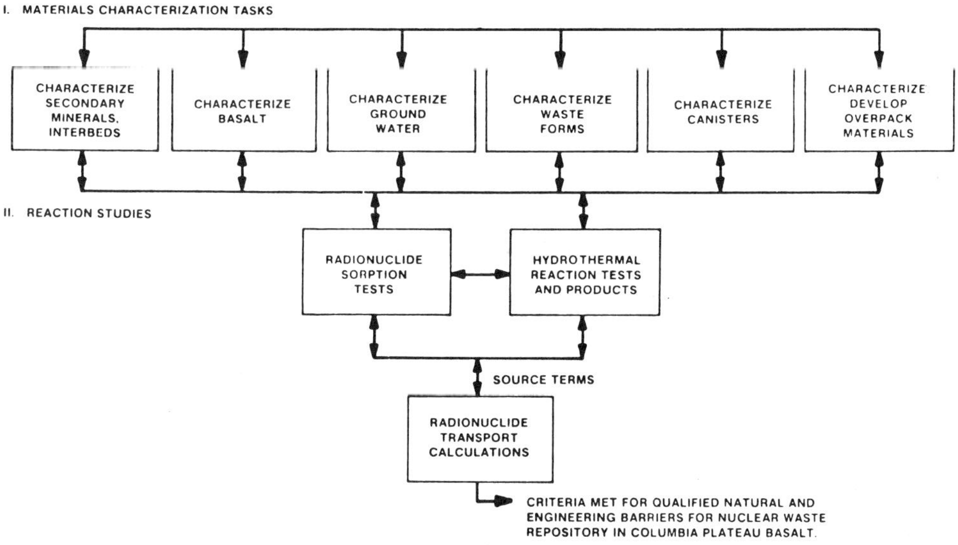

Fig. 3. Simplified Description of Engineered Barrier Activities

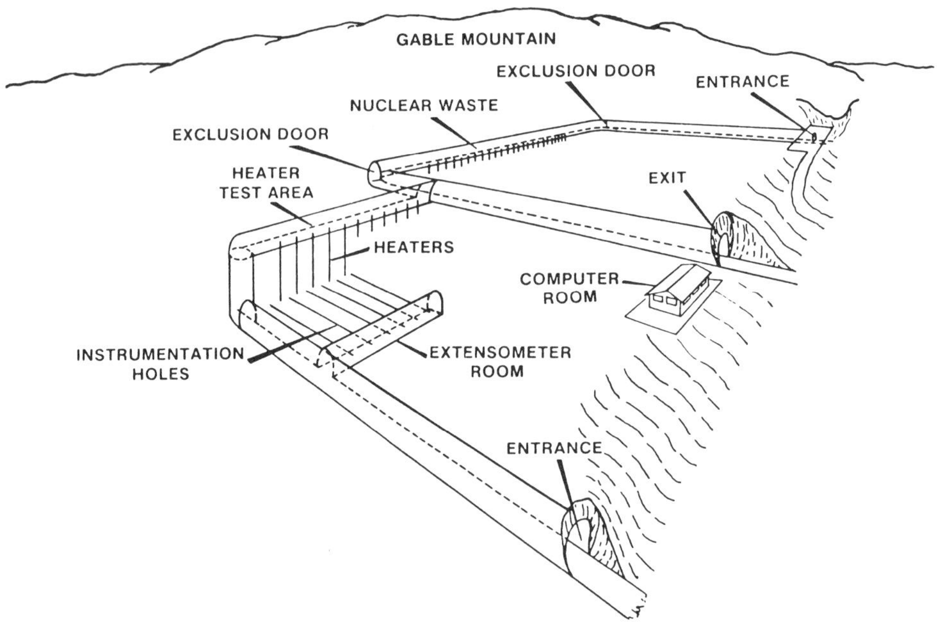

Fig. 4. Near-Surface Test Facility

basalt. This allows a sufficient portion of basalt to remain undisturbed below the test room for the conduct of the test. The facility consists of three entrance tunnels and two test areas; one for heater simulation and one for spent fuel tests (Fig. 4.). Results from the heater tests will provide data on borehole decrepitation, thermal stability, structural integrity, temperature, and displacement fields, and the influence of fractures and joints upon in situ basalt properties. These studies will provide the basis for the design of key repository elements such as canister storage, borehole criteria, borehole liner performance, acceptable waste canister power levels, storage borehole array criteria, and repository step-loading evaluation. In addition, the in situ testing program should aid in the verification of mathematical models of the thermomechanical response of a repository. The spent fuel testing will serve to study the combined effect of radiation and thermal response due to decay. In addition, a combined heater-spent fuel array will be used to heat a large room and examine scale effects and develop room stability considerations.

Tunneling work is now complete. The heater test area required the excavation of approximately 21,000 cubic yards of material from the 2 portal areas and the development of approximately 1,800 feet of underground workings. Tunnels in the heater test area vary in size from an 8-foot diameter for the east access tunnel to 23-foot diameter for the accelerated heater test room and the extensometer room.

The nuclear waste test area required the excavation of an additional 25,000 cubic yards of material from the portal area and another 1,200 feet of underground excavation. The test room in this area is 28 feet high. All test areas have unsupported roofs. A flash coat of shotcrete, however, has been added to the roof for rock stability.

ENGINEERING TESTING

The Engineering Testing Program includes the determination of engineering properties of basalt needed for conceptual engineering design studies and qualification of basalt as a medium for storage of nuclear waste. The program began with an extensive literature search of the engineering properties of basaltic rocks. After this literature search was completed and existing data were tabulated, samples from numerous core holes corresponding to many depth horizons from wells within the Hanford Site were subjected to extensive thermal and mechanical tests. In addition, the test program has been designed to gain basic input data from numerical models, as well as to determine lateral and vertical variations of properties between individual basalt flows and within individual basalt flows.

SYSTEMS INTEGRATION

The Systems Integration Program, as the name implies, is responsible for the use of systems and decision analysis techniques in the integration of the research and development effort. Primary areas of concern involve repository siting: What constitutes feasibility of building a repository in basalt? when is a site qualified? and what are the licensing concerns and requirements? Work in all areas are under way.

PRELIMINARY REPOSITORY ENGINEERING DESIGN

One final area of involvement has been the development of the preconceptual design of a repository in basalt. The preconceptual design was recently completed including five areas: surface facilities; access and isolation shafts; subsurface facilities; waste handling systems; and surface and subsurface service systems.

The preconceptual design also covers the final stage of operations, including the ultimate plugging and abandonment of the facility.

The completed preconceptual design will serve to produce a first-order cost estimate for the repository, schedules for construction, and a preliminary safety assessment.

CONCLUSION

At this time, we have made great progress toward establishing feasibility of using the deep basalts beneath the Hanford Site for underground disposal of commercial nuclear waste. However, some complex issues, such as the nature of the deep hydrology and in situ thermal response of the basalts remain to be fully assessed. These and other related issues are the subject of present and future studies.

GENERAL CRITERIA FOR RADIOACTIVE WASTE DISPOSAL

Margaret N. Maxey*, Laurence I. Moss**, Burdon C. Musgrave+, Goldie B. Watkins++

*University of Detroit, **Energy/Environment Consultant, +Idaho National Engineering Laboratory, ++New York State Department of Health.

INTRODUCTION

Throughout the Department of Energy in-house laboratories and externally funded efforts, there is a series of programs aimed at the development of technology for the conversion of radioactive wastes to appropriate forms and for the placement of these materials into repositories in the earth's surface. To date this methodology development has been conducted largely without the benefit of officially approved or generally accepted criteria, goals or objectives. Agreed upon criteria for the disposal of radioactive waste materials are needed to ensure protection of the biosphere from undue risks.

Several versions of criteria for waste disposal have been offered. Most of these have been limited through one or more problems: (1) lack of a defined objective, (2) difficulties with categories of waste, and (3) lack of a basis or reference by which to judge the success of the standard; that is, to judge how much protection the standard provides. The lack of well defined objectives leads to questions such as, "Are we concerned about management or disposal?" and "Are we to provide criteria for technical specifications?" A waste classification system is needed to provide the rationale for selecting the relative amount of protection from the potential hazards to be offered by the disposal system chosen. Decisions about these first two problems will eliminate the third problem, the selection of a reference for judging the adequacy of the protection provided by the system chosen.

A working group was gathered to develop a draft ANSI standard. Members were chosen from a variety of perspectives and disciplines. Their personal values cover a wide spectrum. The membership draws from a variety of affiliations, including government (state and federal), research, industry, academic institutions and the private sector.

APPROACH

As a starting point, we reviewed several published or proposed criteria. The majority of principles and criteria contained in the literature examined are more accurately defined as statements of concern rather than as criteria for performance. In our early discussions, special attention was given to assuring that the many concerns displayed by the various groups were understood and addressed. To guarantee this, our criteria were written with the following conditions in mind:

> Significant quantities of radioactive materials currently designated as waste do exist. These must be classified as to level of relative toxicity and must be disposed of in a manner that ensures that the biosphere is adequately protected with relation to its supply of food, water and air. The disposal site selected for each class of waste must be characterized adequately to permit prediction of the waste material behavior in that site and to allow the impact on the environment to be established.
>
> This standard will provide the criteria that shall be met for safe disposal and isolation of radioactive waste in the earth's crust until the relative toxicity of the waste is below that for which concern is no longer justified. At all times the applicable radiation protection regulations, including the ALARA principle (1), will be met both during operations and throughout the period of concern for the repository's existence.

We reviewed these two conditions repeatedly in the discussions of the specific criteria. Constant attention was paid to how the principle of ALARA was to be applied. In this regard we decided that discounting radiation risk to the present generation cannot be justified. We feel that the ALARA principle must and can be met during the operational period and throughout the ultimate period of concern.

---

(1) ALARA is defined as: "as low as reasonably achievable, taking into account the state of technology and the economics of improvement in relation to benefits to the public health and safety and other societal socioeconomic consideration" (as defined in 10 CRF Section 50.34a).

The standard's scope was established to apply to the final disposal of radioactive waste materials in a manner ensuring that the characteristics of the site, the method of disposal and the properties of the waste form are jointly considered so as to meet all applicable safety and environmental protection regulations. This is intended to cover all types of radioactive wastes, including transuranic wastes, mill tailings, low beta gamma wastes and the solidified high level wastes. In consideration of conditions acceptable for long term disposal of high level radioactive waste, the questions of institutional integrity and continuity are always raised. These questions, often irrespective of technical problems, require assurances not otherwise asked of societal decisions. While these assurances are technically predictable in nature, they can be answered only with technical experience plus projections into the future. To some who request this type of assurance, this carries far too great an uncertainty to be acceptable.

Because of the possible operation of "Murphy's Law," assurances are sought about all aspects of the system. These include, for example, assurance that 1. the waste forms are stable, 2. the geology is stable, 3. the operational requirements will be met and 4. the caretaker will continue to take care. Technically defensible answers can be provided for 1. and 2. However, concerns related to 3. and 4. are due to a misconception of the nature of the way the institution must function to provide the required control. The question of institutional control is a question of the continuity of important functions. Among these functions is the operation of energy conversion systems of which waste emplacement is a subtask. Successful waste emplacement after the period of beneficial use of nuclear energy does not in itself justify the extensive use of nuclear power. The cost of ultimate disposal is a small fraction of the value of the energy produced on any time scale.

Man is the most probable cause for the loss of functions necessary to operate energy conversion systems. The most probable loss would be caused by nuclear war which would create considerably more hazard than unattended nuclear waste ever could. However, as long as any form of society continues, adequate functions will exist to provide the control required for operation of the repository. After the repository is sealed in any reasonable geologic location, the only institutional function required is human intelligence to recognize the deposit if it is ever encountered.

A question often posed is how much protection is enough. This requires that a judgment be made about when further risk reduction does and does not warrant significant expenditures. This is a difficult judgment to make and certainly is not one for which universal concurrence is likely to be attained. Comparison among different types of risks also requires some weighing of relative comprehension of risks. The comparison is eventually made, if at all, by a regulatory body, which may not act according to well analyzed risk but may act in a political manner to reflect an impression of the public's collective apprehension about that risk. This leads, for example, to the absurd position that more effort per health effect should be spent to reduce risks from radiation from nuclear power than from medical uses of radiation. In this standard, we allow a probabilistic risk comparison as one approach to the evaluation of a repository. We offer as well the alternative of comparison with other risks of the same kind.

Concern is often expressed about the intergenerational problem of the "legacy" of hazardous wastes left for future generations without their consent. A different emphasis is stated by William Mills (2), "I have no desire to provide extreme unwarranted protection against radiation risk and in doing so pass on to my offspring and future generations a legacy of foolish spending and unnecessary debt." Relative risk-benefit analysis can provide a basis for judgment. Similarly, as is often pointed out, people are willing to accept a much higher probability of harm from active (voluntary) risks than from passive (involuntary risks to which people are subjected but have no choice) risks.

Comparison with naturally occurring toxic materials as used in the draft standard accommodates both of these concerns. The net risk to future generations can be established when waste disposal repositories are compared with other very similar passive risks.

WASTE CATEGORIZATION SYSTEM

A waste categorization system is vital to the objective of matching the disposal method and location with the characteristics of the waste. Each class of waste demands specific methods and locations to ensure that the biosphere is adequately protected with relation to its supply of food, water and air and that inadvertent intrusion of humans or animals in the earth's crust. Classification of waste is as follows:

> Class 1: Radioactive wastes under this classification include radioactive materials that clearly constitute a hazard if not disposed of with great care. Therefore, these shall be assigned to an engineered deep geologic repository or disposed of in a manner such as to ensure an equivalent degree of isolation.
>
> Class 2: Radioactive wastes in this category are shown by material properties, specific radioactivity and nuclide distribution

(2) Comments at the Eleventh Annual National Conference on Radiation Control, Oklahoma City, OK, May 6, 1979.

to be adequately contained in shallow geologic burial or disposed of so as to ensure an equivalent degree of isolation.

Class 3: Wastes in this classification contain radioactivity in quantities or concentrations so small that continued control or evaluation for radiation protection is unjustified.

DISCUSSION

Three approaches were developed for judging the adequacy of disposal concepts.

The first alternative ("acceptable risk") reflects, in a conservative manner, society's implied willingness to expend resources to avoid the statistical occurrence of injury, early death or adverse environmental impact. The risk of an occurrence is the consequence-times-the-probability. The variation in the willingness of members of society to allocate resources depends upon these perceived probabilities and consequences.

The second alternative ("ore body comparison") provides high confidence that the hazards are at least as low as the risks from the reference natural uranium ore body. This comparison requires acceptable models.

To simplify the analytical problems, a third approach ("three-stage ore body comparison") was developed in which a comparison with a reference natural uranium ore body would be made from each of three aspects: stability of the waste form/ore, integrity of the host medium and isolation from the biosphere. The third alternative is similar to the second. However, here a comparison is made of each of the three aspects instead of comprehensive probabilistic models being used. If the proposed waste form and disposal site were judged to be safer than the reference natural uranium ore body in each of the three respects, then a more comprehensive analysis would not be required. This third alternative provides high confidence that the contribution to the risks arising from (1) the dispersibility of the waste form in relation to its toxicity, (2) the lack of integrity of the host medium, and (3) the lack of integrity of surrounding media are each less than the contribution to the risk arising from the corresponding aspect of the natural uranium ore body.

Specific criteria were developed in the areas of protection, waste forms, demonstration of site integrity, site suitability, retrievability, and monitoring and identification. A summary of the specific criteria follows.

Protection criteria require the assignment of properly classified waste to appropriate repositories consistent with the potential hazard. The significance of the potential hazard from the classes of waste requiring control can be established through comparison with other hazardous waste disposal-risk reduction actions society undertakes and understanding of the relative hazard of naturally occurring materials in the environment.

When the mobility and the toxicity index (per unit mass) of the Class 1 waste disposed of in any deep geologic formation is comparable to or less than the mobility and the toxicity index of the reference ore body (3), then no further consideration of hazard from the waste shall be deemed necessary. When the access to aquifers and the toxicity index of the Class 2 waste disposed in any shallow site become equal to that of naturally occurring toxic materials in the undisturbed earth's crust in the region (80 km radius) surrounding the site, the hazard from the site shall be considered negligible.

For the second alternative, the ore body comparison, the estimated hazard to mankind from any waste disposal site for any generation during the period of concern shall not be greater than that from the reference natural uranium ore body. The potential hazard to any individual member of the general public shall be no greater than ten times the hazard from the reference natural uranium ore body or restricted to the then-current requirements, whichever is lower.

The chemical and physical forms of the radioactive waste provide the waste with its barrier to dispersion and should be engineered to match the toxicity of the waste. Class 1 wastes should be constrained in a solid matrix; the waste form shall be selected so that it and its reaction products do not induce excessive stress through continued reaction in the host medium. In application of the third approach for wastes that are higher in toxicity than is a comparable mass of the reference natural uranium ore body, the waste's self-containment parameters must be increased proportionately.

The integrity of the selected disposal site shall be assessed based on accepted predictive models which evaluate the potential release of hazardous material and the characteristics of the site host media. The computer codes used to describe these models and test the site characteristics shall employ realistic physical properties and material performance characteristics rather than apply factors of conservatism to these values. The repository site shall be shown to be free of significant viable faulting. Conservative factors should

(3) In the full standard document as it is being reviewed through the standards development system, the reference natural uranium ore body is fully defined. This definition has been developed from the known characteristics of uranium ore bodies in the Western United States.

not be used in calculating expected values for ingestion and inhalation quantities, probabilities or total populations exposed.

The model described by the codes shall be founded on physical principles accepted by consensus of experts in the appropriate fields. It is recognized that predictive models require estimation of certain parameters and the probability of deviations from the estimated values. Uncertainty in modeling and estimating values of parameters shall be treated explicitly, and this quantification of uncertainty shall be carried through the calculations. The models should be designed if practical for field testing to verify the engineering margins of safety in the repository design.

During assessment of the potential repository site and disposal mode, evaluations of the existing natural containment features shall be made. Where applicable, potential radionuclide migration from waste disposal sites should be predicted and quantified. Engineered barriers may be used to reduce the probability of potential release of radionuclides from the repository sites. However, expenditures of effort to increase the integrity of the barriers shall be in proportion to the significance shown appropriate to them in the pathway evaluations.

The geologic medium shall be stable enough not to be significantly degraded by exploration, construction, emplacement of waste and post-emplacement sealing by the radiation, chemical or thermal properties of the waste.

Suitable waste repository sites shall be selected to avoid use of valuable natural mineral resources or environmentally sensitive areas. The site shall be selected where the probability of inadvertent human intrusion in the search of useful resources is low. The extremities of the repository shall be marked at the surface. Retrievability of the waste during the operational period for any repository should be maintained until safe repository performance is demonstrated. However, the long term integrity of the repository and its contained waste shall not be degraded by any consideration of retrievability. Therefore, the ease of retrievability must be weighed against the possible reduction in repository integrity.

CONCLUSIONS

We have presented the approach and the rationale behind the criteria we have developed. Supplementary numerical criteria and technical guides and regulations will be required for execution of a waste disposal approach and to achieve a balanced and technically valid approach to waste disposal criteria. Our criteria are an attempt to avoid the pitfalls of other criteria offered. We sought to achieve a balanced and technically valid approach by providing application of the nuclear safety defense-in-depth approach to a passive (that is, not mechanically operating) system. Through our criteria, we also sought to achieve this balance by ensuring that requirements for performance of the engineering design are specified, sensitive or key parts of the system are strengthened, and conservatism in the design is specified to avoid the consequences of major failures. Finally, an additional conservatism is provided for should it be needed to protect against the possible results of any failure.

CHARACTERIZATION OF WEATHERED GLASS BY ANALYZING ANCIENT ARTIFACTS

Maureen F. Kaplan

The Analytic Sciences Corporation
Six Jacob Way
Reading, MA 01867

ABSTRACT

Ancient artifacts, having been buried for significant periods of time, can be viewed as outcomes in long term durability experiments. This study focuses on ancient man-made glass -- a material which came into use about 3,500 years ago. The weathering products found on these glasses are described and a method for correlating them with glass composition is presented. Relevance to current work on vitrified nuclear waste forms is discussed.

INTRODUCTION

Ancient glass first appears about 1500 B.C. in the areas of modern-day Egypt and Iraq. It was a luxury material and was often used as a substitute for precious stones. But with the advent of glassblowing (approximately 50-0 B.C.), it became inexpensive, and production centers are found throughout Europe and the Mediterranean. The type of glass manufactured was a soda-lime glass; this did not change until about 1000 A.D. when Northern and Western Europe began to produce a potassium-lime glass. The medieval stained glass windows of Europe are predominantly potassium-lime in composition.

Lead glasses were made in the Far East (beginning about 200 B.C.) and during the Islamic period in the Near East. However, these occurrences are rare and the corrosion products are insufficiently known to be included in this study. This work, therefore, will discuss only the weathering products on ancient alkali-lime glasses.

These glasses have either undergone burial or been exposed to the elements for hundreds to thousands of years. An analysis of these weathering products which have formed during long term exposure can illuminate the changes which vitrified nuclear waste forms might undergo during their burial. The following sections describe the types of weathering products found on ancient glasses and outline a method for correlating them with glass composition.

WEATHERING PRODUCTS

Ancient glasses display a wide variation in weathering products. These may be divided into roughly five categories:

Weeping: Some Renaissance and Baroque goblets appear to exude droplets of water when stored under normal museum conditions, hence the name "weeping."

Crizzling: Crizzling is the term used to describe the formation of a network of tiny cracks covering the surface of the object. The cracks do not continue all the way through the vessel walls, but the network may become so dense that the object appears opaque and tiny scales of glass flake off during handling. Like weeping, crizzling also occurs primarily on Renaissance and Baroque glass. Both phenomena are sometimes discussed as two stages in "glass disease." An example of the swiftness with which crizzling can occur is discussed by P. B. Adams (1).

Layer Formation: Many ancient glasses have an iridescent surface, once thought to be a decorative technique lost to modern-day glassmakers. Investigations, beginning with the latter half of the 1800's, indicate that the weathered surface can be resolved into a laminar structure whose layers range from 0.3 μm to 15 μm in thickness. These layers consist primarily of hydrated silica, and the iridescence is caused by the interference of light between these layers.

There is a great deal of debate concerning the physio-chemical processes which form the layers. Brill (2) suggested that the annual shifts in temperature and humidity lead to layer formation. If so, the layers -- like tree-rings -- can be counted in order to obtain an estimate of the age of the object. Newton (3), however, indicates problems with this: a) layers form even when the glass has been submerged in the sea, where annual temperature fluctuations are minimal, and b) the laminar structure is not always uniform. The weathering layers may proceed deeper into the unaltered glass in some areas. These weathering "plugs" may be related to the presence of inhomogeneities in the glass, but further research is needed to be certain. A further distinction may be made within this weathering product. It has been noted that glass found in the East (e.g., Persia, not China) forms an enamel-like layer, while material in the West is often more flaky.

It should be noted that neither author doubts the ability to obtain samples where the original surface of the glass still exists. The weathered layers are porous, delicate and easily detached from the unaltered glass, but the silica matrix has not completely dissolved.

Pitting: Two types of pitting occur, one associated with weathering layers and one without. When weathering layers are mechanically detached from an object, areas of deeper layer penetration may leave pits. Remnants of the weathering layers may be seen in this type of pit.

The second pit clearly involves the complete dissolution of the silica matrix and occurs on the potassium-lime medieval stained glass windows. X-ray diffraction indicates that the powdery white substance found in the center of the pit is gypsum ($CaSO_4 \cdot 2H_2O$).

Crust Formation: Weathering crusts show no laminar structure, are opaque, and appear to totally alter the glass. X-ray diffraction indicates the crust consists of a mixture of gypsum and syngenite ($K_2SO_4 \cdot CaSO_4 \cdot H_2O$) with no trace of the silica remaining (4, 5). This weathering product, like the second type of pitting, is restricted to medieval windows.

CORRELATION OF WEATHERING PRODUCT AND CHEMICAL COMPOSITION

While the weathering products have been investigated for several decades, it has only been recently that attempts to correlate the weathering products with glass composition have been made. The work has concentrated on medieval glasses because of their poor condition, and the European countries are interested in preserving this part of their heritage. One study (6) points out that chemical analyses of ancient glasses are traditionally published in weight percentages, while the chemistry will act according to mole percentages. Of the several hundred analyses of stained glass, only fifty were found where the glass was fully analyzed and its condition was adequately described. The resultant plot of these glasses is given in Fig. 1. and general patterns can be described. Glasses which form crusts tend to fall at the bottom part of the diagram, while those which pit fall slightly higher and to the right (i.e., they contain more silica). The unweathered glasses tend to have even higher amounts of silica.

The promise of this approach warranted further investigation. It is known, empirically, that the older sodium-lime glasses are more durable than the medieval glass. Crizzling glasses are the least

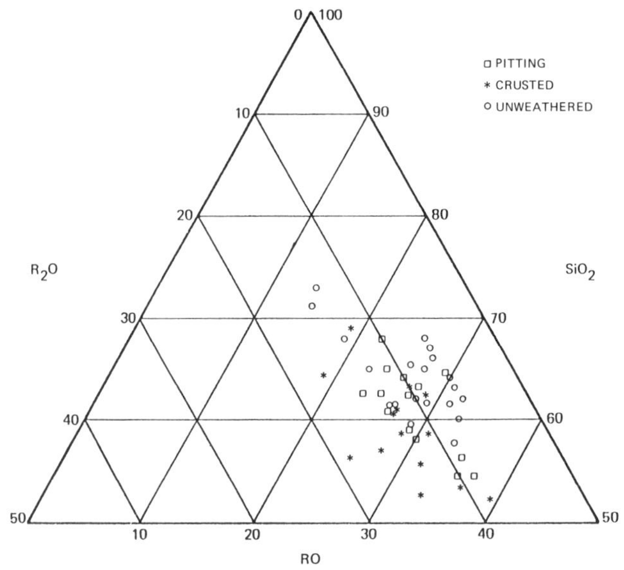

Fig. 1. Chemical Composition of Medieval Glasses, Mole % [After (6)]

durable of the glasses discussed here. These should fall on different areas of the graph if this approach is appropriate. Chemical analyses of crizzling glasses were already available in the literature (7). The spectrographic analysis of several hundred premedieval glasses had been undertaken by Sayre in order to determine whether glass from different manufacturing areas and time periods could be chemically identified. Five compositional categories were distinguished on the basis of weight percentages, and the discriminating elements were magnesium, potasium, manganese, antimony and lead (8). Only the data for these elements were published, but Sayre kindly provided the complete analyses for this investigation.

The chemical data for these samples were transformed into mole percentages and plotted on ternary diagrams. The results are given in Fig. 2.-5. and summarized in Fig. 6. The shaded area delimits the range of chemical composition found in ancient glasses and the weathering products associated with different compositions are marked.

SUMMARY AND COMMENTS

Ancient glasses display a wide range of weathering products, and these products appear to be correlated with the chemical composition of the glass. Glasses which crizzle when brought into a dry museum atmosphere do not necessarily indicate that wet environments may be as or more conducive to glass survival than dry environments. As is

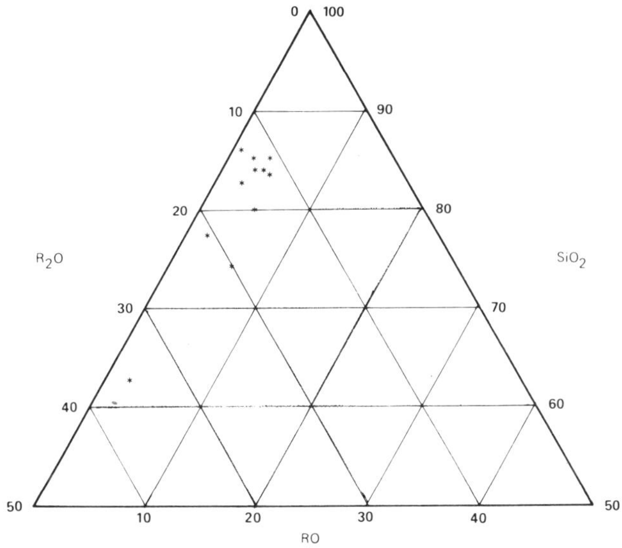

Fig. 2. Chemical Composition of Crizzled Glasses, Mole %

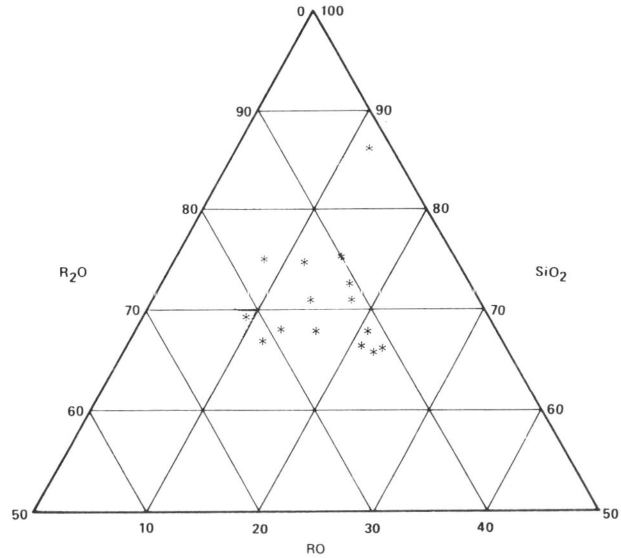

Fig. 3. Chemical Composition of Second Millenium B.C. Glasses, Mole %

evident in Fig. 6, these glasses have far too little lime in them for stability. However, these glasses can be used as an example of compositions to be avoided in nuclear waste forms.

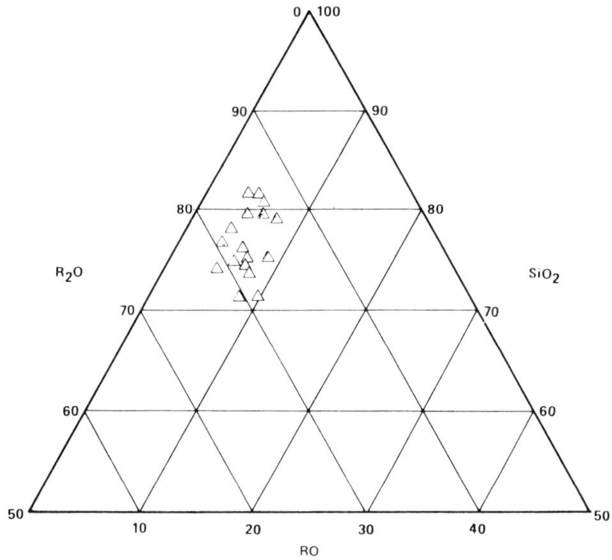

Fig. 4. Chemical Composition of Antimony-Rich (Hellenistic and Eastern) Glasses, 350 B.C.-50 B.C., Mole %

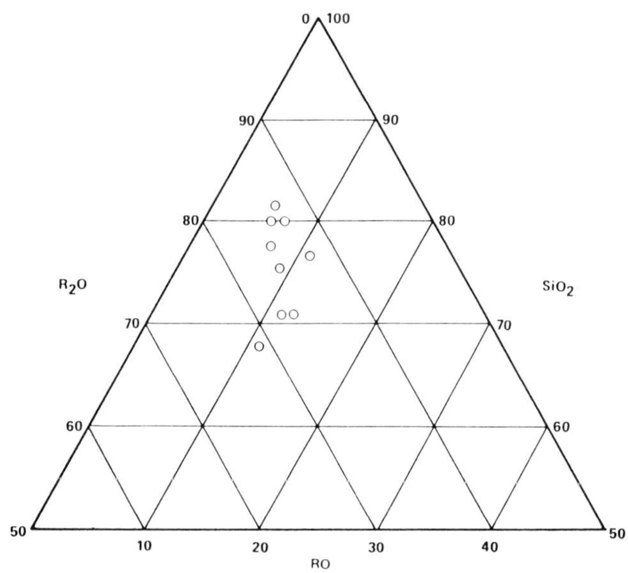

Fig. 5. Chemical Composition of Roman Glasses, 50 B.C.-350 A.D., Mole %

The recent literature on glass corrosion discusses two mechanisms: a) selective leaching or ion exchange and b) network dissolution (9, 10, 11). All of the weathering products discussed above

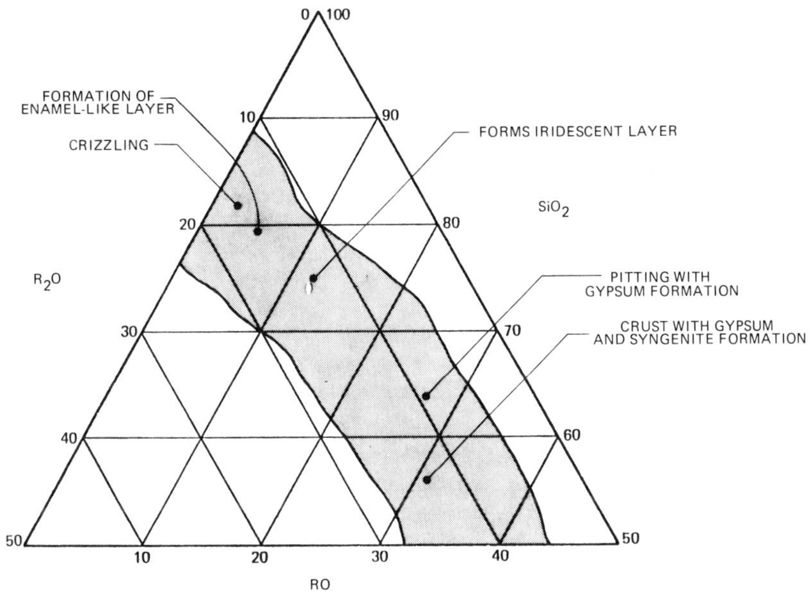

Fig. 6. Chemical Composition of Ancient Glasses and Associated Weathering Products, Mole %

appear to occur with the first mechanism, i.e., leaching of the alkalis. It is interesting to note that the same mechanism can result in such visibly different end products. In addition, network dissolution is apparent in some cases, but not all. For this reason, it is necessary to determine whether network dissolution of nuclear waste forms would occur under actual repository conditions.

Although the nuclear waste glasses are different in composition than ancient glasses, the latter can provide the following information:

1. Illustrate the less durable compositions, i.e., these should be avoided.

2. Document types of corrosion possible under actual burial conditions.

3. Illustrate the range of weathering products which occur even though the same physical mechanisms underlie each situation.

4. Illustrate that the complete dissolution of the silica matrix does not necessarily occur in all cases.

In this work, we have found that the formation of corrosion products is a complex process and that some of the ancient glasses are very durable.

It must be admitted that ancient glasses differ from those designed for the disposal of nuclear waste, but it must also be conceded that the laboratory cannot exactly duplicate burial conditions because of the time factor. The analysis of actual corrosion formed on ancient glasses through centuries or millenia of burial can provide adjunct information on the topic, and its potential should be neither over- not underestimated.

ACKNOWLEDGEMENTS

This work was supported by Battelle Pacific Northwest Laboratories under Subcontract No. B-22596-A-E.

REFERENCES

1. P. B. Adams, "Chemistry of Nuclear Waste Glass Reactions: Problems and Potential of Prediction," Scientific Basis for Nuclear Waste Management 1 (G. J. McCarthy, ed., Plenum Press, New York, 1979).
2. R. H. Brill, "The Record of Time in Weathered Glass," Archaeology 14, 18 (1961).
3. R. G. Newton, "Some Problems in the Dating of Ancient Glass by Counting the Layers in the Weathering Crust," Glass Technology 7, 22 (1966).
4. M. Perez y Jorba, G. Tilloca, D. Michel and J. P. Dallas, "Quelques aspects du phénomène de corrosion des vitraux anciens des églises francais," Verres et Réfractaires 29, 53 (1975).
5. R. Collonques, M. Perez y Jorba, G. Tilloca and J. P. Dallas, "Nouveaux aspects de phénomène de corrosion des vitraux anciens des églises francais," Verres et Réfractaires 30, 45 (1976).
6. C. J. Illife and R. G. Newton, "Using Triangular Diagrams to Understand the Behavior of Medieval Glasses," Verres et Réfractaires 30, 30 (1976).
7. R. H. Brill, "Crizzling--A Problem in Glass Conservation," Conservation in Archaeology and the Applied Arts (London, 1975).
8. E. V. Sayre and R. W. Smith, "Compositional Studies of Ancient Glass," Science 133, 1824 (1961).
9. R. W. Douglas and T. M. el-Shamy, "Reactions of Glasses with Aqueous Solutions," J. Amer. Cer. Soc. 50, 1 (1967).
10. D. E. Clark, E. Lu Yen-Bower and L. L. Hench, "Corrosion Behavior of Zinc-Borosilicate Simulated Nuclear Waste Glass," Symposium on Ceramics in Nuclear Waste Management, Cincinnati, OH, Technical Information Center, Oak Ridge, TN, CONF-790420 (1979).
11. A. Paul, "Chemical Durability of Glasses, A Thermodynamic Approach," J. of Materials Science 12, 2246 (1977).

AN INVESTIGATION ON THE VALENCY STATE OF MOLYBDENUM IN GLASSES WITH AND WITHOUT FISSION PRODUCTS

B. Camara, W. Lutze* and J. Lux

Institut für Werkstoffwissenschaften III der Friedrich-Alexander-Universität Erlangen-Nürnberg
Martensstraße 5, D-8520 Erlangen
*Hahn-Meitner-Institut für Kernforschung Berlin GmbH
Dept. of Chemistry and Reactor
Glienicker Straße 100, D-1000 Berlin 39

INTRODUCTION

Transition elements appear in glass and ceramics in several oxidation states, which depend on composition, temperature and atmosphere. Structure and properties of these elements depend on the oxidation states. Investigations have shown that borosilicate glasses, which contain fission products and are melted in an oxidizing atmosphere, sometimes show separation of alkali molybdates. If these glasses are melted under reducing conditions (1), the solubility of the molybdenum increases. Knowledge of state and structure of molybdenum in glass is therefore highly significant. As yet, investigations in this direction are limited. References 2.-7. deal solely with phosphate glasses with the exception of 3. In the present work, the state of molybdenum in silicate glass is studied by ESR and optical spectroscopy, aiming to clarify the following:

1. Oxidation states of molybdenum and its reactions a) under oxidizing, b) under reducing conditions and c) as a function of molybdenum concentration.
2. Reaction of the molybdenum oxidation states with iron.

EXPERIMENTAL

In order to systematically study the state of molybdenum and its interaction with iron, simple silicate glasses were chosen. Later, glass DC-14, developed at HMI, a borosilicate glass contain-

Table 1. Composition and Color of the Glasses (in mol%)

SiO_2	CaO	Na_2O	MoO_3	Fe_2O_3	color oxidized	color reduced
73.85	6.46	19.48	0.21	-	colorless	colorless
73.56	6.48	19.54	0.42	-	colorless	yellowish
73.27	6.50	19.60	0.63	-	colorless	brown
72.98	6.52	19.65	0.85	-	colorless	dark brown
73.23	6.50	19.61	0.63	0.03	greenish	brown
73.04	6.52	19.05	0.63	0.16	green	green brown
72.82	6.53	19.70	0.63	0.32	green	blue green
72.35	6.57	19.80	0.64	0.64	dark green	blue green
71.88	6.60	19.91	0.64	0.97	dark green	blue

ing waste oxides, was examined. The glasses, which were to be reduced, were homogeneously mixed with 2 g of tannic acid before melting. All glasses were melted in Al_2O_3 crucibles in an electric oven at 1623 K for 2 hours. Table 1 shows compositions and colors of the glasses. For ESR analysis, crushed and sieved samples were used and the same initial weight was used for all measurements. The spectra were recorded with the X-band (9.5 GHz) 414 Bruker-Physik spectrometer. The resonance field was determined with the nuclear resonance probe B-NM 12 (Bruker-Physik); the frequency with the frequency counter 5340 A (Hewlett-Packard). All measurements were made at 298 K. For optical spectroscopy, glass samples were used and measured in transmission.

RESULTS AND DISCUSSION

Mo Glasses

Fig. 1. shows a representative ESR spectrum of the glasses melted under oxidizing conditions. The g-value is 1.906. The spectra for various MoO_3 concentrations are shown in Fig. 2. The intensity increase of the spectrum with the increase of the MoO_3 concentration proves that the spectrum is caused by molybdenum. Also, the intensity proportion of the weak resonances, (~ 1.547), is in accordance with the natural abundance of the isotopes Mo^{97} and Mo^{95}. It is known that molybdenum can appear in several oxidation states, e.g., as VI, V, IV, III. Of these states, Mo(VI) is the most stable. With the configuration $4d^0$, Mo(VI) neither results in an ESR absorption nor can an optical absorption in the VIS-NIR region be expected. This ion is therefore not responsible for the spectrum. Mo (IV) as a d^2 ion can also be excluded (9, 10). Only Mo(V) (d^1, s=1/2) and Mo(III) (d^3, s=3/2) can be considered for the ESR spectrum. However, also with systems with s=3/2, only the main transition -- $1/2 \rightarrow + 1/2$ -- has often been observed. This makes the practical

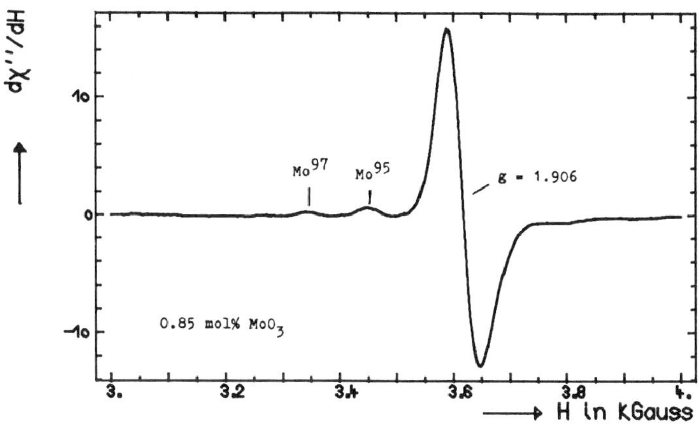

Fig. 1. ESR Spectrum of Mo in a Silicate Glass.

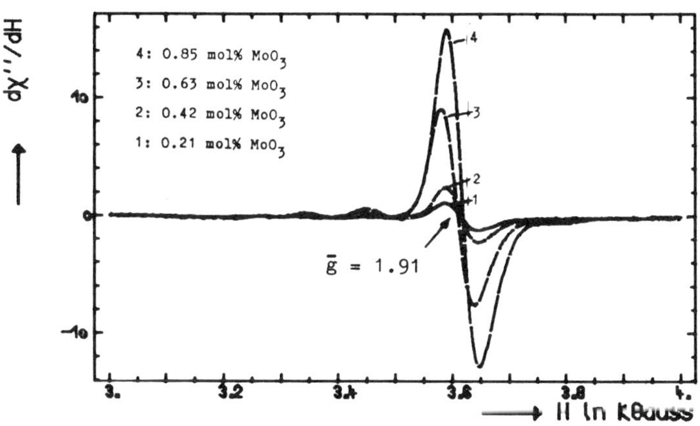

Fig. 2. Dependency of the Mo ESR Spectrum on the MoO_3 Concentration.

differentiation between Mo(V) and Mo(III) in glass especially difficult. All the same, much points to Mo(III) as being the cause of the ESR spectrum: g-values of 1.90 are found in general for Mo(III) (11). As already theoretically expected, larger g-values for Mo(V) are experimentally evaluated (5, 6, 11, 13). Our measurements on $MoCl_5$ and phosphate glass resulted in a g-factor of $\sim$1.945 and $\sim$1.922. A further argument for the existence of Mo(III) is also the brown color (2, 8) of the glasses, while Mo(V) very often colors blue (2, 7, 13). Fig. 3. shows the transmission spectra of the glasses melted under oxidizing conditions. Despite the increasing MoO_3 concentration, no absorption was observed. That means that the Mo oxidation state determined by the ESR (Mo(III)) appears in low concen-

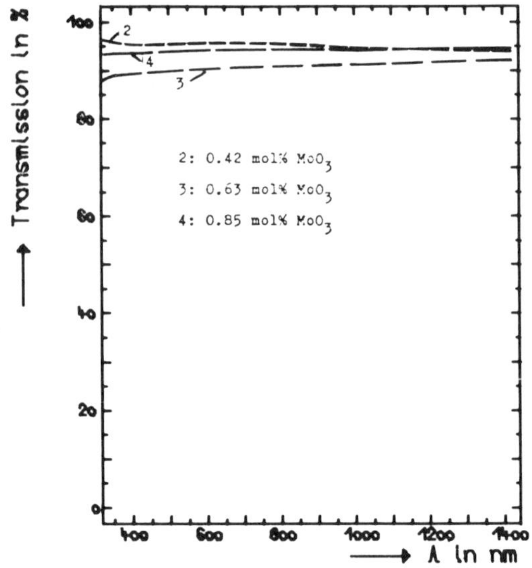

Fig. 3. Optical Spectra of Oxidized Silicate Glasses Containing Mo

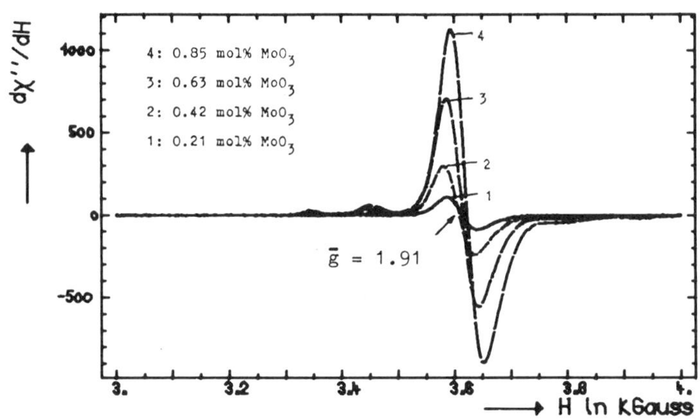

Fig. 4. ESR Spectra of Reduced Mo Glasses

tration. Its quantity is not sufficient to be measured by optical spectroscopy. The main quantity of molybdenum appears as Mo(VI). This appearance was confirmed by the reduced glasses which showed an ESR resonance of very high intensity. To keep the chosen initial weight constant, the signal height had to be reduced by the factor 100. Fig. 4. shows the ESR spectra as a function of concentration of molybdenum. Qualitatively, these spectra are equal to those of the oxidized glasses. In both cases, the same paramagnetic oxidation state of molybdenum is present. An increase of the reducing agent in the glass melt (more tannic acid) resulted in a signal of higher intensity but of the same quality. This result supports the supposition of Mo(III) aside from Mo(VI), also in the oxidized glasses. Fig. 5. shows the optical spectra of the reduced glasses. Absorption bands appear at 940 nm (10,638 cm^{-1}) and 435 nm (22,988 cm^{-1}). Both bands increase with the molybdenum content. The ligand field absorption energy Δ of Mo(III) in octahedral position was assigned by Parke (4) to 22,200 cm^{-1}. If the absorption energy of 22,988 cm^{-1} measured in this work is used for Mo(III) as the ligand field energy Δ of Mo(III) in the silicate glass together with the spin orbit coupling constant λ = 269 cm^{-1} (12), then the according g-value can be evaluated from $g = 2.0023 - 8\lambda/\Delta$. This amounts to 1.909 and is in very good agreement with the experimentally specified 1.906. This is another indication for the presence of Mo(III) instead of Mo(V). The second absorption band at 10,638 cm^{-1} is probably Mo(III) in tetrahedral coordination, as this energy is approximately 4/9 of the absorption energy 22,988 cm^{-1} for the octahedral configuration of Mo(III).

Glasses with Mo and Fe

The question was: How does the Mo react in the presence of iron? This point was examined for a constant content of molybdenum but for different iron concentrations. Fig. 6. shows the results for the oxidized Mo-Fe glasses; in comparison, Fig. 7. shows the iron resonances only. The interpretation of the iron resonances in terms of structure has already been discussed elsewhere by Camara (14).

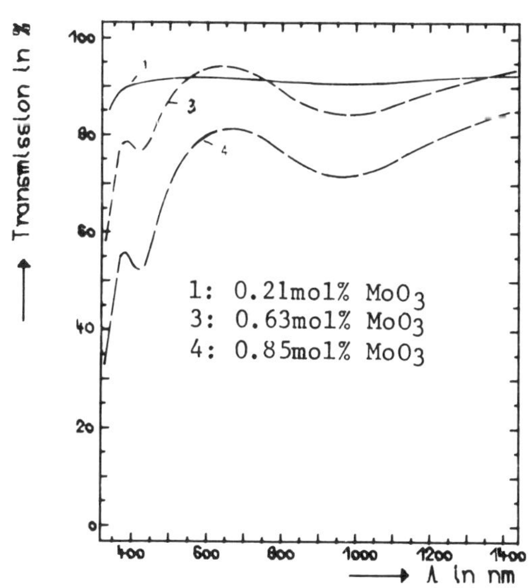

Fig. 5. Optical Spectra of Reduced Mo Glasses

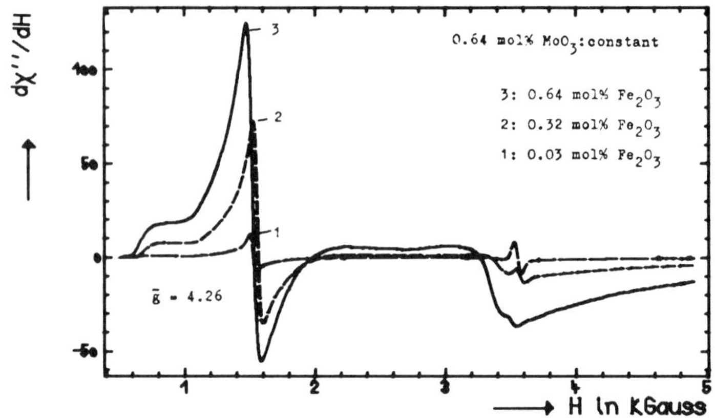

Fig. 6. Behavior of Mo in the Presence of Fe (Oxidizing atm)

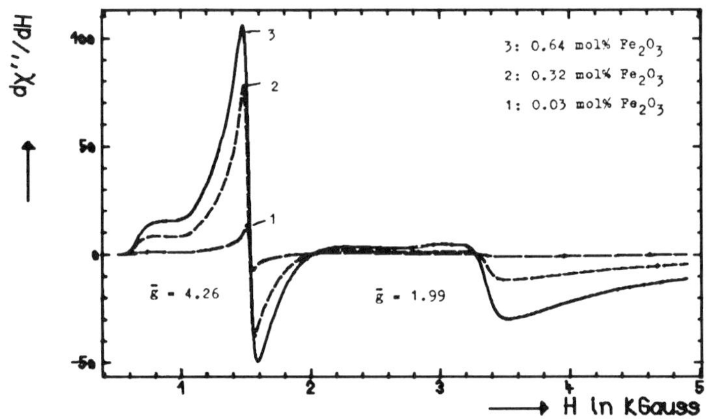

Fig. 7. ESR Spectra of Iron

Thereafter, the resonance of g = 4.26 belongs to Fe(III) in a network former position with the coordination number 4 and the resonance g= 1.99 to Fe(III) in a modifier position with the coordination number 6. The ratio of the occupancy of both structural positions changes with the iron concentration. As shown in Fig. 6., the resonance of the molybdenum changes when adding iron. With increasing iron content, the molybdenum resonance decreases, i.e., Mo(III) decreases. This effect was also examined with the reduced glasses. Fig. 8. shows the results of some chosen Fe_2O_3 concentrations. It is also seen here that with an increasing iron content in the glass, Mo(III) decreases. This result can also be clearly seen in the optical spectrum Fig. 9., especially regarding the absorption band 435 nm of the

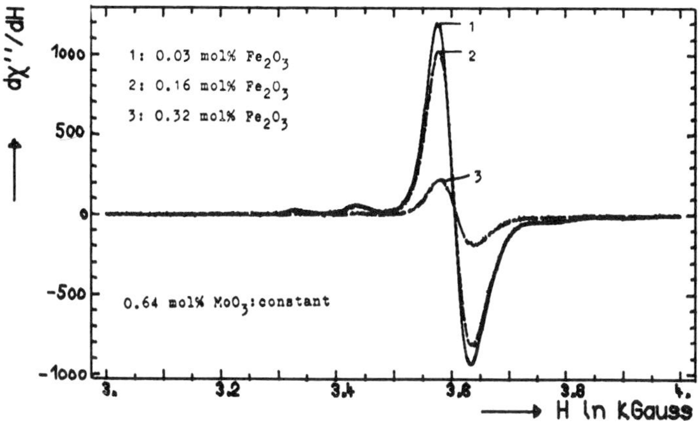

Fig. 8. Behavior of Mo in the Presence of Fe (Reducing atm)

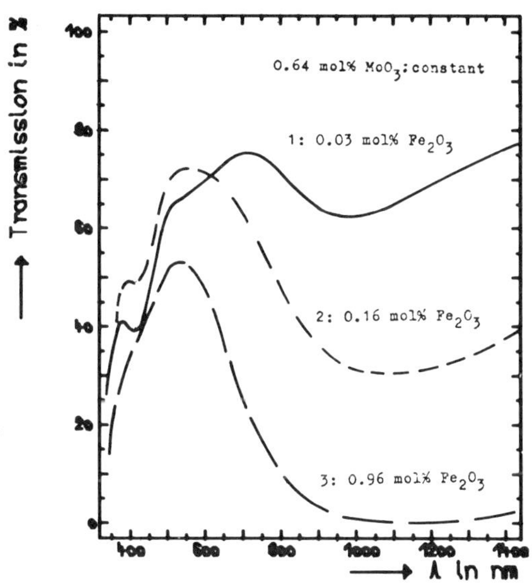

Fig. 9. Dependency of the 435 nm Mo Band on Fe Concentration

octahedrally coordinated Mo(III). At 0.03 mol% Fe_2O_3, this band can be seen well; whereas at 0.16 mol% Fe_2O_3, a decrease is clearly visible. This band behaves similar to the molybdenum resonance, clearly demonstrating that the same molybdenum is responsible for the ESR signal as well as for the optical band 435 nm. The results on the interaction of Mo and Fe show that there is a strong oxidation-reduction reaction:

$$Mo^{3} + 3\,Fe^{3+} \rightleftharpoons Mo^{6+} + 3\,Fe^{2+}$$

The system, which is still being studied, does considerably depend not only on temperature, atmosphere and concentration of the species but also on the presence of other polyvalent ions. Parallel to these reactions, there is an interaction of the "structures" (coordinations) of various polyvalences. Evidence of Fe(III) displacement into a modifier position by Mo(III) has been found.

Waste Oxide Glasses

After examining the silicate glasses, the DC-14 glass was analogously investigated. Fig. 10. shows the ESR spectra of an oxidized resp. reduced glass with the mixture ratio glass frit/waste oxides of 87.40 : 12.60. In both cases, the iron dominates with its known resonances at 1 - 2 KG (g=4.26) and 3 - 4 KG (g=1.99). The Fe(III) content in these glasses is so high that there should be no Mo(III). The spectrum obtained for a larger dilution of the waste products (99:1) for the oxidized glass is shown in Fig. 11. In this Fig., the spectrum of Mn(II) consisting of 6 lines (I-5/2) can be seen in addition to Fe(III). In the reduced glass, the manganese spectrum is not only more distinct, but also the formed Mo(III) can be seen (Fig. 12.). In the optical spectrum of the glass DC-14 (87.4 : 12.6), Nd^{3+}, Co^{2+} were identified but, as yet, no other transition metals. Overlapping bands and concentration effects of the individual species are the probable causes.

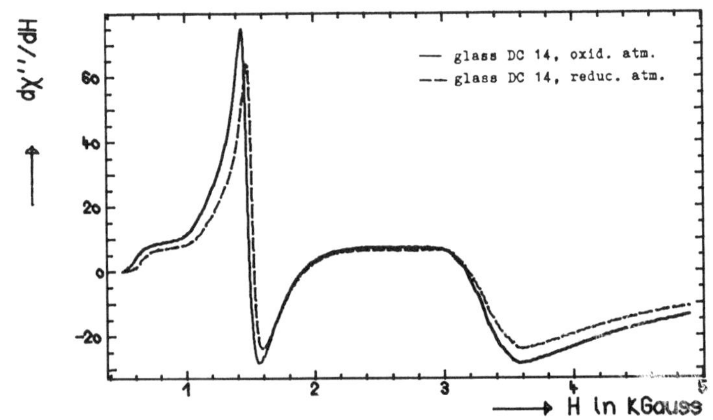

Fig. 10. ESR Spectrum of Glass DC-14

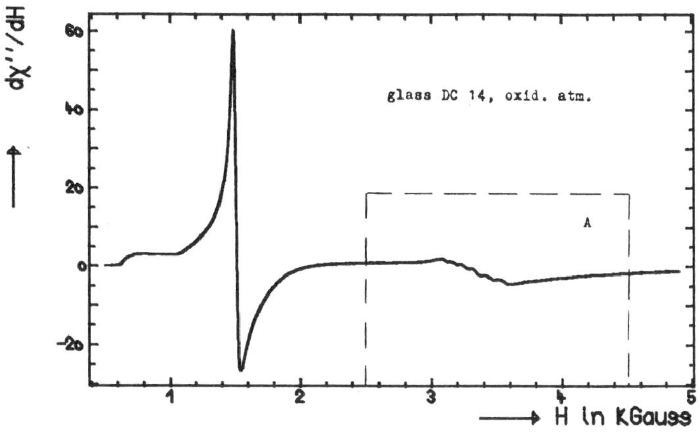

Fig. 11. ESR Spectrum of Glass DC-14, Oxidizing Atmosphere, Waste Oxides Highly Diluted

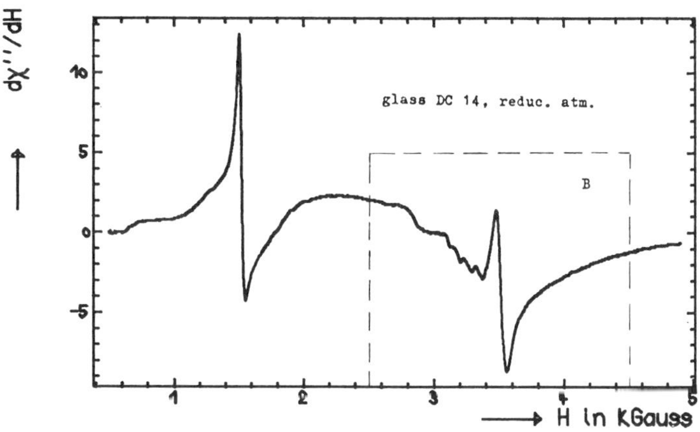

Fig. 12. ESR Spectrum of Glass DC-14, Reducing Atmosphere, Waste Oxides Highly Diluted

CONCLUSIONS

The combination of ESR and optical spectroscopy provides an excellent method of observing redox reactions with molybdenum. The investigation of a silicate glass containing molybdenum closes a gap in literature. The results gained from simply composed silicate glasses will help further examination of the behavior of glasses containing fission products.

REFERENCES

1. C. C. Chapman, Experience with a Joule Heated Ceramic Melter While Converting Simulated High Level Waste to Glass, BNWL-2071.
2. J. Landry and J. T. Fournier, ESR and Optical Absorption Studies of Transition Metal Ions and Color Centers in Glass, Arpa No. 306, Research Center, American Optical Company, Southbridge, MA.
3. J. A. Duffy, "Spectroscopic Study of Mo(V) in a Bisulphate Melt and Glass," Phys. and Chem. Glasses 9, 190 (1968).
4. S. Parke and A. C. Watson, "Absorption Spectra of Mo(III) in Alumino-Boro-Phosphate Glasses," Phys. and Chem. Glasses 10, 37 (1969).
5. J. Baucher and S. Parke, "ESR and Optical Studies of Mo(V) in Phosphate Glasses," Amorphous Material (Douglas and Ellis, ed.; J. Wiley, 1971).
6. F. Lynch and M. Sayer, "Impurity Effects in Tungsten-Phosphate Glasses I," J. Phys. C. Solid State Phys. 6, 2661 (1973).
7. S. Parke, S. Gomolka and J. N. Sandoe, "Effect of Composition and Temperature on Absorption Spectra of Molybdenum in Glasses," J. Non. Crys. Solids 20, 1 (1976).
8. H. Hartmann and J. J. Schmidt, "Über Adsorptionsspektren Komplexer Verbindungen des Dreiwertigen Molybdäns," Z. Phys. Chem. N. F. 11, 234 (1957).
9. J. S. Griffith, "Spin Hamiltonian fòr Even Electron Systems Having Even Multiplicity," Phys. Rev. 132, 316 (1963).
10. J. H. E. Griffiths, J. Owen and I. M. Ward, "Paramagnetic Resonance in Pd and Pt Group Compounds," Proc. Soc. (London) 219, 526 (1953).
11. K. D. Bowers and J. Owen, "Paramagnetic Resonance II," Rep. Prog. Phys. 18, 304 (1955).
12. Handbook of Atomic Data (S. Frage, J. Karwowski and K. M. S. Saxena, ed.; Elsevier Scientific Publ. Co., 1976).
13. K. S. Seshadri and L. Petrakis, "Electron Spin Resonance Spectra of MoO_3 on Various Support," J. Phys. Chem. 74, 4102 (1970).
14. B. Camara, "Einbau von Eisen in Glas," Glastechn. Ber. 51, 87 (1978).

RESISTANCE OF HIGH-LEVEL WASTE MATERIALS TO DISSOLUTION IN AQUEOUS MEDIA*

K. F. Flynn, L. J. Jardine and M. J. Steindler

Argonne National Laboratory
9700 South Cass Avenue
Argonne, IL 60439

ABSTRACT

Results from the studies of the rate of destruction by aqueous media of proposed barrier materials as well as the rate of leaching from proposed matrix materials are reported. A neutron activation analysis technique has been used exclusively for these measurements. Modest thicknesses of barrier material are adequate for "thousand year" security against water intrusion. The effect of temperature on the leach rate of the glass matrices studied is shown to be the predominant variable associated with the leaching medium. Conclusions based on the observed results are drawn.

INTRODUCTION

Transport of hazardous radionuclides from waste to the biosphere depends, among other things, on the rate at which these radionuclides are transferred from an inert solid phase to a mobile aqueous phase. This transfer phenomenon is commonly called leaching.

Radioactive waste will be incorporated into an inert solid matrix surrounded by one or more barriers to the environment. For the purposes of containment, the rate of destruction of the barrier material is of prime concern since the penetration of the barrier allows the aqueous phase to contact the waste matrix. When aqueous contact is made, the destruction of the inactive matrix material is of secondary interest. Rather, the rate of dissolution of the

*Work performed under the auspices of the U.S. Dept. of Energy.

specific hazardous radionuclides (e.g., ^{137}Cs, ^{90}Sr, ^{99}Tc, ^{239}Pu, etc.) is of prime concern.

A neutron activation analysis (NAA) technique has been employed in the present studies in order to tag the specific elements of interest in situ and then to measure their leach rates without physically altering the waste matrix. The details of this technique, together with a critique of its advantages and disadvantages, have been previously reported (1). Briefly, the technique consists of inducing radioactivity in the solid matrix by neutron activation, measuring the quantity of this radioactivity, leaching the solid matrix, and measuring the fraction of the radioactivity leached.

Leach rates (L) are calculated from the ratio (F) of the activated species found in the leachant (A_t) to the same species in the solid matrix (A_o), the time interval of leaching (t), the surface area of the solid matrix (S), and the weight of the solid matrix (W) according to the equation:

$$L = A_t W_o / A_o S t = (F/t) \div (S/W_o). \qquad 1)$$

The term (F/t) identifies the fraction leached per unit time (i.e., the incremental leach rate) and (S/W_o) represents the specific surface area (i.e., cm^2/g) of the solid matrix. The units of L from this equation are grams per square centimeter-day (g/cm^2d). These units allow comparisons to be made among (or extrapolations to) waste forms of significantly different specific surface areas (S/W_o). The S/W_o ratio is related to the surface area to volume ratio (S/V) by the density of the material.

Since a very low total rate of leaching is most desirable for any radioactive waste form, it is desirable to minimize the surface area available to the leaching medium (i.e., minimize the surface to volume ratio). The total surface area of the final waste matrix is a major concern because it is the interface where the radioactive species are transported from the solid phase to the aqueous phase. The surface to volume ratios (S/V) for a series of geometric shapes are given in Table 1. The cylindrical shape reported here (0.3m D x 3.0m) represents an often proposed canister size for glass monoliths. The data in Table 1 demonstrate that a sphere of about 1 cm diameter has about a factor of forty larger surface to volume ratio, compared to an "ideal" glass monolith. Measurements made using large cast cylinders (2) indicate that the surface area increases by a factor of 10 or more (over "ideal") due to cracking. This surface area is further increased if internal fins are used to improve heat transfer. Hence, surface to volume ratios in the range of several hundred $meters^{-1}$ may be realistic for the final waste form regardless of the method used for canister filling (e.g., glass ingot, glass beads in a metal matrix, super-calcine pellets, etc.). Untreated calcine has a substantially larger S/V ratio because of the fine particles pro-

Table. 1 Calculated Surface Area to Volume (S/V) Ratios

Shape	Diameter (m)	S/V (m^{-1})
Spheres	0.005 (0.5 cm)	1200
	0.01 (1 cm)	600
	0.02 (2 cm)	300
	0.04 (4 cm)	150
	0.10 (10 cm)	60
Cylinder	(0.3m D x 3.0m)	14

duced in the preparation. This waste form is not being considered in this work. The sample sizes chosen for these studies were within an order of magnitude of this S/V ratio (200 m^{-1}), so extrapolations to terminal waste storage conditions can be made with some confidence.

EXPERIMENTAL RESULTS

Leach rates for a proposed borosilicate waste glass using a variety of leaching conditions have been determined. Samples of waste glass (PNL code 76-68) were provided by Battelle Pacific Northwest Laboratories (3). These 76-68 glass samples had a density of about 3 g/cm^3 and consisted of uranium doped wafers (discs of 2.3 cm^2 surface area weighing 0.3 grams) and plutonium-neptunium doped beads (1.5 cm^2 surface area weighing 0.35 grams). The surface areas identified above are based on calculations from the geometric shape. Justifications for using this surface area have been given in previous publications (1,4). Leach rates for as many as twenty different chemical elements (e.g., references 1 and 4) using four different leaching media (5) at three different temperatures [ambient room temperature (25°C), boiling aqueous (100°C), and high pressure autoclave (∿280°C)] have been determined as a function of time. The effects of some variables associated with the chemical composition of the leaching medium (e.g., pH and salinity) have been reported previously (5). These effects, while measurable, were not large (i.e., less than an order of magnitude). However, the effect of temperature on the leach rate for this waste glass is pronounced.

Results from an initial seven-day leach test, (using distilled water leachant) for a dozen prominent elements are summarized in Table 2. The leach rate data for ^{134}Cs and ^{152}Eu are plotted as a function of temperature in Fig. 1. The leach rate for cesium increases by four orders of magnitude as the temperature increases from 20°C to 290°C. The effect of temperature on the leach rate for europium was much less (i.e., a change of about one order of magnitude over the same temperature range). These two elements (Cs and Eu) represent the extremes for this temperature effect among the haz-

Table 2. Leach Rates for PNL (76-68) Glass at Indicated Temperatures

Isotope Determined	Incremental Leach Rate $(g/cm^2d)^a$ U Spiked Glass 25°C	100°C	286°C	Pu-Np Spiked Glass 25°C	100°C	254°C
^{60}Co	4.8E-6	3.4E-5	6.6E-5	1.5E-6	1.1E-5	6.8E-5
^{65}Zn	4.9E-6	1.6E-5	4.5E-5	1.6E-6	1.6E-5	b
^{95}Zr	4.5E-7	2.4E-6	1.1E-5	1.5E-7	2.8E-6	1.0E-4
^{103}Ru	1.2E-6	7.4E-6	b	1.7E-6	4.8E-6	b
^{110m}Ag	1.9E-6	2.9E-5	b	1.3E-6	3.2E-5	b
^{131}I	1.5E-6	b	1.3E-3	7.1E-6	3.9E-5	2.6E-3
^{134}Cs	8.4E-7	8.9E-5	3.5E-3	1.7E-6	3.6E-5	1.7E-3
^{140}Ba	1.8E-6	1.5E-5	b	5.5E-7	5.9E-6	b
^{141}Ce	6.5E-7	1.2E-6	2.6E-6	2.5E-7	2.3E-6	b
^{152}Eu	3.5E-7	1.0E-6	3.3E-6	2.9E-7	1.3E-6	3.2E-6
^{237}U	5.9E-7	b	b	7.2E-7	3.1E-6	b
^{239}Np	b	b	b	7.1E-6	2.4E-5	2.7E-4

[a]All leach tests were for seven days in distilled water.
[b]These data were not resolvable from the gamma spectra.

ardous waste elements determined in these experiments. Because of their chemical similarity, the trivalent actinides ought to behave in a manner analogous to europium. The results from uranium and neptunium (cf Table 2) indicate that the temperature effect for these elements would be somewhat greater than that observed for europium. Plutonium was not analyzed for in this series of experiments. If the leach rates at the higher temperatures, as exemplified in Table 2, are considered unacceptable, then additional barrier material would be necessary for as long as the temperature remains high.

Sequential leach tests as a function of time have been performed on both the samples and under all the conditions identified in Table 2. These measured leach rates seem to decrease slowly with time (e.g., a factor of about 2 over a period of 60 days). This phenomenon is commonly observed in leach studies and is attributed to the effects of diffusion in the solid phase. The effect is being quantified by long term studies of various waste forms.

Experiments to compare solubility effects with the measured leach rates, for the elements studies, are in progress. Preliminary determinations, based on material balances, indicate that within experimental error (e.g., an order of magnitude in the case of Eu) the solubilities are not the controlling mechanism.

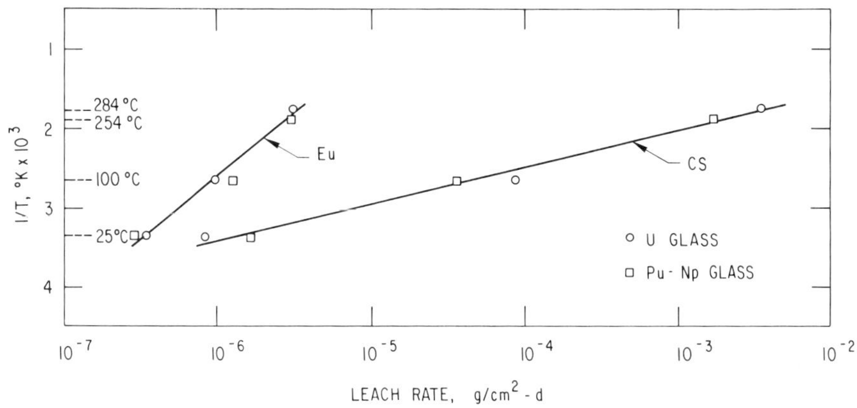

Fig. 1. Effect of Temperature on Leach Rate

CONCLUSIONS

Practical experiments that can help produce answers to specific questions have been invoked in this work. While it is desirable to minimize the surface area subject to potential leaching, it is probably unrealistic to expect to achieve a surface to volume ratio significantly lower than about $100m^{-1}$.

Among the variables associated with the leaching of glass matrices, the effect of temperature is, by far, the most severe. Cesium (and iodine) were found to leach, from the glass matrices studied, three orders of magnitude faster than europium at 290°C, whereas the difference in leach rate at 25°C was significantly less than one order of magnitude. Ten other elements studied were found to behave intermediately between these extremes. If the leach rates at the higher temperatures are considered unacceptable, then additional barrier material would be necessary for as long as the temperature remains high.

ACKNOWLEDGEMENT

The authors wish to acknowledge the efforts of Ed Huff of the ANL Analytical Laboratory for the measurement and analysis of the gamma ray spectra used in this work and Battelle Pacific Northwest Laboratory (PNL) for their cooperation in supplying us with the experimental waste glass.

REFERENCES

1. K. F. Flynn, L. J. Jardine and M. J. Steindler, "Method for Determining Leach Rates of Simulated Radioactive Waste Forms," American Chemical Society Symposium on Radioactive Waste in Geologic Storage, ACS Symposium Series 100, 115 (1979).
2. J. E. Mendel, Annual Report on the Characteristics of High Level Waste Glass, Battelle Pacific Northwest Laboratory Report, BNWL-2252, June (1977).
3. J. L. McElroy, Quarterly Progress Report Research and Development Activities, Waste Fixation Program, Oct.-Dec. 1976, Battelle Pacific Northwest Laboratory Report, PNL-2264, Nov. (1977).
4. K. F. Flynn, R. E. Barletta, L. J. Jardine and M. J. Steindler, "Leach Rate Characterization of Solid Radioactive Waste Forms," Scientific Basis for Nuclear Waste Management, Vol. 1 (G. J. McCarthy, ed., Plenum Press, 149-155, 1979).
5. K. F. Flynn, R. E. Barletta, L. J. Jardine and M. J. Steindler, Trans. Am. Nucl. Soc. 32, 394 (1979).

CHEMICAL STABILITY OF A PHOSPHATE GLASS UNDER HYDROTHERMAL CONDITIONS

B. Grambow and W. Lutze

Hahn-Meitner-Institute für Kernforschung Berlin GmbH
Glienicker Strass 100, D-1000 Berlin 39, Germany

INTRODUCTION

Phosphate glasses have some properties superior to those of borosilicate glasses. Among these are lower melting temperature, higher solubility for molybdenum and better chemical durability for some alkali-low glasses. The high corrosion of metallic containers by the melt was overcome by using refractory-lined melters. The tendency for the phosphate glass to crystallize was suppressed by rapidly cooling small beads. The beads were then embedded in a lead matrix to yield a composite nuclear waste form (1).

The leachability of an alkali-free phosphate glass has been tested with the Soxhlet method and was found to be even lower than that of borosilicate glasses (2). The leach rate for the phosphate glass, derived from weight loss measurements was found to be $(4.9 \pm 0.5) \cdot 10^{-5}$ g/cm^2/d in comparison to an average value of $1.4 \cdot 10^{-3}$ g/cm^2/d for five different borosilicate products. It was, therefore, interesting to test this phosphate glass under hydrothermal conditions that may occur under accidental conditions, e.g., in a rock salt repository. This is also of importance since some results have been published recently indicating a relatively low stability of phosphate glass against concentrated salt solutions (3).

If such tests are to be performed, it is of importance to quantify worst case conditions for the waste/rock interaction in order to perform realistic tests in the laboratory. These conditions have been assumed to be:

(a) the peak temperature at the waste/rock interface shall not exceed 470 K;

(b) a bitter brine, saturated with carnallite, $KCL \cdot MgCl_2 \cdot 6\ H_2O$, having access to the waste form for an indefinite time and being heated to 470 K.

EXPERIMENTAL

The chemical composition of the phosphate glass, consisting of 50 P_2O_5, 15 Fe_2O_3, 5 Al_2O_3% by weight and 30% waste oxides was taken from the literature (2). This composition was developed to solidify LWR waste free of sodium. The glass was melted and dropped on a rotating disc to form beads, having a greatest diameter of 0.5 cm and weighing about 0.25 g each. Some beads were spiked with Cs-137, Sr-85 and Ce-144 for subsequent testing of the leachability of these radionuclides.

The as-prepared samples were analysed by X-ray diffraction and optical microscopy. The hydrothermal experiments were performed in autoclaves containing 200 mL solution and 1 to 8 glass beads in a Teflon beaker. The temperatures ranged from 420 to 510 K. The solutions were pure water, rock salt and a naturally occurring bitter brine found in the Asse II salt mine in West Germany. Its approximate composition was 0.7 Na, 0.6 K, 0.01 Li, 24.0 Mg, 68.3 Cl, 6.4 SO_4, all in weight percent.

The overall hydrothermal attack of the samples was measured by their weight loss. Additionally, optical and scanning electron microscopy, microprobe analysis, X-ray diffraction and γ-activity measurements were used to investigate glass/brine interactions.

RESULTS AND DISCUSSION

The results of a series of experiments to determine the weight loss, the release of phosphorous and of some radionuclides after three days at various temperatures and in various leachants are shown in Fig. 1. Fig. 1a. shows the weight losses in different solutions increasing from rock salt, the least corrosive brine, via pure water to bitter brine. However, weight loss measurements are of low significance when determining the leach resistance of any kind of nuclear waste form if not accompanied by more specific investigations of the leach process. Therefore, several analytical techniques have been applied to understand the substantial attack of the phosphate glass. In the case of pure water the dissolution of phosphorous, the network former, was measured and found that its release is proportional to the weight loss of the sample (Fig. 1., curve 2). The release of Cs-137, Sr-85 and Ce-144, however, show that the corrosion process must not be interpreted as congruent dissolution of the glass by pure water, which is also true for the attack of bitter brine and rock salt solutions. In all cases, cesium is released from the glass

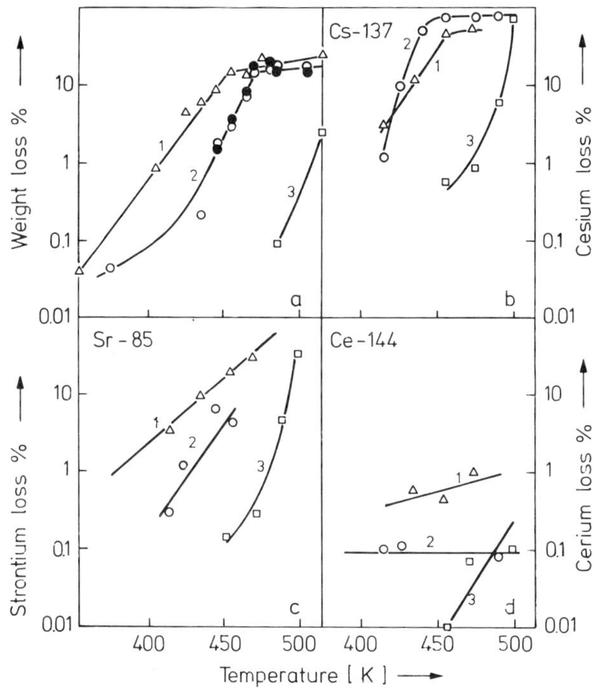

Fig. 1. Weight Loss, Release of Phosphorous (•) and of Some Radionuclides From a Phosphate Glass After 3 Days (1 Bitter Brine, 2 Pure Water, 3 Rock Salt Brine).

at a considerably higher degree than estimated from weight loss. This is also true for strontium (Fig. 1c.), whereas the loss of cerium is in all cases considerably less than would be indicated by the weight loss (Fig. 1d.). The latter is extremely low in water and in rock salt brine, and an order of magnitude higher in bitter brine. The different releases of the radionuclides give evidence for a rather complex reaction behavior of the glass.

The hydrothermal experiments in water and in bitter brine at 460 K and 3 days duration showed a complete alteration of the glass phase under the microscope. No further changes could be seen when the experiments were continued. This may explain the flat part of curves 1 and 2 in Fig. 1a, where the weight loss reached a value of about 20%. At lower temperatures the same results were obtained after longer periods of time. This disappearance of the temperature dependency of weight loss can be understood if either the whole product reacted with the solution to form a mixture of soluble and non-soluble products, or if saturation occurred due to the limited volume of the solution in the autoclave.

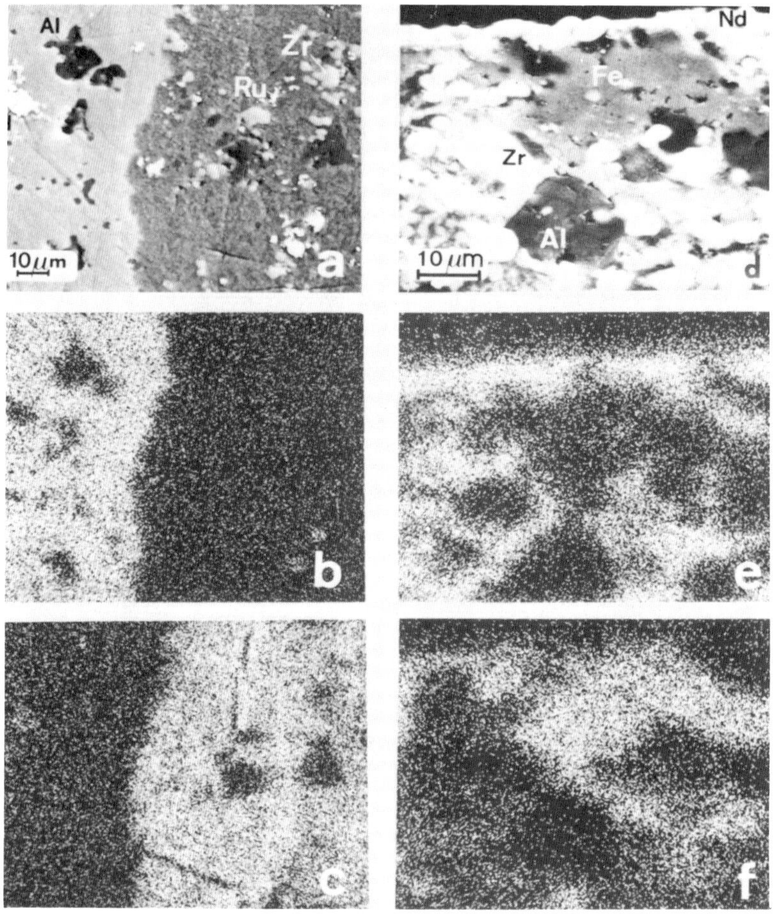

Fig. 2. SEM and MPA Micrographs of Phosphate Glass Sample During and After Reaction with Bitter Brine. (a) Dark Area: Reacted; Bright Area: Before Reaction; Reaction Front in the Middle (Ru=RuO_2, Pd=Pd, Al=$FeAl(PO_4)_{SS}$, Zr=ZrP_2O_7), (b) Cs X-ray, (c) Mg Mapping, (d) SEM Micrograph After Completion of Reaction (Fe=$FeFe_2(PO_4)_2(OH)_2$, Nd=RE_2O_3-rich phase), (e) Nd and (f) Mo Mapping, All Taken From the Same Sample.

Some results of scanning electron microscopy (SEM) and microprobe analysis (MPA) investigations are shown in Fig. 2. Fig. 2a. shows an SEM picture of a reaction front in the glass at an early state of the reaction with bitter brine, the left part showing unattacked glass. Several phases can be seen in both the original glass and in the reaction zone. These contain Al, Pd, Ru and Zr according to microprobe analysis. Fig. 2b. and 2c. are X-ray mappings of Fig. 2a., showing cesium to be almost completely released from the reacted zone, whereas magnesium migrated inwards (Fig. 1c.). As yet, it is not clear for Mg whether ion exchange occurred or whether the signal comes from finely dispersed MgO. The bitter brine undergoes a reaction with water above 420 K: $MgCl_2 + H_2O \longrightarrow MgO + 2\ HCl$. Fig. 2d. shows a region of the sample close to the surface (top) after the hydrothermal reaction has ceased. Some new phases have appeared which were not detectable in Fig. 2a. Due to the comparatively high content of neodymium in the waste, a related X-ray mapping (Fig. 2e.) could be obtained. Cluster-like areas rich in neodymium can be seen throughout in the sample. The same was found for Mo (Fig. 2f.). Both elements were homogeneously distributed in the glass phase of the as-prepared sample.

With the help of X-ray diffraction analyses, several phases could be identified which had formed during the leach experiments when the glass was in contact with the different solutions. The results are summarized in Table 1. As can be seen, some of the phases were already contained in the as-prepared glass. It was checked that new phases did not form upon thermal treatment in the absence of a solution nor did the formation continue when the solution was removed.

A solid solution, $Fe,Al(PO_4)_{SS}$, was identified according to data reported by Kobayashi (4). The formation of Lipscombit, $FeFe_2(PO_4)_2(OH)_2$, was observed after 3 to 4 days above 440 K in water. At 470 K rare earth oxides and Berlinit, $AlPO_4$, crystallized. Three rare earth phases were identified. Using bitter brine, Lipscombit was observed at 440 K; the rare earth oxides were identified only above 510 K reaction temperature. At this temperature, Berlinit was the phase with the highest yield. In no case was a phase containing cesium identified and no enrichment of Cs was found by MPA in any of the observed phases. Only some strontium was found in a molybdenum-barium rich phase. This is consistent with the observed high release of cesium and strontium (Fig. 1b., 1c.).

The neodymium-rich phase forms a phase boundary for the sample shown in Fig. 2d. This phase remained unchanged even after 10 days of contact with the solution and appears to be very insoluble. It is rather likely that this phase hosts the radiactive Ce-144, which is, therefore, strongly bound in the solid material independent of reaction conditions, as seen in Fig. 1d. The leachability of the fission product molybdenum shows a particular behavior: at 440 K, 6% molybdenum was found after 3 days in solution; at 455 K, 21% and at 530 K, 11%. The corresponding values for Al are 1.8, 5.9 and 0.2%. This can be interpreted in terms of competition between dissolution and crystallization. At temperatures above 470 K, the crystallization is very fast; and the molybdenum, which was dissolved at lower temperatures, is now being retained in the solid phase (Fig. 2f.), and an accelerated formation of Berlinit prevents Al from being dissolved.

Table 1. Crystal Phases in a Phosphate Glass After Hydrothermal Reaction (*Also Contained in the As-Prepared Glass)

Phase	Structure	D Spacings Å
ZrP_2O_7*	Cubic	4.74, 4.12, 3.67, 3.37, 2.91, 2.48, 1.84
RuO_2*	Tetragonal	3.17, 2.56, 1.68
Pd*	Cubic	2.22
$Fe,Al(PO_4)*_{SS}$	Tetragonal	4.10, 3.17, 2.91, 2.52
$Fe''Fe'''_2(PO_4)_2(OH)_2$	Tetragonal	3.30, 3.17, 2.03, 2.01, 1.65, 1.59
$AlPO_4$	Trigonal	4.28, 3.37, 1.83
RhO_2	Tetragonal	3.17, 2.56, 1.68
RE_2O_3 (mostly Nd)	Hexagonal	3.01, 2.91, 1.91, 1.71, 1.59
RE_2O_3 (mostly La)	Hexagonal	3.07, 2.98, 2.28, 1.94, 1.75
RE_2O_3 (mostly Sm)	Monoclinic	3.18, 3.04, 2.98, 2.85, 2.79, 2.14, 1.91

There are some unidentified traces left in the X-ray pattern which might be attributed to a crystalline phase containing molybdenum.

Having the results of the hydrothermal experiments in mind, one could try to understand the reaction behavior of the glass in terms of the theory of phosphate glass structure. The phosphate glass has a metal to phosphorous ratio of about 2. It was inferred from other data (5) that the structure of the phosphate glass under investigation contains a lot of "end" groups $(O{=}PO_3)^{2-}$, some "middles" $(O{=}PO_3)^-$ and essentially no "branches" $(O{=}PO_3)$. Thus, the product is a polyphosphate glass with anions of different but short mean chain lengths. The presence of $P_2O_7{}^{4-}$ in the structure is evident due to the presence of ZrP_2O_7. Thus, the leaching of short polyphosphate groups, i.e., the hydrolysis of the oxygen metal bonds in PO-M groups, determines the rate of disintegration of the glass structure. The hydrolysis of polyphosphate groups to orthophosphate is another process contributing to the leaching, whereby the pH value of the solution is being lowered. The leach rate increases with increasing H^+ concentration and, therefore, is highest in the bitter brine which contains HCl. The most polar bond in the glass structure is the PO-M bonding if M = Cs, Rb and will be broken first. Therefore, the leaching leads to the high release of Cs which remains in solution as there is no appropriate mineral host phase. The orthophosphate groups form Lipscombit which is insoluble and, therefore, very little iron was detected in solution. The crystallization of the rare earth compounds is frequently observed in phosphate glasses even from the melt (6). The present glass dissolved the rare earth elements homogeneously in the melt and subsequently showed crystallization of respective phases (Table 1), not only upon hydrothermal reaction but also upon annealing above the transformation temperature (820 K). Here, X-ray traces of lanthanum rich phases were detected.

Sodium ions may also enhance the hydrolysis of polyphosphate ions. The comparatively low corrosivity of the rock salt solution was surprising, but it could be understood when taking the reduced activity of water into account which is due to the high sodium concentration. Above 490 K, considerable ion association takes place in NaCl solutions (8) and the number of free water molecules increases. At and above this temperature, the rock salt solution was found to become very aggressive (Fig. 1b., 1c.).

A comparison of the behavior of phosphate glass with that of borosilicate glass indicates a much higher stability of the borosilicate glass under the hydrothermal conditions applied here and give evidence for a different leaching mechanism. The attack of the brine leads to a weight loss of 2% even at 100 bar and after several weeks (9).

CONCLUSIONS

Phosphate glasses free of sodium may be a suitable nuclear waste form in terms of leach resistance at temperatures below 370 K. At higher temperatures, the leach resistance decreases rapidly. Concentrated carnallite brine was found to be more corrosive than rock salt brine. It is, therefore, worthwhile to test the stability of any nuclear waste form in carnallite if it shall be disposed of in a salt formation and if the presence of carnallite cannot be excluded.

Hydrothermal reactions as they may occur under accidental conditions in a rock salt repository led to complete disappearance of the glassy state and substantial remobilization of solidified radionuclides, such as cesium and strontium. Only rare earth elements and probably the typical trivalent actinides may be almost unaffected and remain immobile, independent of brine composition and temperature.

REFERENCES

1. J. van Geel, H. Eschrich, W. Heimerl and P. Grziwa, Solidification of High Level Liquid Wastes to Phosphate Glass-Metal Matrix Blocks, IAEA.SM-207/83 Vd I S.341-359, Vienna (1976).
2. G. Malow, Testing and Evaluation of the Properties of Various Potential Materials for Immobilizing High Activity Waste, to be published as EUR Report (1979).
3. W. Lutze, J. Borchardt and A. K. Dé, "Characterization of Glass and Glass Ceramic Nuclear Waste Forms," Scientific Basis for Nuclear Waste Management, (G. J. McCarthy, ed., Plenum Press, New York, 1979).
4. T. Kobayashi, "Solid Solubilities of $FePO_4$ and $CrPO_4$ in $AlPO_4$," J. Ceram. Assoc. Japan 74 (6), 190-195 (1966).
5. J. R. van Wazer, Phosphor and Its Compound 1, Kap. 12 S. 731 (1961).
6. J. E. Mendel, Melt Formation From Highly Radioactive Waste, BNWL-1073(4) (1968).
7. C. Y. Shen and D. R. Dryoff, "Hydrolytic Segregation of Sodium Tripolyphosphate in Concentrated Solutions and in Presence of Foreign Ions," I + EC Product Research and Development 5 (2), 97-100 (1966).
8. J. M. Wright, W. T. Lindsay and R. R. Druja, WPAD-TM-204, 365, N.S.A. (1961).
9. W. Lutze, "Glassy and Crystalline High Level Nuclear Waste Forms - An Attempt at Critical Evaluation," International Symposium on Ceramics in Nuclear Waste Management, Cincinnati 30.4-2.5.79 (1979).

THE BEHAVIOR OF ACTINIDES IN α- DOPED GLASSES AS REGARDS TO THE LONG TERM DISPOSAL OF HIGH LEVEL RADIOACTIVE MATERIALS

R. A. Bonniaud, N. R. Jacquet-Francillon, C. G. Sombret

Commissariat A L'Energie Atomique
France

INTRODUCTION

Investigations about the characteristics of glasses involved in the solidification of the fission products solutions are carried out as regards to their mechanical and chemical properties. Since the glasses will contain actinide elements, it is necessary to evaluate the possible modifications of these properties versus time. The major problem in doing so is the simulation of a long term storage to make the effects of α- emitters discernable in a relatively short time span. Several methods can be considered: Irradiation by heavy ions in order to match the energy input which would have been dissipated by α- emitters, neutron irradiation to induce either (n, α) or fission reactions, and finally to doping the glasses with various contents of actinides. C. E. A. (Commissariat a l'Energic Atomique) chose the latter. This method is indeed expected to ensure a more realistic simulation.

BASIC ASSUMPTIONS

To set up the simulation, the following assumptions were taken into account:

(a) Reprocessing operated 3 years after discharge (this is the actual French feature); (b) Subsequent storage of the fission products solution for 1 year; (c) Glass yield: 94 liters per metric ton of spent fuel; (d) Fuel burn up: 33 000 $MWdt^{-1}$; (e) Glass specific gravity: 2.7 $g.cm^{-3}$; (f) Extraction of Pu and U by reprocessing: 99.5%.

FABRICATION OF DOPED GLASSES

^{241}Am, ^{244}Cm and a mixture of Pu isotopes were used as doping elements. Tables 1 and 2 give the main characteristics of the glasses. Pu and Am glasses were made at MARCOULE Center in the cell VULCAIN (1) by a pot vitrification technique with a maximum temperature of 1150°. Samples were poured from the melt into 20 mm diameter 50 mm high cylindrical graphite molds. Glass blocks 80 mm diameter, 100 mm high, were also poured.

Table 1. Characteristics of Actinide Doped Glasses

Actinides	Weight of Actinides per Sample (g)	Alpha Activity (Ci)	Weight of Sample (g)	Cumulated Energy per Liter of Glass for 1 Year (Mean Value in kWh)
Blend of Pu at 0.17% ^{238}Pu	60-100	10-20	1500-2000	5.5
Blend of Pu at 30% ^{238}Pu	50	245	1500-2000	96
^{241}Am	50	160	1500-2000	66
^{244}Cm	0.6	50	50	800

Table 2. Theoretical Chemical Composition of Actinide Doped Glass

Reference	SiO_2	Na_2O	B_2O_3	Al_2O_3	Fe_2O_3	MgO	NiO + CrO_3	OxPF
Blend of Pu* at 30%	45.4	9.5	19.4	----	----	----	----	23.14
^{238}Pu glass**	47.1	9.8	20.1	----	----	----	----	19.6
^{241}Am glass^{+}	50.8	8.2	13.4	----	----	----	----	24.8
^{244}Cm glass^{++}	39.7	18.1	18.1	14.4	5.0	2.5	0.9	----

*2.78% PuO_2; **3.26% PuO_2; $^{+}$2.83% $^{241}AmO_2$; $^{++}$1.3% $^{244}CmO_2$.

Owing to neutronic considerations, the Cm loaded glass was made in Fontenay aux Roses Nuclear Center where suitable facilities were available. The fabrication method was to some extent different from the above mentioned one, and the size of the glass piece was 21 mm diameter and 72 mm high and the glass was cast in an inconel canister. Any tendency for phase segregation when using high PuO_2 concentrations has been discussed (2), and investigations about the quality of these glasses were carried out.

ACTINIDE DISTRIBUTION

Using samples taken during the pouring of Am and Pu glasses or cut from the Cm glass canister, the homogeneity of actinide distributions were determined by α-autoradiography, optical microscopy and scanning electron microscopy.

Plutonium

It appeared (see Fig. 1.) that undissolved small particles of a diameter less than 1μ and made of PuO_2 are rather evenly distributed through the matrix although there is some particle agglomeration. The chemical form of Pu as PuO_2 was determined identified by X-ray pattern. In one of the samples investigated, a complex agglomerate composed of PuO_2, CeO_2, Cr, Ni, Fe and Mn was identified.

Autoradiography showed the occurrence of bubbles (white spots on the photographs) and metallic particles free of actinides.

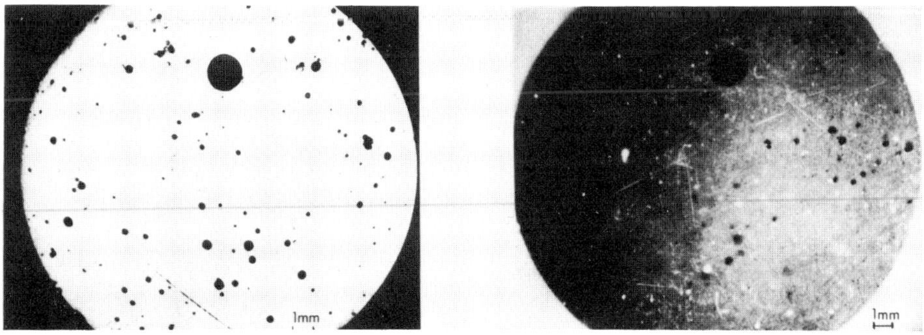

(a) (b)

Fig. 1. ^{238}Pu (Core), (a) Microphotograph, (b) α-Autoradiograph

Americium

The americium distribution is very similar to plutonium, but it was not possible to identify the X-ray pattern.

Curium

A multiphase structure was set up in the matrix. Autoradiography shows (see Fig. 2., the white area) an actinide depleted zone whose composition has not yet been determined. Nevertheless, a vertical γ-scan of the canister pointed out an even distribution. X-ray diffraction did not yield a diffraction pattern, but a non-radioactive specimen, free of Cm, contained some microcrystals of $NiCr_2O_4$.

To verify the theoretical quantities of actinides in the glass, a calorimetry method was used. Table 3 indicates a loss of Pu and Cm during the fabrication.

HELIUM BUILD UP

Helium is produced by α-decay. In order to measure the quantity of helium kept in the glass, samples were melted under vacuum at various times. The released gas was analysed by mass spectrometry

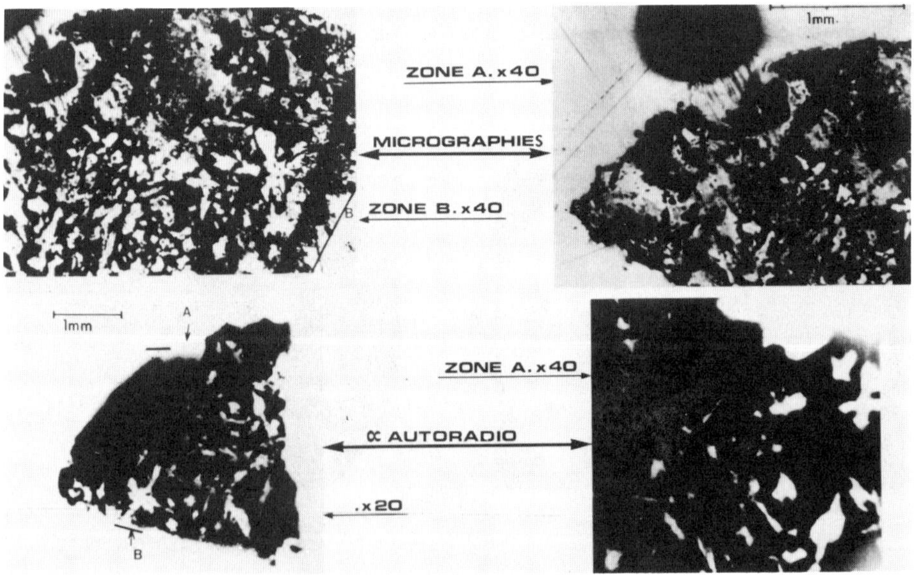

Fig. 2. ^{244}Cm Photomicrographs and Corresponding α-Autoradiographs

Table 3. Comparison of Theoretical and Actual Actinide Content of Glass Samples by Calorimetry (Specific Power - mWg^{-1})

	Pu Glass	Am Glass	Cm Glass
Theoretical	4.06	2.85	29.19
Actual	2.17	2.44	14.5, 12.0 (2 samples)

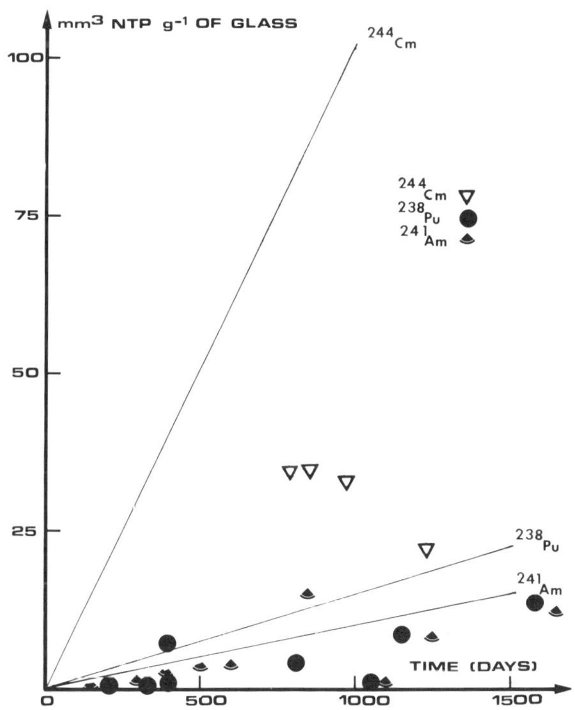

Fig. 3. Comparison of Maximum Theoretical and Actual Build Up of Helium Released from Alpha Doped Glass

and gaseous chromatography. The results correlated well with the content of Pu and Am determined by calorimetry. The data relating to Cm glass were, on the other hand, lower than expected and rather variable. See Fig. 3.

DENSITY AND MICROHARDNESS

Only Pu and Am glasses were investigated. Alpha-irradiation does not seem to have any effect upon either density or microhardness. Variations in the microhardness values are of minor importance because of the difficulty of the measurements and sample surface preparation.

LEACHING

Curium Glass

The leach tests were carried out under static conditions with tap water at room temperature. The weight of the glass sample was 4.2 grams and its geometrical area 6.9 cm^2. The leachant change intervals were generally one day, but they were sometimes increased to find out their effect upon the leaching rate. Various volume of

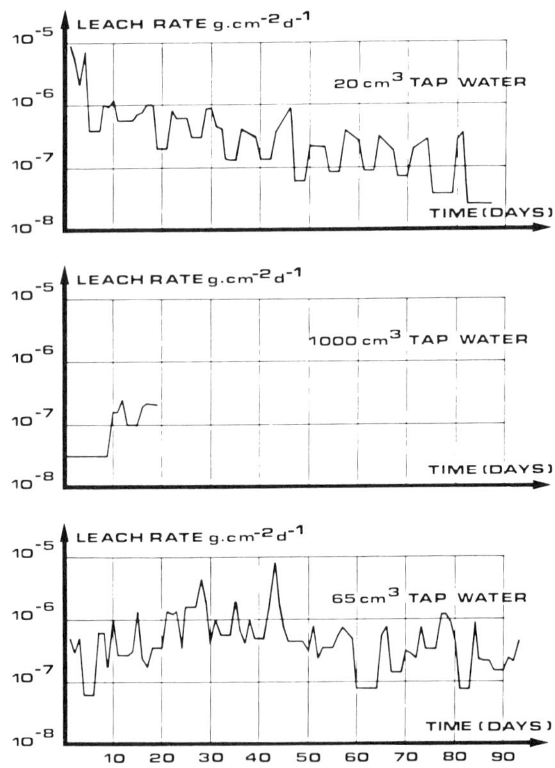

Fig. 4. Leach Test on ^{244}Cm Doped Glass Room Temperature; Static Run

water were used: 20, 65 and 1000 cm^3. The pH behavior was determined during the last day of the leach test and carried out with 65 and 20 cm^3 of solution. The test started when the integrated dose of the sample reached $1.6x10^8$ disintegrations per gram. The leach rate (Fig. 4.) does not show a clear relationship with the ratio of the geometrical area/volume of the leachant. The leach rate is in the range of $5x10^{-7}$ $g.cm^{-2}d^{-1}$*. The increment of pH is, for 1 day, proportional to the logarithm of time (the pH increases from 8.0 to 8.9). Chemical analysis of the leachant indicated a loss of a borate compound containing Na but very little Cm. The leaching is associated with an imporant radiolysis reaction since 10 cm^3 of gas were trapped every day (H_2 66%, O_2 33%, He $25x10^{-4}$% in volume). The volume of the radiolysed water corresponds to a 10.5 μ thick layer destruction. The helium content matches the values previously obtained from the measurement of the helium build up.

*Expressed as $\frac{a}{A} x \frac{P}{S}$ with a: activity of the leachant
A: activity of the glass
P: weight (in gram)
S: surface (in cm^2).

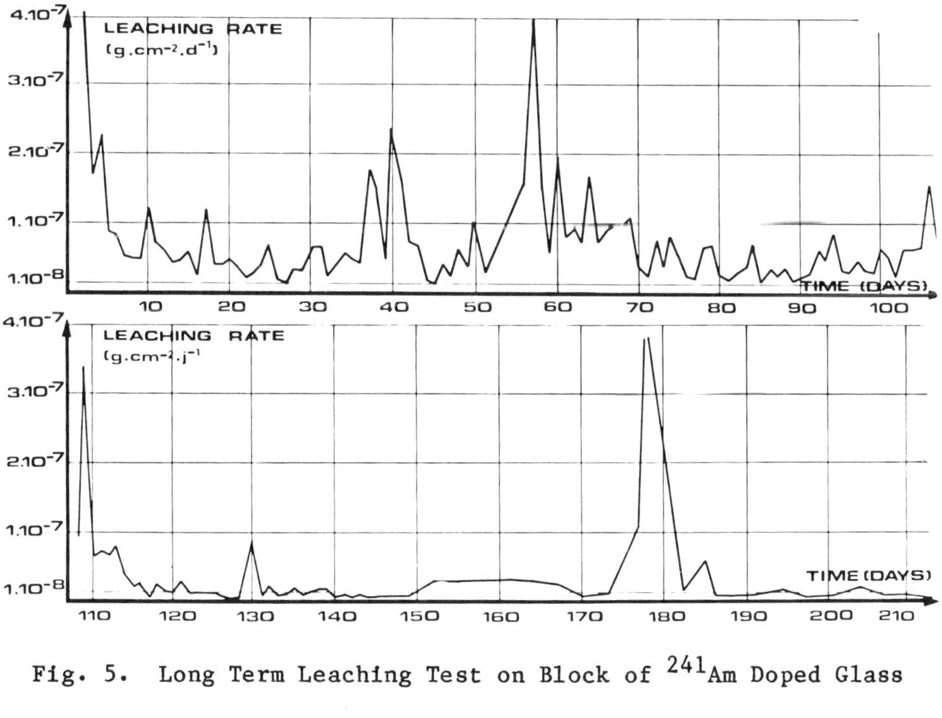

Fig. 5. Long Term Leaching Test on Block of ^{241}Am Doped Glass

AMERICIUM GLASS

The 1900 gram block was leached in the VULCAIN cell in 700 cm^3 of tap water removed daily at room temperature under static conditions. The average leach rate (1 to $2x10^{-8}$ $g.cm^{-2}d^{-1}$) is very low (see Fig. 5.), but some peaks occur almost periodically (3 to $5x10^{-7}$ $g.cm^{-2}d^{-1}$).

PLUTONIUM GLASS

Samples taken during the pouring were used for the leach test. The samples were leached with 120 cm^3 of water under static conditions at 23°, 50°, 70°C and using a variable water renewal rate. The behavior of the leach rate is plotted on Fig. 6. The mean values at the steady state were $3.5x10^{-6}$, $3.5x10^{-6}$ and $1.5x10^{-6}$ $g.cm^{-2}d^{-1}$ for 23°, 50°, 70°C, respectively. It is possible that an insoluble layer located in the water/glass interface acts as a barrier for the diffusing Pu. As for the Cm glass, it is different in regard to its physicochemical properties. The released activity is not related with the water renewal rate for pH 7 to 8.

The leachant recovered from the leach test was acidified and filtered successively on millipore filters grade 8, 0-8, 0.05 μ. It

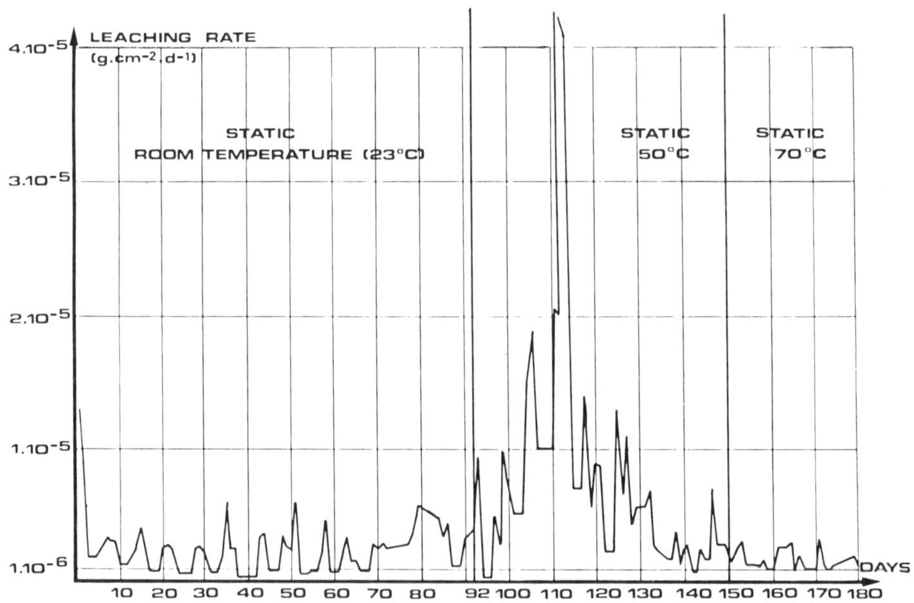

Fig. 6. Leaching Rate Evolution for French ^{238}Pu Glass Samples with Temperature

Table 4. Filtration of Leach Water from ^{238}Pu Glass Sample

Size Distribution/Percentage	23°C	50°C	70°C
$d > 8\mu$	27-46	89-95	90-95
$0.8\mu < d < 8\mu$	5-20	0.2-1	0.2-0.3
$0.05\mu < d < 0.8\mu$	4-12	0.2-5	0.1-3
$< 0.05\mu$	30-50	4-10	2-5

was found that more than 50% of Pu is in the form of a colloidal compound bigger than 500 Å. This quantity increases when temperature increases (see Table 4). Formation kinetics for this phase are fast since 4 successive filtrations of the same liquor with 8 μ filter decreased the activity by a factor 10 to 30.

A leach test with pH 1 liquid at room temperature under static conditions gave rise to values about 100 times more than the ones measured for pH 7 to 8 in the same conditions. The effect of the liquid renewal rate upon the leach rate indicates, at least for several days, the leach process is merely a dissolution of the PuO_2 phase distributed in the glass matrix.

CONCLUSION

Until now, investigations did not point out any marked effect of the irradiation upon the retention efficiency of the glasses or their chemical inertness. It remains to determine why Cm glass does not behave as the others in terms of He build up. The works relating to these topics must be emphasized. But it will be necessary to wait for a larger dose rate to detect a possible modification of the properties of the glasses.

REFERENCES

1. R. Bonniaud, F. Pacaud, C. Sombret, "Quelques Aspects du Comportement de Radioéléments Confinés à Long Terme sous Forme de Verre ou de Produit à Forte Phase Vitreuse," Verres et Refractaires 29 (1), Janvier-Février (1975).
2. F. Laude, R. Bonniaud, C. Sombret, G. Rabot, "Confinement de la Radioactivité dans les Verres," IAEA Symposium SM 207/36 (1976).

ATOMIC DISPLACEMENTS AND RADIATION DAMAGE IN GLASSES INCORPORATING HLW

M. Antonini*, P. Camagni, F. Lanza, and A. Manara

Joint Research Centre
21020 Ispra, Varese
Italy

ABSTRACT

Recent data of radiation effects in borosilicate glasses and amorphous SiO_2 are presented and discussed in terms of atomic displacements produced by direct nuclear collisions and ionization damage produced by interactions of the bombarding particles with the bonding electrons. It is found that density changes and stored energy can properly be represented as depending on the calculated relative number of displaced atoms. Conversely, the comparison of optical absorption spectra of vitreous SiO_2 specimens irradiated by high energy heavy ions and electrons indicates that in some cases it might be of relevance to evaluate the yield of ionization damage, basically related to the total amount of energy dissipated in the glass matrix.

INTRODUCTION

High level waste incorporated in glassy matrices contains various sources of radiation: $\beta - \gamma$ emitters, α emitters of different energies and recoiling heavy ions (1). The relative importance of these particles in long-term damaging of the glass has to be evaluated in terms of parameters appropriate to the specific kind of damage produced. At microscopic level, the interaction of radiation with insulating glasses may result primarily either in direct atomic collisions followed by atomic displacement, or in electronic colli-

*Also: Gruppo Nazionale Struttura della Materia and Istituto di Fisica dell'Universita, 41100 Modena, Italy.

sions followed by bond rupture and heat production within the pattern of the collision cascade. These basic processes may successively cause transformations in the local structure of the amorphous materials with consequent changes in quantities of both scientific and technological interest, such as density (2) and energy content (3). For instance, it is well known that irradiation of pure silica glasses induces a characteristic compaction effect up to some percents of increase in density (4); but the detailed origin of this compaction has never been quite clear, although direct calculations in term of modification of atomic bonding have recently been attempted by means of semi-empirical molecular orbital models (5). The situation is even more complex in more complex glasses, where the experimental data are rather scattered and might be strongly dependent on the glass composition (6).

The amount of energy stored during irradiation is also one of the most extensively studied quantities, because of its twofold interest: (i) to provide a direct assessment of the total damage produced by displacing radiations (7), and (ii) for an intrinsic reason concerning the evaluation of temperature rise during waste storage (3). However, accurate measurements of the stored energy are needed since a small amount of energy is usually released over a large temperature range.

Another advantageous method for the monitoring of radiation effects in insulators is the study of colour centres by means of optical techniques, which provides a specific information on defect properties at the atomistic scale. This often allows more direct insight into the basic damage processes than bulk macroscopic measurements. However, this approach cannot be applied in a straightforward manner to the waste glasses, owing to the experimental and conceptual difficulties brought about by: (i) existence of strong and complex natural absorption, and (ii) uncertainty in the interpretation of optical results, due to the complex multicomponent structure and to the uncontrolled level of pre-existing defects. In this respect, the study of simpler systems such as pure SiO_2 can be useful in the modeling of a basic picture of damage for silica-based glasses.

In the following, we present some recent results of investigations on density changes, stored energy and optical absorption in borosilicate and SiO_2 irradiated glasses.

EXPERIMENTAL

The adopted samples were highly transparent pure SiO_2 glasses and borosilicate glasses representative of those commonly taken into account in Europe for HLW vitrification. Typical dimensions for SiO_2 samples were 5x5 mm with thickness of 0.2 and 1 mm. Some of the samples were coated with a 100 Å Ag-layer to reduce thermal gradients and to prevent space-charge and electrical break-down during irradiation. They were water cooled in order to keep the unexposed surface at room temperature.

The borosilicate glasses were in the form of cylinders, 1 cm diameter and 1 cm thick. They had a weight content of 48% of SiO_2 and incorporated 20% of simulated waste oxide. The detailed composition is given in a previous paper (8).

Heavy ion irradiations were performed at the Variable Energy Cyclotron, AERE, Harwell, using a collimated beam of Ni^{+6} ions at an energy of 46.5 MeV. The calculation of the number of displaced atoms and that of the depth distribution of energy deposition by ion bombardment was made by means of a modified version of the computer program of Manning and Muller (9). Total doses ranged from 0.85×10^{14} to 0.85×10^{16} ions/cm^2, equivalent to 0.01 and 1 dpa, respectively. The last figures are comparable to a number of displacements taking place in about 1 and 10^6 years of waste storage.

Electron irradiations at 1.5 MeV were also made in 3 mm thick pure silica specimens, for a comparison with heavy ions. Part of these irradiations were performed at 150°C. The number of atoms displaced by electrons was calculated by means of semiempirical expressions of the energy loss on the actual range of 3 mm (10).

Borosilicate glasses were charged with different levels of ^{235}U and irradiated with fast and epithermal neutrons, so that fission fragments could damage the samples (8). A maximum amount of 0.36 dpa was obtained at an irradiation temperature around 68°C. Density measurements in these materials had to be made in hot cells, owing to some radioactivity gained by these samples during neutron irradiation. The traditional Archimedean method could be employed, using as a reference fluid tetrochloroethylene, with a density of 1.67 g/cm^3.

Stored energies have been measured by thermal annealing of SiO_2 in a differential scanning microcalorimeter. More details about this instrument and the adopted operational mode are given elsewhere (7).

Optical measurements consisted of recording the optical absorption spectrum with a Cary 14 spectrophotometer in the wavelength interval between 190 and 2400 nm. Alternate positioning of the irradiated sample and of a reference unirradiated sample within the exploring beam allowed to take net coloration spectra, thus minimizing experimental uncertainties.

RESULTS AND DISCUSSION

Density Measurements in Borosilicates

The results obtained are reported in Table 1, which gives density changes versus dose in dpa units. We can see that the glass density decreases rather sharply at the beginning and then reaches a quasi-equilibrium. The scattering of the data is rather high. The observed decrease in density is in agreement with preliminary theoretical predictions based on the difference between the densities of the various components present in our samples and that of the final compound (11).

Stored Energy

A continuous release of energy up to about 50 cal/g is found between 100 and 500°C in a SiO_2 sample, heavy ion irradiated at a total dose of 1 dpa and a dose rate of 10^{-4} dpa/s. This value is only slightly higher with respect to those obtained by other authors (3, 12) after neutron irradiation at lower doses, i.e. 0.3 dpa, thus confirming the tendency towards saturation already mentioned by these authors. At lower dose rates, i.e. at 10^{-5} dpa/s, a higher level of stored energy, is obtained (7). It is possible that this increase marks the onset of structural modifications occurring at an extended scale of distances in the sample during irradiations and successive thermal recovery. Recent results of high-resolution electron microscopy in silica indicate in fact that small crystallites may be present in this compound even when it is prepared in the vitreous state (13). It is well known that a structural transformation takes place during irradiation of crystalline SiO_2 (14).

Table 1. Variation of the Glass Density After Fission Fragments Irradiation

Dose (dpa)	$\Delta\varrho/\varrho_o$(%)
1.2×10^{-3}	-0.50
3.2×10^{-2}	-0.14
1.3×10^{-2}	-0.40
3.6×10^{-2}	-0.80

Optical Measurements

As shown in Fig. 1., the typical optical absorption spectrum of heavy ion irradiated silica presents two maxima: (i) a band B_2 at 5.06 eV, and (ii) a band E_1' at 5.76 eV. Both these maxima were also found after irradiation with neutrons (15) and low energy ions (16). Both maxima were superimposed on a strong absorption tail probably originated from a peak located in the more distant ultra-violet region.

The results of electron bombardment are reported in Fig. 2. It can be noticed that the B_2 band is totally absent. The same figure shows the effect of the temperature of irradiation upon the intensity of the E_1' band: at 140°C the strength of the band is much lower than at 40°C.

The absence of the B_2 band in the electron irradiated samples indicates that this colour centre is generated by direct atomic displacements, whose number is much lower during electron bombardment.

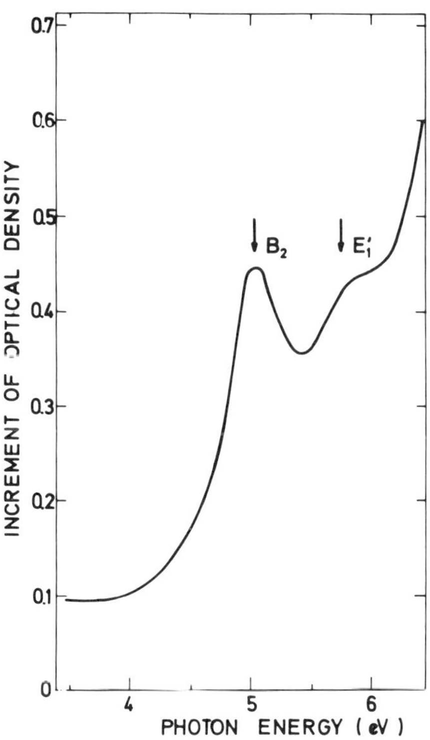

Fig. 1. Increment of Optical Density in Heavy Ion Irradiated Vitreous SiO_2; Dose Is 0.5 dpa.

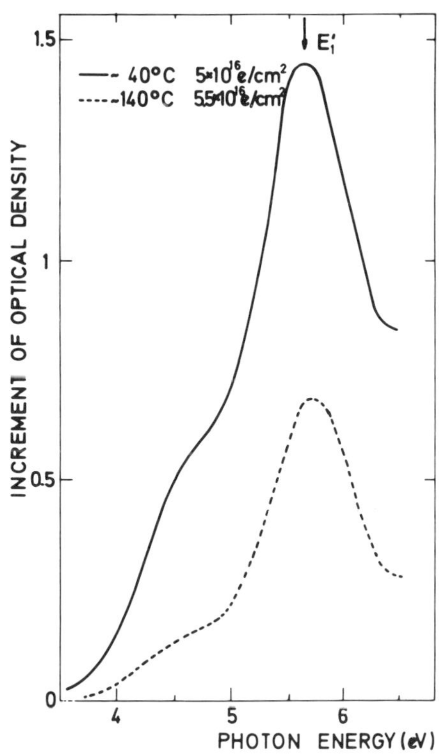

Fig. 2. Increment of Optical Density in Electron Irradiated Vitreous SiO_2.

Table 2. Optical Absorption in Irradiated Silica

Radiation	Displaced Atoms	B_2 Centers	E_1' Centers
Heavy ions (Ni^{+6}, 46.5 MeV)	$7.3x10^{18}$	$7.4x10^{15}$	$1.6x10^{16}$
	$3.6x10^{19}$	$7.9x10^{15}$	$2.1x10^{16}$
	$7.3x10^{19}$	$6.1x10^{15}$	$1.4x10^{16}$
Electrons (1.5 MeV)	$3.1x10^{16}$	none	$5.9x10^{16}$
	$1.2x10^{17}$		$2.0x10^{17}$

The observed data were fitted to computed spectra to obtain the results reported in Table 2. From examination of this table it can be argued that saturation is present in all cases. Moreover, the number of E_1' centres produced by electron bombardment is systematically higher than the correspondent number of displaced atoms and of the number of E_1' centres produced by heavy ion bombardment. It seems therefore that the E_1' band cannot be originated by direct atomic displacements, but its origin may be due to bond ruptures and rearrangements as suggested by some theories (17,18). The number of displaced atoms does not constitute in this case a reliable unit for the damage, but ionization effects have to be considered, although their contribution to density changes and stored energy cannot yet be assessed. In this regard it is recalled that the yield of ionization is related to the total energy dissipated into the damaged volume of the sample. This suggests that a simple evaluation of this quantity is a means of predicting the relevance of ionization effects in complex borosilicates, where a direct observation of the E_1' band is not possible. Proton and electron irradiations having the same dissipated energy are now in progress to control the validity of this assumption.

ACKNOWLEDGEMENTS

We wish to thank Mr. E. Parnisari for vigorous technical assistance during density measurements in a hot cell and collaboration in the preparation of the manuscript.

REFERENCES

1. F. Lanza, A. Manara and F. Van Rutten, Simulation Possibilities of Radiation Effects in Glasses used for Conditioning High Activity Waste, EUR 5560 e (1976).

2. P. Le Clerc, "Action de Rayonnements sur les Verres," Bull. Inf. Sci. Techn., Commissariat a l'Energie Atomique 98, 7 (1965).
3. F. P. Roberts, G. H. Jenks and C. D. Bopp, Radiation Effects in Solidified High Level Waste, BWL-1944, UC-70 (1976).
4. W. Primak, The Compacted States of Vitreous Silica (Gordon and Breach, New York, 1975).
5. P. W. Tasker, A Semi-Empirical Molecular Orbital Model of Silica: Application to Radiation Compaction, AERE-R 9319, (1978).
6. J. E. Mendell, W. A. Ross, F. P. Roberts, J. B. Katayama, J. H. Westsik, Jr., R. P. Turcotte, J. W. Wald and D. J. Bradley, Annual Report on the Characteristics of High Level Waste Glasses, BNWL-2252, UC-70, 75 (1977).
7. M. Antonini, A. Manara and P. Lensi, "Ion Irradiation and Stored Energy in Vitreous SiO_2," The Physics of SiO_2 and its Interfaces, (S. T. Pantelides, ed., Pergamon, New York, 316, 1978).
8. M. Antonini, F. Lanza and A. Manara, "Simulations of Radiation Damage in Glasses," Proc. Int. Conf. on Ceramics in Nuclear Waste Management, Cincinnati, OH, April 30-May 2 (1979).
9. M. D. Matthews, Calculations of the Depth Distribution of Energy Deposition by Ion Bombardment Using the Computer Program of Manning and Muller (1973), AERE R 7805 (1974).
10. J. W. Corbett, "Electron Radiation Damage in Semiconductors and Metals," Solid State Physics, Suppl. 7, (Pergamon, New York, 1964).
11. P. W. Tasker and G. Tinivella, private communication.
12. A. Roux, Energie Emmagasinee dans les Oxydes BeO, MgO, Al_2O_3 et SiO_2 Irradies aux Neutrons, CEA-N-1171 (1969).
13. Y. Bando and K. Ishizka, "Study of the Structure of Silica Glass by High Resolution Electron Microscopy," Jour. non Cryst. Solids 33, 375 (1979).
14. F. Lell, N. J. Kreidl and J. R. Hensler, "Radiation Effects in Quartz, Silica and Glasses," Progr. Ceram. Sci. 4, 1 (1966).
15. E. W. J. Mitchell and E. G. S. Paige, "The Optical Effects of Radiation Induced Atomic Damage in Quartz," Phil. Mag. 1, 1085 (1956).
16. W. Arnold, "Ion-Implantation Effects in Non-Crystalline SiO_2," I.E.E.E. Trans. Nucl. Sci. NS-20, 220 (1973).
17. A. R. Ruffa, "Model for the E_1 Center on SiO_2," Phys. Rev. Lett. 25, 650 (1970).
18. G. N. Greaves, "Intrinsic and Modified Defect States in Silica," Jour. Non Cryst. Solids 32, 295 (1979).

A MICROSCOPIC APPROACH FOR THE SIMULATION OF RADIOACTIVE WASTE STORAGE IN GLASS

J. C. Dran*, Y. Langevin*, M. Maurette*, J. C. Petit**

*Laboratoire Rene Bernas, 91406, Orsay, France
**Service de Radiochimie et Phénoménologie, CEA, 92852, Montrouge Cedex, France

ABSTRACT

Leaching of samples externally irradiated with low energy ($\sim$ 1keV/nucleon) heavy ions shows that the accumulation of α-recoils in glass can drastically reduce the life-time of waste storage glasses in geological environments against complete dissolution by ground water solutions.

INTRODUCTION

The long term stability of glasses, used for the storage of highly radioactive wastes, represents one of the most important constraints on the safety of waste disposal. This stability is likely related in great part to various irradiation effects, due to the decay of fission products and actinides trapped in these materials. For about 20 years, numerous authors have studied such effects by relying on two types of simulation experiments: 1) external irradiations intended to simulate the effects of β and γ decays. 2) incorporation of relatively short-lived α-emitters (^{238}Pu, ^{244}Cm, etc...) in glass samples that simulates in a few years the effect of α-decay of wastes stored for periods of up to several thousand years (1). The stability of storage glasses against radiation damage has been evaluated, mostly by measuring the increase of leach rate of glass immersed in aqueous solutions of various compositions upon irradiation.

Such experiments seem to show that irradiation with high doses of α, β or γ particles, simulating storage time up to a few thousand years, does not noticeably affect the glass leach rate. However, the interpretation of these experiments, including those based on the doping of glass with α-emitters, has been mostly done in terms of the effect of α-particles. Consequently, the specific effect of α-recoil damage has been so far overlooked. In addition this classical approach for investigating α-decay effects presents two other important limitations: 1) the study of the samples has to be performed within a hot cell, and this in turn considerably hampers the detailed analysis of irradiation effect; 2) the simulation of very long storage periods ($\gtrsim$ 1000 years) cannot be done within a reasonable time (a few years).

MICROSCOPIC APPROACH FOR EVALUATING THE EFFECTS OF α-RECOILS

In order to minimize such basic limitations, we have developed a new "microscopic" approach, which is related to our previous studies of radiation damage effects in terrestrial and extraterrestrial minerals (2). In this approach, we investigate the alteration of the long term stability of storage glasses that results from the slow accumulation of α-recoil tracks (3) in the glass matrix by relying on both the artificial irradiation of glass surface with low energy ($\sim$ 1keV/nucleon) heavy ions and the measurement of the etching rate of the irradiated material (the only previous experiment using implantation of lead ions on glasses for simulating α-recoils dealt with radiation induced textural changes) (4). For the ion implantation, we use either U or Pb ions with energies ranging from 0.5 to 3 keV/nucleon. The doses can vary from 10^9 to 10^{14} ions, thus simulating the internal implantation of α-recoils over a storage period extending up to 10^6 years for typical PWR wastes. The targets include polished sections covered with an electron microscope grid which gives a regular succession of irradiated and non-irradiated areas, thus considerably easing the observation and the measurement of irradiation effects. We can also irradiate micro-sized grains firmly fixed on special carbon substrates for electron microscope observations.

The leaching runs are performed within an autoclave in order to choose temperature and pressure conditions corresponding to different burial depths in geological sites. We currently use various etching solutions (sea water, ground waters, NaCl solutions, etc.) in order to simulate different geological environments.

The microscopic measurement of the etching rate of the irradiated glass is performed in two different ways: in the case of the grid-covered polished sections, it is performed with a diamond stylus device (Talystep) which permits the evaluation of depth differences between irradiated and unirradiated areas with an accuracy of $\sim$10 Å, reflecting the increase of the etching rate, $V = V^* - V_o$, where V^* and V_o are the etching rates of these two types of areas, respectively. In the case of irradiated micron-sized grains, the etching rate is directly determined by observing the shrinkage of given grains with a high voltage electron microscope.

This original approach allows the study of non-active samples and the easy simulation of very long storage periods ($\gtrsim 10^5$ yrs.) It thus avoids the two limitations inherent to previous approaches. However, our technique has the limitation of not integrating other effects associated to α-decay such as glass fracturation, helium bubble formation, blistering, etc.

PRELIMINARY RESULTS

So far we investigated three types of glasses, mostly differing with respect to their SiO_2 content, as it is currently thought (5) that the leach rate of glass is mostly related to this parameter: the low silica glass (64% SiO_2, 12% B_2O_3, 24% Na_2O) is similar to those presently proposed for the storage of high level wastes, while the intermediate (75% SiO_2, 1% Al_2O_3, 3% MgO, 9% CaO, 11% Na_2O, 0, 12% K_2O) and high silica (100% SiO_2) glasses look promising for our investigation of fundamental radiation damage processes in glass and because they could possibly represent future technological improvements of storage glasses.

We have implanted doses of $\sim 10^{13}$ Pb/cm^2 (equivalent to $\sim 2\times10^{18}$ internal α-recoils per gram), simulating $\sim$5000 years of disposal for typical PWR wastes. The lead nuclei had an energy of 1keV/nucleon corresponding to a projected range in glass of $\sim$500Å. The irradiated glasses have been leached in sea water or NaCl solutions at atmospheric pressure and temperature of 100°C for leaching times varying from 1mn to 1 day.

Our major findings are the following: 1) In all cases, after a sufficient leaching time, we observed a print of the grid on the glass target, easily visible by phase contrast optical microscopy (Fig. 1.), that reflects the enhanced etching rate of the irradiated glass. With the Talystep device (Fig. 2.) we could measure the increase of the etching rate ΔV. From the value of ΔV we then deduced the maximum lifetime against dissolution (Table 1, column 4) assuming simple, linear reaction rates. This time appears to be particularly short for the storage glass currently used. It can be noted that such a lifetime could be drastically reduced in case of the glass fracturing; however, this was not taken into consideration in the present work. The effect of α-recoil aging on the long-term stability of glass may have been overlooked in previous studies, mainly because they did not combine the necessary conditions to easily observe it. In particular, no highly "irradiated" glass ($\gtrsim 10^{13}$ ions/cm^2) has been leached with realistic solutions (such as a carnallite solution) at sufficiently high temperature ($\sim$100°C).

Fig. 1. Phase Contrast Optical Micrograph of a Typical Glass Surface Irradiated Through a Grid (10^{13} Pb/cm^2) and Subsequently Leached with a NaCl Solution. The Square Zones ($\sim$200μm x 200μm) Correspond to Cavities Resulting from the Preferential Etching of the Irradiated Material.

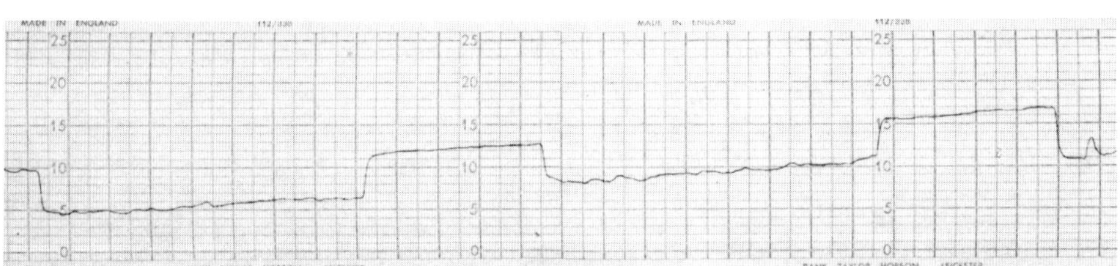

Fig. 2. Depth Profile Obtained with the Talystep Device that Reflects Depth Difference of ~500 Å

2) Upon thermal treatments conducted under vacuum at temperatures varying between 50°C and 600°C, the ΔV values for the microscope slide glass steeply decreases for temperatures ≥400°C as a result of α-recoil track annealing. This high temperature annealing of radiation damage, which is in good agreement with the predictions

Table 1. Effects of α-recoil Nucleii on Glass

Type of Glass	Composition (%)	V (Å/hr)	Life time of a 0.5 m diameter rod (yr)
Silica	SiO_2 100	0.2	$1.4x10^6$
Microscope slide (Fisher)	SiO_2 75		
	Na_2O 11	200	$1.4x10^3$
	CaO 9		
	MgO 3		
Borosilicate glass	SiO_2 64	2700	105
	B_2O_3 12		
	Na_2O 24		

of our new model of nuclear particle tracks in insulators (6), probably excludes the possibility of a natural "curing" of α-recoil effects produced in storage glasses buried in geological formations.

CONCLUSIONS AND FUTURE PROSPECTS

The microscopic approach that we have developed appears to be a very powerful tool for revealing the effects of α-recoils which have been overlooked by previous authors. These effects could drastically affect the lifetime of storage glasses disposed in geological formations where they could be subjected to leaching by natural waters.

Compared to classical doping techniques, our approach presents additional interesting features: its simplicity allows a very rapid evaluation of a great number of possible combinations between glasses of different compositions and disposal conditions in geological sites. Moreover, such studies can be extended to all types of storage material candidates so far proposed.

We are currently performing leaching experiments intended to decipher the respective influences of various environmental parameters (composition of leaching solutions, pH, temperature, etc.) on the α-recoil effects. As materials irradiated in high ion fluence are etched very quickly, we can expect to use pre-irradiated glass targets as probes of the "leaching power" of various natural environments (salt mines, granites, shales, etc.)

in order to identify the most favorable geological formation for waste disposal.

This approach could also be applied with minor changes to a reassessment of α, β and γ irradiation effects on the long term stability of storage materials. In this evaluation we have only taken into account the radiation induced enhancement of the etching rate after a storage period of a few thousand years, corresponding to a dose of 10^{13} lead nuclei/cm^2.

ACKNOWLEDGEMENTS

We thank Drs. R. Klapisch and G. Goutiere for their enthusiastic support and interest. We are indebted to G. Vidal for the use of the Talystep instrument at ONERA, Chatillion sous Bagneux, France. We are grateful to F. Lalu and B. Vassent for their assistance. Financial support was kindly provided to us by "Commissariat a l'Energie Atomique."

REFERENCES

1. A. R. Hall, J. T. Dalton, B. Hudson and J. A. C. Marples, "Development and Radiation Stability of Glasses for Highly Radioactive Wastes," Proc. of the Symposium on Management of Radioactive Wastes from the Nuclear Fuel Cycle, IAEA-ENA (1976).
2. M. Maurette and P. B. Price, "Electron Microscopy of Irradiation Effects in Space," Science 187, 121 (1975).
3. W. H. Huang and R. M. Walker, "Fossil Alpha Particle Recoil Tracks: A New Method of Age Determination," Science 155, 1103 (1967).
4. A. K. De, B. Luckscheiter, W. Lutze, G. Malow and E. Schiewer, "Development of Glass Ceramics for the Incorporation of Fission Products," Ceramic Bulletin 55, 500 (1976).
5. R. H. Doremus, Glass Science (John Wiley and Sons, New York, 1973).
6. E. Dartyge, J. P. Duraud, Y. Langevin and M. Maurette, "A New Method for Investigating the Past Activity of Ancient Solar Flare Cosmic Rays Over a Time Scale of a Few Billion Years," Proc. Lunar Planet Sci. Conf. 9th, 2375 (1978).

DEVITRIFICATION OF NUCLEAR WASTE GLASSES*

R. P. Turcotte, J. W. Wald and R. P. May

Pacific Northwest Laboratory
P. O. Box 999
Richland, WA 99352

Devitrification studies of waste glasses are undertaken for a variety of reasons, but mainly to determine processing conditions through which significant crystallization can be avoided. The first detailed kinetic study by Turcotte and Wald (1) demonstrated that although behavior is complex, the kinetics follow expected trends, based on simple theoretical models of simple glasses.

The present work briefly considers behavior in a variety of glass compositions, using the general experimental approach described previously. One important variation in the X-ray diffraction work has been the use of vitreous waste glass/Al_2O_3 mixtures as "universal" intensity standards. The crystallinity (concentration) data presented in the present paper are, therefore, not precise on an absolute scale, but allow adequate comparison of one glass to another. The compositions of the various glasses examined have been reported elsewhere as follows: glass 72-68 (2), glass 76-68 (3), glass 77-260 (4), glass SRL-C-411 (5), and glass SRL-T-211 (6). Crystallization trends are established by isothermal and constant cooling rate experiments, but only the isothermal results are reported here.

Fig. 1a. shows the ingrowth of Zn_2SiO_4 as a function of time and temperature in glass 72-68. (In this case glass/Zn_2SiO_4 standards are used to evaluate crystal yield.) Well behaved, sigmoidal fractional ingrowth curves are obtained (Fig. 1b.), giving a reasonable (1/T) temperature dependence (Fig. 1c.), with the activation energy calculated to be 331 kjoules/mole.

*This work supported by the U. S. Department of Energy under Contract EY-76-C-06-1830.

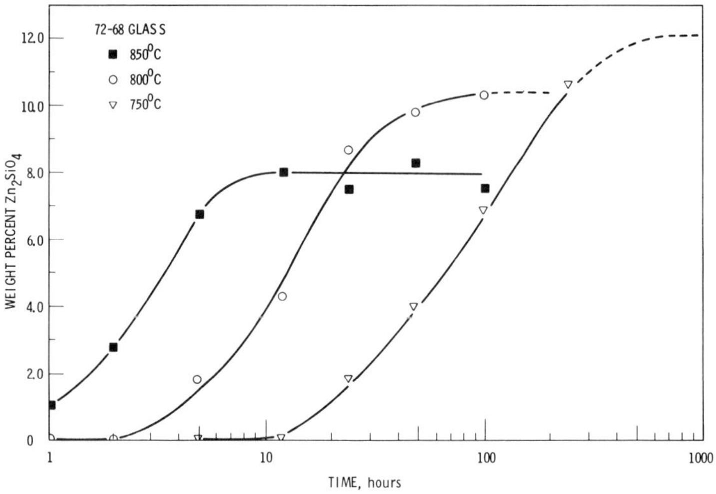

a) Ingrowth of Zn_2SiO_4 in Glass 72-68 at 750°, 800° and 850°C as a Function of Time

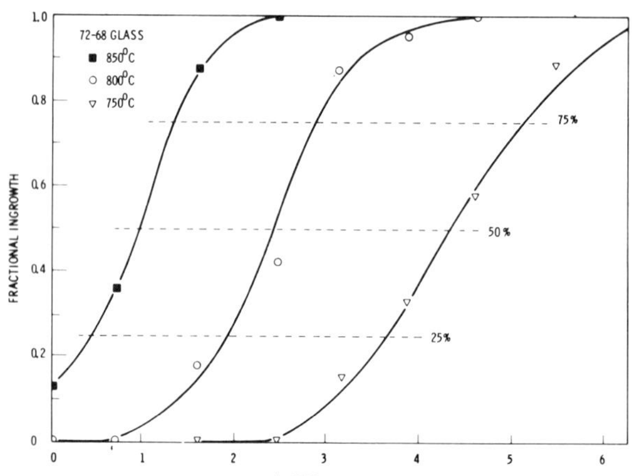

b) Fractional Ingrowth of Zn_2SiO_4 Versus ℓn Time

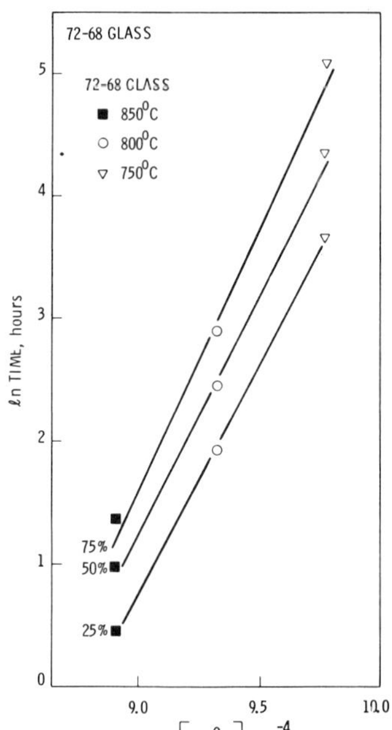

c) Natural Log Time Versus Inverse Temperature for 25%, 50%, and 75% Ingrowth of Zn_2SiO_4

Fig. 1. Crystalline Ingrowth of Zn_2SiO_4 in Glass 72-68 as a Function of Time for Three Different Temperatures

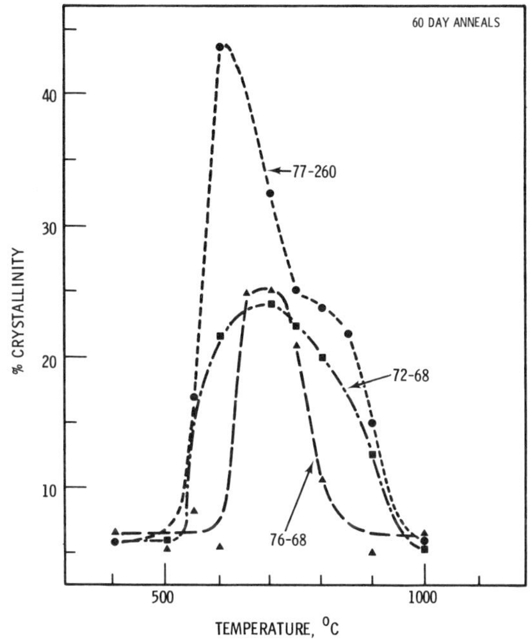

Fig. 2. Percent Crystallinity Curves as a Function of Temperature for Glasses 72-68, 76-68 and 77-260, Annealed for 60 Days

Fig. 2. shows "percent crystallinity" curves as a function of temperature for three glass compositions after annealing for 60 days. Table 1 lists the phases present in each of the glasses. The least stable glass (77-260) shows a bimodal crystallinity envelope arising from rapid formation of a gadolinium titanate phase at high temperatures, whereas formation of a rare-earth apatite (silicate) phase becomes important at longer times and lower temperatures, being the major contributor to a 45% crystalline content after 60 days at 600°C. Composition 76-68, which contains high iron and sodium concentrations, is nevertheless the most devitrification-resistant glass studied--both with respect to crystal yield and to the temperature region over which significant changes occur (600° to 800°C). However, it too becomes extensively crystallized after annealing at 650°C for one year with a slow-forming silicate ($NaFeSi_2O_6$) becoming the predominant phase.

Fig. 3. shows similar data for two glasses developed for defense wastes, after annealing 4 days at the temperatures indicated. In this case, the individual phase concentrations have been evaluated as also summarized in Table 1. Highly devitrified specimens of both SRL glasses contained large crystals of Li_2SiO_3 or $NaFeSi_2O_6$ which result in extensive microcracking. A common feature in some glass systems is the rapid formation of refractory oxides such as CeO_2

Table 1. Principal Melt Insolubles and Devitrification Products for Five Glass Compositions

72-68	76-68	77-260	SRL-C-411	SRL-T-211
RuO_2	RuO_2	RuO_2	$(Ni,Mn)Fe_2O_4$	$(Ni,Mn)Fe_2O_4$
$NiFe_2O_4$	$NiFe_2O_4$	$NiFe_2O_4$	$NaAlSiO_4$	$NaAlSiO_4$
Pd	Pd	Pd	Li_2SiO_3	$NaFeSi_2O_6$
CeO_2	CeO_2	CeO_2		
Zn_2SiO_4	$NaFeSi_2O_6$ [a]	$GdTi_2O_7$		
$SrMoO_4$		$Ca_3Gd_7[SiO_4]_5(PO_4)O_2$		

(a) Appears only after 1-mo anneal.

or $NiFe_2O_4$ (shown by glass T-211), at temperatures not much below the melting temperature, whereas more complex silicate phases form at lower temperatures in all five glasses reported.

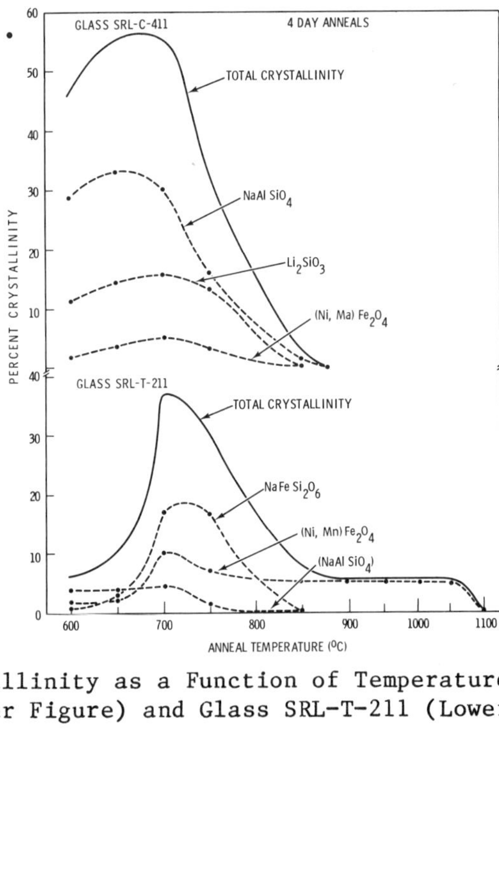

Fig. 3. Crystallinity as a Function of Temperature for Glass SRL-C-411 (Upper Figure) and Glass SRL-T-211 (Lower Figure), Annealed 4 Days

By evaluating yield-curves such as those shown in Fig. 2. and 3., but as a function of anneal time, it is possible to define the devitrification kinetics in the form of time-temperature-transformation (TTT) curves. Fig. 4 shows data based on 5% crystalline ingrowth for two of the glasses. These TTT curves reflect the growth of new crystals in the glass, which is in addition to the 2-5% crystalline content present as melt insolubles. The choice of a 5% ingrowth level is arbitrary except that it indicates minimum conditions that must be met to avoid crystallization. The choice of either 2% or 10% criteria to define extent of crystallization results in only a small effect on the shape and position of the curves. Most of the data available for other glass compositions falls near or within the region bounded by the two curves given. It should be noted that the curves are overall representations of rather complex processes, since 2 or 3 major phases are crystallizing with different time-temperature kinetics--each with their own "liquidous" temperatures. The relatively narrow band describing crystallization kinetics is likely a result of a glass development criteria that viscosity should be ∿200 poise or less at 1050°C. Hence, most waste glass compositions studied to date have similar ion mobilities in the melt, giving rise to relatively narrow property ranges. If the curve for glass (76-68) is examined as an example, ∿5% crystalline ingrowth is achieved in one day (or slightly less) at between 700° and 800°C. At 650°C, 10 days is required, and at 590°C one year is needed. At 875°C, 5% devitrification is not achieved even though the kinetics are rapid, because the glass is near the "liquidous" temperature (∿900°C).

Finally, it should be pointed out that there have been other recent studies of secondary effects caused by devitrification. These include a radiation effects evaluation by Weber, et al. (7) and a

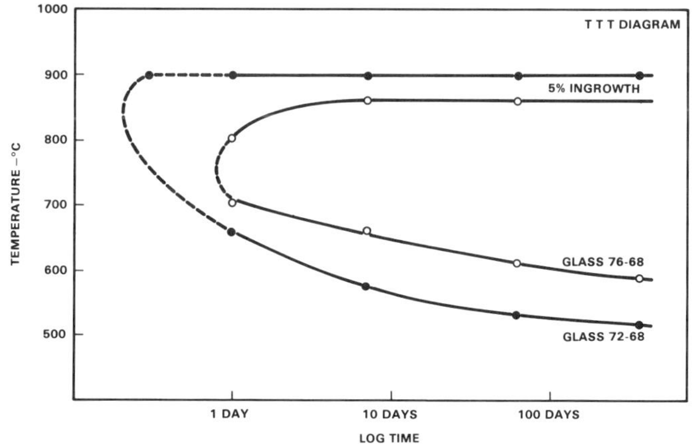

Fig. 4. Time-Temperature-Transformation Behavior for Glasses 72-68 and 76-68. Curves represent 5% crystalline ingrowth.

study of behavior in actual high-level-waste glasses by Wald and Westsik (8). Both of these studies show that the effect of devitrification on leach rate is minimal. With respect to physical integrity, formation of large crystals has been observed to cause microcracking, which, however, can be entirely avoided by maintaining temperatures at not more than 100°C above the dilatometric softening point. For present waste glasses, long time annealing at temperatures beetween 500° to 650°C results in a fine dispersion of micron-sized crystals similar to glass-ceramic textures. No crystallization has been observed below 500°C.

REFERENCES

1. R. P. Turcotte and J. W. Wald, Devitrification Behavior in a Zinc Borosilicate Nuclear Waste Glass, PNL-2247, Pacific Northwest Laboratory, March, (1978).
2. J. L. McElroy, Quarterly Progress Report, Research and Development Activities, Waste Fixation Program, July-September 1975, BNWL-1949, Pacific Northwest Laboratory, January (1976).
3. J. L. McElroy, Quarterly Progress Report, Research and Development Activities, Waste Fixation Program, October-December 1976, PNL-2264, Pacific Northwest Laboratory, Nov. (1977).
4. J. L. McElroy, Quarterly Progress Report, Research and Development Activities, Waste Fixation Program, July-September 1977, PNL-2265-3, Pacific Northwest Laboratory, Oct. (1973).
5. J. L. McElroy, Quarterly Progress Report, Research and Development Activities, Waste Fixation Program, January-March 1979, PNL-3050-1, Pacific Northwest Laboratory Nov. (1979).
6. J. L. McElroy, Quarterly Progress Report, Research and Development Activities, Waste Fixation Program, July-September 1979, PNL-3050-3, Pacific Northwest Laboratory (in preparation).
7. W. J. Weber, R. P. Turcotte, L. R. Bunnell, F. P. Roberts and J. H. Westsik, Jr., "Radiation Effects in Vitreous and Devitrified Simulated Waste Glass," Proceedings of International Symposium on Ceramics in Nuclear Waste Management, CONF-790420, Technical Information Center, Oak Ridge, TN (1979).
8. J. W. Wald and J. H. Westsik, Jr., "Devitrification and Leaching Effects in HLW Glass--Comparison of Similated and Fully Radioactive Waste Glass," Proceedings of International Symposium on Ceramics in Nuclear Waste Management, CONF-790420, Technical Information Center, Oak Ridge, TN (1979).

HIGH-LEVEL WASTE VITRIFICATION PRODUCT CHARACTERIZATION

G. H. Bryan, W. J. Bjorklund and W. L. Kuhn

Pacific Northwest Laboratory
Richland, WA 99352

INTRODUCTION

The Nuclear Waste Vitrification Project (NWVP) was conducted to provide a demonstration of the vitrification of high-level liquid waste (HLLW) from spent fuel discharged from an operating LWR.

The objective of the NWVP was to demonstrate the solidification (vitrification) as a borosilicate glass, of actual high-level liquid waste from spent LWR fuel.

The scope of NWVP included the design, installation, and operation of a small-scale (1 tU LWR fuel/month) plant preparing actual HLLW for vitrification, and waste solidification by operation of a previously used 1 tU/day spray calciner in-can melter (SC/ICM) vitrification system (1). The HLLW was vitrified in two 20.3-cm-(8-in.-) diameter by 2.44-m-(8-ft.-) long canisters. (See Fig. 1).

WASTE VITRIFICATION

Feed Preparation

The HLLW chemical composition was adjusted to that of typical waste by addition of uranium and non-radioactive chemicals (2). The waste was then concentrated to the required volume and acid concentration for feed to the vitrification process.

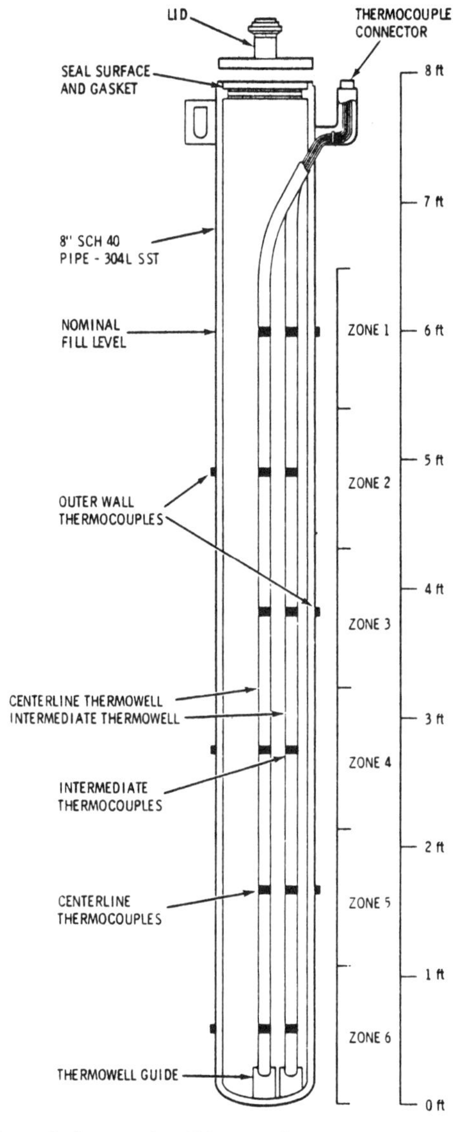

Fig. 1. Schematic View of a Waste Canister

Vitrification

Concentrated waste from the feed preparation runs was used for the two batch operations of the spray calciner/in-can melter system (SC/ ICM). Operating conditions for both vitrification runs are summarized in Table 1. Run #1 used feed prepared from the lower burnup waste, while Run #2 used a mixture of both lower burnup and higher burnup waste feed.

Table 1. Operating Conditions for the Nuclear Waste Vitrification Project

Conditions	Run #1	Run #2
Feed Concentration (L/tU)	833	760
Feed HNO_3(M)	4.4	6.6
Solids Content (g oxide/L)	66	86
Feed Processed (L)*	806	430
Average Feed Rate (L/h)	15.7	12.7
Heat Content (W/L)**	1.12	2.05
Activity (Ci/L)**	313	526
Run Time (h)	53	33.6
Average Fuel Burnup (MWd/tU)	20,400	25,900
Onstream Factor (%)	99	96

*Not all of the calcined feed was vitrified.
**Calculated from ORIGEN code.

The vitrification runs went smoothly, including operation of the effluent cleanup system. One equipment component failed during the first vitrification run, but this was not discovered until after the run was completed. A ceramic cap on the feed nozzle fell off during Run #1, resulting in a smaller glass volume in Can #1 than was expected. The material was removed from the calciner before Run #2. A small amount of this material, plus some added frit, was in the bottom of the second canister before Run #2 and formed an initial layer of calcine. All other equipment operated well.

The waste-glass canisters used in the in-can melting process were maintained at 1000 to 1050°C during the entire filling of the can. Once the cans were full, they were force-cooled with air while still in the furnace, until the wall temperatures were 550°C for all zones; this process took about two hours. The canisters were held for six hours at a wall temperature of 550°C. The forced air was then turned off and the canisters allowed to cool naturally until the walls reached 200°C (30 hours).

Non-Destructive Testing

Characterization of the Nuclear Waste Vitrification Project's canisters of high-level waste glass will be done in a variety of ways. The material presented in this paper covers only the non-destructive aspects; however, the results of these tests will then be used to recommend a choice of canisters to be destructively characterized. Non-destructive examination will utilize methods already available. The non-destructive tests along with their expected benefits are shown in Table 2.

Table 2. Non-Destructive Testing

Test	Benefit
1. Canister measurement	
a) Weight and height of fill.	a) Determination of bulk density.
b) External canister measurements.	b) Evaluation of canister rigidity during melting.
c) Wall thickness.	c) Determination of combined internal corrosion and external oxidation.
2. Radiation	
a) Gross gamma measurement.	a) Direct determination of external gamma field.
b) Axial gamma scan.	b) Determination of radially-averaged homogeneity of gamma emission; height of radioactive fill.
3. Thermal	
a) Canister internal thermocouple readings.	a) Determination of radial temperature distribution.
b) External temperature scan by infrared camera.	b) Evaluation of homogeneity of heat loading near surface.
c) Calorimetry	c) Direct determination of canister heat loading.

Canister Measurement

Weight and Height of Fill. Weight and height of fill measurements on the HLW glass put into each of the two canisters are shown in Table 3. The frit to calcine ratio and glass density values are also shown for comparison purposes.

External Canister Measurements. The NWVP canisters were measured for roundness prior to filling by in-can melting. Canister #1 has been measured to check for deformation that may have occurred during the process. Diameters were measured at 0, 45, 90, and 135° orientations at 40.6 (16 in.), 119.4 (47 in.), and 228.6 (90 in.) centimeters below the canister top. The maximum change is 0.2% and none of the changes are considered significant.

Wall Thickness Measurements. Canister #1 was fabricated from welded pipe. The thickness was exceptionally uniform except along each side of the weld. A variation of about 7% was noted in the wall thickness at this point. The typical variation in wall thickness is less than 1%.

Table 3. Canister Measurements

Conditions	Run #1	Run #2
Glass weight (Kg)	116	145
Glass volume (L)	41.3	52.9
Fill height (cm)	137	175
Frit to calcine ratio	4.2:1	2.2:1
Glass density (g/cc)	2.8	2.7

Radiation

Gross Gamma Measurement. The result of the gross gamma measurement of Canister #1 are shown in Table 4. The measurement was made at a height of 0.6m (2 ft.) from the bottom of the canister at the distances indicated. Very little beta contamination on the external surfaces of the canisters was noted. These measurements have not been made on Canister #2.

Axial Gamma-Scan. NWVP canister #1 and #2 were gamma-scanned while lifted out of the furnace. Relative concentrations of ^{137}Cs, ^{154}Eu, and ^{106}Ru were measured as a function of axial position, starting at the top glass surface and working down each 6 inches to the bottom. The results for ^{137}Cs, ^{154}Eu, and ^{106}Ru are plotted in Fig. 2.

The results show a radiation peak for all species at the very bottom of Canister #2. This indicates a vestigial layer of calcine not successfully mixed with frit prior to the start of Run #2. Fifteen centimeters (6 in.) below the top of fill for either canister there is increased ^{106}Ru and ^{154}Eu activity and decreased ^{137}Cs activity. Evidently, residual insolubles involving ruthenium and europium, but not cesium, floated to the top to form a porous layer. The increased ^{137}Cs activity at the very top is probably due to redeposition of cesium volatilized during vitrification. Aside from the top 15 centimeters (6 in.) and very bottom of Canister #2, both canisters appear homogeneous. The fluctuations in ^{106}Ru are thought to be due to poorer counting statistics than for other species.

Table 4. Gross Gamma Measurement of Canister #1

Distance (m)	R/hr
2	1500
1	4080

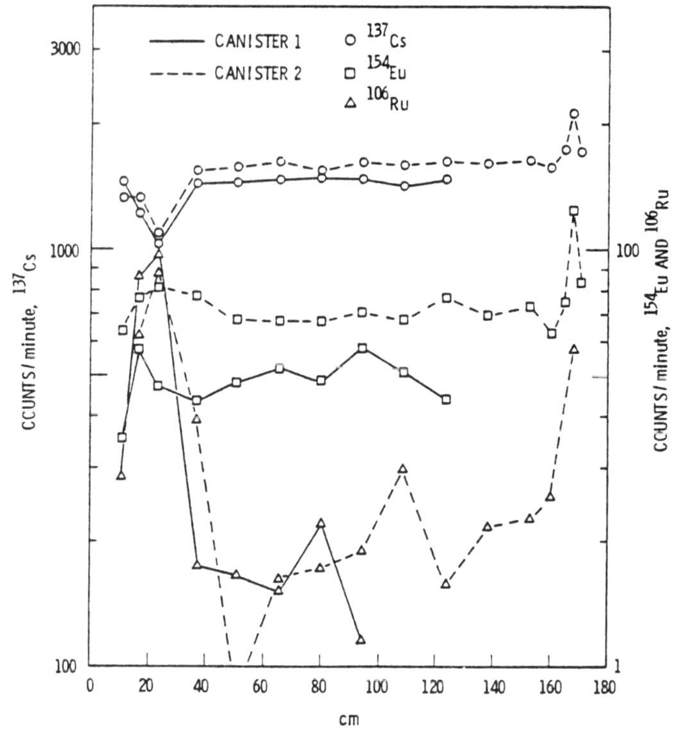

Fig. 2. Axial Gamma Scans of Canisters #1 and #2

Thermal

Canister Internal Thermocouple Readings. Eighteen thermocouples were placed in each canister to measure temperature. They were spaced at 33 centimeter (13 in.) intervals on the vertical axis beginning at 61 centimeter (24 in.) from the top of the canister in groups of six to measure centerline, intermediate (6.85 centimeters off center) and shell temperatures. The actual measured temperatures for Canister #1 are shown in Table 5.

External Temperature Scan by Infrared (IR) Camera. The results of an IR scan of the canister indicate heat output to be uniform. Some anomalies are noted; however, they are believed to be due to non-uniformity of the oxide layer on the surface of the canister, causing a variation in the emissivity.

Calorimeter. The heat generation rate due to radioactive decay was measured in an insulated calorimeter. The calorimeter is a tank of water in which cooling coils can circulate water to adjust the

Table 5.

Centerline TC's		Intermediate TC's		Shell TC's	
1	53.8°C	7	53.7°C	13	55.2°C
2	64.3	8	*	14	*
3	110.4	9	99.8	15	N/A
4	129.9	10	111.8	16	102.2
5	125.0	11	113.5	17	104.1
6	113.4	12	104.1	18	88.9

*Not working

temperature. The heat load can be calculated by measuring the flow rate and temperature change of cooling water required to maintain the calorimeter temperature constant upon insertion of a canister. The water volume is adjusted to keep the same level after inserting the canister.

The cooling is automatically controlled using a cascaded temperature/flow controlling system. The temperature set point is taken to be the calorimeter temperature without the canister inserted. After inserting the canister, a controller adjusts the set point of a flow controller to control the cooling water flow to maintain the temperature set point. A heat generation rate of $\sim$ 350 watts/hr for Canister #1 has been determined. From the heat generation rate and centerline-to-surface temperature differences, the apparent thermal conductivity of the glass is 0.82 W/m°C. This compares favorably with laboratory data (3).

CONCLUSIONS

Non-destructive testing of the first canister of high level waste glass has indicated very little in the way of unexpected behavior. The glass appears to be uniform with no large anomalies noted. Non-destructive testing of the second canister is incomplete at this time; however, tests have not indicated any unexpected or unusual results.

Further testing by destructive methods (e.g., core drilling for an interior view of the glass within the canister, core sampling for determining density, crystallinity, microscopic homogeneity, mechanical conditions, etc., and leach tests to provide fundamental information about leach kinetics and equilibria, etc.) will yield more specific information on the glass. Until such time as these tests are complete, a full comparison of expected and obtained product cannot be made.

REFERENCES

1. J. L. McElroy, K. J. Schneider, J. H. Hartley, J. E. Mendel, G. L. Richardson, R. W. McKeen and A. G. Blasewitz, Waste Solidification Program Summary Report, Vol. II, BNWL-1667, Battelle-Pacific Northwest Laboratory, Richland, WA 99342, (1972).
2. E. J. Wheelwright, W. J. Bjorklund, L. M. Browne, G. H. Bryan, L. K. Holton, E. R. Irish, D. H. Siemens, Technical Summary-Nuclear Waste Vitrification Project, PNL-3038, Battelle-Pacific Northwest Laboratory, Richland, WA 99352, (1979).
3. J. E. Mendel, Annual Report on the Characteristics of High-Level Waste Glass, BNWL-2252, Battelle-Pacific Northwest Laboratory, Richland, WA 99352, (1977).

LONG-TERM EXTRAPOLATION OF LABORATORY GLASS LEACHING DATA FOR THE PREDICTION OF FISSION PRODUCT RELEASE UNDER ACTUAL GROUNDWATER CONDITIONS

F. B. Walton* and W. F. Merritt**

*Atomic Energy of Canada Limited, Whiteshell Nuclear Research Establishment, Pinawa, Manitoba, Canada
**Chalk River Nuclear Laboratories, Chalk River, Ontario, Canada

ABSTRACT

Release and migration of ^{90}Sr and ^{137}Cs from nepheline syenite-based glass buried since 1960 below the water table in sandy soil at Chalk River have been predicted by models using laboratory glass leaching data. Model predications of ^{90}Sr release and migration show good agreement with field measurements. ^{137}Cs concentration profiles in the soil suggest that a simple equilibrium ion exchange model is inadequate to predict ^{137}Cs migration under present test conditions. Soil matrix support of a friable, corrosion-product layer is believed to be largely responsible for observed glass leach rates of the order of 8×10^{-14} kg/m^2·s after exposure to groundwater for 17 years.

INTRODUCTION

In June 1960, twenty-five nepheline syenite-based glass hemispheres containing the fission products ^{137}Cs, ^{90}Sr, ^{144}Ce and ^{106}Ru were buried below the water table in sandy soil at the Chalk River Laboratories of Atomic Energy of Canada Limited (see Fig. 1.). The experiment was designed to provide data for a study of groundwater leaching of the glass and subsequent transport of the fission products. In October 1978, one of these glass blocks was retrieved and brought to WNRE for examination. An accurate assay of the glass fission-product inventory has been made, thus enabling a re-evaluation of the field data obtained since 1960. The objectives of the present study are to determine whether:

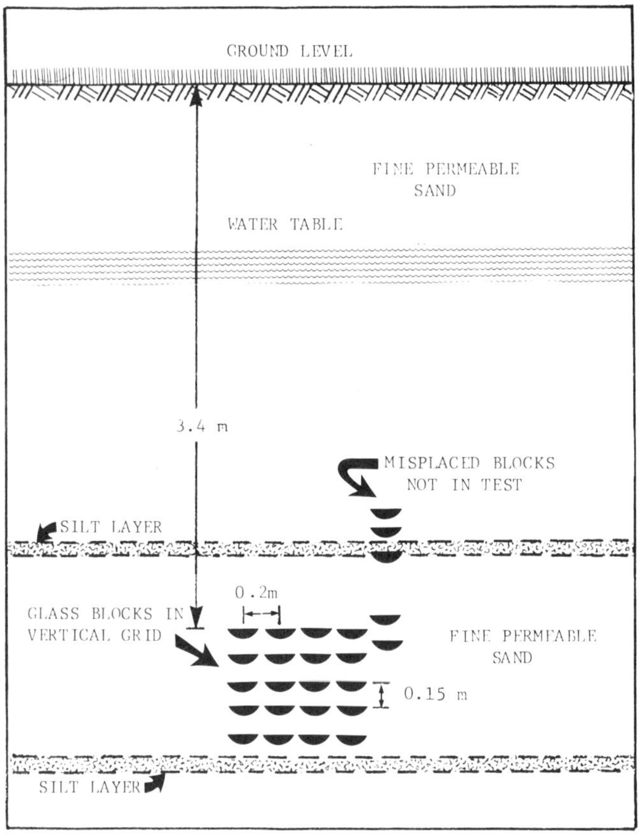

Fig. 1. Grid of Glass Blocks in Soil. Only twenty-two were correctly positioned. One of the misplaced blocks was retrieved in 1978 for examination.

- the release of the fission products ^{90}Sr and ^{137}Cs can be predicted from the extrapolation of laboratory leaching data;
- the groundwater and/or sandy soil have any effect on glass leaching;
- the chemical interaction between dissolved ^{90}Sr and ^{137}Cs and soil can be modelled as simple equilibrium ion exchange;
- a simple one-dimensional porous flow (dispersion) model is adequate to describe the hydraulic transport of the fission products in water-saturated, sandy soil.

MODEL DESCRIPTION AND PARAMETER ESTIMATION

The differential equation describing convective-dispersive, one-dimensional solute transport in an isotropic medium is:

$$\frac{\partial C}{\partial t} = D \frac{\partial^2 C}{\partial X^2} - V \frac{\partial C}{\partial X} + \frac{\rho(1-\varepsilon)}{\varepsilon} \frac{\partial S}{\partial t} \qquad 1)$$

where
- C is the concentration of the solute in the liquid phase
- S is the concentration of the solute adsorbed on the solid phase
- D is the dispersion coefficient
- V is the Darcy water velocity
- ρ is the soil density

and ε is the soil porosity.

If a linear adsorption isotherm of the form $S = k_d C$ is assumed to exist between the solute and the soil matrix, then equation 1) reduces to:

$$\frac{\partial C}{\partial t} = D_o \frac{\partial^2 C}{\partial X^2} - V_o \frac{\partial C}{\partial X} \qquad 2)$$

where $D_o = D/(1 + \rho(1-\varepsilon)k_d/\varepsilon)$ and $V_o = V/(1 + \rho(1-\varepsilon)k_d/\varepsilon)$.

Equation 2) was solved by a finite-difference method assuming:

$$C = 0,\ t < 0;\ C = 0,\ X \to \infty,\ t > 0;\ C = C(t),\ X = 0,\ t > 0.$$

DATA SELECTION

The data required for these calculations can be divided into three areas: hydraulic, soil-water interactions and glass leaching.

HYDRAULIC PARAMETERS

An effective groundwater velocity across the disposal area of 17.8 cm/d has been measured using tritiated water (1). The hydraulic dispersion coefficient was calculated using two methods. For very small velocities, Neretnieks (2) gives the following correlation:

$$D = 5.3 \left(\frac{V d_p}{1-\varepsilon}\right) \qquad 3)$$

where d_p is the particle diameter. For measured soil parameters of $\varepsilon = 0.38$ and $d_p = 0.02$ cm, D is 3.5×10^{-5} m^2/s. This is of the same order of magnitude as would be expected for molecular diffusion.

A correlation based on a theoretical model by Simpson (3) and fit to the crushed quartz sand data of Harleman et al. (4) gives:

$$\frac{D}{Vd_p} = -\frac{0.9\ R}{R + 0.05} \qquad 4)$$

where the Reynolds number $R = d_p V/\nu$ and ν is the kinematic viscosity of the fluid. Using this method, the dispersion coefficient was calculated to be 1.6×10^{-7} m^2/s which is less than molecular diffusion. Since $D \rightarrow$ molecular diffusion as $V \rightarrow 0$, the value of 3.5×10^{-5} cm^2/s, which is close to that of molecular diffusion, was used in these calculations.

SOIL-WATER INTERACTIONS

A cation exchange capacity of ~ 1 meq/100 g soil has been measured by Evans (5) for the Chalk River soil. Based on soil-water pH and the work of Rhodes and Nelson (6) with soil of the same cation exchange capacity, a k_d of 4 - 7 L/kg for ^{90}Sr and > 200 for ^{137}Cs was estimated. Recent measurements for Chalk River soil by Jackson et al. (7) give a k_d for ^{90}Sr of 4.5 to 10 L/kg and a k_d for ^{137}Cs between 80 and 250 L/kg, depending on the method of determination. For illustrative purposes model predictions presented here are for a ^{90}Sr k_d of 4.6 to 7 L/kg and a ^{137}Cs k_d of 80 to 230 L/kg.

GLASS LEACHING

Bancroft (8) had measured the leach rate of glass blocks of the same composition as those used in this experiment using distilled water at room temperature in a flowing system (9). His reported leach rates were based on the following assumptions:

- leaching is by matrix dissolution, and hence measured fission-product concentrations in the leachate reflect the glass leach rate;
- the composition of fission products in the glass can be calculated from the composition of the solution used to fabricate the glass;
- there is no difference between the leach rate of the glass in distilled or soil water.

From a logarithmic plot of the laboratory data, a theoretical model with slope t^{-1} seemed to best fit the longer term leaching behavior of the glass. This time dependence represents a modified form of diffusional control where the corrosion product layer swells during formation and gradually closes off the diffusion pores. Equation 5) gives the leach rate of the glass as a function of time based on the original data.

$$R(t) = 2.1 \times 10^{-4}/t \quad (kg/m^2 \cdot d) \qquad 5)$$

where t is in days.

Equation 5) required further adjustments to take into account soil-water conditions and new information obtained from the glass block which was brought to WNRE for analysis. Surface analysis reported by Strathdee et al. (10) indicated essentially no difference between bulk and surface layers; thus matrix dissolution is a reasonable assumption. Fission-product composition measurements by Chen (11), however, indicated that ^{90}Sr was lower than originally estimated by a factor of approximately five while ^{137}Cs was essentially the same. Because the original leach rate was determined from the total $\beta\gamma$ count in solution, the difference in fission-product inventory requires an increase of 1.16 times the original rate. Leaching experiments using samples of the retrieved glass by Johnson (12) indicated that leach rates in soil waters were lower than those in distilled water by a factor of about two. The effect of groundwater temperature (6°C) was compensated for using an activation energy of 8 - 9 kcal/mole derived from the leaching studies of similar glasses by Westsik and Turcotte (13). This should reduce the in situ leach rate by another factor of 2.5.

The adjusted leach-rate equation becomes:

$$R'(t) = 4.8 \times 10^{-5}/t \quad (kg/m^2 \cdot d) \qquad 6)$$

Assuming perfect mixing of the water flowing through the matrix of glass blocks, 2.1 L/d of water flows past each block. Given a surface area of 0.0462 m^2/block and a measured (11) ^{90}Sr specific activity of 88.8 GBq/kg, the ^{90}Sr concentration in the water at X = 0 is:

$$^{90}Sr(t) = (9.4 \times 10^4/t)\exp(-6.5 \times 10^{-5}t) \quad (Bq/L) \qquad 7)$$

Similarly, for a ^{137}Cs specific activity of 0.83 TBq/kg (22.5 C/kg), the concentration at X = 0 is:

$$C_s(t) = (3.5 \times 10^5/t)\exp(-6.3 \times 10^{-5}t) \quad (Bq/L) \qquad 8)$$

COMPARISON OF MODEL PREDICTIONS AND FIELD MEASUREMENT

Both liquid and soil samples were obtained using techniques described in previous publications (14, 16). The detection of ^{137}Cs was done by γ-spectrometry. ^{90}Sr concentrations were calculated by difference from total $\beta\gamma$-activity.

The concentration of ^{90}Sr in water as a function of time at 46 cm from the blocks and ^{90}Sr concentration in the soil as a function of distance from the block after three years' burial are compared to model predictions in Fig. 2. and 3., respectively. Average yearly concentrations of ^{90}Sr in water at a distance one metre from the blocks are compared to model predictions for the period 1960 to 1977 in Fig. 4. In the three comparisons made here, estimates of data confidence levels are compared to model predictions shown as an envelope encompassing the estimated range of model parameters. Considering the simplicity of the model and the complexity of the actual water movement*, model predictions and ^{90}Sr field measurements agree well with respect to:

- release rate of ^{90}Sr as a function of time from the glass block;
- concentration of ^{90}Sr in the water and the soil at a given point as a function of time;
- velocity of the ^{90}Sr front through the soil relative to the average groundwater movement.

A field sampling program was undertaken in the summer of 1979 with the objective of establishing the extent of the ^{137}Cs plume. Emphasis was placed on measuring concentrations close to the glass blocks. Analysis of soil adhering to the glass block retrieved in the fall of 1978 was also undertaken. Measured concentrations of ^{137}Cs in the soil as a function of distance from the block are compared to model predictions in Fig. 5. These measurements indicate that $> 90\%$ of the ^{137}Cs released from the glass blocks over the last 19 years is still within the first 30 cm of soil downstream of the blocks. The remaining ^{137}Cs is spread out in a low concentration tongue (< 100 Bg/kg) extending up to ~ 7.5 m downstream of the blocks. In contrast, the equilibrium ion exchange model predicts a major concentration peak somewhere between 120 cm ($k_d = 230$) and 360 cm ($k_d = 80$). A comparison between calculated cumulative ^{137}Cs release and the ^{137}Cs inventory measured in soil shows order of magnitude agreement. A more precise comparison is not possible without detailed knowledge of the ^{137}Cs concentration gradient next to the block.

DISCUSSION

The most significant aspect of this experiment is the leaching performance of the glass blocks under actual burial conditions. The leach rate can be estimated from concentrations of ^{90}Sr in groundwater as follows:

*Measured (14) seasonal groundwater velocities varied between 15 cm/d after freeze-up to 20 cm/d during spring run-off. Flow direction varied in direction over a range of 60°.

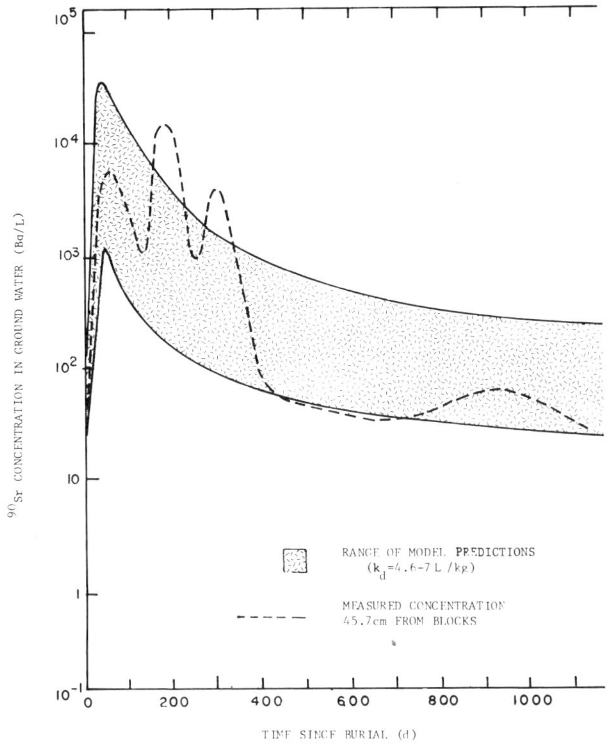

Fig. 2. Concentration of ^{90}Sr in Groundwater 46 cm From Blocks for Period 1960-1963 (14)

Leach rate

$$= \frac{[^{90}Sr \text{ in } H_2O \text{ corrected to time of burial}][H_2O \text{ vol. flow rate}]}{[\text{Glass } ^{90}Sr \text{ specific activity}]\ [\text{Surface area of block}]}$$

For example, in 1977 the average ^{90}Sr activity 1 metre from the glass blocks was 9.8 Bq/L giving a glass leach rate of $\sim 8 \times 10^{-14}$ kg/m^2·s. This agrees well with the prediction of 9×10^{-14} kg/m^2·s using equation 6). The observed leach rate is about 226 times lower than the lowest leach rate reported after 120 days of exposure under laboratory conditions (7). One fifth of this decrease can be accounted for by the lower aggressiveness of groundwater over distilled water and its low temperature (6°C average). The remaining decrease in leach rate (factor of 45) over the first 17 years of burial is thought to be the result of the mechanical support and/or reinforcement by the soil matrix of an otherwise friable protective corrosion-product

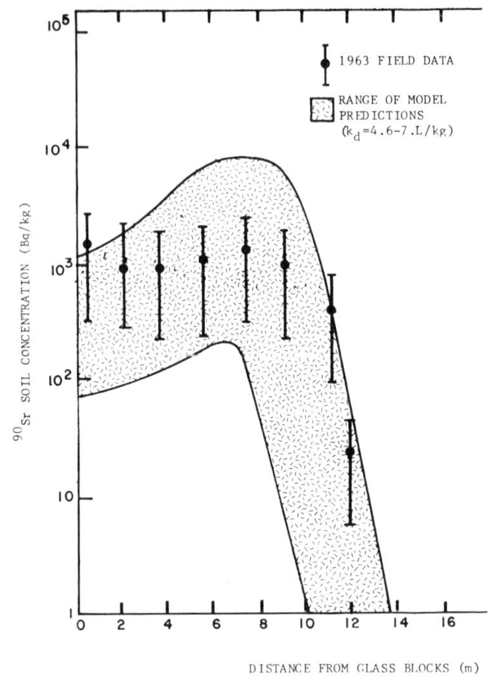

Fig. 3. Concentration of ^{90}Sr in the Soil in 1963 (14) as a Function of Distance From Blocks

layer which allowed the t^{-1} leach rate dependence to continue. This is supported by two further observations. The original laboratory leach tests (8) were performed by a continuous flow method (9) which did not disturb the blocks and produced a t^{-1} leach rate dependence. It has been noted that handling of the glass blocks destroyed the decreasing leach rate trend and restored the rate to near that of a fresh surface (8). In addition, during this summer's soil sample program, one of the glass blocks was inadvertently disturbed by the soil sampler. Subsequent measurements downstream of this block indicate that this slight movement of the block has restored the leach rate to that of a fresh surface for the whole block. Had the soil sampler disturbed only the area of contact, a much smaller increase in concentration would have been noted.

The existence of a corrosion-product layer could not be substantiated from inspection of the glass block retrieved in October 1978 (10). This is not surprising since the block was retrieved by a mechanical grappler thrust down through the soil. The upward movement

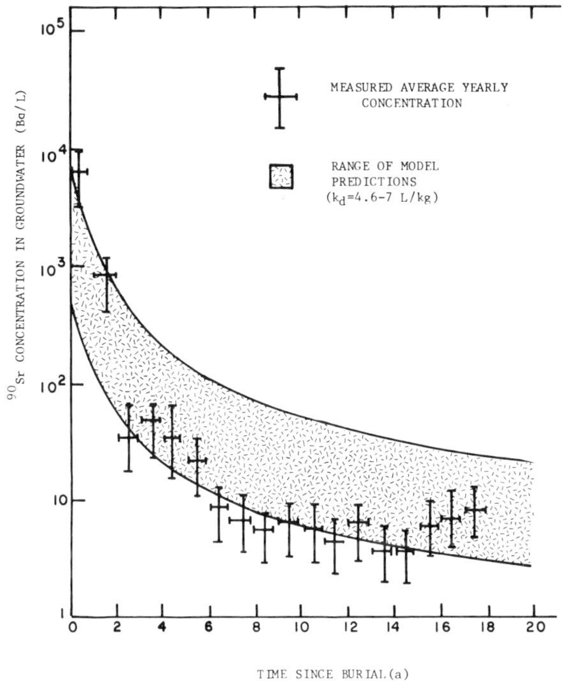

Fig. 4. Yearly Average of ^{90}Sr Concentration in Groundwater One Metre From Blocks

of the block through the soil would have removed any friable surface layers. Since the existence of friable corrosion layers is not unique to nepheline syenite-based glasses, it is possible that the low-temperature resistance to leaching of other types of glass may also show significant improvement if the glass is embedded in a suitable support matrix. This aspect of glass performance is presently being examined at a variety of temperatures at WNRE.

The final point to be made with respect to the glass leaching is that the rate equation predicts a continual decrease with time. This ignores the fact that stresses within a swelling corrosion-product layer may eventually cause its separation from the glass substrata and a subsequent increase in leach rate. Until the corrosion-product layer can be adequately characterized, extrapolation of corrosion rates beyond the present data base is not recommended.

To predict ^{90}Sr and ^{137}Cs concentrations in soil and water, mathematical descriptions of fission-product release from the glass blocks, water movement through the soil and partitioning of the re-

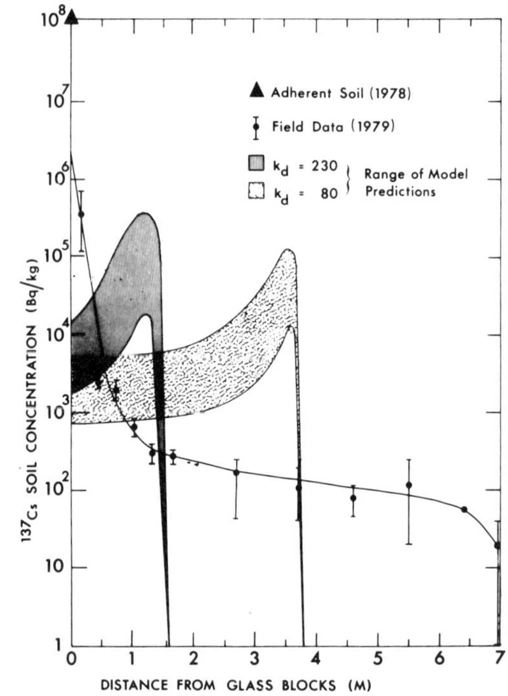

Fig. 5. Concentrations of ^{137}Cs in the Soil as a Function of Distance From Glass Blocks

leased fission products between the soil and water phases are required. The simple, one-dimensional, convective-dispersion model is least accurate in the immediate vicinity of the blocks where the blocks themselves offer an obstacle to flow, producing substantial radial dispersion (14). The radial dispersive component would also be expected to assume increasing importance the farther the front moves away from blocks. This is illustrated in Fig. 3. where the amplitude of the frontal peak has been attenuated almost an order of magnitude by the time it has progressed eight metres. The use of a one-dimensional, convective-dispersion model, therefore, has a range of applicability limited to less than 10 metres under these test conditions.

The relative success of the equilibrium ion exchange model in predicting ^{90}Sr and ^{137}Cs migration is consistent with data available in the open literature. Given sufficient data on groundwater conditions and soil cation-exchange capacity, it would appear possible to predict ^{90}Sr migration at low water velocities in an unconsolidated medium. On the other hand, ^{137}Cs shows considerable deviation from an idealized cation exchange equilibrium. There are at least three phenomena which in whole or part could explain the present observations. Soil analyses (5, 17) indicate the presence of layered silicates. Slow desorption from these silicates could contribute to the type of ^{137}Cs soil concentration profile observed in Fig. 5. Partition coefficients (k_a) for a $^{137}Cs/H_2O$/gneiss (the likely soil precursor) system have been measured (17). These measurements indicate a concave sorption isotherm which would also produce the type of ^{137}Cs soil profile shown in Fig. 5. The third possibility is the migration of ^{137}Cs fixed to fine particulates. Analysis of contaminated soil (7) associated with active waste disposal in the same general area as the glass block experiment has shown that about 5 - 10% of the ^{90}Sr-^{137}Cs activity is associated with soil particles less than 0.074 mm in diameter. An estimate of soil pore diameter based on mean particle size ($\sim$ 0.215 mm) indicates that transport of smaller soil particulates may be possible. Although migration of cesium on suspended solids in both surface waters (18, 19) and through soil columns (20) has been observed, experimental verification would be required to support this hypothesis.

CONCLUSIONS

The following conclusions can be drawn from this study:

- the low temperature leaching performance of nepheline syenite-based glass is improved by mechanical support and/or reinforcement of the corrosion-product layer;
- the leach rate for the first 17 years of glass burial can be predicted from extrapolations of leach rate measurements taken under conditions of continuous flow in the laboratory;
- ^{90}Sr migration through sandy soil at Chalk River can be adequately predicted for distances up to ten metres by a one-dimensional, convective-dispersion model assuming an equilibrium, linear adsorption isotherm. Prediction of migration over longer distances will require at least a two-dimensional model to account for radial dispersion;
- migration of ^{137}Cs under similar conditions cannot be predicted assuming an equilibrium, linear adsorption isotherm.

ACKNOWLEDGEMENTS

The authors wish to acknowledge the assistance of Marsha Sheppard for her evaluation of a particulate transport mechanism as well as others of our co-workers whose support and constructive criticisms have aided in the preparation of this document.

REFERENCES

1. W. F. Merritt, Canad. J. Chem. 36, 425 (1958).
2. I. Neretnieks, KBS-30 (1977).
3. R. J. M. DeWiest, Flow Through Porous Media (Academic Press, 201-214, 1969).
4. D. R. F. Harleman, P. F. Mehlhorn and R. R. Rumer, J. Hydraulic Div., Am. Soc. Civil Engrs. 89, 67 (1963).
5. E. J. Evans, CRER-792 (1958).
6. D. W. Rhodes and J. L. Nelson, HW-54721 (1957).
7. R. E. Jackson, K. J. Inch, R. J. Patterson, K. Lyon, T. Spoel, W. F. Merritt and B. A. Risto, "Adsorption of Radionuclides in an Aluvial-Sand Aquifer: Measurement of the Distribution Coefficients K_d^{Sr} and K_a^{Cs} and Identification of Mineral Absorbents," presented at ACS Division of Environmental Chemistry, April 6 (1979).
8. A. R. Bancroft, CEI-109 (1960).
9. A. R. Bancroft and J. D. Bancroft, Atomic Energy of Canada Limited Report, AECL-718 (1958).
10. G. G. Strathdee, N. S. McIntyre and P. Taylor, Paper No. 51-SI-79, International Symposium on Ceramics in Nuclear Waste Management, American Ceramic Soc., Cincinnati, Ohio, U.S.A., May 2 (1979).
11. J. D. Chen, 23rd ORNL Conference on Analytical Chemistry in Energy Technology.
12. L. H. Johnson, private communication.
13. J. H. Westsik, Jr., and R. P. Turcotte, PNL-2759 (1978).
14. W. F. Merritt and P. J. Parsons, Health Physics 10, 655 (1964).
15. W. F. Merritt, Atomic Energy of Canada Limited Report, AECL-5317 (1976).
16. P. J. Parsons, Atomic Energy of Canada Limited Report, AECL-1325 (1961).
17. T. T. Vandergraaf, private communication.
18. G. G. Eichholz, T. F. Craft and A. N. Galli, Geochim. et Cosmochim. Acta 31, 737 (1967).
19. P. Magno, T. Reavey and J. Apidianakis, BRH-NERHL-70-2 (1970).
20. B. L. Carlile and B. F. Hajek, BNWL-CC-995 (1967).

LEACHING CHARACTERISTICS OF ACTINIDES FROM SIMULATED REACTOR WASTE, PART 2*

H. C. Weed, D. G. Coles, D. J. Bradley**, R. W. Mensing, J. S. Schweiger and J. H. Rego

Lawrence Livermore Laboratory, Livermore, CA 94550 (LLL)
**Battelle Pacific Northwest Laboratories, Richland, WA 99353 (PNL)

ABSTRACT

Leach rates for ^{237}Np and ^{239}Pu were investigated with a single-pass leaching system. The factorial experimental design uses several combinations of solution composition and flow rate and two temperatures, 25°C and 75°C. The 25°C results are compared with those from a modified IAEA procedure. At 25°C, leach rates decrease with time. Agreement between results from the single-pass and modified IAEA methods is fair with WIPP brine leachant, good with $NaHCO_3$ and good with distilled H_2O. Leach rates are approximately independent of flow rates at room temperature but increase with flow rates at high temperature. Rates for ^{237}Np increase with temperature, but those for ^{239}Pu either decrease or do not change with temperature.

INTRODUCTION

This investigation is part of the Waste Isolation Safety Assessment Program (WISAP) conducted by PNL for the Department of Energy. One of the important goals of the WISA program is to be able to calculate the rate of release and subsequent migration of radionuclides in geologic formations surrounding repositories.

*Work performed under the auspices of the U. S. Department of Energy by the Lawrence Livermore Laboratory under Contract No. W-7405-ENG-48.

The present study has the following objectives in support of that goal:

1. To provide information on the source term to be used in radionuclide migration and safety assessment calculations. This includes measuring the effect of solution composition, flow rate, temperature and time on the leach rate of simulated high level reactor waste glass.

2. To compare results obtained from a dynamic one-pass leaching method developed at LLL with those from the modified IAEA method used at PNL.

This interim report includes results to 420 days; a final report will be issued with PNL after the conclusion of the study.

EXPERIMENTAL WORK

As shown in Fig. 1., LLL uses a statistically designed factorial experiment with unequal replication (1). Leach rate is the dependent variable; and solution composition, flow rate, temperature and time are the independent variables. The duration of the experiment is 420 days; each channel is sampled at 11 times during this interval. The PNL experimental design is a modification of the IAEA method (2) in which monthly sampling continues indefinitely and experiments are triply replicated. Both PNL and LLL use the same waste form, leaching solutions and low temperature (25°C) so that the low temperature results can be compared directly. The PNL simulated reactor waste is a sodium zinc borosilicate glass (76-68) in the form of hemispherical beads about 8 mm in diameter. The composition is like that of fully radioactive waste, except for the substitution of non-radioactive fission products. The radionuclides are:

$$^{238}U_3O_8\ (4.2\ w/o),\ ^{237}NpO_2\ (0.46\ w/o),\ \text{and}\ ^{239}PuO_2\ (0.046\ w/o).$$

The leachant solutions are distilled H_2O, 0.03$\underline{M}$ $NaHCO_3$, and synthetic WIPP brine. The final (420-day) effluent sample was collected in June. We are continuing with radiochemical and chemical analyses of the samples, data reduction and interpretation, and post-run examination of the sample cells and beads.

For each combination of temperature, flow rate and solution, the leach rate (R) is given as a function of time (t) by the model:

$$R = \alpha + \beta t^{-\gamma} \qquad 1)$$

The use of least-squares procedures determines estimates of the parameters α, β, and γ, which do not depend on the time but may depend on the temperature, flow rate and solution composition. Analysis of

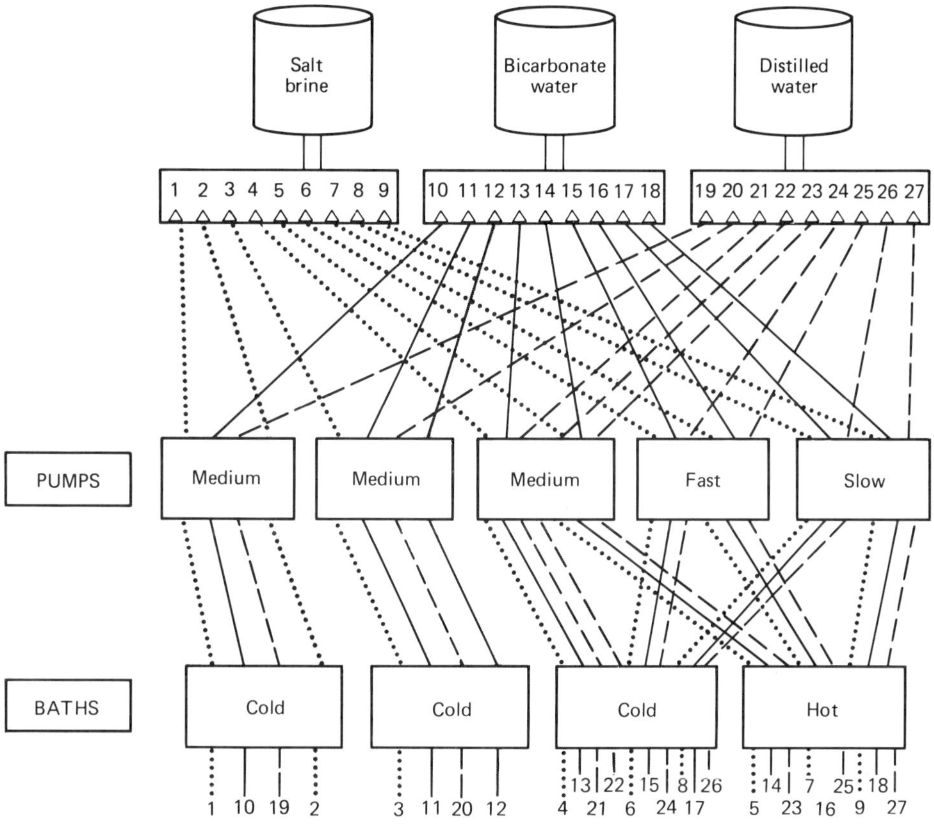

Fig. 1. Experimental Design, One-Pass Leaching Study

variance methods will then be used to determine the nature of this dependence.

RESULTS

Preliminary results are presented here without detailed statistical analysis. The rate R vs. t is shown in Fig. 2.-7. for the complete experiment to 420 days. In the figures, the full lines correspond to 25°C, the dashed lines to 75°C, and F, M, S to fast ($300cm^3/d$), medium ($43cm^3/d$) and slow ($10cm^3/d$) flow rates, respectively. Blank results are labeled B; the horizontal dot-dashed lines are mean values. PNL results are shown as filled circles.

Leach rates vary from $2x10^{-4}$ to $2x10^{-7}$g/(cm^2.d) for ^{237}Np (Fig. 2.-4.) and from $2x10^{-5}$ to $2x10^{-9}$g/(cm^2.d) for ^{239}Pu (Fig. 5.-7.). At high temperature the leach rates for ^{237}Np show little

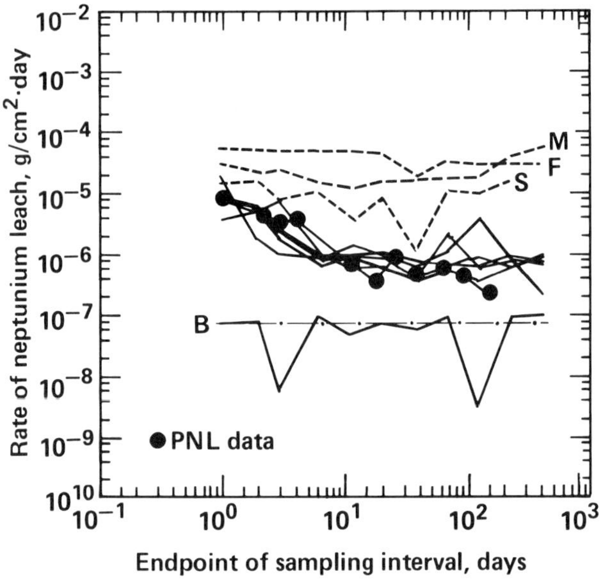

Fig. 2. ^{237}Np Leach Rate in Distilled Water. Dashed lines are 75°C; solid lines are 25°C; F, M, S are fast, medium and slow flow rates. B is the blank.

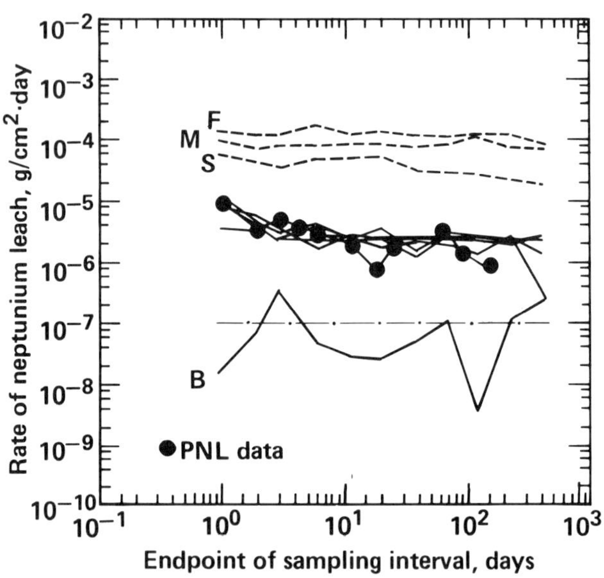

Fig. 3. ^{237}Np Leach Rate in .03M $NaHCO_3$

dependence on time. The ^{239}Pu leach rates decrease with time but in a very irregular fashion. Some of the low flow, high temperature ^{239}Pu leach rates appear smaller than those of the blanks (Fig. 5.). The reason for this is that the radionuclide concentrations for both leach samples and blanks are near the lower detection limit of the counting system, so that differences between them are not significant. Leach rates for both nuclides increase with flow rate for all solutions. High temperature leach rates are greater than those at room temperature for ^{237}Np, but less than or equal to those at room temperature for ^{239}Pu. We plan to look for ^{239}Pu in various parts of the sample cells and effluent tubing in order to determine whether or not it has been leached from the beads.

At room temperature, leach rates are approximately independent of flow rate; this will be checked for statistical significance during variance analysis. Direct comparison of the results from PNL and LLL at 25°C shows that agreement is fair in WIPP brine (Fig. 4. and 7.), good in distilled H_2O (Fig. 2. and 5.) and good in $NaHCO_3$ (Fig. 3. and 6.). The PNL leach rates are consistently lower than those of LLL for ^{239}Pu in WIPP brine (Fig. 7.).

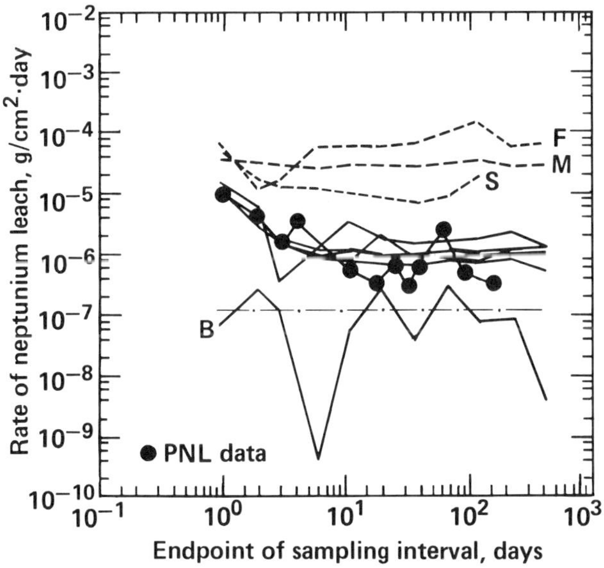

Fig. 4. ^{237}Np Leach Rate in WIPP Brine. Dashed lines are 75°C; solid lines are 25°C; F, M, S are fast, medium and slow flow rates. B is the blank.

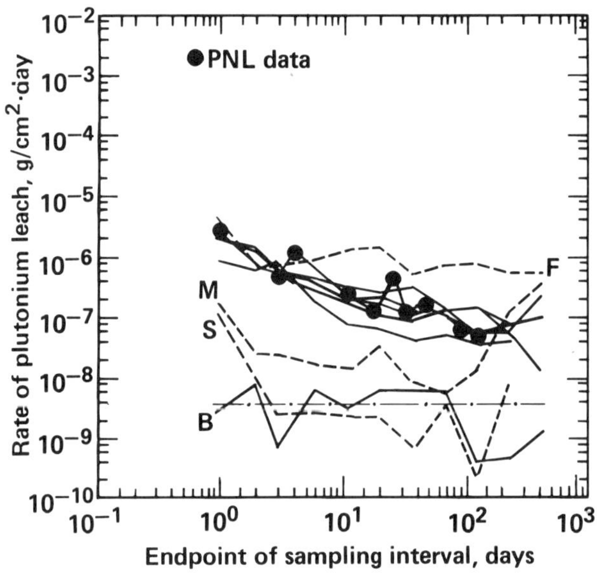

Fig. 5. ^{239}Pu Leach Rate in Distilled Water

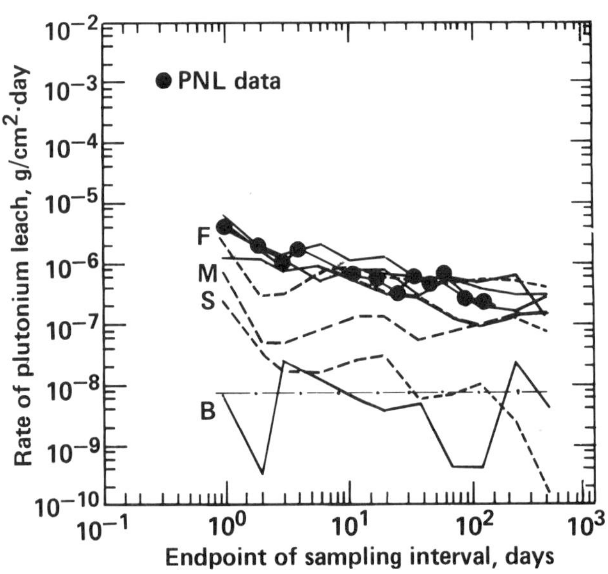

Fig. 6. ^{239}Pu Leach Rate in .03M $NaHCO_3$. Dashed lines are 75°C; solid lines are 25°C; F, M, S are fast, medium and slow flow rates. B is the blank.

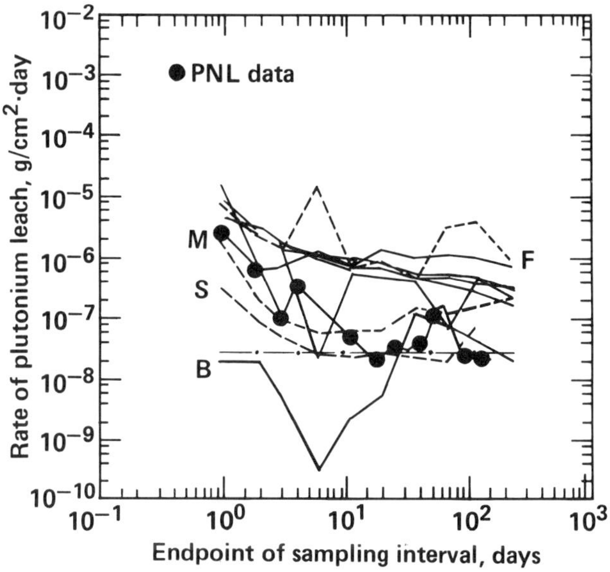

Fig. 7. ^{239}Pu Leach Rate in WIPP Brine

CONCLUSIONS

The following general trends can be seen in the results:

1. Leach rates increase with flow rate at high temperature but are approximately independent of it at room temperature.

2. Agreement between the results from the one-pass method and those from the IAEA method is fair in the case of WIPP brine solution and good in the case of the others.

3. The ^{237}Np leach rates increase with temperature, but the ^{239}Pu leach rates either decrease with temperature or do not change.

REFERENCES

1. H. C. Weed, D. G. Coles, D. J. Bradley, R. W. Mensing and J. S. Schweiger, "Leaching Characteristics of Actinides from Simulated Reactor Waste Glass," Scientific Basis for Nuclear Waste Management, 1 (G. J. McCarthy, ed., Plenum Press, NY, 1979).
2. E. D. Hespe, "Leach Testing of Immobilized Radioactive Waste Solid, A Proposal for a Standard Method," Atomic Energy Review 9, 195 (1971).

STATISTICALLY DESIGNED STUDY OF A NUCLEAR WASTE GLASS SYSTEM

L. A. Chick, G. F. Piepel, W. J. Gray, G. B. Mellinger
and B. O. Barnes

Pacific Northwest Laboratory
P. O. Box 999
Richland, WA 99352

INTRODUCTION

Most commercial glasses and glasses subjected to scientific study contain fewer than eight oxide components. Glasses for the immobilization of nuclear waste are much more complex, usually containing between 15 and 40 components. The wide variety of nuclear waste compositions necessitates development of a similar variety of glass additive compositions, each tailored to immobilize a specific waste. In the past, nuclear waste glasses were developed in an iterative manner, adjusting the composition in successive steps for optimization of desirable processing and immobilization properties. Only limited attempts have been made to understand property behavior over relatively narrow compositional fields.

We have recently broadened our approach to the problem of developing nuclear waste glasses in a generic study which has the objective to develop predictive approximation models for the major glass properties as functions of composition. The experimental text matrix covers the composition range currently foreseen for nuclear waste glasses. Twenty-six major oxide glass components will be included. This paper discusses studies performed on a somewhat restricted matrix with 11 oxide components selected for the purpose of verifying the applicability of newly developed statistical design and analysis techniques.

Properties measured and modeled for this study include those of the melt at high temperature (shear viscosity, electrical conductivity and volatility) and those of the final product in solid form (devitrification products; weight loss by chemical attack in distilled water, buffered pH 9 solution and buffered pH 4 solution; and chemical analysis of ions leached into distilled water). This paper discusses one processing property (volatility) and one immobilization property (weight loss in pH 4 solution).

PROCEDURE

Table 1 defines the composition matrix studied. The upper and lower allowed mole percentages of the components are listed along with additional restrictions placed on the sums of several components. These restrictions define a ten-dimensional hyperpolyhedron from which 70 widely spaced test melt compositions were selected so as to minimize the prediction variance of the approximating models. The approximate centroid of the experimental region is also given in Table 1.

Batches were formulated using reagent grade oxides, carbonates and nitrates. Melting was performed in platinum crucibles in electric resistance furnaces for two hours at temperatures ranging from 1150°C to 1500°C, as required to fully react each batch. Melts were air quenched, crushed, coned and quartered to separate out samples

Table 1. Experimental Composition Matrix

Oxide Component	Low Study Level mole%	High Study Level mole%	Level In Centroid mole%
SiO_2	41	60	49.19
B_2O_3	5.5	15	8.99
Al_2O_3	0	16	6.82
CaO	0	14	6.61
MgO	0	9	3.81
Na_2O	9	17	13.42
ZnO	0	6.5	2.95
TiO_2	0	8	3.73
Cr_2O_3	0	3.5	1.49
Fe_2O_3	0	3.5	1.49
NiO	0	3.5	1.49

Further restrictions on test compositions:

$54 \leq SiO_2 + B_2O_3 + Al_2O_3 \leq 80$
$13 \leq CaO + MgO + Na_2O \leq 35$
Sum of all components = 100

for the various property measurements. Chemical analyses of the test melts were performed by an inductively coupled plasma emission spectrometer. The analysed compositions were used for property modeling (1).

In the pH 4 leach test, one gram of -40 + 60 mesh granules was weighed and enclosed in a 350 mesh stainless steel envelope. The sample was placed in a buffered pH 4 solution and agitated for 19 hours at 25°C. The percent weight loss was calculated after washing, drying and final weighing. Two measurements were performed on each test melt.

In the volatility tests, two grams of each glass were placed in a 1.3 cm high by 1.6 cm diameter platinum crucible and weighed. The sample was suspended in a vertical tube electric resistance furnace for three hours at 1300°C with a controlled updraft of approximately one liter/min. of air. After final weighing, the weight loss was reported as mg.

The investigated nuclear waste glass properties were modeled as functions of composition using the partial cubic polynominal form:

$$P = \sum_{i=1}^{q} B_i X_i + \sum_{ij}\sum B_{ij} X_i X_j + \sum_{ijk}\sum\sum B_{ijk} X_i X_j X_k$$

where: P = property value
q = number of components in glass
X_i = mole% of ith component
B_i = approximation coefficient.

Data were fit by least squares multiple regression analysis. First, all linear (B_i) terms were calculated. Then the most effective quadratic (B_{ij}) and cubic (B_{ijk}) terms were added as necessary to achieve good fit. Statistical quantities RA^2 and F were used to determine the degree of fit. RA^2 measures the fraction of the variability in the data explained by the model. RA^2 approaches one as the fit improves. The F statistic is the ratio of the variability not explained by the model divided by the experimental error as determined by replication of measurements. For this study, values less than six for the F statistic were indicative of satisfactory fit.

Other methods of checking the validity of approximation models are to compare models fit to independent sample melt groups or to compare models fit to original data versus logarithms of the data.

The models are best understood and compared graphically through effects plots. An effects plot is generated by using the approximation model to predict the properties of various compositions. No experimental data are displayed on the plots. A single glass composition within the field of study (usually the centroid of the field) is chosen as the basis of the plot. The model is used to predict its property value. Effects curves are then generated by predicting the property values of glasses with increased or decreased levels of each component. As each component level is changed from the level in the base glass, all other components are adjusted proportionally. For example, when Na_2O is increased by 10 mole%, the sum of all remaining components decreases by 10 mole%, each one decreasing by an amount proportional to its level in the centroid. The slopes of the effects curves indicate the influence of component variation on the properties of interest. In this manner, an empirical understanding of the effects of composition on properties is achieved. The statistical design, modeling techniques and effect analysis have been discussed more rigorously (2).

RESULTS AND DISCUSSION

The pH 4 leach and volatility results for all test melts are reported in Reference 1. The pH 4 solution leaching results ranged from 0 to 98.6 weight percent with a standard error of 0.23 weight percent. Weight loss from volatility tests ranged from 1.8 to 151 mg with a measurement standard deviation of 3.1 mg.

The 70 melts were tested in three groups (A, B and C). A change in pH 4 leach washing and drying procedures occurred between group A and groups B and C tests. As a result, group A pH 4 data was modeled independently from that of groups B and C.

The model of pH 4 data logarithms for the 59 melts in groups B and C achieved an RA^2 of 0.988 and an F of 2.77. This was obtained using the 11 linear terms plus eight quadratic and four cubic terms. Fig. 1. is the resulting effects plot. Prediction error bars were calculated for each one mole percent increment on the plot and the largest and smallest are shown in the figure. If this experiment were repeated many times, the effects lines are expected to fall within one standard error two thirds of the time. Differences in effects of components such as Al_2O_3 and B_2O_3, as indicated in Fig. 1., were not significant.

Due to the changes in washing and drying procedures, pH 4 data for group A were uniformly higher than for groups B and C. Independent modeling of the group A data resulted in $RA^2 = 0.959$ and $F = 38.3$. The high F indicates a significant lack of fit. However, the effects plot is quite similar to Fig. 1. Only NiO and Fe_2O_3 effects changed significantly.

Original data for pH 4 did not model well. Logarithms probably fit better because of the wide range in the values and possibly because of extreme curvature of functions describing original data.

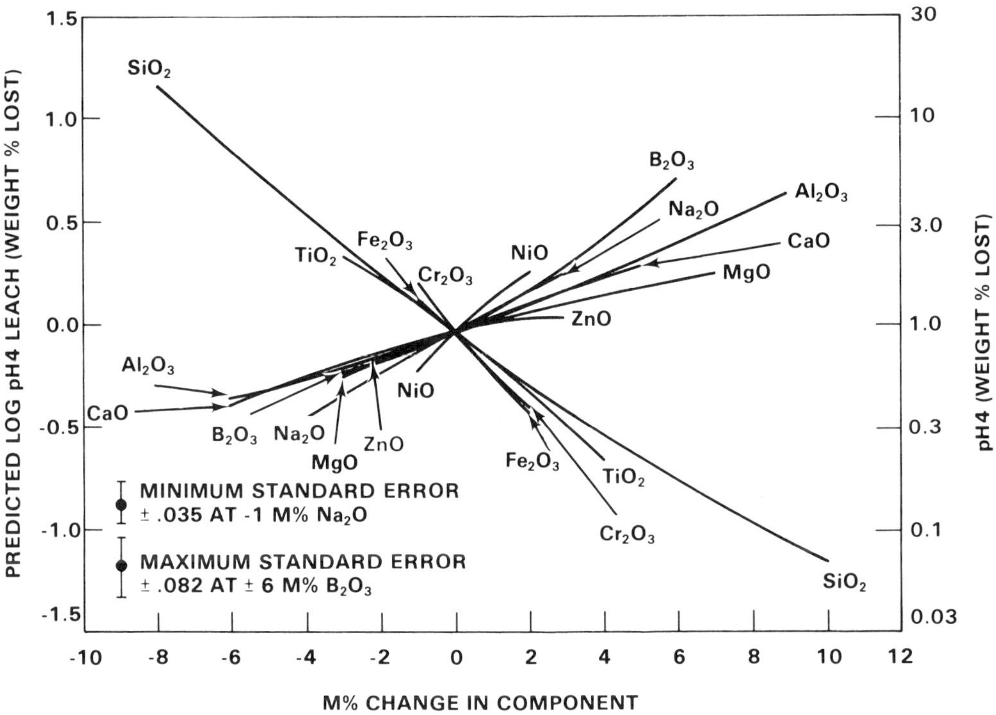

Fig. 1. Effects of Component Concentration on Leaching (Buffered at pH 4)

Fig. 1. shows that increasing the concentrations of SiO_2, TiO_2, Fe_2O_3 or Cr_2O_3 will result in decreasing the pH 4 leach loss of the glass. Increasing the levels of any of the other components will increase the pH 4 leach loss. The reason that Al_2O_3 increases the leach rate in acid solution is not understood. Modeling of other leaching data shows that alumina decreases leach rates in neutral or basic solutions. The causes of the NiO, Cr_2O_3, Fe_2O_3 and ZnO effects are also not understood. However, these components participate in crystalline phases found in some samples, so it is expected that transformation of the compositions to include the crystal phases as components may result in clarification of their effects. This work is being performed using X-ray diffraction and scanning electron microscopy.

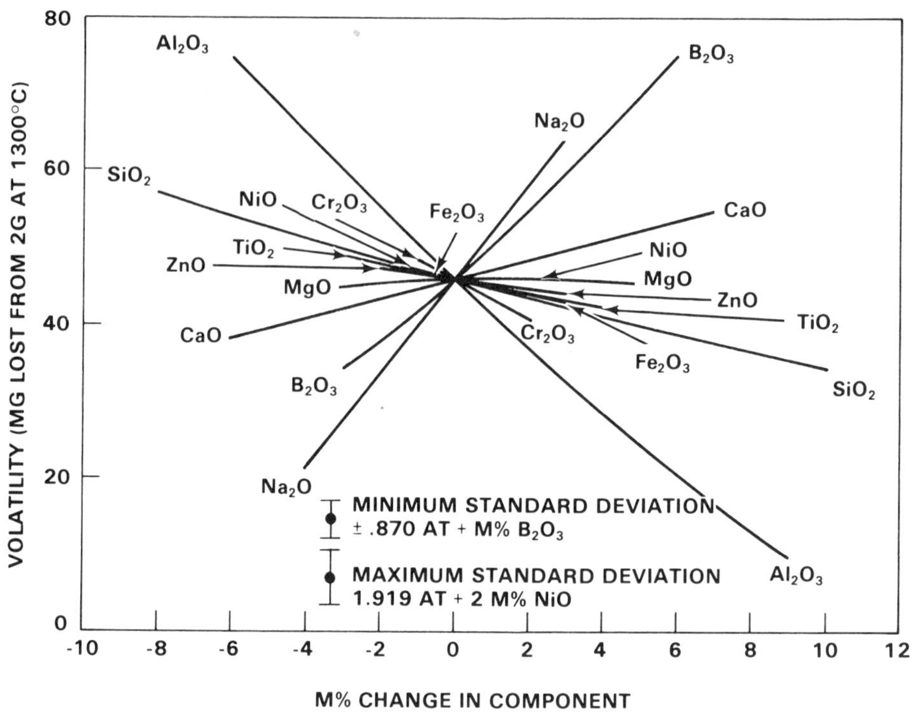

Fig. 2. Effect of Component Concentration on Volatility

Volatility data was consistent between data groups A, B and C so they could be combined for modeling. Fig. 2. shows the effects plot for mg loss at 1300°C. The model had $RA^2 = 0.969$ and $F = 4.61$ using 11 linear, 5 quadratic and 4 cubic terms. For verification, independent fits were made of logarithms of the combined data of original group A data and of original data from groups B and C. All component effects in the resulting plots are in excellent agreement with those in Fig. 2.

Previous (3) and continuing work (1) at Pacific Northwest Laboratories show that of the 11 components in this study, the major volatile elements are sodium and boron. This is consistent with Fig. 2. which shows Na_2O and B_2O_3 to have the most significant positive effects on volatility. SiO_2 and Al_2O_3 have the major negative effects on volatility, probably because they increase the glass viscosity. A strong negative correlation was found (1) between viscosity and volatility data. Furthermore, modeling of viscosity data (1) has confirmed that SiO_2 and Al_2O_3 have large positive effects. Viscosity may correlate with volatility because convection currents replenish volatile species at the melt surface. Alternatively, the correlation could result from the influence of viscosity on diffusion rates of the volatile species toward the surface. The rate controlling mechanism of volatility from these glasses is not yet understood.

CONCLUSIONS

The statistical experimental design and approximation modeling techniques employed in this study have been demonstrated to be effective in yielding empirical understanding and predictive capability of the effect of composition on properties in complex glasses. General agreement between the component effects predicted by modeling of independent sets of data lends credence to the results.

Further work must include crystalline phases as part of the defined compositions so that the effects of components which form the crystals can be better understood.

REFERENCES

1. L. A. Chick, et al., The Effects of Composition on Properties in an Eleven Component Nuclear Waste Glass System, PNL-3188 (in progress, ~March 1980).
2. G. F. Piepel, "Component Effects in Mixture Experiments," Proceedings of 1979 Department of Energy Statistical Symposium, Gatlinburg, TN, Oct. 24-26 (1979).
3. W. J. Gray, Volatility of a Zinc Borosilicate Glass Containing Simulated High Level Radioactive Waste, BNWL-2111, Pacific Northwest Laboratories, October (1976).

DISSOLUTION OF BOROSILICATE GLASSES UNDER REPOSITORY CONDITIONS OF PRESSURE AND TEMPERATURE

Neil A. Chapman and David Savage

Institute of Geological Sciences
Building 151, Harwell Laboratory
Harwell, Oxfordshire, United Kingdom

INTRODUCTION

The projected style of the UK's high level radioactive waste disposal program envisages a period of cooling of the vitrified waste form, prior to disposal in a deep repository. The length of time during which the waste blocks must be kept in store depends on many factors, including social and waste management considerations, and on the limiting admissible temperature which may be attained in a final repository. This temperature value, in turn, depends on the maximum thermal load capacity of the total repository system, i.e. the near-field rock matrix, the canister and backfill materials, and the vitrified waste form itself. One of the factors that these load capacities are dependent upon is the hydrothermal behavior of the various components under deep repository conditions; in other words, their ability to resist significant geochemical changes or corrosion by reaction with warm, pressurized groundwaters.

This paper described laboratory experimental work in progress in the UK to examine the mechanisms of fluid buffered interactions of materials at possible repository temperatures and pressures and aims at proposing limiting temperature values for a disposal facility from the geochemist's viewpoint. During the next few years it is intended to examine the hydrothermal behavior of a range of repository components under relevant induced pressure and temperature conditions, but at present work is concentrated on reconnaissance experiments on dissolution rates of the vitrified waste form and the changes in near-field granite surface chemistry and fissure properties which will have a bearing on nuclide adsorption immediately after release.

RATIONALE

When a repository is eventually backfilled and sealed the physico-chemical environment of the waste will slowly adjust to ambient conditions at depth, superimposed on which will be locally enhanced temperature, stress and geochemical transients caused by the heat output of the waste, largely during the first thousand years.

Any experimental evaluation of materials behavior must thus aim to replicate these conditions of pressure (P), and temperature (T) and induced geochemical fluxes. Once the behavior is quantified, limiting thermal load values can be placed on the disposal system. The physical and chemical behavior of repository materials must be studied at likely conditions of temperature and pressure and perhaps to a degree, over a wider PT spectrum in order to obtain the rates at which processes occur. Reaction kinetics are particularly significant in modelling the effects of nuclide release. Limiting temperature values for any given repository component might be considerably different from those of another, or the total system. The role of the present experimental program is thus to arrive at a set of limiting values valid for the total repository system.

Ambient conditions in rock at depth are reasonably well understood. For example, in granite at an arbitrary depth of 1000 meters ambient PT conditions will be about 35-40°C, 10-30MPa. The value of T depends simply upon the regional geothermal gradient, whereas the value of 'P' is more complex. Simple interconnected hydrostatic fissure or 'pore' fluid pressure is about 10MPa, while lithostatic load pressure is nominally about 10MPa. The latter value is however subject to stress anisotropy and may have a horizontal component up to five times the vertical component. Thermally induced near-field 'P' may reach values as high as 200 MPA in the early life of a repository (1).

The effects of pressure on solid-fluid reactions may be considered as follows. The pressure dependance of the Gibb's free energy of any reaction is given by:

$$\left(\frac{\delta \Delta G}{P}\right)_T = \Delta V$$

where V is the difference in volume between products and reactants and G is the free energy change of the reaction. Therefore, the effect of pressure on the dissolution process will be dependent upon any solid volume change associated with the reaction. The borosilicate glasses studied dissolve incongruently cations such as Na and Li being leached from the glass in solution precipitating oxides such as SiO_2 and Fe_2O_3. Boult, et al. (2) noted that this process resulted in a small solid volume decrease . One would therefore expect higher pressures to increase the amount of glass dissolution. This is, in fact, quite the opposite from the observed effect noted below.

EXPERIMENTAL TECHNIQUE

All hydrothermal experiments were performed in static, cold-seal, argon pressurized reaction vessels with external heating jackets. Pressure was monitored on tube gauges; temperature being measured by Cr/Al thermocouples situated centrally in the reaction cells. Samples were contained in welded Pt or Au capsules, 5 mm O.D. and 4 cm long.

Reconnaisance experiments on dissolution rates and leaching of borosilicate glasses utilzed 3mm cubes of non-active glass of 0.5 cm^2 mean surface area. Two glass compositions, 209 and 189, with different formation temperatures (1000°C and 950°C respectively) were used (Table 1) as these were the reference standards previously studied in atmospheric pressure leach tests (2). From this work it was known that glass 189 has a higher dissolution rate than 209. Prior to each run, the glass cube dimensions were measured accurately and each sample weighed.

Fluid compositions used were a) de-ionized water and b) granite groundwater taken from a 300 m deep borehole at Altnabreac, Caithness, Northern Scotland. Capsules were loaded with a glass sample and 300μl of fluid in an approximate fluid/solid ratio of 5:1, sealed, and leak tested prior to loading in the vessels. The procedure was to pump to just below the required pressure, bring to temperature and make a final pressure adjustment. Vessels took about 150 minutes to reach temperature and run duration was varied between 20 and 228 hours. At the termination of each run the furnace assemblies were slid off the vessels which were quenched with compressed air.

The capsules were opened under 40 ml of deionized water to totally extract the fluid. The diluted fluid was then acidified and analysed for SiO_2, Na, Li, Sr and Cs. Retrieved glass samples were dried at 105°C, examined, scraped and reweighed.

This simplified reconnaisance technique will be modified for further experiments using larger volume capsules with a greater fluid/glass ratio. The pH of solutions will be measured before and after runs, and further elements will be analyzed in solution.

GLASS DISSOLUTION RECONNAISSANCE RESULTS

The borosilicate glasses studied dissolve incongruently. The alkali metals are replaced in the glass structure by hydrogen ions, other elements tending to be deposited on the glass surface as hydro-

Table 1. Chemical Compositions of Glasses Used in This Study.

	Glass 209			Glass 189		
X_yO_z component	from additives	from waste	wt% total	from additives	from waste	wt% total
SiO_2	50.88		50.88	41.51		41.51
B_2O_3	11.12		11.12	21.87		21.87
Al_2O_3		5.11	5.11		5.03	5.03
Li_2O	3.99		3.99	3.69		3.69
Na_2O	8.30		8.30	7.68		7.68
Rb_2O		0.11	0.11		0.10	0.10
Cs_2O		0.77	0.77		0.76	0.76
MgO		6.34	6.34		6.23	6.23
SrO		0.32	0.32		0.32	0.32
BaO		0.38	0.38		0.38	0.38
Y_2O_3		0.17	0.17		0.17	0.17
La_2O_3		0.44	0.44		0.43	0.43
Pr_6O_{11}		0.43	0.43		0.42	0.42
Nd_2O_3		1.82	1.82		1.78	1.78
Sm_2O_3	replaced	by Nd_2O_3				
Gd_2O_3	"	" "				
CeO_2		0.99	0.99		0.98	0.98
ZrO_2		1.43	1.43		1.40	1.40
PO_4		0.24	0.24		0.23	0.23
Cr_2O_3		0.56	0.56		0.55	0.55
MoO_3		1.77	1.77		1.75	1.75
Fe_2O_3		2.73	2.73		2.68	2.68
RuO_2		0.68	0.68		0.67	0.67
RhO_2	replaced	by PdO				
NiO		0.36	0.36		0.36	0.36
PdO		0.44	0.44		0.42	0.42
ZnO		0.44	0.44		0.44	0.44
U_3O_8		0.06	0.06		0.06	0.06
SO_4		0.10	0.10		0.09	0.09

xides. Consequently, dissolution rate was measured by a weight loss technique (after removal of the surface layer) and also by monitoring the concentration of various elements in solution with time.

Weight loss

Results are presented as a rate function ($g.cm^{-2}day^{-1}$) plotted against temperature (Fig. 1) at fixed pressures of 60MPa and 100 MPa and as weight loss per surface area ($g.cm^{-2}x10^3$) as a function of time at fixed P and T in Figs. 2 and 3. Examination of Fig. 1 shows the following features:

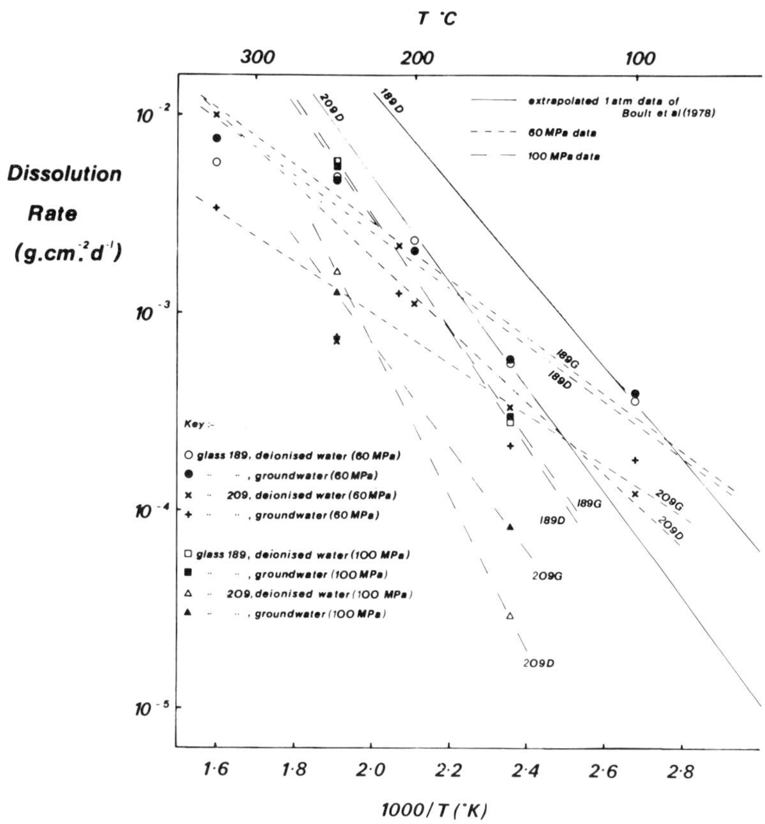

Fig. 1. Glass Dissolution Rate as a Function of 1000/T. Average Dissolution Rates at a Particular Temperature are Plotted for the 60 MPa Data

a) the highest dissolution rates at 200°C, 60MPa were about $1x10^{-3}$ and $2x10^{-3}$ (g.cm^{-2}day^{-1}) and the lowest values (at 100°C, 60MPa were $1x10^{-4}$ and $3x10^{-4}$ (g.cm^{-2}day^{-1}) for 209 and 189 respectively.

b) dissolution rate increases between one and two orders of magnitude between 100°C and 350°C for both glasses.

c) at 100°C glass 189 dissolves faster than glass 209 by a factor of three.

d) increased pressure effectively decreases the rate of dissolution of both glasses.

The physical state of the glasses after dissolution is worth comment. Up to 200°C the cubes remain intact at both pressures,

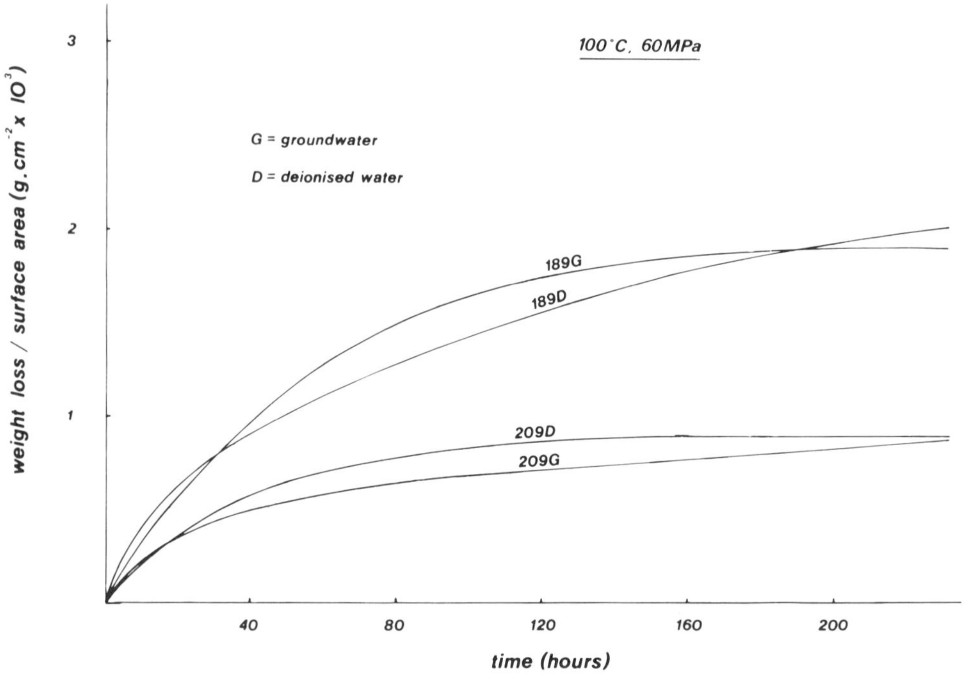

Fig. 2. Weight Loss per Initial Glass Surface Area as a Function of Run Duration at 100°C and 60 MPa

exhibiting only enhanced development of the surface layer (when dry, a porous white coating) and a 'crazing' of the surface in the case of 189. At 250°C, 209 exhibits brittle fracture, although 189 retains its cubic integrity. The leachate is often pale green, containing dispersed gel. At the highest temperature (350°C) the glasses disintegrate completely into blue-gray, angular fragments (apparently formed by swelling along fracture planes) with thick surface coatings of hydroxides. Dissolution rates presented at these temperatures must be treated as minimum values only, as it was not possible to remove the surface layer from these degraded samples. Thus the curves presented in Fig. 1. might be expected to steepen considerably as temperature exceeds 200°C. Figs. 2. and 3. show dissoution rates at 100°C and 150°C, 60MPa. The main feature displayed in these diagrams is the initial rapid dissolution rate as the exposed glass interacts with the fluid, followed by either a steady rate function or decrease in rate, presumably occurring as the surface layer develops and dissolution becomes controlled by ionic diffusion through the hydrated layer. It can also be seen that dissolution rate is generally lower in granite groundwaters than in pure water, being a function of the initial ionic concentrations in the fluid.

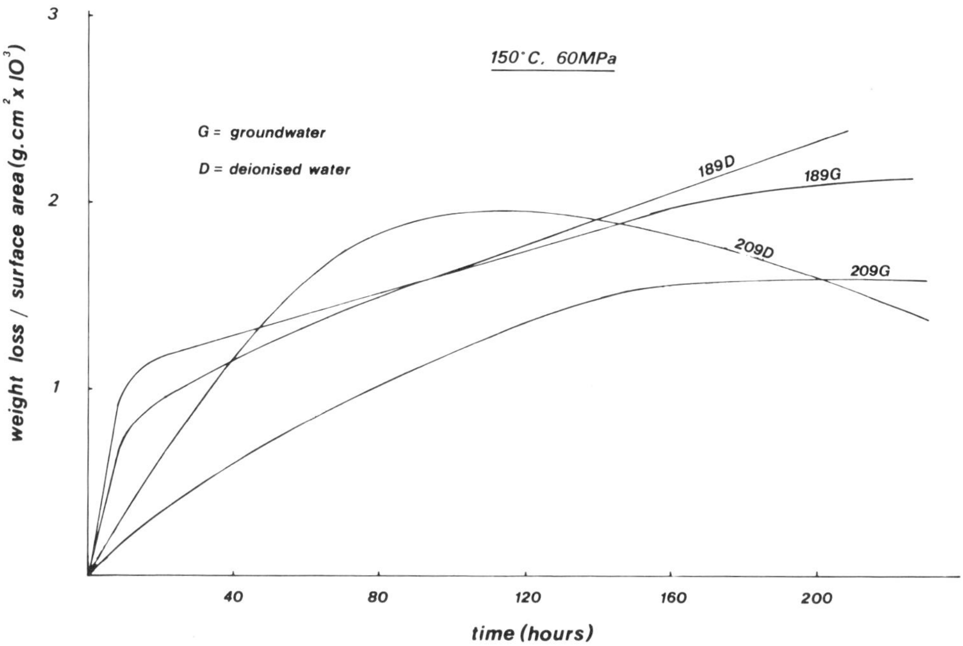

Fig. 3. Weight Loss per Initial Glass Surface Area as a Function of Run Duration of 150°C and 60MPa

Fluid chemistry

From the limited data we have, it is apparent that the concentrations of SiO_2, Na, Li, Sr, and Cs parallel the T-dependent weight-loss data discussed above. There is an initial rapid increase of concentration in solution which increases gradually with time (at fixed P and T). This slowing down of rate may be due to rate control by ionic diffusion through the surface layer or by build-up of concentration of the leached ions in solution.

CONCLUSIONS

The major conclusions are listed below:

1) In all of the experiments the fluid state remained as liquid since temperatures never exceeded 374°C. Under these sub-critical conditions pressure decreases dissolution rate.
2) The ability to extrapolate dissolution rate from values of 100°C to about 200°C, together with the maintenance of physical integrity indicate that the glasses studied are likely to be stable and predictable in leach behavior to

temperatures well in excess of those currently being considered for disposal (generally less than 100°C).

3) Glass 209, as already found by Boult, et al. (2) working at lower temperatures, dissolves more slowly over a wide PT spectrum than does glass 189, although the latter is easier to fabricate.
4) Dissolution rate decreases with increasing time. This is either due to retardation of diffusing ions due to the buildup of a surface layer on the glass or by inhibition due to the increasing concentration of ions in solution.

The principal conclusions of this study so far is that experimental application of realistic repository hydrothermal PT conditions indicate for the first time that current formulations of borosilicate glass would appear to provide for an adequate waste disposal medium which would maintain stability and predictable behavior over a wider pressure and temperature spectrum than previously realized, allowing some latitude in both pre-disposal storage period and eventual loading at the time of disposal.

ACKNOWLEDGEMENTS

The authors thank Mr. J. A. C. Marples and Mr. J. T. Dalton of the UKAEA for providing samples of the glasses, and Dr. I. G. McKinley for constructive discussion during the course of the work. The study was carried out by the Institute of Geological Sciences under contract to the United Kingdom Atomic Energy Authority and is published by permission of the UKAEA, the CEC, and the Director, IGS.

REFERENCES

1. D. P. Hodgkinson, Deep Rock Disposal of High-level Radioactive Waste; Initial Assessment of the Thermal Stress field, AERE-R8999 (1978).
2. K. A. Boult, J. T. Dalton, A. R. Hall, A. Hough, and J. A. C. Marples, The Leaching of Radioactive Waste Storage Glasses, AERE-R9188 (1978).

MECHANISM FOR ELEVATED TEMPERATURE LEACHING*

B. T. Kenna and K. D. Murphy

Sandia Laboratories**
Albuquerque, New Mexico 87185

ABSTRACT

Long-term, elevated temperature leaching and subsequent electron microprobe analysis of simulated waste glass and ceramic materials have been completed. A cyclic leaching pattern was found in all systems over a 20-month period. It appears that the leaching of mobile ions by simple diffusional processes is modified by more complex chemical interactions. The release of immobile ions is primarily a function of their chemical interactions in the matrix which suggests that these ions may be complex species when released into solution. A mechanism is proposed which incorporates these ideas and the cyclic phenomenon observed.

INTRODUCTION

The leaching process of a simple soda-silica glass can be represented by Equations 1-3 (1). The removal of Na^+ from the

$$\equiv Si{-}O{-}Na + H_2O = \equiv Si{-}O{-}H + Na^+ + OH^- \qquad 1)$$

$$\equiv Si{-}O{-}Si\equiv + OH^- = \equiv Si{-}O{-}H + {}^-O{-}Si\equiv \qquad 2)$$

$$\equiv Si{-}O^- + H_2O = \equiv Si{-}O{-}H + OH^- \qquad 3)$$

glass surface results in a depletion layer containing a Na^+ concen-

*This work supported by the United States Department of Energy (DOE), under Contract DE-AC04-76-DP00789
**A DOE facility.

tration gradient. This provides the impetus for diffusion of Na^+ from the interior of the glass thru the depleted zone. Note Equations 1-3 suggest that the silicate matrix is included in all leaching processes of glass - the hydrated depth is termed the gel or leach layer and is itself solubilized at some rate.

The inclusion of additional elements in a glass compounds the leaching processes, e.g., in soda-lime-silica glasses, the formation of calcium silicate can act as a barrier to inhibit the diffusion of Na^+. (1) In a complex system, such as vitrified nuclear waste, the leaching behavior of elements may be quite complicated due to interactions of the elements with each other, with the matrix, with the gel layer, with H^+ ions or with OH^- ions. These interactions can include ion-exchange, metathetical reactions, precipitation, complexation and polymer formation. The relative importance of the interactions depends on the physical form of the materials, leachant, temperature, exposure time, composition, and other physical variables. Thus, long-term leaching studies are of paramount importance since they can indicate the relative importance of specific mechanisms and mechanism changes as a function of time. The present work is an initial step toward gaining understanding of the basic leaching processes.

EXPERIMENTAL

Monolithic forms of two zinc borosilicate glass samples, one prepared at Battelle Northwest (BNW) and one prepared at Sandia Laboratories (SLA) using 73-1 glass frit (3), and one hot pressed sodium titanate waste pellet (STW) containing ca. 10 w/o zeolite, underwent Soxhlet leaching tests for 20 months. The STW pellet was primarily a ceramic, but did contain a small amount of a dispersed amorphous silica phase due to the inclusion of zeolite. All three samples contained nominally 25 w/o simulated high-level calcined waste with a composition similar to that proposed for the Allied General Nuclear Services fuel reprocessing plant planned in Barnwell, SC (4). Technetium, promethium, and transuranics (TRU) were not included. Each sample was leached in a quartz Soxhlet apparatus (2) using deionized water (300 ml), maintained at boiling temperature. The leachate was changed monthly. An identical Soxhlet apparatus containing only deionized water was run in parallel as a blank. The total amounts of Cs, Sr, Na, Mo, Zn and Si in the leachate were determined by atomic absorption spectroscopy.

Initial homogeneity of the samples was assured by scanning several times the total widths of unleached, freshly polished surfaces of identical samples with an electron microprobe for the elements Cs, Sr, Gd, Pd and U. Except for Pd which had agglomerated into islands of the metal throughout the samples, less than 5% variation was found within each sample for these elements.

At the conclusion of the leaching, the samples were sectioned, mounted and polished and the distribution of selected elements in the gel layer was determined by electron microprobe analysis. Both quantitative analysis and pictorial distributions were obtained.

RESULTS AND DISCUSSION

From about the second month of leaching on, the glass samples (SLA, BNW) displayed a reddish-brown "skin" on their surfaces which periodically sloughed inhomogeneously, generally reexposing the glass surface. The primary elemental composition of this layer in order of decreasing concentration as determined by emission spectroscopy is: Al~RE>Zr>Fe>U>Mo>Cr~Ca. These elements generally are considered as immobile elements, i.e. those which do not diffuse or diffuse very slowly. There was no "skin" formation or sloughing observed on the ceramic sample (STW).

The cyclic leaching pattern is not a new phenomenon and has been reported previously (5-8). Figs. 1. and 2. serve to demonstrate its characteristics in this study. Concentration maxima of immobile elements, e.g. U, RE, Si, Mo, in the leach solution generally precede the maxima of mobile elements, e.g., Na, Cs, Sr, by up to one month whereas the concentration minima are temporally coincident. This cyclic phenomenon was observed for all three samples although it was somewhat subdued for the ceramic sample (STW). This might be due to the much lower silicate content (4 w/o) in the ceramic sample as opposed to the glass samples and/or because the silicate phase is not continuous in the ceramic sample, but is dispersed throughout the ceramic phase.

Electron microprobe studies of the leached samples identified the presence of a discrete gel (leach) layer in the glass samples, but none for the ceramic sample. Because of this, a definitive leaching mechanism for the ceramic sample has not been formulated and further studies are in progress. Therefore, the following discussion concerns only the glass samples (SLA, BNW).

Elemental distributions were determined across the gel layers several times at different locations by the electron microprobe with consistent results. When compared to unleached glass, several conclusions can be made. Figs. 3 thru 8 are from the BNW glass which is used as an example. (a) Fig. 3: the gel (leach) layer appears to be porous in nature, (b) Fig. 4: there are distinct, continuous linear zones where Na, Fe, Zn, and perhaps P are concentrated. Since Na, Zn, and P have some diffusion mobility, these layers suggest that there are diffusion barriers formed in the gel layer. The Na-Fe-Zn-P layers, once formed, may themselves become diffusion barriers, (c) Fig. 5: mobile elements (Na, K, Cs, Sr) are depleted in the leach layer (Na is shown in Fig. 4 and is depleted except in the Na-Fe-Zn-

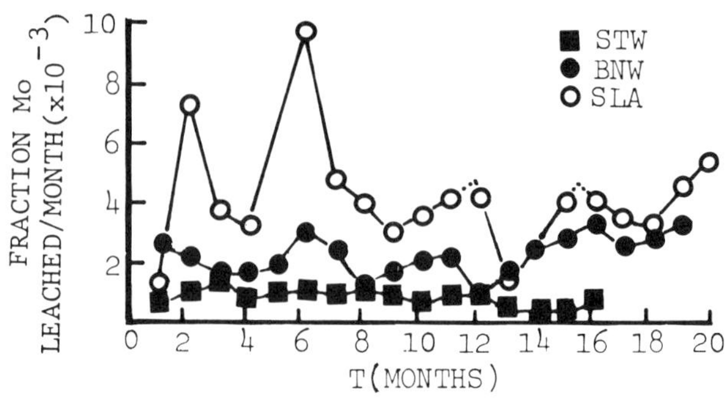

Fig. 1. Soxhlet Leaching of Mo from Glasses (SLA, BNW) and Ceramic (STW) Loaded with 25 w/o Waste Oxide Simulant

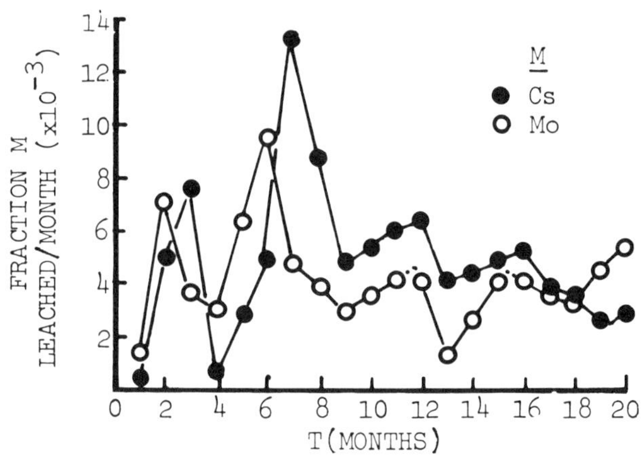

Fig. 2. Comparison of Soxhlet Leaching of Mobile (Cs) and Immobile (Mo) Elements from Glass (SLA) Loaded with 25 w/o Waste Oxide Simulant

P layers), (d) Fig. 6: the relative concentrations of immobile elements (Zr, U, RE) are enhanced in the leach layer, (e) Fig. 7: matrix elements do not exhibit identical behavior, i.e., the relative concentration of Al is decreased but approximately constant in the leach layer, Si definitely decreases toward the outer edge of the gel layer, and Zn generally decreases except in the thin barrier zones as shown in Fig. 5. Note that these results indicate the solubilization of the silicate matrix as suggested in Equations 1-3, (f) Fig. 8: in the leach layer, Mo is depleted, probably due to the formation of soluble MoO_4^{-2} species, and Cr is depleted only in the outer half of the leach layer. This latter result may indicate a species change from an insoluble chromium compound to a soluble

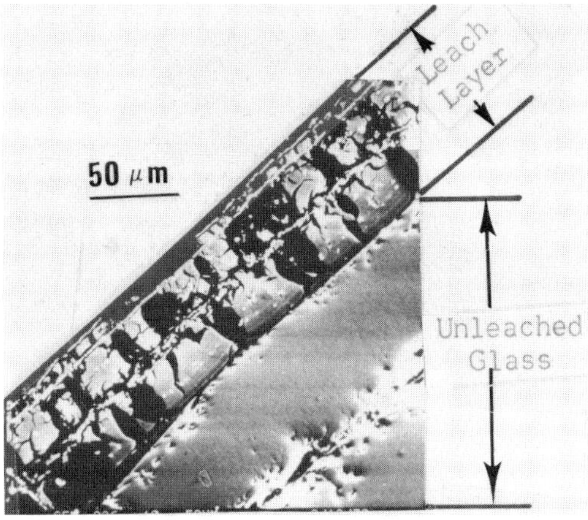

Fig. 3. Scanning Electron Microscopy Photomicrograph of Glass After 20 Months Soxhlet Leaching. Leach Layer and Unleached Glass are Denoted

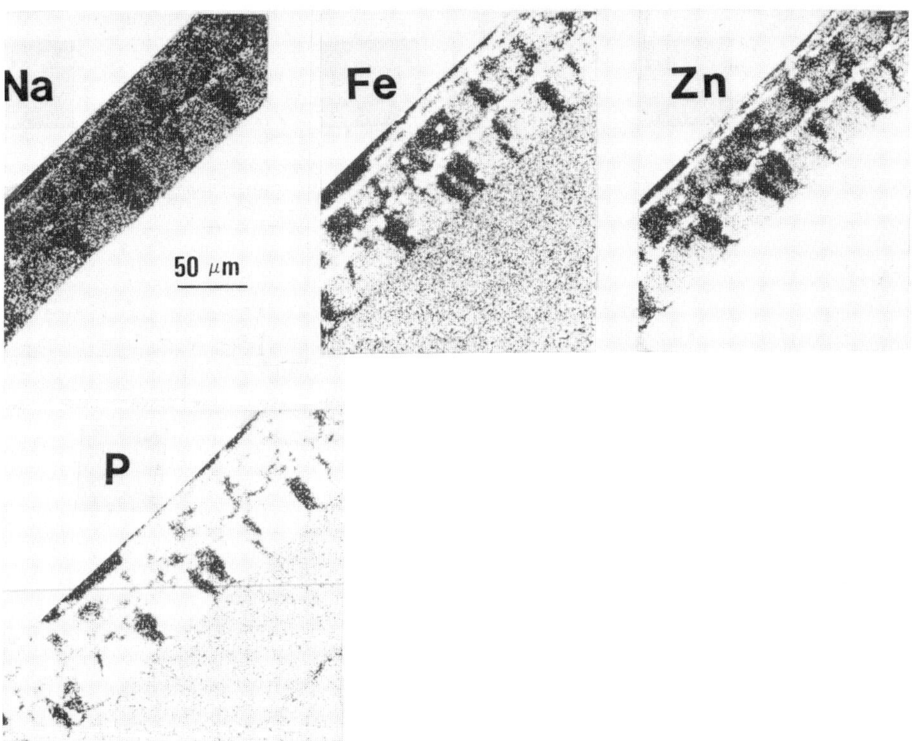

Fig. 4. Elemental Distributions of Na, Fe, Zn, and P in Glass as Determined by Electron Microprobe

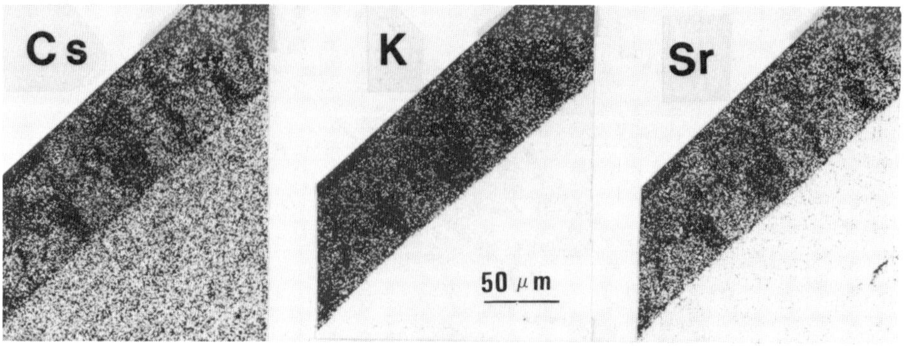

Figure 5. Elemental Distributions of Mobile Ions in Glass as Determined by Election Microprobe

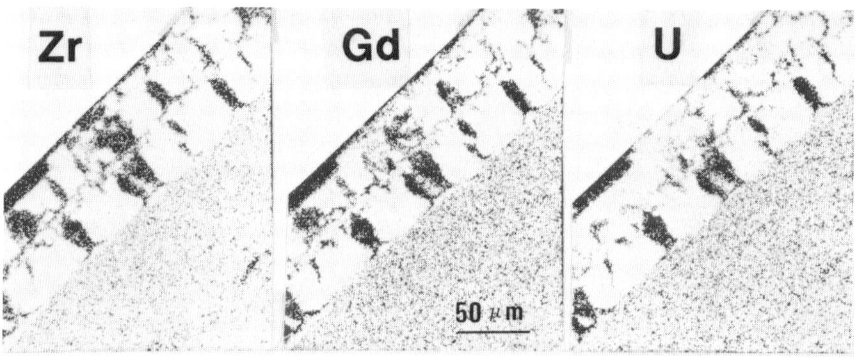

Fig. 6. Elemental Distributions of Immobile Elements in Glass as Determined by Electron Microprobe.

CrO_4^{-2} species. The fact that this change occurs at a Na-Fe-Zn-P barrier may or may not have significance.

The following leaching mechanism is consistent with the experimental results. The leachant reacts with the fresh glass surface to form siliceous gels initially releasing both mobile and immobile ions into solution. The immobile ions form hydroxides or other insoluble species in the gel layer which prevents their rapid diffusion or solubilization. They remain to produce a porous framework as the siliceous gels dissolve away. Mobile ion diffusion becomes increasingly inhibited because the effectiveness of diffusion barriers in the gel layer is continually enhanced by the formation of contiguous secondary diffusion barriers, e.g., the Na-Fe-Zn-P layers. Eventually, the leach layer becomes strained as the siliceous matrix is dissolved

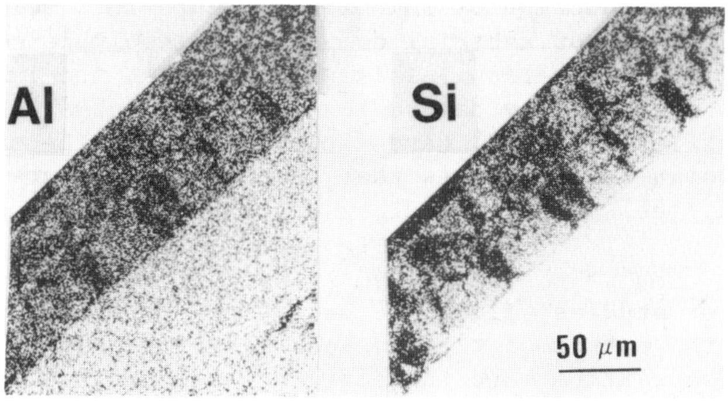

Fig. 7. Elemental Distributions of Matrix Elements in Glass as Determined by Electron Microprobe

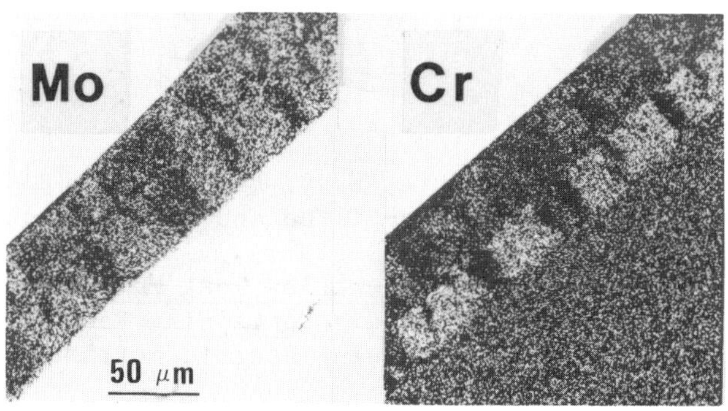

Fig. 8. Elemental Distributions of Possible Anion Forming Elements in Glass as Determined by Electron Microprobe

and the immobile ions accumulate in insoluble form. The leach layer begins to lose physical integrity with the matrix, possibly at a Na-Fe-Zn-P interface and begins to slough off. Thus, the total immobile ion concentration in the leachant begins to increase and reaches a maximum approximately when the sloughing process is complete and then begins to decrease as a "new" gel layer begins to form. In the sloughing process, the diffusion barriers are removed, or at least decreased, and mobile ions can diffuse more freely, thereby increasing the concentration of mobile ions in the leachant. After the sloughing process is complete, new diffusion barriers are being established and the concentration gradient is being decreased as a new gel layer is formed, thereby causing a decrease in the mobile ion leaching. The cycle then is repeated.

The introduction of immobile elements into solution is by a sloughing process of material composed of insoluble compounds with subsequent solubilization. This suggests that the immobile ion species in solution may be complex in nature rather than a simple ion. Scheffler et al (4) have shown this to be the case for plutonium. The point to be made is that the chemical nature of these species must be known in order to describe their transport through geomedia.

Further studies will be necessary to characterize the diffusion barriers, the effect of time between leaching changes, the effect of leachant composition, and the effect of temperature. Characterization of the species produced through leaching of a wasteform should also be considered.

ACKNOWLEDGEMENT

The authors express their appreciation to P. F. Hlava, Sandia Laboratories, for the electron microprobe work and to J. K. Johnstone, Sandia Laboratories, for informative discussions.

REFERENCES

1. D. E. Clark, M. F. Dilmore, E. C. Ethridge and L. L. Hench, "Aqueous Corrosion of Soda-Silica and Soda-Lime-Silica Glasses," J. Amer. Ceram. Soc. 56, 373 (1973).
2. J. E. Mendel, A Review of Leaching Test Methods and the Leachability of Various Solid Media Containing Radioactive Waste, USDOE Rept. BNWL-1765, Battelle Pacific Northwest Labs, Richland, WA, 44 (1973).
3. Report BNWL-1841, 29, Table VIII, Battelle Northwest Laboratories, Richland, WA, (1974).
4. R. G. Dosch, "Ceramic Forms for Nuclear Waste," Radioactive Waste in Geologic Storage (S. Fried, ed., ACS Symposium Series 100, 1979)
5. B. T. Kenna, K. D. Murphy and H. S. Levine, "Long-Term Elevated Leaching of Solid Waste Forms," Scientific Basis for Nuclear Waste Management 1 (G. J. McCarthy, ed., Plenum Press, NY, 157, 1979).
6. K. Scheffler, U. Riege, K. Louwrier, Hf. Matzke, I. Ray and H. Thiele, Long Term Leaching of Silicate Systems: Testing Procedure, Actinides Behavior and Mechanism, EUR-5509e, Kernfurschugezentrum Karlruhe, Germany, 60 (1977).
7. H. Born, The Mechanism of Dissolution of Silica in HCl-HF Mixtures, PhD Thesis, University of Texas, Austin, TX (1976).
8. M.N. Elliot and D. B. Auty, "The Durability of Glass for the Disposal of Highly Radioactive Waste," Glass Tech. 9,5 (1968).

BRINE CHEMISTRY EFFECTS ON THE DURABILITY OF A SIMULATED NUCLEAR WASTE GLASS*

J. W. Braithwaite

Sandia Laboratories**
Albuquerque, NM 87185

ABSTRACT

The effects of various solution chemistry parameters on the alteration and leaching characteristics of a copper borosilicate simulated waste glass have been determined under hydrothermal conditions. Results are presented which demonstrate that leachant salinity, volume to glass surface area ratio, pH, cation content and dissolved SiO_2 concentration effects are important. A brief explanation of these results is given which is based on ion exchange and solution saturation equilibria.

INTRODUCTION

The leaching process generates the source for radionuclide dispersal. This makes the study of the chemical durability of nuclear waste forms in geologic isolation environments an essential portion of the technological basis of radioactive waste management. It has been shown in previous studies that a number of experimental and environmental parameters affect the waste form leaching process (1, 2, 3, 4). Of particular importance to a bedded salt and subseabed site specific study is the effect of possible wide variations in leachant chemistry on chemical durability. Chemical variations in the leachant may result from any or all of the following four sources:

---

*This work supported by the United States Department of Energy (DOE), under Contract DE-AC04-76-DP00789.
**A DOE facility.

1. Repository site: Leachant compositions can range from a saturated NaCl brine solution to a number of saturated Na-K-Mg chloride brines in bedded salt to the unsaturated Na-K-Mg chloride containing seawater in subseabed sediments (5).

2. Leachant volume to waste form surface area ratio (V/SA): The mechanism of leachant intrusion (i.e., thermal migration versus a flooding accident in bedded salt) can affect the V/SA ratio; this, in turn, influences the concentration of solubilized products.

3. Dissolved SiO_2 concentration: The additional presence of geologic and canister backfill materials (clays and quartz) will change the concentration of dissolved SiO_2 in solution.

4. Radiation: The steady state concentration of gamma radiolysis products (H_2, H_2O_2, Cl_2^-, OCl^-) is a function of solute concentration (6).

The results of a study investigating the effects of the above four solution chemistry factors on the leaching characteristics of a copper borosilicate simulated waste glass in hydrothermal brine environments will be presented in this paper. An explanation of results, which is based on ion-exchange and solution saturation equilibria, will be given. It should be noted that the maximum anticipated interface temperature for waste glass under inundated bedded salt repository conditions is less than 150°C (7). A temperature of 250°C was used exclusively in this study in order both to overtest the waste glass durability and to accelerate the leaching reactions. A more complete description of current repository environments is given in Reference 5.

EXPERIMENTAL

A copper-titanium borosilicate glass frit (#199), obtained from the Battelle Pacific Northwest Laboratory, was melted with a simulated Barnwell-type oxide waste to form the waste glass studied. The waste glass composition was 30% waste oxides, 37% SiO_2, 9.8% B_2O_3, 8.4% Na_2O, 2.1% K_2O, 2.1% CaO, 6.3% TiO_2, 1.1% Al_2O_3 and 3.1% CuO.

The majority of the experiments were conducted in high pressure one liter 4130 steel autoclaves. The solution and 3 mm cubes of the waste glass were sealed inside a gold capsule. When a large solution excess was desired, one liter and four liter stainless steel and Hastelloy C-276 autoclaves were used.

Two saturated chloride brines, denoted as Brine A and Brine B, along with seawater and deionized (DI) water were used as the basic leach solutions in this study. The two saturated brines are representative of solutions which may be encountered in bedded salt environments, and their detailed composition and origin are given in Reference 5. Basically, Brine A contains high concentrations of Na-K-Mg chloride (107 g/l, 57 g/l, 138 g/l, respectively), whereas Brine B contains predominantly NaCl (290 g/l). Seawater is an unsaturated chloride brine containing a moderate concentration of $MgCl_2$ (5 g/l). Several variations of the above four solutions were also used.

Waste glass durability was assessed using a combination of weight loss measurements, solution analyses and a leached layer characterization. The sample weight loss was successfully used as a method of determining the extent of glass matrix alteration by measuring the final weight after the leached layer had been mechanically removed. Solution analyses were performed using atomic absorption, flame emission and induction-coupled plasma emission spectroscopy. Selected leached layers were characterized on the electron microprobe (EMP).

RESULTS

The solution volume to waste glass surface area (V/SA) ratio is a very important experimental parameter. The dramatic dependency of matrix alteration and cesium solubility on V/SA ratio is shown in Table 1. Cesium solubilization was monitored because previous studies have shown cesium to have both the highest leach rate of all the simulated radionuclides and to be least involved in the solid alteration product reactions (1, 3, 4). Data for the solubilization of other simulants will be given in the Discussion section. From the data shown in Table 1, one can conclude that reaction products, if allowed to concentrate, will significantly retard both the corrosion and leaching reactions. In an effort to understand both this retardation and the effect the presence of geologic and backfill materials will have, experiments were conducted with varying amounts of colloidal silica added. The colloidal silica served as a source for silicic acid in the high temperature water (solubility of amorphous silica in 250°C water is 2 g/l). The effect of dissolved SiO_2 on waste glass durability is shown in Table 2. The solution composition was varied by changing the chloride cations used. The influence of cation additions and leachant composition is shown in Table 3. To determine the effect of radiation, glass waste was leached at 90°C for 79 days in the four standard leach solutions in the presence of gamma radiation from a 10^7 rads/hr ^{60}Co source. Identical experiments were performed outside the gamma source for comparative purposes. The production of radiolysis products in the brine solutions increases the oxidizing potential of the solution; this might increase the extraction of species such as uranium which need to be oxidized. It was found, however, that the presence of radiation did not cause a detectable change in either matrix alteration or cesium and uranium solubilization. The effects of radiolysis represent the chief unknowns presently and will continue to be studied.

The following leachant chemistry observations can be made based on the data given above:

Table 1. Effect of Solution Volume to Waste Glass Surface Area (V/SA) Ratio on Matrix Alteration and Cesium Solubilized T = 250°C, t = 28 days)

Leachant	V/SA (cm)	Percent Altered	Percent Cs Solubilized
DI Water	0.5	3.6	1.3
DI Water	4.4	16.4	8.6
DI Water	44.0	70.0	38.4
DI Water	750.0	100.0	89.5
Brine A	4.2	6.6	0.8
Brine A	109.0	12.6	<1.9

Table 2. Effect of SiO_2 Concentration on Waste Glass Durability (T = 250°C, t = 28 days)

Leachant	V/SA (cm)	SiO_2 Added (g/l)	Percent Altered	Percent Cs Solubilized
DI Water	4.4	0.0	16.4	8.6
DI Water	4.4	0.2	15.5	9.0
DI Water	4.4	1.0	6.0	4.6
DI Water	4.4	2.0	3.0	1.3
DI Water	750.0	0.0	100.0	89.5
DI Water	750.0	2.0	0.1	3.6
Brine A	4.2	0.0	6.6	0.8
Brine A	4.2	2.0	1.0	0.4

Table 3. Effect of Solution Composition on Waste Glass Durability (T = 250°C, t = 28 days, V/SA = 4-5 cm)

Leachant	Percent Altered	Percent Cs Solubilized
DI Water	16.4	8.6
Seawater	11.5	10.3
Brine A	6.6	0.8
Brine B	4.6	1.1
100 g/l NaCl	3.0	1.7
300 g/l NaCl	3.0	0.8
57 g/l KCl + 243 g/l NaCl	2.0	0.6
138 g/l $MgCl_2$	15.0	5.5
138 g/l $MgCl_2$ + 145 g/l NaCl	9.0	2.0
pH 0 H_2SO_4	100.0	102.
pH 0 300 g/l NaCl	25.0	8.4
pH 12 NaOH	100.0	84.
pH 12 300 g/l NaCl	100.0	29.4

1. Salinity: As the salt concentration is increased, matrix alteration and Cs solubilization decrease. This effect occurs over a pH range of 0 to 12. The pH of the non-magnesium containing brines remains near neutral during testing, whereas the pH of the magnesium containing brines drops to 3-4 after being heated to 250°C (5).

2. pH: Both unsaturated low and high pH solutions produce significant leaching and alteration. However, the addition of 300 g/l NaCl retards both cesium extraction in this pH range and alteration only at low pH.

3. Presence of SiO_2: The presence of dissolved SiO_2 (as Na_2SiO_3, H_2SiO_4) results in an increase in the waste glass durability.

4. Cation Additions: The effects of Na and K chloride on alteration and cesium release are similar and affect leaching by producing changes in solution salinity. However, additions of magnesium chloride result in significant increases in both matrix alteration and cation release (including Cs, Ca and Sr) by lowering the solution pH and by entering into the glass alteration reaction (1).

5. Radiation: The leaching process is not affected by the presence of high intensity gamma radiation or radiolysis products.

From the above results, solution corrosivity can be expected to increase in the order: saturated KCl-NaCl brine, unsaturated KCl-NaCl brine, saturated $MgCl_2$-NaCl brine, unsaturated $MgCl_2$-NaCl brine. The various site specific leachants can then be categorized into one of the above basic chemistries.

Leachant chemistry and corrosivity could be altered by the presence of solids in the geologic formation itself. Possible solids which will be encountered in the sites being considered in this study include engineered backfill materials, bedded salt, clays dispersed in bedded salt and sea sediments. It was shown previously that excess bedded salt does not affect waste glass durability (1). A mixture of quartz and hectorite (a lithium montmorillonite $((Mg,Li)_6\ Si_8\ O_{20}(OH)_4)$ is a leading candidate backfill material. The clay mixture found in bedded salt and the deep sea sediments is composed of mixtures of montmorillonites, illites, kaolinites and other clays. The effect of these solids on the alteration of waste glass in various solutions in shown in Table 4. It was expected that these complex silicate clays will equilibrate with the hydrothermal solutions and retard the alteration reactions (similar to additions of SiO_2). This expectation was confirmed only for the sea sediments. The presence of bedded salt clay and hectorite had a deleterious or accelerating effect on the alteration mechanism. Magnesium is present in all three clays; yet it was shown in Table 3 that even extremely high concentrations of $MgCl_2$ could not produce the alterations caused by the additions of hectorite and bedded salt clay. Additionally, the solution pH remained around 5 to 6 in all of the clay experiments. Thus, the differences in alteration produced by clays and sea sediments must be due to the release from the impuri-

Table 4. Influence of the Additions of Various Geologic Clays on Waste Glass Alteration (T = 250°C, t = 28 days, V/SA = 4.2 cm, solution/clay - 4:1)

Clay	Leachant	Percent Alteration Without Clay	Percent Alteration With Clay
Hectorite	DI Water	16.4	70.0
Bedded Salt Clay	DI Water	16.4	46.0
Sea Sediments	DI Water	16.4	1.7
Bedded Salt Clay	Brine A	6.6	55.0
Sea Sediments	Seawater	11.5	5.9

ties in the clays of an effective silica complexing species or even a silica network attacking ion such as flouride. It should be noted that the accelerating effect of hectorite and bedded salt clay has only been observed at low V/SA. These results point out the existence of a potential waste form problem in an actual repository if proper backfill materials are not used. This effect is currently being studied in more detail and at more probable lower temperatures.

DISCUSSION

The mechanism of the dissolution of simple glasses has been extensively studied and need not be presented here except to note a few salient points (8). First, glass cations (especially alkali ions) are extracted by an ion exchange reaction with H_3O^+ leaving a silica rich phase. The rate of exchange is controlled by the diffusion rates of the ions involved through the leached layer. Hydroxide ions present or produced when H_3O^+ ions are consumed during ion exchange attack the SiO glass bond resulting in dissolution of the leached layer. The resistance of glass to leaching is then a function of the thickness and integrity of the diffusion limiting SiO_2-rich leached layer and the rate of hydration of the silica. It is believed the chemical durability of glasses increases with divalent oxide additions because these oxides fill microvoids in surface films to form a more impermeable barrier (9).

The glass used in this study is a very complex mixture of various oxides. For discussion purposes, assume the glass is a binary sodium silicate glass which is leached by the following reaction sequence in near neutral solutions:

$$2H_2O \rightleftarrows 2H^+ + 2OH^- \qquad 1)$$

$$Na_2SiO_{3(glass)} + 2H^+_{(aq)} \rightleftarrows H_2SiO_{3(glass)} + 2Na^+ \quad 2)$$

$$H_2SiO_{3(glass)} + 2OH^-_{(aq)} \rightleftarrows SiO^{-2}_{3(aq)} + 2H_2O \quad 3)$$

From the mass action expression for equation 2), it can be shown that:

$$\log a_{H_2SiO_3} = \log(K_{eq}) - 2 \log a_{Na^+} - 2pH \quad 4)$$

If it is assumed that the above processes are applicable and the system is at equilibrium, then the results discussed in the previous section can be qualitatively explained as follows:

1. Increases in the salt content in the leachant result in less glass cation ion exchange and less silica dissolution by the retardation of reaction 2 [see equation 4)]. This result was confirmed with both electron microprobe (EMP) mapping, which showed higher silica, cesium and sodium content in leach layers produced in brine than in DI water, and by product layer density measurements. The original glass density of 2.8 decreased to 2.0 for a seawater leached layer to 1.4 for a DI water leached layer. Thus, the silica layer forming during DI water leaching is not as compact and protective as the one formed during brine leaching. The solution pH could have a contributing effect here because prior studies have shown that significant SiO_2 dissolution occurs at high pH (8). DI water leachants maintained a pH of 8-9 and brine leachant pH was 5 to 6.

2. Increases in silica content in solution retard the glass matrix dissolution [reaction 3)] which results in a more effective diffusion barrier.

3. Lower solution volume to surface area ratios allow both dissolved alkali and silica concentrations to increase. Increases in Na, Cs and SiO_2 solution concentrations have been observed with decreases in V/SA ratios. Thus, a combination of the effects of Nos. 1 and 2 above retard the reactions as V/SA ratios decrease.

4. The accelerating effect of Mg additions in the leachant is due to the substitution of Mg^{+2} in the glass lattice for divalent cations like Ca, Sr, Cu. These divalent ions are not leached in DI water or Brine B. This substitution has been confirmed using the EMP and occurs because it is thermodynamically favorable due to magnesium's high charge to size ratio. Magnesium content in a Brine A leached layer reached 10-15% of the total constituents. The impermeability of a Mg containing leach layer is not as high as a leached layer still containing Ca and Sr, and therefore the alteration and leaching rates increase as Mg is substituted. The lack of an effect of time on glass durability which has been noticed in other studies is probably due to the saturation and equilibrium effects discussed

above (1, 4). Finally, the elements Gd, Ti, Pd, U, La and Mo are not detectably solubilized into high temperature brine solutions.

SUMMARY

The solution chemistry parameters salinity, pH, cation content, volume/exposed surface area and dissolved SiO_2 concentration have been shown to have a significant effect on waste glass durability.

1. Solution corrosivity increases in the order: saturated KCl-NaCl brine, < unsaturated NaCl-KCl brine, < unsaturated $MgCl_2$-NaCl brine.
2. Additions of alkali ions to the solution retard glass cation ion exchange.
3. Dissolved SiO_2 in solution improves the impermeability of the leached layer.
4. The presence of Mg decreases glass durability by substituting for divalent oxides like CaO, SrO and CuO in the glass lattice.
5. The effect of gamma radiolysis products on waste glass durability appears minimal.

REFERENCES

1. J. W. Braithwaite and J. K. Johnstone, "Chemical Durability and Characterization of Nuclear Waste Forms in a Hydrothermal Environment," Paper VI-7, Proceedings of the Symposium on Ceramics in Nuclear Waste Management, American Ceramics Society, Cincinnati, May (1979).
2. J. E. Mendel, Annual Report on the Characteristics of High Level Waste Glasses, BNWL-2252, June (1977).
3. J. H. Westsik and R. P. Turcotte, Hydrothermal Reactions of Nuclear Waste Solids, PNL-2759, September (1978).
4. G. J. McCarthy, Reaction of Water with a Simulated High Level Nuclear Waste Glass at 300°C, 300 Bars, DOE Report SA-35, Rockwell Hanford Operations (1978).
5. J. W. Braithwaite and M. A. Molecke, "Nuclear Waste Canister Corrosion Studies Pertinent to Geologic Isolation," Chemical and Nuclear Waste 1(1), in press (1980).
6. J. W. Braithwaite, N. J. Magnani and J. W. Munford, "Titanium Alloy Corrosion in Nuclear Waste Isolation Environments," Paper 213, Corrosion 80, NACE, Chicago, March (1980).
7. J. R. Wayland, Private Communication, July (1979).
8. A. Paul, "Chemical Durability of Glasses -- A Thermodynamic Approach," J. Mat. Sci. 12, 2246-2268 (1977).
9. D. E. Clark, "Aqueous Corrosion of Soda-Silica and Soda-Lime-Silica Glass," J. Amer. Cer. Soc. 59(1-2), 62-65 (1976).

HYDROTHERMAL INTERACTION OF SIMULATED NUCLEAR WASTE GLASS IN THE PRESENCE OF BASALTS

B. E. Scheetz, S. Komarneni, D. K. Smith*, C. A. F. Anderson, S. D. Atkinson and G. J. McCarthy**

Materials Research Laboratory and *Department of Geosciences, The Pennsylvania State University, University Park, PA 16802

INTRODUCTION

One of the leading candidate waste forms for disposal of commercial high level nuclear waste is glass. This waste form has received a great deal of attention over the past twenty years of development. Alteration of this waste form in the presence of mild hydrothermal solutions, which might be anticipated during the first several hundred years after emplacement of the waste into a geological repository, was shown (1) to be extensive at 300°C. These stability experiments, however, were conducted in a closed static hydrothermal system with the waste form and deionized water alone.

Basalt is one of the geomedia which is receiving significant consideration (2) as a repository. The mineral assemblage commonly encountered in the basalts from the Columbia River flows contain, as dominant phases, labradorite, feldspar, pyroxene and glass. The ample ferrous/ferric iron in these phases can act as a buffer of the chemical system that may be encountered in a flooded repository. Preliminary results that support this observation have been reported (3). In these experiements, the hydrous alkali uranyl silicate, weeksite, which forms extensively from glass in a closed hydrothermal system, was reacted with excess basalt. One product from this interaction was U_4O_9.

The present study was initiated to examine the effects of the presence of basalt on the alteration of a simulated nuclear waste glass under the static, closed-system hydrothermal conditions.

**Now at the University of North Dakota, Fargo, ND.

EXPERIMENTAL

Materials

Basalt. Two different basalts from the Columbia River Basalt Flows were used in these studies. The first was a generic basalt represented by the USGS standard, BCR-1, while the other was a sample obtained from the targeted repository location nearly 1000m beneath the Hanford Reservation. This sample was designated DDH-3/A2120/3320 and represents a site specific sample from the Umtanum Formation. Detailed compositional data for BCR-1 are presented by Flanagan (4), while compositional data for DDH-3 are contained in Deju, et al. (5).

Glass. The simulated nuclear waste form used in these studies was a formulation developed at Battelle Pacific Northwest Laboratories, designated PNL 76-68, which was designed to incorporate PW-7a calcine into a monolithic borosilicate-based glass. Detailed characterization of the glass was performed at Pacific Northwest Laboratories (6).

Procedures

All of the experiments were conducted in sealed gold capsules which were subjected to temperatures up to 300°C at 300 bars of pressure. The temperatures and pressures were developed and maintained in conventional cold-seal hydrothermal vessels with resistance heating element furnaces.

The solid charges for these experiments were of two forms: one employed intimate mixtures of -200 mesh powder of basalt and glass, and the other employed chunks (-6+10 mesh) of both solids in approximately equal masses. The solid charge was sealed into the gold capsule with either five or ten times the mass of deionized water relative to the total mass of the solids.

Characterization Methods

Solids. All reaction products from these experiments were examined optically with a 40X binocular microscope to ascertain gross spatial relationships before being examined on a scanning electron microscope (SEM). The SEM used in these studies was a JOELCO 50A equipped with a Kevex energy dispersive X-ray spectrum analyzer. In addition, samples of the solid reaction products larger than 5 μm were hand-picked and examined by means of Gandolfi microsampling X-ray diffraction. A combination of the qualitative chemistry and the structure type based on matches with data in the Powder Diffraction File (7) allowed for the identification of reaction products.

Liquids. The liquid phase of these experiments was carefully collected (1) for quantitative chemical determinations. Eighteen to twenty elements were analyzed in each sample with the aid of atomic emission spectrometry, atomic absorption spectrophotometry and fluorometry. A Spectraspan III instrument was employed for analysis of all elements except Na, Rb, Cs and U. Na, Rb and Cs were determined using a PE403 atomic absorption spectrophotometer. U concentrations were determined by a fluorometric technique.

RESULTS AND DISCUSSION

Solid Reaction Products

Characterization of solid reaction products from experiments conducted at 100°C did not reveal any interaction and/or alteration products. These results were anticipated, based upon the observations of the individual components of the system, which also exhibited very little alteration (8).

At 300°C, however, extensive changes had occurred. The dominant products that were characterized were similar to those identified for the alteration of glass alone at this temperature (4). Extensive pyroxene formation was observed. The pyroxenes were characterized by SEM and X-ray diffraction as being an acmite-augite composition with a green body color. This dark body color is more typical of natural pyroxenes of this composition that contain partially reduced iron which allows for charge transfer between the divalent and trivalent iron through a common oxygen ligand. The partially reduced pyroxenes were rarely observed in the alteration products of glass alone.

In those experiments where the solid charges were composed of the larger fragments (small surface area), the mineral weeksite ($K_2(UO_2)_2Si_6O_{15} \cdot 4H_2O$) was observed to form extensively developed crystal clusters. This is contrasted to the experiments in which the charges consisted of intimate mixtures of powders (large surface area). In the latter experiments, weeksite was never detected as an alteration product. These observations suggest that the effective oxygen fugacity established by the basalt lies below the stability limits for weeksite, thus preventing its formation. The finer particle size of the latter experiments allows the equilibrium oxygen fugacity to be more closely approached than when coarse particles were used. Further, the relative changes of the two solid forms in these experiments suggest that in experiments using charges of smaller surface area, the glass waste form alters fast enough to dominate the effective oxygen fugacity of the closed system.

In those experiments in which weeksite did not form, an unidentified phase was always observed. This phase, which is designated as "unknown A," exhibits hexagonal morphologies and has a qualitative chemistry composed of Ca, Zn, Fe and Si. Weeksite and this unidentified phase were observed together in only one experiment at 300°C after four months of hydrothermal treatment.

Doubly terminated crystals with lengths of up to 25 μm were also isolated in the alteration products of these experiments at 300°C. Characterization of these crystals by SEM/EDX and X-ray diffraction identified them as quartz. Quartz was never observed as an alteration product from the 300°C experiments when glass alone was treated with deionized water. Although at this point the controlling factors for the formation of quartz in these experiments are not known with certainty, the data suggest a possible modifying influence of basalt upon the pH of the fluids established by the alteration of glass. The high alkali content of the solutions in contact with the altered glass in the presence of basalt may be further evidence to substantiate the controlling effect of pH on the formation of quartz relative to the pyroxenes.

Grains of a phase morphologically and chemically similar to pollucite, $CsAlSi_2O_6$ (PDF #25-194), were isolated and identified as such by X-ray diffraction. This observation in the 300°C, 4-month experiment marks the first time that pollucite has been identified as a product in these experiments.

Two additional solid phases were isolated and identified in the reaction products. A zinc silicate, probably willemite (Zn_2SiO_4), has been identified by SEM/EDX but has not yet been confirmed by X-ray diffraction. The second phase was picked from a white crust in the reaction products and identified as orthoclase, $KAlSi_3O_8$ (PDF #19-931), by X-ray diffraction. These latter two phases, which were isolated in the products of the 300°C, 4-month experiments, were not identified in products from shorter runs.

Liquid Phase Characterization

Results of the chemical analyses of solutions are expressed as concentrations (μg/ml) and as weight percentages of the amount of each element present in the solids. The concentration data cannot be directly compared between the glass alone and glass + basalt experiments because the glass-to-deionized water ratios are 1:10 and 1:20, respectively, even though they have a constant total solids-to-water ratio of 1:10. When the solid charge is a mixture of glass + basalt, the elements are expressed as percentages of glass only for two reasons: a) the elements Cs, Rb, Sr, Ba, Mo and B are present in glass, and b) the alteration of glass (noncrystalline solid) is expected to be much faster than the basalt, which is largely crystalline. Therefore, even those elements which are common to both glass and basalt will be extracted mainly from glass.

Quantitative determination of the constituents of the liquids in contact with the solid charge in these experiments takes on much more significance at low temperatures where the solid alterations are minimal.

Data for 1- and 4-month runs at 100°C indicate the percentages of Cs, Rb, Mo and B taken into solution were higher in the presence of basalt (Tables 1 and 2). The value of Cs at 1 month is approximately 2.5% of the total available Cs in the original solid for the glass-basalt experiments, contrasted to no detectable Cs in the experiments with glass alone. The values do not appear to change significantly as a function of the duration of the runs. For example, only 2% of the Cs was detected in the glass-basalt experiments, even after 4 months (Table 2).

The amounts of detected cations in solutions increased sharply with the enhanced alteration at 300°C (Table 3). Alteration of glass alone and alteration of glass in the presence of basalt were investigated as a function of time at 300°C (Table 3). After 1 week of treatment, the percentages of Cs and Rb released into solution stayed the same with or without the presence of basalt. However, at 2, 3 and 4 weeks, the percentages of these two elements in solution increased by a factor of 2-4 in the presence of basalt. The alkaline earth elements Sr, Ba and Ca could not be detected in solution in the presence of basalt but could be detected in the absence of basalt. These observations indicate that these elements are probably reacting

Table 1. Element in Solution at 100°C[a] for 4 Weeks

	deionized water +					
	DDH-3		glass		DDH-3 + glass	
	DD 232		average of GD 203 + GD 204		DGD 229	
element[b]	μg/ml	%	μg/ml	%	μg/ml	%
Cs	<0.05[c]	-	<0.05	-	13	2.6
Rb	<0.005	-	<0.005	-	2	3.2
Na	35	1.5	155	1.7	170	1.8
Sr	<0.001	-	<0.001	-	1	0.4
Ba	<0.001	-	<0.001	-	<0.001	-
Ca	70	1.4	<0.001	-	25	1.7
Mo	-	-	19	1.2	20	2.5
B	-	-	33	1.1	30	4.0
Si	110	0.4	105	0.6	250	1.3

[a]P = 300 bars; solid-to-deionized water ratio = 1.10.

[b]Al, Fe, Cr, Ni, Zn, Nd, La, Ce, Gd, Ti, Zr, P, and U were not detected in solution.

[c]The cumulative experimental errors are within ±5% of the reported values.

Table 2. Element in Solution at 100° C[a] for 4 Months

	deionized water +					
	BCR-1		glass		BCR-1 + glass	
	BD 271		GD 273		BGD 270	
element[b]	μg/ml	%	μg/ml	%	μg/ml	%
Cs	<0.05[c]	-	6	0.6	10	2.0
Rb	<0.005	-				
Na	6	0.32	80	2.9	160	1.7
Sr	1	2.1	<0.001	-	<0.001	-
Ba	<0.001	-	<0.001	-	<0.001	-
Ca	40	0.8	50	3.4	25	1.7
Mo	-	-	40	2.4	20	2.5
B	-	-	92	3.0	70	4.2
Si	210	0.8	330	1.7	270	1.4

[a]P = 300 bars; solid-to-deionized water ratio = 1.10.

[b]Al, Fe, Cr, Ni, Zn, Nd, La, Ce, Gd, Ti, Zr, P, and U were not detected in solution.

[c]The cumulative experimental errors are within ±5% of the reported values.

Table 3. Percent of Element in Solution at 300°C

	1 week		2 weeks		3 weeks		4 weeks	
element[a]	glass	glass + BCR-1 basalt[c]	glass	glass + BCR-1 basalt	glass	glass + BCR-1 basalt	glass	glass + BCR-1 basalt
Cs	4.3	4.3	5.6	14.8	4.6	15.6	5.0	20.3
Rb	ND[d]	ND	1.4	ND	3.4	6.6	8.7	14.0
Na	46	38	45	34	44	34	45	38
Sr	0.62	ND	0.55	ND	0.22	ND	0.15	ND
Ba	1.47	ND	1.67	ND	0.04	ND	0.1	ND
Ca	0.51	ND	0.55	ND	0.62	ND	1.2	ND
Mo	72	78	72	85	72	72	72	85
B	84	87	87	81	89	84	93	91
Si	2.7	1.3	3.7	1.0	6.4	0.9	4.6	1.0
U	0.15	0.06	0.07	0.15	0.03	0.05	0.03	0.05

[a]Zn, Ni, Cr, Fe, La, Nd, Ti, and Zr were not detected in solution.

[b]1:10 glass-to-deionized water ratio.

[c]1:10 glass-to-deionized water ratio, but total solids-to-water ratio is 1:5.

[d]ND = None Detected, based upon the detection limits for the experimental design. (See Tables 1, 2, and 4.)

with the basalt to form new solid phases. The Na and Si contents decreased in solution when glass altered in the presence of basalt.

Table 4 contrasts the solution chemistry of basalt, glass and glass-basalt experiments that were conducted under identical conditions at 300°C for 4 months. These data suggest that the extraction

Table 4. Element in Solution at 300°C[a] for 4 Months

	1:10 solid-to-deionized water ratio					
	BCR-1		glass		BCR-1 + glass	
	BD 260		GD 262		BGD 259	
element[b]	μg/ml	%	μg/ml	%	μg/ml	%
Cs	<0.05[c]	-	136	14	95	20
Rb	<0.005	-	25	2	10	17
Na	70	2.8	4750	51	2000	21
Sr	<0.001	-	<0.001	-	1	0.4
Ba	<0.001	-	0.6	0.1	<0.001	-
Ca	25	0.5	51	3.4	25	1.7
Mo	-	-	1040	68	440	59
B	-	-	2370	80	1300	88
Si	310	1.2	430	2.3	290	1.5

[a] P = 300 bars.

[b] Al, Fe, Cr, Ni, Zn, Nd, La, Ce, Gd, Ti, Zr, P, and U were not detected in solutions.

[c] The cumulative experimental errors are within ±5% of the reported values.

of Cs into solution increased at least by a factor of 2 in those experiments in which glass altered in the presence of basalt, compared to glass alone. Rb followed a similar pattern but with nearly an 8-fold increase. Results for all the other elements excepting Mo parallel those at shorter durations, as discussed above. Molybdenun in solution decreased after 4 months of treatment in the presence of basalt. This decrease may be due to the formation of powellite by reaction with Ca from basalt.

CONCLUSION

One important result of these experiments is the verification of the oxygen buffering capacity of the ferrous/ferric content of the basalt and its effect on the uranium component of the waste glass. Where the glass was the most reactive material, as in the experiments with solids of low surface area, weeksite was formed, indicating that the U^{6+} content was not effectively reduced. However, where the surface area was large, the U^{6+} component was reduced, preventing the formation of weeksite.

The solid phase characterization has revealed several compounds from the long duration runs that were not observed in products of the shorter runs. New phases such as quartz, orthoclase, pollucite and willemite were found. The presence of these phases may be indicative of a closer approach to equilibrium, or at least an indication of phases which are more stable in the presence of basalt than in its absence.

Chemical analyses of the solutions indicated that the percentages of Cs released from glass into solution at 100° or 300°C increased by a factor of 2 to 4, while Sr, Ba, Ca, Na and Si seem to have decreased to various extents in the presence of basalt. The principal conclusion which results from these studies is that basalt has a marked influence on the hydrothermal alteration behavior of the glass waste form. Not only does it cause new solid phases to form, but it also modifies the solution chemistry significantly. In particular, it seems to increase the amount of Cs which is taken into solution, a result which requires more careful investigation.

ACKNOWLEDGEMENTS

This research was sponsored by the U.S. Department of Energy through the Rockwell Hanford Operations under Contract No. EY-77-C-06-1030.

REFERENCES

1. G. J. McCarthy, B. E. Scheetz, S. Komarneni, D. K. Smith and W. B. White, "Hydrothermal Stability of Simulated Radioactive Waste Glass," Solid State Chemistry: A Contemporary Overview, Advances in Chemistry Series, American Chemical Society (in press).
2. Staff, Basalt Waste Isolation Program, Basalt Waste Isolation Program Plan, RHO-CO-132, Rockwell Hanford Operations, Richland, WA, May 15 (1979).
3. B. E. Scheetz, D. K. Smith, M. W. Barnes, S. Komarneni, L. W. Stull and C. A. Smith, Simulated High Level Waste-Basalt Interaction Experiments, Annual Progress Report for the Period Oct. 1, 1977, to Sept. 30, 1978, RHO-BWI-C-33, Rockwell Hanford Operations, Richland, WA, Nov. (1978).
4. F. J. Flanagan, "1972 Values for International Geochemical Referenced Samples," Geochimica et Cosmochimica Acta 37, 1189-1200 (1973).
5. R. A. Deju, R. K. Ledgerwood and P. E. Lang, Reference and Waste Form Basalts and Ground Water Systems for Waste Interaction Studies, RHO-BWI-LD-11, Rockwell Hanford Operations, Richland, WA, Sept. (1978).
6. W. A. Ross, D. J. Bradley and L. R. Bunnell, Annual Report on the Characterization of High Level Waste Glasses, PNL-2625, Battelle Pacific Northwest Laboratories, Richland, WA, June (1978).
7. Powder Diffraction File, International Center for Diffraction Data, Swarthmore, PA (W. F. McClane, Mgr. Ed., JCPDF, 1979).
8. B. E. Scheetz, S. Komarneni, D. K. Smith, A. F. Anderson and S. D. Atkinson, Interactions Between Glass/Basalt and Supercalcine/Basalt, RHO-BWI-CI, Rockwell Hanford Operations, Richland, WA, submitted Sept. 30 (1979).

VITRIFICATION OF HIGH-LEVEL ALUMINA NUCLEAR WASTE

J. R. Brotzman

Exxon Nuclear Idaho Company, Inc.
P. O. Box 2800
Idaho Falls, ID 83401

ABSTRACT

Borophosphate glass compositions have been developed for the vitrification of a high alumina calcined defense waste. The effect of substituting SiO_2 and P_2O_5 for B_2O_3 on the viscosity and leach resistance was measured. The effect of the alkali to borate ratio and the $Li_2O:Na_2O$ ratio on the melt viscosity and leach resistance was also measured.

INTRODUCTION

Vitrification of high level nuclear waste is one of the waste management alternatives being investigated at the Idaho Chemical Processing Plant (ICPP). Currently the nuclear waste solutions are calcined to a mixture of granules and powder by a fluidized bed calcination process (1). The calcined waste has been produced since 1963 and is currently stored in near-surface stainless steel bins within concrete vaults. The two most abundant wastes produced are zirconia calcine, consisting of approximately 25% ZrO_2, 20% Al_2O_3 and 55% CaF_2, and alumina calcine, consisting of approximately 90% Al_2O_3, 3% Na_2O, 3% B_2O_3, 0.5% fission products and 3.5% miscellaneous. Glass compositions have been developed for the zirconia calcine (2). The flux developed for the zirconia waste does not readily vitrify the alumina calcine nor does it vitrify mixtures of the two wastes with >15% wt % alumina calcine. Because of this, formulations to vitrify the alumina calcine were developed. Although both borophosphate and borosilicate frits were evaluated, only the borophosphate frits reported on here resulted in practical alumina waste glasses.

Each of the glass compositions was evaluated on the basis of melt viscosity and leach resistance. These two criteria were used because glasses with melt viscosities of 50-200 poise at temperatures of 1050°-1200°C are desired for the proposed nuclear waste vitrification processes of in-can melting (3) or joule heated melting (3,4) and glasses with high leach resistance are desired for environmental safety.

EXPERIMENTAL METHOD

Glass Preparation

Experimental glasses were melted at 1100°C in either 100 mL platinum or high fired alumina crucibles. Glasses were made using reagent grade oxides and carbonates and a simulated alumina calcine (5). A 20-24 h fining period was used after the charging of the glass additives was completed. The glass samples were poured into graphite molds, cooled and then ground to a -16+30 mesh size for leach testing.

Upon pouring, each glass was judged on a relative viscosity scale as follows: 1=25-75 poise, 2=75-150, 3=150-300, 4=300-500 and 5=not pourable, as estimated from standard viscosity fluids. The actual viscosity of selected melts was measured with a Brookfield viscometer.

Numerous compositions developed a meltable borophosphate glass containing 23 mol% calcine which was then used as a base glass for further study (Table 2). The glass was evaluated in the four leach solutions shown in Table 1. The solutions were designed to evaluate the glass over a wide range of pH conditions. The Soxhlet test was used because it measured the corrosion rates of the glasses in a neutral pH region at an elevated temperature. Preliminary experiments showed that the borophosphate glasses varied most in the Soxhlet leach test and had good leach resistance in the other leach solutions. Thus, the Soxhlet leach test was used to compare the modifications made to the base borophosphate glass. The modified glasses were checked periodically for both acid and Soxhlet leach resistance.

Leach Tests

Experimental procedures previously used at the ICPP (2) were followed in leaching the glasses. Four grams of the ground glass sample were weighed and placed in either a 100 mL leachant solution (Table 1) at 25°C or in a Soxhlet apparatus at 95°C. After 19 hours on a magnetic stirrer or 72 hours in the Soxhlet apparatus, the sample was removed, dried and then weighed. The gross weight percent lost was then calculated. All leach rates referred to in this report

Table 1. Leaching Solutions and Conditions

Solution	pH	Temperature, °C	Time, h	Description
A	3.6-3.8	25	19	1M acetic acid, buffered
B	7	95	72	distilled water, Soxhlet test
C	9	25	19	1M NH_4OH, buffered
D	7.5	25	19	WIPP brine water, 306 g/L dissolved solids

Table 2. Base Borophosphate Glass Composition and Properties

	Composition, mol%						
Glass No.	SiO_2	B_2O_3	P_2O_5	Li_2O	Na_2O	CuO	Al_2O_3 Calcine
178	4	40	10	10	10	3	23

Properties	
Viscosity (poise) at 1100°C	25-75
Acid Leach, wt% lost/19 h (25°C)	0.5
Soxhlet Leach, wt% lost/72 h (95°C)	6

are expressed as weight percent lost in either a 19-hour or a 72-hour period.

BOROPHOSPHATE GLASS DEVELOPMENT

Single Substitutions of SiO_2 or P_2O_5 for B_2O_3

The base borophosphate glass (No. 178) has a high Soxhlet leach rate and a low melt viscosity. Individual substitutions of SiO_2 and P_2O_5 were made for B_2O_3 in this base glass (Table 3) to improve the Soxhlet leach rate. The substitution of SiO_2 (8-20 mol%) decreased the Soxhlet leach rate to only 1.5 wt% while only moderately increasing the melt viscosity. The substitution of (15 mol%) P_2O_5 also lowered the Soxhlet leach rate without significantly raising the melt viscosity. At levels above 15 mol% P_2O_5, the glasses became very viscous, melted with heavy surface scums and had high Soxhlet leach rates.

Table 3. Substitution of SiO_2 or P_2O_5 for B_2O_3

Base Glass 178

A. Substitution of SiO_2

Glass No.	Mol% SiO_2-B_2O_3	Relative Viscosity	Soxhlet Leach Rate, wt% lost/72 h
187	0-44	1	16.5
178	4-40	1	6.0
188	8-36	1	5.1
211	12-32	2	3.2
212	16-28	2	2.5
271	18-26	2	1.4
272	20-24	2	1.3

B. Substitution of P_2O_5

Glass No.	Mol% P_2O_5-B_2O_3	Relative Viscosity	Soxhlet Leach Rate, wt% lost/72 h
176	0-50	1	27
178	10-40	1	6
186	12.5-37.5	1	6
185	15-35	1	3
192	17.5-32.5	3	18
193	30-20	4	not reproducible
194	27.5-22.5	5	-

Having observed a useful range of 10-15 mol% P_2O_5 in the base glass, it was decided to substitute SiO_2 for B_2O_3 in glass 185 which contained 15 mol% P_2O_5 (Table 3). Substitutions of 4-14 mol% SiO_2 had little effect on the Soxhlet leach rate and only moderately increased the melt viscosity. It was not until substitutions of 14-16 mol% SiO_2 were made that the leach rate decreased to 1.5% and the viscosity increased to 75-150 poise. The Li_2O to Na_2O ratio and alkali to borate ratio were then varied in the selected glasses (Table 4) to improve the melt viscosity and Soxhlet leach resistance.

Variation of the Alkali Ratio

The alkali ratio was varied in a low silica glass (No. 178) and a high silica glass (No. 271). The effect of variation in the alkali ratio on the viscosity and Soxhlet leach rate of glass 178 was markedly different from that of glass 271 (Tables 4 and 5).

Table 4. Selected Low and High Silica Borophosphate Glasses

	Composition, mol%						
Glass No.	SiO_2	B_2O_3	P_2O_5	Na_2O	Li_2O	CuO	Al_2O_3 Calcine
178	4	40	10	10	10	3	23
185	4	35	15	10	10	3	23
271	18	26	10	10	10	3	23
272	20	24	10	10	10	3	23
274	16	23	15	10	10	3	23

Table 5. Effect of Variation of Alkali Ratio on Viscosity and Leach Rate

	Glass 271		
Glass No.	Mol% Li_2O-Na_2O	Relative Viscosity	Soxhlet Leach Rate, wt% lost/72 h
301	20-0	1	1.0
302	5-15	1	1.3
271	10-10	2	1.4
303	15-5	2	2:0
304	0-20	2	1.5

All of the low silica glasses (not shown) with an Li_2O:Na_2O ratio of 1:1 or greater had leach rates of 6-9 wt% while glasses with less Li_2O present had leach rates of 25-30 wt%. The variation of the Li_2O:Na_2O ratio had no effect on melt viscosity. This glass was not improved by changing the alkali ratio.

All of the high silica mixed alkali glasses (Table 5) had leach rates of 1-2 wt% but the viscosity of the glasses decreased with increasing Li_2O content. As an example, glass 301, which contains only Li_2O, had the lowest viscosity (50 poise) and leach rate (1 wt%).

Variation of the Alkali to Borate Ratio

The alkali to borate ratio was varied in two low silica glasses (Nos. 178 and 185) and two high silica glasses (Nos. 272 and 274)

Table 6. Effect of Variation of Alkali to Borate Ratio on Viscosity and Leach Rates

Base Glass 178

Glass No.	Mol% Alkali-B_2O_3	Relative Viscosity	Soxhlet Leach Rate, wt% lost/72 h
248	16-44	1	43.3
249	18-42	1	23.6
178	20-40	1	6.0
250	22-38	1	22.0
251	24-36	1	18.9

Base Glass 185

Glass No.	Mol% Alkali-B_2O_3	Relative Viscosity	Soxhlet Leach Rate, wt% lost/72 h
267	16-39	1	8.3
268	18-37	1	2.7
185	20-35	1	3.0
269	22-33	1	1.8
270	20-31	1	2.0

Base Glass 272

Glass No.	Mol% Alkali-B_2O_3	Relative Viscosity	Soxhlet Leach Rate, wt% lost/72 h
272	20-24	2	1.3
305	22-22	2	1.4
306	24-20	2	1.4

Base Glass 274

Glass No.	Mol% Alkali-B_2O_3	Relative Viscosity	Soxhlet Leach Rate, wt% lost/72 h
281	16-27	3	4.7
282	18-25	2	1.3
274	20-23	2	1.3
283	22-21	1	.85
284	24-19	1	1.8

(Table 6). It was observed that the viscosity of the glasses decreased as the alkali content increased and that the effect became more pronounced as the SiO_2 and P_2O_5 content in the glass increased. For example, the most significant decrease in viscosity occurred in glass 274, a high SiO_2, high P_2O_5 glass.

The low silica glass 178 had acid leach rates of 25 to 40 wt% when the alkali content was other than 20 mol% (alkali:borate=1:2). The low P_2O_5 and SiO_2 contents of the glass are the cause for the high leach rates. The other glasses showed no major changes in leach rate if the alkali content was between 18-24 mol%. Thus, improvements of the glasses (other than 178) can be made by increasing the alkali to borate ratio of the glass.

Brookfield Viscosity Measurements

Actual 1000°-1100°C viscosities, as measured by a Brookfield viscometer, were determined for four glasses containing 10% P_2O_5 and 20% alkali and for two glasses containing 15% P_2O_5 and 20% alkali. In addition, the reduction in viscosity achieved in these glasses by increasing the alkali to borate ratio and the Li_2O to Na_2O ratio was measured (Table 7).

SUMMARY AND RECOMMENDATIONS

Substitutions of SiO_2 and P_2O_5 for B_2O_3 were made using a base borophosphate (10 mol% P_2O_5) composition. Both of the additives increased the melt viscosity and decreased the Soxhlet leach rate.

Table 7. The Viscosity-Temperature Profile of Selected Borophosphate Glasses

Condition	Glass No.	Glass Viscosity (poise)		
		1000°C	1050°C	1100°C
Comparison of 10%, 15% P_2O_5 Glasses	185	175	105	30
	212	110	60	40
	272	275	150	90
	274	580	250	120
Alkali to Borate Ratio	272	275	150	90
	306	210	120	70
	274	580	150	120
	283	180	100	70
Use of Li_2O	271	210	130	80
	301	160	90	50

Substitutions of 4-15 mol% SiO_2 in a 15 mol% P_2O_5 base glass also lowered the leach rate and increased the viscosity, but these glasses were more viscous when compared at equal leach rates. The viscosity of the low silica glasses, such as Nos. 178 and 185, was slightly decreased by increasing the alkali to borate ratio. These modifications produced variable effect on the leach rates.

The viscosity of the high silica glasses, such as Nos. 272 and 274, was decreased by increasing the alkali to borate ratio and by increasing the Li_2O to Na_2O ratio. These modifications cause only slight changes in the Soxhlet leach rate. The acid leach rates of these high and low silica glasses were quite low (0.5 wt%).

Based on Soxhlet leach rate, melt viscosity and glass meltability, there are a number of borophosphate glasses suitable for vitrification of the alumina calcine. High silica glasses with 10-15 mol% P_2O_5 and 20-24 mol% alkali (with an Li_2O to Na_2O ratio of 1:1 or greater) have Soxhlet leach rates approximately one half that of the low silica glasses. All glasses meet the viscosity requirements of joule heated melting or "in-can" melting. These experiments demonstrate that leach resistant glasses containing alumina calcine can be made at temperatures compatible to the proposed vitrification processes.

REFERENCES

1. G. F. Offutt and B. R. Wheeler, AEC-IN-1091, March (1968).
2. D. Gombert, H. S. Cole, J. R. Berreth, DOE-ICP-1177, Feb. (1979).
3. H. T. Blair, DOE-BNWL-2061, May (1976).
4. J. H. Valentine and B. J. Newby, DOE-ICP-1163, Oct. (1978).
5. S. A. Birrer, DOE-ICP-1148, April (1978).

IMPROVED GLASS COMPOSITIONS FOR IMMOBILIZATION OF SRP WASTE*

M. J. Plodinec

Savannah River Laboratory
E. I. du Pont de Nemours and Company
Aiken, SC 29801

INTRODUCTION

Radioactive waste produced from reprocessing nuclear materials for defense programs at Savannah River Plant (SRP) is stored in large underground tanks on the plant site. This alkaline waste is made up of water soluble salts and insoluble "sludges." Most of the waste actinides and fission products are contained in the insoluble sludge, consisting mainly of hydroxides and hydrous oxides of iron, aluminum, and manganese. The rest of the waste is either in the form of a soluble crystalline salt cake or a nearly-saturated supernatant salt solution. This fraction contains about 95% of the fission product cesium and traces of other radionuclides.

Methods to immobilize SRP waste for long-term storage are being developed at Savannah River Laboratory (SRL). In the current reference process (Fig. 1), the supernatant liquid is pumped from the tanks and decontaminated by ion exchange. The sludge is slurried with water removed from the tank, and washed with hot caustic to remove about 75 wt % of the aluminum. The sludge is then water-washed to remove soluble salts. The cesium-loaded zeolite from ion exchange are then mixed with the sludge.

The radioactive sludge, cesium-loaded zeolites, and residual radionuclides from ion exchange are dried by spray calcination, mixed with pre-melted glass frit, and melted to form a borosilicate glass

* The information contained in this article was developed during the course of work under Contract No. AT(07-2)-1 with the U. S. Department of Energy.

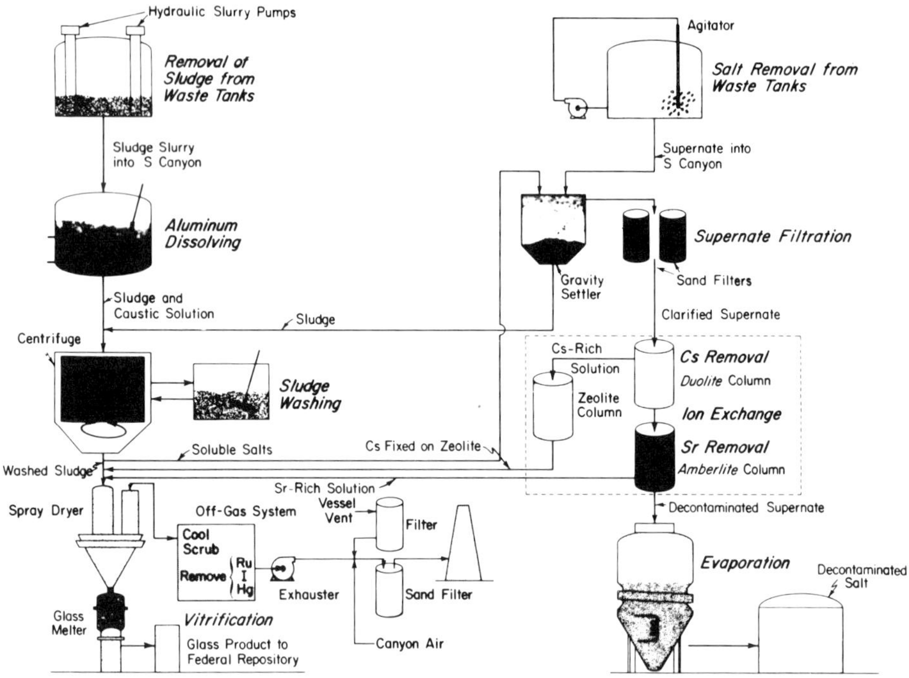

Fig. 1. Reference Process for Immobilization of SRP Waste

in a joule-heated melter. The molten glass is poured into a steel canister and allowed to cool. Table 1 shows the composition of the average calcine and extreme compositions expected.

Initial glass composition development was directed toward demonstrating the feasibility of immobilizing SRP waste in glass. A large amount of experimental data was obtained on the properties of simulated and actual waste immobilized in glass made from Frit 21 (Table 2). In laboratory tests, this frit formed durable glasses containing 35 wt % sludge which spanned the range of compositions found in SRP wastes. Therefore, Frit 21 was chosen for large-scale testing.

However, results with large-scale tests with simulated SRP waste indicated several problems:

- Melting rates were low.
- A persistent foam formed during melting, which slowed melting.
- A ferrite-spinel slag formed at the bottom of the melter.
- A stable foam was generated when a cooled melt was reheated (reboil).

Table 1. Major Components of SRP Waste Calcines, wt % (after 75 wt % Al removal)

	High Aluminum	High Iron	Average
Fe_2O_3	14.80	55.90	47.16
Al_2O_3	50.77	1.29	9.24
MnO_2	12.16	3.74	12.98
U_3O_8	3.57	12.98	4.25
CaO	0.97	3.80	3.51
NiO	2.17	9.54	5.84
SiO_2	1.57	--	1.12
Na_2O	5.41	4.75	6.63
Na_2SO_4	--	1.21	--
Zeolite	8.58	8.00	8.05

Experiments at SRL have indicated that these results can be alleviated by the following methods:

1. Increasing melt temperature.
2. Reducing sludge loading.
3. Adequate mixing of frit and waste.
4. Improving glass compositions.

The first three methods are costly and might cause other processing problems. Higher melt temperatures would increase the volumes of off-gas. Reduced loading would increase glass production. Longer mixing would cause excessive dusting. Therefore, the effects of frit composition on both process and product quality were studied in detail.

Discussion

The processing variables studied were viscosity and waste solubility (primarily ferrite solubility); the product quality variables studied were leachability, and devitrification during cooling.

Viscosity

As reported previously (1,2) the viscosity is mainly controlled by the ratio of the moles of alkali oxide to moles of silica in the frit, and by the mole fraction of Li_2O in the alkali oxide. However, increasing the B_2O_3, alkaline earth, or TiO_2 content of the frit also reduced viscosity slightly.

Table 2. Frit composition, wt %

	Frit 21	Frit 211	Frit 131
SiO_2	52.5	58.3	57.9
B_2O_3	10.0	11.1	14.7
Na_2O	18.5	20.6	17.7
Li_2O	4.0	4.4	5.7
CaO	5.0	5.6	--
MgO	--	--	2.0
TiO_2	10.0	--	1.0
ZrO_2	--	--	0.5
La_2O_3	--	--	0.5

Waste Solubility

Ferrite solubility is affected most by the B_2O_3 and the TiO_2 content of the frit. As Fig. 2 and 3 show, B_2O_3 increases, and TiO_2 decreases ferrite solubility. However, small additions of TiO_2 do increase actinide solubility. Also, the effect of B_2O_3 was not as great on high-alkali frits as on lower alkali frits. Additives to improve leachability, such as ZrO_2 or La_2O_3, also greatly reduced ferrite solubility; 1 wt % of either in the frit increased the temperature at which the average sludge composition was soluble (liquidus temperature) by about 50°C.

Leachability

Leachability is sensitive to frit composition. Variations in all components can have large effects. Substitution of Li_2O for Na_2O on a molar basis dramatically reduces the leach rate of the glass in a variety of media. Increasing the B_2O_3 content of the frit increases the leach rate in acid, but decreases the leach rate in alkali. Substitution of MgO for CaO also reduces the leach rate, especially in acid. Small additions of ZrO_2, TiO_2, or especially La_2O_3 reduce leach rate.

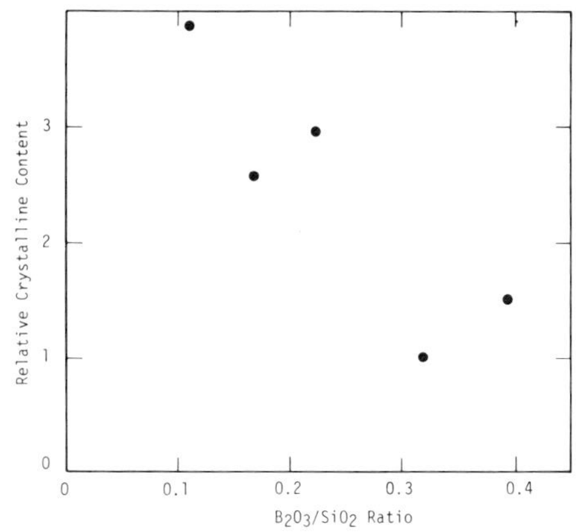

Fig. 2. Effect of B_2O_3 on the crystalline content at 1000°C

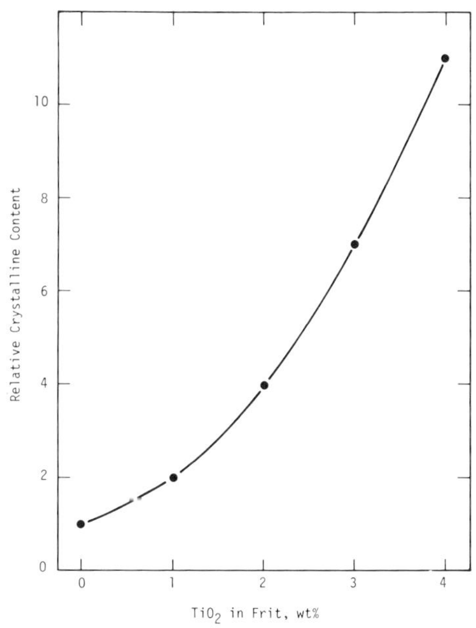

Fig. 3. Effect of TiO_2 on the crystalline content at 1000°C

Devitrification During Cooling

Li_2O or TiO_2 in the frit promotes devitrification; B_2O_3 inhibits it. Table 3 shows an almost linear correspondence between the extent of devitrification and the TiO_2 content of the frit. The effect of Li_2O is probably kinetic because the viscosity of the glass between 600 and 900°C is very low.

Table 3. Effect of TiO_2 on Devitrification During Cooling

TiO_2 Content of Frit, wt %	0	5	10
Temperature to get to 5% Crystals, °C	930	1000	1030
Devitrification After Cooling, vol %	10	50	90

Improved Frits

Based on these tests, two improved frits were designed, Frits 211 and 131 (Table 2). Frit 211 was derived from Frit 21 simply by eliminating TiO_2. Table 4 summarizes preliminary tests of the simulated waste glasses. The leach rates of glass from Frit 211 are comparable to glass from Frit 21, but the waste solubility, tendency to devitrify, and viscosity of Frit 211 glasses are better. In tests in a small-scale continuous joule-heated melter, Frit 211 dissolved simulated waste twice as fast as Frit 21. Frit 211 is now being used in large-scale tests using simulated waste, and small-scale tests using actual waste.

Frit 131 contains more B_2O_3 and less TiO_2 to increase waste solubility. A small amount of TiO_2 has been retained to increase actinide solubility and to reduce leachability. The molar ratio of Li_2O to total alkali has been increased to further improve leach-

Table 4. Frit 211 Compared to Frit 21

Property	Frit 21	Frit 211
Leach Rate at pH 4.4, wt %	0.31	0.23
Leach Rate at pH 11.5, wt %	0.23	0.46
Tendency to Devitrify	High	Moderate
Viscosity, poises	22	13
Waste Solubility at 1150°C, Max. Loading - Calcine	25	30
Melting rate, g/min	4.1	9.8

ability. MgO has been substituted for CaO to improve leachability and waste solubility in the glass. ZrO_2 and La_2O_3 increase leach resistance.

REFERENCES

1. M. J. Plodinec, "Development of Glass Compositions for Immobilization of SRP Waste," USDOE Report DP-1517 (1979).
2. M. J. Plodinec, "Development of Glass Compositions for Immobilization of SRP Waste," Scientific Basis for Nuclear Waste Management I (G. J. McCarthy, ed., Plenum Press, New York, 31-35, 1979).

MULTIBARRIER STORAGE OF SAVANNAH RIVER PLANT WASTE

M. D. Dukes, W. C. Mosley, W. N. Rankin, M. H. Tennant
J. R. Wiley

Savannah River Laboratory, E. I. du Pont de Nemours
& Company, Inc., Aiken, SC 29801

INTRODUCTION

The Savannah River Plant (SRP), operated by Du Pont for the U.S. Department of Energy, has produced nuclear materials for national defense programs since 1953. Acidic waste from the fuel reprocessing facilities on the plant site are neutralized and made alkaline for storage in large carbon steel tanks. Most fission products and waste actinides along with large amounts of nonradioactive waste (Fe, Mn, Al, etc.) precipitate from the alkaline solution to form an insoluble sludge on the bottom of the waste tanks. Supernatant liquid is evaporated to form salt cake and residual liquor that are also stored in the waste tanks. About 20 million gallons of waste are now stored at SRP. To provide for long term management of this waste, a conceptual process for incorporating waste in borosilicate glass is being developed at the Savannah River Laboratory. Solidified waste will be shipped to a federal repository.

The current approach to repository storage of solidified waste emphasizes a multibarrier system to prevent hazardous amounts of radioactivity from entering the biosphere. Elements of the multibarrier system include the waste form, canister, overpack, backfill material and the host geology. A conceptual multibarrier system for SRP waste is shown schematically in Fig. 1.

Experimental work is in progress at Savannah River Laboratory to determine how borosilicate glass will perform as an element in such a multibarrier system. Work completed to date includes: 1) calculating typical storage temperatures of SRP waste in a salt repository, 2) measurement of leach rates in typical repository brines, and 3) corrosion tests for candidate canister materials.

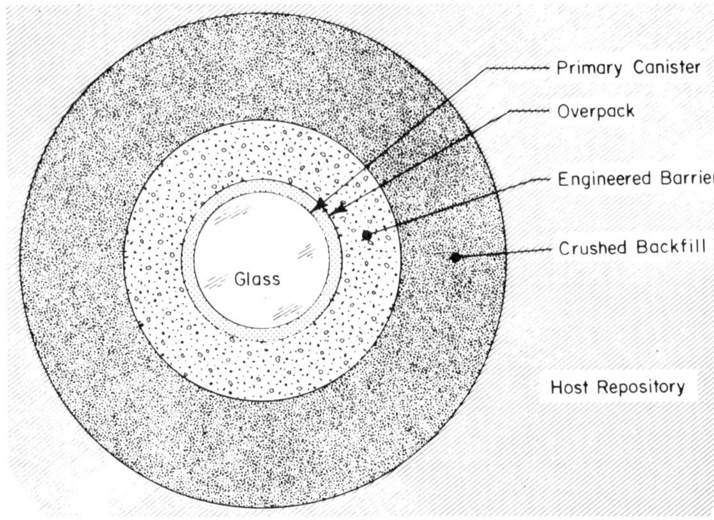

Fig. 1. Conceptual Multibarrier Waste Storage

DISCUSSION

Waste Solidification

The reference process for SRP waste solidification is shown in Fig. 2. Salt cake will be redissolved in the waste tanks and pumped out. Sludge will be removed from the tanks by hydraulic slurrying. This method of waste removal was recently demonstrated in a SRP waste tank. The salt solution is allowed to settle, is filtered to remove entrained solids, and is decontaminated by an ion exchange process that removes ^{137}Cs and the traces of ^{90}Sr and Pu that remain in solution (1, 2). The insoluble sludge is then treated with NaOH to remove about 75% of the aluminum, washed, dried and incorporated in borosilicate glass. This glass has been tested extensively as a matrix for solidifying SRP waste (3-8). The composite composition of SRP waste glass is shown in Table 1.

The Waste Package

The waste glass will be poured into stainless steel containers 2 feet in diameter and 10 feet tall. The glass within this primary container is the nucleus of the multibarrier system designed to ensure that harmful amounts of radioactive isotopes will not be released to the environment. The container of glass can be placed in an overpack prior to shipment to a federal repository or at the repository itself. The primary container and shipping cask (plus possibly an overpack) will protect the glass in the event of an accident

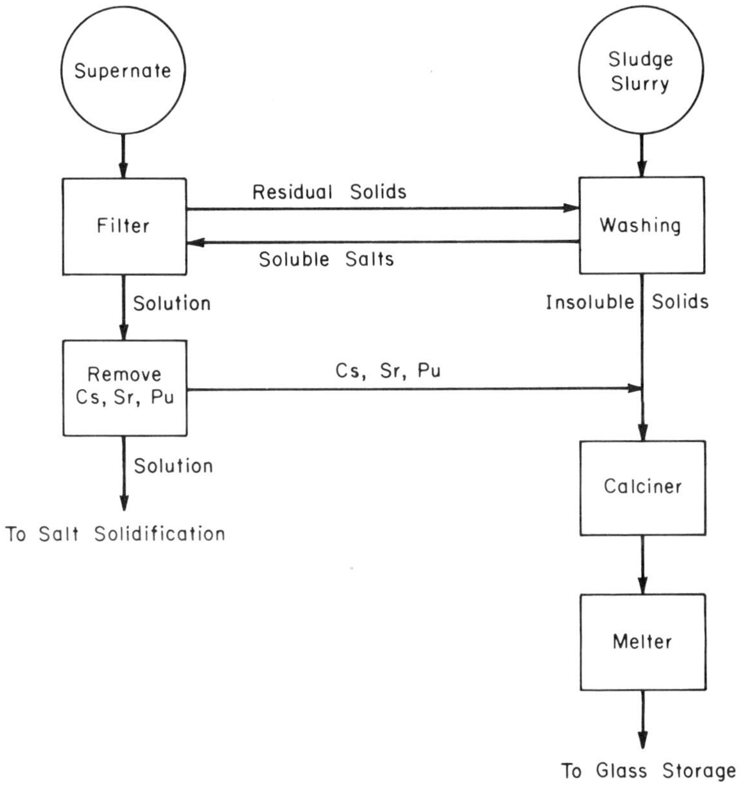

Fig. 2. Reference Waste Solidification Process

Table 1. Average Glass Composition

Component	
Glass Frit	**Wt. Percent**
SiO_2	40.4
Na_2O	16.2
B_2O_3	7.2
TiO_2	7.2
CaO	4.5
Li_2O	2.9
Waste Oxides	
Al_2O_3	3.1
Fe_2O_3	12.1
MnO_2	3.4
U_3O_8	1.1
NiO	1.5
Na_2SO_4	0.4

during shipment to the repository. After emplacement in the repository, the overpack provides a secondary barrier against release of radionuclides and should prevent intrusion by water into the primary container for 10's to 100's of years. A diagram of the reference package for SRP waste is shown in Fig. 3.

Canister Temperature (Near-Field Model)

Temperature and storage environment will have the most significant effects on the lifetime of the waste package. Detailed calculations were made to estimate repository temperatures to serve as a basis for laboratory tests of the components of the waste package. Details of the near-field model used for the calculations are being presented in a separate paper at this conference (9).

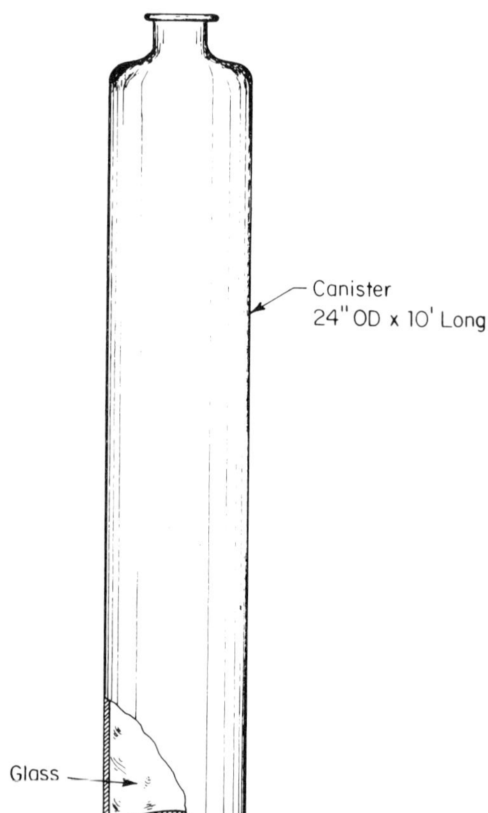

Fig. 3. Reference Canister for SRP Waste Glass

The heat produced by a reference canister of SRP waste depends on the age of the waste it contains. The first SRP waste to be vitrified will be nearly 35 years old and will produce only about 375 watts/canister. The freshest waste to be vitirified according to present plans will have been out of the reactor five years and will produce about 1050 watts/canister. Using these as limiting cases, Fig. 4. shows that in a salt repository, canister temperatures can be varied over a wide range depending on the canister spacing. For SRP waste, Fig. 4. shows that the canister temperature would be 100°C for a spacing of 20-25 feet. A temperature of 120°C has been used as a reasonable maximum temperature for glass durability tests.

Glass Durability

Because glass is the primary solidification matrix, a series of salt repository-specific leach tests were conducted at 20°-25°, 90° and 120°C. The 20°-25° and 90°C tests were extended term tests in which samples of crushed glass containing ^{137}Cs tracer were leached

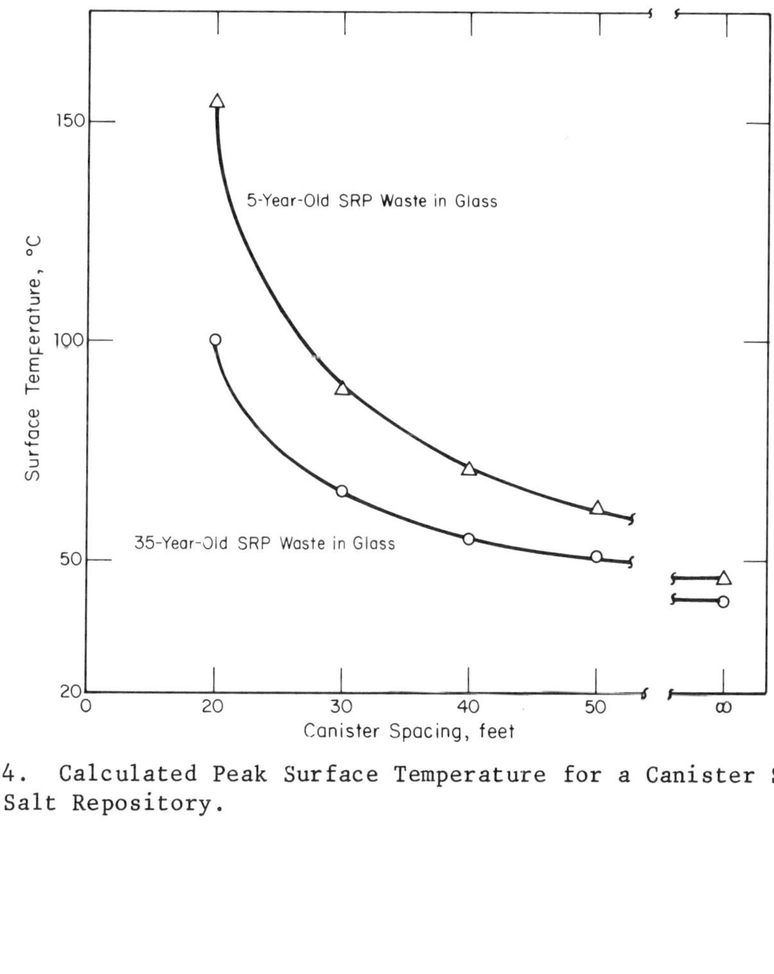

Fig. 4. Calculated Peak Surface Temperature for a Canister Stored in a Salt Repository.

in brine solutions. Solutions were changed on a regular schedule. The representative brine used (WIPP B) has the composition of that made by saturating distilled water with rock salt (10). The 120°C tests were done in an autoclave using WIPP B and a bittern brine composition, NBT-6 (11). The NBT-6 brine represents that which might be trapped in pockets in a salt repository.

In the 20°-25°C tests, glass containing simulated SRP waste has been exposed to room temperature brine solutions for over a year. The initial leach rate was 1.3×10^{-6} g/cm^2-day. After 30 days, the rate had decreased to 6.6×10^{-8} g/cm^2-day; and after 370 days, the rate had decreased to 1.5×10^{-9} g/cm^2-day (Fig. 5.). This corresponds to a penetration of 3.6×10^{-6} cm into the glass.

Glass was also exposed to WIPP B brine at 90°C for 60 days. The initial leach rate was 8.4×10^{-6} g/cm^2-day. The leach rate decreased rapidly to 1.2×10^{-7} at the end of 30 days and 5.6×10^{-8} after 60 days. The cumulative amount leached in 60 days was 2.6×10^{-5} g/cm^2 (Fig. 5.). These are a factor of 3-5 higher than the leach rates at room temperature.

Glass buttons have been exposed to distilled water, WIPP B brine and NBT-6 brine at 120°C for 1 month. At the end of this period, the glasses leached in each solution had a leached layer estimated to be about 1 μm thick. The composition of the surface layer could not be

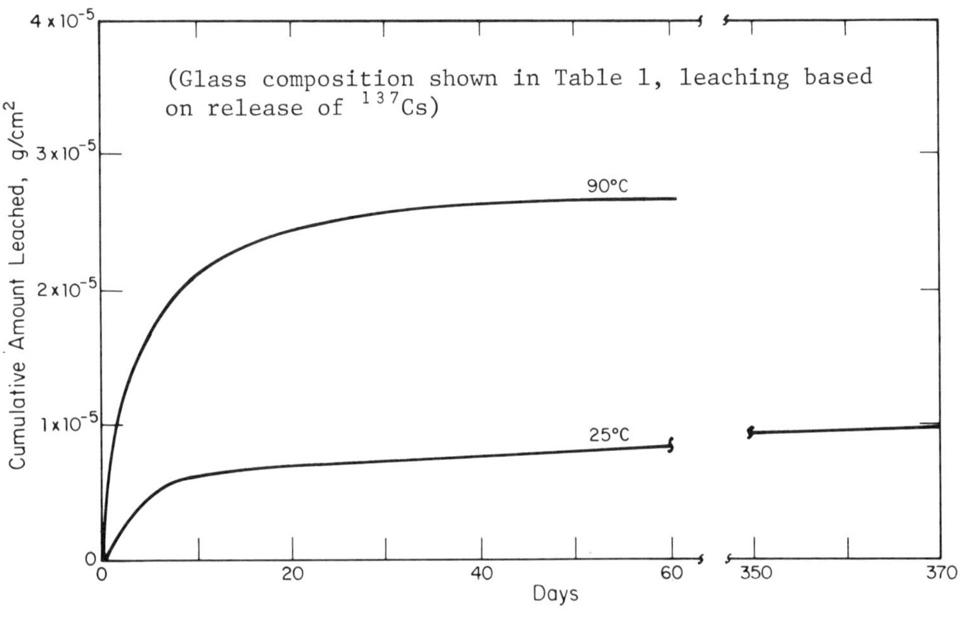

Fig. 5. Cumulative Leaching of Typical SRP Glass in Brine

determined quantitatively with the Scanning Electron Microprobe Quantometer because the layer was so thin. ESCA/Auger spectroscopic analyses of this surface will be used to determine the composition. These tests, as well as those at lower temperatures, will be continued with the goal of establishing glass leaching mechanisms.

Canister Corrosion

Candidate canister materials, Cor-Ten* A and type 304L stainless steel, heated with salt for 10,000 hours showed no significant reaction with dry salt. Details of the corrosion tests are being presented in a separate paper at this conference (12). Wet salt and brine solutions are expected to be much more corrosive than the dry salt. Wet corrosion tests are just beginning.

Additional Barriers

Beyond the primary container and overpack, engineered barriers to radionuclide migration may include adsorbent materials such as bentonite clay (Fig. 1.). Such material would retard movement of water toward the waste form and, in the event radionuclides were released, retard their migration. Backfill material could also be chosen to act as a pH buffer to prevent pH excursions that could be caused by small amounts of water contacting the glass. Such pH excursions may be the most severe leaching conditions that would be encountered in a repository. Backfill materials are an important part of the multibarrier system but are as yet unspecified. Ion exchange properties, buffering capacity and absence of interference with the functioning of other parts of the system will be the primary criteria for material selection.

The massive geologic repository itself provides the final barrier between waste and the biosphere (Fig. 1.). Choice of the repository will be based on technical requirements and public acceptance. The reference SRP waste vitrification process is essentially independent of the choice of the repository location because proper use of engineered barriers between the primary container and the rock type can ensure compatibility.

CONCLUSIONS

Borosilicate glass will perform well as part of a multibarrier system for SRP high level waste. Storage temperatures can be kept to ≤100°C with appropriate spacing of the reference canisters. Under

*Registered Trademark of U. S. Steel.

dry conditions, the waste package should remain intact for very long periods of time. Even if the package is breached and water contacts the glass, the radionuclide release rate will be very low.

ACKNOWLEDGEMENT

The information contained in this article was developed during the course of work under Contract No. DE-AC09-76SR00001 with the U. S. Department of Energy.

REFERENCES

1. J. R. Wiley, "Decontamination of Alkaline Radioactive Waste by Ion Exchange," I & EC Process Design and Development 17, 67 (1978).
2. P. K. Baumgarten, D. Whitehurst, R. M. Wallace and J. M. Steed, "Development of an Ion Exchange Process for Removing ^{137}Cs From High Level Radioactive Waste," These Proceedings.
3. J. A. Kelley, "Evaluation of Glass as a Matrix for Solidification of Savannah River Waste, Nonradioactive and Tracer Studies," USDOE Report DP-1382, E. I. du Pont de Nemours & Company, Inc., Savannah River Laboratory, Aiken, SC, (1975).
4. J. A. Kelley, "Evaluation of Glass as a Matrix for Solidification of Savannah River Plant Waste, Radioactive Studies," USDOE Report DP-1397, E. I. du Pont de Nemours & Company, Inc., Savannah River Laboratory, Aiken, SC, (1975).
5. M. J. Plodinec, "Development of Glass Compositions for Immobilization of Savannah River Plant Wastes," Scientific Basis for Nuclear Waste Management 1, 31 (G. J. McCarthy, ed., Plenum Press, New York and London, 1979).
6. J. R. Wiley, "Leach Rates of High Activity Waste from Borosilicate Glass," Nucl. Tech. 43, (1979).
7. J. R. Wiley, "Long Term Leach Rates of Actual Waste Glasses," International Symposium on Ceramics in Nuclear Waste Management, American Ceramic Society, Nuclear Division, April 30-May 2, Cincinnati, Ohio, (1979).
8. W. N. Rankin and J. A. Kelley, "Microstructure and Leachability of Vitrified Radioactive Wastes," Nucl. Tech. 41, (1978).
9. M. H. Tennant, "Temperature Generated by Underground Storage of Defense Waste Canisters," These proceedings.
10. Composition provided by M. A. Molecke, Sandia Laboratories, Albuquerque, NM.
11. Composition provided by R. W. Potter, USGS, Menlo Park, CA.
12. W. N. Rankin, "Canister Compatibility with Carlsbad Salt," International Symposium on the Scientific Basis for Nuclear Waste Management, Boston, MA, Nov. 26-29, (1979).

TEMPERATURE DEPENDENCE FOR HYDROTHERMAL REACTIONS OF WASTE GLASSES AND CERAMICS

J. H. Westsik, Jr., J. W. Shade, and G. L. McVay

Pacific Northwest Laboratory
Richland, WA 99352

INTRODUCTION

While leach tests to date on simulated high level waste (HLW) forms have been conducted at temperatures ranging from 20° to 350°C, little has been done to determine the actual temperature dependence for the interaction between the waste form and the leachate. This report discusses the results of a study to determine the effect of temperture on the leaching behavior of several HLW forms.

Three simulated HLW glasses, a supercalcine, SYNROC-B and a natural glass, were leached three days in deionized water at 25°, 50°, 100°, 150°, 250° and 350°C. Actual geologic repository environments will probably include only the lower end of the temperature spectrum used in this test; we have used the higher temperatures (250° and 350°C) to accelerate the leaching reaction and to look for changes in release mechanisms at the higher temperatures. Deionized water was used as a reference leachant in these experiments because it provides a tie line to the WISAP/WRIT programs in which a HLW glass is being leached in deionized water and in a variety of simulated geologic solutions (1). From analyses of the leachate solutions and examination of the reacted waste forms, preliminary information on the temperature effects on the complex leaching reactions were determined.

STARTING MATERIALS

The compositions and characteristics of the waste glasses 72-68, 77-107 and 77-260 and the supercalcine SPC-4 have been described previously (2,3,4). The natural glass is a pyroxene rhyolite obsidian estimated to be 670,000 years old; it has a composition in weight percent of: Na, 3.3; Al, 11.6; Si, 36.7; K, 4.1; Ca, 0.6; Ti, 0.1; and Fe, 0.8 with oxygen comprising the remaining weight. The right circular cylinders of each glass were obtained by core-drilling larger samples. The supercalcine samples were pressed and sintered into their cylindrical forms.

The SYNROC-B composition is shown in Table 1 and is identical with that reported by Ringwood (5), except for the addition of 9.1% PW-4b-6 calcine (2). The SYNROC was prepared from an acid nitrate solution containing Al_2O_3, titanium hydrous oxide previously precipitated from Ti-isopropoxide, the calcine powder and the other components. The solution was boiled to dryness and nickel powder was added. The resulting powder was cold-pressed and sintered for one to two hours at 1300°C in purified nitrogen.

The three primary SYNROC titanate phases -- hollandite, zirconolite ($CaZrTi_2O_7$) and perovskite ($CaTiO_3$) -- were identified by X-ray diffraction and SEM/EDX. However, the hollandite phase appeared to be the tetragonal $BaAl_2Ti_5O_{14}$ (6) rather than $BaAl_2Ti_6O_{16}$ reported by Ringwood (5). The sintered pellets were somewhat porous and had a maximum density of 3.4 g/cc. Crystal sizes ranged from about 4 to 20 μm. Chemical analyses of the sintered pellets indicated that about 70% of the Cs and 20% of the Mo and Te were lost during firing, but there were negligible losses of the other components. It may be that in the hot pressing procedure used by Ringwood, the volatility losses are reduced.

Table 1. Synroc-B Composition

Constituent	PNL	Ringwood (5)
TiO_2	54.6	54.9
ZrO_2	9.0	9.0
Al_2O_3	10.0	10.0
CaO	12.6	12.6
BaO	3.8	3.8
Waste	9.1	10.0
Ni	0.5	0.5
	99.6	100.8

EXPERIMENTAL

Samples of simulated HLW glasses, 72-68, 77-107 and 77-260, SPC-4 supercalcine, SYNROC-B and the natural glass were leached three days in deionized water. The test temperatures were 25°, 50°, 100°, 150°, 250° and 350°C, except that the SYNROC was not tested at below 100°C. Each sample was placed in a Hastelloy autoclave with 100 ml of deionized water, giving solution-volume-to-sample-surface-area ratios of between 12 and 17 cm. The autoclaves were pressurized with nitrogen so that the operating pressure at temperature was 13.8 MPa (2000 psi) -- except at 350°C, when the pressure was 16.9 MPa (2450 psi).

After the tests, the solutions were chemically analyzed using the inductive-coupled plasma (ICP) spectrometer and the atomic adsorption spectrometer. Each waste form sample was sectioned and examined optically and with the scanning electron microscope (SEM) with an energy-dispersive X-ray (EDX) attachment.

EXPERIMENTAL RESULTS

Waste Form Behavior

Typical cross sections of the test samples after leaching at 250°C are shown in Fig. 1. The supercalcine and all the glass samples developed visible reaction layers at this temperature. The reaction layer on the supercalcine has more porosity than does the bulk of the sample. Cesium was identified by SEM/EDX as the major constituent lost from the reaction layer. The SYNROC pellets did not have a visible reaction layer. However, SEM/EDX measurements showed high Al and Ti concentrations and lower concentrations of metallic phases of Mo, Pd, Fe and Ni at the edges as compared with the interior of the samples.

Leach Solution Analyses

The weight fraction of individual components leached per square centimeter of surface area for all test specimens is shown in Fig. 2. through 5. for temperatures from 25° to 350°C. In general, leaching activity in glasses shows the greatest temperature dependence; supercalcine is next, and SYNROC leaching displays the least temperature dependence. There also seems to be a grouping of elemental releases that is independent of waste form. The alkali metal elements, Mo and Si, seem to have the greatest fractional releases; Al and the transition metals are intermediate, and the alkaline earths and rare earths are the lowest. Some elements, such as the alkalies, Mo, Al, La and Si, exhibit an increase in releases with temperature; while others, such as Ca, Sr, Fe and Ba, show a leach rate decrease as temperature

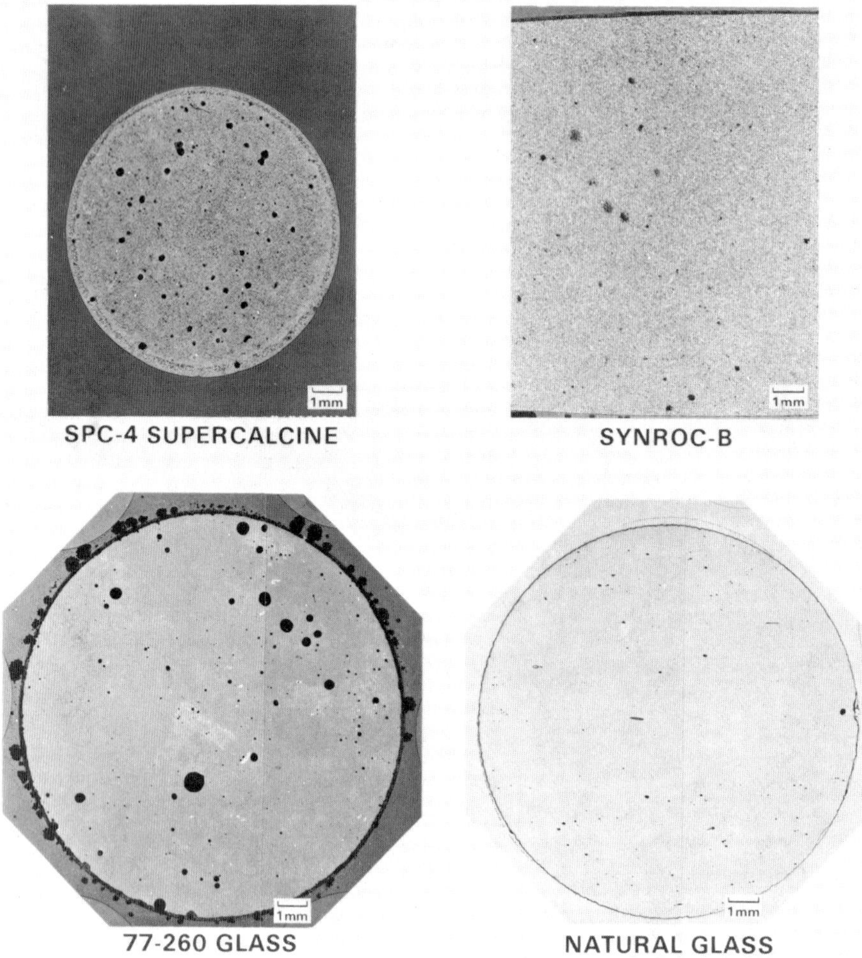

Fig. 1. Cross Section of Samples Leached 3 d in 250°C Deionized Water

increases. There also appears to be a release maximum for some elements near 250°C as compared with 350°C exposure. This may be the result of co-precipitation or secondary phase formation.

In comparing these leach data, it appears that the leach rates are similar at below 200°C, with glass and supercalcine leaching somewhat lower than SYNROC for Cs, Mo and Na, but with SYNROC slightly lower for Ba and Sr. At higher temperatures, SYNROC may have better element retention capabilities. The leach results for selected elements in all three types of waste forms are compared in Fig. 6. These results illustrate that for temperatures of up to 150°C, many elements exhibit Arrhenius-type behavior and that all waste forms are similar in leaching characteristics. On the basis

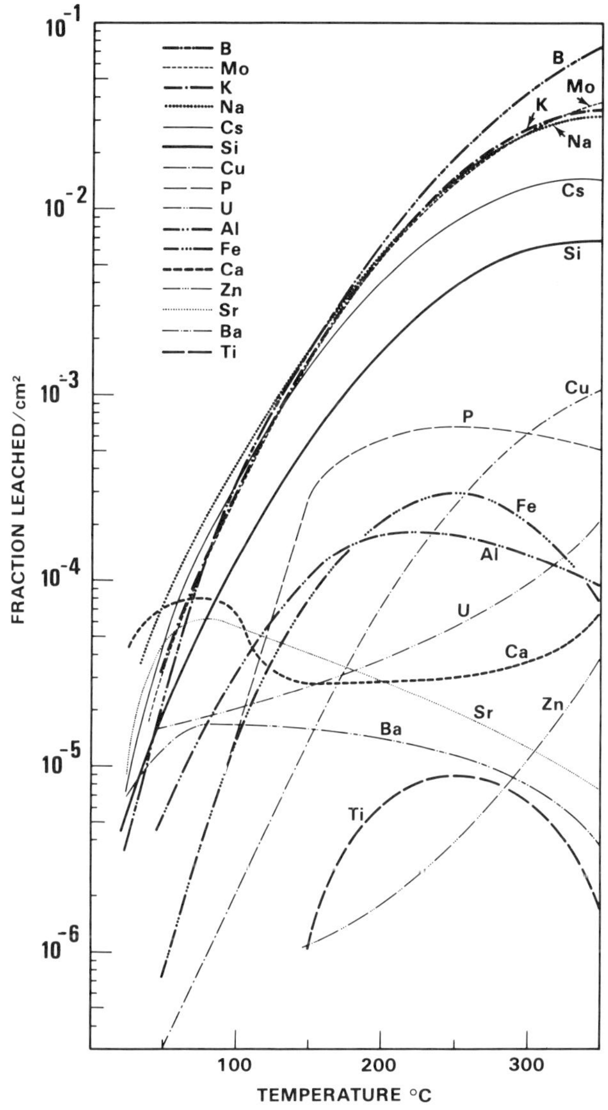

Fig. 2. Elemental Releases From 77-260 Glass

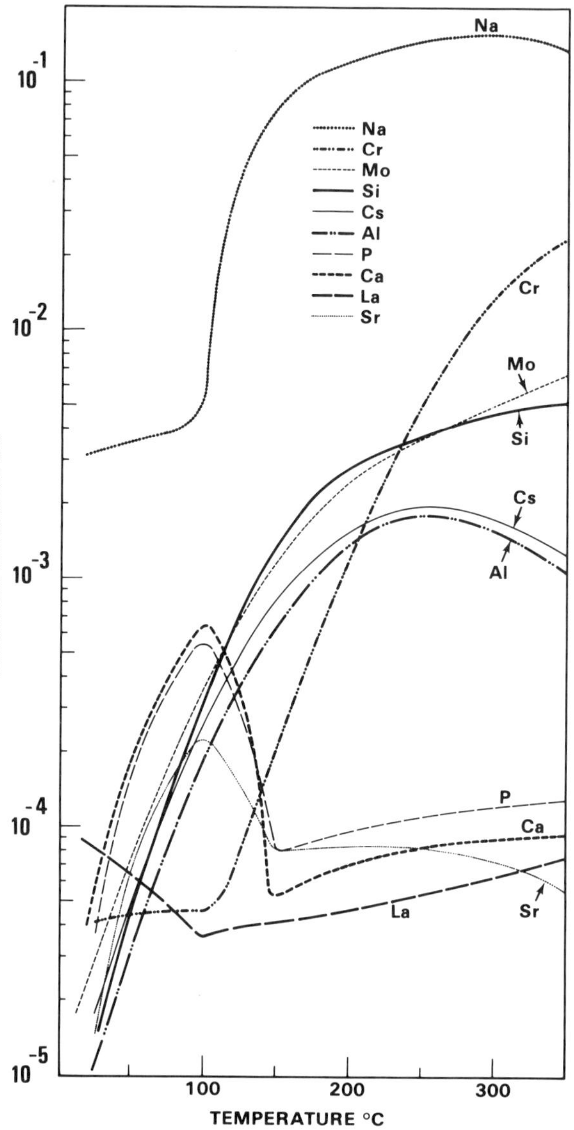

Fig. 3. Elemental Releases From SPC-4 Supercalcine

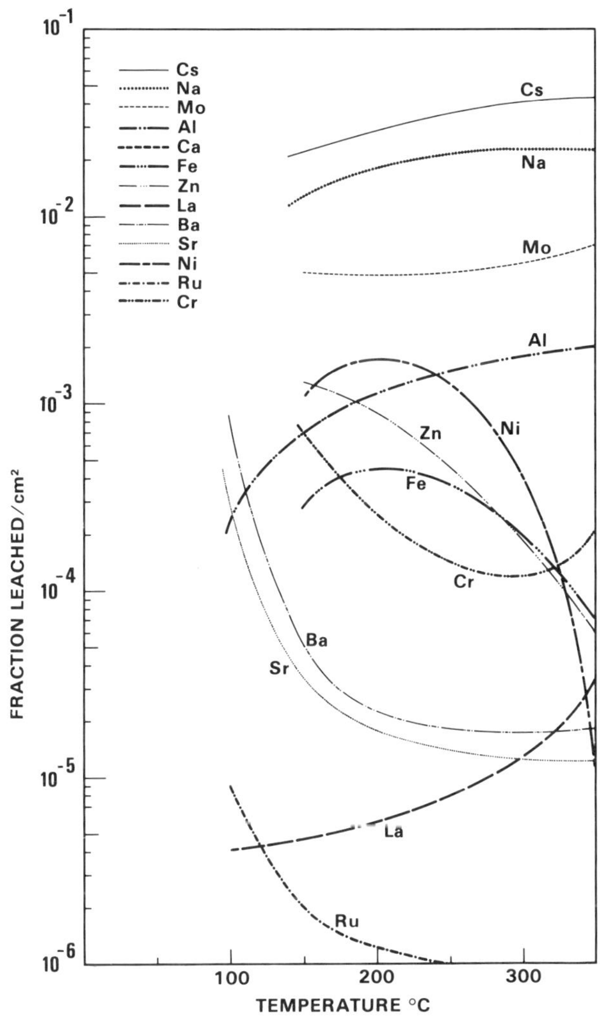

Fig. 4. Elemental Releases From SYNROC-B.

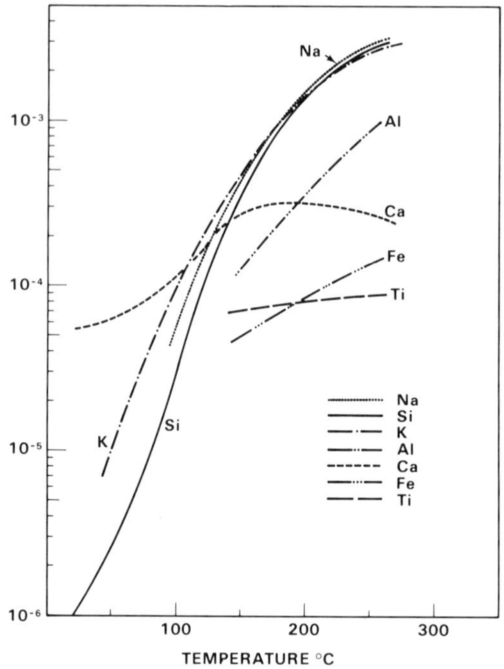

Fig. 5. Elemental Releases From a Natural Glass.

of this preliminary investigation, it does not appear that one waste form is clearly superior to the others. The general similarity of the temperature dependent leach rate curves for many elements, regardless of waste form type, implies that solvent properties may be equally as important as is the actual waste form type. In the temperature range considered, the density of liquid water changes from about 1.0 g/cc to 0.6 g/cc. Also, along the liquid vapor or boiling curve, the ion product of liquid water ($Kw = (H^+)(OH^-)$) can change nearly three orders of magnitude from 1E-14 to 1E-11.5 (7). These solvent properties affect the stability of soluble complex species (e.g., Al^{+++}, $Al(OH)^{++}$ and $Al(OH)_4^-$) which may lead to increased solubility for some aluminosilicates. Soluble complex species stability and secondary phase compositions incorporating elements capable of multivalent states are further affected by redox conditions during leaching.

The waste form type and composition are important in controlling the release of elements to solution, the surface alteration and the secondary phase composition. It is conceivable that even for the same waste form, the redox conditions during fabrication that affect the distribution of elements among different phases may limit the release rate of some elements. For some elements, however, the role of

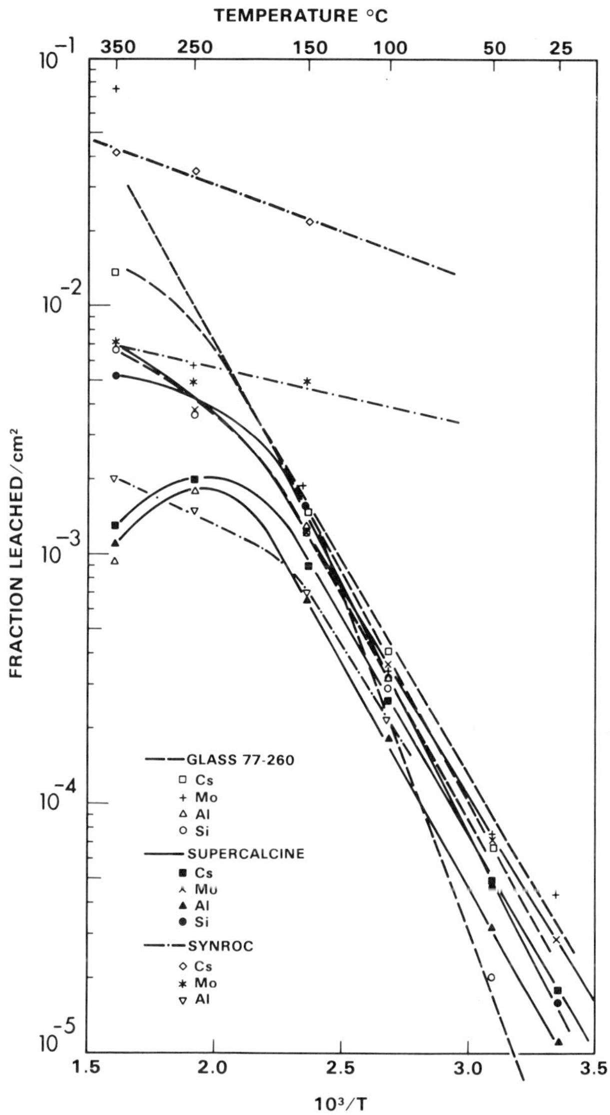

Fig. 6. Arrhenius Plot of the Elemental Release From 77-260 Glass, SPC-4 Supercalcine, and SYNROC-B.

the waste form may be primarily to control the release of an element by influencing secondary phase formation.

CONCLUSIONS

This study shows that simulated high level waste glasses, supercalcine and SYNROC exhibit similar leaching behaviors in hydrothermal environments. The ceramics show a smaller temperature dependence than do the glasses. At temperatures above 250°C, the ceramics show lower releases; but below 250°C, where we expect actual repository temperatures to lie, releases from the waste glasses are lowest. The actual temperature relationship observed depends on the elements studied. The chemistry of the element, changes in the solution properties and secondary reactions all influence the release of elements from the waste forms.

ACKNOWLEDGEMENT

This work was performed at the Pacific Northwest Laboratory operated for the U. S. Department of Energy by Battelle Memorial Institute under contract EY-76-C-06-1830.

REFERENCES

1. J. H. Westsik, Jr., and C. O. Harvey, PNL-3172, Pacific Northwest Laboratory, Richland, WA 99352 (1980).
2. J. E. Mendel, BNWL-2252, Pacific Northwest Laboratory, Richland, WA 99352 (1977).
3. J. L. McElroy, PNL-2265-3, Pacific Northwest Laboratory, Richland, WA 99352 (1979).
4. J. M. Rusin, PNL-2668-1, Pacific Northwest Laboratory, Richland, WA 99352 (1978).
5. A. E. Ringwood and S. E. Kesson, Ceramics in Nuclear Waste Management, CONF-790420, Technical Information Center, USDOE (1979).
6. J. P. Guha and D. Kolar, Jour. Solid State Chem. 16, 49-54 (1976).
7. J. R. Fisher and H. L. Barnes, Jour. Phys. Chem. 76, 90-99 (1972).

MICROSTRUCTURAL CHARACTERIZATION OF GLASS AND CERAMIC SIMULATED WASTE FORMS*

Thomas J. Headley, John T. Healey and Paul F. Hlava
Sandia Laboratories**
Albuquerque, NM 87185

Michael J. Kupfer and Denis M. Strachan
Rockwell Hanford Operations
Richland, WA 99352

INTRODUCTION

Glass is currently the primary waste form candidate for immobilization of Hanford high level radioactive wastes from three generations of nuclear fuel reprocessing. Several chemically varied waste types must be immobilized, including sludges and highly concentrated liquids, and each may require a chemically different but optimum glass composition to fix the appropriate set of radionuclides. Efforts are also underway to evaluate the application of ceramic forms as an alternative to glass for the fixation of Hanford defense waste. Results of laboratory scale studies for fixation of these wastes into glass (1) and ceramic (2) forms have been reported and the studies are continuing.

The microstructures of three nonradioactive glass samples, which simulate three Hanford process waste forms, were characterized in this study. In addition, two samples of iodine sodalite which simulate the fixation of radioactive iodine were characterized. These five samples were intended only to demonstrate the potential for fixation of waste in a high integrity solid form and are not necessarily representative of the final products to be used in the immobilization of Hanford defense waste.

*This work supported by the U.S. Dept. of Energy (DOE) under Contract DE-AC-4-76DP00789.
**A U.S. DOE facility.

The samples were analyzed at Sandia Laboratories utilizing the techniques of X-ray diffraction, scanning electron microscopy (SEM) coupled with X-ray energy dispersive spectrometry (EDS), transmission electron microscopy (TEM) coupled with EDS, and electron microprobe analysis (EMP). Thorough microstructural analysis is considered essential to understanding isolation mechanisms and to interpreting leach rate and other test data on such solid waste forms. A similar microstructural characterization study was reported at the first symposium last year. (3)

One of the three glass samples, Glass 1, demonstrates the potential fixation of sludge generated from the Hanford bismuth phosphate ($BiPO_4$) process which is a first generation process for separating plutonium and uranium from other radionuclides. This sludge typically consists of a mixture of insoluble $BiPO_4$ and other hydrated metal oxides and hydroxides. The principal radionuclides in sludge are ^{90}Sr and ^{239}Pu.

Alkaline salt solutions stored in large underground tanks at Hanford are presently being evaporated to salt cake using large evaporator-crystallizer units. A highly alkaline solution called residual liquid remains when the salt solution cannot be further evaporated. The principal radionuclides in residual liquid are ^{137}Cs, ^{90}Sr and ^{99}Tc. This residual liquid could be treated to remove the radionuclides resulting in a small volume of highly radioactive waste suitable for conversion to glass and a large volume of "decontaminated" solids salt which need not be converted to a high-integrity, high-cost solid such as glass. Glass samples 2 and 3 demonstrate the potential for fixation of the concentrated residual liquids.

The two ceramic iodine sodalite ($Na_4Al_3Si_3O_{12}I$) samples demonstrate the potential for fixing radioactive iodine which volatilizes during fuel reprocessing and is removed from the process off-gas. Sodalite is a naturally occurring halide containing mineral. Iodine sodalite is being investigated at Rockwell Hanford to determine its acceptability as a material for the long term storage of ^{129}I.

SAMPLE PREPARATION

The final compositions of the three glasses, calculated from initial batch mixtures, are given in Table 1 expressed as wt% oxides. Glass formation temperatures were 1573K for Glass 1 and 1473K for Glasses 2 and 3. The melts were held for one hour in air atmosphere and then air-quenched. Glasses 1 and 2 were then reheated to 1273K and cooled at 3K/h to 673K to determine the devitrification behavior.

Table 1. Calculated Compositions of Glass Samples (Wt%)

Oxide	Bi-phos. Glass 1	Res. Liq. Glass 2	Res. Liq. Glass 3
SiO_2	58.1	63.0	63.0
B_2O_3	10.5	10.0	10.0
Na_2O	7.1	17.3	11.1
Li_2O	2.0	2.0	2.0
Fe_2O_3	7.5		
Cr_2O_3	1.1		
ZrO_2	0.1		
Bi_2O_5	10.4		
CeO_2	0.1		
P_2O_5	3.0		1.4
SO_3	0.05		
F	0.05		
Al_2O_3		5.8	5.5
TiO_2		0.9	0.6
SrO		0.5	6.0
Cs_2O		0.5	0.4

The two iodine sodalite samples were prepared by mixing 19M NaOH, NaI and a kaolinite clay (nominally $Al_2Si_2O_5(OH)_4$ with several percent impurities) in a mass ratio of 1.0:1.25:3.22, based on the dry solids. After drying, one sample was hot pressed in graphite dies at 14 MPa and 1253K. A portion of this sample was then reground and hot pressed in graphite dies at 14 MPa and 1353K. (See Reference 4. for a more complete discussion on iodine sodalite preparation.)

RESULTS OF MICROSTRUCTURAL CHARACTERIZATION

Glasses

A. Bismuth Phosphate Glass 1: Optical, SEM and TEM microscopy revealed that this glass did undergo devitrification during the slow cool from 1273K to 673K. The resulting microstructure was composed of four separate crystalline phases dispersed in a glassy matrix, as shown in Fig. 1. Phase 1 is present as large plate-like crystals which comprise approximately 30-40 vol. % of the sample. Both electron diffraction analysis of these crystals in the TEM and X-ray diffraction analysis of the bulk sample identified this most prevalent crystalline phase as tridymite (SiO_2). Quantitative EMP analysis indicated that the platelets are almost pure SiO_2 but contain about 0.25 wt% iron oxide.

Fig. 1. SEM Micrographs of Bismuth Phosphate Glass 1 Showing Four Crystalline Phases (1-4) Dispersed in Glassy Matrix

Phase 2 was present in plane sections as long stringers of small crystallites. This geometry is interpreted in 3-dimensions as cross sections through planar arrays of small crystallites. This phase comprised only about 1-2 vol. % of the total sample. EDS spectra obtained from these crystals in the TEM contained only phosphorous peaks (elements with $Z \leq 11$ are not detected by EDS). EMP analyses suggested that the particles are a phosphorous oxide containing some additional undetected component(s). Li and/or B oxide, neither of which is detected by EDS, are likely possibilities. D-spacings which were measured from electron diffraction patterns of this phase did not match any listed oxide of phosphorous with Li and/or B. Therefore, it was concluded that this phase is probably a Li- and/or B-containing phosphorous oxide not reported in the X-ray Powder Diffraction File.

Phases 3 and 4 were found to be two Cr-Fe oxides with different morphologies. Phase 3 was present as small blocky-shaped particles while Phase 4 had a plate-like morphology. It was determined from electron diffraction analysis that Phase 3 has a spinel structure with $a_0 = 0.812$ nm, but EMP analysis indicated that it has a much higher Cr:Fe ratio than the 2:1 ratio of the normal $FeCr_2O_4$ spinel. EMP analysis also suggested the presence of a small amount of Li and/or B oxide in solution. The structure of phase 4 was not determined, but morphological characteristics suggested that it is not the Cr-Fe spinel.

Quantitative EMP analyses of these 4 crystalline phases and the glass matrix indicated that all Bi from the $BiPO_4$ addition went into the glass while P partitioned between the P-rich stringers (phase 2)

and the glass. All other elements not found in the four crystalline phases were retained in the glass matrix.

B. Residual Liquid Glass 2: Although this glass was subjected to the 1273-673K slow cool test, it did not undergo any devitrification. The glass was found to be completely amorphous when examined by SEM and TEM. Thus all elements of the mix were retained in the glassy matrix.

C. Residual Liquid Glass 3: The increased Sr and P in this glass caused devitrification, even though it was not subjected to the slow cool test. The microstructure, shown in Fig. 2, was a two-phase structure composed of approximately 5-10 vol. % of small crystallites dispersed in the glass matrix. Crystallinity was confirmed by the presence of spots in electron diffraction patterns. The particles were highly spherical with an average diameter of about 80 nm. Comparison of EDS spectra obtained from the particles and from the adjacent glass matrix in the TEM indicated that the particles were high in Sr and P and low in Al and Si compared to the matrix. A positive identification of this phase was not obtained from electron diffraction analysis, but it is probably a Sr-phosphate. EDS analysis indicated that Sr partitioned between these crystallites and the glass matrix, while the Cs content, albeit low, was retained in the glass.

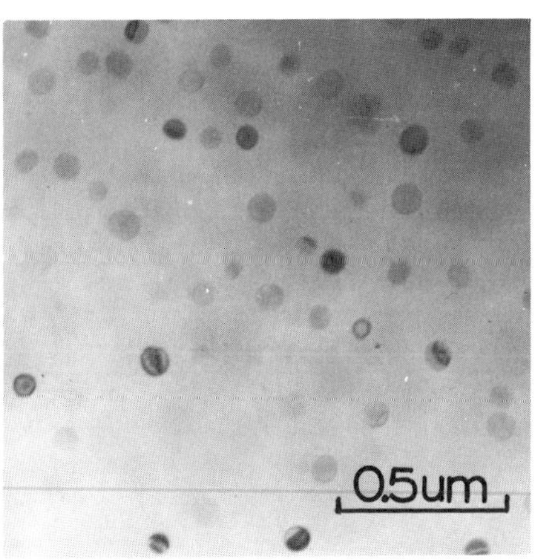

Fig. 2. TEM Micrograph of Small Crystallites in Residual Liquid Glass 3

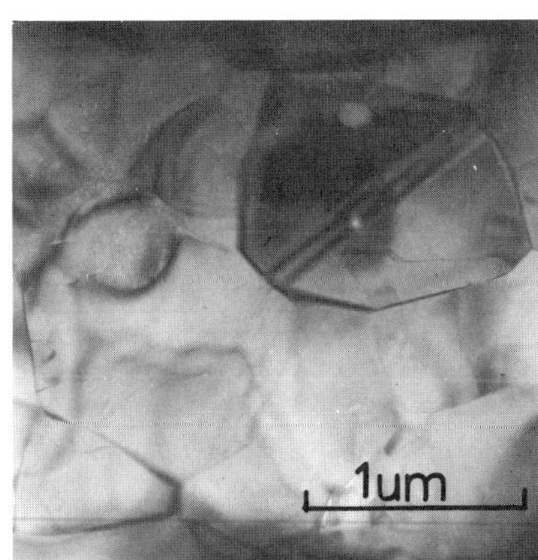

Fig. 3. TEM Micrograph of the Iodine-Loaded Ceramic Which was Reground and Repressed

Ceramics

The two iodine sodalite samples were found to be similar in chemistry and crystalline phase structure. However, the sample which was not reground and repressed contained up to approximately 50% interconnected porosity. This indicated that insufficient consolidation had occurred. Conversely, the reground and repressed sample was essentially fully dense as shown in Fig. 3. The ceramic microstructure consisted primarily of an equiaxed grain structure of a single matrix phase. EDS spectra obtained from these grains contained large Si, Al and I peaks and a lower Na peak. Both X-ray diffraction analysis of the bulk sample and electron diffraction analysis of individual grains identified the matrix as iodine sodalite, $Na_4Al_3Si_3O_{12}I$. Results of quantitative EMP analyses of the matrix indicated that it was close to the ideal chemistry for stoichiometric $Na_4Al_3Si_3O_{12}I$.

In addition, several of the analysis techniques revealed the presence of particles of both a Ti-rich and an Fe-rich phase scattered throughout the matrix. These two phases comprised an estimated 1-2 vol. % of the sample and are believed to result from impurities in the clay.

REFERENCES

1. M. J. Kupfer, Vitrification of Hanford Radioactive Defense Wastes, RHO-SA-89, Rockwell Hanford Operations, Richland, WA, May (1979).
2. D. M. Strachan, Crystalline Materials for Immobilization of Hanford Nuclear Defense Waste, RHO-SA-13, Rockwell Hanford Operations, Richland, WA, May (1978).
3. John T. Healey, Thomas J. Headley, Paul F. Hlava, Denis M. Strachan and Michael J. Kupfer, "Microstructural Characterization of Solidified Simulated Reactor Waste Forms," Scientific Basis for Nuclear Waste Management, 1 (G. J. McCarthy, ed., Plenum Press, NY, 1979).
4. D. M. Strachan and H. Badad, Iodide and Iodate Sodalites for the Immobilization of Iodine-129, RHO-SA-83, Rockwell Hanford Operations, Richland, WA, April (1979).

ALTERNATIVE WASTE FORMS -- A COMPARATIVE STUDY*

J. M. Rusin, J. W. Wald and R. O. Lokken

Pacific Northwest Laboratory
P. O. Box 999
Richland, WA 99352

INTRODUCTION

Several alternative process and waste form options exist for the immobilization of nuclear wastes. These options can be categorized into seven groups (1) including the reference monolithic glass process: glass, sintered ceramic, glass ceramics, hot-pressed ceramics, coatings, concrete and metal matrix. Although data exists on the characterization of these alternatives, it is difficult to conduct comparisons due to lack of standard testing procedures.

A characterization study utilizing comparative tests was intitiated at PNL to assess product inertness of alternative waste forms. A parallel study is also underway to assess process complexity. Final assessment of alternative waste forms will be made by comparison of product integrity and process complexity. Process complexity has been previously presented on a preliminary basis (2). Specific waste form options were selected which were representative of each basic waste form category. The simulated nuclear waste loading varied from 5%-100% depending upon waste form type.

This paper discusses the characterization of selected simulated nuclear waste form options. Characterization test results include bulk properties, phase analysis, microstructure, compressive strength, volatility and leachability.

*This work supported by the U. S. Department of Energy under Contract EY-76-C-06-1830.

WASTE FORM DESCRIPTIONS

Eight simulated waste forms were prepared for property comparisons. Seven of the waste forms were tailored from a calcined liquid waste high in sodium; i.e., a PW-9 waste stream; and the glass form, PNL composition 76-68, was prepared from a PW-8a waste stream which has a higher sodium and iron content (3). The four product types investigated are sintered ceramics, glass ceramics, hot-pressed ceramics and concrete.

The sintered ceramic products were cold-pressed cylinders of powder which were subsequently fired in air to densify. Materials included PW-9 calcine, PW-9 calcine plus 50 wt% glass frit* and SPC-5B supercalcine (4). In addition, PW-9 calcine pellets were produced by hot pressing. Formulation of glass ceramic was based upon a celsian type (5) and was prepared containing 20 wt% PW-9 calcine. Two concrete products were produced from Type II Portland cement** and either 10 wt% PW-9 calcine or 10 wt% crystallized SPC-5B supercalcine.

WASTE FORM FABRICATION

76-68 Simulated Waste Glass

Glass containing 76-101 frit and PW-8a-4 calcine (2:1) was melted at 1200°C for 2 hours in a platinum crucible and cast into graphite molds which were preheated to 500°C (3). After air cooling to ∿500°C, the glass was annealed and slow cooled. Samples were core drilled from the glass for characterization and testing.

Sintered Ceramics

Spray-calcined PW-9 powder was pressed into green compact cylinders at 7800 psi in a hardened steel die 0.57-inch diameter. Butyl stearate was used as a die lubricant. The green pellets were sintered in air at 975°C for 2 hours with a heating and cooling rate of 200°C per hour. Frit/PW-9 calcine pellets (1:1 mixture) were prepared as above by pressing powder at 9200 psi and sintering for 2 hours at 750°C. SPC-5B supercalcine pellets were pressed at 19,500 psi from powder ball milled for 16 hours and were sintered at 1125°C for 2 hours.

Hot-pressed PW-9 calcine pellets were prepared by cold pressing as-calcined powder in graphite dies at 1500 psi. The samples and

*O. Hommel Co., Code OHCO 3GF29E.
**Lehigh Portland Cement Co., Type II, ASTM C-150.

dies were outgassed under vacuum for 45 minutes at 350°C and then hot pressed at 3500 psi for 20 minutes at 900°C.

Glass Ceramics

Celsian glass ceramic was prepared by mixing 20 wt% PW-9 calcine with glass frit and melting at 1200°C for 2 hours. The glass was cast into pre-heated graphite molds, allowed to air cool to ∿550°C, annealed and furnace cooled to room temperature. Crystallization was effected by heating at 2°C/minute to 600°C, hold for 3 hours to precipitate nucleation centers and heating at 2°C/minute to 800°C, for a 12-hour hold for crystal growth.

Concrete

Concrete waste forms were prepared by mixing Type II Portland cement with 10 wt% PW-9 calcine powder and enough water to form a smooth consistency paste (water/solid ratio was 0.34 to 0.37). The paste was cast into a mold, vibrated to release entrapped air, sealed and cured. As-calcined SPC-5B supercalcine requires heat treatment to form the desired crystalline phases; therefore, the powder was heated in air to 1125°C for 2 hours, ball milled for 20 hours and screened to -100 mesh. The resultant powder was mixed with Type II Portland cement and prepared as above.

WASTE FORM CHARACTERIZATION

The crystalline phases in the samples as determined by X-ray diffraction were identified in Table 1 along with representative bulk properties. Simulated 76-68 waste glass is vitreous except for some undissolved RuO_2. The sintered PW-9 calcine product contains a number of crystalline phases including the water soluble Na_2MoO_4 phase. With the addition of glass frit to the PW-9 calcine, a $BaMoO_4$ phase is formed in place of Na_2MoO_4. Sintered SPC-5B supercalcine is the most complex waste form analyzed. The formation of pollucite, apatite and tetragonal phases are observed; however, it appears that complete reaction of nepheline and sheelite to form the sodalite solid solution has not occurred. Unreacted nepheline $[(Na,K)AlSiO_4]$ is seen to exist along with a sodalite type analog Hauynite $[(Ca,Sr,Ba)_2(NaAlSiO_4)_6(MoO_4)_2]$. Three major crystalline phases were present in the celsian glass ceramic product: cymrite ($BaAl_2Si_2O_8$), $BaMoO_4$ and a rare earth titanate ($RE_2Ti_2O_7$). Both cement waste forms were characteristic of the cement matrix materials, calcium hydroxide and calcium silicate, rather than the waste constituents.

Table 1. Crystalline Phases Observed in Comparative Studies Materials

Waste Form	Waste Load	Bulk Density	Approximate Porosity	Phase Observed
Sintered PW-9				
Hot Pressed PW-9	100%	3.55 gm/cm^3	10%	CeO_2 (Zr, Ce, ...)O_2 Na_2MoO_4 RuO_2
SPC-5B	~50%	2.4 gm/cm^3	20	$(Ca,Sr,Ba)_2(NaAlSiO_4)_6(MoO_4)_2$ $(Na,K)AlSiO_4$ $(Cs,Na)AlSi_2O_6$ RuO_2 $Ca_2(RE)_8(SiO_4)_6O_2$
PW-9+50% frit	50%	2.21 gm/cm^3	35	$BaMoO_4$ RuO_2 (Unknown)
Glass 76-68	33%	3.0 gm/cm^3	0	RuO_2 Pd
Celsian Glass Ceramic	20%	3.10 gm/cm^3	0	$BaAl_2Si_2O_8$ $BaMoO_4$ $RE_2Ti_2O_7$
Cement + PW-9	10%	1.71 gm/cm^3	15	$Ca(OH)_2$
+ SPC-5B	5%	1.85 gm/cm^3	15	Ca_3SiO_5

WASTE FORMS TESTING

Description

Three basic tests have been performed on all the comparative studies materials to assess product inertness: (1) static leaching, (2) elevated temperature volatility and (3) mechanical strength through impact and load-to-failure compression tests.

Static leach testing of alternative waste forms was performed to evaluate the effects of time, temperature, solution, sample geometric surface area and volume of leach solution on leach behavior. The leach test was designed to provide kinetic information as well as steady state or "saturation" values by measuring ion concentrations in solution by the Induction Coupled Plasma Spectrometry technique. Analysis of pre-leached solid materials after fabrication was also conducted by "ICP" to give a base line by which to evaluate the percent of available ion leached. Leach testing covered a time-temperature span of 1-9 days between 30°C and 200°C at a surface area to solution volume range of 0.6:1 to 0.006:1cm^{-1}. The principle leachant was deionized water; however, salt brine as well as an acid buffer solution were also evaluated on a limited basis.

Initial volatility evaluations on all comparative studies materials were conducted at evalated temperature, typical of previous work (6), and were performed on right circular solid cores with a nominal surface area measurement of 472 mm^2. Four temperatures were chosen for the test: 700°, 900°, 1100° and 1300°C, with a hold time at temperature selected at 3 hours. Volatile loss was measured on all samples by determining the weight change before and after heating. Materials are placed in platinum crucibles, slowly lowered into the furnace preheated to temperature, and heated with a dry air flow adjusted to a range where weight loss is independent of flow rate. Even though the temperature range of this test may represent the mid to upper end of the fabrication temperatures for most of the materials studied, its significance will be illustrated in the releases that were measured.

Comparisions of the mechanical strength of the wide range of glass and ceramic waste forms examined in this study requires a test that can be applied meaningfully throughout. Standard procedures have been developed and used previously on the more brittle materials which involve an impact type loading and particle size fractioning after loading (3). This test has not been found adequate for comparative purposes on the entire range of materials studied here since product forms such as cement and sintered calcine do not tend to "fracture" but rather impact-compress after loading. Meaningful particle size analysis is, therefore, not possible. As a result, a secondary load to failure type test has been selected to evaluate comparative mechanical properties. The test employed is called the Brazilian Diametral compression test and has been described at some length elsewhere (7, 8). Right circular cylinders, 1/2 in. dia. x 1/2 in. l., are loaded on a tensile test apparatus at a crosshead rate of 0.02 in/min in a diametral fashion until failure occurs. Failure loads in pounds per square inch are then calculated using standard equations (7, 8).

RESULTS

Leach data for five of the materials studied are given for 90°C-9 day (SA/V = 0.6 cm^{-1}) in Fig. 1. and Table 2. From this data, it can be clearly seen that the sintered calcine waste form is not successfully retaining any of the ions of concern. In fact, after this period of time and temperature, the sintered calcine waste form had completely broken apart. When the same waste calcine is added to a matrix former such as cement or a glass forming frit, the leach resistance is improved significantly. Retention of sodium and cesium are still quite low with the cement matrix. When waste ions are incorporated in a crystalline or glassy matrix such as supercalcine or glass, another increase in leach resistance is noted. Glass and supercalcine show similar trends under both hydrous and hydrothermal conditions with glass exhibiting a greater retention of sodium and

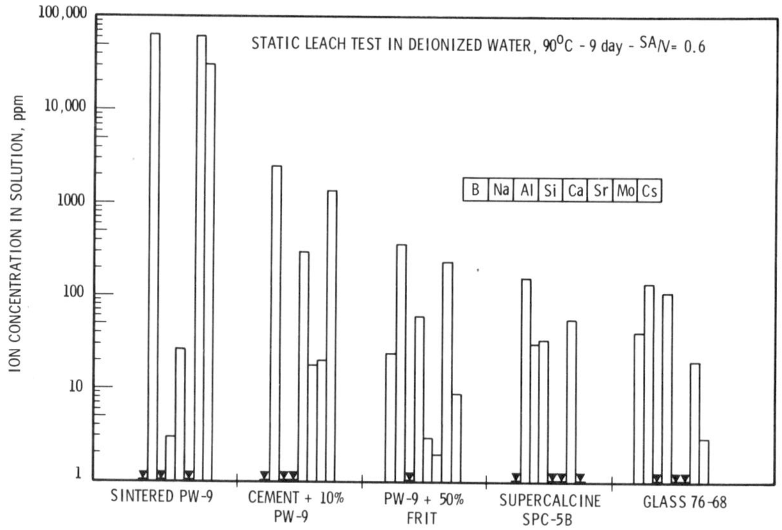

Fig. 1. Static Leach Test of Alternative Waste Forms Under Hydrous Conditions in Deionized Water (90°C-9 day-SA/V = 0.6 cm^{-1}).

Table 2. Static Deionized Water Leach Test of Alternative Waste Forms: Percent of Available Ion in Solution (90°C-9 day, SA/V = 0.6 cm^{-1})

Element	Sintered PW-9	Cement + 10% PW-9	PW-9 + 50% Frit	Supercalcine SPC-5B	Glass 76-68
Na	100%	62.8	2.1	0.8	0.4
Si	N.A.	0.01	0.2	0.1	0.1
Sr	< 0.01	2.4	0.05	< 0.01	< 0.01
Mo	100	0.8	1.7	0.8	0.4
Cs	56.0	47.5	0.1	< 0.01	0.1

molybdenum, supercalcine having a greater retention of cesium, and both forms approximately equal in strontium retention. As found previously in high temperature tests (9), elemental leach rates are not very different in comparing glass and supercalcine.

In the waste forms examined, there appears to be a greater effect of surface area to volume on the leaching of any species than either time or temperature. In general, for a given leached ion, the larger the surface area to volume ratio, the greater concentration of ion in solution. Temperature seems to have the second most pronounced effect on the leaching behavior of these waste forms. Generally, at the lower surface to volume ratios (0.06:1, 0.006:1), the effect of temperature was minimal after one day's leaching, since concentrations in solution for 30°C were approximately the same as at 90°C. At the larger surface area to volume ratio (0.6:1), however, the higher the temperature, the greater the ion concentration in solution. For the most part, it would appear that ion concentration in solution is not changing after one day's time at surface area to volume ratios below $0.06 cm^{-1}$. Slightly greater increases are noted with time at the larger surface area to volume ratios; and in this case, even 9 days may not represent a steady state level; but at the lower surface area to volume ratios, only a small percentage (< 5%) increase is noted with time out to 9 days.

Volatility measurements for three materials are illustrated in Fig. 2. as weight loss per surface area versus the inverse of the absolute temperature. Sintered PW-9 calcine is seen to be the poorest performer with the largest weight loss at all temperatures. As in the other tests, it would seem to follow that without a host matrix former, mechanical agglomeration of particles is not adequate to keep dispersion, i.e., volatility, down. An order of magnitude decrease in volatility is realized when the waste stream is tied to a crystalline host, as in the case of supercalcine. The rate of loss, judged by the slope of the curves, would appear to be similar to that of the calcine alone, while overall losses are about one tenth as great. The volatile losses in the glass waste form are about two orders of magnitude lower than sintered PW-9 calcine at the lower temperature regime (700-900°C); however, they increased to only about one-half as much at the upper temperature regime (1100-1300°C). This indicates that as the glass approaches or exceeds its melting point, volatile losses will increase significantly but are still below that measured with sintered PW-9 calcine.

Volatile measurements for glass 76-68 agree closely with previous work (10); however, volatile losses for the supercalcine SPC-5B waste form are much higher than previous supercalcine types. These higher losses may be attributed to higher sodium concentration in this supercalcine type, as sodium has been found to be a volatile species in the supercalcine system.

Some difficulties were encountered in volatility measurements on both hot-pressed PW-9 calcine and PW-9 plus 50% frit. Hot-pressed PW-9 samples fractured spontaneously when lowered into the furnace, apparently as a result of trapped gasses within the sample. The PW-9 plus 50% frit samples were found to melt and foam over the crucible when heated above 800-900°C.

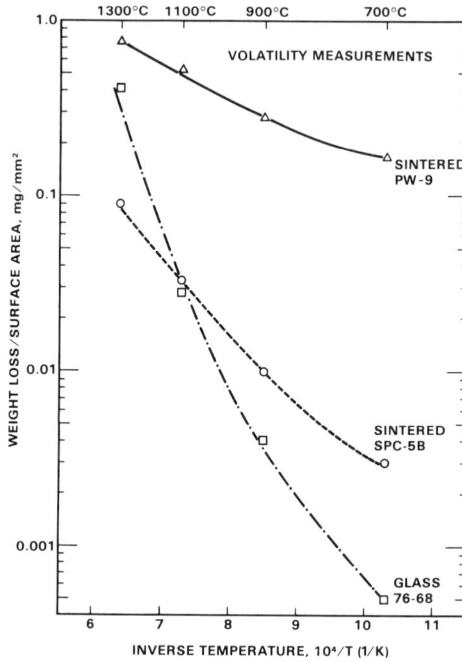

Fig. 2. Weight Loss Per Surface Area Versus Inverse Absolute Temperature for Comparative Studies Materials.

Compressive strength test results are illustrated in Fig. 3. Although the diametral compression test does not assess the impact type "risk analysis" questions, it does give a meaningful measure of comparative mechanical strength as the test is applicable to the entire range of materials studied. Deviations were found to be approximately ±10%.

Sintered PW-9 and cement matrix materials have the lowest strength values of the materials measured; however, through the application of hot pressing or the addition of 50% of a glass forming frit, the average strength values for the calcine waste form can be increased by about 60%. This would apparently result from an enhanced mechanical bonding realized by increased particle agglomeration. The supercalcine waste form shows an increase in average strength over the next lowest material by approximately 120%. In this case, this increase might be due to increased physical bonding resulting from the ingrowth of crystalline phases and the interaction with an amorphous phase. This represents an improvement over the simple mechanical bonding of discrete particles as in cement or simple calcine systems. The glass waste form represents a matrix containing less defects than the ceramic materials, with the continuum of vitreous material providing the main basis of the strength.

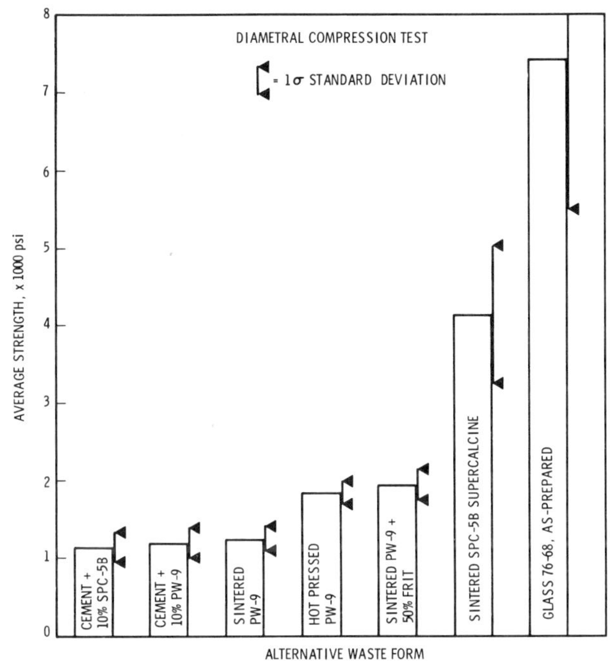

Fig. 3. Average Strength of Seven Comparative Studies Materials by Diametral Compression Testing.

CONCLUSIONS

A characterization study utilizing comparative tests has been conducted to assess product inertness of alternative waste form materials, having evaluated at this point four basic product types: sintered ceramics, glass ceramics, glass and concrete. The seven specific waste form materials studied represent simulated nuclear waste loading of 5% to 100%, processed between room temperature and 1200°C and subjected to characterization tests including phase analysis, microstructure, compression testing, volatility and leach testing.

Significant conclusions based upon the results obtained to date are:

1. Sintered calcine waste form PW-9 does not retain Na, Mo and Cs when leached @90°C and, in fact, does not remain a solid.

2. Glass and supercalcine are alike under both hydrous and hydrothermal leach conditions with glass exhibiting a greater retention of sodium and molybdenum, supercalcine having a greater retention of cesium, and both forms approximately equal in strontium retention.

3. Volatility measurements indicate that an order of magnitude decrease in volatility occurs when a calcine waste form is incorporated in a crystalline or glassy host.

4. Glass 76-68 is superior to supercalcine SPC-5B in retention of volatiles below 1100°C because of the high release of Na from SPC-5B; however, as the temperature approaches or exceeds the glass melt temperature, volatile losses of the glass equal or exceed that of SPC-5B.

5. Glass 76-68 and supercalcine SPC-5B have high compressive strengths when compared to sintered PW-9 and cement products. This is apparently due to a stronger continuum bond resulting from a glassy matrix or crystalline ingrowth over a simple mechanical agglomeration of particles.

REFERENCES

1. J. M. Rusin, PNL-2775, Pacific Northwest Laboratory, Sept. (1978).
2. W. A. Ross, J. M. Rusin and J. L. McElroy, Proceedings of the Symposium on Waste Management, Tucson, Arizona, Feb. (1979).
3. W. A. Ross, PNL-2625, Pacific Northwest Laboratory, June (1978).
4. J. L. McElroy, PNL-2999-1, 2, 3, 4, Pacific Northwest Laboratory, May (1979).
5. W. Lutze, J. Borchardt and A. K. De, Scientific Basis for Nuclear Waste Management, I, (G. J. McCarthy, ed.; Plenum Press, New York, 1979).
6. J. M. Rusin, W. J. Gray and J. W. Wald, PNL-2668-2, Pacific Northwest Laboratory, Aug. (1979).
7. R. H. Marion and J. K. Johnstone, SAND-75-0347, Sandia Laboratories, July (1975).
8. R. M. Spriggs, L. A. Brissette and T. Vasilos, Materials Research and Standards, 218, May (1964).
9. J. H. Westsik, Jr., and R. P. Turcotte, PNL-2759, Pacific Northwest Laboratory, Sept. (1978).
10. W. J. Gray, "Volatility of Some Potential High Level Radioactive Waste Forms," to be published in Nuclear Waste Management (1980).

IMMOBILIZATION OF U.S. DEFENSE NUCLEAR WASTES USING THE SYNROC PROCESS

A. E. Ringwood, S. E. Kesson, and N. G. Ware

Research School of Earth Sciences
Australian National University
Canberra, A.C.T. 2600, Australia

INTRODUCTION

United States defense nuclear wastes are presently in tank storage, partly as sludges comprising Fe, Mn, Ni, U and Na oxides and hydroxides, together with 0.5 to 5 percent of fission products and actinides (exclusive of uranium). The relative proportions of Al, Fe, Mn, Ni, U and Na in the sludges from different tanks vary considerably, except that (Fe+Al+Mn) are by far the major components and Fe is more abundant than Mn. (Typical compositions of some calcined sludges from Savannah River (Garvin, personal communication) are given in Tables 1, 2, 3). In this paper we will briefly describe how the SYNROC process, utilizing straightforward technology, can be readily adapted to the problem of defense waste immobilization, yielding a dense, inert ceramic waste form, SYNROC-D.

This objective can be achieved as follows: about 30-50 wt.% of additional inert components (TiO_2 + ZrO_2 + CaO $\pm$ Al_2O_3 $\pm$ SiO_2 $\pm$ BaO) would be intimately mixed with the sludges, followed by heating under controlled oxygen fugacity conditions to form a ceramic synthetic rock composed of a specific equilibrium mineral assemblage. This assemblage comprises the "original" SYNROC phases: zirconolite, perovskite and hollandite, together with certain other refractory and leach-resistant minerals such as spinel, pseudo-brookite, ilmenite and nepheline. In keeping with the SYNROC strategy, the immobilization of fission products and actinides would be achieved by incorporating them as dilute solid solutions forming integral parts of the crystal lattices of hollandite, zirconolite and perovskite. The additional thermodynamically compatible phases (spinel, etc.) serve to immobilize the "inert" sludge components, mainly Fe, Al, Mn, Ni and Na oxides.

The sodium contents of sludges influence the selection of an optimum immobilization assemblage. The sludges whose compositions are given in Tables 1, 2 and 3 were calcined after two or three stages of dispersal in water at ambient temperatures followed by centrifuging, filtering and drying. It seems possible that the sodium content could be substantially or greatly reduced by more intensive treatments, e.g., by washing in hot aqueous solutions containing ammonium or calcium ions. Accordingly, we will discuss two classes of processes - one designed to immobilize sludges containing "normal" amounts of sodium (cf. Tables 1, 2, 3) and the other designed for otherwise similar sludges which are, however, strongly depleted in sodium as a result of more efficient washing procedures. The processes which are discussed in the following sections are based upon the results of over 250 experimental hot-pressing runs followed by detailed phase chemistry investigations using the electronprobe microanalyser and X-ray diffraction. The number of full analyses of individual phases by electronprobe exceeds 2000.

IMMOBILIZATION OF LOW-SODIUM SLUDGES

In this process, it is assumed that the sodium content of the (dried) sludge can be reduced to less than one weight percent. The treatment of sludges of this type was described by Ringwood et al. (1). About 30-50 wt.% of TiO_2 + CaO + ZrO_2 + BaO in appropriate proportions would be intimately mixed with the sludge, followed by sintering or hot-pressing under mildly reducing conditions around 1100-1200°C. High-alumina sludges treated in this manner crystallize to a dense ceramic composed principally of the minerals hercynite-rich spinel ($FeAl_2O_4$), perovskite ($CaTiO_3$), zirconolite ($CaZrTi_2O_7$) and "hollandite" ($BaAl_2Ti_6O_{16}$). High-iron sludges behave similarly, except that the spinel phase consists dominantly of the ulvospinel component, Fe_2TiO_4, whilst an additional barium-aluminum-iron-titanate phase "BAIT" may also be formed.

IMMOBILIZATION OF HIGH-SODIUM SLUDGES

It is difficult to immobilize sludges containing several percent of sodium by the above technique without adding excessive quantities of TiO_2. To avoid this, it was found preferable to add sufficient SiO_2 so that nearly all sodium was immobilized in nepheline ($NaAlSiO_4$). In this variant of the SYNROC process, about 30-35 wt.% of TiO_2 + CaO + ZrO_2 + SiO_2 $\pm$ Al_2O_3 in appropriate proportions (Tables 1, 2, 3) is intimately mixed with the sludges. The dried mixture is then sintered or hot-pressed under mildly reducing conditions (near the Ni-NiO oxygen fugacity buffer) around 1100°C to 1200°C, thereby forming strong ceramic synthetic rocks of high density. Under these conditions, high-alumina sludges are converted to an equilibrium assemblage of hercynite spinel + perovskite +

Table 1. Immobilization of High-Alumina Sludges, Savannah River

	High-Al Sludge Calcine	Inert Additives
SiO_2	-	6.5
TiO_2	-	16.7
ZrO_2	-	4.2
UO_2	2.3	-
Al_2O_3	33.8	-
Fe_2O_3	17.7	-
MnO	5.3	-
NiO	0.6	-
CaO	2.1	5.5
Na_2O	3.4	-
FP,A*	2.0	-
Sum wt.%	67.0	33.0

Typical Examples of Coexisting Minerals[+]

	Zirconolite	Perovskite	Hercynite Spinel	Nepheline
Estimated Proportions, wt.%	16	13	55	16
SiO_2	-	-	-	41.5
TiO_2	30.1	53.2	5.8	0.2
ZrO_2	38.5	0.7	0.3	-
UO_2	14.2	4.8	-	-
Al_2O_3	1.1	0.3	48.3	35.9
FeO•	4.2	2.4	37.4	0.8
MnO	0.9	0.7	7.2	0.2
NiO	-	-	0.4	-
CaO	12.5	38.1	-	-
Na_2O	0.4	-	-	21.5
Sum wt.%	101.9	100.2	99.4	100.1

* Fission products plus actinides excluding U.

[+] Run conditions: Fe capsule, 5 kb, 1200°C, 1 hr.

• Total iron assumed as FeO.

Table 2. Immobilization of High-Iron Sludges, Savannah River

	High-Fe Sludge Calcine	Inert Additives
SiO_2	1.4	4.8
TiO_2	-	16.8
ZrO_2	-	4.0
UO_2	2.4	-
Al_2O_3	3.7	5.4
Fe_2O_3	36.8	-
MnO	6.1	-
NiO	6.6	-
CaO	1.9	5.0
Na_2O	3.2	-
FP,A*	2.0	-
Sum wt.%	64.0	36.0

Typical Examples of Coexisting Minerals[+]

	Zirconolite	Perovskite	Ferrite Spinel	Nepheline
Estimated Proportions, wt.%	15	11	59	15
SiO_2	-	-	-	40.6
TiO_2	35.1	52.4	7.9	0.5
ZrO_2	25.2	1.4	-	-
UO_2	15.5	2.6	-	-
Al_2O_3	0.4	1.5	8.1	34.4
Fe_2O_3	7.8[++]	1.6[++]	43.5	5.0[++]
FeO	-	-	20.7	-
MnO	1.8	0.8	9.0	-
NiO	-	0.2	9.7	-
CaO	14.6	39.8	-	-
Na_2O	0.2	-	-	20.1
Sum wt.%	100.6	100.3	98.9	100.7

* Fission products plus actinides excluding U.

[+] Run conditions: Pt capsule, 5 kb, 1150°C, 1 hr.

[++] Total iron assumed as Fe_2O_3.

Table 3. Immobilization of Average Savannah River Sludge Composition

	Average Sludge Calcine	Inert Additives
SiO_2	0.6	7.5
TiO_2	-	15.0
ZrO_2	-	4.0
UO_2	2.3	-
Al_2O_3	21.2	-
Fe_2O_3	27.1	-
MnO	6.1	-
NiO	3.4	-
CaO	2.0	4.9
NaO	3.8	-
FP,A*	2.1	-
Sum wt.%	69.0	31.0

Typical Examples of Coexisting Phases^{+}

	Zirconolite	Perovskite	Spinel	Nepheline
Estimated Proportions, wt.%	16	11	55	18
SiO_2	-	-	-	41.0
TiO_2	32.5	53.6	14.3	0.5
ZrO_2	32.7	<.1	0.2	-
UO_2	11.2	3.0	<.1	-
Al_2O_3	1.1	0.1	17.2	35.2
Fe_2O_3	7.1^{++}	2.7^{++}	20.5	1.1^{++}
Sm_2O_3	2.5	6.8	-	-
FeO	-	-	31.0	-
MnO	0.9	<.3	10.8	-
NiO	2.1	0.2	5.7	-
CaO	9.8	34.1	0.2	0.8
Na_2O	0.7	<.5	<.1	21.4
Sum wt.%	100.6	100.5	99.9	100.1

* Fission products plus actinides excluding U.

$^{+}$ Experimental conditions: Ni capsule, 5 kb, 1150°C, 1 hr. Starting material doped with 1-2% Sm_2O_3.

$^{++}$ Total iron assumed as Fe_2O_3.

zirconolite + nepheline (Table 1). High-iron sludges behave similarly, except that the spinel is a complex ferrite solid solution (Table 2). The behavior of average Savannah River sludge is intermediate between these two cases (Table 3). Compositions of coexisting minerals and their relative proportions in the SYNROC-D waste forms are given in Tables 1, 2, 3.

Despite the very wide range of chemical compositions between high-Al and high-Fe sludges, inspections of Tables 1, 2, 3 show that the amounts and proportions of additives are similar in all of these cases. Moreover, this entire compositional spectrum can be satisfactorily immobilized by heat treatment under the same redox conditions (near the Ni-NiO oxygen buffer). This is a most felicitous situation with respect to the development of a relatively simple and economically efficient industrial scale process for treating the wastes. To a large extent, it is a consequence of the fact that the principal inert components in the waste are oxides of Al, Fe, Mn and Ni. When heated under appropriate conditions, these components become immobilized in a suitably inert spinel-type mineral phase. Apart from a small amount of SiO_2 needed to immobilize sodium, the remaining additives (TiO_2, ZrO_2, CaO $\pm$ Al_2O_3) are dedicated primarily to the objective of immobilizing the very small proportion (0.5 to 5%) of fission products and actinides (exclusive of U) present in the waste. Thus the basic SYNROC concept according to which the fission products and actinide components of the wastes are immobilized by incorporating them as dilute solid solutions in titanate minerals (mainly zirconolite and perovskite) is retained; the principal difference being that the large excess of "inert" components in the waste crystallize as the additional phases, spinel and nepheline.

Distributions of fission products and actinide elements among SYNROC minerals have been determined by electronprobe microanalysis. Rare earths (and trivalent actinides) preferentially enter perovskite, whereas tetravalent actinides enter zirconolite. Uranium becomes immobilized in the latter mineral as the $CaUTi_2O_7$ component. Strontium is strongly partitioned into perovskite whilst Cs, Rb, Na and Ba enter nepheline. Phosphorous forms $Ca_3(PO_4)_2$ whilst sulphur, bismuth, ruthenium, rhodium and palladium alloy with nickel to form minor accessory minerals.

TECHNOLOGICAL ASPECTS

After the removal of the sludges from the tanks, followed by washing and blending, they would be dispersed and intimately mixed as a slurry with the required components (mainly TiO_2 + ZrO_2 + CaO $\pm$ SiO_2). After centrifuging, washing and drying, the SYNROC feedstock would be devolatilized by heating to 600-900°C, for example, in a spray calciner. There are several alternative further courses of treatment which might be envisaged. The first, and simplest, would be to pelletize and sinter the SYNROC powder. Strong pellets possessing low porosity can be readily produced by sintering at 1100-1200°C in a controlled atmosphere, e.g., in a stationary heated kiln. The pellets could then be encapsulated in a continuous metallic matrix of copper or lead to provide a sealed waste form possessing high mechanical strength, thermal conductivity and corrosion resistance. Techniques for incorporating ceramic pellets in metallic matrices are well advanced (2). A variant of the above sintering process might aim to produce much larger integral sintered bodies, e.g., short cylinders, which could be deposited in hollow steel canisters and the relatively small proportion of remaining void space then filled with molten lead.

Alternatively, more sophisticated processing could involve large scale isostatic hot-pressing (3). The SYNROC oxide mixture, incorporating a small amount of dispersed metallic nickel powder, would be hot-pressed at 1100-1200°C, thereby producing a dense, mechanically strong waste form possessing negligible porosity. Although this requires a more elaborate technology than the sintering operation described above, the ceramic product may well be significantly superior in its properties to an extent which justifies the additional expense, as discussed by Ringwood et al. (1).

The SYNROC-D waste forms produced as described above are vastly more stable than borosilicate glasses when buried in appropriate geochemical-geological environments. They are immune from devitrification and are composed of thermodynamically compatible phases which possess crystal structures identical to those of natural minerals which are known to have survived geologically for periods of 500 to 2000 million years and to have retained radioactive elements for these periods. It is this evidence, provided directly by Nature herself, which can demonstrate to the community that the much shorter times required for radwaste immobilization under the much less extreme P,T conditions in a suitable repository can be successfully achieved.

Moreover, in addition to the key issues of safety and public acceptability, there is yet a further area in which the SYNROC strategy possesses major advantages over immobiliization in borosilicate glasses. Because it is possible to immobilize about three times the weight of sludge in a given volume of SYNROC-D as compared to borosilicate glass, the very large costs of fabrication, transport and ultimate geological disposal are significantly reduced; moreover, radiological hazards to technical personnel are correspondingly diminished.

REFERENCES

1. A. E. Ringwood, S. E. Kesson, N. G. Ware, W. B. Hibberson, and A. Major, "The SYNROC Process: A Geochemical Approach to Nuclear Waste Immobilization," Geochemical Journal, in press (1979).
2. L. J. Jardine and M. J. Steindler, "Metal Encapsulation of Ceramic Nuclear Waste," Scientific Basis for Nuclear Waste Management, 1 (G. J. McCarthy, ed., Plenum Press, 1979).
3. H. Larker, "Hot Isostatic Pressing for Consolidation and Containment of Radioactive Waste," Scientific Basis for Nuclear Waste Management, 1 (G. J. McCarthy, ed., Plenum Press, 1979).

THE EFFECTS OF RADIATION DAMAGE ON SYNROC

A. E. Ringwood, V. M. Oversby, and W. Sinclair

Research School of Earth Sciences
Australian National University
Canberra, A. C. T. 2600, Australia

INTRODUCTION

The basic objective of the SYNROC method of immobilizing high level nuclear reactor wastes (HLW) is to incorporate the fission products and actinides as dilute solid solutions in the crystal lattices of synthetic minerals. These minerals are closely related to natural minerals known to have been stable in a wide range of geological and geochemical environments for periods of more than 10^9 years. The minerals employed in SYNROC formulations designed to immobilize HLW from reprocessed fuel rods of power reactors are hollandite ($BaAl_2Ti_6O_{16}$), zirconolite ($CaZrTi_2O_7$), and perovskite ($CaTiO_3$). It has been demonstrated that these minerals have the capacity to accept nearly all HLW fission products and actinides into solid solutions in their crystal lattices (1).

It is now essential to demonstrate that these minerals are capable of immobilizing the HLW safely for periods sufficient to allow the radioactive fission products and actinides to decay to safe levels. The required minimum immobilization period is commonly estimated to lie between 10^5 and 10^6 years. Because this timescale is beyond human experience, there has been a great deal of concern within society as to whether safe immobilization of HLW can be firmly guaranteed by existing technologies. One of the major advantages of the SYNROC approach is that natural minerals themselves provide key evidence that safe immobilization is possible for these long periods. This factor may well prove to be of critical importance in determining the public acceptability of waste immobilization technologies. Indeed, in some countries, the public concern with the long term safety of nuclear waste disposal is so intense that the very future of nuclear power may rest upon an acceptable answer to this problem.

A key technical question is the effect of radiation damage on the structures of minerals proposed for waste immobilization. How much accumulated radiation damage is necessary to cause metamictization of these minerals? Do radiation damage and metamictization impair the capacity of the minerals to immobilize wastes?

ALPHA DOSE AS A FUNCTION OF AGE FOR SYNROC

By far the most important source of lattice damage in SYNROC minerals would arise from recoil of the parent nucleus during alpha decay. We have used the fuel compositions calculated by Cohen (2) to estimate the actinide composition of the waste material. The fuel type is U-only with 3% of ^{235}U, a burnup of 33,000 MWd/MTHM, and a cooling time of 150 days before reprocessing. Reprocessing of one metric ton of this fuel would generate approximately 35 kg of HLW calcine. We plan to incorporate this waste calcine as 10% of the SYNROC mixture.

In SYNROC, the quadrivalent and trivalent actinides are partitioned into zirconolite and perovskite phases. These represent about 60% of the total SYNROC and will absorb most of the alpha recoil damage. To evaluate the effect of alpha decay on mineral structures, we have calculated the cumulative alpha dose per gram of mineral as a function of SYNROC age. The dose rate is initially very high due to decay of ^{242}Cm and ^{244}Cm. After about 50 years, decay of americium isotopes dominates until 10^3 years. From 10^3 to 10^4 years, ^{243}Am produces half the total activity and ^{239}Pu and ^{240}Pu make up most of the remainder. Total alpha dose per gram for zirconolite and perovskite would be 5×10^{17} at 100 years, 1.1×10^{18} at 1000 years, and 2.2×10^{18} at 10,000 years. During this period, the cumulative dose doubles for each factor of 10 increase in age of SYNROC.

The period from 10^4 to 10^5 years is transitional. ^{243}Am decreases in importance while ^{239}Pu (produced from ^{243}Am) decay dominates most of the period. Decay of ^{237}Np and its short-lived daughters begins to make an important contribution. Cumulative dose absorbed by 10^5 years is 4.2×10^{18} alphas per gram of zirconolite plus perovskite. The dose rate is still roughly a factor of two increase for a factor of 10 increase in age.

From 10^5 to 10^7 years, ^{237}Np and its daughters completely dominate the alpha activity. Because the ^{237}Np daughter products are all short lived, the alpha dose increases during the time while Np decays. For each factor of ten increase in SYNROC age from 10^5 to 10^7 years, the cumulative alpha dose increases by roughly a factor of three. At 10^6 years the total is 1.5×10^{19} alphas per gram, and at 10^7 years the total is 4.9×10^{19}. After 10^7 years the system is essentially zirconolite and perovskite with an average U content of

2.4% and inert hollandite. The activity is then nearly constant over periods of 10^8 years at 7.4 x 10^{10} alpha events per year per gram of zirconolite or perovskite. For a SYNROC age of 10^8 years, total alpha dose received would be 5.8 x 10^{19} per gram. This is an increase of less than 20% for a factor of ten increase in age.

X-RAY STUDIES

The first question which we must ask is, "What is the effect of radiation damage on the crystal structures of zirconolite and perovskite?" To answer this question, we have examined a series of naturally occurring zirconolite specimens which have an accumulated alpha dose ranging from 5 x 10^{17} to 1 x 10^{20} alphas per gram. The uranium and thorium content of all specimens were first determined by electron microprobe analyses after which they were investigated by X-ray diffraction methods applied to single crystals or powder.

The Kaiserstuhl (Germany) carbonatite complex contains traces of zirconolite in a late stage vein. The estimated age of these samples is 16 my (3). The grains of zirconolite vary in their U and Th contents from a low of 0.8% UO_2, 1.7% ThO_2 to a high of 1.5% UO_2, 2.9% ThO_2. The age of SYNROC which would have received an alpha dose equivalent to the high U sample would be about 10^3 years. X-ray diffraction of powders and single crystals shows this zirconolite to be well crystallized. The zirconolite structure is closely related to that of fluorite. The X-ray data for these samples show that the fluorite subcell reflections remain constant while the reflections from the zirconolite superstructure decrease slightly in intensity with increasing alpha dose. This effect is especially noticeable at high 2θ values where slight broadening of spots is also seen. Based on these observations, we would expect that the zirconolite superstructure lines would still be present at twice the alpha dose. This would correspond to a SYNROC age of 8 x 10^3 years. After annealing at 1200°C, a slight increase ($\sim$2%) in density is noted.

A small grain of zirconolite from Jacupiranga, Brazil, with an age of 133 my contains 0.4% UO_2 and 6.6% ThO_2. The cumulative alpha dose for this sample is 7.6 x 10^{18} alphas per gram, which corresponds to a SYNROC age of 4 x 10^5 years. Preliminary X-ray photographs from single crystal chips of the Brazilian zirconolite show numerous, quite sharp reflections characteristic of a fluorite-type lattice. However, the zirconolite superstructure reflections have completely disappeared. Thus, it appears that the basic effect of that high radiation dose has been to cause a degree of disorder of the cations, leaving the fluorite-type anion sublattice essentially intact.

Several Russian zirconolite samples were also examined. Our unpublished isotopic measurements indicate an age of about 650 my. Single crystal oscillation and rotation X-ray photographs showed that they were largely metamict; however, a few faint elongated reflections could be seen. The strongest reflection occurs at approximately 3.0 Å, and corresponds to the 111 reflection of the fluorite-type subcell. Taking the age as 650 my, we find that this sample has accumulated a dose of 1.7×10^{19} alphas per gram. This is somewhat more than the total alpha dose received by SYNROC in 10^6 years. Thus, in order to cause zirconolite to become truly metamict, the alpha dose would need to exceed that received by SYNROC over this period.

Small samples of loparite, a perovskite mineral containing large amounts of cerium and niobium, have also been studied by X-ray and electronprobe. The samples come from the Lovozero alkaline massif of the Kola Peninsula. Samples of loparite from this intrusion dated by Komlev et al. (4) gave a lead isotopic age of 300 my. The U and Th contents of our samples range from U = 700 ppm, Th = 1300 ppm to U = 1600 ppm, Th = 3400 ppm.

For an age of 300 my, an alpha dose of 1 to 2.3×10^{18} would have accumulated for these samples. The corresponding ages for SYNROC at this dose would be 10^3 to 10^4 years. Powder X-ray photographs reveal that the loparite crystals have not suffered any perceptible radiation damage to their crystal structure. The powder lines are sharp, and at angles of high 2θ, the α_1 and α_2 peaks are clearly resolvable. Comparison of these X-ray observations with those on Kaiserstuhl zirconolite indicates that the perovskite mineral is even more resistant to radiation damage than zirconolite. On annealing loparite at 1200°C, a slight decrease in "d" spacings occurs, equivalent to an increase in density of 1.7%. This is quite a small change in volume in relation to the alpha dose received and further demonstrates the stability of perovskite to radiation damage.

ISOTOPIC STUDIES

At very high alpha dosage ($> 2 \times 10^{19}$ alphas/gram), zirconolite becomes completely metamict. We have studied three samples from Sri Lanka which have received alpha doses in excess of 1×10^{20} alphas per gram (5). This is more than twice the dose for 10^7-year-old SYNROC. In fact, it would take SYNROC 750 my to accumulate this high dose.

For all of the minerals studied, it is important to know how well they are able to retain HLW. This involves the resistance of the minerals to leaching and the retention of daughter products from alpha decay processes. To examine both of these aspects, we have analyzed two specimens of zirconolite from the gem gravels of Sri Lanka for U and Pb concentrations by isotope dilution, Pb isotopic composition, and Th concentration by electron microprobe. These samples should represent a severe test since the accumulated alpha dose is very high, the samples are metamict, and the minerals have been weathered out of their original matrix and deposited by stream sedimentation. As such, the samples have been subjected to extensive attack by ground water.

The lead isotopic ages are given in Table 1. Sample 83800 has a relatively large uncertainty in the $^{207}Pb/^{206}Pb$ age because of a small amount of common lead in the sample. This specimen contains 18% ThO_2, 1.8% UO_2 and 0.56% Pb.

There is no detectable loss of any of these elements since the three calculated ages agree within error limits, and the Th content is appropriate to the measured $^{208}Pb/^{206}Pb$ and age of the sample.

Sample B20392 contains less common lead than 83800; the measured $^{206}Pb/^{204}Pb$ was 11,800. This produces a more precise $^{207}Pb/^{206}Pb$ age which is in excellent agreement with that obtained for sample 83800. The uranium-lead ages for this zirconolite are in excellent agreement and are within error limits of the 207/206 age. Uranium, thorium and lead concentrations are similar to sample 83800.

The concordance of lead ages for Sri Lanka zirconolites is far better than that normally obtained for zircon samples. We believe that this is because the decay product, lead, can be readily accepted into the zirconolite structure, whereas it does not fit into the zircon lattice. A further proof of the ability of zirconolite to retain daughter products was provided by Blake and Smith (5). They found a large amount of helium (equivalent to an "internal pressure" of 1 atmosphere) retained in Sri Lanka zirconolite. Their analysis corresponds to approximately 36% retention of the total helium produced over the crystal's lifetime. This is remarkable in view of the readiness with which helium diffuses through most glasses and silicate minerals.

Table 1. Lead Isotopic Ages of Sri Lanka Zirconolites

Sample Number	83800	B20392
$^{207}Pb/^{206}Pb$ age	570 $\pm$ 20	565 $\pm$ 3 my
$^{207}Pb/^{235}U$ age	567 $\pm$ 5	560 $\pm$ 2 my
$^{206}Pb/^{238}U$ age	567 $\pm$ 5	559 $\pm$ 2 my

Zirconolite is also found as small (30-100 micron) grains in the Jimberlana mafic complex in Western Australia, which is about 2500 my old [(6) and Lovering, personal communication]. Wark (personal communication) has studied several samples and carried out careful electronprobe analyses for U, Th and Pb. The grains range from Th 2.0%, U 0.59% to Th 8.38%, U 2.27%. Wark has shown that the lead abundances in these grains are closely consistent with those expected to result from radioactive decay of the contained U and Th in 2500 my. Since it is unlikely that these grains contain any signifcant amount of common lead, this implies closed system behavior for U, Th and Pb over a period of 2500 my. More recently, Lovering (personal communication) has dated two Jimberlana zirconolite grains by ion microprobe. He obtained lead 206/207 ages of 2500 my, thus establishing the closed system behavior. This is remarkable in view of the enormous amount of radiation damage (sometimes exceeding 3 x 10^{20} alphas/gram) received by these small grains and the demonstrated fact that water obtained access to the intrusion as shown by partial serpentization of olivines (7,8).

CONCLUSION

Concern has been expressed in several quarters that the integrity of SYNROC as a waste form might be compromised by the effects of alpha radiation damage on its constituent minerals. The evidence discussed herein addresses that concern. Radiation damage to crystal lattices is caused dominantly by alpha emitters (actinide elements) present in zirconolite and perovskite phases. We have seen that the zirconolite phase in SYNROC containing 10% of HLW would require about 10^6 years before becoming completely metamict, i.e., losing the capacity to diffract X-rays. By this time, the alpha activity in SYNROC would have declined to a value smaller than that present in a natural uranium ore body. This level of activity can be considered effectively safe for a waste form buried in a suitable geologic repository. Moreover, even when metamict, the zirconolite continues to demonstrate a remarkable capacity to immobilize U, Th and their decay daughter products. The isotopic measurements on Sri Lanka and Jimberlana zirconolites demonstrated that these metamict minerals had behaved as closed systems for U, Th and Pb for periods of 560 and 2500 million years, respectively, despite intense radiation damage (SYNROC equivalent ages, $\gtrsim 10^9$ years). Moreover, the Sri Lanka zirconolite was sufficiently "tight" to immobilize a considerable proportion of radiogenic helium, one of the most mobile of elements, for 560 million years. Our data on perovskite are limited and more work remains to be done. Nevertheless, present results indicate that perovskites are even more resistant to radiation damage than zirconolite.

The suite of natural zirconolites and perovskites studied here have experienced complex geological histories, including elevated temperatures and pressures, contact with ground waters, weathering and erosion over periods hundreds or even thousands of times longer than are needed for decay of high level waste to safe levels. They have been subjected to a range of geochemical stresses, far more intense than are likely to be found in any geological waste repository. Despite this, they have retained their waste elements (alpha emitters) quantitatively. These observations provide strong grounds for confidence that the integrity and safety of SYNROC as a waste form will not be adversely affected by radiation damage before the high level waste have decayed to safe levels. The kind of evidence which we have described is important in another context. It provides a means of demonstrating to the general public in a readily understandable way that safe immobilization of radioactive waste elements over periods which are enormously long in comparison with human experience _can_ indeed be achieved. This factor is likely to exert a strong influence on public acceptability of the feasibility of waste disposal. It should be noted that borosilicate glass waste forms, preferred by the nuclear power establishment, are intrinsically incapable of providing this kind of reassurance.

ACKNOWLEDGEMENTS

We wish to record our appreciation of the persons and institutions who made it possible to obtain the samples of rare minerals used in this investigation. In this respect, we are particularly indebted to Jörg Keller, Gene Kaula, Carl Francis, Brian Mason, Andrew Graham, Andrew Clark, and to the British Museum (Natural History), the Smithsonian Institution, Harvard University, University of Freiburg, and the USSR Academy of Science. We are also indebted to Professor J. F. Lovering and Mr. D. Wark of the Department of Geology, Melbourne University, for permission to quote from unpublished results. The researches described in this paper were supported by grants provided by the National Energy Research Development and Demonstration Program of the Commonwealth Department of National Development and by the Australian Atomic Energy Commission. This support is gratefully acknowledged.

REFERENCES

1. A. E. Ringwood, S. E. Kesson, N. G. Ware, W. Hibbesson, and A. Major, "Immobilization of High Level Nuclear Wastes in SYNROC," _Nature_ _278_, 219 (1979).
2. B. L. Cohen, "High Level Radioactive Waste from Light Water Reactors," _Rev. Mod. Phys._ _49_, 1 (1977).
3. Keller, personal communication.
4. L. V. Komlev, A. D. Mikhalevskaya, and S. I. Danilevich, "The Age of the Alkaline Intrusions of the Chibina and Lovozero

Tundras (Kola Peninsula)," Dokl. Akad. Nauk. SSSR (Earth Sciences Trans.) 136, 31.

5. G. S. Blake and G. F. H. Smith, "On Varieties of Zirkelite from Ceylon," Mineral. Mag. 16, 309 (1913).
6. A. Turek, "Rubidium-Strontium Isotopic Studies in the Kalgoorlie-Norseman Area, Western Australia," Aust. Natl. Univ., Ph.D. Thesis.
7. I. H. Campbell, "A Study of Macro-rhythmic Layering and Cumulative Processes in the Jimberlana Intrusion, Western Australia, Part I: The Upper Layered Series," J. Petrol. 18, 183 (1977).
8. A. P. Vinogradov and A. I. Tugarinov, "Problems of Geochronology of the Pre-Cambrian in Eastern Asia," Geochim. Cosmochim. Acta. 26, 1283.

URANIUM AND THORIUM MINERALS: NATURAL ANALOGUES

FOR RADIOACTIVE WASTE FORMS

Richard F. Haaker and Rodney C. Ewing

Department of Geology
University of New Mexico
Albuquerque, NM 87131

INTRODUCTION

SYNROC (1) and Supercalcine (2) are proposed crystalline radiowaste forms that may be regarded as highly radioactive "artificial rocks." In these crystalline radioactive waste forms, rare-earth elements (REE), U and transuranium (TRU) elements are strongly partitioned into five phases. These phases have U- and Th-bearing mineral analogues. In SYNROC, the actinide bearing phases are perovskite ($CaTiO_3$) and zirconolite ($CaZrTi_2O_7$). "BAIT" ($BaAl_2Fe_8Ti_{13}O_{38}$) is a phase which may occur in SYNROC formulations for wastes with large amounts of Fe and Al. "BAIT" may be structurally similar to davidite ($Fe_6Ti_{15}O_{36}$). Supercalcine actinide phases include silicate-apatite ($CaREE_4(SiO_4)_3O$), monazite ((Ce, La) PO_4 and uraninite (UO_{2+x}). The purpose of this paper is to summarize the relevant geologic literature on these phases. Emphasis is given to: (1) solid-solution chemistry, (2) mode of occurrence, (3) alteration and (4) radiation damage effects. A special effort is made to identify stable and unstable mineral associations, as this may be important if hydrothermal conditions occur in the waste repository. The idea of devising a radioactive waste form which is in thermodynamic equilibrium with its repository is appealing but impossible. This is apparent from the conditions for equilibrium between two or more phases, namely, for each chemical species, i, the chemical potential, U_i, must in each phase be equal. This condition cannot be met since there are no appreciable amounts of ^{137}Cs, ^{90}Sr, Tc, Pu, Cm, Np, etc. in the wall rock. At best it would be possible to pick a waste form-wall rock combination such that a "near equilibrium" situation with high kinetic stability exists. At lower temperatures (i.e. lower heat content) or in the absence of water, thermodynamic stability is not necessarily important. There are two basic scenarios for performance of a waste form under hydrothermal conditions: (i) If the waste form and wall rock are chemically "compatible" then alteration effects such as hydration and perhaps only superficial corrosion of the waste form may be expected, (ii) If the waste form and wall rock are not a stable assemblage then some (or all) of the waste phases may be degraded or replaced altogether. In such a case specific alteration patterns can be expected, the extent of which will be determined by reaction kinetics. This paper is an extended abstract of a report which is being prepared for Battelle Pacific Northwest Laboratories (PNL) (EY-76-C-06-1830). That report also summarizes the behavior of natural analogues to the nonactinide phases in Supercalcine and SYNROC.

APATITE

A great number of natural and synthetic phases are isostructural with apatite ($Ca_5(PO_4)_3(OH,F)$). The rare apatite-group mineral, britholite ($((Ca,Ce,Y,Th)_5((Si,P,Al)O_4)_3(OH,F))$, is a close natural analogue to the REE-silicate apatite phase in supercalcine. Britholite can contain up to 56 wt % REE. The heavy REE and Y can be present in amounts up to 19 wt %. Britholite commonly contains 6 wt % ThO_2 and can contain 19 wt % ThO_2. Major amounts of P or Al can substitute for Si. CaO is present in amounts of 10-17 wt % but SrO rarely exceeds 1 wt %.

Occurrence and Alteration

Britholite occurs in nepheline syenites, alkali syenites, granites and pegmatites. Associated minerals include quartz, feldspars ($(Na,K,Ca)AlSi_2(Al,Si)O_8$), nepheline ($NaAlSiO_4$), amphiboles, pyroxenes ($(Ca,Mg,Fe)SiO_3$), natrolite (a zeolite, $Na_2Al_2Si_3O_{10}\cdot 2H_2O$), fluorite ($CaF_2$) and niobate minerals. Descriptions of britholite occurrences indicate that it is a late stage mineral, often of metasomatic origin (i.e. formed by replacement of older rock under hydrothermal conditions). Britholite is oxidized under supergene conditions (i.e. contact with near surface ground waters). Oxidation of britholite results in the development of an alteration rind of cerianite $(Ce,REE)O_{2-x}$. Britholite is frequently hydrated, probably occurring after volume changes caused by metamictization. REE-carbonates, -phosphates and silicophosphates have been cited as alteration products of britholite. Thorian varieties of britholite with as little as 0.80 wt % ThO_2 may show radiation damage effects. Completely metamict britholite with 11 wt % ThO_2 is known (3). Radiation damage in Nd-silicate apatite ($CaNd_4(SiO_4)_3O$) doped with 1 wt % ^{244}Cm is being studied at Battelle PNL.

MONAZITE

The orthophosphates of the REE, La through Gd, have the monazite structure. Monazite ($(Ce,La)PO_4$), a supercalcine phase, has a dense and tightly packed structure without large vacant interstitial voids. It is monoclinic and isostructural with huttonite ($ThSiO_4$). Minor amounts of Y and the REE, Tb through Ly, are present. Monazite with 15.6 wt % UO_2 is known, but it usually does not have more than 0.5 wt % UO_2. Monazite is the major source of Th. It commonly contains 5 wt % ThO_2 and can contain up to 35 wt % ThO_2. The common mechanisms of Th substitution are $Th^{+4} + Ca^{+2} = Ce^{+3} + Ce^{+3}$ and $Th^{+4} + SiO_4^{-4} = Ce^{+3} + PO_4^{-3}$. There is major solid-solution in the system $CePO_4$-$ThSiO_4$-$ThCa(PO_4)_2$.

Occurrence and Alteration

Monazite occurs throughout the world as a constituent of detrital "black sand" along with other resistant minerals such as magnetite, zircon and garnet. Monazite also occurs in clay-containing and sand-containing metamorphosed sediments. In igneous rocks monazite occurs in granites, pegmatites, nepheline syenites and carbonatites. Some of its igneous mineral associations include quartz, feldspars ($(Na,K,Ca)AlSi_2(Al,Si)O_8$), muscovite ($K_2Al_6Si_6O_{20}(OH,F)_4$), nepheline ($NaAlSiO_4$), pyroxenes ($(Mg,Ca,Fe)SiO_3$), zircon ($ZrSiO_4$), apatite ($Ca_5(PO_4)_3F$), xenotime ($YPO_4$), thorite ($ThSiO_4$), magnetite ($Fe_3O_4$), ilmenite ($FeTiO_3$) and titanite ($CaTiSiO_5$) (4). Monazite is sufficiently stable to occur in detrital beach sands and in ancient placer deposits. Monazite occurrences over 2 x 10^9 years old are known. It is crystallized from hydrothermal solutions of H_3PO_4 - $(Ce,La)Cl_3$ - H_2O (300°C and 95 atm). Metamorphism increases the thorium content of monazite. Despite the durability of monazite, it is seldom found without outwardly altered surfaces. It does appear subject to oxidation in weathering zones, as rare CeO_2 (cerianite) pseudomorphs after monazite are known (5). Natural monazite with 20 wt % ThO_2 has an unusually high stability to radiation damage.

URANINITE

The fluorite structure phase in supercalcine is expected to contain U, REE, Zr, TRU elements and possibly Ca. Important natural analogues to this phase include uraninite and pitchblende (UO_{2+x}), thorianite (ThO_2) and cerianite (CeO_2). There is complete solid-solution among ThO_2, stoichiometric CeO_2 and stoichiometric UO_2, but limited miscibility of these phases with ZrO_2. Uraninite is always oxidized beyond UO_2, and uraninite of near $UO_{2.6}$ is known. Major amounts of REE_2O_3 or CaO can be incorporated into uraninite (UO_{2+x}), thorianite (ThO_2) and cerianite (CeO_2).

Occurrence and Alteration

Uraninite (UO_{2+x}) occurs as a minor mineral in pegmatites. Associated minerals include K-feldspar, quartz, muscovite (a mica), zircon ($ZrSiO_4$), garnet, allanite ($(Ca,REE)_2(Al,Fe,Mg)_3(SiO_4)_3(OH)$), REE-phosphates, REE-silicates, niobate-titanate minerals, spodumene ($LiAlSi_2O_6$), topaz ($Al_2SiO_4(F,OH)_2$), beryl ($Be_3Al_2Si_6O_{18}$) and lepidolite (a mica). There are major deposits of uraninite of hydrothermal origin which occur along cracks or fissures in rock. Some associated minerals in these deposits include Co-, Ni-, Bi-, As- and Ag-minerals, quartz, carbonate minerals, fluorite and sulfides. Uraninite also occurs in important amounts in quartz-rich sandstone interbedded with organic material or mudstone(4). In general, felsic igneous rocks are enriched in uranium relative to nonfelsic rocks. Under reducing conditions, uraninite is sufficiently stable to form placer deposits. Thorianite (ThO_2) occurs as a detrital mineral and cerianite (CeO_2) occurs as an alteration (weathering) product of REE-minerals including monazite. Under reducing conditions the weathering of uraninite is dominated by mechanical agents. Soluble uranyl carbonates and secondary uranium minerals are formed when uraninite is exposed to surface weathering or supergene conditions at low temperatures. Uraninite, thorianite and cerianite have unusually high stabilities to radiation damage.

PEROVSKITE

A large number of ABO_3 compounds including some titanates, niobates, zirconates and aluminates are isostructural with perovskite ($CaTiO_3$). The perovskite phase in SYNROC would be expected to contain Ca and minor amounts of REE, Na, Sr, U^{+4}, and TRU elements in the A-cation site. The B-cation site contains Ti with minor amounts of Mo^{+4}, Fe^{+3}, Al and Tc. In natural perovskite, Nb is the only element which substitutes for Ti to any major extent. There are varieties of perovskite with atom % Nb > Ti. Natural perovskite commonly contains up to a few wt % Na, REE and Sr, but usually less than 1 wt % ThO_2. Irinite and loparite are nearly Ca-free varieties of perovskite which can contain up to 27 wt % REE and 13 wt % ThO_2.

Occurrence and Alteration

The occurrence of perovskite (except for Ca-poor varieties) is restricted to rocks which are undersaturated with silica (e.g. carbonatites, ultramafics, phonolites, basic pegmatites, metamorphosed limestones or kimberlites). Minerals associated with perovskite include nepheline, leucite ($KAlSi_2O_6$), forsterite (Mg_2SiO_4), ilmenite ($FeTiO_3$), calcite, biotite (a mica), apatite ($Ca_5(PO_4)_3F$, pyrochlore ($(Na,Ca)_2(Nb,Ta)_2O_7$), magnetite, pyrrhotite ($Fe_{1-x}S$), dolomite ($CaMgSi_2O_6$), spinel ($MgAl_2O_4$), titanite ($CaTiSiO_5$) and the pyroxenes, diopside ($CaMgSi_2O_6$) and augite ($(Ca,Na)(Mg,Fe,Al)(Si,Al)_2O_6$). The absence of perovskite in granites, nepheline syenites and most basalts is due to the thermodynamic instability of perovskite + saturated silicate assemblages. Titanite replaces perovskite when silicate minerals such as quartz, feldspars or enstatite ($MgSiO_3$) are present. Hydrothermal studies indicate that perovskite is reactive with alkali-feldspars at 700°C and 1 kbar. The association of perovskite with quartz, alkali-feldspars and enstatite is unstable at STP as well (6). Kinetics of these reactions should be slow at low temperatures.

Perovskite (variety loparite) has been reported as a placer mineral and occurrences older than 5×10^8 years are known. At high temperatures perovskite will alter to titanite ($CaTiSiO_5$) if it is exposed to solutions in contact with quartz, alkali feldspars or enstatite. Pyrochlore, "hydropyrochlore", REE oxides of the pyrochlore type, hydrated loparite, anatase (TiO_2) and monazite are alteration products of or pseudomorphs after REE-containing varieties of perovskite. Perovskite may alter to anatase (TiO_2) + ilmenite ($FeTiO_3$) (7) or leucoxene (an ill-defined mixture of titanates). Most of these alteration effects appear to be due to metasomatism (e.g. corrosion by hydrothermal solutions) rather than surface weathering. Natural perovskite with 0.2 wt % $(Th,U)O_2$ shows only minor radiation damage effects. Irinite with 13.0 wt % ThO_2 and loparite with 3.0 wt % ThO_2 are susceptible to metamictization (8), which results in decreased density, loss of X-ray diffraction pattern and glass-like physical properties.

ZIRCONOLITE

Zirconolite ($CaZrTi_2O_7$), is one of the three major SYNROC phases. The REE, actinides and perhaps Na may substitute for Ca and Zr. Small ions present in the waste stream may substitute for Ti. Terrestrial zirconolite with 8.3 wt % ThO_2 and 4.6 wt % U_3O_8 has been reported. Up to 17 wt % REE_2O_3 may be present. TiO_2 is known to vary between 18 and 32 wt %. Major amounts of Nb_2O_5 (up to 24 wt %) may substitute for TiO_2. Approximately 6 wt % ($FeO + Fe_2O_3$) is typical (9). Lunar zirconolites have more ZrO_2, similar amounts of REE and Ti, and less ThO_2 and U_3O_8 (10).

Occurrence and Alteration

Terrestrial zirconolite occurs as an accessory mineral in ultrabasic and alkalic rocks. Associated minerals include titanite, perovskite ($CaTiO_3$), baddeleyite (ZrO_2), zircon, pyrochlore, forsterite, magnetite, apatite ($Ca_5(PO_4)_3(OH,F)$), phlogopite (a mica), Ti-magnetite, nepheline, calcite and dolomite ($CaMg(CO_3)_2$) (11). That zirconolite may be unstable in the presence of quartz and alkali feldspars is suggested by its absence in granites, syenites. In these rocks zirconium silicate minerals occur in place of zirconolite. Kinetics of zirconolite + "silica" reactions should be sluggish at low temperatures or in the absence of H_2O. Alteration of zirconolite increases the contents of SiO_2 and H_2O (9). Zirconolite occurs as a durable mineral in placer deposits (12). Terrestrial zirconolite is usually metamict. The original structure is recovered when metamict zirconolite is annealed at 1100 - 1200°C.

DAVIDITE

Davidite is a complex titanite, $A_2B_5(O,OH)_{12}$, where A = U, REE, Fe^{+2}, Th, Ca, Mg, Mn, Na and B = Ti, Fe^{+3}, Cr, V and Al. According to Ringwood (1) it is structurally similar to the "BAIT" phase, $BaAl_2Fe_8Ti_{13}O_{38}$, which may occur in some SYNROC formulations. Ba and Al do not seem to occur in significant amounts in davidite but "BAIT" has several weight percent of these components. Fe and Ti contents appear to be similar. Natural davidite can contain up to 10 wt % REE_2O_3 and up to 10 wt % UO_2. Davidite usually contains less than 1 wt % ThO_2.

Occurrence and Alteration

Unlike perovskite and zirconolite, davidite occurs in granitic rocks. It appears to be a metasomatic mineral that can contain inclusions or intergrowths of quartz, hematite, rutile and ilmenite. Minerals associated with davidite include titanite, magnetite apatite, tourmaline, molybdenite (MoS_2), quartz, calcite, dolomite, Fe-sulfides and Cu-sulfides (4). In weathering zones, the Fe^{+2} and U^{+4} in davidite can be oxidized. Weathering of davidite results in formation of alteration rinds of carnotite (a uranyl-vanadate) (4). Davidite is known to occur in placer deposits. It is subject to radiation damage and metamictization. There are rare occurrences of nonmetamict davidite, usually with less than 1 wt % $(U,Th)O_2$ (13).

CONCLUSIONS

It is inappropriate to conclude that one radioactive waste form "mineral assemblage" is "better" than another unless it is with reference to a specific set of conditions. The compatibility of waste form and wall rock are of key importance in determining the long-term stability of any crystalline waste form at high temperatures. Unfortunately, much of the thermodynamic and kinetic data needed to make long term predictions are not yet available. Nonetheless, it is sometimes possible to predict whether a phase will be stable in a particular set of conditions. Without kinetic data it is not possible to predict whether or not thermodynamic instability will be of any significance. One can, from the geologic literature, only predict what alteration trends may be expected. The most easily predicted alteration trends result from oxidizing and high silica activity conditions. Considering only actinide phases and "BAIT", the UO_2 supercalcine phase is the most sensitive to E_h, as it can form soluble uranyl carbonate complexes at low tempertures under oxidizing conditions. Ferrous iron in the SYNROC "BAIT" phase, ($BaAl_2Fe_8Ti_{13}O_{38}$) which is related to davidite will probably be subject to oxidation if placed in a near surface environment. To help insure permanent fixation of uranium in either Supercalcine or SYNROC oxidizing conditions in the repository should be avoided. Generally the actinide phases in Supercalcine are stable in rocks with higher silica concentrations than the SYNROC actinide phases. Supercalcine monazite, uraninite and apatite should be stable in rocks ranging from granites (quartz-containing rocks) to nepheline syenites and certain basalts. The Supercalcine actinide phases may also be stable in low silica rocks. Compositional differences between SYNROC "BAIT" ($BaAl_2Fe_8Ti_{13}O_{38}$) and davidite ($Fe_6Ti_{15}O_{36}$) make the assumption of similar reactivity with SiO_2 and alkali feldspars unwise. "BAIT" and davidite are probably not very similar in their geochemistry. The occurrence of perovskite and terrestrial zirconolite appears to be restricted to low silica rocks which have no feldspar.

Kinetics are probably more important than thermodynamics but kinetic data are not yet available for wasteform-wallrock interactions. On the basis of thermodynamic considerations, the SYNROC actinide phases should be suitable for burial in quartz-free and feldspar-free rocks. In either case, oxidizing conditions should be avoided. If unfavorable waste-rock interactions are a problem, they could be reduced by using a suitable canister overpacking material, engineered barriers and waste with a lower heat content.

REFERENCES

1. A. E. Ringwood and S. Kesson, "Immobilization of High Level Nuclear Wastes in Ceramic Materials - A New Approach," Proceedings of International Symposium on Ceramics in Nuclear Waste Management, American Ceramic Society, Cincinnati, April 30-May 2 (1979) (in press).
2. G. J. McCarthy, "High-Level Waste Ceramics: Materials Considerations and Product Characterization," Nuclear Technology 32, 92 (1977).
3. A. A. Gluschchenko and A. F. Li, "Britholite from an Alkaline Massif in the Northern Baikal Area," Nauch. Tr. Irkutsk. Gos. Nauch.-Issled. Inst. Redk. Tvsvet. Met. 14, 100 (1966) (Russ) [Chemical Abstracts 67:119051r (1967)].
4. Clifford Frondel, "Systematic Mineralogy of Uranium and Thorium," Geological Survey Bulletin, 1064, (1958).

5. B. B. Jensen, "Distribution Patterns of Rare Earth Elements in Cerianite," Norsk Geologisk Tidsskrift 47, 1 (1967).
6. R. D. Schuiling and B. W. Vink, "Stability Relations of Some Titanium Minerals (Sphene, Perovskite, Rutile, Anatase)," Geochim. Cosmochim. Acta 31, 2399 (1967).
7. V. I. Lebedev and O. M. Rimskaya-Korsakova, "Ilmenitization of Perovskite," Doklady Akad. Nauk SSR 66, 257-60 (1949). [Chemical Abstracts 43, 7379 (1949)]
8. S. A. Gorzhevskaya and G. A. Sidorenko, "Loparite Alteration Products," Mineral. Sb. (Lvov) 23, 270 (1969) (Russ). [Chemical Abstracts 73, 89938Q (1970)]
9. K. A. Vlasov, ed., Mineralogy of Rare Elements, Vol. II, (Israel Program for Scientific Translations, Jerusalem 1966).
10. F. D. Busche, M. Prinz, K. Keil and G. Kurat, "Lunar Zirkelite: A Uranium Bearing Phase," Earth and Planetary Science Letters 14, 313 (1972).
11. A. G. Bulakh, G. A. Il'inskii and A. A. Kukharenko, "Zirkelite from deposits of the Kola Peninsula," Zapiski Vesesoyuz. Mineral. Obschchestva 89, 261 (1960) [Chemical Abstracts 55, 9167 (1961)].
12. G. S. Blake and G. F. Herbert Smith, "On Varieties of Zirkelite from Ceylon," Mineral. Mag. 16, 309 (1913).
13. L. Van Wambeke, "A Second Occurrence of Nonmetamict Davidite," Mineral Deposita (Berl.) 3, 178 (1968).

MONAZITE AND OTHER LANTHANIDE ORTHOPHOSPHATES AS ALTERNATE ACTINIDE WASTE FORMS

L. A. Boatner, G. W. Beall,* M. M. Abraham,
C. B. Finch,** P. G. Huray[+] and M. Rappaz[++]

Solid State Division
Oak Ridge National Laboratory[+++]
Oak Ridge, TN 37830

INTRODUCTION

The recent results of leach tests performed under hydrothermal conditions have shown that borosilicate glasses are physically and chemically unstable at elevated temperatures and pressures (1). These results emphasize the need for investigations of alternate high level waste forms which offer the potential of both increased resistance to leaching and stability under various possible metamorphic geological conditions. Although many of the previous (and present) studies of high level waste forms emphasize the containment of fission products, in fact, the dominant long term (i.e., after 300 to 500 year) radiation hazard is posed by the α-emitting actinides. Accordingly, the object of the present investigation is to evaluate an alternate means of actinide containment which is potentially superior to encapsulation in borosilicate glass. The mineral monazite, which is a mixed lanthanide-actinide orthophosphate [i.e., (Ce, La, Nd, Th, U)PO_4] of varying composition, is a crystalline material with established chemical and physical stability when subjected to natural geological conditions over extremely long time periods. Additionally, since monazite is a natural source of the actinides Th and U, it

*Chemistry Division, ORNL.
**Metals and Ceramics Division, ORNL.
[+]Guest Scientist from the University of Tennessee, Knoxville, TN.
[++]Guest Scientist from the Swiss National Fund for Scientific Research.
[+++]Operated by Union Carbide Corporation with the U.S. Department of Energy under Contract W-7405-eng-26.

has been subjected to radiation damage (including α-particle damage) during its natural lifetime, and the continued maintenance of its physical and chemical integrity is evidence of its intrinsic resistance to radiation. In view of these obviously desirable characteristics, a series of investigations of the physical and chemical properties of both natural monazite and the related, synthetically produced lanthanide orthophosphates has been undertaken. These investigations have included the usual applied studies such as leach testing under various conditions and the consolidation of precipitated orthophosphate powders by hot pressing as well as more fundamental determinations of the valence states and site symmetry of the actinide impurities. The details of this work are presented in the following discussion.

LANTHANIDE ORTHOPHOSPHATE CRYSTAL GROWTH

Single crystals of the lanthanide orthophosphates were grown in order to perform X-ray, electron paramagnetic resonance, Mössbauer, radiation damage, and chemical stability studies. In going through the lanthanide series, the orthophosphates from La to Gd are characterized by the monoclinic (monazite) structure while the orthophosphates from Dy to Lu have the tetragonal (zircon) structure. Terbium orthophosphate is dimorphic but crystallization at elevated temperatures produces only the tetragonal form. Orthophosphate crystals of every member of the lanthanide transition series (excepting $PmPO_4$) have been grown using the flux technique described by Feigelson (2). Crystals of YPO_4 and $ScPO_4$ (zircon structure) were also grown. Fig. 1a. shows several crystals of $LaPO_4$ doped with 0.5 wt% americium 241 oxide. These crystals are shown inside the growth crucible and are still entrained by the solidified $Pb_2P_2O_7$ flux which is removed by boiling in nitric acid. Natural monazite ores are known to contain relatively high concentrations of uranium (i.e., up to 15 wt% UO_2), and it was possible to grow single crystals with similar uranium concentrations. Crystals of $LaPO_4$ with 10 wt% $^{238}UO_2$ are shown in Fig. 1b. Additionally, plutonium-doped $LaPO_4$ crystals have been grown from a starting composition of 90 wt% La_2O_3 and 10 wt% $^{242}PuO_2$. The maximum crystal size obtained for the monoclinic orthophosphates was approximately 4 mm on an edge, but significantly larger crystals of the tetragonal structure materials were obtained. As an example, an exceptionally large crystal of $LuPO_4$ is shown in Fig. 1c.

LANTHANIDE ORTHOPHOSPHATE PRECIPITATION AND CONSOLIDATION

Although the lead pyrophosphate flux process noted above represents a reliable method for obtaining single crystal specimens for leach tests and more fundamental research investigations, it is not considered to be a practical process for the synthesis of actual waste material. The method of precipitation in molten urea as em-

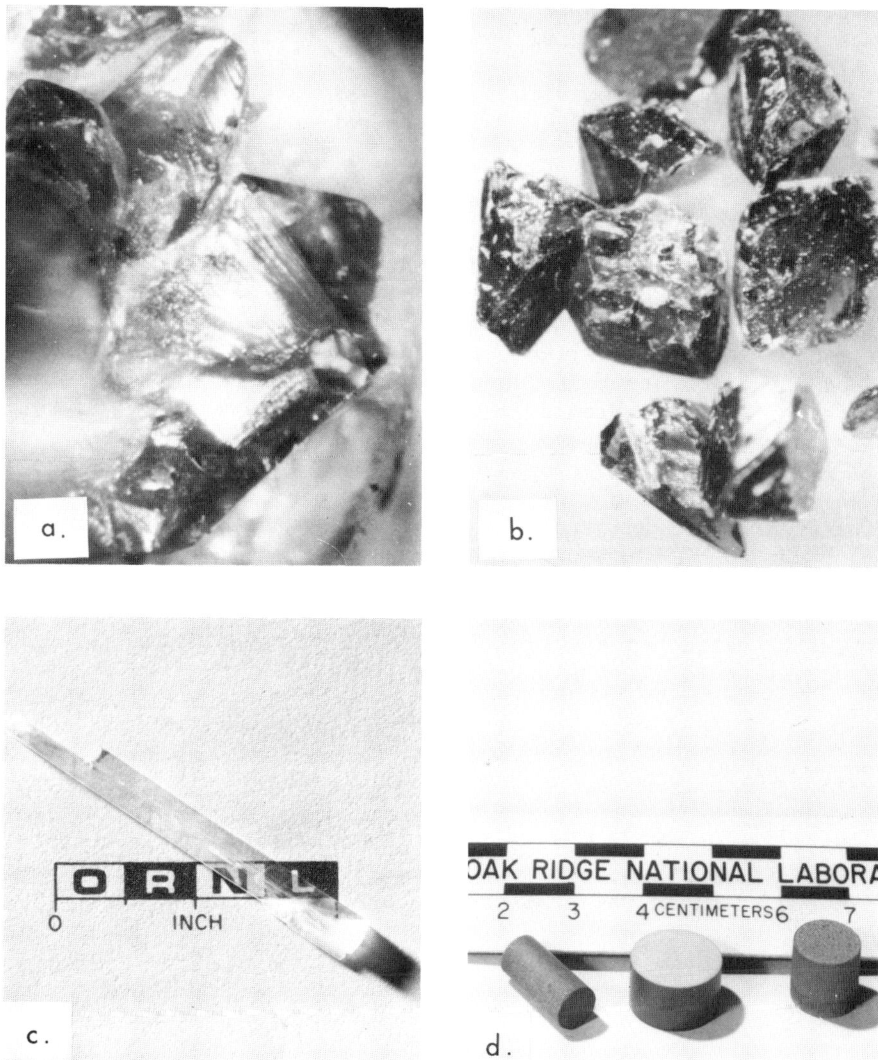

Fig. 1. a. Single Crystals of La (^{241}Am)PO_4 in the Pt Growth Crucible. Starting composition 99.5 wt% La_2O_3, 0.5 wt% Am_2O_3. b. Single crystals of La(^{238}U)PO_4. Starting composition 90 wt% La_2O_3, 10 wt% UO_2. c. Large single crystal of $LuPO_4$ grown with a cooling rate of 0.5°C/hr between 1360° and 900°C. d. Hot pressed pellets of $CePO_4$ formed from precipitated powders.

ployed by W. S. Aaron, T. C. Quinby and E. H. Kobisk in the formation of their cermet material does, however, represent a promising process for the practical synthesis of large quantities of actinide containing rare-earth orthophosphates (3). Figure 1d. shows several pellets formed by hot pressing $CePO_4$ powders. The phosphate particle size can be controlled in the urea precipitation process and consolidation via hot pressing of powders with a suitably small particle size has resulted in bodies with 98% of the theoretical density.

SINGLE CRYSTAL X-RAY STRUCTURE ANALYSIS

Earlier single crystal determinations of the monazite structure were conducted using natural materials and were not carried out to the precision obtainable using current crystallographic methods. A redetermination of the structure using pure $CePO_4$ single crystals yielded more accurate bond distances and indicated nine coordinations rather than the eight found in the earlier measurements. These crystals were monoclinic (space group $P2_1/n$ with lattice constants of a=6.777(1), b=6.993(1), c=6.445(1) and β=103.5°. Fig. 2. shows a stereo projection of the cerium and phosphorus coordination. Each cerium atom has eight oxygen atoms at a distance of approximately 2.5 Å with a ninth oxygen 2.7 Å away. The polyhedron is irregular and may reflect the ability of this site to accommodate such varied ions as the lanthanides, tetravalent uranium, thorium and plutonium, along with charge compensation ions. The phosphate group is a distorted tetrahedron that links the cerium atoms.

OPTICAL SPECTRA

The visible spectra for $LaPO_4$ single crystals doped with various actinides are shown in Fig. 3. $LaPO_4$ was selected as the host orthophosphate since it is totally transparent in the region studied. The ^{238}U-doped crystals whose absorption spectrum is shown at the top of

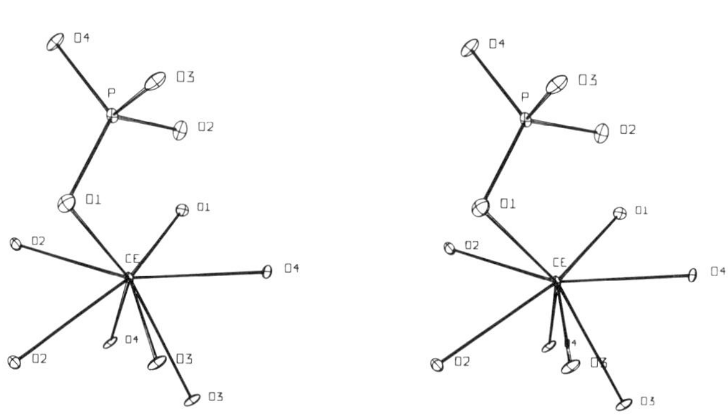

Fig. 2. Stereographic View of the $CePO_4$ Structure

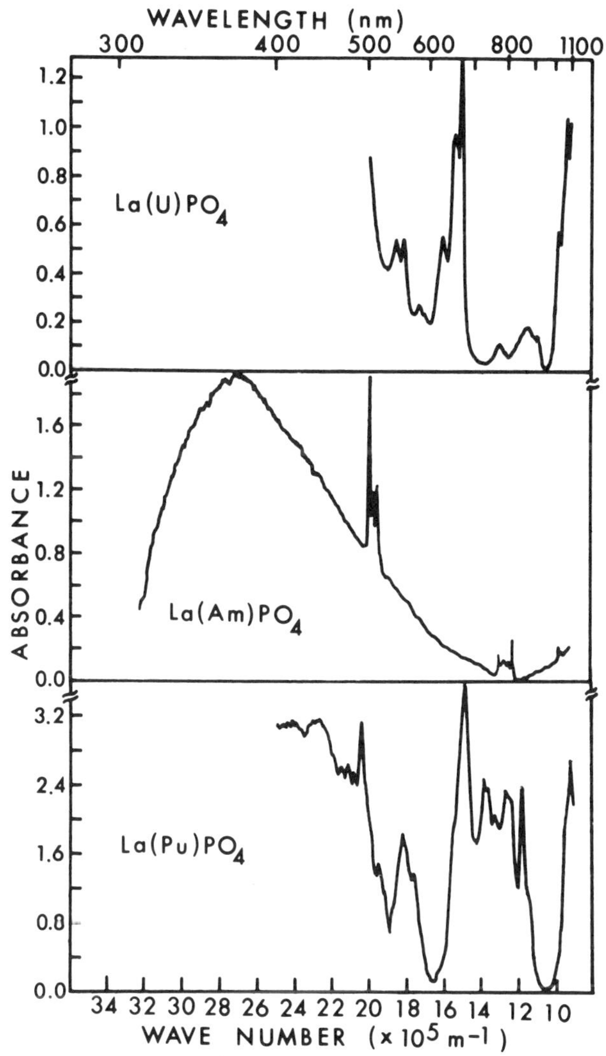

Fig. 3. Top: Optical Absorption Spectrum of $LaPO_4$ Grown with 10 wt% UO_2. Middle: Optical spectrum of $LaPO_4$ doped with 0.5 wt% Am_2O_3. Bottom: Optical absorption of $LaPO_4$ grown with the addition of 10 wt% $^{242}PuO_2$.

Fig. 3. contain 10 wt% uranium dioxide and have a deep green color indicative of tetravalent uranium. When the spectrum is examined in detail, the 4+ state is confirmed (4). Crystals of $CePO_4$ and (45 at% Sm, 45 at% Ce)(PO_4) have also been grown with a 10 at% uranium dopant. In both of these crystals, the ^{238}U tetravalent state was also confirmed. (The cerium in these crystals appears to be a mixture of Ce^{4+} and Ce^{3+}, with Ce^{3+} representing the predominant species.) These results are in agreement with observations of tetravalent uranium in natural monazite. The middle spectrum in Fig. 3. was obtained for $LaPO_4$ containing 0.5% americium 241. This spectrum shows that the americium is in the trivalent state, an observation that is consistent with the known redox chemistry of americium (5). These crystals have a pronounced amber color after cooling from the growth process and darken to a deep brown after several days at room temperature. This deepening color appears to be associated with radiation damage due to alpha decay and atom recoil from the americium 241. The color can be annealed out at approximately 500°C to produce perfectly clear material but returns in less than 24 hours at room temperature. The spectrum shown at the bottom of Fig. 3. was obtained for $LaPO_4$ containing 10 wt% plutonium dioxide. These crystals have a deep brown color which can be identified as due to tetravalent plutonium. The brown color darkens slightly following an anneal at 1000°C. This color change cannot be explained at this time.

MÖSSBAUER INVESTIGATIONS

Mössbauer effect studies were performed using La(Am)PO_4 as a source and NpO_2 as an absorber at 4.2 K. The observed Mössbauer spectrum (see Fig. 4.) shows that an absorption with a small splitting has occurred with an isomer shift in the neighborhood of +17 mm/sec and that a second absorption with a larger splitting has occurred with an isomer shift in the neighborhood of -35 mm/sec. The observed splitting is believed to result from a small electric field gradient (efg) for the absorption at 17 mm/sec (consistent with Np^{5+}) and a larger efg for the absorption at -35 mm/sec (consistent with Np^{3+}). The absorption at -35 mm/sec is probably a result of the recoiling ^{237}Np having found a normal $NpPO_4$ site prior to its recoilless γ de-excitation. The large 5+ Mössbauer absorption is somewhat surprising since the parent Am is trivalent. A permanent change in valence of this type following α-decay is very unusual and could have far-reaching implications concerning the leachability of the actinides.

ELECTRON PARAMAGNETIC RESONANCE STUDIES

The technique of electron paramagnetic resonance (EPR) was used to determine the local site symmetry of Gd^{3+} impurity ions in single crystals of the lanthanide orthophosphates, and these results were compared with the EPR spectra obtained from powders formed using the

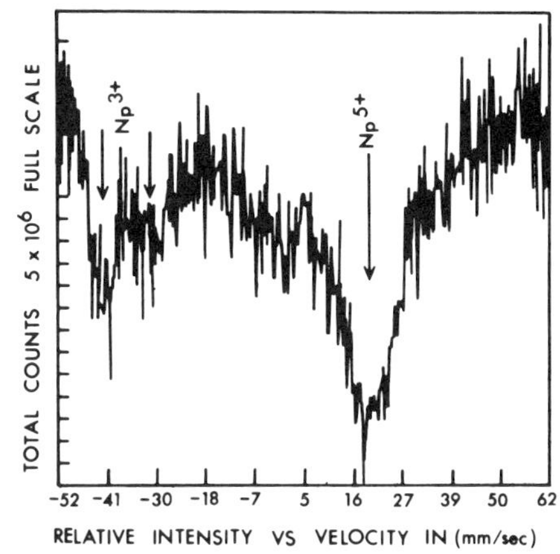

Fig. 4. Mössbauer Spectrum of the ^{237}Np Daughter of ^{241}Am in $LaPO_4$

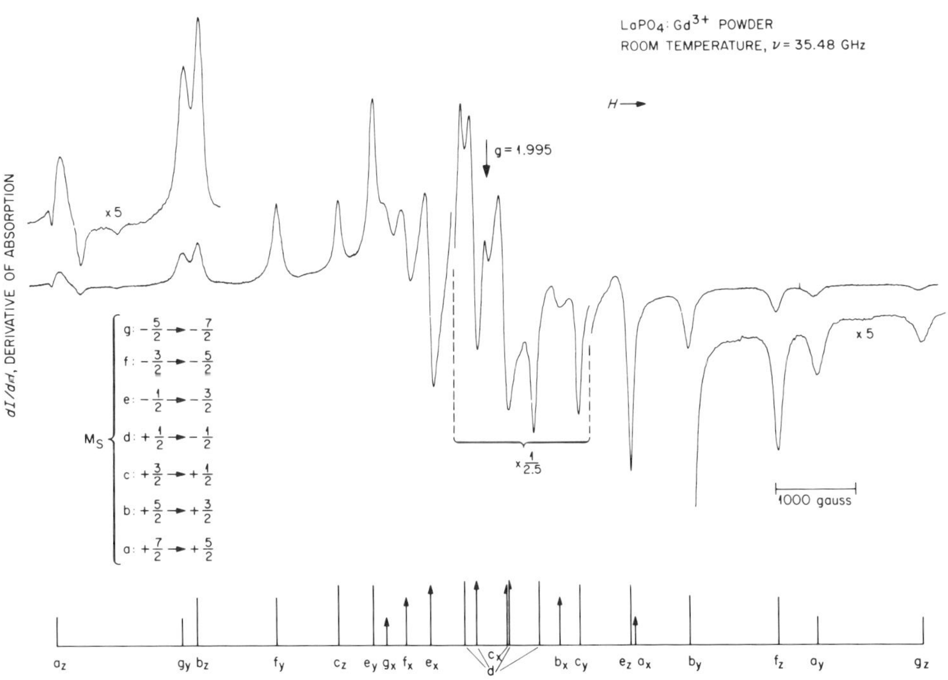

Fig. 5. EPR Powder Spectrum of Gd^{3+} Doped $LaPO_4$. Positions of the shoulders and divergences (arrows) are indicated by the scale.

urea process (3). An example of a Gd^{3+} powder spectrum is shown in Fig. 5. By comparing powder spectra such as this with the spectra obtained from single crystals, it was possible to show that the Gd^{3+} impurities occupy identical substitutional sites in both the single crystal and urea-precipitated powder specimens. Since the crystalline electric field splitting of the Gd^{3+} $^8S_{7/2}$ ground state can be used as a sensitive probe of the local environment, experiments are underway to exploit this sensitivity in investigations of α-particle damage in the lanthanide orthophosphates.

LEACHING STUDIES

The leachability of a waste form is one of the most important factors in determining its suitability for use in a nuclear waste repository. Leach tests of orthophosphate crystals resulted in IAEA leach indices that were smaller than those reported for grouts or borosilicate glasses by a factor of 16 for $CePO_4$ into 4 M NaCl at 200°C and 250 psi and a factor of 295 for $La(Am)PO_4$ into distilled water at 200°C and 250 psi (6). These results are even more striking since the leaching studies on the glasses and grouts were conducted at room temperature and atmospheric pressure.

REFERENCES

1. G. J. McCarthy, W. B. White, R. Roy, B. E. Scheetz, S. Komarneni, D. K. Smith and D. M. Roy, "Interactions Between Nuclear Waste and Surrounding Rock," Nature 273, 316 (1978).
2. R. S. Feigelson, "Synthesis and Single-Crystal Growth of Rare-Earth Orthophosphates," J. Am. Ceram. Soc. 47, 257 (1964).
3. W. S. Aaron, T. C. Quinby and E. H. Kobisk, "Cermet High Level Waste Forms," U.S. DOE Report ORNL/TM-6404, June (1978).
4. D. M. Gruen and R. L. McBeth, "Oxidation States in Complex Ions of Uranium in Fused Chlorides and Nitrates," J. Inorg. Nucl. Chem. 9, 290 (1959).
5. C. B. Finch and G. W. Clark, "Gamma Irradiation Effects in Ferroelectric KDP," J. Phys. Chem. Solids 34, 921 (1973).
6. J. G. Moore, H. W. Godbee and A. H. Kibbey, "Leach Behavior of Hydrofracture Grout Incorporating Radioactive Wastes," Nucl. Tech. 32, 39 (1977).

CRYSTAL CHEMISTRY AND PHASE RELATIONS IN THE SYNTHETIC MINERALS OF CERAMIC WASTE FORMS: I. FLUORITE AND MONAZITE STRUCTURE PHASES

Gregory J. McCarthy, John G. Pepin[+] and Dennis D. Davis[+]

Departments of Chemistry and Geology, North Dakota State University, Fargo, ND 58105; [+]Materials Research Laboratory, The Pennsylvania State University, University Park, PA 16802.

INTRODUCTION

McCarthy (1-4) has suggested that synthetic minerals of the fluorite and monazite types are potentially ideal hosts for the lanthanide*-actinide** elements in nuclear wastes. He has synthesized numerous compositions of these structure types relevant to nuclear waste chemistries (1-5). Their geologic stability, radiation stability and high solid solution capacity for actinides have been previously discussed (1-10). The Ln and An elements are very important constituents of high level nuclear wastes (HLW) and transuranic wastes (TRU), and thus are receiving high priority consideration in the design of ceramic nuclear waste forms. The oxides of the Ln's and An's can constitute more than 50 weight percent (wt%) of some HLW formulations. Also included among the actinides are long-lived hazardous radionuclides such as Np-237, Pu-238 and Am 241. We discuss here crystal chemistry and recently determined phase relations data on synthetic fluorites and monazites in the context of designing synthetic minerals for HLW ceramics.

FLUORITE PHASES

Crystal Chemistry

The prototype mineral for this structure type is fluorite, CaF_2. Several oxide minerals crystallize in this structure type. These are

*Ln = La, Ce, Pr, Nd, Pm, Sm, Eu, Gd, Tb, Dy.
**An = Ac, Th, Pa, U, Pu, Am, Cm.

listed in Table 1 along with their mineral names, ionic radii and unit cell parameters. In these fluorite structure oxides, metal ions are coordinated by eight oxygens in an approximately 5.5Å unit cell having four formula units per cell and space group Fm3m.

Zirconium dioxide, ZrO_2, has been included in Table 1 because it is also a major constituent in commercial HLW and has a fluorite structure when heated to a very high temperature. The size of the Zr^{4+} ion (0.84 Å)* is somewhat small for regular 8-coordination, so the cubic fluorite structure distorts to a tetragonal form at intermediate temperatures and a monoclinic form at low temperature. The mineral baddeleyite occurs in this monoclinic form. Incorporation of larger ions such as Ca^{2+} (1.12Å) or Y^{3+} (1.02Å) into solid solution with ZrO_2 stabilizes the cubic phase.

The trivalent oxides of the lanthanides (Ln_2O_3), some of which are abundant in commercial HLW, crystallize in one of three fluorite-related structure types and form solid solution with the dioxides of Ce, Th and U. Because of their valence, solid solution of Ln^{3+} in MO_2 is accompanied by oxygen defects according to the model $(M_{1-2x}Ln_{2x})O_{2-x}\square_x$.

Another type of solid solution mechanism is observed when UO_2 is oxidized. Conversion of a portion of the U^{4+} to U^{6+} leads to incorporation of interstitial oxygen (O_i) into the fluorite structure according to the model $(U^{4+}_{1-x}\ U^{6+}_x)O_2O_{x}^{i}$. The limiting value of x for the fluorite structure is $x \sim 0.25$ ($UO_{2.25} = U_4O_9$). The stable oxide of uranium at typical ceramic processing temperatures (800°-1400°C) is $U_3O_8(UO_2{\cdot}2UO_3)$ in which 2/3 of the U is hexavalent. As will be shown below, solid solution with other fluorite structure oxides can stabilize the cubic fluorite structure in high uranium content phases during crystallization of HLW ceramics in air.

The solid solution range among oxides is greatest when the size of cations is close (within $\sim$15% according to Reference 12) and the charges of the cations differ by no more than ± 1. Thus, extensive solid solution is to be expected among CeO_2, ThO_2 and UO_2. Trivalent lanthanide radii in 8-coordination vary from 1.16Å for La to 1.03Å for Dy. Thus, the extent of solid solution of Ln_2O_3 in CeO_2, ThO_2 and UO_2 should increase as the Ln^{3+} ion becomes smaller. These predicted trends are observed in the phase behavior and crystal chemistry of current supercalcine ceramics (9, 10).

---

*Ionic radii for 8-fold coordination after Reference 11.

Table 1. Fluorite Structure Minerals

MO_2 Oxide	Mineral Name	M^{4+} Radius[a]	Cubic Cell Parameter
UO_2	Uraninite	1.00	5.468
CeO_2	Cerianite	0.97	5.411
ThO_2	Thorianite	1.05	5.600
ZrO_2	(b)	0.84	5.07 [c]

[a]Ionic radius in 8-fold coordination after Reference 11.
[b]Mineral name for the monoclinic phase of ZrO_2 is Badelleyite.
[c]Stable only above 2400°C; cell parameter approximate.

Phase Relations in the System CeO_2-ThO_2-UO_{2+x}-ZrO_2

Phase relations in the ternary or pseudoternary subsystems of this system for a firing temperature of 1200°C have been investigated. Additional systems with the lanthanide oxides Nd_2O_3 and Gd_2O_3 added are currently being studied. A summary of results to date is presented here. Detailed discussions of the experimental procedures, data reduction and interpretations will be published elsewhere.

Phase relations in the four subsystems are shown in Fig. 1-4. They were determined by firing large numbers of compositions from each subsystem in air at 1200°C for one to two weeks. The compositions were pelletized oxide mixtures that had been prepared by drying and then denitrating solutions of the metal nitrates. Phase identification was by X-ray powder diffraction. The presence of a steady state phase assemblage was ascertained by monitoring of the phase composition, cubic fluorite structure solid solution (F_{ss}) cell parameter and crystallinity by X-ray diffraction of sequentially fired specimens. An internal standard (Er_2O_3 calibrated to NBS-SRM-640 silicon) was used in all of the cell parameter determinations. Phase boundaries were fixed by plots of these F_{ss} cell parameters and are expected to be accurate within ±2 mole %. Lines or curves of "iso-cell parameter" (cell parameter isopleths) are plotted in those one and two phase regions involving F_{ss}.

After being fired at 1200°C, the specimens were removed from the furnace and quenched to below "red heat" in 2-3 seconds. During the cooling, the tetragonal ZrO_2 phase, stable at 1200°C, converts to monoclinic ZrO_2. There is also the possibility of oxygen diffusion into the sample on cooling and perhaps some exsolution from the F_{ss} phase at composition near the limit of F_{ss} solid solution. Strictly speaking, then, the diagrams represent phase relations observed at room temperature for compositions that had been fired at 1200°C in air rather than 1200°C isotherms.

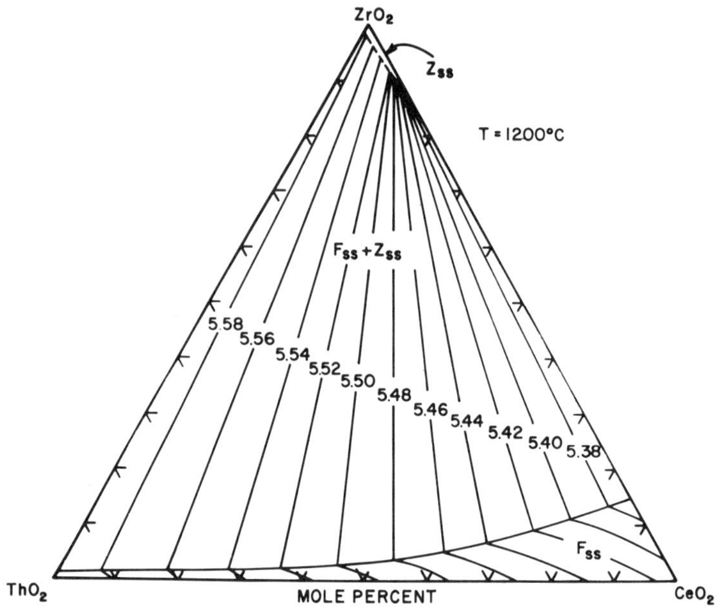

Fig. 1. Phase Relations in the System ThO_2 - CeO_2 - ZrO_2 for a Firing Temperature of 1200°C. Note the change in cell parameter with variations in composition in the two phase region.

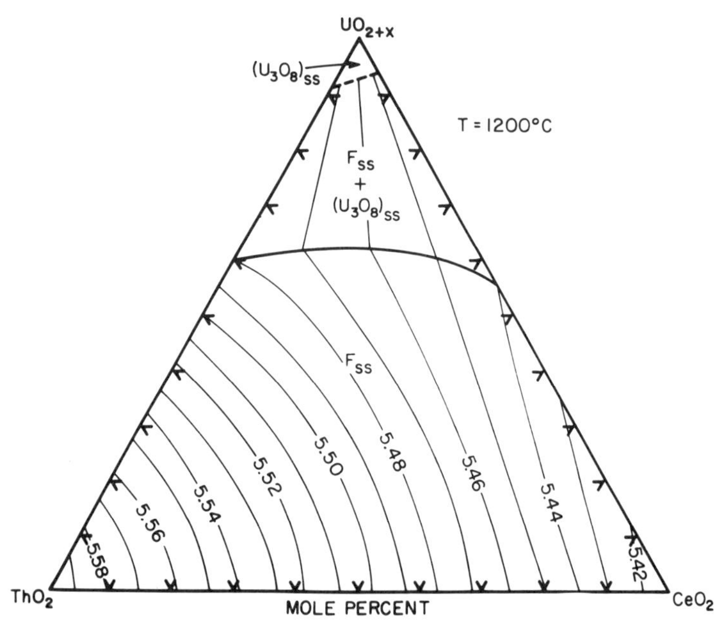

Fig. 2. Phase Relations in the System CeO_2 - ThO_2 - UO_{2+x} for 1200°C Firings in Air

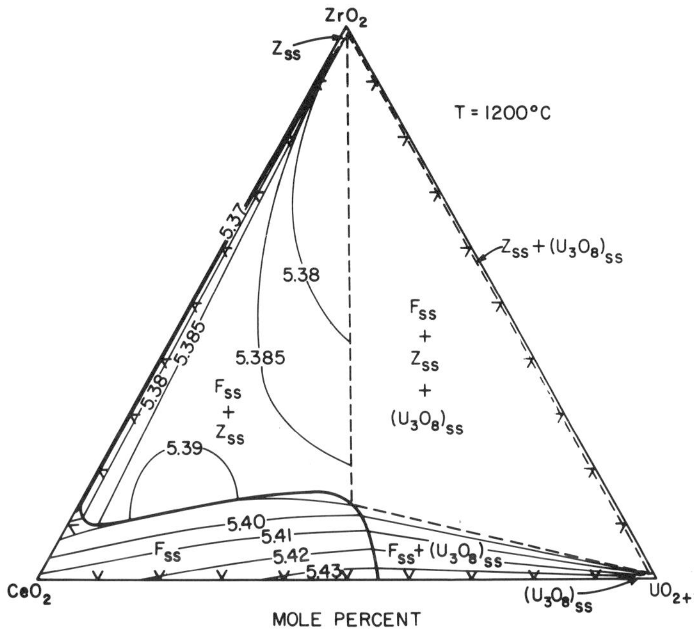

Fig. 3. Phase Relations in the System CeO_2 - UO_{2+x} - ZrO_2 for 1200°C Firings in Air

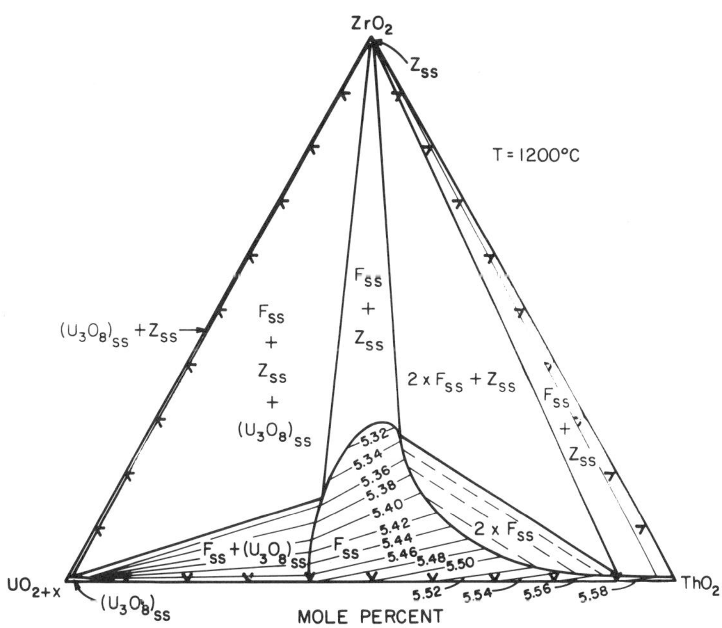

Fig. 4. Phase Relations in the System ThO_2 - UO_{2+x} - ZrO_2 for 1200°C Firings in Air

The systems containing U oxide are not true ternaries because the compositions of all phases cannot be expressed in terms of a unique U oxide component. Near the U oxide apex, U_3O_8 is the appropriate U oxide component, but in the U-poor F_{ss} phase fields, the appropriate U oxide component has a composition closer to $UO_{2.0}$. The fact that the system was open to air and contained a variable valence oxide makes this a "pseudoternary" system. The variability of the oxygen concentration as a function of composition results in some unusual looking tie-"lines" in the 2-phase $F_{ss} + Z_{ss}$ region of Fig. 3. Tie-lines in a true ternary system, such as CeO_2-ThO_2-ZrO_2, must be straight.

Only three phases occur in these diagrams, F_{ss}, zirconia solid solution (Z_{ss}), and U_3O_8 solid solution [$(U_3O_8)_{ss}$]. The extent of solid solution in the F_{ss} phase is consistent with the crystal chemical preditions. There is complete miscibility of CeO_2 and ThO_2 (Fig. 1., 2. and 4.), but there is only a small amount of miscibility between Z_{ss} and F_{ss}. There is somewhat more miscibility between Z_{ss} and CeO_2 than Z_{ss} and ThO_2 because the radii of Zr^{4+} and Ce^{4+} are closer than those of Zr^{4+} and Th^{4+}. There is also very little miscibility between $(U_3O_8)_{ss}$ and Z_{ss} (Fig. 3. and 4.).

In Fig. 2. it is shown that the cubic CeO_2-ThO_2 F_{ss} can incorporate up to 60 mole % of U oxide without formation of $(U_3O_8)_{ss}$. In Fig. 3. it is seen that the F_{ss} phase formed by solid solution of U oxide in CeO_2 can further incorporate 10-15 mole % of ZrO_2. Finally, the phase relations in Fig. 4. show that as much as 30 mole % of ZrO_2 is miscible in the UO_{2+x}-ThO_2 F_{ss} phase when the Th to U ratio is near unity.

MONAZITE

Crystal Chemistry

Monazite is a solid solution mineral based on the larger lanthanide (La→Gd) orthophosphates, $LnPO_4$ (Table 2). In nature and in synthetic phase syntheses it has been shown that this structure type can accomodate a wide range of solution substitutions accompanied by charge compensation (4, 13). This is an important consideration in tailoring of synthetic mineral ceramics. Monazite is a common host for Th in nature and is its primary ore. Uranium occurs in lower concentrations than Th in natural specimens, but some monazites are known to contain up to 16 wt% UO_2 (4). The greater size difference between U^{4+} and the larger Ln°s compared to Th^{4+} and these Ln's is one explanation for the lower U concentrations in monazites.

Monazite is a sedimentary resistate, as shown by its typical occurrences in stream beds, beach sands and clastic sedimentary rocks. Such minerals are prime candidates for nuclear waste hosts (1, 2, 4,

14-17). Also, Ewing (7, 8) has pointed out that monazites rarely, if ever, occur in the metamict state in spite of high concentrations of radioactive elements.

Monazite is an ABO_4 structure type in which AO_8 polyhedra and BO_4 tetrahedra form a dense, space filling network (13). The type composition, $CePO_4$, has an $A^{3+}B^{5+}$ charge model. In order to accomodate tetravalent actinides, charge balance must be made by other substitutions such as Ca^{2+} for A^{3+} or Si^{4+} for P^{5+}. Both of these are observed in nature. Cheralite is a monazite structure solid solution between $CePO_4$ and $(Ca_{0.5}Th_{0.5})PO_4$. Huttonite is $ThSiO_4$ in which there has been complete replacement of Ce^{3+}/P^{5+} by Th^{4+}/Si^{4+}. Coffinite ($USiO_4$) and thorite ($ThSiO_4$) are two zircon structure minerals that could show substantial miscibility in monazite.

Nature, then, provides ample evidence for extensive solid solution substitutions in monazites. However, our ability to duplicate nature with ceramic preparation techniques and the phase relations in relevant compositions under these synthesis conditions is in need of investigation.

Phase Relations

A study of the phase relations among $NdPO_4$* (monazite), $(Ca_{0.5}Th_{0.5})PO_4$ (cheralite), $(Ca_{0.5}U_{0.5})PO_4$, $ThSiO_4$, (thorite/huttonite) and $USiO_4$ (coffinite) was recently initiated. Starting mixtures are prepared by drying and decomposing nitrate-phosphate-silica sol solutions, precipitates or gels to form a reactive oxide mixture. Synthesis temperatures range from 800° to 1200°C. Routine pressureless sintering in air and hot pressing are being employed.

Table 2. Selected Monazite Structure Phases[a]

$A^{3+}B^{5+}O_4$	$A^{2+}B^{6+}O_4$	$A^{2+}_{0.5}A^{4+}_{0.5}B^{5+}O_4$
$LnPO_4$ (Ln=La→Tb) (Monazite)	$PbCrO_4$ (Crocoite)	$(Ca_{0.5}Th_{0.5})PO_4$ (Cheralite)
$LnAsO_4$ (Ln=La→Nd)	$SrSeO_4$	$(A_{0.5}Th_{0.5})PO_4$ (A=Cd, Sr, Pb)
$LaVO_4$	$SrCrO_4$	
$LaCrO_4$	$PbSeO_4$	
$PuPO_4$		$(A_{0.5}Th_{0.5})AsO_4$ (A=Ca, Cd, Sr, Pb)
$PuAsO_4$	**$A^{4+}B^{4+}O_4$**	
$AmPO_4$	$ThSiO_4$ (Huttonite)	
$AmAsO_4$		
	$PaSiO_4$	

[a]Data from Reference 13.

To date, it has been found that there is complete solid solution between $NdPO_4$ and $(Ca_{0.5}Th_{0.5})PO_4$ and thus, that the natural monazite-cheralite series can be duplicated in the laboratory. This solid solution series will accept approximately 20 mole % of $(Ca_{0.5}U_{0.5})PO_4$. When the concentration of $(Ca_{0.5}U_{0.5})PO_4$ is increased beyond 20 mole %, a small amount of a second, and as yet unidentified, phase appears. It is suspected that this is a metastable phase and that with different synthesis conditions, complete solid solution among these three end members will be found.

APPLICATIONS TO NUCLEAR WASTE CERAMICS

Purex Waste Supercalcine Ceramics

A class of supercalcine ceramics was developed during the mid-1970's for typical Purex process reprocessing wastes by McCarthy et al. (3, 5, 9, 18). In the Purex waste PW-7 (18), the molar ratio of tetravalent elements is approximately 1 Ce:2U:2Zr. This composition falls within the three phase region, $[F_{ss} + Z_{ss} + U_3O_8]$ in Fig. 3. The F_{ss} and Z_{ss} were found to be major constituents of these ceramics, but U_3O_8 was definitely not present in any measurable (by X-ray diffraction) amount. Thus, there must be other constituents that promote U incorporation in F_{ss} (Z_{ss} shows very little solubility for U) rather than U_3O_8. The Ln's are the logical choice for these constituents based on crystal chemistry, and they have been observed to be major solid solution components of F_{ss} by electron microprobe and scanning transmission electron microscopy (9, 10).

It should be noted that while supercalcine ceramics are crystallized for typically two hours at 1200°C, the phase relations are determined after two weeks of firing. Tests of selected compositions for phase formation after only two hours indicated that the same phase assemblage is observed, but the F_{ss} cell parameter may differ by a few parts per thousand from the steady state values. Thus, comparison of phase relations in these dioxide systems with phase behavior in these ceramics is appropriate.

Thorex Waste Supercalcine Ceramics

One typical formulation of reprocessing waste from a Th-U cycle fuel (TW-3) has the tetravalent elements present in approximately the ratio 5Ce:10Th:1U:45Zr (19). If the small amount of U is omitted, the dioxide phase relations after crystallization at 1200°C are represented by Fig. 1. The diagram predicts that Ce+Th+Zr oxides would

---

*$NdPO_4$ was selected over $CePO_4$ because Nd^{3+} is a more appropriate simulation of the Ln's in HLW (18).

react to form Z_{ss} plus an F_{ss} with a cell parameter near 5.55Å. Based on the observed phased behavior in Purex waste supercalcine ceramics and on Fig. 2., the small amount of U should be incorporated into the CeO_2-ThO_2 F_{ss} along with some Ln oxide.

Synthetic Mineral Ceramics for Defense Waste Sludges

Defense waste sludges can contain substantial U concentrations. For example, Savannah River Plant sludges have been found to contain from 4-10 wt% U_3O_8 (20). If air firing is used to form a ceramic, the stable phase of U would be U_3O_8. It is well known that uranyl (U^{6+}) containing oxides are 10^2 to 10^3 more soluble than stoichiometric UO_2 (21). Under reducing conditions, the UO_2 phase could be formed, but there may be advantages in air atmosphere processing. Thus, it would be desirable to modify the U phase behavior to provide for stabilization as U^{4+} in some synthetic mineral.

Fig. 3. indicates that the addition of equivalent Ce to the available U will promote formation of an F_{ss}. While the exact valence of U in Ce-U F_{ss} phases is not presently known (such studies have been initiated), it should be closer to U^{4+} than is the case with U_3O_8.

ACKNOWLEDGMENTS

This research was supported in its initial stages by the Department of Energy (DOE) through the Battelle Pacific Northwest Laboratory and is currently supported by the DOE through the Rockwell Energy Systems Group. J. G. Pepin and D. D. Davis acknowledge fellowship support from the U.S. Department of Education.

REFERENCES

1. G. J. McCarthy, Solid State Chemical Aspects of Radioactive Waste Disposal, Final Report, NSF Contract GK-27781, 32, The Pennsylvania State University, (1973).
2. G. J. McCarthy, "Atomic Level Isolation," Quarterly Progress Report, Waste Fixation Program, Jan.-March, DOE Report BNWL-1826, Pacific Northwest Laboratory, 43 (J. L. McElroy, ed., 1974).
3. G. J. McCarthy, Nucl. Technol. 32, 92-104 (1977).
4. G. J. McCarthy, W. B. White and D. E. Pfoertsch, Mat. Res. Bull. 13, 1239-1245 (1978).
5. G. J. McCarthy and M.T. Davidson, Am. Ceram. Soc. Bull. 54, 782-786 (1975).
6. G. J. McCarthy, Management of Commercially Generated Radioactive Waste, Sect. 3.2 and Appendix P, DOE-EIS-0046-D, April (1979).

7. R. C. Ewing, High Level Radioactive Solid Waste Forms, NUREG/CP-0005, 623-649, Dec. (1978).
8. R. C. Ewing, Ceramics in Nuclear Waste Management, CONF-790420, 305-309, DOE Tech. Info. Cr. (1979).
9. G. J. McCarthy, J. G. Pepin, D. E. Pfoertsch and D. R. Clarke, Ceramics in Nuclear Waste Management, CONF-790420, 315-320, DOE Tech. Info. Cr. (1979).
10. G. J. McCarthy, J. G. Pepin and D. E. Clarke, The Rare Earths in Modern Science and Technology, 2, (G. J. McCarthy, ed., Plenum Press, New York, in press).
11. R. D. Shannon, Acta Cryst A32, 751 (1976).
12. V. M. Goldschmidt, J. Chem. Soc., 655-672 (1937).
13. O. Muller and R. Roy, The Major Ternary Structural Families, (Springer-Verlag, New York, 1974).
14. R. E. Isaacson and L. E. Brownell, Management of Radioactive Wastes from Fuel Reprocessing, OECD Proceedings, Paris, 953-986 (1972).
15. S. Forberg and T. Westermark, Scientific Basis for Nuclear Waste Management, 1, 201-206 (G. J. McCarthy, ed., Plenum Press, New York, 1979).
16. A. E. Ringwood, Nature 278, 219-223 (1979).
17. R. Roy, High Level Radioactive Solid Waste Forms, NUREG/CP-0005, 353-356, Dec. (1978).
18. J. M. Rusin, Multibarrier Waste Forms, Part I: Development, DOE Report PNL-2668-1, Pacific Northwest Laboratory (1978).
19. J. M. Rusin (Pacific Northwest Lab.), private communication.
20. G. J. McCarthy, Nucl. Techn. 44, 451-452 (1979).
21. D. Langmuir, Geochim. Cosmochim. Acta. 42, 545-569 (1978).

CHARACTERIZATION OF IMPACT FRACTURE OF BRITTLE SOLID WASTE FORMS

W. J. Mecham, L. J. Jardine, R. H. Pelto and M. J. Steindler

Argonne National Laboratory
9700 S. Cass Avenue
Argonne, IL 60439

INTRODUCTION

The fracture of solid waste forms resulting from mechanical impacts that could occur during in-plant handling and storage and during off-site transportation needs to be known to evaluate the increased risks of potential dispersion of radioactive wastes. Two physical properties of the waste form which may be useful to characterize the increased dispersion risks from fracture are the increase in surface area and the (mass) fraction of particles less than the respirable size (i.e., $< 10\,\mu m$) for the waste form. An analysis and evaluation of the effects of impacts have been initiated for metal-encapsulated reference glass and ceramic waste forms. Descriptive and calculational models were developed for relating impact energy to deformation effects, both for brittle and ductile materials and for characterizing the surface area resulting from impact fracture of brittle materials (1).

This paper presents a methodology for analyzing particle size distributions obtained in impact testing of brittle waste form materials. The methodology includes (a) a linear two-parameter lognormal correlation of the weight fraction less than any given size, (b) a mathematical function of the two lognormal parameters to determine the total surface area in terms of a dimensionless shape factor, and (c) a surface-energy constant to predict the anticipated increase of surface area from the known energy absorbed in the impact of the brittle material. Preliminary measurements were made on impacted simulated glass specimens. The analysis is used as an example of the methodology. These results are also compared with reanalyzed impact data reported by others.

DEFORMATION OF BRITTLE MATERIALS BY MECHANICAL IMPACT

Brittle fracture of typical glass or ceramic solids under an impact results in fragmentation and formation of new surface area (ΔS_F). Fracture energy (W_F) absorbed in the brittle material is proportional to ΔS_F over a considerable range of energy input for a given specimen and mode of impact. The proportionality constant is termed (1) the fracture surface energy, γ_F. Thus,

$$W_F = \gamma_F \ \Delta S_F \qquad 1)$$

In equation 1), W_F is generally smaller than the gross impact-energy input (W_I) due to deformation in the test equipment. Even for "bare" glass specimens impacted by hard steel surfaces, deformation of the ductile metal typically absorbs more than 50% of the gross impact energy (W_I). Variations of γ_F can be expected for different materials, sizes and shapes, and for variations of input energy concentrations in specimen volumes (V_o).

In the work on brittle fracture of Piret and associates (2, 3, 4), both single impacts and slow compression crushing were applied to small (~ 1.5 cm^3) solid specimens (spheres and cylinders) of Pyrex glass and of prisms of natural quartz at input energy densities up to 17 J/cm^3. Multiple impacts and slow compression crushing were applied to particulate samples (10-14 mesh) of natural quartz. The principal result was a measurement of the net energy absorbed (W_F) in the brittle materials in producing the new surface area, as measured by a gas adsorption technique. Calorimetric methods were used to establish the net energy absorbed. Results were presented as plots of the net energy density (W_F/V_o) applied in impacting specimens versus the net fracture-surface energy, γ_F. A representative selection of the total number of data points is given in Fig. 1. There is a great deal of scatter in the slow compression data for a given test material, e.g., 10-fold variations of γ_F (4 to 80 J/m^2) for the range of values of W_F/V_o. However, γ_F is nearly constant (~ 100 J/m^2) for impact fracture while W_F/V_o varied over two orders of magnitude. This implies that equation 1) may be used to estimate ΔS_F for an impact input energy W_F, since γ_F is reasonably constant at most energy densities.

THE LOGNORMAL PARTICLE SIZE DISTRIBUTION

It is shown below that the particulate solids formed in brittle fracture exhibit a particle size distribution that can be described by the lognormal probability function. (Details as to why this is the fundamental relationship to be expected based on the physical laws will be given in a separate paper) (5). This function has two parameters obtainable from graphical analysis, namely the geometric mean diameter D_g and the geometric standard deviation σ_g. D_g is the

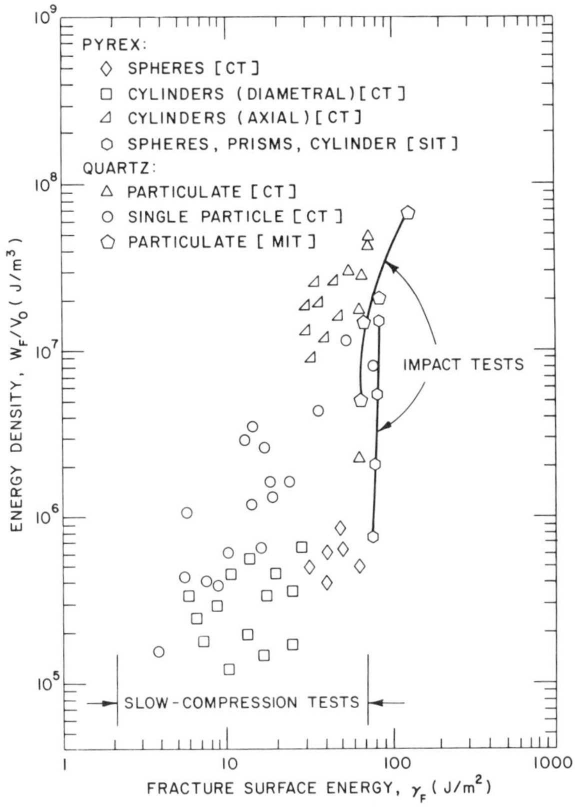

Fig. 1. Results of measurements (2, 3, 4) of the fracture surface energy, γ_F, of pyrex and quartz at various specimen net impact energy densitites for compression tests [CT], single impact tests [SIT] and multiple impact tests [MIT].

diameter at the 50% mass fraction, and σ_g is the ratio of the diameters plotted at 50% and 16% mass fractions. The ratio (S_F/V_o) of total particulate surface area to solid volume can be expressed (6) as a particulate "equation of state" in terms of the lognormal parameters as

$$\frac{S_F}{V_o} = \frac{\alpha \ \sigma_g^{0.5 \ln \sigma_g}}{D_g} \tag{2}$$

where α is a dimensionless shape factor. The minimum value of α is 6 (for spheres and cubes). For brittle fragments of other shapes, α is typically 2 to 3 times higher. The total particulate surface area

S_F is related to the original specimen surface area S_o and to the surface area formed in fracture, ΔS_F, as

$$\Delta S_F = S_F - S_o. \tag{3}$$

The relations expressed by the three above equations provide quantitative parameters which can serve as a basis for systematic test programs for characterizing the results of impact on various composite bodies of brittle and ductile materials, such as canistered glass waste forms or metal encapsulated composites.

APPLICATION OF THE ENERGY AND SIZE RELATIONSHIPS TO IMPACT TESTING

Two 5.3 g samples consisting of 19 Pyrex spheres (6.2 mm) were each impacted by a drop weight device at two energy densities 6.9 x 10^6 and 3.3 x 10^6 J/m^3, based on gross impact energy input. The net energy input was not measured. Size distributions were measured by sieve analyses and are labeled ANL-1 and ANL-2 in Table 1. The fracture surface area was also measured for one test (ANL-2) by (Kr) gas BET adsorption. The cumulative lognormal plot of these particle size data are given in Fig. 2. The linearity of the plots is important for estimating the cumulative mass fraction of respirable sizes (< 10 μm). The lognormal parameters from the graphical analysis and the calculated fracture-surface energies are also shown in Table 1. The fracture surface area for test ANL-1 was calculated from the lognormal surface/volume ratio (S_F/V_o in Table 1) with the value of the shape factor $\alpha = 12$ as determined from test ANL-1.

For comparison, also reported in Table 1 and Fig. 2. are data selected from the reported literature on the impact fracture of brittle materials representative of solid waste forms. Impact tests on two materials at two different energy densities made at the Savannah River Laboratory (SRL) (7) were: (1) a simulated borosilicate waste glass as a cylindrical specimen 0.50 inch diameter and 1.0 inch long and (2) a concrete cylinder consisting of "neat" high alumina cement 1.0 inch diameter and 0.50 inch long. Two tests made at Battelle Pacific Northwest Laboratory (PNL) (8) used a reference commercial soda-lime silica glass cylinder 0.44 inch diameter and 0.50 inch long. Engineering studies of impact effects on fuel capsules (for aerospace applications) were made at Monsanto's Mound Laboratory (ML) (9); two impact tests of PuO_2 microspheres sealed into a stainless steel (6 1/2" long x 7/8" O.D. with 3/8" walls) capsule were reanalyzed. Only data from the glass specimens are reported in Table 1. However, all particle size data, when replotted lognormally as in Fig. 2., formed straight lines and all values of σ_g (from graphical analysis) were in the range shown for the glass specimens.

The analysis of the above data shows that for a given material, σ_g is nearly constant over a range of energy densities. The D_g

Table 1. Selected Test Data on Impact Fracture of Brittle Specimens

KEY	⊙	⊙	⊡	⊡	◈	◈
INDEX	ANL-1	ANL-2	SRL-1	SRL-2	PNL-1	PNL-2
MATERIAL	Pyrex Beads	Pyrex Beads	B-Glass Cyl	B-Glass Cyl	SL Glass Cyl	SL Glass Cyl
VOLUME V_o (cm^3)	2.38	2.38	3.33	3.33	1.25	1.25
.SURFACE S_o (cm^2)	23	23	12.7	12.7	6.4	6.4
SURFACE S_F (cm^2)	424 (calc'd)	1000 (meas'd)	354 (calc'd)	1320 (calc'd)	1540 (meas'd)	2870 (meas'd)
ENERGY $\frac{W_I}{V_o}$ $\frac{J}{m^3}$	$3.3x10^6$	$6.9x10^6$	$5.9x10^6$	$2.4x10^7$	$2.2x10^7$	$8.9x10^7$
Lognormal parameters:						
D_g(m)	2100	930	7000	750	470	225
g	4.4	4.6	6.7	3.8	3.8	3.8
Wt % D 10 m	0.02	0.32	0.03	0.06	0.20	0.90
$\frac{S_F}{V_o}$ $\frac{1}{m}$	$1.46x10^3$	$3.44x10^3$	$8.72x10^2$	$3.25x10^3$	$5.19x10^3$	$1.08x10^4$
F (J/m^2)	1.8 x 10^2	$1.6x10^2$	$5.6x10^2$	$6.0x10^2$	$1.8x10^2$	$3.9x10^2$

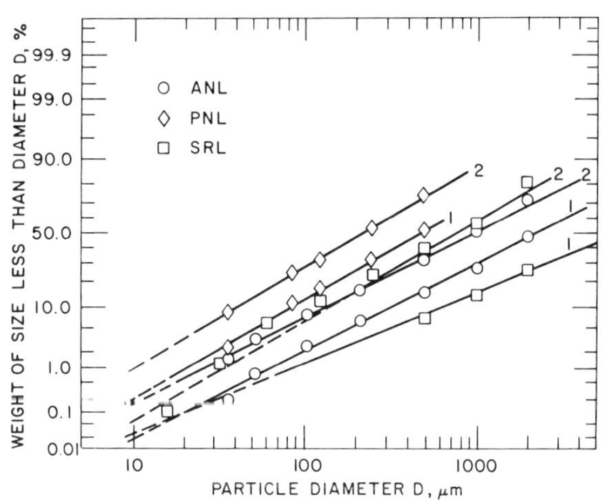

Fig. 2. The Cumulative Lognormal Plot of Measured Particle Size Distributions for Various Impact Tests of Glass Materials.

varies inversely with the energy density. The surface area as determined directly by gas adsorption or as calculated from the lognormal parameters and the shape factor α = 12 allows calculation of a fracture-surface energy γ_F (J/m^2) if the net energy is known. For impacts, γ_F appears to be nearly constant for a given material and to increase at higher energy densities (W_F/V_o). The energies reported here are the total input energies; the net energy absorbed in brittle fracture is expected to be lower by a factor of about two for "bare" specimens and a factor of more than four for canistered specimens. The considerable regularities in lognormal relationships of these tests of different specimens under quite varied test conditions indicate promise as a basis of further systematic tests and analyses.

Three different glass compositions of small specimens were impacted (8) at four different energies. Particle sizes were measured by sieving. To test the methodology on different glass compositions, the particle size data were replotted on lognormal graphs to determine the two lognormal parameters and the surface area. These lognormal plots are shown in Fig. 3., and the lognormal parameters, calculated surface areas and fracture-surface energies are shown in Table 2. From reported data (8) on surface areas of typical particle size distributions of soda-lime-silica glass, a mean shape factor of $\alpha = 21 \pm 3$ was calculated. This value of α was used to calculate surface areas from the lognormal S_F/V_o ratio given in equation 2). These parameters show the comparative fracture properties of the materials. The constancy of the values of σ_g is striking, as shown by the similar slope in Fig. 3.

To obtain a range of impact energies, multiple impacts on a single specimen have been used at SRL and PNL. For example, tests of the same type of specimen were made with the same total energy-input density, but with that total energy applied in either 1, 2, 4 and 8 blows (8). The size distributions have been plotted lognormally in Fig. 4. From the lognormal parameters, the highest energy density (1 blow) showed the smallest σ_g, the largest γ_F and the smallest S_F/V_o. The principal difference was the respirable fraction

Table 2. Lognormal Parameters of Impact Fracture of Three Different Glass Compositions (Fig. 3.)

Glass Type	W/V (J/m^3)	D_g (μm)	σg	Wt% $D<10\mu m$	$\frac{S_F^a}{V_o}$ m^{-1}	Calc γF (J/m^2)
Soda-lime-silica Glass	$1.70x10^8$	192	3.7	1.2	$3.75x10^5$	$4.5x10^2$
	$8.9x10^7$	220	3.7	0.8	$2.47x10^5$	$3.6x10^2$
	$4.6x10^7$	370	4.1	0.6	$1.65x10^5$	$2.8x10^2$
	$2.2x10^7$	490	4.4	0.4	$1.19x10^5$	$1.8x10^2$
Frit	$1.7x10^8$	290	4.3	1.0	$2.1x10^5$	$8.1x10^2$
	$8.9x10^7$	490	4.4	0.4	$1.3x10^5$	$6.8x10^2$
	$4.6x10^7$	530	4.4	0.4	$1.2x10^5$	$3.8x10^2$
	$2.2x10^7$	720	4.2	0.2	$8.3x10^4$	$2.7x10^2$
Fused Silica (vitreous)	$1.7x10^8$	135	4.1	3.2	$4.2x10^5$	$4.0x10^2$
	$8.9x10^7$	150	4.0	2.8	$3.7x10^5$	$2.4x10^2$
	$4.6x10^7$	220	4.0	1.4	$2.5x10^5$	$1.8x10^2$
	$2.2x10^7$	300	4.0	0.6	$1.9x10^5$	$1.2x10^2$

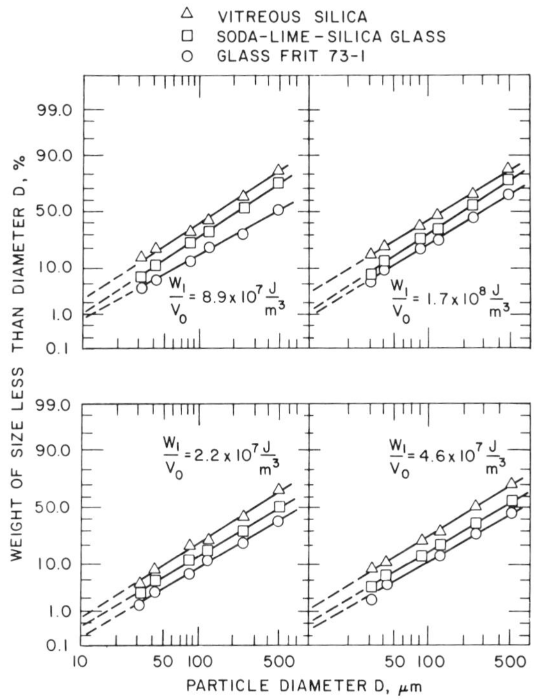

Fig. 3. Lognormal Plots of Data Reported (8) for Three Types of Impacted Glass Compositions at Four Energy Densities.

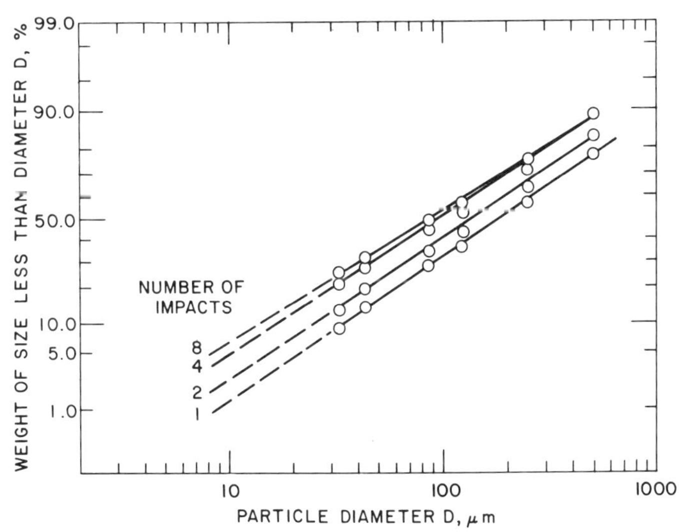

Fig. 4. Lognormal Plot of Data Reported (8) for Impacted Glass Specimens Receiving the Same Total Energy but with a Different Number of Impacts.

(< 10 μm diam.) which was 1.3% for 1 blow and 6.6% for 8 blows. This effect is consistent with other brittle-fracture data.

CONCLUSIONS

Data obtained from impact fracture of brittle materials can be correlated with a linear lognormal particle size distribution and with a nearly constant brittle-fracture surface-energy factor. This methodology allows estimation of total surface area of particles from impact fracture. Reliable extrapolations of particle size distributions to any lognormal size range seem achievable. Confirmations of this methodology are subjects of current study.

REFERENCES

1. W. J. Mecham, L. J. Jardine and M. J. Steindler, Proceedings of the International Symposium on Ceramics in Nuclear Waste Management, American Ceramic Society, April 30 - May 2, Cincinnati, OH (1979), CONF-790420, p. 327.
2. J. W. Axelson and E. L. Piret, Ind. Eng. Chem. 42, 665 (1950).
3. W. J. Kenny and E. L. Piret, AIChE Journal 7, 199 (1961).
4. R. A. Zeleny and E. L. Piret, I&EC Process Design and Development 1, 37-41 (1962).
5. W. J. Mecham, L. J. Jardine and M. J. Steindler, American Association for the Advancement of Science Annual Meeting, San Francisco, CA, Jan. 3-8 (1980).
6. G. Herdan, Small Particle Statistics, (Academic Press, London, 1960).
7. R. M. Wallace and J. A. Kelley, Savannah River Laboratory Report DP-1400 (1976).
8. J. L. McElroy, Quarterly Progress Report, Oct.-Dec. 1976, Battelle Pacific Northwest Laboratory Report PNL-2264 (1977).
9. P. H. Bonnell, Mound Laboratory Report MLM-1626 (1969).

DEVELOPMENT AND CHARACTERIZATION OF CERMET FORMS FOR RADIOACTIVE WASTE*

W. S. Aaron, T. C. Quinby, and E. H. Kobisk

Oak Ridge National Laboratory
Post Office Box X
Oak Ridge, TN 37830

INTRODUCTION

Cermets designed to isolate high level wastes in a solid form are a composite consisting of various ceramic phase particles uniformly dispersed in and microencapsulated by an iron-nickel base alloy matrix. The metal matrix provides this waste form with many advantageous features including excellent thermal conductivity and mechanical strength. These cermets are formed by first dissolving the waste in molten urea, precipitating and calcining all the constituents, compacting the calcine, and sintering and reducing to form the final product. The exact formulation of cermets through additions to the waste is designed to fix most of the fission products in stable leach-resistant ceramic phases which are subsequently microencapsulated by an alloy matrix. The alloy matrix, which is derived primarily from the waste itself and includes the reducible fission and activation products from the waste, can be compositionally adjusted through additions to optimize its corrosion resistance under conditions existing in various disposal environments.

The processes by which cermets are formed include several new and unique materials preparation options that are being developed to permit engineering scale-up and to be compatible with remote operations. Cermets formed by alternate processing methods are being characterized. Simulated wastes, radiotracer-containing simulated wastes and actual high level wastes have been studied. Until

*Research sponsored by the Division of Waste Products, U.S. Department of Energy, under contract W-7405-eng-26 with the Union Carbide Corporation.

recently, investigations have been focused on NFS Thorex waste and SRP dried sludge and fresh acid wastes; however, the applicability of cermet processing to a widening variety of other waste types is also being investigated.

CERMET PROCESSING DEVELOPMENT

Initially, cermet samples were prepared using a laboratory scale, batch process developed for the preparation of special ceramics having high compositional uniformity and excellent sinterability (1,2). The first step in this process involved the dissolution of the waste and additives in a molten urea solution. Upon heating, dehydration and denitration of the solution occurred, and eventually a precipitate was formed containing the waste and additive ions distributed homogeneously throughout the precipitate. Calcination of this precipitate in air at 800°C decomposed the precipitate into its component oxides. Hydrogen reduction of this submicron size calcine powder at 800°C converted the reducible metal oxides to metal (Fe, Ni, Cu, etc.) from which the final alloy matrix was derived. Subsequent densification of this metal and oxide powder mixture by hot pressing or cold compaction and sintering formed the cermet samples for property evaluation purposes (3,4).

The modification of this batch process to one suitable for scale-up and remote operation is the subject of this paper. Significant improvements in the precipitation and calcination steps have resulted from the development and use of a spray calciner. A 10-cm i.d. spray calciner with an 800°C wall temperature using simulated waste-additives-molten urea solutions has combined time-consuming precipitation and calcination process steps into one very rapid unit operation with additional advantages. With spray calcination, a 1:1 molar ratio of urea to nitrate ion is used rather than the 5:1 to 40:1 ratio commonly employed in batch processing. This not only drastically reduces urea consumption, but also the effluent gases from the process are decreased and the exothermic nature of the urea-nitrate reaction improves the calcination of the material as it falls through the spray calciner. The off-gas resulting from spray calcination of a 1:1 urea-to-nitrate ratio is also more chemically inert than that generated over the course of comparable batch processing due to the greatly reduced NH_3 production and the elimination of HCN from the off-gas. Spherical calcine particles, composed of an intimate mixture of the oxides present in the waste-additives system, are produced in sizes up to approximately 10 μm in diameter under these operating conditions. Further development and optimization of spray calcination operating parameters will be pursued following the completion of a larger 20-cm i.d. spray calciner which will permit greater operating versatility and improved process monitoring.

Although facilities do not currently exist for waste component volatility loss studies using radiotracers in these calciners, it is expected that these losses would be even lower than those encountered during batch processing since exposure times of the material to elevated temperatures are greatly reduced in the spray calcination process. Maximum volatility losses encountered in batch processing were 0.27%, 0.23% and 0.0001% for Cs, Ru and Sr, respectively.

A second major improvement in the process has been the development of liquid phase sintering as a densification technique. The calcine powder, consisting of an intimate homogeneous mixture of oxides, is first cold compacted into the desired shape using water as the binder. Under proper temperature conditions, a liquid phase can be developed by the simultaneous reduction and sintering of the cold compacted material, which greatly enhances its sinterability. Bulk densities comparable to those obtained by conventional sintering under a reducing atmosphere for periods up to 24 hours at temperatures of 1200°C or higher appear achievable by the liquid phase sintering method in less than one-half hour at a temperature of approximately 1100°C. More important, the microporosity typically produced by conventional sintering is greatly reduced by the liquid phase sintering technique. Micrographs in Fig. 1a. and 1b. of samples prepared from actual high level waste clearly illustrate the microstructural differences resulting from conventional and liquid phase sintering. In Fig. 1a. the rounded ceramic particles, although uniformly dispersed throughout the metal matrix, tend to be located within micropores without intimate metal encapsulation. This condition is undesirable because maximum density has not been achieved and, more importantly, because the leach and dispersion resistance provided by the metal matrix is of less than optimum quality. On the other hand, the microstructure developed by the presence of a liquid phase during sintering, as shown in Fig. 1b., has very little microporosity and excellent microencapsulation of the larger angular particles of the various ceramic phases by the metal matrix. Refinements in the liquid phase sintering technique remain to be made along with its application to the sintering of cylinders larger than 2 cm in diameter; however, this densification technique appears to be very promising since it not only provides significant advantages in processing but also improves the properties of the resulting cermet waste form. The development of this densification technique along with spray calcination has resulted in the improved basic flowsheet, shown in Fig. 2., for cermet processing of sludge and acid wastes.

CERMET CHARACTERIZATION

General properties and characteristics which were determined for cermets formed primarily from SRP and NFS wastes using various formulations and processing methods are shown in Tables 1 and 2. Presented are typical ranges for cermet materials which have been found

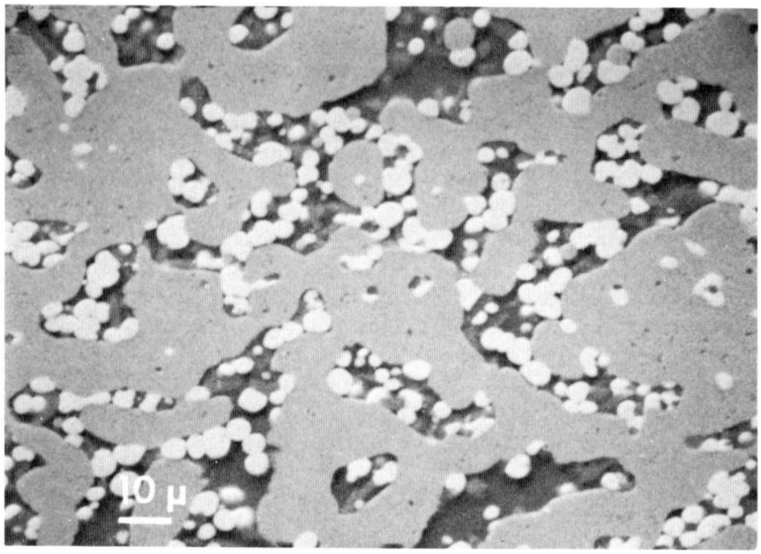

Fig. 1a. Microstructure Resulting From Conventional Sintering Methods

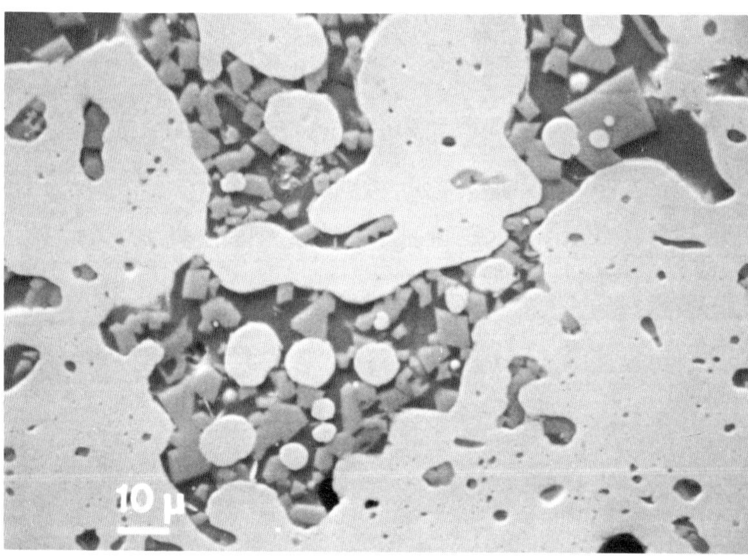

Fig. 1b. Improved Microencapsulation and Minimized Microporosity Resulting from Liquid Phase Sintering

Fig. 1. Microstructural Comparison of Cermets Densified by Conventional and Liquid Phase Sintering

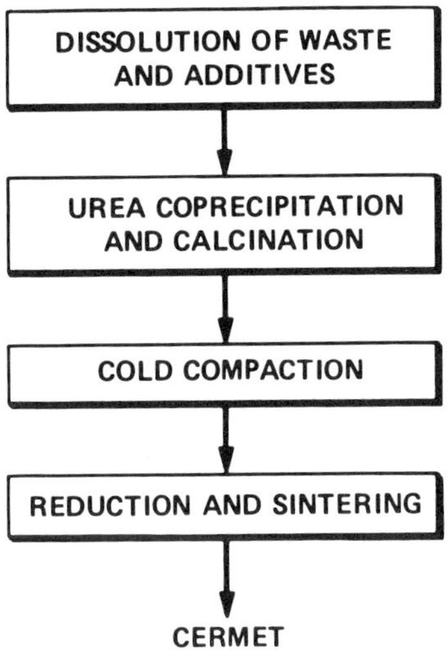

Fig. 2. Improved Basic Flowsheet for Cermet Processing of High Level Wastes.

to be representative of a wide variety of compositions and processes and are expected to remain valid in spite of future changes in the process or the wastes considered. The distinction between cermet characteristics related to processing methods and those related to the cermet formulation is somewhat ambiguous; however, characteristics related most closely to cermet formulations can vary over a wide range while intrinsic cermet properties are less variable and essentially determined by the general ceramic-to-metal ratio.

In Table 1, the waste loading levels are primarily determined by the waste composition, more specifically, the metal and sodium contents of the waste. As the metal content increases, waste loading levels can be increased since this metal is used in the formation of the alloy matrix requiring lower additions of metals from other sources. On the other hand, as sodium content increases, greater additions of aluminosilicate formers are required to fix the sodium and thus maintain the integrity of the cermet body. If it is desired to maintain a particular metal-to-ceramic ratio, increased metal additions may be required to compensate for the aluminosilicate additions, thus further reducing the maximum possible waste loading. The waste volume reduction factor is primarily dependent on the waste loading level and, to a smaller extent, on the final cermet density. As greater additions of materials are made, especially those forming lower density ceramic phases, the waste volume reduction factor is reduced.

Table 1. Processing Related Cermet Characteristics

Waste loading	2-75 wt.%
Waste volume reduction	1:1-100:1
Ceramic phase loading	20-50 wt.%
Metal phase loading	50-80 wt.%
Metal phase composition	70% Fe, 20% Ni, 5% Cu, 5% Misc.

Table 2. Typical Cermet Property Ranges.

Density	6.0-8.2 g/cc
Thermal conductivity	3-20 W/m · °C
Decay heat loading	0.02-0.2 W/cc
Oxidation	
8 hr - 500°C air	∿2 μm thick oxide layer
2 hr - 800°C air	∿5-7 μm thick oxide layer
Thermal shock (2 cm diameter cylinders)	
3 cycles: 900°C ⟶ water quench	No damage detected metallographically
Leach rate	
Soxhlet test (72 hr)	7.1×10^{-6} g/cm^2 · d, 0.024% Cs loss

As would be expected, the ceramic and metal phase loadings are dependent on the waste composition and the amount and type of additives used in each particular case. The upper limit for ceramic loading while maintaining continuity of the metal and satisfactory microencapsulation has not yet been determined. A trade-off does exist between the amount of ceramic phase in a cermet and the level of retention of the advantageous properties supplied by the metal matrix, such as thermal conductivity, mechanical strength, etc. Finally, although little investigation of alternative metal matrices has been performed, it appears that an iron-nickel base alloy will most likely be chosen for cermets because of the amount of these metals in typical wastes. The metal maxtrix is composed of those metals which can be reduced during processing by hydrogen or carbon-monoxide-containing atmospheres. These include reducible fission or activation products in the waste which result in some degree of variability in the alloy composition as the waste composition (type) changes. In cases examined up to this point, no detrimental effects from this variation have been observed.

In Table 2, property measurements taken from various simulated and/or actual waste cermets are shown. The composite cermet densities achieved are dependent upon the densification technique used and the relative amounts of low density ceramic phases and high density metal present. Thermal conductivity is likewise dependent on the density (more properly, the porosity) and the ceramic phase loading. As both porosity and ceramic phase loading of the cermet increase in magnitude, the bulk thermal conductivity is decreased. Decay heat loading of cermet waste samples is strictly determined by the type, age and loading level of the waste in a particular cermet sample. The 0.02 W/cc (cermet) was obtained from a sample prepared from from actual NFS acid Thorex waste at a waste loading level of approximately 28 wt.%. The value of 0.2 W/cc or 31 W/kg of cermet was determined for a sample formed from fresh SRP acid waste at a loading level of only 2.1 wt.%.

Some tests on the oxidation resistance and thermal shock performance of cermets have been performed. These properties are almost exclusively dependent on the alloy matrix. As shown in the table, upon exposure to air at elevated temperatures, thin oxide layers were formed on the surfaces of the cermet samples. Further testing for longer periods and over wider temperature ranges will be required to more definitely ascertain the oxidation behavior of cermets.

Again in limited testing of the thermal shock performance of cermets, no detrimental effects (except oxidation) could be found in samples cycled three times from 900°C to room temperature by water quenching. Here also, further testing will be required to better define the resistance of cermets to thermal shock degradation.

Significantly more leach testing of cermets is required to determine the performance of this waste form under the variety of disposal environments that might be encountered. A variety of leach tests, reported in detail elsewhere (5), have been performed both at ORNL and PNL. Testing difficulties were encountered that, in most cases, prevented the quantification of leach rates for cermets. Galvanic coupling effects between the cermet samples and test apparatus, except the Soxhlet test (see Table 2), were chiefly responsible for the problems encountered in these radiotracer leach tests. Indications are, however, that the cermet waste form possesses a leach resistance comparable to or greater than various alternate waste forms currently under evaluation.

Since significant changes have occurred in the process by which cermets are formed and subsequently are expected to change a few of the properties of the final cermet, detailed studies of these properties, especially leach resistance, have been delayed until samples representative of the products from these new processes can be produced repetitively. At this point, quantitative in-depth characterization of cermets will be undertaken.

FUTURE DIRECTIONS

The cermet program is of recent origin, and more development of the process and characterization of the product must be performed to permit quantitative evaluation of this alternate for nuclear waste disposal. As indicated above, such studies are being actively pursued. A previously unconsidered disposal option, space disposal, is now being considered as an application for cermet waste forms. Engineering and economic feasibility analyses have been initiated.

REFERENCES

1. T. C. Quinby, "Neutron Dosimetry," U.S. Patent 3,971,944, July (1976).
2. T. C. Quinby, "Method for Producing Mixed Metal Oxides and Metal-Metal Oxide Mixtures," U.S. Patent 4,072,501, Feb. 7 (1978).
3. W. S. Aaron, T. C. Quinby, and E. H. Kobisk, Cermet High Level Waste Forms, ORNL/TM-6404, June (1978).
4. W. S. Aaron, T. C. Quinby, and E. H. Kobisk, "Development of Cermets for High Level Radioactive Waste Fixation," to be published in the Proceedings of the International Symposium on Ceramics in Nuclear Waste Management, USDOE CONF 790420, sponsored by the American Ceramic Society, May (1979).
5. R. E. Blanco and A. L. Lotts, High Level Waste Program Progress Report for October 1, 1978, through March 31, 1979, ORNL/TM-6866, May (1979).

LONG-TERM LEACHING OF IRRADIATED SPENT FUEL

Y. B. Katayama
D. J. Bradley

Pacific Northwest Laboratory
Richland, WA 99352

ABSTRACT

Spent light-water reactor (LWR) fuel with burnups of 9, 28 and 54 MWd/kgU were leach tested at 25°C in deionized water in a Paige apparatus. No discernible differences in long term leach rates were observed relative to burnup. An accelerated leaching period was observed during the Paige leach test of the 54 MWd/kgU spent fuel. Additionally, the 28 MWd/kgU fuel was IAEA leach-tested in five different leach solutions. Deionized water gave the highest leach rates, and calcium chloride solution gave the lowest.

Comparisons between spent fuel and borosilicate waste glass leach rates are made. In sodium bicarbonate solution, the leach rates are nearly equal and the glass becomes increasingly more resistant in calcium chloride solution, followed in order by sodium chloride solution, WIPP "B" brine and deionized water--in which the glass is two to three orders of magnitude more leach-resistant than is spent fuel.

INTRODUCTION

There are many types of leach tests being used to evaluate solidified waste forms (1). At Pacific Northwest Laboratory (PNL), the Paige procedure (2) was chosen for leaching radioactive samples of solidified product from the Waste Solidification Engineering Prototype (WSEP) Program (3). The first leach tests of spent LWR fuels (1975) were in these Paige leach systems. Two years later in 1977, spent fuel leach tests were started using a modification of the International Atomic Energy Agency (IAEA) procedure. The modifications included more sampling (continuous at monthly intervals) and the use of "free form fragments" of waste materials. In 1979, spent fuel leach tests were started using a static leach test procedure. In this paper, we will present spent fuel leach data from the Paige tests and IAEA tests. The contents are extracted from two detailed publications (4, 5).

EXPERIMENTAL PARAMETERS

Two leaching procedures (Paige and IAEA) were used to generate the data presented in this paper. In the Paige system, the leach solution was recirculated with an airlift pump at a flow rate of 75 liters/day. The leach solution is changed at the time of sampling. Sampling was done daily for the first week, weekly for three weeks, and monthly thereafter. Samples to measure container adsorption (plateout) were not taken from the Paige apparatus. However, analytical samples are acidified to prevent plateout in the sample bottles.

In the IAEA system, the leachant is static during the leaching interval between sampling periods. The modified IAEA sampling period is daily for the first week, weekly for eight weeks, and monthly thereafter. Plateout samples were taken and analytical samples were acidified to prevent plateout in the bottles.

All the leach tests to date have been with unclad fuel fragments. The fuels used in the Paige leach tests were from three different reactors and had different average burnups; the Zorita Reactor fuel (burnup of 54.5 MWd/kgU), the H. B. Robinson-II Reactor fuel (burnup of 28.0 MWd/kgU), and the Quad Cities-I Reactor fuel (burnup of 9.0 MWd/kgU). Only one fuel (H. B. Robinson-II) was leached in the IAEA apparatus. The IAEA tests were done in the following five leach solutions: 1) deionized water, 2) 0.03 $\underline{M}$ NaCl, 3) 0.015 $\underline{M}$ $CaCl_2$, 4) 0.03 $\underline{M}$ $NaHCO_3$, and 5) saturated WIPP "B" salt brine. One Paige leach test was made with a synthetic sea water and 54.5 MWd/kgU fuel.

RESULTS AND DISCUSSION

Paige test leach rate data for the 54.5 MWd/kgU fuel in deionized water at 25°C are shown in Fig. 1. for duplicate samples Z-6 and Z-7. The curves for the period up to approximately 550 days appear as continuations of the data reported after 140 days (6), and the cesium and curium leach rates are up to four orders of magnitude apart. From 600 days to 800 days, the curves converge and the leach rates for all the radionuclides become nearly equal. This period is characterized by a hump in the leach rate curves and is referred to as the "accelerated leaching period" in this paper. After 750 days, the leach rates start to decrease in value.

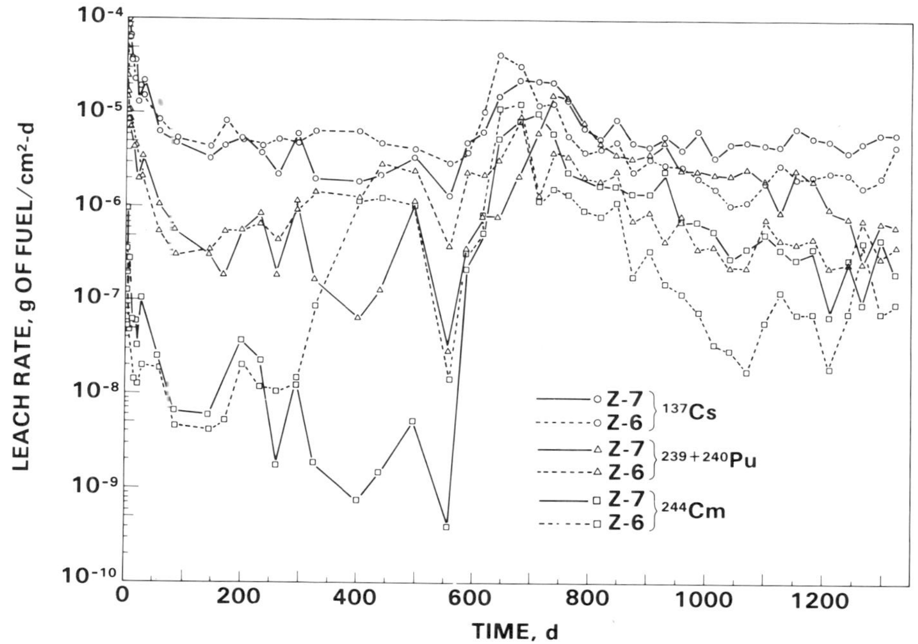

Fig. 1. Paige Leach Rate of 54.5-MWd/kgU Spent Fuel in Deionized Water at 25°C (Duplicate Samples are Designated Z-6 and Z-7)

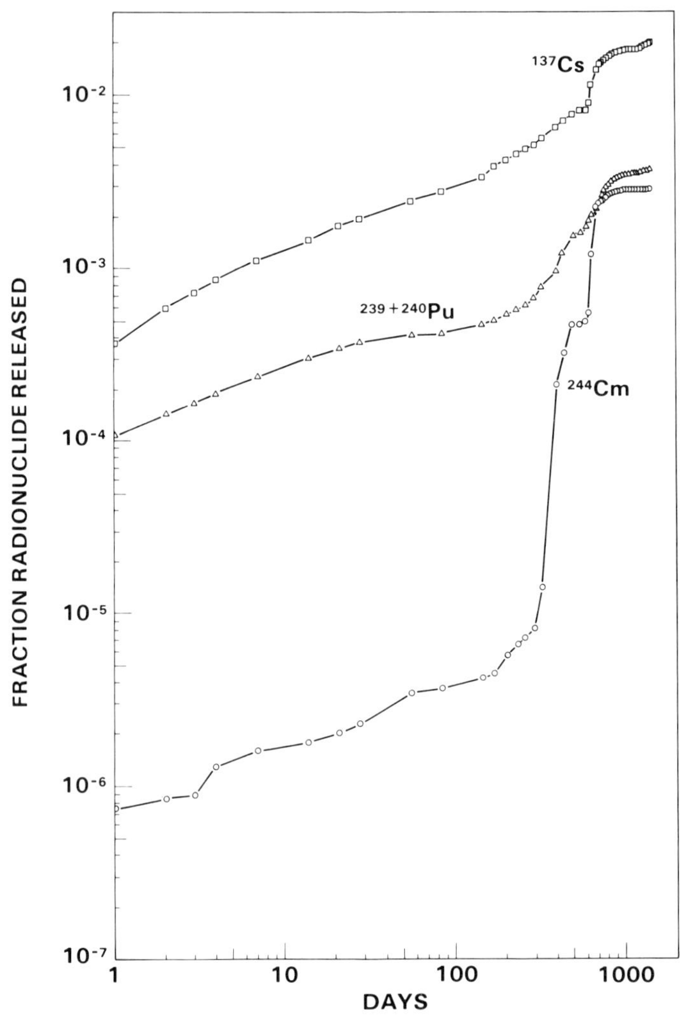

Fig. 2. Cumulative Release of Cs, Pu and Cm from 54.5-MWd/kgU Spent LWR Fuel in Deionized Water at 25°C (Sample Z-6)

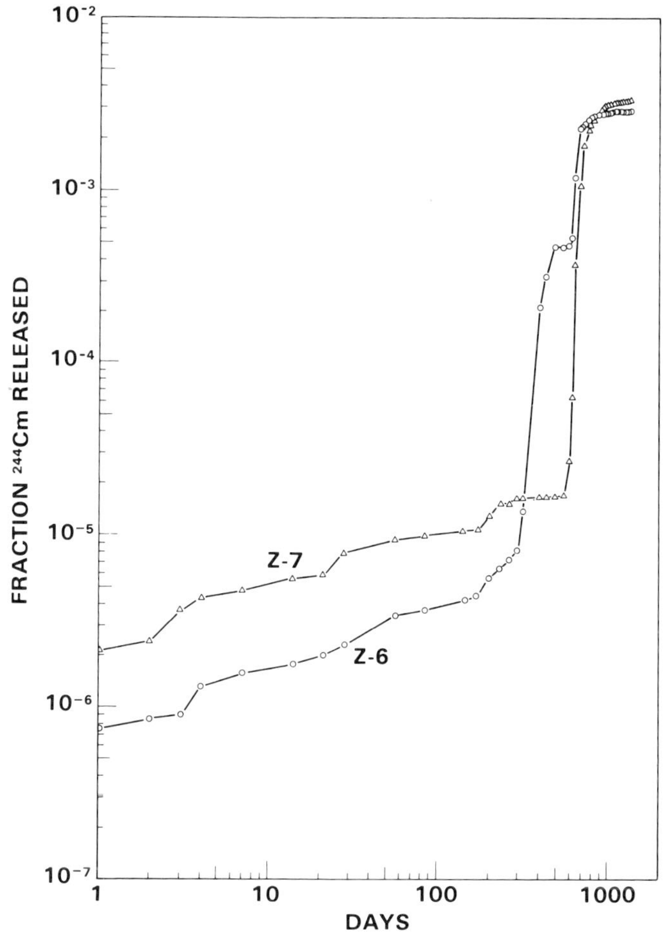

Fig. 3. Cumulative Fraction of ^{244}Cm Released from Spent LWR Fuel, 54.5 MWd/kgU, in Deionized Water at 25°C

Not shown in Fig. 1. are the leach rate curves for ^{144}Ce and ^{154}Eu, which were not continuous leach curves. These radionuclides were analytically detected during the first 200 days of the leach test and then were not radiochemically detectable until the start of the accelerated leaching period. The reappearance of ^{144}Ce and ^{154}Eu during the accelerated leaching period indicates that new sources of these radionuclides became available--that new fuel surfaces were probably exposed.

Graphical representations of cumulative fractional release of ^{137}Cs, $^{239+240}Pu$ and ^{244}Cm, as a function of the days leached, are shown in Fig. 2. for sample Z-6. After 1378 days of leaching, 2.0% of the cesium, 0.38% of the plutonium and 0.29% of the curium were released to the deionized water. The plutonium and curium fractional releases differ by about 2 orders of magnitude for the first 300 days and then become nearly equal during the accelerated leaching period--indicating the disappearance of a preferential leaching mechanism at 300 days. Fig. 3. is a graph of the cumulative fractional release of curium for the duplicate samples Z-6 and Z-7. Although major fractional releases of curium from the 2 samples occurred about 300 days apart, the fractional release became nearly equal during the accelerated leaching period--indicating similar chemical composition and distribution between the duplicate samples.

Spent LWR fuel with a burnup of 54.5 MWd/kgU was leach tested (Paige) in a synthetic sea brine. Cesium-based leach rate data for 1017 days of cumulative leaching in brine and deionized water are shown in Fig. 4. Based on the release of ^{137}Cs, the leach rate of spent LWR fuel is a half order of magnitude lower in sea brine than in deionized water. The accelerated leaching period was not observed in the sea brine leach test.

Spent LWR fuels with three different burnups (54.5, 28.0 and 9.0 MWd/kgU) were leached by the Paige procedure in deionized water at 25°C. Fig. 5. shows the leach rate curves for these fuels based on the release of ^{137}Cs. The leach curves are similar except for the accelerated leaching period observed for the 54.5 MWd/kgU fuel.

The IAEA leach rate data for the 28.0 MWd/kgU fuel in deionized water at 25°C is shown in Fig. 6. Cesium has the highest initial leach rate and tends to stay at the top of the leaching trend for all the measured radionuclides. The leach rate curves for the other solutions (NaCl, WIPP "B" Brine, $NaHCO_3$ and $CaCl_2$) start at values similar to those in deionized water, but the general trend with cumulative time of leaching is lower than in deionized water. For comparison purposes, if we choose the midpoint of the leaching curves in deionized water at 467 days of leaching as 5×10^{-6} g/cm^2-d, then the midpoints for the five solutions are as follows:

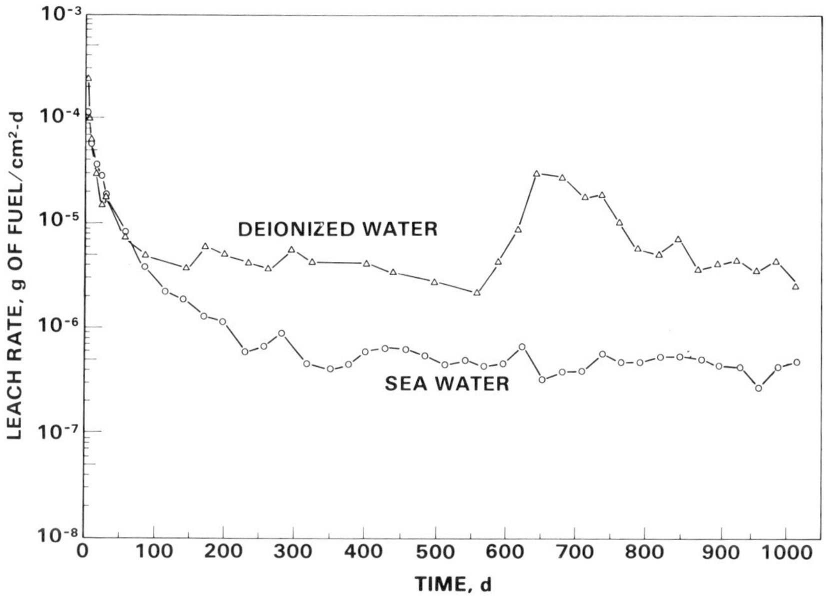

Fig. 4. Paige Leach Rate of 54.5 MWd/kgU Spent LWR Fuel Based on Release of ^{137}Cs in Sea Brine and Deionized Water at 25°C

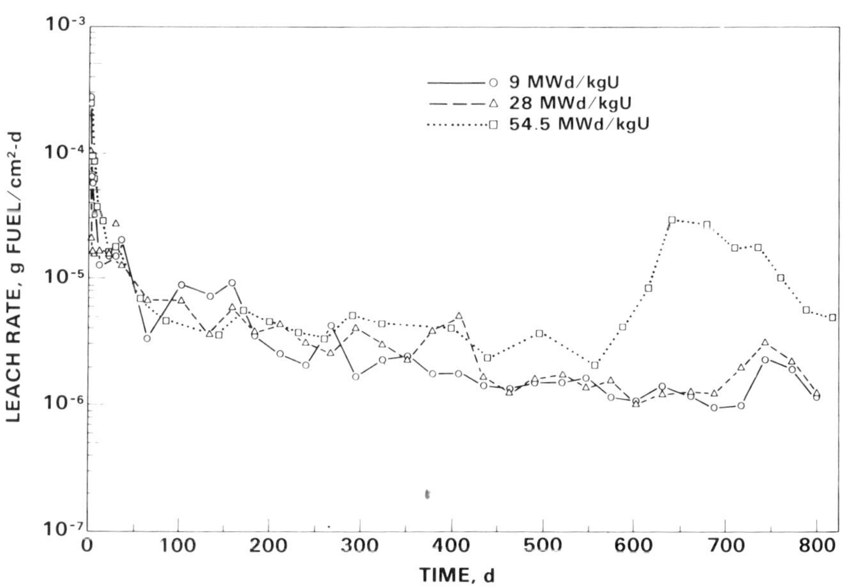

Fig. 5. Paige Leach Rate of Spent Fuels with Burnups of 9.0, 28.0, and 54.5 MWd/kgU Based on Release of ^{137}Cs in Deionized Water at 25°C

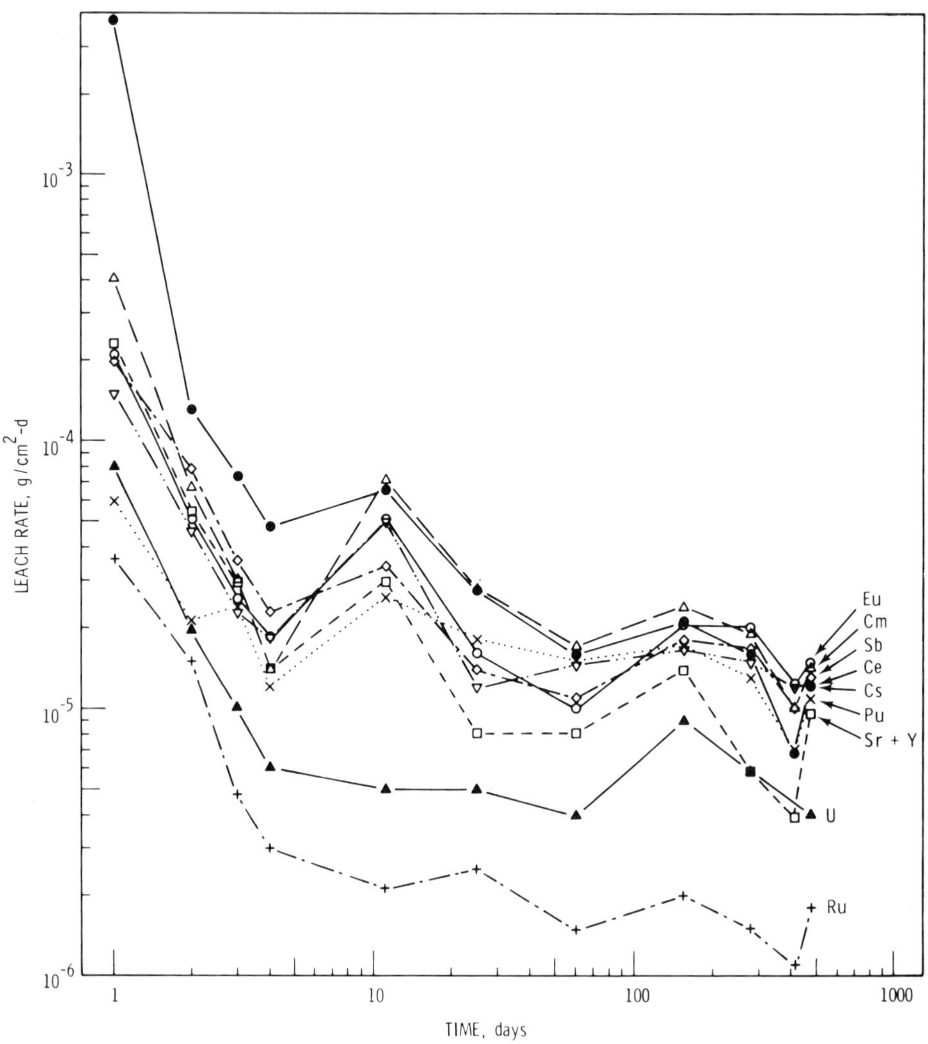

Fig. 6. IAEA Leach Rate for 28-MWd/kgU Spent Fuel in Deionized Water at 25°C

Solution	Midpoint at 467 Days (g/cm^2-day)
Deionized Water	$5x10^{-6}$
0.03 M NaCl	$2x10^{-6}$
Saturated WIPP "B" Brine	$1x10^{-6}$
0.03 M $NaHCO_3$	$8x10^{-7}$
0.015 M $CaCl_2$	$2x10^{-7}$

The leach rates in deionized water are the highest and the leach rates in $CaCl_2$ solution are a factor of 25 lower. Also, in Fig. 6., the relative leachability of the measured radionuclides after 467 days of leaching is listed for deionized water. For the other leach solutions (NaCl, WIPP "B", $NaHCO_3$ and $CaCl_2$) the relative leachabilities are similar except for uranium which was seen to vary between the highest and lowest leachable element during the 467 days of leaching. Fig. 7. shows the leach rate curves based on the release of uranium for the five leach solutions. The release of uranium to the leach solution is not at a uniform rate. This behavior is being investigated at PNL in a task directed toward the understanding of the mechanism and kinetics for UO_2 spent fuel release in aqueous solutions.

Comparison of IAEA and Paige leach rates at a cumulative time of 467 days shows about one order of magnitude difference (IAEA higher than Paige) in the values of releases based on ^{137}Cs, $^{239+240}Pu$, and ^{244}Cm. The difference may be due to radionuclide plateout on the Paige apparatus, which was not measured.

Paige leach rates for spent fuel (deionized water at 25°C) were found to be comparable to the first-generation borosilicate glasses vitrified during the WSEP demonstration at PNL (6). Comparison of spent-fuel and glass leach rates using the IAEA procedure are tabulated in Table 1. The ^{244}Cm- and $^{239+240}Pu$-based glass leach rates are for a second-generation borosilicate glass coded 76-68 (see Ref. 7). In deionized water leach solution the spent-fuel leach rate is two to three orders of magnitude higher than that of glass. These differences decrease as tests progress from deionized water to the WIPP "B" brine to the NaCl solution to the $CaCl_2$ solution, and nearly disappear in $NaHCO_3$ solution.

CONCLUSIONS

The spent-fuel leach tests continue. A few conclusions may be drawn from information obtained to date. The choice of leach-test procedures may introduce differences in leach-rate values (up to one

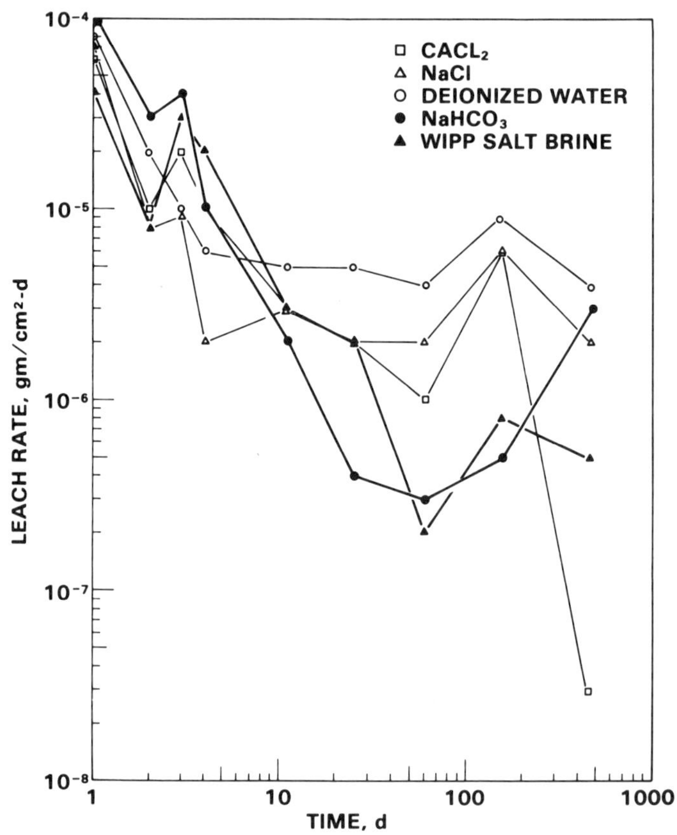

Fig. 7. IAEA Leach Rate for 28-MWd/kgU Spent Fuel at 25°C Based on Release of Uranium

Table 1. Comparison of IAEA Leach Rates at 25°C

Leach Solutions	$^{239+240}$Pu Leach Rates[a] g/cm^2-d		^{244}Cm Leach Rates[b] g/cm^2-d	
	76-68 Glass	Spent Fuel	76-68 Glass	Spent Fuel
Deionized Water	$5x10^{-8}$	$2x10^{-5}$	$4x10^{-7}$	$2x10^{-5}$
WIPP "B" Brine	$2x10^{-8}$	$2x10^{-6}$	$1x10^{-8}$	$3x10^{-6}$
NaCl	$7x10^{-8}$	$3x10^{-6}$	$2x10^{-7}$	$4x10^{-6}$
$CaCl_2$	$2x10^{-8}$	$3x10^{-6}$	$1x10^{-7}$	$9x10^{-8}$
$NaHCO_3$	$2x10^{-7}$	$1x10^{-7}$	$2x10^{-7}$	$2x10^{-7}$

(a) 151 d
(b) 454 d

CONCLUSIONS

order of magnitude between Paige procedure and IAEA procedure). Comparison of leach rates for spent fuels with burnup of 9, 28, and 54 MWd/kgU showed no discernible effect of burnup on long term leach rate. The leach rates are highest in deionized water and lowest in 0.015 M $CaCl_2$ solution. Based on the release of plutonium and curium, spent fuel has a leach rate comparable to 76-68 glass in sodium bicarbonate solution and is two to three orders of magnitude less leach resistant than is 76-68 glass in deionized water and WIPP "B" brine solution. The observed accelerated leaching period indicates the need for long term leach tests for spent LWR fuels.

ACKNOWLEDGEMENTS

This research was supported by the Waste Isolation Safety Assessment Program and the Waste/Rock Interaction Technology Program being conducted by Pacific Northwest Laboratory. These programs are sponsored by the Office of Nuclear Waste Isolation, which is managed by Battelle Memorial Institute under contract EY-76-C-06-1830 with the Department of Energy.

REFERENCES

1. J. E. Mendel, A Review of Leaching Test Methods and the Leachability of Various Solids Media Containing Radioactive Wastes, BNWL-1765, Pacific Northwest Laboratory, Richland, WA (1973).

2. B. E. Paige, Leachability of Alumina Calcine Produced in the Idaho Waste Calcining Aluminum Waste, IDO-14672, Phillips Petroleum Company, Idaho Falls, ID 83401 (1966).
3. J. L. McElroy, K. J. Schneider, J. N. Hartley, J. E. Mendel, G. L. Richardson, R. W. McKee, and A. G. Blasewitz, Waste Solidification Program Summary Report, Volume II, Evaluation of WSEP High Level Waste Solidification Processes. BNWL-1667, Pacific Northwest Laboratory, Richland, WA 99352 (1972).
4. Y. B. Katayama, Spent LWR Leach Tests, PNL-2982, Pacific Northwest Laboratory, Richland, WA 99352 (1979).
5. Y. B. Katayama, D. J. Bradley and C. O. Harvey, IAEA Leach Test of Spent LWR Fuel - A Progress Report, PNL-3713, Pacific Northwest Laboratory, Richland, WA 99352 (1979).
6. Y. B. Katayama, Leaching of Irradiated LWR Fuel Pellets in Deionized and Typical Groundwater, BNWL-2057, Pacific Northwest Laboratory, Richland, WA 99352 (1976).
7. D. J. Bradley, C. O. Harvey and R. P. Turcotte, Leaching of Actinides and Technetium from Simulated High Level Waste Glass, PNL-3152, Pacific Northwest Laboratory, Richland, WA 99352 (1979).

LEACHING OF IRRADIATED CANDU UO_2 FUEL

T. T. Vandergraaf, L. H. Johnson, and D. W. P. Lau

Atomic Energy of Canada Limited
Whiteshell Nuclear Research Establishment
Pinawa, Manitoba, Canada ROE 1L0

INTRODUCTION

An assessment of the concept to dispose of spent, irradiated nuclear fuel in an underground repository requires information on the rates of radionuclide leaching from the fuel matrix and of fuel matrix dissolution.

The CANDU-PHW* reactor is fueled with natural uranium, sheathed in zirconium. Heavy water is used as moderator and as coolant. Each fuel bundle is made up of a cluster of 50-cm-long fuel elements and contains approximately 22 kg uranium as UO_2. The fuel is irradiated to a design burnup of 500 to 680 GJ/kg U (160 to 190 MWh/kg U) at a linear power rating of 50 to 63 kW/m. At discharge the fuel contains 0.4% Pu and 0.8% fission products, while the ^{235}U content is 0.2%.

The study reported here was initiated in November 1976. Sections of sheathed, irradiated fuel are leached in a static system. A minimum of chemical separation of the leachates is performed to remove interferences in the radiometric assay. The leach rates of the various radionuclides are expressed in Bq/day and compared to the rate of UO_2 dissolution.

EXPERIMENTAL

Four 5-cm-long sections of fuel, with attached sheath, were cut from an outer element of a 28-element fuel bundle irradiated in the

*CANada Deuterium Uranium - Pressurized Heavy Water

Pickering Generating Station near Toronto. This fuel had been irradiated to a burnup of 610 ± 20 GJ/kg U at a linear power rating of 53 kW/m for 13 months. It was discharged from the reactor on 1973-03-07 and cooled for 2.5 years in a water-filled storage bay. Table 1 lists the major fission product and actinide inventories as obtained from a CANIGEN-based compilation (1).

Two leaching solutions were selected:

(a) distilled, deionized water (pH 5.9) and
(b) filtered and slightly chlorinated river water (pH 7.7).

The major species in the river water with their concentrations in μg/g are: Na (2.3), K (0.9), Mg (3.8), Ca (11.8), Si (2.9), HCO_3 (64), Cl (4.1), and SO_4 (4.9).

Leaching is done in duplicate, at room temperature, in polyethylene bottles with 100 mL of air-saturated solution. The fuel is periodically transferred to new bottles containing fresh solutions, and the leachates acidified with 5 mL concentrated nitric acid. The acidified solutions are left to equilibrate for at least two weeks to allow any sorbed species to desorb from the container walls, and aliquots are removed for radionuclide assay and uranium determination.

^{134}Cs and ^{137}Cs concentrations are determined on aliquots of the leachate using a Ge(Li) detector. Other γ-emitting radionuclides are determined after scavenging the cesium with ammonium molybdophosphate (2), synthesized in our laboratory. ^{90}Sr is isolated as its carbonate after a double co-precipitation of ^{90}Y with ferric hydroxide. Since ^{89}Sr has decayed to negligible levels, no correction for ^{89}Sr is necessary. A combination of α-counting and α-spectrometry is used to determine α-emitting radionuclides in aliquots of the original leachate. Uranium is determined by thermal neutron activation analysis, measuring the induced ^{239}U activity.

RESULTS AND DISCUSSION

Leach rates, expressed in Bq/day, have been obtained for ^{60}Co, ^{90}Sr, ^{106}Ru, ^{125}Sb, ^{134}Cs, ^{137}Cs, ^{144}Ce, ^{154}Eu, $^{239+240}$Pu, ^{241}Am and ^{244}Cm. Since the UO_2 fuel is fractured, it is difficult to assign a creditable surface area to it. The cross-sectional area of a fuel pellet is 1.6 cm^2 and, as the fuel sections have both ends exposed to the leaching solutions, conservative leach rates can be calculated by assuming that all leaching takes place at these exposed surfaces and by dividing the leach rates by 3.2 cm^2.

Table 2a lists the leach rates and the fractional release rates for the above radionuclides, and Table 2b gives the dissolution

Table 1. Inventory of Major Radionuclides in 5-cm-Long Sections of Irradiated UO_2 Fuel as Obtained From CANIGEN (1) (In Bq/sample*).

ISOTOPE	Days After Start of Experiment		
	10	100	1000
$^{90}Sr/Y$	3.9E10	3.8E10	3.6E10
$^{106}Ru/Rh$	4.1E10	3.4E10	6.2E09
^{125}Sb	2.3E09	2.2E09	1.2E09
^{134}Cs	1.1E10	1.0E10	4.5E09
^{137}Cs	5.7E10	5.7E10	5.4E10
$^{144}Ce/Pr$	4.4E10	3.4E10	3.8E09
^{147}Pm	6.9E10	4.6E10	3.6E10
^{154}Eu	2.1E09	2.1E09	1.6E09
^{239}Pu	4.2E08	4.2E08	4.2E08
^{240}Pu	5.6E08	5.6E08	5.6E08
^{241}Pu	4.0E10	3.8E10	3.5E10
^{241}Am	2.8E08	2.9E08	4.3E08
^{244}Cm	2.5E07	2.5E07	2.2E07

*Average sample weight: 84.4 ± 1.5 g UO_2

rates and fractional dissolution rates of the UO_2 matrix. Fig. 1., 2. and 3. show representative plots of leach rate against time while Fig. 4. is a plot of the UO_2 dissolution rate as a function of time. In addition, all four figures show a curve denoting the removal of 10^{-6} of the radionuclide inventory or matrix per day as an arbitrary reference.

No leach rates were obtained for the two other major radionuclides listed in Table 1: both ^{147}Pm and ^{241}Pu are pure β-emitters and require complete separation from other radioisotopes. On the basis of leaching behavior, the nuclides fall into two groups: (1) the cesium isotopes, ^{125}Sb, and to a lesser extent ^{90}Sr, show a marked decrease in fractional leach rates over the first 40 days; (2) the actinides and rare earth fission products have much lower initial fractional leach rates which decrease less with time. ^{106}Ru, and presumably ^{147}Pm and ^{241}Pu, also fall into this category.

These differences in leaching behavior can be explained by the migration of certain radionuclides, such as the cesium isotopes, along the temperature gradient existing across the fuel element during irradiation (3). The extent of migration is governed by the chemistry of the individual fission product and by the slope of the gradient, which in turn is a function of the linear power rating of the fuel. In a separate experiment, we have observed a 50-fold decrease in ^{137}Cs leach rates with a 20% decrease in power rating of the fuel. This agrees with similar trends which have been observed elsewhere (4).

Table 2a. Leach Rates and Fractional Release Rates of Radionuclides From UO_2 Fuel (Burnup 610 GJ/kg U)

Isotope	Leaching Solution	10 days		30 days		100 days		300 days		900 days	
		LR	FRR	LR	FRR	LR	FRR	LR	FRR	LR	FRR
^{60}Co	Distilled	1.3E4		4.2E3	-	1.4E3	-	8.2E2	-	2.0E2	-
	River	2.6E3		1.0E4	-	1.9E3	-	8.1E2	-	1.6E2	-
^{90}Sr	Distilled	8.2E5	2.1E-5	5.6E5	1.4E-5	1.3E5	3.3E-6	3.6E4	9.2E-7	1.1E3	2.9E-8
	River	7.6E5	1.9E-5	2.8E5	7.2E-6	1.3E5	3.3E-6	4.6E4	1.2E-6	8.8E2	2.3E-7
^{106}Ru	Distilled	1.3E4	3.2E-7	2.8E3	7.2E-8	8.8E3	2.7E-7	2.5E3	1.2E-7	-	-
	River	1.6E4	3.9E-7	4.0E3	1.0E-7	8.8E3	2.7E-7	2.5E3	1.1E-7	8.7E1*	1.2E-8*
^{125}Sb	Distilled	3.3E5	1.4E-4	1.4E4	6.1E-6	1.4E3	7.0E-7	1.7E3	9.4E-6	4.7E1	3.9E-8
	River	2.4E5	1.0E-4	3.1E4	1.3E-5	8.3E3	4.2E-6	1.8E3	2.0E-6	1.2E3	1.4E-6
^{134}Cs	Distilled	5.4E6	4.9E-4	2.1E6	1.9E-4	1.9E5	2.1E-5	7.2E4	1.0E-5	7.4E3	1.6E-6
	River	5.7E6	5.2E-4	1.8E6	1.6E-4	4.6E5	5.1E-5	7.9E4	1.1E-5	8.4E3	1.9E-6
^{137}Cs	Distilled	3.0E7	5.3E-4	1.2E7	2.1E-4	1.2E6	2.2E-5	5.2E5	9.8E-6	9.1E4	1.7E-6
	River	3.1E7	5.4E-4	1.0E7	1.8E-4	3.0E6	5.3E-5	5.6E5	1.0E-5	1.1E5	2.0E-6
^{144}Ce	Distilled	3.1E4	7.0E-7	2.4E4	5.5E-7	1.1E4	3.2E-7	1.8E4	9.0E-7	1.4E1*	2.3E-8*
	River	5.7E4	1.3E-6	1.8E4	4.1E-7	4.2E4	1.2E-6	1.8E4	8.8E-7	1.8E2*	3.1E-8*
^{154}Eu	Distilled	-	-	8.1E2	3.9E-7	3.3E2	9.2E-8	1.0E3	3.1E-7	1.9E1*	6.2E-9*
	River	-	-	3.4E2	1.6E-7	2.0E2	6.2E-8	4.9E2	1.5E-7	3.2E1*	1.0E-8*
$^{239+240}Pu$	Distilled	1.3E3	2.2E-6	1.7E3	2.9E-6	-	-	7.8E2	7.8E-7	8.1E0	8.1E-9
	River	1.8E3	3.1E-6	1.6E3	2.7E-6	1.1E3	1.1E-6	5.3E2	5.3E-7	1.2E1	1.2E-8
^{241}Am	Distilled	4.7E2	1.6E-6	6.7E2	2.4E-6	-	-	3.7E2	1.1E-6	5.0E0	1.1E-8
	River	7.2E2	2.5E-6	6.6E2	2.3E-6	4.3E2	1.5E-6	2.4E2	7.1E-7	5.8E1	1.3E-7
^{242}Cm	Distilled	4.6E1	1.8E-6	4.2E1	1.7E-6	-	-	2.9E1	1.2E-6	9.1E-1	1.6E-7
	River	4.8E1	1.9E-6	3.2E1	1.3E-6	2.5E1	1.0E-6	1.8E1	7.9E-7	4.9E-1	7.8E-8

Table 2b. Dissolution rates and Fractional Dissolution Rates of U From UO_2 Fuel (Burnup 610 GJ/kg U)

	DR	FDR	DR	FDR	DR	FDR	DR	FDR	DR	FDR
Distilled	8.0E0	1.1E-7	1.0E1	1.3E-7	1.0E0	4.1E-8	1.3E1	1.8E-7	1.1E0	1.5E-8
River	1.0E1	1.3E-7	1.0E1	1.3E-7	3.9E1	5.4E-7	2.6E1	3.6E-7	6.4E0	8.8E-8

LR = Leach rate in Bq/day; FRR = Fractional release rate in fraction/day; DR = Dissolution rate in μg/day; FDR = Fractional dissolution rate in fraction/day; * = Rates after 800 days.

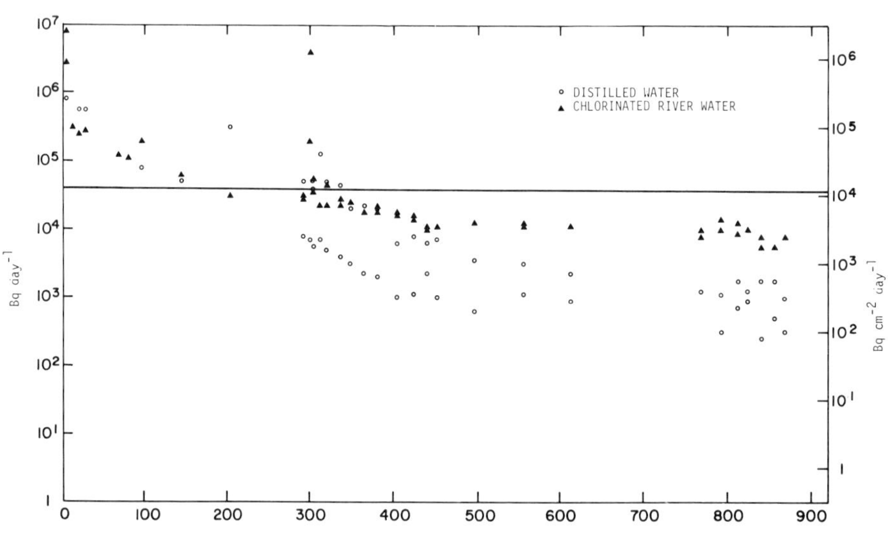

Fig. 1. ^{90}Sr Leach Rate as a Function of Time (Solid Line Denotes Arbitrary Reference of 1 ppm of ^{90}Sr Inventory)

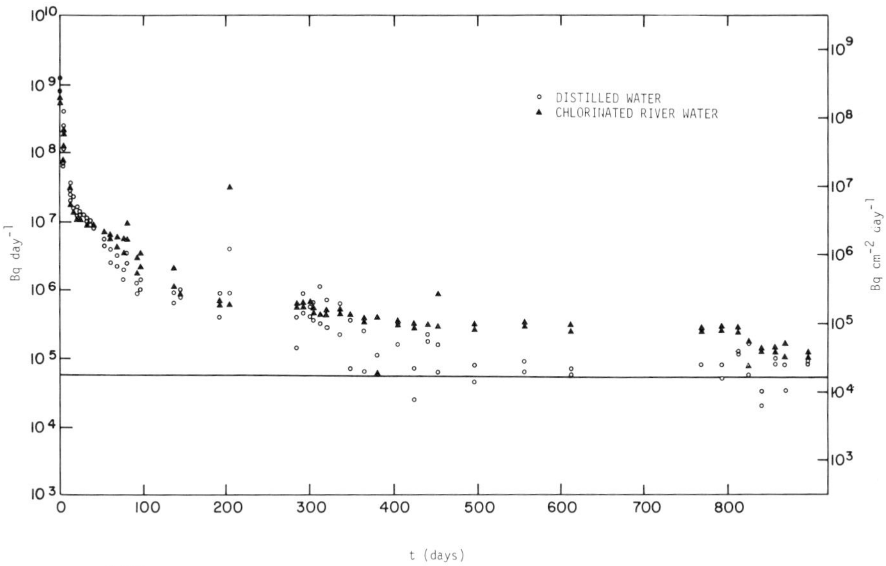

Fig. 2. ^{137}Cs Leach Rate as a Function of Time (Solid Line Denotes Arbitrary Reference of 1 ppm of ^{137}Cs Inventory)

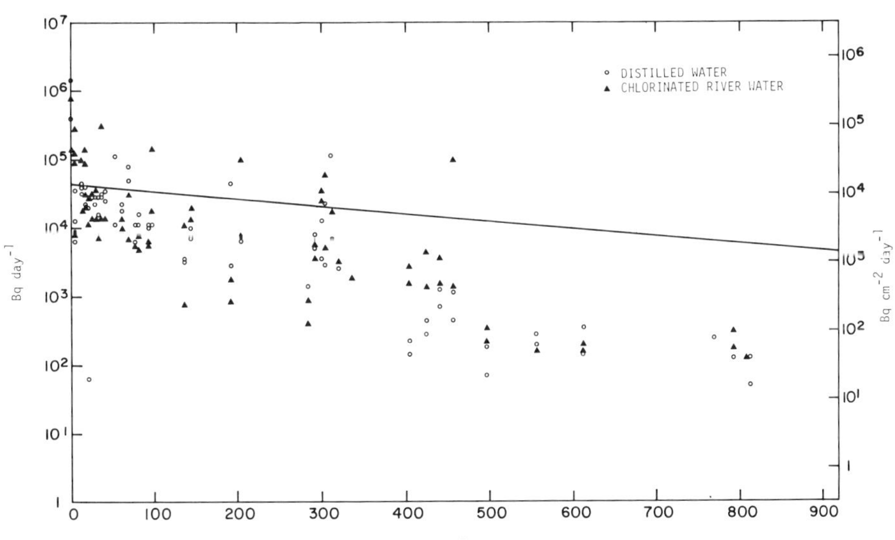

Fig. 3. ^{144}Ce Leach Rate as a Function of Time (Solid Line Denotes 1 ppm of ^{144}Ce Inventory as an Arbitrary Reference Line). Slope of the line is due to radioactive decay of ^{144}Ce

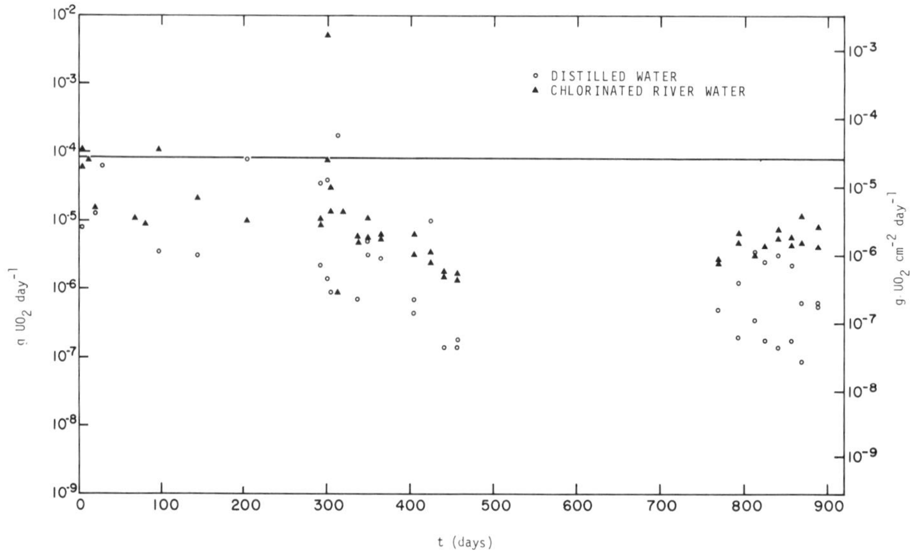

Fig. 4. Rate of UO_2 Dissolution as a Function of Time (Solid Line Denotes 1 ppm of Matrix as an Arbitrary Reference)

The rare earths and actinides, on the other hand, form solid solutions with UO_2 and tend not to migrate; and their leach rates are controlled by the dissolution rate of the fuel matrix. Alteration of the surface of the matrix structure by cutting of the fuel, accompanied by some inevitable oxidation of the UO_2 to U_3O_7 or U_3O_8 (5), helps to explain the initially high fractional release rates of the rare earths and actinides compared to those for uranium. With time, these leach rates approach the matrix dissolution rate and the rare earths and actinides will be leached congruently with the fuel matrix.

If matrix dissolution rates must be calculated from the leach rates of radionuclides from the fuel, they must thus be based on those nuclides that form solid solutions with the matrix. Rates based on cesium will result in erroneous values, as can be seen from the calculated and experimental dissolution rates shown in Table 3. The calculated rates were obtained from the fractional release rates of the various radionuclides (Table 2) and are based on an exposed surface area of the fuel of 3.2 cm^2. Compared to results obtained by Katayama (6), our values are generally higher. It is conceivable that we have considerably underestimated the surface area of the fuel and that much more than 3.2 cm^2 of fuel surface per sample is exposed to the leaching solutions. On the other hand, by removing the vessel with the leachates, we avoid the error of radionuclide sorption on the walls which would lead to low results.

Table 3. UO_2 Matrix Dissolution Rates In 10^{-6} g cm^{-2} day^{-1}

Based On	Leaching Solution	Days Leaching			BNWL Results
		100	300	900	140 d
^{90}Sr	Distilled	86	24	0.75	1
	River	86	30	6.0	0.4^{+}
^{106}Ru	Distilled	7.0	3.0	-	-
	River	7.0	2.8	0.30	-
^{125}Sb	Distilled	18	24	1.0	2
	River	110	52	37	1^{+}
Cs Av.	Distilled	550	260	45	7
	River	1300	280	52	5^{+}
Rare Earth Av.	Distilled	5.4	16	0.38	-
	River	16	13	0.53	-
Actinide Av.	Distilled	-	27	1.4	0.15*
	River	31	17	1.9	0.007^{+}
U^{++}	Distilled	1.0	4.1	0.33	-
	River	12	8.0	2.0	-

+ Results obtained in Hanford groundwater.
* Average of $^{239+240}Pu$ and ^{244}Cm results.
++Dissolution rates obtained from U analyses in the leachates.

The total amount of material leached from the fuel varies from ~4% for ^{137}Cs after 900 days to < 1% for the actinides. The largest fraction of the cesium is removed over the first few days.

The composition of the leaching solution has an effect on the leach and dissolution rates. An increase in these rates of up to an order of magnitude is seen with the chlorinated river water. This may be due to increased solubility resulting from formation of complexes. The effect of organic species (7) on the dissolution rate of UO_2 has been considered. However, analysis of the leaching solution has shown no detectable organic carbon (< 0.02 μg/g).

FUTURE WORK

Leaching studies similar to those reported here are underway at elevated temperatures (< 150°C), and leaching of whole, intentionally defected fuel elements under a variety of conditions is planned for the near future. The effect of chlorine is also being investigated.

SUMMARY

Irradiated fuel, leached at room temperature with distilled water and with slightly chlorinated river water, releases ∿4% of its cesium inventory over a comparatively short period of a few days but releases its actinides and rare earths more slowly. The matrix itself dissolves at a rate conservatively calculated to be less than $\sim 2 \times 10^{-6}$ g UO_2/cm^2 day and, with time, the leach rates of the various nuclides approach this value.

ACKNOWLEDGEMENTS

We would like to acknowledge the contribution by C. J. Provencal who developed the cesium separation procedure; the efforts of L. M. Carefoot who adapted the ^{90}Sr separation and carried out most of the chemical separations; W. A. Boivin, E. H. K. Lau, and R. E. Irwin who performed the radiometric analyses; and B. J. Komadowski and W. Tiede who assisted in sampling the leaching solutions in the hot cells.

REFERENCES

1. L. J. Clegg and J. R. Coady, "Radioactive Decay Properties of CANDU Fuel, Volume 1," The Natural Uranium Fuel Cycle, Atomic Energy of Canada Limited Report, AECL-4436/1 (1977).
2. J. van R. Smit, J. J. Jacobs, and W. Robb, "Cation Exchange Properties of the Ammonium Heteropolyacid Salts," J. Inorg. Nucl. Chem. 12, 95-103 (1959).
3. W. B. Lewis, J. R. MacEwan, W. H. Stevens, and R. G. Hart, "Fission Gas Behavior in UO_2 Fuel," Proc. Third International Conference on Peaceful Uses of Atomic Energy, Geneva 1964 11, 405-415 (1965).
4. U. B. Ekland and R. Forsyth, Leaching of Irradiated UO_2 Fuel, KBS Report 70, February (1978).
5. D. G. Boase and T. T. Vandergraaf, "The Canadian Spent Fuel Storage Canister: Some Materials Aspects," J. Nucl. Mat. 49, 105-107 (1973).
6. Y. B. Katayama, "Leaching of Irradiated LWR Fuel Pellets in Deionized and Typical Groundwater," BNWL-2057 (1978).
7. D. E. Grandstaff, "A Kinetic Study of the Dissolution of Uranite," Econ. Geol. 71, 1493-1505 (1976).

LOW TEMPERATURE CERAMIC WASTE FORM: CHARACTERIZATION OF MONAZITE CEMENT COMPOSITES

D. M. Roy, B. E. Scheetz, L. D. Wakeley and S. D. Atkinson

Materials Research Laboratory
The Pennsylvania State University
University Park, PA 16802

INTRODUCTION

Central to the concept of geologic waste isolation is the design of a stable radioactive waste form. This paper describes research on an alternative low-temperature ceramic waste form. The concept is to utilize modified calcium silicate, aluminate and other cements as matrix solidification utilizing a hydration bonding or other low temperature bonding mechanism. Previous workers have shown that conventional cements are suitable for solidifying a variety of wastes (1). Advantages of cement-based solidification include relative ease of processing (2) and potentially low cost; indeed, intermediate level wastes have been incorporated in cementitious grouts at ORNL for a number of years (3).

While conventional cements promise to serve well in many applications, further optimization was desired in order to generate a durable low-temperature ceramic waste form. One possibility involves low-temperature "hot-pressing"; and studies of simulated Purex wastes PW-4b and PW-4c treated at temperatures in the range 423-523K (150-250°C) were made earlier (4, 5). The products had very high strengths (up to 345 MPa [50,000 psi] compressive strengths) and were dense, impermeable and characteristically leach resistant. Related processing at modestly elevated temperatures and low pressures (6) is also expected to have much potential. The most recent work has extended the studies of "hot-pressed" formulations to those based upon multiphase supercalcine-ceramic (7) cement combinations (8) and has also involved elevated temperature, atmospheric pressure studies. Products which were resistant to accelerated hydrothermal leaching were found.

In order to predict the behavior of a multiphase material such as a supercalcine-ceramic in combination with a matrix material such as cement, it is desired to understand the behavior of the major component phases. This paper examines the behavior of one component, a monazite structure utilizing $NdPO_4$, in combination with various types of cement, to be followed by future studies of other components. Two types of cementitious composite matrix were studied. The first composite was composed of three readily available as-received cements -- two silicate based cements, a Class C and H, and an aluminate based cement, Secar. Bulk chemical analyses for these three cements are presented in Table 1. The second set of composite cements used the initial three cements but with adjusted and more potentially desirable bulk chemical compositions. The formulations investigated in this study are presented in Table 2. Phase pure $NdPO_4$, monazite, was prepared for this study by first achieving a chemical mixing in the aqueous solutions of stoichiometric amounts of 1 molar $Nd(NO_3)_3$ and

Table 1. Bulk Chemical Analyses for the Cement Samples

oxide	PSU Code C2 class C wt/o	PSU Code H3 class H wt/o	PSU Code S2 Secar 71 wt/o
SiO_2	21.2	22.7	0.3
Al_2O_3	2.72	3.8	70.8
Fe_2O_3	5.17	4.8	0.25
MgO	0.90	1.6	0.5
CaO	65.7	65.0	27.2
SO_3	2.62	1.5	---
P_2O_5	0.05	----	---

Table 2. Bulk Compositions for Individual Samples

	% of total solids			added	
sample number	Class C	Class H	Secar	SiO_2	$NdPO_4$
79C200-001	100	---	---	---	---
79C200-002	61.5	---	---	38.5	---
79C2M1-003	83.5	---	---	---	16.6
79C2M1-004	51.3	---	---	32.0	15.8
79H300-005	---	100	---	---	---
79H300-006	---	62.3	---	37.7	---
79H3M1-007	---	83.4	---	---	16.6
79H3M1-008	---	52.1	---	31.3	16.6
79S200-009	---	---	100	---	---
79S200-010	---	---	61.8	38.2	---
79S2M1-011	---	---	83.4	---	16.6
79S2M1-012	---	---	51.5	31.8	16.7
79H300-013	---	100	---	---	---
79H3M1-014	---	83.4	---	---	16.6

$(NH_4)H_2PO_4$. This step was followed by 378K (105°C) dehydration and denitration for 4 hours at 873K (600°C). The final phase pure monazite was prepared by firing the denitrated powder at 1473K (1200°C) for 24 hours. Phase purity was determined by X-ray diffraction. The product from this preparation procedure matched very well to the PDF #25-1065 standard pattern. All samples used in this study were prepared following API* RP-10B procedures. The individual samples were prepared with the minimum w/c** ratio, approximately 0.3, and cast into polyethylene vials, compacted on a vibrating table for 60 seconds and initially cured for 24 hours >95% relative humidity at 333K (60°C). After the initial curing period, the molds were removed and the samples were coated with a polymer and returned to the 333K (60°C) curing chamber.

Chemistries of the initial silicate based cements were adjusted to approximate the tobermorite composition; and in the aluminate based, silica was added to achieve the desired bulk composition. All of the different modifications were prepared both with and without the monazite waste form component. These formulations are detailed in Table 2.

Each of the individual samples was designed to follow a curing schedule of 28d, 56d, 6m, 12m and store. For this paper, results of the 28-day sampling period are reported. Hydrothermal accelerated leach testing of these formulations was performed on samples that had cured for 28 days. The leaching was conducted in sealed noble-metal capsules with a 10-fold excess of deionized water to the mass of sample. The hydrothermal leaching was conducted at 473K (200°C) and 30.4 MPa (300 bars) of overpressure for 28 days. After the hydrothermal treatment, all solutions in contact with the cementitious composite were collected and analyzed for Ca, Al, Fe, Mg, Si, Nd and P by means of an atomic emission spectrophotometer. Detailed discussions of the treatment procedures and liquid sampling procedures are presented in Reference 9. The solid components of the hydrothermal experiments were characterized by bulk X-ray diffraction and by scanning electron microscopy.

RESULTS AND DISCUSSION

X-ray Diffraction Characterization of the 333K (60°C) Cured Samples

Class C Cement Samples. Bulk X-ray diffraction characterization of the 28-day cured samples of the as-received cement indicated the formation of portlandite, CSH gel and calcite along with some unre-

*American Petroleum Institute procedure.
**w/c - weight water/cement.

acted β-C_2S.* This assemblage was contrasted to the chemically adjusted formulation which indicated the formation of CSH gel and calcite with unreacted β-C_2S. Although crystalline tobermorite was not formed with 333K (60°C) curing, the absence of the deleterious portlandite phase suggests that all of the available calcium is being used for the formation of the tobermorite precursor, CSH gel. The presence of monazite does not alter the phases that were observed. In both cases, the same phase assemblage was identified.

Class H Cement Samples. Although the bulk chemistry of the Class C and H cements are similar, the differences in coarseness of the grinding between the two products produce differences during hydration. The coarser Class H cement hydrated to form CSH gel and calcite after 28 days. Unreacted β-C_2S was also identified in this hydration product. This behavior is contrasted to the finer grind Class C cement which contained appreciable amounts of portlandite at the same curing interval. The Class H cement adjusted to tobermorite contained the same hydration products, with calcite, β-C_2S, plus detectable amorphous silica. When $NdPO_4$ was incorporated into the as-received cement, the formation of CSH gel did not occur in detectable quantities, with the resultant formation of portlandite to accommodate the excess calcium oxide. Both residual β-C_2S and calcite along with $NdPO_4$ were detected. The adjusted compositions also failed to indicate the presence of much CSH gel; however, no portlandite was formed. The phases calcite, β-C_2S and monazite were identified in the products along with the possible presence of α-C_2SH and C_4AF.

Secar Cement Samples. In the presence of excess water, the bulk chemical composition of this cement is very nearly that of the hydrogarnet. Therefore, it is not surprising that the as-received cement readily formed C_3AH_6 along with possible $C_4A_3H_3$. The anorthite adjusted formulation also contained these two hydration products with amorphous silica. The presence of monazite did not alter the hydration products in either the as-received cement or the compositionally adjusted one.

Scanning Electron Microscopy

The samples cured 28 days all appeared as dense competent ceramic bodies. Characterization of these products revealed typical cementitious morphologies (10), while monazite appeared as discrete grains dispersed throughout the cementitious matrix. Careful examination did not reveal any monazite-cement interactions across the

*Cement chemistry abbreviations: C = CaO, A = Al_2O_3, S = SiO_2, F = Fe_2O_3, H = H_2O.

grain boundaries, although the monazite grains were very tightly bonded into the matrix.

X-ray Diffraction Characterization of the Hydrothermally Leached Samples

Class C Cement Samples. CSH gel that was formed at 333K (60°C) curing conditions was crystallized to tobermorite in minor amounts after the hydrothermal treatment. Unreacted β-C_2S was detected as was portlandite. For the formulation that was adjusted to tobermorite, extensive formation of both tobermorite and xonotlite were observed with only minor amounts of unreacted β-C_2S remaining. In the presence of monazite, the as-received cement contained the same phase assemblage -- portlandite, minor tobermorite and β-C_2S. However, the adjusted formulation with monazite indicated only tobermorite and monazite with minor α-C_2SH.

Class H Cement Samples. Tobermorite, portlandite, calcite and possibly C_4AF were recognized in the reaction products of the as-received cement, along with unreacted β-C_2S. The compositionally adjusted sample without monazite crystallized to a mixture of tobermorite and xonotlite. For those samples that contained monazite, the above observation of individual phases was the same, suggesting that very little if any interaction between the cementitious matrix and $NdPO_4$ occurred.

Secar Cement Samples. Extensive formation of hydrogarnet, C_3AH_6, and γ-AlOOH, boehmite, with a trace of calcite were characterized in the reaction products of the as-received Secar cement. On the other hand, the compositionally adjusted formulation crystallized to tobermorite, boehmite and α-C_2SH instead of anorthite. The as-received Secar plus monazite formed the same set of reaction products. The adjusted formulation with monazite, however, produced boehmite, hydrogarnet and hexagonal anorthite (11).

Scanning Electron Microscopy

Characterization of the reaction products of these hydrothermally leached samples typically revealed enhanced crystal growth. Surface characterization of these samples generally did not indicate monazite because of the extensive hydrous calcium silicate formation. Tobermorite was readily recognized as fibrous crystals. The as-received samples of the silicate based cements typically exhibited crystal growth of the normal phases described above. The aluminate based cement exhibited extensive development of a ground mass of spherical clusters of blade-like crystals of boehmite with euhedral hydrogarnet trapezohedra commonly twinned.

Solution Characterization after Hydrothermal Leaching

Duplicate experimental runs for each sample in Table 2 were hydrothermally treated at 473K (200°C) and the solution phase analyzed. Only Ca and Si were routinely detected in solution for the silicate based cements while Ca, Si and Al were detected in the aluminate based cement. For those samples in which monazite was incorporated, no detectable Nd nor P was observed in the solutions. This observation, coupled with the results of the SEM, strongly suggests that the monazite component of a tailored ceramic waste form should be stable in the presence of a cementitious matrix material. This statement is in harmony with the anticipated behavior of monazites found in nature (12), which suggest excellent phase stability. Ancillary to this observation, an examination of the data in Table 3 suggests that the increased amount of Ca and Si in solution in the silicate based cements is a function of the larger CaO/SiO_2 ratios designed into the individual formulations.

CONCLUSIONS

Experiments were designed to study the solidification of monazite, in which Nd simulates rare earths and actinides, in a cementitious matrix. The initial results of these experiments are encouraging. Detailed SEM examination of the composite waste form suggests that the monazite phase is being bound into the matrix. No interactions across the grain boundaries were observed, but some exchange might be anticipated with the minor amount of calcium phosphate present in the cement with prolonged curing at elevated temperature and pressure. No detectable Nd or P was found in the solution analyses for these experiments after a severe accelerated leach treatment at 473K (200°C) for 28 days, which is also very encouraging.

ACKNOWLEDGEMENT

This research was partially supported by the U.S. Department of Energy, Prime Contract DE-AC09-79ET41900.

REFERENCES

1. D. M. Roy, Proc. Conf. on High Level Radioactive Solid Waste Forms, U.S. Nuclear Regulatory Commission, Dec. 18-21, Denver, CO, NUREG/CP-0004 (Leslie A. Casey, ed.; 1978).
2. J. A. Stone, Evaluation of Concrete as a Matrix for Solidification of Savannah River Plant Waste, DP-1448, June (1977).
3. J. G. Moore and E. W. McDaniel, Cerm. Bull. 57, 324 (1978).

4. D. M. Roy and G. R. Gouda, Hot-Pressed Cement in Radioactive Waste Mangement, Report to Battelle Pacific Northwest Laboratories, BSA Subcontract 841, August (1974).
5. D. M. Roy and G. R. Gouda, Nucl. Tech. 40, 214-219 (1978).
6. S. O. Oyefesobi and D. M. Roy, Cem. Concr. Res. 7, 95 (1977).
7. G. J. McCarthy, Trans. Am. Nucl. Soc. 23, 168-169 (1976).
8. D. M. Roy, B. E. Scheetz, M. W. Grutzeck, A. K. Sarkar and S. D. Atkinson, Proc. Intl. Symp. Ceramics in Nuclear Waste Management, (T. D. Chikalla and J. E. Mendel, ed.; Nucl. Div. Am. Ceramic Soc. and U.S.D.O.E., April 30-May 2, 1979a).
9. G. J. McCarthy, W. B. White, S. Komarneni, B. E. Scheetz, W. P. Freeborn and D. K. Smith, Proc. Intl. Symp. Ceramics in Nuclear Waste Management (T. D. Chikalla and J. E. Mendel, ed.; Nucl. Div., Am. Ceramic Soc. and U.S.D.O.E., April 30-May 2, 1979a).
10. C. A. Langton, M. W. Grutzeck and D. M. Roy, Am. Ceramic Soc. Bull. 57 (3), 324 (1978).
11. D. M. Roy, E. L. White, C. A. Langton and M. W. Grutzeck, Proc. 1979 SPE Intl. Symp. Oilfield and Geothermal Chemistry, Houston, TX, 153-161, January (1979b).
12. G. J. McCarthy, W. B. White and D. E. Pfoertsch, Mat. Res. Bull. 13, 1239-1245 (1978).

STRUCTURE AND STABILITY OF INCINERATED α-WASTE PRODUCTS*

R. De Batist,[+] E. De Grave,[++] W. Timmermans,[+] J. Vangeel[+++] and P. Van Iseghem[+]

[+]Materials Science Department, S.C.K./C.E.N., B-2400 Mol, Belgium; [++]Laboratory of Magnetism, University of Gent, B-9000 Gent, Belgium; [+++]Waste Department, S.C.K./C.E.N., B-2400 Mol, Belgium

INTRODUCTION

The management of α-bearing wastes is determined to a very large extent by their plutonium content, i.e., by technico-economic considerations about recoverability of the contained fissile material. Supposing the plutonium concentration has been reduced to a sufficiently low level, the volume of the remaining waste material has to be reduced as much as possible and insolubilized (e.g., in concrete, bitumen or glass) so that it can be disposed of safely (1). Depending on the techniques used for volume reduction, the waste has to be separated into, e.g., combustible and non-combustible materials. This additional step can be avoided in an integrated system allowing incineration of a mixture of combustible and non-combustible waste. A further advantage of such an integrated system is that the resulting product may already be sufficiently insoluble for disposal. Such a one-step incinerator has been installed at S.C.K./C.E.N. and is presently being prepared for operation under α-conditions. The incinerator is a FLK type furnace with a capacity of 100 to 150 kg. h^{-1}, operating at 1400° to 1500°C and has been described in more detail earlier (2). The most significant feature for the present context is, in addition to the relatively high working temperature, the very short melting time: the feed material drips down the feed cone as soon as it is sufficiently fluid and it is collected and very rapidly solidified in a water bath below the furnace. The resulting

*Part of this work has been supported by the Commission of the European Communities.

product takes the form of small, porous, amorphous granules which may be hot pressed for further volume reduction. In this contribution, we will first discuss some of the properties of the FLK product, both in granular form and following hot pressing, as obtained in a number of typical furnace runs under nonactive and slightly $\beta-\gamma$ active conditions. A second part contains preliminary results of a supporting basic study carried out with reference glasses designed to simulate the FLK product. This supporting research aims at clarifying the glass-forming abilities of, e.g., iron, aluminum and copper in oxide mixtures as appearing in the FLK product and at investigating the microstructural behavior of uranium and plutonium.

FLK PRODUCTS

Typical compositions of the waste materials fed to the FLK furnace and of the slags produced in the incinerator are given in Table 1. The slags resulting from two representative FLK runs have been further characterized in terms of granulometry, microstructure, density, microhardness (Vickers test) and leach rates. Results are summarized in Table 2. Also included in this table are results of hot-pressed compacts. Of the two selected FLK runs, one has been performed with non-radioactive feed material, whereas the other concerns slightly $\beta-\gamma$ active material (about 1 $\mu Ci.kg^{-1}$). The hot-pressing tests were carried out with non-active granules at various temperatures (700°, 800°, 900°C) and various pressures (0.16, 0.33, 0.5, 1 GPa). In addition to the data given in Table 2, the following observations are worth mentioning:

-- the morphology of the granules is rather irregular;
-- except for the lowest temperature (700°C) and pressures (0.16 and 0.33 GPa), a macroscopically monolithic block is obtained by hot-pressing;
-- the granules are mainly amorphous, have a basalt-like appearance and contain small amounts of crystalline spinel-type precipitates consisting of Fe_2O_3, $NiFe_2O_4$ and $(Fe,Mg)(Cr,Fe)_2O_4$; hot-pressing usually yields a higher fraction of crystallinity;

Table 1. Composition of FLK-Feed and of FLK-Slags (Wt%)

Feed	wt %	Slags	wt %
Paper	20 - 50	SiO_2	50 - 70
Plastics	10 - 40	$FeO + Fe_2O_3$	10 - 30
Ion Exchangers	1 - 5	Al_2O_3	5 - 20
Chemical Sludges	5 - 10	CuO	0 - 10
Metals	5 - 10	Alkali Oxides	5 - 10
Glass	5 - 10	Alkaline Earth Oxides	5 - 10
Water	5 - 20	Others (mainly oxides of Mo,Cr,Ti,Ni...)	0 - 5

Table 2. Characteristics of FLK Granules and Compacts

Characteristics	Granules	Hot-Pressed Compacts
Granulometry (%)		
0.5 - 5 mm	75	
5 - 10 mm	15	
Density (10^3 kg m^{-3})		
tap	1.70-1.80	
air	3.40-3.70	3.55-3.60
Microhardness (10^7 N m^{-2})		
glass matrix	520-570	590-630
BET specific surface area (m^2 g^{-1})	0.10-1.00	0.20-0.30
Leach rate (g cm^{-2} d^{-1})		
- at 100°C		
first 72 h	$(2.8\text{-}4.3)\times10^{-4}$	$(1.4\text{-}1.7)\times10^{-4}$
after 21 d	$(0.6\text{-}2.0)\times10^{-4}$	$(1.0\text{-}1.3)\times10^{-4}$
- at 25°C (demineralized water)*		
for Si, Al	$(1\text{-}10)\times10^{-11}$	
for Ca, Sr	$(1\text{-}6)\times10^{-8}$	

*Leaching tests at 25°C were performed over a period of 18 months with granules from earlier FLK runs. Leach rates were measured for specific elements, using a BET specific surface area of 0.2 m^2/g in each case.

-- for any given run, the composition is fairly homogeneous, although the larger granules have a tendency to contain larger amounts of iron oxide and copper oxide;

-- both wet chemical analysis and Mössbauer spectroscopy indicate the iron to be for about 85% in the divalent state; furthermore, roughly 60% of the ferrous ions appear with tetrahedral coordination, while the rest of the ferrous ions as well as the ferric ions show octahedral coordination;

-- the geometrical specific surface area ranges from 6.10^{-4} m^2 g^{-1} for the larger fraction to 19.10^{-4} m^2 g^{-1} for the smaller fraction; since the pore diameter, as determined in BET experiments, is found to be less than 3 nm (for the largest fraction of pores), it seems preferable to use this geometrical surface area for evaluating leaching experiments rather than the much larger BET values incorporating these very fine pores which can be considered to be unaccessible for the water in the leach test; this would increase the 25°C leach rates with a factor of about 200.

Table 3. Data for the Reference Glasses (Compositions are Given in Mole %)

Series	nr	SiO_2	Al_2O_3	Fe_2O_3	Li_2O	Na_2O	K_2O	MgO	CaO	BaO	UO_2	melting temp. (°C)	melting time (h)	Soxhlet leach rate ($\times 10^{-5} gcm^{-2}d^{-1}$)	Leach rate after 14 days ($\times 10^{-5} gcm^{-2}d^{-1}$)
I	1	71.5	2.0	3.0	-	6.6	2.2	5.1	9.7	-	-	1400 1500 1600	3	4.1	1.3
	2	70.0	2.0	4.5	-	6.6	2.2	5.1	9.7	-	-	1400	3	14.8	3.5
	3	69.2	2.0	5.8	-	6.6	2.2	5.1	9.7	-	-	1400	3	6.5	2.8
	4	66.4	2.0	8.1	-	6.6	2.2	5.1	9.7	-	-	1400	3	27.6	6.0
	5	64.5	2.0	10.0	-	6.6	2.2	5.1	9.7	-	-	1400	3	43.4	15.0
	6	62.3	2.0	12.0	-	6.6	2.2	5.1	9.7	-	-	1400	3	36.7	39.0
II	1	79	-	5	4	-	4	4	4	-	-	1550	3	1.3	0.2
	2	74	-	10	4	-	4	4	4	-	-	1400 1550	1 - 3	2.4	1.5
	3	69	-	15	4	-	4	4	4	-	-	1550	3	10.9	1.0
	4	64	-	20	4	-	4	4	4	-	-	1400 1550	1 - 3	2.8	1.7
	5	54	-	30	4	-	4	4	4	-	-	1400 1550	1 - 3	8.1	3.0
III	1	70.0	2.0	2.9	-	4.2	4.2	5.0	9.5	-	2	1500	3	12.4	4.0
	2	70.0	2.0	2.9	4.2	4.2	-	5.0	9.5	-	2	1500	3	10.6	6.5
	3	70.0	2.0	2.9	4.2	-	4.2	5.0	9.5	-	2	1500	3	3.6	0.6
	4	70.0	2.0	2.9	-	4.2	4.2	7.25	7.25	-	2	1500	3	28.2	2.5
	5	70.0	2.0	2.9	-	4.2	4.2	-	7.25	7.25	2	1500	3	9.1	2.0
	6	70.0	2.0	2.9	-	4.2	4.2	7.25	-	7.25	2	1500	3	6.4	2.0

REFERENCE GLASSES

Since one of the unavoidable characteristics of this waste is its variability in composition, it is very important to understand the influence of the various constituents on the final properties of the products. Therefore, a number of reference glasses have been prepared, characterized firstly by a relatively small number of constituents and secondly by a quantitatively well-defined composition. Based upon the components found in the actual FLK material, the glass compositions given in Table 3 were selected. The following considerations were applied in arriving at these compositions:

-- for series I, the iron oxide content is varied in order to investigate its influence on glass formation;

-- for series II, the iron oxide content is still further increased, whereas the alkali oxides and alkaline-earth oxides are more evenly distributed over two different M_2O's and MO's; furthermore, differences in melting procedure and in cooling rate are investigated;

-- in series III, the effects of mixing M_2O's and MO's are further studied (3);

-- the role of the actinides is to be investigated by means of the behavior of UO_2 and PuO_2 (simulated by CeO_2 in preliminary experiments).

Table 4. Ionic Properties of Iron and Aluminum (4)

Ion	Electronegativity	Ionic radius (nm)
Fe^{2+}	1.8	0.077
Fe^{3+}	1.9	0.065
Al^{3+}	1.5	0.053
Glass Former	1.5 - 2.1	0.031 - 0.058

In the present contribution, however, we will be concerned primarily with the behavior of iron oxides in silica glasses representative for FLK material.

Considering the ionic properties of both Fe^{2+} and Fe^{3+} in comparison with those of Al^{3+} and with those expected for a good glass former (Table 4), one concludes that neither is a real glass former, although they might be relatively close. Aluminum is known to be able to act as glass former, leading to fourfold coordination in the silica-glass network provided each Al^{3+} ion is compensated by an alkali ion (5). If the ratio of aluminum ions to alkali ions exceeds one, the excess aluminum cations become glass modifiers. Following this line of thought, it is conceivable that iron ions, too, might be incorporated in the glass network provided a sufficient amount of glass modifiers is present for compensation of the glass-forming irons. It is obvious that the way iron is incorporated in the glass matrix may strongly influence the chemical stability of the glass. Results reported thus far in the literature concerning the behavior of iron ions in glass structures are far from being conclusive (6-12). In the present work, the structural behavior of iron ions in glass is studied by correlating data obtained by means of Mössbauer spectroscopy (valence state, coordination) with the chemical stability as studied by Soxhlet experiments.

Experimental Conditions

The glasses are prepared by melting in glazed porcelain or Pt-10% Rh crucibles thoroughly mixed fine powders of SiO_2, Fe_2O_3, (UO_2), alkali and alkaline earth carbonates, sodium or potassium feldspars. ^{57}Fe-Mössbauer spectroscopy is done at the State University of Gent, using a time-mode spectrometer with 1024 channels and a channel width of 0.017 mm/s. Leaching tests at 100°C are performed using a Soxhlet technique with a quartz vessel and specimen holder. The specimens are polished with 600 grit SiC paper. Corrosion rates are evaluated by means of weight loss measurements with periodicities of about 1 day (without disturbing the surface layer).

Experimental Results

The results of Mössbauer spectroscopy are summarized in Fig. 1. and 2. The Soxhlet leach rates (averages over the first 3 days) as well as the leach rate after 14 days at 100°C are given in Table 3.

Discussion

From the preliminary experimental results available thus far, the following characteristic features can be extracted:
-- The Mössbauer data for Fe^{2+} lead to values for the isomer shift between 0.78 and 0.83 mm s^{-1} and for the quadrupole splitting of $\lesssim$ 2.0 mm s^{-1}, suggesting fourfold coordination; the Fe^{3+} ions, on the other hand, show both fourfold and sixfold coordination. As also observed by others, this suggests that Fe^{3+} can be either glass modifier or glass former, in which case the $(Fe^{3+}O_{4/2})^-$ tetrahedron should, for electrical neutrality, be compensated by combination with an interstitial monovalent cation (6, 8-11). The fourfold coordination of Fe^{2+}, which contrasts with results reported in References 7-9

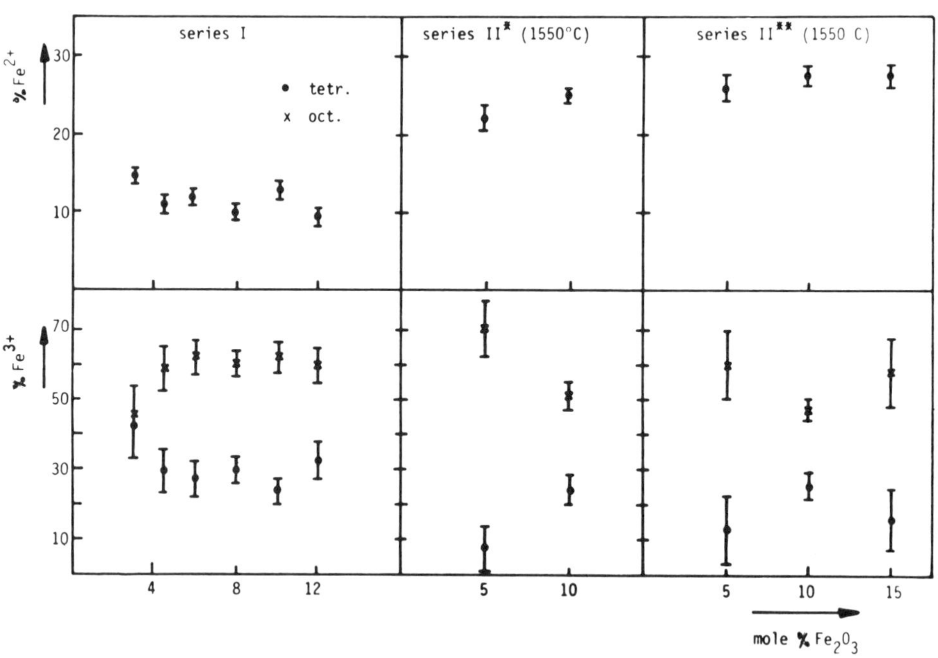

Fig. 1. Fe^{2+}, Fe^{3+}-Distribution as a Function of Fe_2O_3 Content (*Air Cooling; **Water Quenching)

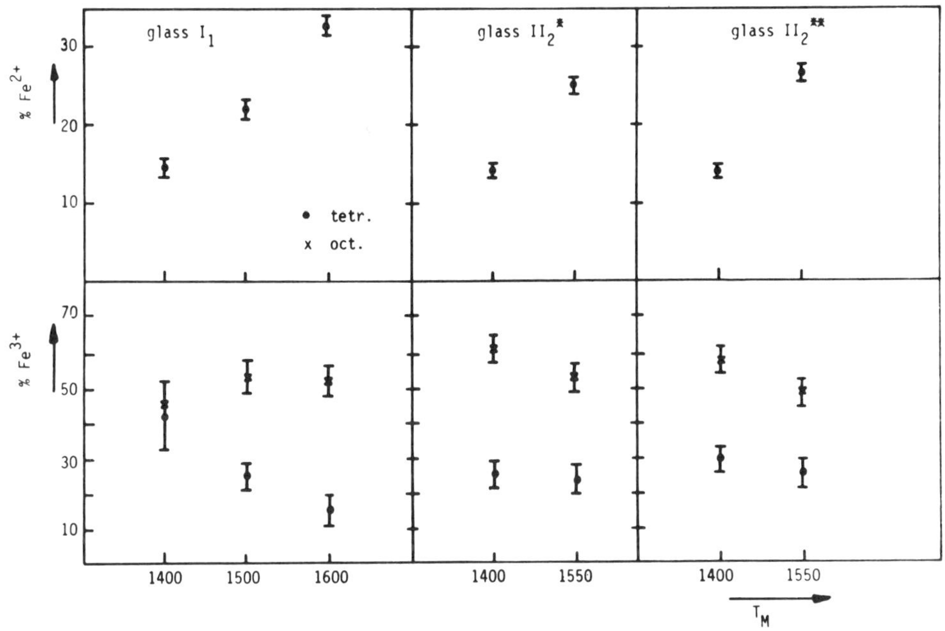

Fig. 2. Fe^{2+}, Fe^{3+}-Distribution as a Function of Melting Temperature (*Air Cooling; **Water Quenching)

and 12 but agrees with Labar and Gielen (10), combined with its ionic radius of 0.08 nm and with the covalent character of FeO, suggests that it too could be incorporated within the glass network. This implies that the $(Fe^{2+}O_{4/2})^{2-}$ tetrahedron requires, for electrical neutrality, interaction with two interstitial monovalent cations (or possibly with one divalent cation). Besides considerations of electrical neutrality, size effects clearly have also to be taken into account. This may explain why, for the compositions investigated thus far, the amount of tetrahedral coordination for the iron ions appears to be limited at 30-50% in spite of the fact that enough glass modifier is present.

-- As expected, melting conditions have a pronounced effect on the valence state of the iron. Thus, the ferrous content increases with increasing melting temperature (Fig. 2.), whereas the proportion of ferrous iron is even larger in the FLK granules as a result of the reducing atmosphere used in the FLK furnace (in view of volatility problems). One might also expect the amount of tetrahedral iron ions to increase with increasing melting temperature due to the decrease in density and the correspondingly increased probability for accommodating a large Fe^{2+} or Fe^{3+} ion within an oxygen tetrahedron. This expectation appears to be borne out, provided one takes into account the above mentioned limitation of fourfold coordination to about 50% of the iron. Compare, e.g., glass II 2 (Fig. 2.), where the ferric content in octahedral sites is found to decrease with increasing temperature, with glass I 1, where the 50% level is reached already at the lowest temperature. It is interesting to notice that also for the FLK granules, a 50-50 ratio between fourfold and sixfold coordination is found.

-- Whereas melting time and total iron content appear to have very little if any effect on the valence and coordination states of the iron, the cooling rate is found to strongly influence the glass formation ability. Increasing iron content above 10 mole % leads to crystalline precipitates of Fe_3O_4 in air-cooled material, whereas 20 mole % also leads to partially crystalline material in water-quenched melts.

-- The average leach rates for the reference glasses presented in Table 3 vary between $2x10^{-6}$ g $cm^{-2}d^{-1}$ and $4x10^{-4}$ g $cm^{-2}d^{-1}$, as compared with a value of about 10^{-4} g $cm^{-2}d^{-1}$ for the FLK granules. The following preliminary comments can be made:

-- For series I, leachability can be correlated with the concentration of glass forming cations (Fig. 3.).

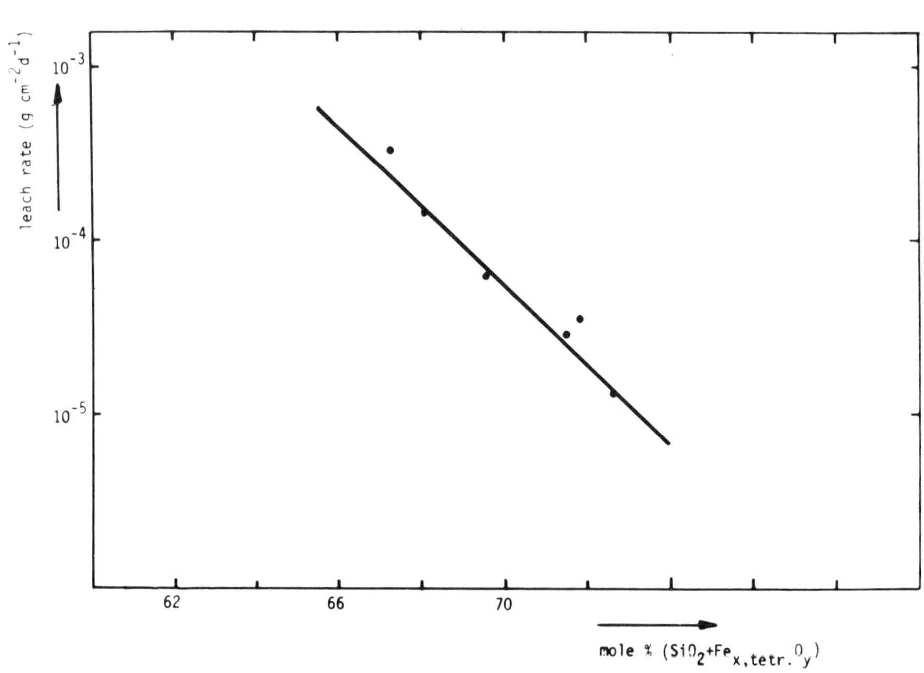

Fig. 3. Leach Rate at 100°C as a Function of the (SiO_2 + $Fe_{x,tetr.}$ O_y)-Content

-- For series II, the occurrence of Fe_3O_4 crystallites does not appear to lead to an increase in leachability.

-- The mixed-alkali effect leads to a pronounced improvement in chemical durability; this is also indicated by a comparison of the results for glasses II 2 and I 5, containing both 10 mole % of Fe_2O_3 (3). A similar observation can be made for mixing alkaline earth oxides.

-- SEM observations of glass surfaces following 14 days of Soxhlet testing indicate the formation of an iron rich surface layer which appears to cover an underlying layer which is also enriched in iron with respect to the bulk glass.

CONCLUSIONS

The preliminary experimental results discussed in this contribution indicate the strong influence of the melting conditions upon the structural behavior of iron in silica glasses. It has been shown that a considerable fraction of the iron can be incorporated within the glass network. Some indications about the effects of iron on the chemical durability are also derived.

REFERENCES

1. J. E. Mendel, BNWL-2117 (1976).
2. N. Van de Voorde, G. Dumont, J. Claes, A. Taeymans, J. Vangeel and P. Dejonghe, IAEA-SM-207/7, Proc. IAEA-NEA Intern. Symp. on the Manag. of Rad. Wastes from the Nucl. Fuel Cycle, Vienna, Austria (1976).
3. J. O. Isard, J. Non-Cryst. Sol. 1, 235-261 (1969).
4. W. D. Kingery, H. K. Bowen and D. R. Uhlmann, Introduction to Ceramics, (John Wiley & Sons, 1976).
5. D. E. Day and G. E. Rindone, J. of Am. Cer. Soc. 45, 579-581 (1962).
6. N. Iwamoto, Y. Tsunawaki, H. Nakagawa, P. Yoshimura and N. Wakabayashi, J. Non-Cryst. Sol. 29, 347-356 (1978).
7. C. R. Kirkjian and D. N. E. Buchanan, Phys. & Chem. of Glasses 5, 63-70 (1964).
8. L. Pargamin, C. H. P. Lupis and P. A. Flinn, Met. Trans. 3, 2093-2105 (1972).
9. R. A. Levy, C. H. P. Lupis and P. A. Flinn, Phys. & Chem. of Glasses 17, 94-103 (1976).
10. C. Labar and P. Gielen, J. Non-Cryst. Sol. 13, 107-119 (1973-1974).
11. C. R. Kirkjian and E. A. Sigety, Phys. & Chem. of Glasses 9, 73-83 (1968).
12. J. A. Boon and W. S. Fyfe, Chem. Geol. 10, 287-298 (1972).

KRYPTON-85 STORAGE IN SOLID MATRICES

G. L. Tingey, E. D. McClanahan, M. A. Bayne, W. J. Gray and *C. A. Hinman

Pacific Northwest Laboratory, Richland, WA 99352
*Hanford Engineering Development Laboratory, Richland, WA

INTRODUCTION

Of the several radioactive inert gas isotopes generated by the fission process in nuclear reactor fuels, only Kr-85 has a long enough half-life (10.73 years) to be considered for long term storage. Kr-85 decays to Rb-85 releasing a beta particle with a maximum energy of 0.66 MeV and a very low flux of 0.52 MeV gamma rays. The chemical inertness of this isotope reduces markedly the biological effects of the gas but contributes to its long lifetime as a component of the earth's atmosphere. In the past, the small quantities of Kr-85 generated and subsequently released from the fuel have been dispersed in the atmosphere, but federal regulations to take effect in 1983 limit the release to 50,000 Ci/gigawatt-yr of electrical power generated by a nuclear reactor (1). This amount is about 15% of that generated in a typical light water reactor (LWR) fuel. To meet this requirement, studies are underway to develop methods of separating the Kr from other reactor gases and storing it for periods of about 100 years until decay is largely complete.

Several storage methods have been suggested previously. In this paper, we will describe two of these which are under investigation in our laboratory; namely, ion implantation of Kr in sputter deposited metal and dissolution in low density glasses.

ION IMPLANTATION DURING SPUTTER DEPOSITION

Atoms can be implanted into solids by accelerating highly energetic ions toward the solid surface. The depth of penetration of the atoms is a function of ion kinetic energy, ion mass and density of the solid. For Kr^+ with energies in the keV range, this amounts to only a few atom layers. Thus, to produce bulk solids containing Kr, it is necessary to perform the implantation while building up the solid by some deposition technique. Sputtering of solids appears to be particularly well suited for gas incorporation where the sputtering ion is generated from the gas which is to be implanted in the deposit.

In our system, a dense thermionically supported plasma is generated in the annulus between a target and a substrate as shown in Fig. 1. Positive Kr ions are formed by collisions between energetic electrons and krypton gas atoms and are accelerated toward a negatively charged target surface. These highly energetic ions transfer momentum to the target atoms and "sputter" them from the solid with a cosine spatial distribution where they are deposited on any solid surface within their path. In most applications of sputter deposition, attempts are made to minimize gas incorporation in the deposit to optimize its physical and electronic properties. However, in our case, the gas content is maximized by applying a negative bias voltage to the substrate as well as the target and by maintaining the substrate temperature at below 25°C. Of course, the substrate voltage/target voltage ratio must be maintained at a value which will permit a net deposit on the substrate. The gas trapping rate for this process depends on the sputtering yield for the target, bombarding ion current and ion energy, the sputtering geometry and the sticking coefficient for the Kr ion in the material being deposited.

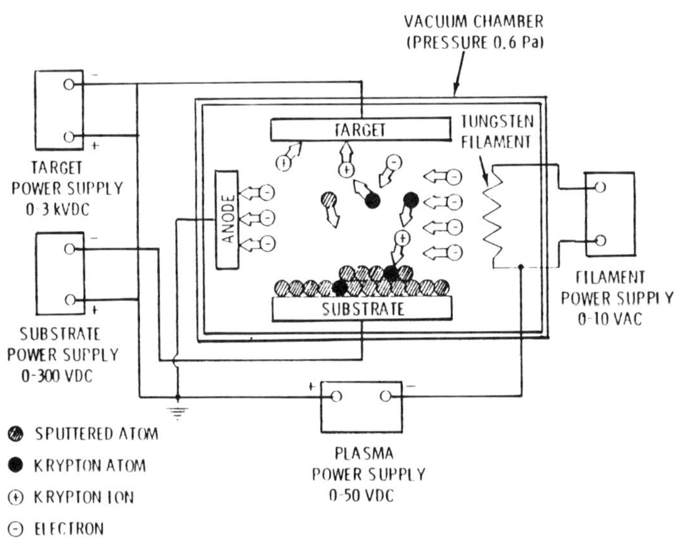

Fig. 1. Thermionically Supported Plasma Sputtering/Ion Implantation System

The details of the sputtering apparatus and the experimental procedures and parameters have been published earlier (2, 3). The initial phases of this work involved an attempt to adjust the sputtering parameters to yield a deposit with the highest gas content. Table 1 is a summary of the results of sputtering in several metallic substrates. Whenever possible, the targets were fabricated from commercially available materials. For the multicomponent alloys where no commercial materials were available, the targets were fabricated either by inserting plugs made of the minor metallic component into small holes drilled into the major target metal or by spot welding small wafers onto the surfaces of the target (4). The structure of the sputter deposited materials, as determined by X-ray diffraction, was found to be crystalline, glassy, or a mixture of both. In general, crystalline and mixed crystalline and glassy deposits contain Kr up to 5 atomic % and require a substrate bias of -1300 to -2500 volts which must be pulsed to maintain a useful net deposition rate. The metal alloys which produced glassy deposits were found to trap approximately twice as much Kr with a negative bias of only 250 volts applied continuously without reducing the net deposition rate appreciably. The gas pressure equivalent of the Kr concentration is given in Table 1 to show the comparison with gas bottle storage. Kr-85 is generally stored in gas bottles at 3.4 MPa (500 psi); thus the gas concentrations in the sputtered metals exceed that of gas bottles by up to a factor of six.

Gas release measurements from sputter deposited crystalline metals were performed by heating the sample while increasing the temperature at a linear rate of about 3°C/minute. During the heating cycle, the gas was continually removed for gas analysis. At lower temperatures, practically immeasurable quantities of Kr were released but some H_2 and CO was observed. Significant Kr release occurred at about 600°C for Fe and Ni samples and at about 300°C for Al. Gas release rates were also measured for the most promising glassy metals. An Arrhenius plot of the release of Kr from $Fe_{0.79}Y_{0.12}Kr_{0.09}$ is given in Fig. 2. At lower temperature, the release rate deviates from the linear plot, yielding rates up to three orders of magnitude higher than the extrapolation of the high temperature line. The low temperature rates decrease with time and approach a steady state rate about equal to that expected from the high temperature data. This steady state rate is approached early in the cycle well before 0.1% of the Kr is released. A similar plot is obtained for the $Fe_{0.76}Zr_{0.19}Kr_{0.05}$ but release occurs at about 100°C higher temperature.

Extrapolation of these data for release indicates that less than 2% of the Kr would be released in 10 years from $Fe_{0.79}Y_{0.12}Kr_{0.09}$ at 300°C and less than 5% from $Fe_{0.76}Zr_{0.19}Kr_{0.05}$ at 400°C. Long term release data are required to more accurately determine the rates of release in a storage environment, and tests have been initiated to obtain these data. The total Kr content was determined by integrat-

TABLE 1. Krypton Loadings in Sputter-Deposited Metals

Target Material	Sputtered* Product	Krypton Content: Cm^3 of Kr(STP) / g of deposit	Krypton Content: Cm^3 of Kr(STP / Cm^3 of Deposit	Pressure Equivalent (MPa)	Substrate Bias Voltage (Volts)	Deposition Rate (nm/sec)
Ni-200	$Ni_{0.95}Kr_{0.5}$ Crystalline	16.9	135	13.8	-1500 Pulsed	6.05
Al	$Al_{0.96}Kr_{0.04}$ Crystalline	30.1	75	7.8	-1300 Pulsed	5.0
A-108 Steel	$Fe_{0.95}Kr_{0.05}$ Crystalline	19.8	140	13.8	-2500 Pulsed	2.7
Ti	$Ti_{0.96}Kr_{0.04}$ Crystalline	17.0	70	6.9	-2300 Pulsed	4.1
Steel/Y Plugs	$Fe_{0.79}Y_{0.12}Kr_{0.09}$ Glassy	30	189	19.3	-225 Continuous	8.0
Steel/Y Plugs	$Fe_{0.97}Y_{0.03}$ Crystalline	$<$ 0.05	-	-	-250 Continuous	7.0
Steel/Zr Plugs	$Fe_{0.76}Zr_{0.19}Kr_{0.05}$ Mixed	17	120	12.1	-240 Continuous	15.0
Zircaloy IV/Fe Plugs	$Zr_{0.68}Fe_{0.24}Kr_{0.08}$ Glassy	22	143	14.5	-160 Continuous	11.0
Steel/Zr/Ta	$Fe_{0.70}Zr_{0.20}Ta_{0.05}Kr_{0.05}$ Glassy	16	110	11.0	-220 Continuous	11.0
316SS/Y Plugs	$Fe_{0.60}Cr_{0.20}Ni_{0.06}Y_{0.10}Kr_{0.04}$ Mixed	15	100	9.7	-250 Continuous	12.
316SS	$Fe_{0.69}Cr_{0.22}Ni_{0.07}Kr_{0.02}$ Crystalline	8	60	6.0	-2500 Pulsed	7.8

*Composition not specifically determined; previous experimental results show that metal deposit composition is approximately the same as the target.
Glassy--no crystals greater than 30A. Mixed--glassy with crystalline phases.

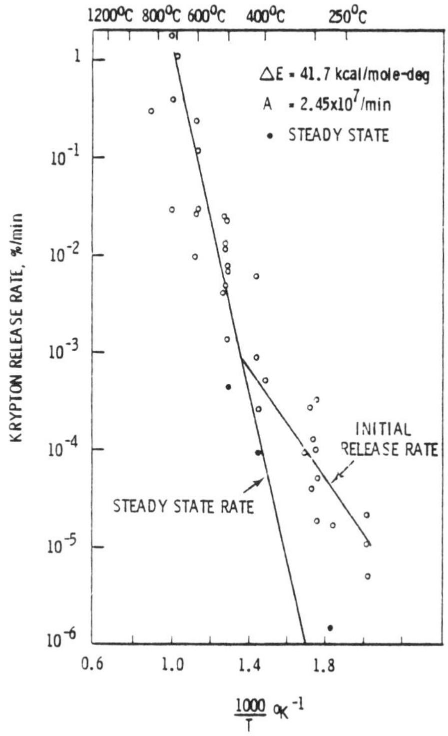

Fig. 2. Krypton Release Rate from Amorphous $Fe_{0.79}Y_{0.12}Kr_{0.09}$

ing the Kr release over the entire temperature range and is given in Table 1.

The mechanism of gas release from sputtered deposits and particularly glassy deposits is not currently understood. On the basis of very preliminary observations, release from crystalline metals appears to involve bubble information followed by bubble diffusion. Gas release from the glassy metals appears to be much more closely tied to the crystallization phenomenon. In the $Fe_{0.79}Y_{0.12}Kr_{0.09}$ deposit, the sample disintegrates to a fine powder during gas release. Apparently, crystallization and/or gas release causes high stress in the solid and fracture occurs. X-ray diffraction measurements after heating at various temperatures have also been made, and the Kr release correlates reasonably well with growth of the 110 α Fe peak.

CONTAINMENT OF KRYPTON IN LOW DENSITY GLASS

The solubility of gas in glass is a well-known phenomenon in glass technology often used as a parameter for tailoring the properties of glass. Shelby has reviewed the solubility, diffusivity and permeability of low molecular weight inert gases (He, Ne, A) in glasses as a function of glass composition (5). These data provided the encouragement to evaluate Kr solubility in glass as a method for long term storage of Kr-85.

Because of the very low expected diffusivity of Kr in glasses, initial studies were conducted on glass powders and porous materials such as silica gels to minimize the diffusion path required. A variety of glass types were studied by subjecting them to 193 MPa (1905 atm) pressure of Kr for 24 hours at 510°C. The Kr content of these glasses varied from 0.3 cm^3 of Kr (STP)/g of glass for fused silica powders to 4.5 cm^3 of Kr(STP)/g of glass for Vicor powders. Kr was released from these materials upon heating at temperatures between 400° and 600°C. The Kr solubility of all of these glasses was so disappointingly low that it was necessary to look for alternate materials. In 1966, Doremus published results of a study in which he showed that the solubility of inert gases in glass is a function of free volume in the glass matrix (6). In an attempt to increase the solubility and diffusion rate of Kr in glass, we obtained low density SiO_2-based glass rods which had about 30% open porosity and a surface area of 100 to 200 m^2/g. The densification rate of these materials measured using a dilatometer showed that some densification occurred at low temperature, but the rates were quite slow below about 800°C. Samples were loaded with Kr by heating them in a Kr atmosphere to 900°C and maintaining the temperature for a period of time sufficient to condense the glass to 80 to 90% of theoretical density. Initial tests were conducted with the Kr pressure at about 1 atmosphere and holding at 900°C for four hours. Under these conditions, Kr loadings of about 0.025 cm^3 of Kr(STP)/g of glass occurred. Several porous SiO_2 samples were then loaded with Kr at a pressure of about 35 MPa (340 atm) while heating to temperatures from 750° to 900°C and holding for various times. The results of these experiments show relatively high Kr content and excellent retention of the gas.

Samples containing up to 7 cm^3 of Kr(STP)/g of glass have been prepared. At a temperature of 900°C, it appears that one hour is sufficient to saturate the glass with Kr and to densify sufficiently to obtain a low Kr release rate. Similar results are also obtained in four hours at 850°C, but the minimum time required has not been established. At 800°C or below, the heating time to load and densify the glass becomes impractical if not impossible. During an effective Kr loading process, the sample densifies to 80 to 85% of its original volume, thus removing from 60 to 85% of the original porosity. Lower temperature heat treatment yields a Kr containing product, but the Kr loading is generally lower and the Kr is much more easily removed.

Gas release measurements have been conducted on the Kr containing glass samples by heating at a rate of about 100°C/hour while monitoring both the gas pressure and composition. Under these test conditions, properly densified glasses release less than 1% of the total

Kr below 900°C. Long term tests of Kr release rate have also been conducted at 420°C for over 400 hours. A plot of the fractional release (R) as a function of time appears to be linear with the square root of time ($t^{1/2}$) and is expressed by $R = 1.0 \times 10^{-5} t^{1/2}$, where t is in hours. Extrapolation of this curve yields a release of 0.3% of the Kr in the first 10 years.

The form of krypton in the SiO_2 matrix has not been conclusively determined. Two possibilities are considered: 1) true dissolution in the glass and 2) trapping of Kr in the very small connected porosity of the sample. Differentiating between the two cases is not an easy task, and perhaps the differences are not significant for radioactive waste storage. We have, however, attempted to shed some light on the subject. If all of the Kr were trapped in the pores of the sample, the pressure in the pores at 900°C would be 50 to 100 MPa (i.e., somewhat higher than the Kr pressure in the system). Only two possibilities are apparent. First, Kr permeates into the pores during cool down or the pores close and then continue to shrink against the gas pressure due to surface tension. To further test the form of the Kr in the solid, we have ground the sample into a fine powder. If the gas is contained in the pores, some should be released during the grinding process due to fracture of the pore. Initial results of such an experiment show essentially no reduction in Kr content with grinding to particle sizes down to 72 μm. However, when the particle size is further reduced to 40 μm, approximately 40% of the Kr was released. Although we do not consider these data conclusive at present, it now appears that both forms may be involved; i.e., some dissolved Kr and some trapped in the remaining pore volume. The $t^{1/2}$ release appears to suggest a diffusion mechanism for release. Work is currently underway to determine the effective diffusion coefficients for Kr in the glass matrix.

CONCLUSIONS

Storage of Kr-85 will be required in support of nuclear power reactors beginning in 1983. Both approaches described here appear to meet the requirements for such a storage medium. Entrapment of the Kr during sputtering has several rather obvious safety advantages. The operation of the process at low pressure and at or below room temperature should reduce markedly the potential for significant Kr-85 release to the environment during processing of the waste stream. It also appears that adaptation of this process for handling radioactive materials would also be simpler than the large high pressure, high temperature apparatus required for loading the glass sample. Furthermore, a significantly higher Kr loading is possible in the sputtered metals, thus reducing the volume required for storage by as much as a factor of 50 to 100. On the other hand, the low density loaded glass process takes advantage of a very inexpensive starting material and existing commercial technology for high temperature, high pressure processes. The volume of the Kr loaded glass matrix could obviously be reduced by going to still higher pressures. This concept will be tested as part of our continuing program, but we presently believe that the possible trade-offs favor a much simpler pressure chamber with the resultant lower loading obtainable at a pressure of 34 MPa.

ACKNOWLEDGEMENT

The authors gratefully acknowledge the U.S. Department of Energy, Office of Nuclear Waste Management, Contract No. EY-76-C-06-1830, for financial support of this work. We also acknowledge P. J. Raney for analytical work on gas release. The high pressure loadings of glass were done by K. Wheeler at our laboratory and by D. Knecht at Exxon Nuclear Idaho Company.

REFERENCES

1. "Environmental Radiation Protection Standards for Nuclear Power Operation," Federal Register 42 (9), Title 40, Part 190, Jan. 13 (1977).
2. G. L. Tingey, E. D. McClanahan, M. A. Bayne and R. W. Moss, Entrapment of Krypton in Sputter-Deposited Metals--A Storage Medium for Radioctive Gases, PNL-2879, Pacific Northwest Laboratory, Richland, WA (1979).
3. M. A. Bayne, R. W. Moss and E. D. McClanahan, "Krypton Entrapment in Pulse-Biased, Sputter-Deposited Metals," Thin Solid Films 54, 327 (1978).
4. M. A. Bayne, R. W. Moss and E. D. McClanahan, "Krypton Entrapment in Continuous Biased, Sputter-Deposited Glassy Metals," Thin Solid Films 63, 137 (1979).
5. J. E. Shelby, A Comprehensive Review of Gas Permeation, Diffusion and Solubility in Inorganic Glasses, SLL-73-0259, Sandia Laboratories, Livermore, CA, Aug. (1973).
6. R. H. Doremus, "Physical Solubility of Gases in Fused Silica," J. Amer. Cer. Soc. 49 (9), 461 (1966).

TECHNICAL FEASIBILITY OF KRYPTON-85 STORAGE IN SODALITE

R. W. Benedict, A. B. Christensen, J. A. Del Debbio,
J. H. Keller, D. A. Knecht

Exxon Nuclear Idaho Company, Inc.
Idaho Falls, ID 83401

INTRODUCTION

Federal regulations limit the release to the environment of krypton-85 produced by nuclear fission in commercial electric power generation systems (1). These regulations apply to nuclear fuels irradiated after January 1, 1983 (1). To meet the EPA regulation, the ^{85}Kr produced must be collected and stored for sufficient time to allow radioactive decay to reach a level that will allow release consistent with the regulations (2,3). This paper presents experimental results and analyses to demonstrate the technical feasibility of encapsulation in zeolites to immobilize krypton-85 for safe, long term storage (4).

BACKGROUND

Previous work by Barrer and Vaughan shows that krypton can be encapsulated by sodalite to provide a product which might be suitable for long term storage of krypton-85 (5,6). Such a product is illustrated in Fig. 1, where a structural model of sodalite cages (each $\sim$ 6.6 Å free diameter with 2.3 Å cage opening) is shown with a trapped krypton atom ($\sim$ 3.5 Å). If each cage were occupied, a theoretical saturated loading V_s, of 53 cm^3 STP of krypton per gram sodalite can be calculated (5). This would give a storage volume equivalent to that in a pressurized cylinder of krypton gas at 46 atmospheres. Vaughan obtained equilibrium encapsulation loadings in the range 20-35 cm^3 STP/g[6] (equivalent to storage volumes for pressurized cylinders at 16 to 28 atm). Gas release rates measured at about 400 to 560°C could be used to predict low leakage rates at storage temperatures of about 150°C (5,6,7).

Fig. 1. Model of Sodalite Cages. Upper Cage Contains a Krypton Atom. Relative Cage Opening Shown on Right Side of Upper Cage.

THEORY

A Langmuir adsorption model (equations 1) and 2) can be used to describe equilibrium loading of kyrpton on sodalite (8);

$$V_{eq} = \frac{V_s . k(T) \cdot f(P,T)}{1 + k(T) \cdot f(P,T)} \qquad 1)$$

$$k(T) = \exp\left[\frac{H}{RT} + \frac{S}{R}\right] \qquad 2)$$

with V_{eq} and V_s the equilibrium and saturated loadings, respectively, f the fugacity, R the universal gas constant, T the absolute temperature, P the pressure, k the equilibrium constant, H the enthalpy, and S the entropy. The equilibrium loading increases with pressure but decreases with temperature of encapsulation.

The rate of encapsulation or leakage is described by Fick's law. Fick's law can be solved to describe the total krypton uptake during encapsulation as a function of time in equation 3) or to describe the rate of krypton leakage from sodalite in equation 4),

$$V = V_{eq}\left[1 - \frac{6}{\pi^2}\sum_{n=1}^{\infty}\frac{1}{n^2}\exp\left(\frac{-Dn^2\pi^2 t}{R_0^2}\right)\right] \qquad 3)$$

$$\frac{d\left(\frac{Q_t}{Q_{oo}}\right)^2}{d\ t} = \frac{36D}{\pi R_o^2} = \alpha D \text{ For } \frac{Q_t}{Q_{oo}} < 0.3 \qquad 4)$$

with V the average concentration of krypton in sodalite, V_{eq} the equilibrium loading, t the time, D the diffusivity, R_o the average sodalite particle radium Q_t the volume of gas released at time t and Q_∞ the volume release at infinte time (9, 10, 11).

The diffusivity is a strong function of temperature and is described by equation (5),

$$D = D_o \ \exp\left(\frac{-E}{RT}\right) \qquad 5)$$

with E the activation energy and D_o a constant (11). Other factors which may affect the krypton diffusion include the concentration of different-sized cations in sodalite, the adsorbed water content, crystal defects, and the concentration of other gases, such as xenon (4,9,11,12).

EXPERIMENTAL PROCEDURE

Encapsulation

Samples were introduced into a 0.25-L pressure vessel and activated by heating overnight under vacuum at the run temperature. After activation, the vessel was pressurized to a run pressure for a given time. The vessel was then cooled at the run pressure, vented, and the samples removed to a dry glove box. The vented gas was returned to the storage cylinder for reuse. Gas sorption was measured using mass sprectrometry (4).

Leakage Measurement

High-temperature leak-tests (300°-500°C) were made by placing a weighed sample of sodalite loaded with krypton in a quartz tube and evacuating it. The tube was heated at constant temperature and the gas evolution measured by a pressure gauge and a gas chromatograph (4).

Long-term storage tests (50-300°C) were made by placing a sample in a quartz tube, which was evacuated and sealed. The tubes were kept in ovens at constant temperature, and then removed after predetermined storage times. The gas volume and composition released into the tube were measured by mass spectrometry.

RESULTS AND DISCUSSION

Equilibrium loadings of krypton on leached sodalite were measured in the ranges 400-500°C and 1200-1910 atm. The fitted parameters for equations 1) and 2) were very close to those obtained by fitting Vaughan's data at lower pressures (6). The parameters obtained by fitting Vaughan's and our combined data are: V_s = 45.2 cm^3/g; H/R = 2083.7 K; and S/R = -8.54 (4).

When mixtures of krypton and xenon were used, the krypton equilibrium loadings decreased with xenon fraction (Henry's Law), as shown in Fig. 2. Krypton loadings decreased by 25-30% as the gas mixture changed from 0 to 64% xenon. The xenon loadings were 1 to 5 cm^3/g for oven-dried (OD) and spray-dried (SD) sodalites, respectively, indicating that xenon was generally excluded from the sodalite cages.

Krypton loading in oven-dried sodalite powder (OD-2) is shown in Fig. 3. as a function of time. Higher temperatures strongly increase the loading rate, e.g., 20 cm^3/g is reached in 2 h at 575°C or in 10 h at 500°C. Leaching the OD-2 to remove excess NaOH (sample OD-2-L48H) also increased the loading rate. Not shown in Fig. 3., the rate of loading in pilled sodalite was not different from the rate in powdered sodalite.

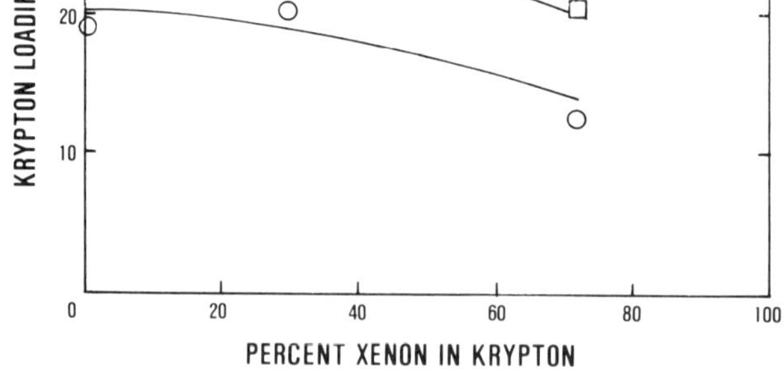

Fig. 2. Equilibrium Krypton Loading in Sodalites as a Function of Gas Composition at 500°C and 1220 atm.

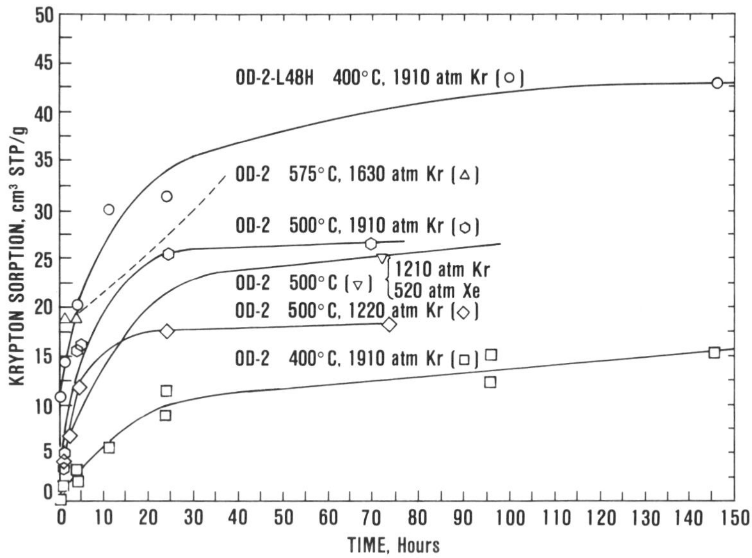

Fig. 3. Krypton Loading on Oven-Dried Sodalite as a Function of Time at Various Temperatures and Pressure

Diffusivities calculated from high temperature leakage rates are compared with Vaughan's results (6) in Fig. 4. Leached spray-dried sodalite leaks faster than unleached oven-dried sodalite. Not shown in Fig. 4. are the effects of pilling, initial krypton loading on leakage and added xenon: pilled sodalite behaves the same as powdered sodalite, while a lower initial krypton loading increases the leakage rate (4). Diffusivities calculated from fitting equation 3) to the rate of krypton loading are identical to diffusivities calculated from krypton leakage from sodalite at low initial loadings (4). Added xenon lowers the rate of leakage of krypton from leached spray-dried sodalite by a factor of ~ 8, resulting in diffusivities lying about half-way between the OD-2 and SD-LG lines.

The results of up to nine months of long term leakage measurement are shown in Fig. 5. for unleached sodalite. The lowest leakage was obtained with samples having a high krypton loading and a low water content. The sample containing 21 cm^3/g Kr and 10 gaseous cm^3/g H_2O was obtained by allowing the material with 21 cm^3/g Kr and 5 cm^3/g H_2O to contact air saturated with water vapor at 25°C for 14 h. Diffusivities calculated from the leakage measurements in Fig. 5. are compared with high temperature values in Fig. 4.

If the leakage rates shown in Fig. 5. are extrapolated, the predicted leakage and its 95% confidence interval of total krypton at ten years varies from $0.05 \pm 0.3\%$ to $1.6 \pm 1.1\%$ with most of the

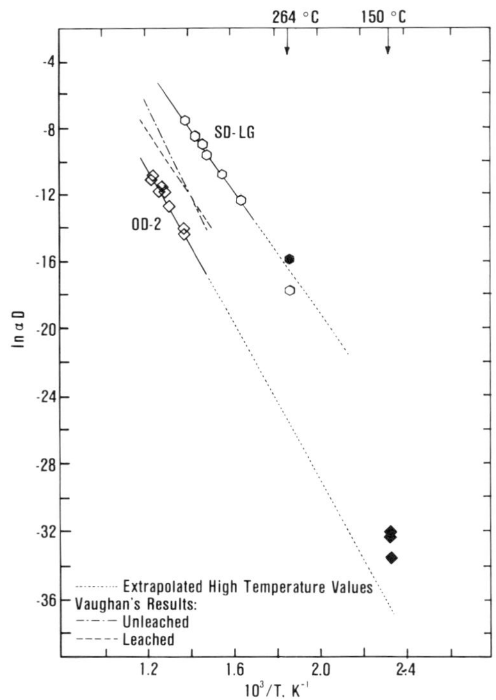

Fig. 4. Comparison of the Temperature Dependence of Diffusion of Krypton in OD-2 and SD-LG with Vaughan's Results and with 264 and 150°C Leakage Measurements. High-Temperature Method Used for Open Points; Low Temperature Method for Filled Points. See Equation 3) for α.

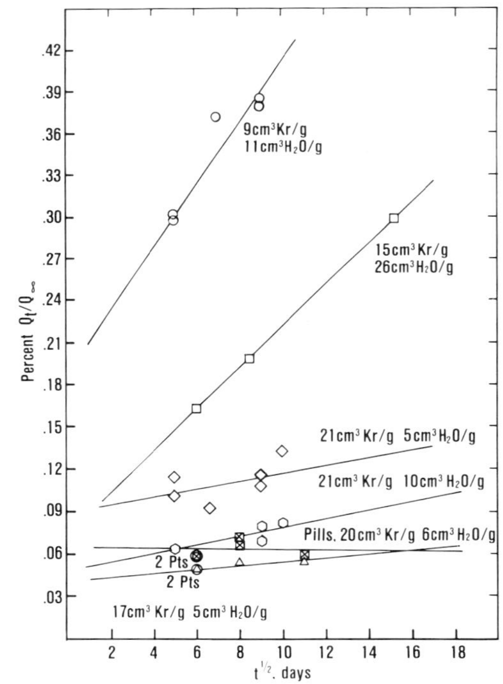

Fig. 5. Rates of Percent Leakage at 150° of Krypton Encapsulated in Unleached Sodalite at Various Kr Loadings and Amounts of Adsorbed Water.

values around 0.3 ± 0.5% (4). The predicted loss of ^{85}Kr is approximately half of these values due to radioactive decay. After ten years, the exponential rate of radioactive decay of ^{85}Kr predominates over the square root rate of leakage, resulting in decreasing amounts of ^{85}Kr which can be found outside the sodalite crystals (7).

Evacuated quartz tubes containing sodalite samples with encapsulated krypton were exposed to $\sim 2 \times 10^9$ rads of gamma radiation at 48-115°C, with a daily average of 60°C. There was no evidence of crystal damage or increased krypton leakage compared with laboratory control samples. However, a samll amount of hydrogen but almost no oxygen was formed, in agreement with other results (13).

TECHNICAL FEASIBILITY EVALUATION

Based on these experimental results, the process that is technically feasible for a reference (14) 2000 metric ton of heavy metal (MTHM) per year reprocessing plant producing $\sim$17 MCi or $\sim 190 m^3$ at STP would encapsulate krypton at approximately 20 cm^3/g from krypton at temperatures greater than 575°C and pressures greater than 1660 atm with one batch a day in a 58-L high pressure vessel (4). Based on preliminary measurements at 500°C, the same process also would be feasible for a 70% krypton and 30% xenon mixture (4).

REFERENCES

1. "Environmental Radiation Protection Standards for Nuclear Power Operation," Federal Register 42(9) Part VII, 2858 (1977).
2. D. A. Knecht and R. A. Brown, "Strategy Analysis for Krypton-85 Waste Management," Trans. ANS (To be published in 1979).
3. Alternatives for Managing Wastes from Reactors and Post-Fission Operations in the LWR Fuel Cycle, ERDA-76-43, Chapt. 14.1, Vol. 2, 14.1-14.7, NTIS, Springfield, VA (1976).
4. R. W. Benedict, Technical and Economic Feasibility of Zeolite Encapsulating for Krypton-85 Storage, ENICO-1011, Idaho National Engineering Laboratory (1979).
5. R. M. Barrer and D. E. W. Vaughan, "Trapping of Inert Gases in Sodalite and Cancrinite Crystals," J. Phys. Chem. Solids 32, 731-743 (1971).
6. D. E. W. Vaughan, Encapsulation of Rare Gases (PhD. Thesis, Imperial College, London 1967).
7. R. A. Brown, M. Hoza and D. A. Knecht, "^{85}Kr Storage by Zeolite Encapsulation," Proc. 14th ERDA Air Clean. Conf., CONF-760822 1, 118-131, NTIS, Springfield, VA (1977).
8. S. Brunauer, The Adsorption of Gases and Vapors, (Princeton University Press, Princeton 1943).

9. J. Crank, The Mathematics of Diffusion, (Clarendon Press, Oxford 1975).

10. H. S. Carslow and J. C. Jaeger, Conduction of Heat in Solids, (Clarendon Press, Oxford 1959).

11. R. M. Barrer, Diffusion In and Through Solids, (Cambridge University Press, Cambridge 1941; Xerox University Microfilms, Ann Arbor, MI 1975).

12. D. M. Ruthven, "Diffusion in Molecular Sieves. A Review of Recent Developments," Molecular Sieves II (ACS Symposium Ser. 40, American Chemical Society, Washington, D. C. 320-333, 1977).

13. N. E. Bibler, Radiolytic Gas Production from Tritiated Water Sorbed on Molecular Sieves, DPST-77-375, Savannah River Laboratories 1977).

14. Management of Commercially Generated Radioactive Waste Draft Environmental Impact Statement, DOE/EIS-0046-D, NTIS, Springfield, VA 1979).

CORROSION-RESISTANT METALLIC CANISTERS FOR NUCLEAR WASTE ISOLATION*

N. J. Magnani and J. W. Braithwaite

Sandia Laboratories**
Albuquerque, NM 87185

ABSTRACT

A program with the goal of estimating the potential of a material to survive 300 years as a nuclear waste canister is underway at Sandia Laboratories. The corrosion and stress corrosion cracking behavior of the leading candidate, TiCode-12***, is contrasted to that of a commonly used engineering alloy, 304 stainless steel. Experimental evidence is presented which shows the inadequacy of 304 stainless steel in potential repository environments and shows that TiCode-12 could endure the desired 300 years. Further work required to qualify TiCode-12 is outlined.

INTRODUCTION

A program based on the premise that a corrosion resistant nuclear waste canister could make an important contribution to the long term containment of nuclear waste is underway at Sandia Laboratories. The object of the program is to assess the ability of a canister alloy to survive more than 300 years in the repository environment.

There are essentially four different barrier lifetimes which could be required for a nuclear waste isolation canister. They are as follows:

*This work supported by the United States Department of Energy (DOE), under Contract DE-AC04-DP00789.
**A DOE facility.
***A trademark of TIMET.

0 Year Lifetime: This type of canister would be required to survive the transportation and emplacement stages of a disposal program. Material with a low corrosion resistance could be used for this category of canister and a corrosion evaluation is unnecessary.

~25 Year Lifetime: If retrievability of wastes is legislated or if the possibility of unanticipated conditions in the repository make it prudent to retain the option of easy retrievability, then a canister which will survive a repository environment for ~25 years will be required. A moderately corrosion resistant material would be expected to survive this length of time, and materials will be identified by Sandia's program as a by-product of the study of materials which are expected to last ~10 times this long.

~300 Year Lifetime: This is the option emphasized by Sandia's program. This type of canister would survive the thermal period of the waste which would eliminate potential hydrothermal waste interactions. Several investigations (1-3) have shown that the chemical durability of nuclear waste forms is low in hydrothermal conditions. For example, the initial leach rate of Cs from borosilicate glass in water increases by over 10^4 when the temperature is increased from 100° to 250°C (3, 4). Containing the heat producing fission products ^{90}Sr and ^{137}Cs would also limit the interaction of heat and radionuclides with the geologic formation to the near field. This option, while not considered to be essential in providing waste containment, could be relied upon as an additional barrier which provides even greater confidence in the isolation system.

$>10^4$ Year Lifetime: This type of option considers the canister to be an ultimate barrier which would last until the radioactivity of the waste decayed to near background levels. Presently, this is not required in any waste disposal program nor is it likely to be because of the extreme difficulty of qualifying a material for such an extended lifetime with present day techniques.

The 300 year option was selected for this program because this should be obtainable with present day technology at a reasonable cost. Furthermore, the benefit of a 300 year can to a total waste program is very significant. The materials for this type of canister would cost ~$3000 (based on 1979 prices for a TiCode-12 canister surrounding an inner 304-L canister) (5). The only important impact would occur if an alloy was selected which contained large quantities of strategic materials. This has been considered in the selection of candidate materials for this program.

Experiments to assess canister lifetimes must be conducted in realistic environments. Therefore, a knowledge of the repository environment is necessary to define laboratory environments for simulation experiments. This includes information on how the environment will change with time due to the perturbations of heat and radiation. Ultimately, confirmatory testing must be done in-situ.

Presently, alloys are being evaluated in simulated environments which might be encountered in bedded salt (Waste Isolation Pilot Plant), including both dry and wet interactions (representing worst case accident conditions) and in seawater and saturated North Pacific sediments (Seabed Disposal). The effects of high temperatures on corrosion rates have been extensively evaluated; however, the effects of radiation and brine radiolysis have only recently been studied. The results of radiation exposure have shown that the alteration of the environment caused by radiation may be the most important factor in evaluating the lifetime of a waste canister.

Limited literature information is available concerning the effects of radiation on concentrated Cl^- solutions such as those utilized in this investigation. A discussion of radiation effects is available elsewhere (6). Suffice it to say that exposure of Cl^- containing aqueous solutions to gamma irradiation results in a solution with an increased oxidation potential.

The results of laboratory experiments conducted on the leading candidate for a 300 year canister, TiCode-12 (Ti-0.8% Ni-0.3% Mo), and on a very common "corrosion resistant" material, 304 stainless steel (Fe-18% Cr-8% Ni), will be presented in this paper. TiCode-12 is an α phase alloy which is similar in mechanical behavior, machinability and weldability to Grade 3 commercial purity titanium (Ti-75A) and is available in a wide variety of finished mill products. Results for both general and localized corrosion (including stress corrosion cracking/hydrogen embrittlement) behavior in simulated high temperature environments will be presented. Also included will be the results of experiments conducted in a high intensity ^{60}Co gamma source (10^7 rads/hr). From this data, it will be concluded that 304 stainless steel does not have sufficient corrosion resistance to serve as anything other than a transportation and emplacement canister and that TiCode-12 has a high probability of surviving 300 years and warrants further study.

RESULTS AND DISCUSSION

General Corrosion

The uniform corrosion rates of TiCode-12 and 304 stainless steel (304SS) in 250°C* deoxygenated brines are shown in Table I. Brine A is a saturated Na/K/Mg chloride brine which is representative of

*250°C represents an overtest. The maximum temperature expected in an inundated bedded salt repository is 150°C.

Table 1. Corrosion Rates in 250°C Deoxygenated Solutions After 28 Days of Exposure

Alloy	Brine A (μm/yr)	Seawater (μm/yr)
304SS	18	6
TiCode-12	3	1

brine inclusions found in bedded salt. One should not conclude from these data that both materials would survive for 300 years as a waste canister. The reason this should not be done can be explained by referring to Fig. 1. In this Fig. are shown the electrochemical polarization responses for TiCode-12 and 304SS in 250°C seawater. The most striking observation to be made from these curves is the difference between the potential which separates the respective anodic and cathodic branches for TiCode-12 and 304SS. This anodic-cathodic potential spread is an indicator of corrosion resistance because to cause appreciable corrosion, the potential of the metal must be driven to the very positive voltages corresponding to anodic dissolution. For example, the addition of 500 ppm of dissolved oxygen at 250°C increased the corrosion potential of TiCode-12 by 0.6 volts and of 304SS by 0.3 volts. As can be seen from Fig. 1., increases such as these will not significantly affect TiCode-12 dissolution but will be extremely detrimental to 304SS corrosion behavior. The other point to be made from these plots is that a hysteresis in the anodic curve, such as that shown for 304SS, indicates a susceptibility for passive film breakdown and rapid pitting. This same problem exists for TiCode-12, but the potentials and currents which must be applied are much higher than those shown (> 2.5 volts and 25 mA/cm^2). Additionally, both of these metals have been shown to support an oxygen current. Localized differential aeration cells can result from this which can lead to crevice corrosion. A differential aeration cell cannot be totally responsible for crevice corrosion in TiCode-12 for reasons pointed out in Reference 7. However, because of its quasi-susceptibility, this problem was investigated by subjecting TiCode-12 crevices to 250°C and 300°C oxygenated seawater environments. Crevice corrosion was not evident. It should be pointed out that similar tests conducted with commercial purity titanium produced crevice corrosion and led to its dismissal as a candidate alloy. These observations have been substantiated in practice, i.e., 304SS does undergo significant pitting and crevice corrosion in environments such as the ones used in this study and TiCode-12 does not (8). Penetration rates of the metal due to these forms of localized attack are unpredictable, but certainly are several times faster than by uniform corrosion of the metal.

The quantity of electrolyte present at the canister interface as a function of time is critical to lifetime assessments for canister materials. For example, it has been shown in both laboratory and in-

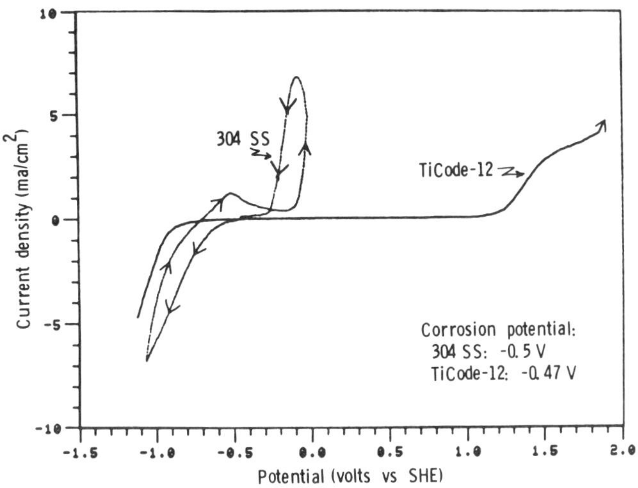

Fig. 1. Polarization Response of TiCode-12 and 304 Stainless Steel in 245°C Deoxygenated Seawater. (Scan rate - 120 mV/min). Anodic currents are positive and cathodic currents are negative.

situ experiments that both 304SS and mild steel will last indefinitely if bedded salt remains absolutely dry. As the water content of the environment is increased up to saturation, the corrosion rates of 304SS and TiCode-12 also increase to those shown previously in Table 1. Another important issue is the effect of gamma radiation on corrosion processes. As mentioned previously, a number of corrosion experiments were conducted with simultaneous intense gamma irradiation. The following observations were made: 1) the production of brine radiolysis products results in an increase in the concentration of solution oxidants; this, in turn, can significantly increase the uniform corrosion rates along with the susceptibility to stress corrosion and localized attack of many alloys including mild steel and 304SS; 2) the effect of radiation is a function of the solute concentration, dose rate and total dose; and finally, 3) the corrosion rates of TiCode-12 are only slightly increased by simultaneous exposure to brines and gamma radiation.

Stress Corrosion Cracking

Slow strain rate stress corrosion cracking tests were conducted in environments representative both of proposed repositories and of worst case accident conditions. These environments included dry salt, saturated brine and seawater. It has been shown that Brine A is more corrosive than seawater (5); therefore, this environment was used extensively and only confirmatory tests were conducted in seawater.

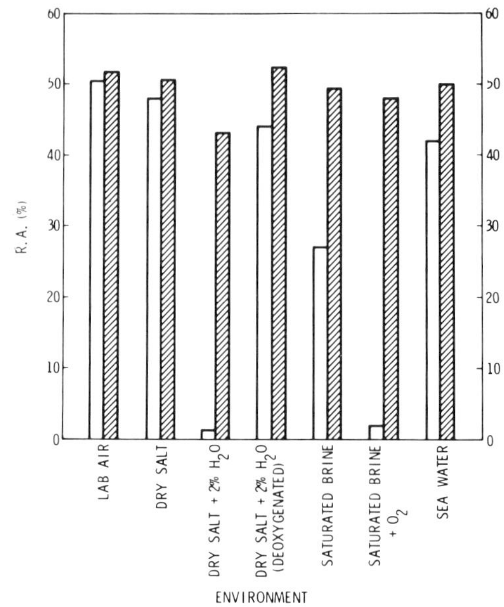

Fig. 2. The Reduction in Area for 304 Stainless and TiCode-12 Slow Strain Rate Specimens Exposed to Selected Environments at 250°C. 304 stainless steel tests (open bars) conducted at $1x10^{-6}$/sec; TiCode-12 tests (hatched bars) conducted at $1x10^{-5}$/sec.

In a slow strain rate test, the reduction of ductility of a material caused by the environment is considered to be related to the susceptibility of the material to stress corrosion cracking or hydrogen embrittlement. Ductility data show that 304SS is not embrittled by dry salt at 250°C (0.2% H_2O) but is embrittled by both dry salt with 2% H_2O added and by saturated brine at 250°C. The data also suggest that the optimal strain rate for further evaluation of the 304SS cracking phenomena is at $1x10^{-6}$/sec or slower. The data for TiCode-12 show that it is not embrittled by either dry salt or brine at 250°C and that its ductility is not a function of strain rate.

The results of tests conducted on both materials in several pertinent environments are shown in Fig. 2. The data show that 304SS is very susceptible to stress corrosion cracking in dry salt when a few percent of water is added. The data also show that if O_2 is removed from the system, the material fails in a ductile manner. The importance of oxygen to 304SS cracking is also illustrated by an experiment where O_2 was added to saturated brine and the R.A.* dropped from 27 to 2%. The TiCode-12 alloy did not exhibit any significant reduc-

*The reduction in area, R.A., is a direct measurement of specimen ductility.

tion in R.A. in any of the environments tested, indicating this material has a higher resistance to stress corrosion cracking than 304SS. However, fractographic analysis of a TiCode-12 specimen which failed in 250°C oxygenated brine showed that the environment caused some degradation in the near surface regions. The extent of the interaction is such that TiCode-12 should not be precluded from consideration as a canister material. Tests are underway to further quantify the magnitude of the interaction.

Slow strain rate tests were also conducted on TiCode-12 specimens which had been exposed to high intensity γ radiation for 31 days (total dose = 8×10^9 rads) while partially immersed in a saturated brine solution. The estimated temperature during exposure was 90°C. The test results, obtained at room temperature at strain rates of 1×10^{-5} and 1×10^{-6}/sec, showed that specimen ductility is not significantly affected by exposure to the radiation source. Fractographic analysis of the specimen, however, did show that the material interacts with the environment. A narrow, $\leq 10 \mu m$, band of material along the surface failed in a brittle manner, indicative of stress corrosion cracking. A test is underway at a very slow strain rate, 1×10^{-7}/sec, to determine if this effect will become more pronounced at slower strain rates. This would be expected to be the case if H_2 redistribution is the rate controlling process.

304SS corrosion specimens cracked due to residual stresses when exposed to irradiated conditions which were similar to those discussed for the TiCode-12 experiments. Therefore, slow strain rate stress corrosion cracking tests were not conducted on 304SS specimens exposed to irradiated environments.

CONCLUSIONS

1. 304 stainless steel is susceptible to both localized corrosion and stress corrosion cracking in Cl^- + H_2O containing environments.

2. TiCode-12 is very corrosion resistant material in bedded salt and seabed environments. No pitting or crevice corrosion problems have been observed for temperatures ≤ 300°C, and uniform corrosion rates are typically 1-10 μm/yr. This alloy may well survive 300 years or more. However, further study is still necessary to qualify the material for such an extended lifetime.

3. High dose, high intensity gamma irradiation increases corrosivity of Cl^- containing environments. This effect is being quantified.

4. TiCode-12 is embrittled in highly oxidizing environments, but the extent of embrittlement does not preclude its use for canisters.

FUTURE STUDY

Although the initial results of TiCode-12 exposures to isolation environments appear encouraging, further extensive investigation is needed in several areas before the potential of this alloy can be completely assessed. The effect of radiation on corrosion conditions in a repository probably represents the largest unknown at the present time, and a program is underway to define both how the environment is altered and how the altered environment affects the performance of canister materials. The extent of embrittlement in simulated environments must be quantified and the rate controlling process(es) must be determined. Very accurate and sensitive corrosion monitoring techniques which will enable general corrosion mechanisms to be determined for highly corrosion resistant alloys need to be perfected. This is necessary for confidence in the extrapolation of rates to very long times. The titanium alloy family, and especially TiCode-12, are relatively new, and therefore, no long term environmental degradation data are yet available. A set of experiments is being implemented which will be continued indefinitely in order to provide this data at some later date. Finally, after the alloy has been adequately assessed based on laboratory tests, confirmatory tests will be conducted in-situ.

REFERENCES

1. J. H. Westsik and R. P. Turcotte, Hydrothermal Reactions of Nuclear Waste Solids, PNL-2759, Sept. (1978).
2. G. J. McCarthy, Reaction of Water with a Simulated High Level Nuclear Waste Glass at 300°C, 300 Bars, Report SA-35, Rockwell Hanford Operations (1978).
3. J. W. Braithwaite and J. K. Johnstone, Chemical Durability and Characterization of Nuclear Waste Forms in a Hydrothermal Environment, SAND79-2259C (1979).
4. J. E. Mendel, Annual Report on the Characteristics of High Level Waste Glasses, BNWL-2252, June (1977).
5. J. W. Braithwaite and M. A. Molecke, "Nuclear Waste Canister Corrosion Studies Pertinent to Geologic Isolation," Chemical and Nuclear Waste 1, (2) (1980).
6. J. W. Braithwaite, N. J. Magnani and J. W. Munford, Titanium Alloy Corrosion in Nuclear Waste Isolation Environments, paper no. 213 to be published and presented at Corrosion 80, NACE, Chicago, March (1980).
7. W. H. Smyrl, L. L. Stephenson and J. W. Braithwaite, Behavior of Candidate Canister Materials in Deep Ocean Environments, SAND77-0243, April (1977).
8. L. C. Covington, "Selection of Titanium Alloys for Corrosion Resistance," Selection of Materials Symposium, AIChE, New York.

CORROSION OF MATERIALS IN A CLAY ENVIRONMENT*

Florent Casteels, Hugo Tas, Jan Naets**, Martin Brabers**

S.C.K./C.E.N., 200 Boeretang, B-2400 Mol, Belgium;
**K.U.L., Leuven, Departement Metaalkunde, B-3030 Heverlee

INTRODUCTION

Disposal of radioactive waste in stable geological formations is presently considered as one of the safest methods for the isolation of radionuclides contained in the waste from the ecosphere. The burial of radioactive waste in sheathed canisters in holes in geologic formations provides different barriers against dispersal of radionuclides, namely, the stability of the waste, the canister, possible chemical and sorptive barriers present in the storage hole or in the geological formation itself. The canister material has to be selected with the intention of good durability and service life under the conditions prevailing in the final repository. In Belgium, the use of a geological clay formation is being studied for the burial of radioactive waste (1). The metallic materials fabricated under well defined rules should be compatible in the different steps in the lifetime of the canister, namely, during reprocessing, interim storage and during final storage. A further aim of this study is to characterize the environment present in holes of a geological clay formation as a function of canister temperature, temperature of the clay formation, and time, both in normal service life and under accidental conditions.

EXPERIMENTAL CONDITIONS

The corrosion resistance of candidate canister materials and construction materials of the galleries and pits to be used for the

*Work performed under contract with the Commission of the European Communities, Brussels.

storage of low, medium and high activity radioactive waste in a geological clay formation must be evaluated in the different atmospheres which can be present in the holes. In ideal conditions, the compatibility of candidate canister and structural materials has to be evaluated after corrosion experiments are carried out in direct contact with clay (with and without backfill with sand), in interstitial clay water, and in a humid atmosphere loaded with corrosive products from the clay. A major source of water intrusion in the geological clay repository located at Mol at a depth of 220 m could be ground water present in the sandy clays just above the Boom clay formation. The alloys were tested in the annealed and heat-treated conditions (temperature - time dependence of heat treatment on candidate canister materials determined by the French AVM process). Welded and stressed specimens (for high level vitrified waste) were also inserted in the corrosion programmes. Austenitic and ferritic stainless steels, nickel, titanium, and super alloys were included in the tests. Copper alloys will be tested during a second series of corrosion experiments. The atmosphere present in holes of a geological clay formation, corresponding to pyrolysis of the clay at 300°C, was simulated using synthetic gas compositions (i.e., air with a well-known humidity and loaded with corrosive agents). The tests were carried out under dynamic conditions either in a corrosion chamber (at 50°C) or in inert furnaces (at 150°C and 300°C). The composition of the moisture is given in Table 1. Besides weight measurements, metallographic techniques were used to determine the extent of general and localized corrosion. The morphology of the as-received and corroded surfaces was evaluated by SEM techniques. The composition of the corrosion products was determined by microprobe techniques.

EXPERIMENTAL RESULTS

Characterization of Atmospheric Clay Environment

The nature of corrosive products present in the atmosphere in holes in a clay formation was studied by exposing representative clay samples (600 g) to an air stream in an inert furnace. The pH and chemical composition of the moisture escaping from the clay and condensed at room temperature or -50°C or chemically fixed on sodium hydroxide solutions was monitored as a function of time and exposure temperature (25°, 50°, 100°, 150° and 200°C). The exposure time lasted two months at each temperature. The pH of the liquid condensed at 25°C and -50°C shifts from 8.2 over 4.1, 2.3 to 1.4 for temperatures of the clay of 25°C, 60°C, 100°C and 150°C. At 25°C the condensed liquid contains mainly chlorides. Chlorides, sulphates, ammonia and fluorides are the major products determining the pH of the moisture condensed after exposure of the clay to a temperature of 50°C. Fluorides and ammonia are liberated at the beginning of the experiment at 50°C. The pH of the condensed moisture extracted at 100°C is mainly conditioned by the presence of chlorides, sulphates

Table 1. Composition of Moisture (ppm)

acid (salt)	HCl	H_2SO_4	HF	NaCl	CaF_2	KF	$MgCl_2$
	4.9	15	0.40	8	0.21	0.44	0.1

and phosphates and to a minor extent by fluorides and ammonia. At 150°C, sulphur dioxide is the main corrosive compound determining the pH of the solution. Other components escaping at 150°C are chlorides and organic products. About 11 ppm SO_2 is found to be present in a flow of air of 2 L/hr. Candidate canister materials were exposed in the air stream in direct contact with the clay. Rather low corrosion rates were obtained; these results were comparable with the data obtained in synthetic (gas) atmospheres carried out in identical inert furnaces.

Weight Changes

Corrosion rates were measured in synthetic gaseous atmospheres as a function of temperature, time and alloy composition (Table 2). The conditions in the corrosion chamber operated at 50°C consist of alternating 30 min heating periods and 30 min vaporization periods of the corrosive moisture in the corrosive chamber. About 3.5 cm^3/hr moisture together with 14.5 l air was injected into the inert furnaces which were operated at 150°C and 300°C. The corrosion rates were found to decrease with decreasing relative humidity (varying temperature). Carbon steel, chromized steel, aluminum and its alloys, possess the lowest resistance in the wet atmosphere. In the series of ferritic and austenitic stainless steels, the high alloyed UHB 904L, AISI 316 and AISI 304 had the lowest weight gains. In the series of nickel alloys and superalloys, Inconel 625 and Hastelloy-C had the lowest weight changes. Among titanium and its alloys, IMI 260 (Ti alloyed with Pd) had the lowest weight changes.

Optical and Electron Microscopy

Metallography, microscan and microprobe techniques have been applied for the determination of localized attack (pitting, grain boundary), the morphology of corroded and cleaned surfaces, and the composition of surface deposits and subsurface scales. The results of the depths of localized attack found on canister materials after exposure during 9 and 16 months at 49°C in the corrosion chamber are reported in Table 3. It can be concluded that only a few alloys are resistant against localized forms of attack in this atmosphere: titanium and its alloys, Hastelloy C, UHB 904 L and to a lesser extent, 1803 MoT, AISI 316, Ni-200 and Inconel 625.

Table 2. Weight Changes as a Function of Alloy Composition, Exposure Time and Temperature ($mg.cm^{-2}$) $\pm$ 0.005 $mg.cm^{-2}$

	50°C				150°C		300°C	
	4 wks	4 mos	9 mos	16 mos	1 mo	6 mos	1 mo	6 mos
Carbon steel	4.80	19.53	destroyed		---	---	---	---
Chromized steel	1.31	5.69	5.39	4.24	0.01	0.08	0.01	-0.05
AISI 304	0.06	0.28	0.25	0.15	0.00	0.03	0.02	-0.04
AISI 304 L	0.09	0.48	0.14	0.15	---	---	---	---
AISI 316	0.07	0.27	0.48	0.12	-0.01	-0.01	0.01	-0.01
AISI 316 L	0.07	0.35	0.36	0.05	-0.02	0.07	0.00	-0.01
AISI 430	0.14	0.76	1.15	1.47	0.01	-0.02	0.00	-0.07
1803 T	0.09	1.04	0.85	0.37	-0.01	0.02	-0.01	-0.06
1803 MoT	0.10	0.44	0.66	0.65	0.03	0.02	-0.01	-0.02
UHB 904 L	0.00*	0.16	0.27	---	---	---	---	---
Inconel 600	0.10	0.48	0.75	0.56	0.06	-0.01	0.00	-0.02
Inconel 625	0.12*	0.15	0.27	---	---	---	---	---
Incoloy 800	0.06	0.37	0.95	0.19	-0.01	0.02	-0.02	-0.46
Hastelloy B	0.17*	0.39°	0.76	0.49	---	---	---	---
Hastelloy C	0.14*	0.24°	0.40	0.14	---	---	---	---
Ni 200	0.18*	0.50°	0.54	1.12	---	---	---	---
Al	0.29	1.47	1.27	2.69	---	---	---	---
AA-5052H-32	0.39*	1.00°	1.13	1.84	---	---	---	---
AA-6061-T-6	0.24*	-0.68	1.60	---	---	---	---	---
Ti	0.06	0.24	0.41	0.15	-0.02	0.01	-0.2	-0.07
Ti-AlV-64	0.07*	0.11	0.31	---	---	---	---	---
Ti-AlSn-52	0.09*	0.14	0.29	---	---	---	---	---
IMI 260 (Ti-Pd)	0.05*	0.15	0.16	---	---	---	---	---

*5 weeks; °6.5 months.

Table 3. Depth of Localized Attack on Candidate Materials (10^{-6}m)

Alloy	9 months	16 months
AISI 304	78	60
AISI 316	20	10
UHB 904 L	no preferential attack	not available
Ni-200	26	20
Inconel 625	36	not yet available
Hastelloy C	no preferential attack	no preferential attack
Al	76	---
Ti and alloys	no preferential attack	---

Aluminum and its alloys. Specimens of aluminum, AA-5052 H-32 and AA-6061-T-8 have been examined after exposure in the corrosion chamber. The formation of Na_2SO_4 deposits was found on the surface of the alloys; the pH (3.65) and the composition of the moisture fixed on the Na_2SO_4 deposits does not allow restoration of the passive oxide layers. The development of pitting corrosion starts underneath the deposits. The localized attack proceeds in depth as well as parallel to the surface and results in the formation of large but rather shallow pits. The extension of the pits proceeds also along grain boundaries and has a nature typical for corrosion induced by S-compounds (Fig. 1a.). AA-6061-T-8 possesses the highest corrosion resistance of the three alloys tested.

Carbon steel and chromized steel. Carbon steel possesses only a limited resistance towards polluted humid atmospheres. This resistance is normally ameliorated by the application of noble or non-noble surface claddings. A Cr-coating was applied on the carbon steel using a vapor deposition technique followed by adequate heat treatments. The carbon steel was attacked in this atmosphere due to the formation of FeO and FeS. The unprotective nature of the corrosion layers results in rapid attack and destruction.

The corrosion attack of the chromized steel proceeds along fine cracks present in the as-received coating. The formation of less-dense corrosion products compared to the density of the starting materials results in a "bulging" effect leading to the formation of protuburances, giving easier access to the corroding atmosphere (Fig. 1b.).

Austenitic and ferritic stainless steels. The behavior of AISI 304, AISI 316 and the low carbon interstitial grades together with the Ni, Cr, Mo and Cu-rich UHB 904 L was compared with the behavior of the martensitic ferritic grade AISI 430 and the high purity ferritic stainless steel grades 1803 T and 1803 MoT. The morphology of the salts, deposits and corrosion products found on the surface of the corroded alloys changes with the nature of the substrate. The surface of AISI 304 steel showed pustule like deposits which contained besides S and Cl, and in minor quantities Al, Si and K, in addition to the main alloying elements of the steel. The surface cleaned ultrasonically revealed the presence of different types of pits. The morphology of the pit is either typical for S or for Cl attack. Microprobe analysis revealed the presence of Na, S, O, Cl, Mg, Si, Al, K, Ca, Cu and Zn. When compared to chloride, sulphur was present in much higher concentrations in the pits. Metallographic examination of a cross section revealed a subsurface attack typical for the sequestration by S-containing compounds (Fig. 1c.). The surface aspect of the specimens which were corroded either at 150°C or at 50°C in an air stream in direct contact with clay is identical to the forementioned case. Si, Al, Ca and S rich deposits have been found on the specimens corroded in the "clay" atmosphere.

The surface and cross section os AISI 304 L steel showed a much better behavior: a smaller number of pits with less penetration. As usual, the pits were surrounded by small satellite pits. The compo-

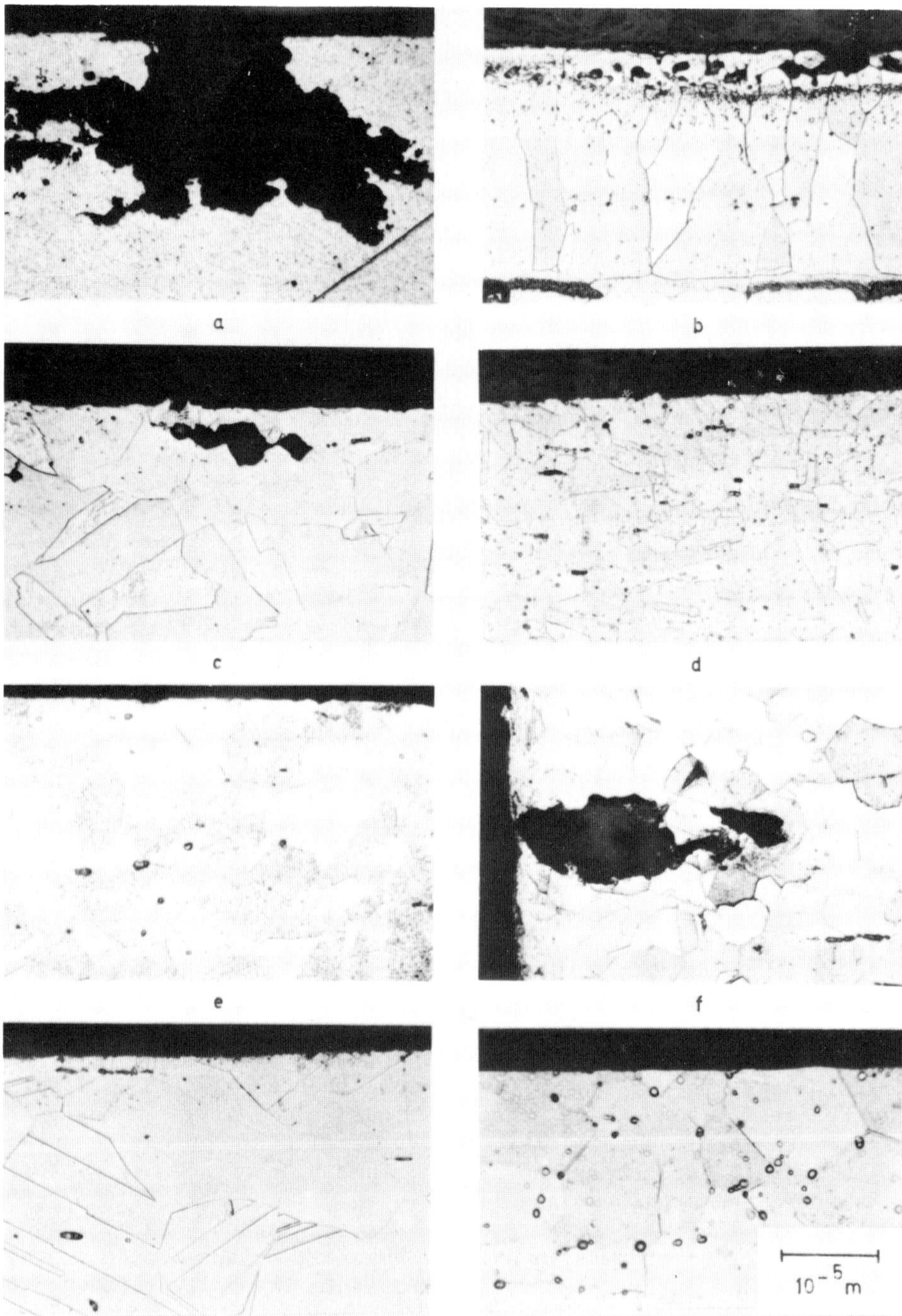

Fig. 1. Metallographic Cross Sections of Alloys After Exposure at 50° C in Corrosion Chamber

sition of AISI 316 steel and particularly the Mo-content makes this steel more resistant than AISI 304 steel. Nevertheless, a much higher number of pits was developed which were very small and had little penetration. Metallographic analyses of the penetration showed that after 16 months, pit-growth was small (Fig. 1d.).

Scanning electron microscope analysis equipped with an EDAX analysis system revealed the presence of needle-like S-rich deposits on the UHB 904 L steel. Besides the alloying elements, Cr, Ni, Fe and Mn, the elements Mg, Al, Si, K and Ca were found to be present. After removal of corrosion and deposition products, the presence of isolated shallow pits was found. These pits are due to leaching the remainder of the deoxidation products ("Misch-Metall") present in the as-received material. Examination of cross-sections after 16 months of exposure did not reveal any trace of preferential attack (Fig. 1e.).

The surface of the AISI 430 steel corroded during 4, 9 and 16 months at 50°C in the corrosion chamber was covered with a brown-black corrosion layer which covered a number of deep pits or areas with grain boundary attack. The layer of the areas having general and localized attack are non-uniform and are rich in S, Cl, Si and the alloying elements Cr and Fe. The elements present in the deep pits are mainly Fe, Cr, S, Na, O, Mg, Ca, Al, Si and Cl. Cr was found to be associated with S.

The ferritic stainless steels 1803 T and 1803 MoT with low carbon content are not immune to the atmosphere of the corrosion chamber. S, Na and Si are the most important constituents present in the continuous corrosion layer and in the pits found in steel 1803 T. The deposits present on 1803 MoT steel are rich in S, Si, Na, Cr, Ti and Fe and possess a completely different morphology compared to the previously examined alloys which were tested under identical conditions. Addition of Mo decreased the intensity of localized attack.

Nickel alloys and superalloys. The surface of Ni-200 samples corroded at 50°C in the corrosion chamber showed a green colored corrosion layer. A large number of deep (V-shape) and shallow pits were present under the corrosion layer which were composed of Ni, S, Na, Si and Cl. The attacked grain boundaries of Ni-200 are decorated by large precipitates.

Inconel 600 was also covered with a S-rich reaction layer alloyed mainly with iron and chromium. Ni was present only in small quantities. The reaction layer showed many crevices which allowed easy contact with the corrosive environment. The alloy suffered from the same subsurface attack as observed for AISI 304. Tunnels and subsurface attacks as shown in Fig. 1f. have been found after 9 months and 16 months exposure at 50°C. This alloy also suffered from pitting corrosion after exposure for 6 months at 150°C in synthetic gas composition.

The beneficial effect of Mo on the overall corrosion and localized attack of Ni alloys has been demonstrated by the post-corrosion analyses of Inconel 625 tested under identical conditions. The zone adjacent to the surface was characterized by the presence of a high density of precipitates on the grain boundaries. Further exposure at 50°C resulted in the formation of subsurface porosities. A large number of pustule-like deposits were present on the surface of specimens corroded for 4 months at 50°C. These deposits were rich in Ni, Cr, Mo and Fe. The S content in the deposits was rather low. The enhanced corrosion resistance of this alloy compared to Inconel 600 can be related to the fact that Mo decreased the reaction kinetics of sulphates with Ni-rich alloys. An enhanced Cr content has the same positive effect, although Cl^- present in the corrosive environment decreases the protective nature of Cr_2O_3 towards S penetration (2).

Incoloy 800 had the lowest resistance of the high-alloyed nickel alloys in the corrosive environment. This fact appeared after only 4 weeks of exposure at 50°C. A large number of S-rich deposits with the morphology of a cross covered only part of the exposed surface; other deposits with the morphology of a pustule completely covered the surface and contained Si, S and Cl. The attack started under the salts obtained after evaporation of the liquid from the surface. The formation of a pit was preceded by grain boundary attack. The Incoloy 800 alloy is rather sensitive for S-induced corrosion.

Excellent corrosion resistance has been found for the two superalloys, Hastelloy B and Hastelloy C. After rather short exposure times at 50°C, the surface of Hastelloy B samples was covered with dense and cracked deposits rich in Ni, S and alloyed with small amounts of Si, Mn and Fe. The local attack of the grain boundaries is limited to a few μm. The rather good corrosion resistance could be expected since this alloy shows a rather good performance in H_2SO_4 solutions (3). The highest corrosion resistance has been found for Hastelloy C. A large number of deposits with different morphologies were present on the alloy surface. Pustule-like deposits mainly contained a S-product. The needle-like deposits mainly contained S and Cr. The continuously creviced surface layer was composed of all elements originally present in the moisture evaporated in the corrosion chamber. Metallographic analysis of cross-sections did not reveal any indication of localized attack even after 16 months of exposure (Fig. 1g.).

Ti and Ti alloys. Titanium and Ti-alloys have been exposed to the environment of the corrosion chamber. The needle-like deposits found on the surface of the exposed alloy contained mainly a S-compound; other deposits contained Si, S, K. Ti and Fe. Although after ultrasonic cleaning, a certain roughening of the exposed surfaces could be seen, metallographic analysis of cross-sections did not indicate any localized corrosion even after 16 months of exposure (Fig. 1h.).

DISCUSSION AND CONCLUSIONS

The analysis of the corrosive products present in an air stream passing over clay revealed that at higher temperatures S and S-compounds are the main products governing the corrosiveness of the atmosphere in holes of a geological clay formation. The presence of SO_2 in the moisture and gases escaping from the clay is mainly due to the presence of pyrite in the clay and the chemical reaction induced in the clay with pyrite and humidity.

Exposure of aluminum and its alloys to the acid environment results in rapid uniform attack and pit formation induced by S^- and Cl^- compounds (4). The corrosion resistance of steels in the acid S and Cl^- containing environment is in accordance with literature data (4). Carbon steel had large corrosion rates in acid environments; the quality of a Cr coating is insufficient to protect the non-noble underlying carbon steel. The results obtained for the austenitic grades are in agreement with published corrosion data in acid SO_2, Cl^- and H_2SO_4 containing environments. Higher molybdenum contents and lower interstitial contents result in enhanced corrosion resistance (4). Sulphur, and to a lesser extent Cl^-, determine the corrosion mechanism of the austenitic and ferritic grades. The result obtained for nickel and superalloys can be explained by the positive influence of higher Cr and Mo contents on general and localized attack by S compounds in the presence of Cl^- ions. The results obtained for Ti and its alloys are in agreement with the excellent behavior observed in different chemical environments. Salts and salt deposits played a dominant role in the corrosion mechanism of all tested material, since the hygroscopy of the salts keeps the aggressive environment in contact with the tested alloys even during the heating period of the corrosion chamber.

REFERENCES

1. A. Bonne, M. D'Allessandra, "Probabilistic Analysis of Geological Segregation of Radioactive Waste in a Clay Formation," paper, this conference.
2. J. F. Nowak, "A Kinetic and Morphological Study of the Sulphidation of Ni-33Cr Alloys in SO_2 and H_2S Vapor", Corrosion Science 18, 971 (1976).
3. L. Beanier, "Sur la Corrosion Intergranulaire des Alliages à Base de Nickel", Métaux 625, 285 (1971).
4. Uhlig, Corrosion Handbook (John Wiley, New York, NY, 1948).

CANISTER COMPATIBILITY WITH CARLSBAD SALT*

W. N. Rankin
To be presented by J. R. Wiley

Savannah River Laboratory
E. I. du Pont de Nemours and Company
Aiken, SC 29801

INTRODUCTION

This paper describes the results of the examination of candidate canister alloy specimens heated with Carlsbad salt in sealed capsules for up to 5000 hours and in unsealed capsules for up to 10,000 hours. The compatibility of candidate alloys, for use in fabricating canisters for solidified radioactive waste, with salt from a potential final storage location for waste forms is being evaluated by long term heating tests (1). In these tests, canister alloys are being exposed to salt in a manner similar to that expected during final storage of canisters of solidified waste forms in a salt deposit. Test specimens are being heated at the temperature expected during waste form storage and at higher temperatures to accelerate any reactions that may occur. Reaction rates are determined by examining replicate specimens periodically removed from the test.

SUMMARY

No significant reaction was found when candidate canister alloys were heated with salt from Carlsbad, New Mexico, for up to 5000 hours in sealed capsules and for up to 10,000 hours in unsealed capsules at temperatures (80 to 225°C) that bracket the maximum temperature calculated for reference Savannah River Plant (SRP) waste containers at 20-foot spacings in salt (2). Additional tests were made at 600°C in sealed capsules to characterize reactions that may occur between

*Work performed under USDOE contract AT(07-2)-1.

candidate canister alloys and any component of the salt that is released when decrepitation occurs. Under these extreme conditions there was no significant attack of Type 304L stainless steel. But, there was up to 20-mils attack of the low-carbon steel.

DISCUSSION

Materials Used in Tests

Candidate Canister Alloys. Canister alloys for these tests were chosen for testing because of their oxidation resistance, low temperature ductility and cost. Cor-Ten A (Cor-Ten is a trademark of U.S. Steel Company) is a low-alloy steel containing 1% chromium, 0.5% nickel and 0.35% copper. This alloy costs only a little more than low-carbon steel, but it has better resistance to air oxidation. ASTM-A-516 (Grade 70) covers a low-carbon steel plate that is suitable for use at low temperatures because of its low nil-ductility transition temperature (NDTT). This specification could be important if low-carbon steel is considered for use as the outer canister (3) and low-temperature impact resistance is required as it is in the case of shipping casks for radioactive material (4). Past experience indicates that low-carbon and low-alloy steels have similar compatibility with vitrified waste. Therefore, the behavior of these two alloys should be representative of any other similar steels that might be selected (1).

Type 304L stainless steel has better resistance to air oxidation than either Cor-Ten A or low-carbon steel [ASTM-A-516 (Grade 70)]. It is more resistant than either of these alloys to accelerated corrosion in a radiation field. In a radiation field nitrogen combines with the oxygen and water vapor in the air to form nitric acid. But, stainless steel costs about three times more than the other alloys. It might not be suitable for final storage in salt because of its susceptibility to chloride pitting and/or stress corrosion cracking.

Carlsbad Salt. The salt used in these tests was from a section of core from Drill Hole AEC-8 in Carlsbad, New Mexico. The core, four inches in diameter and three feet long, was taken from the depth of 2142.7 to 2145.3 feet. Carlsbad is a potential final storage site for canisters of solidified radioactive waste and is being studied in the Waste Isolation Pilot Plant Program (5, 6).

The salt is primarily halite (NaCl)(7) and as in most bedded (stratified) salt, (8) moisture is trapped in small pockets on the grain boundaries and within the grains. When the salt is heated to around 250°C, the steam pressure in these pockets becomes greater than the bonding forces. This results in the salt fracturing (decrepitating) with considerable violence.

Experimental Procedure

Two types of specimens are being used in these tests: unsealed capsules and sealed capsules.

Unsealed Capsules. The first generation compatibility tests were started in unsealed capsules (Fig. 1.). Small capsules of Cor-Ten A and Type 304L stainless steel were tested. In these tests the capsules were filled with crushed salt. The capsules were then covered, but not sealed and placed in holes drilled in a solid piece of this salt. Unsealed capsules are being heated for up to 50,000 hours at 80°C, a reasonable temperature expected during waste form storage, and also at 225°C to accelerate any reactions that might occur.

Sealed Capsules. The second series of compatibility tests were started in sealed capsules (Fig. 2.). Small right circular cylinders of low-carbon steel ASTM-A-516 (Grade 70), or Type 304L stainless steel were partially embedded in a small block of salt. This assembly was then put inside a capsule made of Type 304L stainless steel high-pressure pipe fittings. The capsules were sealed by tungsten-inert-gas (TIG) welding. These tests are designed to characterize reactions that occur between the canister alloys and any volatile components of the salt. Reaction of the portion of the specimen embedded in the salt will be indicative of the compatibility between the canister alloy and solid salt and vapor from the salt. Reaction of the portion of the specimen exposed only to the vapors from the salt will isolate the effect of the salt vapor alone.

Specimens in sealed capsules are being heated for up to 50,000 hours at 80, 225 and 600°C. The amount of attack of the specimens heated in sealed capsules at 80 and 225°C will be compared with the amount of attack of specimens heated in unsealed capsules at these temperatures.

Tests at 600°C in sealed capsules were included in the test matrix to characterize any reactions that may occur between candidate canister alloys and any component of the salt that is released when decrepitation of the salt occurs. Decrepitation of the salt at about 250°C will occur, but this will not destroy the test assembly, as in the case of the unsealed capsules. The salt, the metal specimen, and any material that is released from the salt by decrepitation will remain in contact with the specimen inside the capsule.

Results

Unsealed Capsules. After the unsealed capsules had been heated for 10,000 hours, they were removed from the block of salt, the powdered salt was poured out of the capsules, and the surfaces of

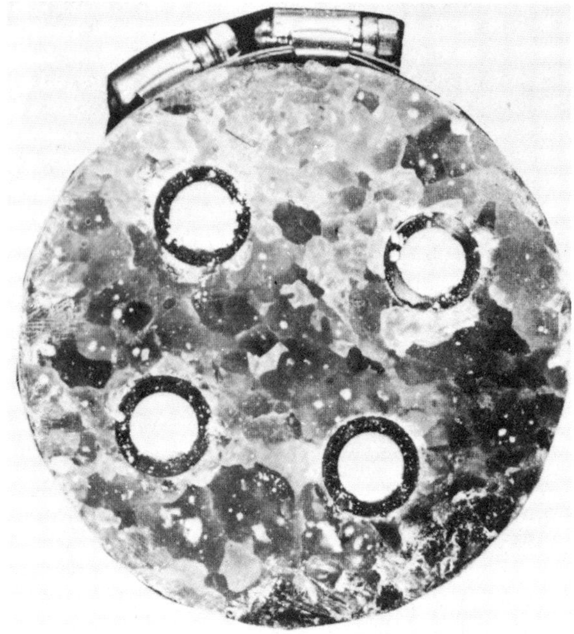

Fig. 1. Unsealed Capsule Assembly with Test Capsules Filled with Crushed Salt. Hose Clamps Were Used to Keep Salt from Breaking.

the capsules were examined visually and at low magnifications. There was no significant change in the appearance of the capsules heated at 80°C for 10,000 hours and the appearance of replicate specimens heated 1000 hours examined previously (9). Corrosion products and some pieces of salt adhered to the OD and ID surfaces of the Type 304L stainless steel capsule heated at 225°C for 10,000 hours. The change in dimensions of the unsealed capsules before and after heating were <5 mils which is within the experimental error of the technique used. A cross section through capsules heated at 225°C was examined on the optical microscope and the scanning electron microscope (SEM). Very little attack was found of the Type 304L stainless steel capsule heated 10,000 hours at 225°C. No oxide film was found on the OD surface. Patches of oxide only 0.1 mil thick were found on the ID surface. The ID surface of the low-carbon steel capsule heated 10,000 hours at 225°C was attacked less than the OD surface. An oxide film 0.6 mil thick was found on the OD surface. An oxide film 0.3 mil thick was found on the ID surface. The capsules heated at 80°C were not examined on the microscope because no significant attack was found of capsules heated at 225°C (9).

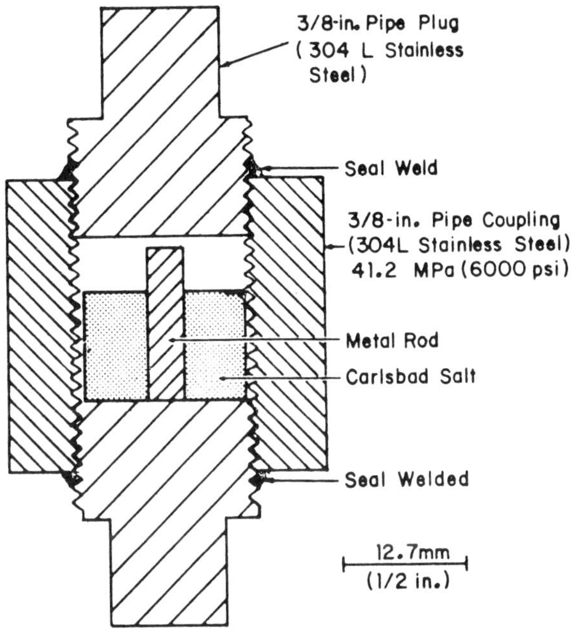

Fig. 2. Sealed Capsule Assembly

Sealed Capsules

The Type 304L stainless steel specimens heated at 80 and 225°C showed no visible sign of oxidation. They had the same appearance as recently machined specimens. The low-carbon steel specimens were covered with a black oxide film. There was no difference in the appearance of either the end of the specimen exposed to the vapor or the end of the specimen in the hole in the block of salt. Both ends of the Type 304L stainless steel specimen heated at 600°C were covered with a thin black oxide film and no salt adhered to the specimen. The end of the low-carbon steel exposed to vapor was covered with a thick black film. Salt was bonded to the oxide on the end of the specimen in the salt. No significant changes in dimensions occurred to specimens in test at conditions expected during waste form storage. Changes in diameter were <2 mils for Type 304L stainless steel and low-carbon steel specimens heated for up to 5000 hours at 80 and 225°C. In specimens exposed to more extreme conditions (600°C), there was no significant attack of Type 304L stainless steel in 5000 hours. But, extensive attack of the low-carbon steel occurred. The end of the specimen exposed to the vapor was attacked 20 mils in 5000 hours. The end of the specimen

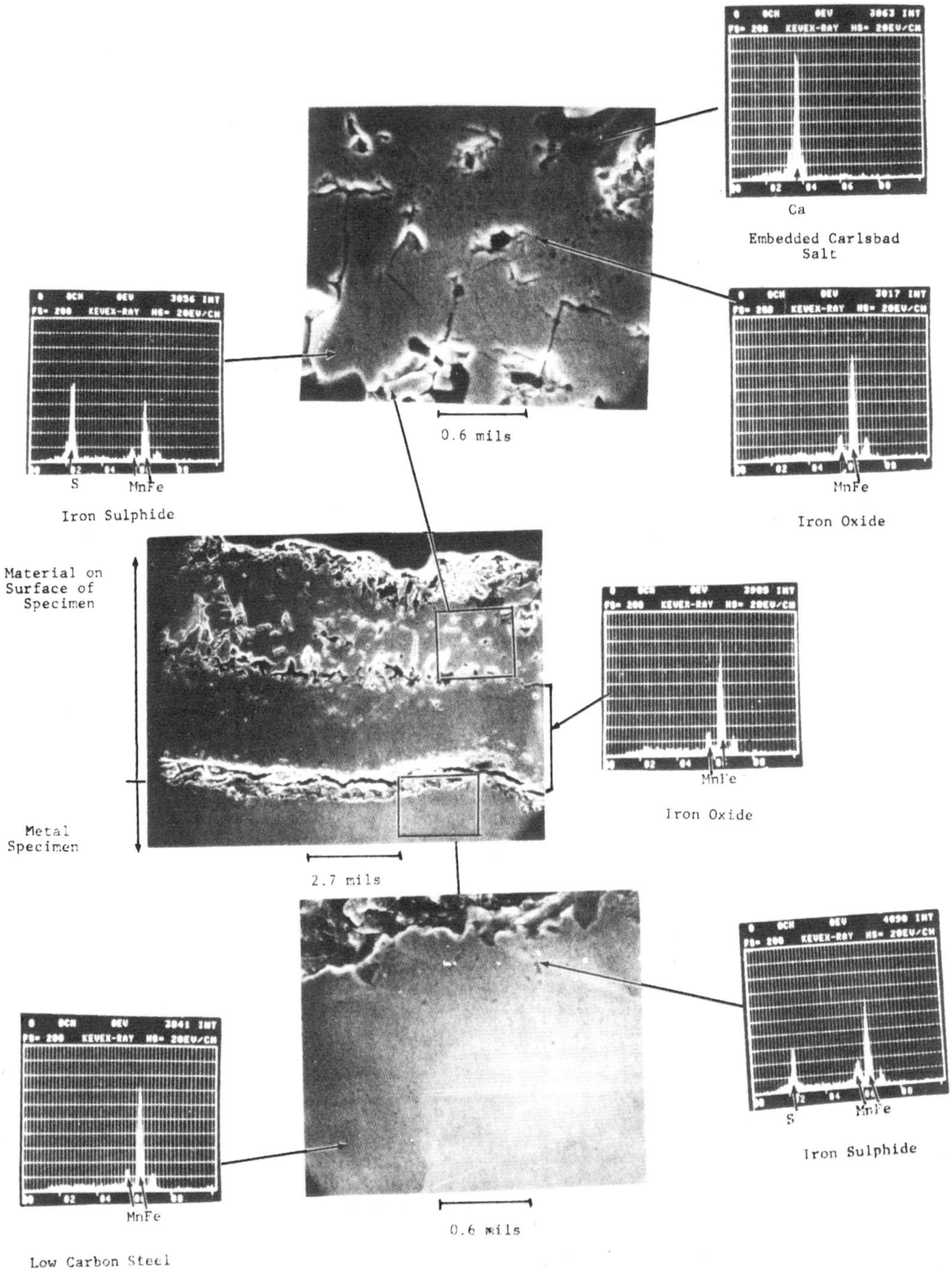

Fig. 3. Longitudinal Section Through the Surface of Low-Carbon Steel Exposed to Carlsbad Salt in a Sealed Capsule (Penetration 9.3 mils in 5000 hours at 6000°C)

embedded in the salt was attacked 9 mils in 5000 hours. X-ray diffraction analysis was performed on a piece of material from the surface of the portion of the specimen exposed to vapor. A longitudinal section through the entire specimen and transverse sections through each end of the specimen were examined on the SEM.

The appearance of the layers of material on both ends of the specimen indicate that the mechanism of attack of both ends was similar (Fig. 3.). There was about 0.8 mil of intergranular penetration of the low-carbon steel specimen. On the grain boundaries there was a material which contained iron and sulfur. Adjacent to the metal was a layer of material composed of the same elements as the low-carbon steel. This material was identified by X-ray diffraction analysis as iron oxide. The iron oxide layer was covered with a layer composed of iron oxide and a small amount of iron sulfide.

APPLICATION OF RESULTS

The results of these tests at 80 to 225°C show that there is no significant attack of Type 304L stainless steel or low-carbon steel under the conditions of the tests and at a higher temperature designed to accelerate any reactions that might occur. The results of the tests at 600°C indicated no significant attack of Type 304L stainless steel occurs at this temperature by the volatile products of the salt released by decrepitation, however, there was extensive attack of the low-carbon steel under these extreme conditions.

ACKNOWLEDGMENTS

The author is grateful to D. W. Powers of Sandia Laboratories for supplying the salt. He is also indebted to G. G. Wicks of Savannah River Laboratory (SRL) for the X-ray Diffraction Analysis and to J. A. Donovan formerly of SRL and P. K. Smith of SRL for their helpful discussions.

REFERENCES

1. W. N. Rankin, "Compatibility Testing of Vitrified Waste Forms," Presented at Corrosion/78, Houston, TX, National Association of Corrosion Engineers, March 6-10 (1978).
2. M. H. Tennant, "Temperatures Generated by Underground Storage of Defense Waste Canisters," to be presented at Corrosion/80, Houston, TX, National Association of Corrosion Engineers, March 6-10 (1980).
3. A. S. Jennings, "Plans for Solidification of Savannah River Plant High Level Wastes," presented at the 70th Annual AIChE Meeting on

Long-Term Management of Defense High Level Waste, American Institute of Chemical Engineers, New York, NY, Nov. 15 (1977).

4. L. B. Shappert, "Cask Designers Guide," USERDA Report ORNL-NSIC-68, Oak Ridge National Laboratory, Oak Ridge, TN (1970).
5. WIPP Conceptual Design Report, SAND 77-0274 (Parts I and II), Sandia Laboratories, Albuquerque, NM (1977).
6. J. W. Braithwaite and M. A. Molecke, "High Level Waste Canister Corrosion Studies Pertinent to Geological Isolation," Proceedings of Conference on High Level Solid Radioactive Wastes, Denver, CO, NUREG CP-0005 (1979).
7. M. A. Molecke, Personal Communication, Sandia Laboratories, Albuquerque, NM, Oct. (1976).
8. R. L. Bradaham and W. C. McClain, Editors, Project Salt Vault: A Demonstration of the Disposal of High-Activity Solidified Wastes in Underground Salt Mines, ORNL-4555, UC-70, Waste Disposal and Processing, Oak Ridge National Laboratory, Oak Ridge, TN, April (1971).
9. Savannah River Laboratory Quarterly Report, Waste Management, DPST-77-125-2, April-June (1977).

THE BACKFILL AS AN ENGINEERED BARRIER FOR NUCLEAR WASTE MANAGEMENT*

E. J. Nowak

Sandia Laboratories **
Albuquerque, New Mexico 87185

ABSTRACT

This paper presents results from an experimental backfill barrier development program. The swelling, plastic flow and relative impermeability of bentonite and hectorite were observed and measured after wetting with concentrated brines. Measurements of stable values of pH > 6.5 for the interstitial brines in wetted bentonite and hectorite confirmed conditions favorable for precipitation and sorption of transuranics. Values of $K_d > 2000$ ml/g were measured for Pu and Am. Calculated estimates of the effectiveness of a one-foot-thick backfill barrier are presented. They show that the breakthrough of Pu and other transuranics ($K_d = 2000$ ml/g) can be delayed for 10^4 to 10^5 years. The breakthrough of most fission products ($K_d = 200$ ml/g) can be delayed for 10^3 to 10^4 years, sufficient time for them to decay to very low concentrations. A backfill barrier can contribute significantly to a radioactive waste isolation system.

INTRODUCTION

Backfill material emplaced and maintained as a continuous layer or blanket surrounding waste containers can be designed to be part of an interacting multiple barrier concept for radioactive waste isolation. Barr and O'Brien (1) proposed that radioactive waste containers can be surrounded by radionuclide adsorbers or "getters" and that they could be emplaced as mixtures with backfilled

*This work supported by the United States Department of Energy (DOE), under Contract DE-AC04-76-DP00789.
**A DOE Facility

solids, as powders, or as monolithic hole liners. Jacobson and Pusch (2) gave their criteria for the properties of suitable "buffer substances" which could be emplaced around waste containers as elements in a Swedish (KBS) conceptual repository design. They suggested the use of bentonite-sand mixtures for that purpose. Guiffre et al.(3) also recognized the potential effectiveness of a backfill as a physical and sorptive barrier to radionuclide migration.

Radionuclide sorption and retention measurements on clays and soils have been underway for many years, and a ponderous body of literature has been generated. However, those data are not readily generalizable to specific backfill barrier applications. Comparatively little has been reported on measurements and calculations specific to the conditions of deep geologic isolation (4,5,6,7,8). It is to the latter area that this paper is directed.

A schematic representation of a backfill barrier is shown in Fig. 1. Groundwater, including brine which may be present in bedded salt, is the vehicle for potential waste migration. Capacity for sorption of waste species, low hydraulic permeability, adequate thermal conductivity, adequate plasticity to resist fracture, and sufficient support strength for waste containers are among the important properties which contribute to the effectiveness of a backfill as a barrier to radioactive waste migration. The backfill may also be designed to modify the groundwater chemically for decreased corrosion of the overpack and canister and decreased leaching of the wasteform. A more detailed description of this system and migration variables is given elsewhere (9).

A backfill barrier is best designed and analyzed as a part of a specific radioactive waste isolation system. The time by which it can delay breakthrough of radionuclides is a measure of the backfill's effectiveness. The time to breakthrough is determined by the sorption and mass transport properties of the backfill and the groundwater. Interactions among the groundwater, wasteform, canister, overpack, backfill and host geologic formation under the thermal, radiation and lithostatic pressure conditions in a specific repository system can determine those pertinent sorption and transport properties.

SORPTION MEASUREMENTS

The capability to sorb waste species is a crucial property of getters or sorbers for use in backfill barriers. Measurements of the batch sorption of ^{152}Eu by clays, soils, and a zeolite have been made. ^{152}Eu dissolved in nearly saturated brines was used to simulate transuranics for the proposed Waste Isolation Pilot Plant (WIPP) near Carlsbad, New Mexico. Brine A is representative of groundwater in contact with potash-containing layers above the proposed waste

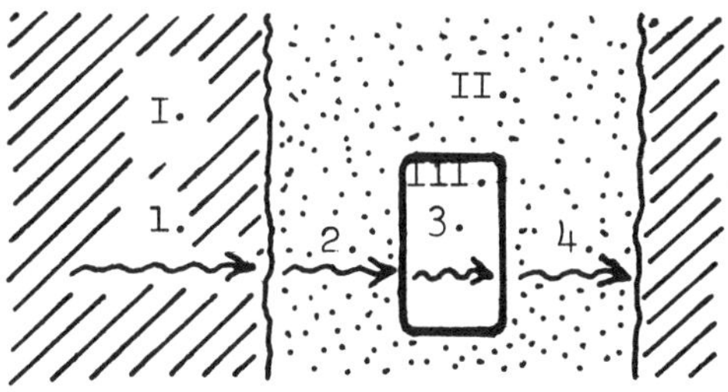

Fig. 1. Schematic Diagram of Emplaced Backfill Barrier and Hypothetical Sequences of Processes in its Function.
1. Groundwater Movement. 2. Retarded Groundwater Movement.
3. Corrosion, Dissolution, Leaching. 4. Retarded Migration.

horizon, and brine B represents groundwater in contact with the halite of the proposed waste horizon itself. Nominal compositions of brines A and B are given in Table 1.

Sorption measurements were made by contacting brine containing ^{152}Eu and a solid sorbent in a continuously agitated vessel. Values of pH were controlled by additions of aqueous HCl or NaOH. Control samples of Eu-containing brine alone (sorbent-free) were included for each set of experimental conditions to correct the results for Eu losses not attributable to the sorbent. All samples, including the controls, were filtered through 0.8 μm pore size filter media before gamma-counting. Some of the getter materials were heated in air at 300°C for 6 hours and cooled prior to the room temperature sorption measurements. This heat treatment was done to test for the effect of a dry thermal cycle on clay properties.

Screening results are given in Table 2 as empirical distribution coefficients, K_d. It was shown previously (10) that the liquid phase precipitation of Eu-containing solids from brines A and B is unlikely at pH $\leq$ 5.5. At pH = 5.5, the values of K_d for all of the materials were nearly the same except for montmorillonite in Brine B. At pH = 6.5, there was greater variability among values of K_d, possibly because small changes near pH = 6.5 caused large changes in precipitation. Nevertheless, the values of K_d were consistently larger at pH = 6.5 than at pH = 5.5, and no large decreases in K_d resulted from heating the getters at 300°C. These results show that K_d values in the approximate range of 10^3 to 10^4 are achievable for all of the materials tested and that exposure to a dry thermal cycle from high level waste would not adversely affect their sorption properties.

Although increased K_d with increased pH is consistent with sorption by a single ion exchange reaction, other evidence is contradictory. Table 3 presents such evidence for sorption at pH = 5.5 in

Table 1. Representative Brine Compositions

	Major Constituents, Molarity							
	Na^+	K^+	Mg^{++}	Ca^{++}	Cl^-	SO_4^{--}	HCO_3^-	BO_3^{---}
Brine A	1.8	0.8	1.4	0.02	5.4	0.04	0.01	0.02
Brine B	5.0	---	---	0.03	5.0	0.04	----	----

dilute and concentrated brines. There was little change in K_d for montmorillonite and hectorite with a two-orders-of-magnitude change in the competing ion concentration (virtually all Na^+ for brine B). It was concluded that the sorption mechanism for montmorillonite and hectorite is complex. Neretnieks (6) reported a similar conclusion for natural zeolites. The same conclusion was reached for the soils in these tests. However, the results for Zeolon are consistent with a single ion exchange reaction mechanism at pH = 5.5.

SELECTION AND PROPERTIES OF BENTONITE AND HECTORITE

The smectite swelling clays (11) montmorillonite and hectorite have physical properties which make them good candidates for backfill barrier components. When wetted, they swell and flow plastically to fill and seal voids, they are essentially impermeable and they have adequate thermal conductivity and support strength when mixed with sand (12). Swelling to nearly double their initial volume and subsequent impermeability were verified experimentally for brines A and B at atmospheric pressure. Measurements of pH $\geq \sim 6.5$ for the supernatant brine confirmed conditions favorable for large values of K_d. Finally, initial measurements yielded ~ 0.5 W/mK for the thermal conductivities of dry clay-sand mixtures at atmospheric pressure.

Because of their favorable properties, commercially available bentonite (containing montmorillonite) and hectorite mixed with sand were chosen as basic backfill materials for further investigation. These materials establish physical properties for the backfill and a pH range for the brine or groundwater. They also act as getters for transuranic nuclides. Some preliminary batch sorption measurements with 10 wt% bentonite, 90 wt % sand mixtures have yielded $K_d > 2000$ ml/g for Pu and Am in brine B. These measurements included corrections for losses of Pu and Am from solution that are not attributable to the presence of the clay-sand mixtures. Other backfill components (13) may be required for getters of fission products in the presence of concentrated brines. Potential getters for pertechnetate, iodide, and iodate anions have been reported (14, 15,16).

Table 2. Batch Equilibration Results for $^{152}Eu^{3+}$ Sorbate at Room Temperature

$C_o = \sim 2 \times 10^{-7}$ $\underline{M}$ Eu^{3+} ($\sim 0.1\ \mu Ci/ml$)

K_d, ml/g

Getter	Brine A			Brine B		
	pH = 5.5	pH = 6.5		pH = 5.5	pH = 6.5	
	----	----	Heated*	----	----	Heated*
DLR(soil)**	200	1600	1800	270	14000	7300
Caliche	140	8000	9000	220		11000
Tuff	200	2500		200	1400	
Zeolon(Zeolite)	50	690	270	60	6000	1400
Montmorillonite (SWy-1)***	100	850	1100	6700	1300	3500
Hectorite (SHCa)***			5500			7200
Kaolin(DGa-1)***	60	1100		200	1600	

*Heated 6 hours in air at 300°C before sorption measurement at room temperature.

**Dewey Lake Redbeds, an outcropping in the Los Medanos area, Carlsbad, NM.

***Samples from Source Clay Mineral Repository, University of Missouri, Columbia, MO.

BACKFILL BARRIER PERFORMANCE ESTIMATES

Estimated breakthrough times for migrating waste species in backfill barriers free of cracks or channels were calculated. Breakthrough was defined as the appearance of a migrating species at $\sim$ 1% of its initial concentration. A fixed bed model with sorption by linear equilibrium ion exchange was used as a first empirical approximation. Calculation techniques given by Vermeulen et al.(17) and Hefferich (18) were used for migration by convective transport. Relationships from Crank (19) were used for migration predominantly by diffusion. Details of these calculations are given elsewhere (9).

Values for K_d, the effective porosity ε, and the interstitial groundwater velocity v_g were chosen to bracket the appropriate ranges. A value of $K_d = \sim 2000$ ml/g has been shown to be achievable for Pu in concentrated brine, and it is realistic for other groundwaters (20). $K_d = 200$ ml/g is attainable for Sr in brine (13) and Cs in dilute aqueous solutions (20). Values of ε in the range of 0.01 to 0.1 are conservative estimates for clays (21). At interstitial groundwater velocities less than 0.1 ft/year, migration is predominantly by molecular diffusion and essentially independent of interstitial velocity. An upper bound of 1000 ft/year includes most of the interstitial velocities predicted or measured for generic or specific repository sites(20,21).

Calculated estimates for breakthrough times are given in Table 4. They are in the range of 10^4 to 10^5 years for Pu ($K_d = 2000$ ml/g) and other transuranics. For $K_d = 200$ ml/g, breakthrough times are in the range of 10^3 to 10^4 years, sufficient time for most of the fission products in high level waste to have decayed to very low concentrations. The significant effects of interstitial groundwater velocity, K_d, and ε are also illustrated. These estimates show that backfill barriers are potentially effective contributors to the isolation of radioactive wastes in the present of concentrated brines and other groundwaters.

Table 3. Sorption at Different Competing Ion Concentrations

pH = 5.5, $C_o = \sim 2\times10^{-8}$ $\underline{M}$ Eu^{3+} ($\sim$ 0.01 μCi/ml)

	K_d, ml/g	
	100% Brine B	1% Brine B in Deionized Water
Montmorillonite	500	1000
Hectorite	3400	2600
Zeolon	12	25000

Table 4. Times to Breakthrough for a One Foot Thick Backfill Barrier Having a Bulk Density of 2 g/cm^3

ε	K_d, ml/g	Interstitial Groundwater Velocity ft/year	Calculated Time to Breakthrough, years
0.1	2000	0.1	$1x10^4$
		1.0	$1x10^4$
		10.0	$2x10^3$
		100.0	$3x10^2$
		1000.0	$3x10^1$
0.01	2000	0.1	$1x10^5$
		1.0	$1x10^5$
		10.0	$2x10^4$
		100.0	$3x10^3$
		1000.0	$3x10^2$
0.1	200	0.1	$1x10^3$
		1.0	$1x10^3$
		10.0	$2x10^2$
		100.0	$3x10^1$
		1000.0	$3x10^0$
0.01	200	0.1	$1x10^4$
		1.0	$1x10^4$
		10.0	$2x10^3$
		100.0	$3x10^2$
		1000.0	$3x10^1$

REFERENCES

1. G. E. Barr and P. D. O'Brien, Personal Communication and Invention Disclosure to Sandia Laboratories and the United States Department of Energy (DOE), (March 1976).
2. A. Jacobson and R. Pusch, "Deposition of High-Level Radioactive Waste Products in Bore-holes with Buffer Substance," KBS Technisk Rapport 03, Stockholm (1977).
3. M. S. Guiffre, C. M. Koplick, R. L. Plum and R. Talbot, "Information Base for Waste Repository Design," Vol. 5: Decommissioning of Underground Facilities, NUREG/CR-0495, TR-1210-1 (1979).
4. B. Allard, H. Kipatski and J. Rydberg, "Sorption of Long-lived Radionuclides in Clay and Rock. Part 1, Determination of distribution Coefficients", KBS Technisk Rapport 55, Stockholm (1977).
5. B. Allard, H. Kipatski and B. Torstenfelt, "Sorption ay Langlivade Radionuclider i Lera och Berg", Del 2, KBS Technisk Rapport 98, Stockholm (1978).

6. I. Neretnieks, "Retardation of Escapng Nuclides from a Final Repository", KBS Technisk Rapport 30, Stockholm (1977).
7. I. Neretnieks and C. Skagius, "Diffusivity Measurements in Wet Compacted Clay Na-lignosolfonate, Sr^{++}, Cs^{+}, KBS Technisk Rapport 87, Stockholm (1978).
8. I. Neretnieks, "Transport of Oxidants and Radionuclides Through a Clay Barrier", KBS Technisk Rapport 79, Stockholm (1978).
9. E. J. Nowak, "The Backfill Barrier as a Component in a Multiple Barrier Radioactive Waste Isolation System", Sandia Laboratories Report SAND79-1109 (1979).
10. E. J. Nowak, "The Migration of Eu Through Geologic Media," Trans. Am. Geophysical Union 59, 1224 (1978).
11. R. E. Grim, Clay Mineralogy (McGraw-Hill, New York, 1968).
12. R. Pusch, "Highly Compacted Na Bentonite as Buffer Substance," KBS Technisk Rapport 33, Stockholm (1977).
13. R. G. Dosch, "The Use of Titanates in Decontamination of Defense Waste," Sandia Laboratories Report SAND78-0710 (1978).
14. S. Fried, A. M. Friedman, D. Cohen, J. J. Hines and R. G. Strickert, "The Migration of Long-lived Radioactive Processing Wastes in Selected Rocks," Argonne National Laboratory Report ANL-78-46 (1978).
15. R. A. Couture, M. G. Seitz, and S. Steindler, "Adsorption of Iodate by Hematite," presented at the ANS Annual Meeting, Atlanta, Georgia (June 1979).
16. E. Akatsu, R. Ono, K. Tsukuechi and H. Uchiyama, "Radiochemical Study of Adsorption Behavior of Inorganic Ions on Zirconium Phosphate, Silica Gel and Charcoal," J. Nucl. Sci, Techn. 2, 141 (1965).
17. T. Vermeulen, G. Klein, and N. K. Hiester, "Adsorption and Ion Exchange," Chemical Engineers' Handbook (5th Edition, R. H. Perry and C. H. Chilton, eds., McGraw-Hill, New York 1973).
18. F. Helfferich, Ion Exchange (McGraw-Hill, New York, 1962).
19. J. Crank, The Mathematics of Diffusion (Oxford Press, London, 1956)
20. I. Y. Borg, R. Stone, H. B. Levy, and L. D. Ramspot, "Information Pertinent to the Migration of Radionuclides in Groundwater at the Nevada Test Site, Part 1, Review and Analysis of Existing Information," Lawrence Livermore Laboratory Report UCRL-52078 Part 1 (1976).
21. R. K. Kibbe and A. L. Boch, "Technical Support for GEIS: Radioactive Waste Isolation in Geologic Formations," Vol. 21 Groundwater Movement and Nuclide Transport, Office of Water Isolation Report Y/OWI/TM-36/21 (1978).

SUPEROVERPACK: TAILOR-MADE MIXTURES OF ZEOLITES AND CLAYS

Sridhar Komarneni and Rustum Roy

Materials Research Laboratory
The Pennsylvania State University
University Park, PA 16802

INTRODUCTION

Cation exchange and selective adsorption of cations by clays, zeolites and gels is well known and has been the subject of intensive research for decades. Numerous equations have been derived to describe the exchange of ions by clay systems (1, 2), and selective sorption of cations by clays has been long studied (3-5). Ion exchange has been established as a standard way of describing molecular sieve zeolites (6), and selectivities shown by a number of zeolites for Cs or NH_4 or K (7-9) illustrate a few of many ion separations which could be performed with zeolite exchangers. Hydrous metal oxides typically as noncrystalline oxyhydroxide gels have also been extensively studied for the selective sorption of cations (10, 11). The vast majority of these studies of cation exchange equilibria and of selective adsorption have been made on single cation exchangers, i.e., the solution from which sorption occurs typically contains only one cation. Moreover, most of the work also involves one solid phase at a time, and little attention was paid as to what extent the sorbed ions are retained or fixed by the solid phase. Very little work has been done on adsorption and fixation of complex <u>mixtures of ions</u> by <u>mixtures of clays and zeolites</u>.

The reason for this is clear. There was an insuperable problem in obtaining the information because of the difficulty in separating the two or more crystalline phases for analysis so that partitioning of ions among the exchangers could be determined. Early work by Denny and Roy (12) using the electron microprobe showed a way of overcoming this difficulty by individual particle analysis. It was, however, tedious work at the limit of instrumental capability.

High level liquid radioactive wastes contain some 40 ions in a particular combination and within a certain range of ratios. The adsorption and retention of these from liquid onto filters, or the adsorption of any ions which may under any circumstances be extracted from a previously solidified waste form, are of great significance for both the science and the future technology of the disposal of nuclear wastes, resulting principally so far from the U.S. weapons program. Thus, such studies to understand the partitioning of many ions among two, three or more cation exchangers are of considerable importance in solving problems connected with radioactive waste disposal. The objective of this research paper, therefore, is to study the mechanisms of adsorption and fixation of complex mixtures of ions by mixtures of clays and zeolites with specific attention to the typical fission products in high level waste. In connection with radioactive waste fixation, these adsorbent mixtures can be used in four different situations: (1) as a "superoverpack" surrounding the canisters containing radioactive waste which would constitute a first level sponge to sop up any ions escaping from the container; (2) as a filter bed to pick up all or some of the ions in liquid waste; (3) as an additive in grout preparation to fix all the high level liquid waste (HLLW) ions; and (4) as a special "firefighting" material to be used as a slurry in the case of leaks either before or after geological disposal. Most of the zeolites and clays retain their basic structure at modest pH_2O and temperatures below 200°C. The stability of the different phases may need to be investigated further if they are to be used as major hosts in geological disposal.

EXPERIMENTAL

Solid Phases Studied

The phases studied were selected to represent a wide variety of mineralogical and crystallographic types with different exchange capacities and different size cavities, etc. For this preliminary study, it was felt that a level of purity of "single phase to X-ray diffraction" was sufficient. All phases used in this study were found here to be monomineralic as determined by X-ray diffraction, except for the "Conasauga shale" used in the Oak Ridge grouting process which consists of illite, quartz, chlorite and albite minerals. These phases were used at 25°C with their natural cation saturation because it was felt that the effects of exchangeable cation (which determines the hydration state) are minor in this exploratory study. The shale and natural zeolites were dry ground gently in a ball mill while the vermiculite sample was wet ground in a blender followed by sieving to obtain the different size fractions (see Table 1). The montmorillonite, oxide and hydroxide samples were sieved to get their grain size fractions (see Table 1), the individual crystal size being much smaller, of course. The size-fractionated artificial zeolites were supplied by the manufacturer.

Table 1. Sorption[a] and Fixation of Ions by Individual Mineral Phases

Sample, state, particle size (-mesh)	mg of Cs sorbed/g of sample	% Cs fixed	mg of Rb sorbed/g of sample	% Rb fixed	mg of Sr sorbed/g of sample	% Sr fixed	mg of Ba sorbed/g of sample	% Ba fixed
Natural Zeolites								
Clinoptilolite, AZ, 200 mesh	5.35	81	0.643	77	1.490	26	2.346	nd
Clinoptilolite, ID, 200 mesh	5.35	82	0.648	82	1.590	75	2.359	nd
Mordenite, AZ, 200 mesh	5.43	92	0.558	61	0.618	53	1.969	nd
Mordenite, NV, 200 mesh	5.37	89	0.558	69	0.410	70	1.985	nd
Erionite, OR, 200 mesh	5.28	89	0.650	63	0.550	7	1.879	nd
Erionite, CA, 200 mesh	5.51	91	0.648	75	1.346	9	2.363	nd
Chabazite, AZ, 200 mesh	5.47	88	0.598	68	0.945	0	2.197	nd
Phillipsite, CA, 200 mesh	5.44	89	0.663	94	0.775	0	2.364	nd
Synthetic Zeolites[b]								
Linde, 3A, 600 mesh	2.40	38	0.073	0	1.778	87	2.368	nd
Linde, 4A, 600 mesh	4.27	70	0.323	40	1.783	70	2.368	nd
Linde, 5A, 600 mesh	5.06	69	0.345	41	1.670	30	2.343	nd
Clay Minerals								
Montmorillonite, AZ, 200 mesh	4.24	33	0.378	21	0.913	0	1.650	nd
Vermiculite, SC, 60 mesh	4.76	65	0.218	55	1.213	0	1.521	nd
Oxides and Hydroxide[b]								
Ferric Oxide (Red), 60 mesh	0.00	0	0.000	0	0.090	24	0.388	0
Ferric Oxide (Yellow), 60 mesh	0.00	0	0.000	0	0.090	94	0.283	6
Manganese (IV) dioxide, 60 mesh	0.00	0	0.000	0	0.075	89	0.073	75
Gibbsite, heated,[c] 60 mesh	0.00	0	0.000	0	0.675	0	1.602	nd
Shale								
Conasauga shale, TN, 200 mesh	1.56	23	0.168	45	0.330	0	nd	nd

[a] La, Nd and U were totally sorbed by these samples because of their small concentrations in the PW-4b solution. The concentrations of La, Nd and U are 200, 180 and 40 μg/ml (ppb) respectively.
nd = not determined

[b] Obtained from commercial vendors.

[c] Heated at 500°C for 3 hours.

Sorption

All the sorption studies to be described here were performed using a solution made by the treatment of 0.8 g of simulated radioactive waste, PW-4b (13) in 5 ml of Conc. HNO_3 for one hour followed by neutralization to pH 7 with NaOH, and the total volume made up was 2 liters. The neutralized solution referred to hereafter as simulated waste solution was found to contain 22.44, 2.75, 7.14, 9.47, 0.20, 0.18 and 0.04 μg/ml (ppm) of Cs, Rb, Sr, Ba, La, Nd and U, respectively. In addition to the above elements, Na^+ and NO_3^- are present in the solution to the extent of ~0.04N. Sorption studies were done either on individual phases or 1:1 by weight of mixtures of phases as follows: a 25 ml aliquot of the simulated waste solution was added to 100 mg solid phase in a glass vial (solution/solid ratio = 250), tightly capped, and the mixture was shaken at room temperature for 24 hours, which was found to be sufficient to attain equilibrium. The solid/solution mixture was then centrifuged to separate the solid and solution phases. The solution was analyzed for Cs and Rb by atomic absorption spectrophotometry (AAS); for Sr, Ba, La and

Nd by atomic emission spectrometry and for U by fluorimetry. The detection limits by these methods are 0.005, 0.002, 0.001, 0.001, 0.001, 0.025 and 0.001 μg/ml for Cs, Rb, Sr, Ba, La, Nd and U, respectively. Analyses of the solutions for these elements give the partitioning between the solid and solution phases because the total amounts of various elements in the initial solution are known. The amount sorbed is simply the amount taken up by the solid phase from the solution.

After the solid and solution phases were separated, the solid phase was washed free of excess salts by water and acetone. "Fixation" or retention of the various ions by these solid phases was determined as follows: an aliquot of 15 ml of 0.1N NaCl + 0.1N $CaCl_2$ (to simulate ground water) was added to 50 mg samples in glass vials, tightly capped, and the solid/solution mixtures were shaken for 24 hours. The solid and solution phases were separated and the solution was analyzed for the amounts of Cs, Rb, Sr and Ba released from solid as described above. The percent of element fixed by the solid sample is simply 100 times the ratio of the amount retained in the solid after one extraction to the amount sorbed by the sample.

RESULTS AND DISCUSSION

Sorption and fixation data will be discussed under three sections: sorption and fixation of ions by individual phases, sorption and fixation of ions by mixtures of minerals with vermiculite, and sorption and fixation of ions by mixtures of zeolite minerals.

Sorption and Fixation of Ions by Individual Mineral Phases

All the natural zeolites used here have high sorption and fixation capacities for Cs and Rb and have high sorption capacity for Ba (Table 1). The two clinoptilolite samples have high Sr sorption and fixation capacity as well, while all the other natural zeolites have moderate Sr sorption capacity. The reason for the big difference in the Sr fixation capacity of the two clinoptilolites is not clear. Mordenite has a high Sr fixation capacity, but not chabazite, erionite and phillipsite, as can be seen from Table 1. Among the natural zeolites used here, clinoptilolite and mordenite have the highest sorption and fixation capacities for Cs, Rb, Sr and Ba from simulated waste solution.

The synthetic zeolites, Linde 3A, 4A and 5A, have high Sr and Ba sorption capacities, but only Linde 3A and 4A have large Sr fixation capacity (Table 1). Linde 4A and 5A zeolites have moderate sorption and fixation capacities for Cs and Rb, while Linde 3A is poor for their sorption and fixation. Thus, no one of these synthetic zeolites is exceptionally good in the sorption and fixation of all the fission products, Cs, Rb, Sr and Ba, from the simulated waste solution.

Among the clay minerals, only the montmorillonite and vermiculite group of minerals have high cation exchange capacities, and that is why only two of these clay minerals are used in these experiments. Vermiculite is superior to montmorillonite in the sorption and fixation of Cs, Rb and Sr (Table 1), but both these minerals are not as good as the natural zeolites. The Cs fixation ability of vermiculite is greater than that of montmorillonite, which can be explained by layer collapse in the former (14). Vermiculite has a complex Cs sorption behavior (15) which depends on the charge density of the layers (16), the concentration of Cs in solution, the effect of contact time, etc. Therefore, it cannot be ruled out as a potential mineral in an overpack. The oxides and hydroxide materials used here show no sorption capacity for Cs and Rb from the simulated waste solution. The sorption capacity of oxides for Sr and Ba is poor, while the heated gibbsite is good for their sorption. The poor sorption capacity for divalent ions of the oxides used here may be attributed to their crystalline nature (as evidenced by X-ray diffraction) because amorphous iron and magnesium oxides have good sorption capacity for divalent and trivalent ions (11, 12).

Conasauga shale, which consists of different fine-grained minerals, has been used for the disposal of low and/or intermediate level radioactive waste (17). The sorption and fixation capacities of Conasauga shale are given here for comparison. As can be seen from Table 1, Conasauga shale is a poor sorber compared to the zeolites or clay minerals, but better than the oxides and hydroxide used in these studies. Thus, when high concentrations of fission products are involved, Conasauga shale has a poor sorption and fixation ability.

Sorption and Fixation of Ions by Mixtures of Minerals with Vermiculite

Vermiculite has a high capacity for sorption and fixation of Cs, depending upon the Cs concentration in solution and the charge density of the layers (16), and that is why it is further explored here in combination with other mineral phases. The addition of vermiculite to Conasauga shale, AZ montmorillonite, heated gibbsite, and Linde 3A and 4A zeolites has a synergistic effect, i.e., in general, the Cs and Rb sorption and fixation capacities of the mixtures are greater than what can be expected from their individual sorption capacities. For an example, only 2.38 mg of Cs/g of sample are expected to be sorbed by a 1:1 mixture of heated gibbsite and vermiculite based on their individual sorption capacities for Cs (Table 1). However, this mixture sorbed about 4.81 mg of Cs/g (Table 2), which indicates that the Cs sorption capacity of vermiculite increased because of a change in the total amount of Cs available for sorption by

Table 2. Sorption and Fixation of Ions by Mixtures of Minerals with Vermiculite

Sample mixtures	mg of Cs sorbed/g of sample	% Cs fixed	mg of Rb sorbed/g of sample	% Rb fixed	mg of Sr sorbed/g of sample	% Sr fixed	mg of Ba sorbed/g of sample	% Ba fixed
1:1 Mixtures with Vermiculite, SC								
Conasauga shale, TN	4.720	66	0.223	54	0.250	0	1.414	nd
Montmorillonite, AZ	4.568	62	0.300	13	1.678	0	1.567	nd
Gibbsite, heated	4.808	64	0.163	56	0.908	0	1.607	nd
Clinoptilolite, AZ	5.058	80	0.603	78	1.055	26	2.339	nd
Phillipsite, CA	5.245	92	0.643	94	0.975	0	2.362	nd
Chabazite, AZ	5.180	86	0.520	71	1.085	0	2.136	nd
Linde, 3A	4.688	89	0.235	78	1.355	32	2.368	nd
Linde, 4A	4.865	73	0.463	59	1.775	25	2.368	nd
Linde, 5A	4.748	73	0.273	64	1.778	14	2.332	nd

nd = not determined

vermiculite in the mixture. The Sr and Ba sorption capacities stayed about the same or slightly decreased in the mixture of the above phases with vermiculite compared to the individual phases (Tables 1 and 2). The addition of vermiculite to phillipsite, chabazite, clinoptilolite and Linde 5A zeolites did not improve the sorption and fixation of any of the ions compared to the individual phases (Tables 1 and 2). These studies point out that mixtures of zeolites may be superior to the mixtures of clay minerals with zeolites.

Sorption and Fixation of Ions by Mixtures of Zeolite Minerals

The data on 1:1 mixtures of synthetic zeolites with naturally occurring zeolites reveal that the sorption and fixation of Cs and Rb increased significantly. This points out that most of the Cs and Rb are taken up by the natural zeolites because all these natural zeolites are excellent sorbers and fixers of Cs and Rb (Table 1). The sorption of Sr and Ba also increased in the 1:1 mixtures because of the synthetic zeolites which are good sorbers of Sr and Ba. However, the Sr fixation capacity of the mixtures decreased, which indicates that Sr is distributed between the natural and synthetic zeolites. If all Sr is sorbed by synthetic zeolites in these mixtures, the fixation of Sr would have been greater than that reported in Table 3.

All the 1:1 mixtures of natural zeolites have good sorption and fixation capacities for the four fission products. It is not surprising that these mixtures are good for Cs and Rb because all the natural zeolites are good sorbers and fixers of Cs and Rb (Table 1). But what is surprising is the fact that Sr fixation is much greater in certain mixtures than can be expected from the individual minerals. These mixtures involve the poor Sr fixers, such as erionite, chabazite and phillipsite. The increased Sr fixation in these mix-

Table 3. Sorption and Fixation of Ions by Mixtures of Zeolite Minerals

Sample mixtures	mg of Cs sorbed/g of sample	% Cs fixed	mg of Rb sorbed/g of sample	% Rb fixed	mg of Sr sorbed/g of sample	% Sr fixed	mg of Ba sorbed/g of sample	% Ba fixed
Mixtures of Natural and Synthetic Zeolites								
Phillipsite, CA + Linde 3A	4.693	75	0.473	70	1.783	38	2.368	nd
Phillipsite, CA + Linde 4A	5.303	84	0.663	88	1.783	44	2.368	nd
Phillipsite, CA + Linde 5A	5.335	84	0.655	89	1.706	12	2.360	nd
Chabazite, AZ + Linde 3A	4.730	71	0.383	35	1.783	47	2.368	nd
Chabazite, AZ + Linde 4A	5.335	83	0.570	58	1.783	59	2.368	nd
Chabazite, AZ + Linde 5A	5.313	83	0.535	58	1.685	10	2.319	nd
Clinoptilolite, AZ + Linde 3A	4.390	66	0.350	37	1.783	47	2.361	nd
Clinoptilolite, AZ + Linde 4A	5.193	77	0.630	68	1.783	62	2.356	nd
Clinoptilolite, AZ + Linde 5A	5.193	77	0.595	66	1.695	15	2.328	nd
Mixtures of Natural Zeolites								
Clinoptilolite, ID + Mordenite, AZ	5.335	85	0.598	63	1.428	80	2.309	86
Clinoptilolite, ID + Mordenite, NV	5.323	83	0.623	66	1.541	80	2.326	85
Clinoptilolite, ID + Erionite, OR	5.385	82	0.605	66	1.445	68	2.303	82
Clinoptilolite, ID + Erionite, CA	5.418	82	0.635	67	1.590	68	2.334	85
Clinoptilolite, ID + Chabazite, AZ	5.325	79	0.620	63	1.482	62	2.309	80
Clinoptilolite, ID + Phillipsite, CA	5.340	83	0.655	84	1.532	72	2.360	100
Erionite, CA + Mordenite, AZ	5.390	89	0.605	62	1.242	47	2.196	64
Erionite, CA + Mordenite, NV	5.440	86	0.628	67	1.330	52	2.211	56

nd = not determined

tures points out that the Sr is mostly partitioned on to either clinoptilolite or mordenite which are good Sr fixers. For example, in a mixture of clinoptilolite and chabazite, the Sr fixation is 62% (Table 3). If Sr had been equally partitioned between clinoptilolite and chabazite, the fixation capacity of their mixture would be about 38%, because clinoptilolite and chabazite fix 75% and 0%, respectively, when they are present alone (Table 1). The increase in Sr fixation by about 24% in the 1:1 clinoptilolite and chabazite mixture is indicative of the great selectivity of clinoptilolite for Sr. The partitioning of Sr can thus be deduced from the mixtures of zeolites. Based on these studies, the best mixtures for greatest sorption and fixation of various fission products are the mixtures of clinoptilolite with mordenite and clinoptilolite with phillipsite.

CONCLUSIONS

Preliminary investigation of the sorption and fixation properties of clay minerals, oxides, zeolites and their mixtures with a simulated waste solution indicated that the mixtures of clinoptilolite with mordenite and clinoptilolite with phillipsite have the highest sorption and fixation capacities for fission products such as Cs, Rb, Sr and Ba. However, further work is needed to clearly establish the interesting results obtained here.

REFERENCES

1. E. Eriksson, "Cation Exchange Equilibria on Clay Minerals," Soil Sci. 74, 103-113 (1952).
2. G. H. Bolt, "Analysis of the Validity of the Gouy-Chapman Theory of the Electric Double Layer," J. Colloid. Sci. 10, 206-218 (1955).
3. T. Tamura, "Development and Applications of Minerals in Radioactive Waste Disposal," Proc. Intl. Clay Conf. I, 425-439 (1966).
4. T. Tamura, "Selective Sorption Reactions of Cesium with Soil Minerals," Nucl. Safety 5, 262-265, 267-268 (1968).
5. S. Komarneni, "Cesium Sorption by Clay Minerals and Shales at Elevated Temperatures," J. Inorg. Nucl. Chem. 41, 397-400 (1979).
6. R. M. Barrer, "Separation of Mixtures Using Zeolites as Molecular Sieves, Part I. Three Classes of Molecular Sieve Zeolite," J. Soc. Chem. Ind. 64, 130-131 (1945).
7. R. M. Barrer and J. Klinowski, "Ion Exchange in Mordenite," J. Soc. Chem. Faraday I 70, 2362-2367 (1974).
8. L. Ames, "Mass Action Relationships of Some Zeolites in the Region of High Competing Cation Concentrations," Am. Mineral. 48, 868-882 (1963).
9. R. M. Barrer, J. A. Davies and L. V. C. Rees, "Thermodynamics and Thermochemistry of Cation Exchange in Chabazite," J. Inorg. Nucl. Chem. 31, 219-232 (1969).
10. D. G. Kinniburgh, M. L. Jackson and J. K. Syers, "Adsorption of Alkaline Earth, Transition, and Heavy Metal Cations by Hydrous Oxide Gels of Iron and Aluminum," Soil Sci. Soc. Am. J. 40, 796-799 (1976).
11. R. R. Gadde and H. A. Latinen, "Studies of Heavy Metal Adsorption by Hydrous Iron and Manganese Oxides," Anal. Chem. 46, 2022-2026 (1974).
12. P. J. Denny and R. Roy, "Cation Exchange Between Mixtures of Clay Minerals and Between a Zeolite and a Clay Mineral", Clays & Clay Miner. 12, 567-589 (1964).
13. D. M. Roy and G. R. Gouda, "High-level Radioactive Waste Incorporation into (Special) Cements", Nucl. Tech. 40, 214-219 (1978).
14. B. L. Sawhney, "Regularity of Interstratification as Affected by Charge Density in Layer Silicates", Soil Sci. Soc. Am. Proc. 33, 42-46 (1969).
15. D. G. Jacobs, "Cesium Exchange Properties of Vermiculite", Nucl. Sci. Eng. 12, 285-292 (1962).
16. S. Komarneni and D. M. Roy, "Effect of Layer Charge and Heat Treatment on Cesium Fixation by Layer Silicate Minerals", J. Inorg. Nucl. Chem. 40, 893-896 (1978).
17. D. G. Jacobs, "Sorption of Cs by Conasauga Shale", Health Physics 4, 157-163 (1960).

SOLUTION INTERACTION OF NUCLEAR WASTE ANIONS WITH SELECTED GEOLOGICAL MATERIALS

G. W. Bird and V. J. Lopata

Atomic Energy of Canada Limited
Whiteshell Nuclear Research Establishment
Pinawa, Manitoba, Canada ROE 1LO

INTRODUCTION

In the Canadian nuclear fuel waste management program, we are considering geological disposal in vaults mined deeply into plutonic rocks of the Canadian Shield (1). In this context ^{99}Tc and ^{129}I are of concern due to their long half-lives and their tendency to go into solution as anions which are largely unretarded by interaction with silicate minerals (2). For these reasons we have been looking for materials which would be stable within the vault environment and which could be used as part of a multi-barrier containment system to retard the movement of Tc and I. As part of this program, we have contacted $SeO_4^{=}$, TcO_4^- and I^- solutions with a wide variety of geological materials to test for sorption and hence retardation of their movement in a natural environment. We have included the study of Se for completeness even though ^{79}Se is present as only a minor component of the waste. Most of our experiments have been carried out under oxidizing conditions as we consider these to be the worst possible circumstances in terms of geological transport. In this paper we report our results for a wide variety of materials of diverse chemical behavior. In many cases, the work is still in progress and a detailed understanding of the chemistry is unavailable.

EXPERIMENTAL

Most of the experiments reported here were carried out at room temperature using natural minerals supplied by the Geological Survey of Canada or purchased from Ward's Scientific Supply, reagent grade chemicals and deionized water. For each experiment, 0.25 g of mineral was weighed into a small plastic or glass vial and 25 mL of radioactive tracer solution was added. The isotopes ^{75}Se, ^{99}Tc and ^{125}I were used as tracers. The selenate and iodide solutions were prepared by the addition of sufficient tracer to solutions of stable Na_2SeO_4 and NaI to permit us to sample 0.1 to 0.4 mL for radiometric counting. The TcO_4^- solutions were made from $KTcO_4$ prepared by recrystallization of an NH_4TcO_4 stock solution.

For each experiment we made up three samples containing the mineral plus anion and one reference sample containing the anion only. The pH of the reference sample was adjusted to be similar to those containing the minerals. All four solutions were then stored in identical vials to standardize the effects of adsorption on the container walls. The sorption of the anions on the minerals was measured by sampling the solutions and counting the amount of radioactive tracer remaining in solution. The ^{125}I and ^{75}Se solutions were pipetted onto small plastic counting vials and γ-scanned. The ^{99}Tc samples were pipetted into a liquid scintillator mixture and counted on an automatic scintillation counter. The solutions were not filtered and small amounts of solid contaminant would have been counted with the solution. The amount sorbed was calculated by averaging the counts of the three solutions containing the mineral and comparing this to the reference solution. The reproducibility of counting a single sample is within $\pm$ 2%, and the overall reproducibility of the average of three samples and reference solution is within $\pm$ 10%. The possibility of slight contamination by the solids means that the measured sorptions are a minimum.

Solutions were sampled shortly after preparation and then daily or every few days until the composition had ceased to change. Rate studies were not attempted. After the solution composition had equilibrated for several days, the solution was carefully decanted and replaced with a new solution free of any radionuclide. Desorption was followed by counting the new solution as before. Since some of the original solution always remains with the mineral and this solution contains some of the radionuclide which is then counted with the material leached from the mineral, the amount of desorption recorded is a maximum. The desorption experiments were repeated a number of times until count rates in solution had dropped below a level suitable for reasonable radiometric counting times.

A wide variety of starting materials was used. The natural minerals were first crushed to less than 40 μm grain size, checked for purity by X-ray diffraction (XRD) and then washed with distilled water to remove fines. The reagent grade chemicals were used largely as from the bottle.

In the lead monoxide experiments, we used chemical reagent grade PbO supplied by Fisher Scientific Company. This material consisted of a mixture of massicot, the yellow orthorhombic high temperature polymorph, and litharge, the red tetragonal low temperature polymorph. To test whether the polymorphic form of the oxide had any effect on the reactions, we recrystallized litharge by the method of Garret et al. (3) and compared its reaction with PbO from the bottle. No difference in chemical behavior was noted and the chemical reagent grade PbO was then used as supplied.

To determine whether dissolved gases had any effect on the reactions of the lead oxide, we purged several samples with nitrogen or argon and compared the results with air-saturated samples. No differences were noted in short duration experiments. In some of the solutions which sat for several weeks in the laboratory, a basic lead carbonate film was precipitated on the surface of the solution. The formation of the carbonate did not affect the sorption of the radionuclides.

RESULTS AND DISCUSSION

The sorption results for the silicates, sulfides and apatite are summarized in Table 1. The results for copper metal, copper oxides and PbO are discussed below. The results cannot be generalized as none of the minerals except PbO showed any affinity for all three anions and these cannot therefore be considered "anion sorbers." Several materials were found to sorb $SeO_4^=$. The silicates augite, hornblende and vermiculite sorbed substantial quantities of $SeO_4^=$ and held it tightly when desorption was attempted. Galena, sphalerite, calcite, smithsonite and apatite also took up $SeO_4^=$ (Table 1). Apatite was found to completely desorb all of the $SeO_4^=$ in three desorptions. The calcite lost approximately 90% in one desorption while desorption from the sphalerite was not measurable. We have not confirmed a mechanism for the sorption of $SeO_4^=$ on any of the materials tested. Our results, however, are compatible with the geochemistry of the minerals tested and with that of selenium compounds. The iron-bearing silicate minerals are known to dissolve in oxygenated water by a mechanism which leads to a hydrated ferric oxide crust on the surface of the grains (4). In soils, selenium is known to be associated with such hydrated oxides and to be held so strongly that the selenium can be unavailable to plants (5). Thermodynamic calculations on the solubility of $SeO_4^=$ and $SeO_3^=$ compounds suggest that sparingly soluble precipitates will form with a number of divalent cations (J. Paquette, personal communication). Such precipitates could be contributing to the sorption of the $SeO_4^=$ anion.

Copper metal was found to react with the $SeO_4^=$ anion to form discrete crystalline products which were examined by XRD. The diffraction pattern was not attributable to any known selenate phase but was very similar to that of basic cupric sulfate hydrate. We conclude that a basic cupric selenate hydrate compound was formed. We have not determined the solubility limits or stability field of this compound. Sorption of $SeO_4^=$ on CuO was also noted (Table 1).

Table 1. Measured Sorptions for $SeO_4^=$, TcO_4^- and I^- on a Number of Minerals. Each mineral was contacted with each anion separately. Results are given for only those minerals which showed measurable takeup of the anions. The ratios of the anions to minerals were 0.1 mole/kg, 1.0×10^{-2} mole/kg, and 1.0×10^{-5} mole/kg for $SeO_4^=$, TcO_4^-, and I^-, respectively, except as noted.

Minerals	Approximate Formula	Anions Attached	Original Concentration Attached (Moles/kg mineral)	Number of Desorptions Attempted	Concentration Remaining on Solid	Equilibrium pH of Solution
Silicates						
Augite	$(Ca,Na,Mg,Fe^{+2},Mn,Fe^{+3},Al,Ti)_2[(Si,Al)_2O_6]$	$SeO_4^=$	4.0×10^{-2}	2	3.2×10^{-2}	8.2
		TcO_4^-	1.9×10^{-3}	1	1.4×10^{-3}	8.1
Hornblende	$(Na,K)_{0-1}Ca_2(Mg,Fe^{+2}Fe^{+3},Al)_5[Si_{6-7}Al_{2-1}O_{22}]OH,F)_2$	$SeO_4^=$	2.2×10^{-2}	2	1.6×10^{-2}	8.2
Biotite	$K_2(Mg,Fe^{+2})_{6-4}(Fe^{+3},Al,Ti)_{0-2}[Si_{6-5}Al_{2-3}O_{20}](OH,F)_4$	I^-	2.4×10^{-6}	0*	-	7.8
		TcO_4^-	9×10^{-5}	0	-	-
			5×10^{-6}**	0	-	
Vermiculite	$(Ca,Mg)_{0.7}(Mg,Fe^{+3},Al)_{6.0}[(Al,Si)_8O_{20}](OH)_4 \cdot 8H_2O$	$SeO_4^=$	3.2×10^{-2}	4	2.0×10^{-2}	6.8
Kaolinite	$Al_4[Si_4O_{10}](OH)_8$	TcO_4^-	3×10^{-4}	0	-	6.8
			8×10^{-6}**			
			8.0×10^{-3}**			
Sulfides, carbonates, Phosphates						
Galena	PbS	$SeO_4^=$	5×10^{-3}	0	-	6.1
		TcO_4^-	2.7×10^{-7}	0	-	-
		I^-	3.0×10^{-4}	0	-	-
Sphalerite	(Zn,Fe)S	$SeO_4^=$	3.9×10^{-2}	1	3.9×10^{-2}	-
		TcO_4^-	2.2×10^{-5}	0	-	-
Calcite	$CaCO_3$	$SeO_4^=$	9.7×10^{-3}	1	1.5×10^{-3}	8.2
Smithsonite	$ZnCO_3$	$SeO_4^=$	9.8×10^{-4}	0	1.5×10^{-3}	-
		TcO_4^-	3.2×10^{-5}	0	-	-
Apatite	$Ca_5(PO_4)_3(OH,F,Cl)_2$	$SeO_4^=$	2.5×10^{-2}	3	0	8.0
Metals, oxides						
Copper		$SeO_4^=$	see text	0	-	-
		TcO_4^-	see text			
		I^-	see text			
CuO		$SeO_4^=$	1.3×10^{-2}	0	-	-
Cu_2O		I^-	see text			
PbO		$SeO_4^=$	see text			
		TcO_4^-	see text			
		I^-	see text			

* Spontaneous desorption occurred, see text.

** 1×10^{-5} moles TcO_4^-/kg mineral.

The pertechnetate anion was found to sorb on augite, biotite, kaolinite, galena, sphalerite and smithsonite. The sorption was always at least an order of magnitude less than that of $SeO_4^=$, and only augite sorbed sufficient TcO_4^- to test for desorption. Although copper metal, Cu_2O, magnetite and graphite were tested for TcO_4^- sorption, none was observed. This suggests that the reduction of TcO_4^- on the surface of these materials is inhibited by kinetic factors, even though the redox potentials are favorable, or the TcO_2 is readily reoxidized under our experimental conditions.

The I^- anion was found to be the most difficult of the three anions to remove from solution. The only silicate found to have any affinity for I^- was biotite (kaolinite was not tested). Although biotite initially took up a small quantity of I^-, spontaneous desorption over a relatively short time released all of it back into solution. The sorption of I^- on galena confirms results reported previously (6), although we have not attempted to measure sorptive capacity. Sorption experiments with I^- on copper metal under Eh conditions suitable for the formation of Cu_2O show that I^- will sorb on copper in amounts controlled by the availability of sorption sites and the solubility of CuI. Results of this work have been submitted for publication elsewhere (7).

Our sorption studies with lead minerals have included lead monoxide (massicot and litharge), basic lead carbonates ($3PbCO_3 \cdot 2Pb(OH)_2$ or $2PbCO_3 \cdot Pb(OH)_2$) and cerussite ($PbCO_3$). With the exception of cerussite, all the lead minerals were found to sorb some I^-, TcO_4^- or $SeO_4^=$. The PbO system was the best sorbent we have tested for anions. PbO (litharge and massicot) in water containing an electrolyte in contact with the atmosphere reacts spontaneously to form either hydrocerussite ($3PbCO_3 \cdot 2Pb(OH)_2$) or plumbonacrite ($2PbCO_3 \cdot Pb(OH)_2$) (8). This reaction buffers the solution pH between 10 and 11. The greatest extent of sorption of the three anions tested occurred under these conditions. Fig. 1A. shows the extent of sorption of the TcO_4^- anion on PbO as a function of the TcO_4^-/PbO molar ratios. We examined the products of the Pb-TcO_4^- experiments by XRD and found that all of the powder patterns contained lines which were not attributable to known phases. We attempted to precipitate Pb-Tc compounds by titration of solutions containing $Pb(NO_3)_2$ and $KTcO_4$ with NaOH. In these experiments two distinct breaks were found in the titration curves. The first break was found around pH 6 and was accompanied by the formation of a white precipitate. X-ray powder patterns of the solids show that the white precipitate is a basic lead nitrate. With continued addition of base to around pH 9, the precipitate changed from white to lemon-yellow color, and the second change in slope of the titration curve was noted. The X-ray powder pattern of the yellow precipitate is much more complicated but appears to indicate a mixture of basic lead-nitrate and a second phase. To confirm that NO_3^- anions were not responsible for the reaction, we have precipitated a similar yellow compound from $Pb(ClO_4)_2$ solution. It appears

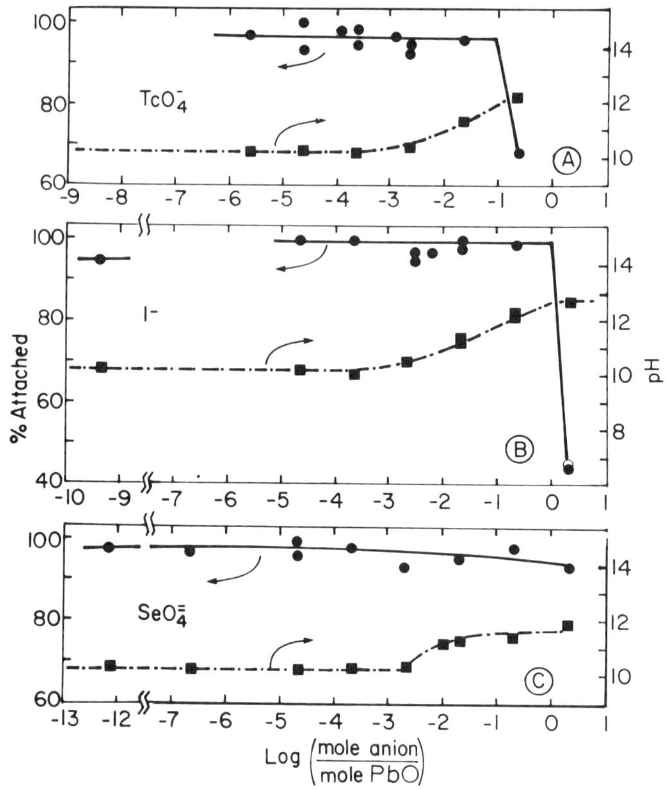

Fig. 1. Sorption of TcO_4^-, I^- and $SeO_4^=$ on PbO for a Range of Molar Ratios. The solid circles are sorptions; the squares are the measured pH of the solution in equilibrium with the solid. The open circle on Fig. 1B. is the calculated I^- attachment assuming all the PbO has been converted to PbI(OH).

that these yellow compounds are basic lead pertechnetates, but we have not found any previous reports of such compounds. We have now successfully isolated single crystals of the compound and intend to determine the molecular composition from single crystal X-ray diffraction.

Removal of I^- by PbO (Fig. 1B.) has been measured over I^-/PbO ratios ranging from $4.5x10^{-10}$ to 3.0. Greater than 95% of the I^- was taken up by the solid at all concentration ratios less than 0.2. At a I^-/PbO molar ratio of 3.0, PbI(OH) was formed and confirmed by X-ray diffraction. At the lower ratios, however, PbI(OH) was not found and the XRD data suggested that one or more previously observed but unidentified phases were present and that these compounds are variable in composition. A lead oxide-iodide phase with the composition $4PbO \cdot PbI_2$ has been previously reported (9), but our XRD data are not compatible with those for this compound. Comparison with lead oxide-chloride and lead oxide-bromide compounds suggests that the ratio of $PbO:PbI_2$ could vary over the range of 1:1 to 4:1, although such I^- compounds have not been described. Similar basic lead compounds of the type $(X)Pb(A^{n-})_{2/n} \cdot (Y)Pb(OH)_2$ where $A = CO_3^=$, NO_3^-, ClO_4^-, etc., and n = 1 or 2 have been found (10). We have measured the pH of the solutions in equilibrium with the solids, and it is compatible with reactions of the type:

$$(X)\ PbO + (Y)H_2O + (2/n)(Y)A^{n-} = (X-Y)PbO \cdot YPb(A)_{2/n} + (2Y)OH^- \qquad 1)$$

$$(X+Y)PbO + (2X/n)A^{n-} + (X+Y)H_2O = (X)Pb(A^{n-})_{2/n} \cdot (Y)Pb(OH)_2 + (2X)OH^- \qquad 2)$$

More work is required to characterize these lead compounds and to identify the reaction by which they form. The $PbO-SeO_4^=$ data are summarized in Fig. 1C. The general chemical behavior of this system is similar to that of the $PbO-I^-$ and $PbO-TcO_4^-$ systems, and we are having similar difficulties characterizing the reaction products. It is clear, however, that the PbO is efficient at removing $SeO_4^=$ from solution over molar concentration ratios ranging from 10^{-12} to 3.

The combined results given above suggest that the mechanisms of reaction are not simply dissolution-precipitation processes. The high sorption at the extremely low radionuclide levels in the I^- and $SeO_4^=$ systems would require that the compounds be extremely insoluble and on the basis of previous work on lead chemistry, this seems unlikely. We suggest that the removal at the lowest concentration ranges is the result of adsorption while the removal at the higher concentrations occurs by precipitation of insoluble phases.

A mixture of the basic lead carbonate plumonacrite and hydrocerussite was prepared by dissolving PbO in HNO_3 and precipitating from this solution with Na_2CO_3. Sorption experiments with the precipitate gave results similar to those with the PbO. A substantial amount of further research is required to fully delineate the chemistry of these systems.

CONCLUSIONS

Experiments with a wide variety of minerals have shown that a number of minerals will retard the movement of the radioactive waste anions $SeO_4^=$, TcO_4^- and I^-. The $SeO_4^=$ anion is retarded by the greatest variety of materials which includes a number of silicates, sulfides and oxides. This retardation and the presence of ^{79}Se in the waste in relatively low quantity implies that it does not represent a hazard, as similar or greater retardation is expected for the more reduced selenium species (5).

The TcO_4^- and I^- anions are more difficult to retard. However, we have shown that a number of geological materials will retard TcO_4^- under oxidizing conditions. Research is in progress on the geochemical behavior of TcO_4^- and I^- in reducing systems, and preliminary indications show that TcO_4^- will precipitate as an oxide. The retardation of I^- is the more difficult, but we have shown that a number of minerals will retard its movement. Two lead minerals and Cu-metal have been shown to retard I^- or at least minimize its solubility. The PbO-system sorbs I^- from very dilute solutions. However, it remains to be shown that this material would be stable under repository conditions.

REFERENCES

1. J. Boulton (ed.), Management of Radioactive Fuel Wastes: The Canadian Disposal Program," Atomic Energy of Canada Limited Report AECL-6314 (1978).
2. R. Rancon and J. Rochon, Workshop on Migration of Long-Lived Radionuclides in the Geosphere, Brussels, Jan. 29-31 (1979).
3. A. B. Garret, S. Vellanga and C. M. Fontana, J. Am. Chem. Soc. 61, 637 (1939).
4. R. Siever and N. Woodford, Geochim et Cosmochim Acta 43, 717-724 (1979).
5. H. Howard I. III, Conf. on Trace Elements in Environmental Health Proceedings, U. of Missouri (1971).
6. S. Fried, A. M. Friedman, D. Cohen, J. J. Hines and R. Strickert, Annual Report of Office of Waste Handling, Project AN 0115A, Argonne National Laboratory (1978).
7. Z. Haq, G. M. Bancroft, W. S. Fyfe, G. W. Bird and V. J. Lopata, submitted to J. Environ. Sc. and Tech.
8. W. Stumm and J. J. Morgan, Aquatic Chemistry (Wiley-Interscience, Inc., New York, 1970).
9. M. A. Bredig, C. E. Bannberger and D. J. Richardson, J. Inorg. Nucl. Chem. 40, 1479 (1978).
10. B. Charreton, Soc. Chim. 5e Serie, Memoires, 323-344 (1956).
11. R. M. Garrels and C. L. Christ, Solutions, Minerals and Equilibria (Harper and Row, Inc., New York, 1965).
12. E. A. Bondietti and C. W. Francis, Science 203, 1337 (1979).

CLAY MINERALS SUITABLE FOR OVERPACK IN WASTE REPOSITORIES: EVIDENCE FROM URANIUM DEPOSITS

D. G. Brookins

Department of Geology
University of New Mexico
Albuquerque, NM 87131

INTRODUCTION

Various clay minerals have been proposed for overpack around radioactive waste canisters. The laboratory studies of the migration of elements from the nuclear waste through proposed overpack have resulted in some disagreement. As an alternative approach to the laboratory studies, a review of data for the uranium deposits of the Grants, New Mexico, mineral belt has been undertaken (1, 2). This paper reviews the evidence from the clay mineralogy, the Rb-Sr systematics and the trace element analyses of the clay size ($< 2\mu$) fraction of rocks from the uranium ore zones and the barren host rocks. The $< 2\mu$ fraction is dominated by clay minerals, but small amounts of other minerals and organic matter are present as well. These deposits are well suited to such a study because (a) they are old, (b) some have been remobilized and reprecipitated, (c) some have been destroyed by subsurface waters and the uranium disseminated, and (d) each type of uranium deposit has a rather diagnostic clay mineral suite and characteristic suite of elements associated with it. The oldest ore is found in argillaceous sandstone in deeply buried sediments which, based on geochronologic study, have remained closed systems since about 135-145 m.y. ago. That the Rb-Sr systematics for these deposits are undisturbed also indicates that Cs, with a greater tendency to be retained in clay minerals than Rb, is also retained. In addition, pronounced enrichment of Y, the REE, Nb, Mo, Sb, Te, Ba and Cd is noted in the mixture of vanadiferous chlorite, an illite and mixed layer illite: montmorillonite with coffinite, pyrite, organic matter. While the organic acids derived from lignin-derived humates have been associated with ore formation, they have not evidently affected the ore in post-formational time due presumably to the low Eh and intermediate (6.5-7.5) pH values. It is interesting to note that this suite of elements, although in different proportions and of normal isotopic composition, contains many elements in the mass range produced by 235-U fission.

Clay Minerals from Barren Rocks and Ore Zones

The clay minerals associated with different uranium deposits are discussed in the next section. The Cretaceous Mancos Shale and Dakota Formation can be summarized as follows: The Mancos Shale is characterized by both variable amounts of kaolinite and illite plus illite-montmorillonite. Illite plus illite-montmorillonite are always enriched relative to montmorillonite. The Dakota Formation can be divided into those samples rich in bentonites and/or shales relative to those rich in sandstone. The former are conspicuously montmorillonite-rich with variable kaolinite. The sandstones, on the other hand, are always rich in kaolinite.

The Late Jurassic Morrison Formation units studied for this report include, in descending stratigraphic order, the Brushy Basin Member (a shale with the ore bearing Jackpile sandstone near the top and the Poison Canyon sandstone near the bottom), the Westwater Canyon Member (argillaceous sandstone) and the Recapture Member (shale). The stratigraphic sequence, number and types of samples studied, percent clay minerals in rocks and abundances of various clay minerals are given in Table 1. Of further interest is the fact that many of the sandstones contain 5-20% argillaceous material, thus giving an approximate sand/shale range from 19-1 to 4-1.

Clay minerals from ore zones are typically enriched in chlorite, illite or mixed-layer illite-montmorillonite relative to the more montmorillonite-rich barren samples. Any particular ore deposit may be characterized by (a) chlorite > illite plus illite-montmorillonite, (b) chlorite $\sim$ illite plus illite-montmorillonite and (c) chlorite < illite plus illite-montmorillonite. Kaolinite is common in ore deposits located near the main redox front of the Grants mineral belt (2) but less abundant in the ore found in reduced rocks (2, 4). Discussion of the clay mineral abundances is given in detail by Adams, et al. (1, 2, 4, 6).

Calcite is common as a cement in many of the deposits; it is thought to be related to ore forming processes in reduced rocks and as secondary to such processes in deposits near or at the redox front (2).

Rubidium-Strontium Studies in the Grants Mineral Belt

For the Grants mineral belt, the following factors must be considered: (a) the Morrison Formation is non-marine; thus, more rapid

Table 1a. Clay Minerals (Both From Barren Rocks and From Uraniferous Rocks

Formation Member Bed/Unit	Age	Number of Samples Included In This Report Barren	 Ore Zone
Mancos	Cretaceous	6	--
Dakoka	Cretaceous	9	--
Morrison			
Brushy Basin	L. Jurassic		
Jackpile Ss.		6	30
Shales		18	--
Poison Canyon Ss.		22	12
Westwater Canyon Ss.	L. Jurassic	24	40
Recapture Shale	L. Jurassic	9	--

Table 1b. Additional Clay Mineral Information

Unit	Clay Minerals (%) in Rock	Avg. Range in Abundance of Clay Minerals Kaol.	 Chlor.	 Ill.	 Mont.	 Ill.-Mont.
Barren						
Mancos	~90	0-5	< 5	10-20	< 20	50-70
Dakota-bentonite	~70	2-40	< 2	< 10	20-50	10-20
-sandstone	~25	72-95	--	< 10	< 10	< 15
Jackpile	2-5	40-95	< 15	5-15	10-20	0-10
Brushy Basin Sh.	70-95	< 20	< 20	< 15	80-90	20-30
Poison Canyon	0.5-3	25-85	< 5	10-20	< 10	30-60
Westwater Canyon	1-20	10-70	< 10	< 20	< 15	30-60
Recapture	80-100	0-25	< 5	5-20	10-30	40-100
Ore Zone						
Jackpile	< 20	> 50	< 30	20-40	< 20	10-40
Poison Canyon	< 10	> 50	< 20		10-40	
Westwater Canyon	< 25	20-60	10-50	10-20	15	10-50

Note: The clay content of the rocks studied are extremely variable (2) as are the abundances of clay minerals in the barren and ore zone samples (1, 2, 4). Abbreviations: Kaol. = kaolinite; Chlor. = chlorite; Ill. = illite; Mont. = montmorillonite; Ill.-Mont. = mixed-layer montmorillonite.

sedimentation than for marine sediments is probable; (b) clay minerals penecontemporaneous with uranium mineralization presumably formed over a short interval of time; (c) chlorite formed in the ore zones is more stable than illite or mixed-layer illite-montmorillonite. Either of these last two may, if mixed with chlorite, be armored from surrounding solutions by the chlorite. In Table 2, it is noteworthy

Table 2. Summary of Radiometric Age Determination by the Rb-Sr Method (From Minus-Two Micron Clay Mineral Fraction)

Formation/Mine	(n)*	Rb-Sr Age (MYBP)	Initial $^{87}Sr/^{86}Sr$	Range in $^{87}Rb/^{86}Sr$
Mancos Shale	6	83 ± 11	0.7235 ± 0.0009	2.0 - 9.5
Dakota Formation	10	93 ± 8	0.7097 ± 0.0003	0.1 - 4.0
Jackpile Sandstone	14	142 ± 14	0.7100 ± 0.0007	0.3 - 7.2
Westwater Canyon (mont.-rich)	10	131 ± 26	0.7071 ± 0.0029	3.0 - 13.2
Sec. 23-30 Mine (chlorite-rich)	9	139 ± 10	0.7091 ± 0.0020	2.8 - 12.0
Smith Lake Ore (chlorite-rich)	9	139 ± 13	0.7102 ± 0.0013	1.9 - 12.5
Jackpile Mine (chlor.-ill.-rich) (-2 micron)	14	115 ± 9	0.7112 ± 0.0016	0.8 - 11.2
(-0.5 micron)	8	110 ± 10	0.7088 ± 0.0016	0.7 - 11.5

Note: n = number of samples for Rb-Sr isochron construction. MYBP = millions of years before present; ages calculated by method of York. See References 1, 2 and 5 for details of clay mineralogy.

that for the six suites of clay minerals analyzed, a minimum of six samples was used for one isochron (Mancos Shale) and as many as fourteen samples for another (Jackpile-Paguate mine ore zone, $< 2\mu$, chlorite-rich material). With sufficient data as in these two cases, the data are subjected to a refined least squares treatment in order to obtain the slope and $(^{87}Sr/^{86}Sr)_o$ intercept. The largest scatter of data about the isochron constructed for any of the clay mineral suites is from the Westwater Canyon Member. Such scatter may reflect open system conditions for several samples at the time of formation or possible post-formational effects. Ore zone clay minerals from Sec. 23-30 and Smith Lake Mines yield dates of 139 ± 10 and 139 ± 13 m.y., respectively, and barren rock montmorillonites from the Jackpile-Paguate mine yield 142 ± 14 m.y. Since the accepted age for the Jurassic-Cretaceous is 136 m.y., the date for the Jackpile-Paguate montmorillonite is compatible with its Late Jurassic age assigned on geologic grounds. More significant, however, is that the Ambrosia Lake and Smith Lake ore zone chlorites give identical dates at 139 m.y. Both suites of clay minerals were taken from trend ore, and the writer proposes that these dates are close to the age of sedimentation. The detailed clay mineralogy for these and other units is given in References 1, 2 and 5.

At the Jackpile-Paguate mine, the ore zone chlorite-rich material was examined for < 2μ and < 0.5μ size fractions. The date obtained for the < 2μ fraction yielded 115 ± 9 m.y., much younger than the montmorillonite date from barren rocks of 142 ± 14 m.y. The < 0.5μ fraction also yielded a young date, 110 ± 10 m.y. X-ray diffraction study of both fractions reveals both suites of samples to be identical. Since Adams and others (6) propose destruction of original ore at the Jackpile-Paguate mine area with subsequent remobilization and reprecipitation of uranium before uplift, tilting, erosion and deposition of the Cretaceous Dakota Formation, these 110 and 115 m.y. dates are consistent with such an interpretation. Recently, R. S. Della Valle and the writer have completed Rb-Sr dating of the Dakota Formation and the Mancos Shale. The results are presented in Table 2. We again used the < 2μ fraction for dating purposes in order to avoid detrital material. The Rb-Sr dates are in excellent agreement with K-Ar dates of 90-94 m.y. for the Dakota and 80-89 m.y. for the Mancos (2).

Trace Elements in Ore Zones

Trace element data are given in Brookins (1). The sample suites are divided into three different groups: outcrop, bleached barren zone and ore zone. This grouping is based on sample availability; it is intended only to establish background factors and is not related to any predetermined concept of ore genesis. The clay-sized fraction is preferentially enriched in most of the elements investigated except for Hf and Rb. This observation is more pronounced for REE, Fe, Cr, Ta, Co and Sb than for others. Whole-rock enrichment of Hf and Rb probably represents concentration of these elements in resistate minerals.

REE enrichment in sandstone type uranium deposits has not been reported, although some La and Y have been noted in the northern Colorado Plateau. The results of the present study indicate that the whole rock fractions of the three groups of samples have similar distribution patterns except for the enrichment of Eu and Tb in ore zone samples. Except for these two elements, the REE content of the ore zone samples is not abundant. The clay-sized fraction of the ore zone samples, however, is different in both the absolute REE abundances and distribution pattern. There is only a slight Ce anomaly, but the Ce content is higher than in other samples. The bleached zone samples contain the least amount of REE, and the distribution is similar to that in the outcrop samples.

Both the whole rock and clay-sized fractions of the ore zone samples show chromium enrichment. The enrichment is by a factor of about two. Chromium is an intrinsic element in sandstone and independent of any enrichment due to uranium mineralization. Lee (4) has shown that the magnetite separates from the Westwater Canyon Sandstone Member contain quite high chromium contents. Along with other unstable detrital minerals, the dissolution of these magnetites would have provided enough chromium for the enrichments noted.

Occasional anomalies of Nb have been noted in uranium deposits; the occurrence of Ta has not been reported. Owing to similar chemical behavior under oxidizing conditions, Nb and Ta could occur together. The amount of Nb and Ta that go into solution as oxy-metal anion complexes is believed to be proportional to that of V. Since V has almost phenomenal association with the U deposits, Ta (and Nb) enrichment would be expected under approximately the same Eh-pH conditions as for V. Ta is indeed enriched in the ore zone samples. The enrichment of the clay-sized fraction is as high as tenfold. Barren samples generally have very low background concentrations, to as much as a few ppm.

The presence of Co and Ni has been reported previously. Co and Ni are estimated to be about 20 times as abundant in ores as in unmineralized sandstones. Nickel enrichment is more pronounced in the clay-sized fraction than in the whole rock samples. The somewhat high nickel content in the outcrop samples is probably due to the small number of samples analyzed. Although the clay-sized fraction contains more scandium than the whole rock, ore zone enrichment was not detected. The overall Sc content appears to correlate with the Fe content. Manganese analyses were determined for the representative samples of the clay-sized fraction. The limited data indicate that the manganese is somewhat enriched in the ore zone clay minerals. The distribution of group IA and IIA elements appears to be controlled by lithology and to bear no specific relation to the uranium deposits. However, occasional enrichment of Ba and Ca is possible near the ore zones owing to the presence of barite and calcite. Cs and Ba are slightly enriched in the clay fraction although their distribution is essentially uniform in all sample groups.

In both hydrolyzate and oxidate sediments, Sb occurrence is associated with As in amounts roughly one-fifth by weight. It is also concentrated in some sulphidic sediments. Owing to similar ionic radius, Sb (III) can substitute for Nb and Ta. Since both As and Ta are enriched in the ore zones, one would expect Sb enrichment as well.

Data for thorium in sandstone type uranium deposits are generally absent. In sandstones it is mainly concentrated in the resistate heavy minerals. In the sedimentary cycle, Th behaves similarly to Sc. Because of the very high ionic potential of Th^{4+}, only very small amounts of Th can be kept in ionic solution in near neutral waters owing to precipitation of highly insoluble $Th(OH)_4$.

In a previous report (1), it was noted that the REE were enriched in many of the ore zone samples studied. Further examination of these samples showed also that this enrichment was due to enrichment in the clay size (< 2 μ) fraction (1, 4). Of further interest from these studies was the observation in ore zone samples from the Grants Mineral Belt that not only were the REE enriched in the clay-sized fraction but that a positive Eu anomaly was commonly noted. Europium is commonly (along with Ce) used as a valuable petrogenetic indicator among the REE because it can occur in nature both as Eu^{+2} and Eu^{+3} ions.

The REE data are of extreme interest because the R_X values (Table 3) are all greater than 1.0. While the mudstones all show values between 1.06 and 1.74, the values for the sandstones vary from 2.06 to 14.3. Most sandstones fall between 2.06 and 2.89; however, the Dakota Sandstone yields an anomalously high (14.3) value.

Older conventional ideas concerning the REE suggested that they should be enriched in resistate fractions of sediments; and thus, values of R_X should, under most circumstances, be less than 1.0. That all units studied to date show values in excess of 1 and sandstones in excess of 2 clearly show that the hydrolyzate fraction plays a major role in abundance and distribution of the REE.

Table 3. Rare Earth Element Data From Barren Rocks (Data in ppm)

		A.	B.	C.	D.
		< 2 μ REE	WR REE	R	R_X
Mancos Shale		68 - 147	75 - 110	0.9 - 1.5	1.22
Dakota Formation		307 - 502	13 - 413	1.2 - 34.4	14.3
Jackpile Sandstone		266 - 3100	91 - 1500	2.1 - 3.3	2.78
Poison Canyon		23 - 498	48 - 186	0.3 8.8	2.89
Brushy Basin		50 - 238	41 - 536	0.3 - 3.0	1.74
Westwater Canyon	a)	82 - 409	44 - 388	1.1 - 3.8	2.06
	b)	101 - 568	47 - 255	1.0 - 5.5	2.28
	c)	55 - 138	55 - 106	0.9 - 1.3	1.06
Recapture Shale		68 - 147	75 - 110	0.9 - 1.5	1.22

Note: Barren rocks refers to those with low uranium contents (i.e., less than 100 ppm). Column A = range in total REE for the < 2 μ; Column B = range in total REE for whole rocks; Column C = ranges in the average < 2 μ REE/WR REE for each rock unit; Column D = mean (R_X) for the ranges shown in Column C. a) = Ambrosia Lake district; b) = San Mateo area; c) = area north of Ambrosia Lake.

The REE are associated with clay minerals in ore zones, but it has not been unequivocally demonstrated that the REE are located in the clay mineral structures or absorbed as surface oxides-hydroxides (2).

DISCUSSION AND CONCLUDING STATEMENTS

Collectively, the argillaceous sandstones and arenaceous mudstones of the Morrison Formation, host formation to much of the uranium ore of the Grants mineral belt, approximate a mixture of quartz (plus feldspar) and clay minerals in a ratio of 80-90:20-10. This mixture is not very different from the quartz:montmorillonite used by the Swedish Government in its overpack studies. Further, the Rb-Sr dating of clay minerals penecontemporaneous with uranium mineralization indicates closed system conditions to Rb and Sr since times of ore formation at 135-140 and 113 MYBP; because of the greater retentivity of Cs than Rb in most clay minerals, an indirect argument can be made for these rocks being closed to Cs since time of formation as well. The clay minerals exhibiting the best approximation of truly closed system conditions are from ore zones and consist of chloride and/or illite-rich material. Due to the low cation exchange capacity (CEC) of chlorites relative to montmorillonites, this is not surprising. However, even the barren samples show ancient Rb-Sr dates suggesting diagenetic factors rather than very young processes to account for the age discrepancies noted. The barren samples are typically more montmorillonite-rich or mixed-layer montmorillonite-illite rich, and both contain appreciable amounts of accessory Ti-magnetite.

The trace element data are useful for several reasons. Most of the trace elements are enriched in the clay-sized ($< 2\mu$) fraction in both barren rock and ore zone samples relative to the whole rock abundances. In zones of even minor uranium mineralization, enrichment of Th and the REE is noted along with enrichment in chalcophile elements and Nb, Ta, Ba. Were data for very rare elements such as Ru, Rh, Pd, Ag, Cd, Tc, In, Sn available, theoretical considerations (3) would argue for fixation of these elements in the same zones as the actinides and lanthanides where Rb-Sr study of clay minerals indicates closed system conditions.

The information from uranium deposits suggests (1, 2) that chlorite will form during the mineralization process from pre-existing montmorillonite or illite or mixed-layered illite-montmorillonite accompanied by fixation of V, Mo, Se, As, REE, Th. Based on this fact coupled with the observation that the barren rocks are magnetite-rich under low Eh conditions, then an argument for the use of the very abundant sand-poor shales of the Morrison Formation as material suitable for overpack can be made. In the event of canister failure, any escaping actinides, actinide daughters or fission products should react efficiently with the intermediate CEC montmorillonite-illite-rich material and promote low CEC chlorite formation and sorption of actirides and lanthanides. Formation of such low CEC chlorite would not only scavenge the earliest radioactive elements released but would also retard fluid flow through the surrounding overpack by fissure filling which in turn would further impede radionuclide escape. The distribution of trace elements around uranium deposits of the Grants mineral belt is entirely compatible with such a scenario.

REFERENCES

1. D. G. Brookins, Uranium Deposits of the Grants, New Mexico, Mineral Belt I, USERDA Report GJO-1636-1, 120 (1976).
2. D. G. Brookins, Uranium Deposits of the Grants, New Mexico, Mineral Belt II, USERDA Report GJBX-76-029E, 411 (1979).
3. D. G. Brookins, "Thermodynamic Considerations Underlying the Migration of Radionuclides in Geomedia: Oklo and Other Examples," Scientific Basis for Nuclear Waste Mangement 1, 355-366 (G. J. McCarthy, ed., Plenum Press, 1978).
4. M. J. Lee, Geochemistry of the Sedimentary Uranium Deposits of the Grants Mineral Belt, New Mexico, unpub. Ph.D. dissert., 241, Univ. New Mexico (1976).
5. M. J. Lee and D. G. Brookins, "Rb-Sr Investigation of Morrison Formation (Late Jurassic), Grants Mineral Belt, New Mexico: Minimum Ages of Sedimentation-Uranium Mineralization and Provenance," Amer. Assoc. Petroleum Geol. Bulletin 62, 1673-1683 (1978).
6. S. S. Adams, H. S. Curtis, P. J. Hafen and H. Salek-Nejad, "Interpretation of Post-Depositional Processes Related to the Formation and Destruction of the Jackpile-Paguate Uranium Deposit, Northwestern New Mexico," Econ. Geol. 73, 1635-1654 (1978).

COMPARISON OF TAILORED CEMENT FORMULATIONS FOR BOREHOLE PLUGGING IN CRYSTALLINE SILICATE ROCKS AND EVAPORITE MINERAL SEQUENCES

B. E. Scheetz, D. M. Roy, E. L. White, and
D. Wolfe-Confer

Materials Research Laboratory
The Pennsylvania State University
University Park, PA 16802

INTRODUCTION

The development of adequate technology for repository sealing, which includes borehole plugging, formation grouting and shaft sealing, is an important aspect of safety assurance in the geologic waste isolation program. Boreholes or shafts which penetrate specific repository formations must be sealed adequately, for these artificial penetrations could provide the most direct pathway for release of radioactive waste (RW) nuclides into aquifers, and subsequently into the biosphere. Based upon results of previous studies (1-5), it is probable that cement-based materials (broadly considered) will fulfill the combination of plugging and sealing requirements most adequately, either alone or in combination with other component materials. Thus, current studies of cementitious composites are in progress by the investigators (6).

The following requirements are considered to be highly important. The plugs should be mechanically adequate, i. e. well bonded to the surrounding rock, and should not undergo significant volume change with long time (5,7,9). The plugging materials must be tailored as closely as possible to the rock. They should be resistant to destructive thermal expansion and contraction (6); and in a combination of properties provide the overall stability which will guarantee impermeability (or extremely low permeability) (8) to circulating fluids, thus minimizing possible transport of RW species. Factors related to the emplacement of large-scale structures have also been summarized in a recent report (10). Furthermore, the properties which control ease and reliability of emplacement are vital in achieving the above requirements, for they affect the final properties of the materials (6,11-14). Further constraints on the choice of cementitious materials considered for the plug are abundance, expertise, and use of existing technology. The goals of this program therefore, include designing potential cement-based plugging materials for use to seal penetrations, which will match the properties and characteristics of the host rock as closely as is consistent with their placement characteristics and potential durability. The following discussion describes investigations of the physical, mechanical and rheological properties, i.e. hardness, density, permeability, setting time and viscosity (volume change was described in Reference 7) of two families of materials: one with compositions adjusted for placement in salt-saturated formations; the second one designed to be compatible with crystalline silicate rocks, e.g. shales, basalts and granites.

SALT-COMPATIBLE CEMENTS

Experimental Procedures

The following experiments were designed to determine the effects of sodium chloride and other typical evaporite minerals on the stability and properties of hydrated cement composites. Experiments were performed both with and without use of water-reducing admixtures of the short-chain polymeric sulfonated naphthalene or melamine-formaldehyde condensate type. Experiments were designed to determine the effect of initial composition as well as curing variables upon a combination of properties, including setting time, mechanical and physical properties (i.e., strength, microhardness, permeability), and phase compositions that are considered important for predicting the long term stabilities of the hydrated products.

The initial samples were prepared as neat pastes with one cement, a Type I cement (I-5), using a w/c of 0.3. The mixing procedures followed ASTM-C305, water used was either deionized water or a saturated NaCl solution (34% at 25°C). The samples, cast in polyethylene molds 1 inch in diameter and 3 inches long, were compacted on a vibrating table for 60 seconds; initially cured at greater than 95% relative humidity for 24 hours; then removed from the cylinder and stored under a saturated solution of $Ca(OH)_2$ or NaCl in sealed glass jars. The curing was accomplished at room temperature and 333K (60°C) in a water bath at constant temperature.

Comparisons were made between the effects of the type of mixing water, the interrelationship of a fixed concentration (1%) of a superplasticizer, the presence or absence of NaCl in the mixing water, and the effect of the type of curing solution [sat. $Ca(OH)_2$ or sat. NaCl]. The nomenclature used throughout the experiments to specify the sample characteristics is as follows:

cement type	:	w/s[†]	:	mixing water	:	% admixture	:	curing water	:	curing temperature
I		0.3		NaCl*		1		$Ca(OH)_2$		RT
I		0.3		DW		0		NaCl		60(°C)

*The concentration of NaCl in the solution is not included as solid in the designated [†]water:solid weight ratio; DW = deionized water.

The phases identified by X-ray diffraction were portlandite [$Ca(OH)_2$], β-C_2S, possibly C_3S, and the calcium chloroaluminate hydrate "Friedel's salt" (15). A low angle diffuse background which usually left the chart, beginning in the range of from 6 to 10Å (16) was found in all of the four experimental mixes both with and without the presence of admixture, for the reference samples prepared with an admixture and cured at room temperature. The phase contribution to the low-angle diffraction is observed only in the samples cured 7 days. The calcium chloroaluminate hydrate, Friedel's salt ($3CaO{\cdot}Al_2O_3{\cdot}CaCl_2{\cdot}10H_2O$) is believed to be responsible for the distinctive diffraction peak at 11.2°2θ [7.9Å] (15). This phase is recognized in all the samples that were associated with sodium chloride either in the mixing water or in the curing water or both, but was not observed at 7 days.

The density of samples of a particular composition was found, for most samples, to remain constant (within $\pm$1.5 percent) with different curing times. Only two compositions showed significant departures from this behavior. This indicates reproducibility in techniques of sample preparation.

Microhardness, which has been shown to be inversely correlated with porosity and indirectly related to strength and density (4), was measured using the Vicker's technique. The hardness increase ranged from 10% to 70%, with the average increase approximately 30%. The reference neat paste (without admixture or NaCl) cured at 60°C showed the most rapid increase in hardness and also possessed the greatest hardness. The samples prepared with deionized water are fully 40% harder than the samples with salt admixture (see Fig. 1).

One of the most effective predictive measures of the potential for mass transport through a particular medium is its permeability. The permeability measurements were all made with nitrogen, and the Klinkenberg extrapolation (8) to infinite pressure, to provide an approximation to liquid permeability. The values obtained for the entire specimen set, range from about 0.1 microdarcy to tens of microdarcys. All those samples with deionized water appear to have permeabilities that were less than or equal to the values for samples made with saturated sodium chloride (Table 1).

The setting time for the cement samples was measured using a Vicat needle and following ASTM-C191 procedures. Samples selected contained a Type I cement (I-5) and a Class C (C-2), with a w/c ratio of 0.27, both with and without admixture and with and without sodium chloride in the mixing water. Fig. 2 compares the data as a function of salt concentration in the mixing water. For both cements, the increased concentrations of sodium chloride act as retarders to setting. The Class C cement paste is typically retarded to a greater extent than the Type I. Fig. 3 presents the data for Type I paste (without NaCl) with varying concentrations of two admixture types. The sulfonated naphthalene causes a nonlinear increased retardation with increasing concentrations of admixture. (This relative increase is partly explained by a higher concentration of organic in the admixture solution of the SN.) Additional details on behavior of Class C cement are given elsewhere (6).

Studies are in progress to compare quantitatively the behavior of the mixes in the plastic or fluid state, because of the complexities due to the chemical combinations of salt, cement and superplasticizing admixture. Quantitative viscometric measurements on some other related materials have been reported elsewhere (14). The behavior of cement pastes and slurries are non-Newtonian, but selection of data at a single shear rate reveal interesting comparisons of some effects. Fig. 4 shows a comparison of "relative viscosities" of mixes with Type I cement measured at 10 rpm paddle speed on a Brookfield viscometer. The two different admixtures cause behavior more closely parallel to each other than to the behavior of mixes without admixture. The anomalous behavior of the single sample of the latter, containing mix water with 2.5% NaCl, is unexplained.

Table 1. Permeability vs. Curing time for Sodium Chloride Treated Samples

		1 week admixture		4 weeks admixture		8 weeks admixture	
		without	with	without	with	without	with
I-0.3NaCl-0-Ca(OH)$_2$RT	3A	ND		4.4×10^{-4}		2.8×10^{-5}	
I-0.3NaCl-1-Ca(OH)$_2$RT	1A		7.0×10^{-4}		9.0×10^{-4}		2.5×10^{-5}
I-0.3NaCl-0-Ca(OH)$_2$-60	3A'	3.7×10^{-7}		ND			
I-0.3NaCl-0-NaCl-RT	3B	9.0×10^{-8}		4.5×10^{-7}		ND	
I-0.3NaCl-1-NaCl-RT	1B		2.2×10^{-6}		ND		7.0×10^{-6}
I-0.3NaCl-0-NaCl-60	3B'	1.03×10^{-7}		ND		ND	
I-0.3DW-0-Ca(OH)$_2$-RT	4A	4.25×10^{-6}		2.9×10^{-6}		1.5×10^{-9}	
I-0.3DW-1-Ca(OH)$_2$-RT	2A		4.0×10^{-4}		3.9×10^{-6}		8.25×10^{-6}
I-0.3DW-0-Ca(OH)$_2$-60	4A'	6.0×10^{-7}		1.2×10^{-6}		ND	
I-0.3DW-0-NaCl-RT	4B	9.8×10^{-6}		5.2×10^{-8}		7.4×10^{-6}	
I-0.3DW-1-NaCl-RT	2B'		6.9×10^{-6}		3.2×10^{-5}		3.2×10^{-5}
I-0.3DW-0-NaCl-60	4B	4.0×10^{-7}		ND		ND	

ND = Not Determined.

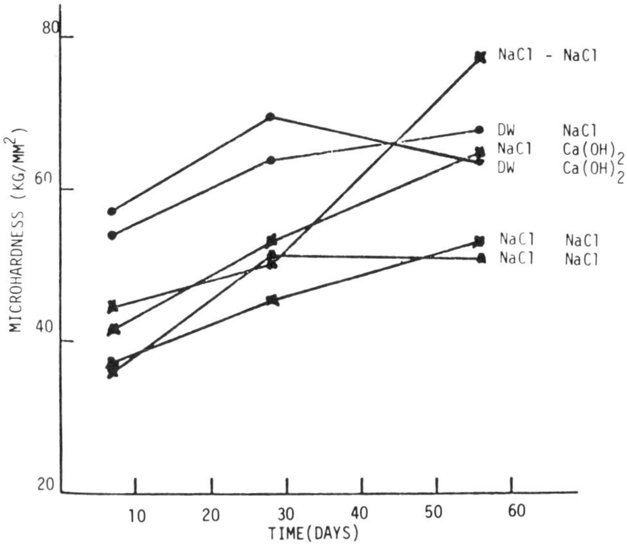

Fig. 1. Microhardness vs. Curing Time

SILICATE CEMENTS

Cements adjusted to compositions approaching that of the mineral tobermorite were prepared, first by the addition of silica as very

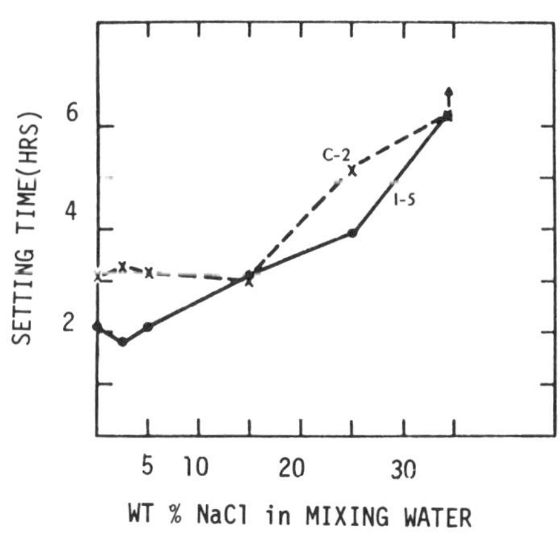

Fig. 2. Comparison of Setting Times for Type I and Class C Cement Pastes in the Presence of Varying Salt Concentrations.

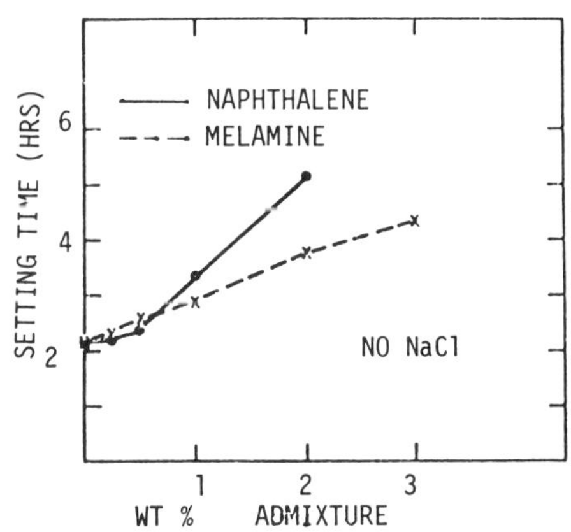

Fig. 3. Effect of Type of Water-Reducing Agent Upon Initial Setting Times of Type I Cement Without NaCl as a Function of Admixture Concentration.

fine grained (5 μm) quartz. The results of this series were reported earlier (4) in which the mechanical and physical properties were demonstrated to be adequate, but not exceptional, probably because of the relatively high w/c ratio necessitated by the fine SiO_2. Experimental results have now been obtained for parallel experiments, but with the use of fly ash as the silica source and a Class C cement.

This measure [approximately proportional to strength (4)] was used extensively. The compositionally adjusted Class C cement with fly ash used as the silica source (also containing much alumina) developed a higher early hardness as compared to the parallel samples which employed fine quartz as the source of SiO_2. This would be expected because of a higher reactivity of the fly ash. The experiments were designed to allow comparisons between specimens prepared with a CaO/SiO_2 ratio of 0.73 vs. 0.77, room temperature curing vs. elevated temperature curing at 60°C, and seeding at 1 wt/o vs. nonseeding. In all cases, the w/s (cement + fly ash) ratio was maintained at 0.33. A general increase in microhardness with curing time was found up to 2 months. In some of the specimens, a slight as yet unexplained decrease was observed with longer time. Bulk densities and X-ray diffraction measurements have also been used for characterization, as have compressive strengths; and the results are reported elsewhere (6).

Measurements were made on a series of eight Class C cementitious composites (mixed with fly ash), with and without tobermorite

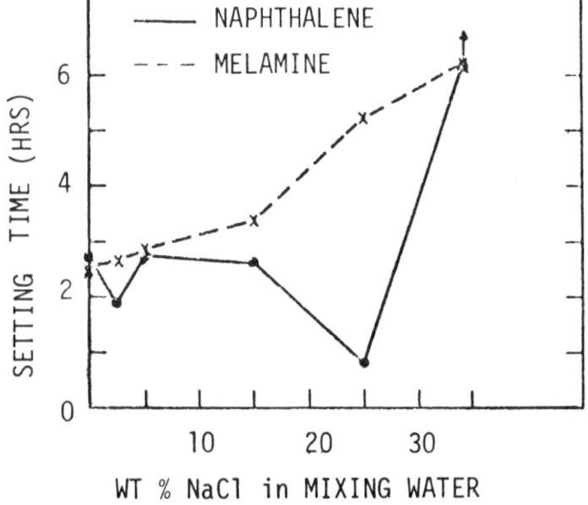

Fig. 4. Effect of Two Superplasticizers Upon Initial Setting Times of Type I Cement as a Function of NaCl Concentration.

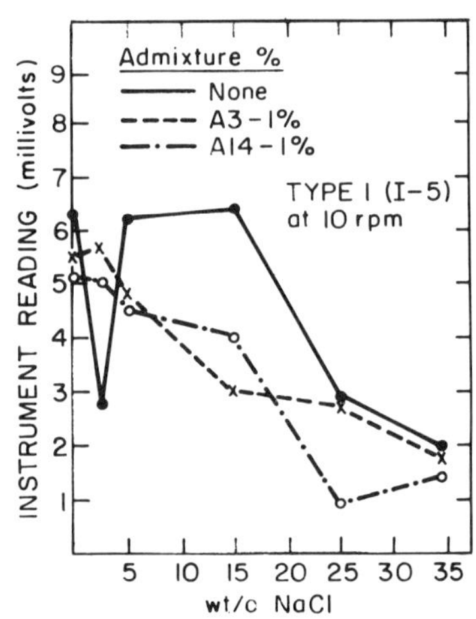

Fig. 5. Relative Shear Stress of a Type I Cement Paste, 0.30 w/c, 1% Water Reducing Agent, as a Function of Sodium Chloride Concentration.

seeding. The gas permeabilities extrapolated to liquid equivalent by the Klinkenberg method (8) were lower for those samples cured at 60°C than those cured at room temperature, as shown in Table 2. The difference may reflect more advanced reaction in the higher temperature specimens. Variations in the permeabilities were minimal after 28 days of curing; after 2 months of curing the permeability appears to be independent of both curing and starting composition. Direct water permeability measurements at 13.8 MPa gave a value of $1x10^{-6}$ darcys for sample #1A cured at room temperature for 28 days. As has been observed previously, the water and extrapolated (gas) permeability values are within an order of magnitude of each other.

The extrapolated gas permeability of another sample from the previous series (4), a tobermorite-seeded sample, Type V cement (V-2) cured at room temperature for 6 months, was $1.7x10^{-7}$ darcys. The direct water permeability was $1.5x10^{-7}$ darcys, consistent with previous observations. The same composition, but without the tobermorite seeding and also cured for 6 months at room temperature had an extrapolated gas permeability of $8x10^{-8}$ darcys.

CONCLUSIONS

Cementitious materials tailored for plugging penetrations through specific types of geologic repository media have been studied. Those designed to be compatible with halite-rich evaporites have yielded acceptable products, having permeabilities generally in the microdarcy range or lower. These are in contrast to much higher permeabilities measured in samples of a 17-year-old plug emplaced in a potash-rich zone in the Carlsbad area (16). The potential interaction of NaCl with water reducing agents was initially an important unknown. The rheological investigation results revealed a mild

Table 2. Fly Ash Samples. Tobermorite Seedings, vs. Non-tobermorite Seeding and Variation in the c/s Ratio of 0.73 to 1.0. Class C (C-1).

sample	14 days	28 days	2 months
	gas permeability		
1A-C-1-0.73-FA-0.33-OT-$Ca(OH)_2$-RT	$1.30x10^{-5}$	$8.50x10^{-6}$	$9.50x10^{-6}$
1B-C-1-0.73-FA-0.33-OT-$Ca(OH)_2$-60	$1.50x10^{-8}$	$3.30x10^{-6}$	$2.45x10^{-6}$
2A-C-1-0.73-0.33-1T-$Ca(OH)_2$-RT	$2.00x10^{-5}$	$8.25x10^{-6}$	$2.30x10^{-6}$
2B-C-1-0.73-FA-0.33-1T-$Ca(OH)_2$-60	$6.00x10^{-7}$	$2.30x10^{-6}$	$8.20x10^{-6}$
3A-C-1-1-FA-0.33-OT-$Ca(OH)_2$-RT	$1.75x10^{-5}$	$1.05x10^{-5}$	$1.15x10^{-5}$
3B-C-1-1-0.33-OT-$Ca(OH)_2$-60	$3.30x10^{-6}$	$2.50x10^{-6}$	$2.75x10^{-6}$
4A-C-1-1-FA-0.33-1T-$Ca(OH)_2$-RT	$1.35x10^{-5}$	$1.15x10^{-5}$	$7.00x10^{-6}$
4B-C-1-1-FA-0.33-1T-$Ca(OH)_2$-60	$1.35x10^{-6}$	$3.40x10^{-6}$	$4.60x10^{-6}$

retardation with increasing salt concentration, in the combinations with two leading classes of superplasticizers, but no interaction of a prohibitive nature was found.

When the properties of the composites formulated for an evaporite horizon are contrasted to those of formulations designed for silicate rocks, many of the physical properties are directly comparable, though some phase compositions are different. The permeabilities of fly ash-modified composites are typically in the microdarcy range, in reasonable agreement with those for the evaporite sequence.

ACKNOWLEDGEMENT

This research was supported by the U. S. Department of Energy, Office of Nuclear Waste Isolation, ONWI Subcontract E512-00500.

REFERENCES

1. D. M. Roy and W. B. White, ORNL/Sub-4091/1, May 31 (1975).
2. R. T. Martin, Prepared for Oak Ridge National Laboratory, Contract ORNL-Sub-3960, June (1975).
3. D. L. Black, Prepared for Union Carbide, Oak Ridge, TN, WANL-TME-2870, June (1975).
4. D. M. Roy, B. E. Scheetz, E. L. White and M. Daimon, ONWI-5, 119, Oct (1978).
5. D. M. Roy, M. W. Grutzeck and P. H. Licastro, Topical Report, ONWI-30, Feb (1979).
6. D. M. Roy, ONWI/SUB/78/E512-00500-3 (Task I), Quarterly Progress Report, June 30 (1979).
7. N. Setter and D. M. Roy, Cem. Concr. Res. 8, 621-632 (1978).
8. E. L. White, B. E. Scheetz, D. M. Roy, K. G. Zimmerman and M. W. Grutzeck, Scientific Basis for Nuclear Waste Management Vol. 1 (G. J. McCarthy, ed. Plenum Press, NY 1979).
9. D. M. Roy, ONWI/SUB/78/E512-04200-1, Quarterly Progress Report, May 1-July 31 (1979).
10. D'Appolonia, ONWI-15, Jan (1979).
11. D. M. Roy, M. Daimon, B. E. Scheetz, D. Wolfe-Confer and K. Asaga, Scientific Basis for Nuclear Waste Management, Vol. 1 (G. J. McCarthy, ed. Plenum Press, NY 1979).
12. M. Daimon and D. M. Roy, Cem. Concr. Res. 8, 753-764 (1978).
13. M. Daimon and D. M. Roy, Cem. Concr. Res. 9, 103-110 (1979).
14. D. M. Roy and K. Asaga, Cem. Concr. Res. 9(6), 731-740 (1979).
15. H. G. Smolczyk, Proc. 5th Intl. Symp. Chem. Cement, Tokyo, Vol. III, 274-280 (1969).
16. B. E. Scheetz, M. W. Grutzeck, L. K. Wakeley, and D. M. Roy, "Characterization of a Cement Borehole Plug in Bedded Evaporite From Southeastern New Mexico." Topical Report, ONWI, July (1979).

CLAY MINERAL-BRINE INTERACTIONS DURING EVAPORITE DIAGENESIS: LANTHANIDE DISTRIBUTION IN WIPP SAMPLES

J. K. Register, D. G. Brookins, M. E. Register, S. J. Lambert*

Department of Geology, University of New Mexico
Albuquerque, NM; *Sandia Laboratories
Albuquerque, NM 87185

INTRODUCTION

Clay minerals are common accessory constitutents of the bedded evaporite sequence at and near the WIPP site, southeastern New Mexico. Many of these are detrital in origin as documented by preservation of pre-emplacement ages, grain size and other criteria (1). Some phyllosilicates disseminated throughout the evaporite sequence have either formed by authigenic processes or by diagenetic reactions between small amounts of detritus and brine (2). In order to study the behavior of clay minerals considered for overpack in salt media, we have examined the detrital clay minerals by scanning electron microscopy (SEM), neutron activation analysis (NAA), X-ray diffraction (XRD) and supporting techniques (3). While special emphasis is placed on lanthanide elements, data for other trace and some major elements are reported as well. Fourteen samples were selected which, from SEM and XRD examination, showed clear evidence for some interaction between clay minerals and brine.

These analyses were undertaken to help establish a data base for clay minerals which have been immersed in an evaporite brine solution. It was also felt that the trace element data would be of use in explaining possible reactions between clay minerals used as overpack and brines formed at or near a canister interface. Further, to understand the geochemical systematics of the evaporite rocks of the Delaware Basin, it is desirable to determine the effectiveness of clay minerals as ion exchange scavengers.

CLAY MINERALOGY OF DETRITAL BEDDED CLAY MINERALS AT THE WIPP SITE

Twenty-two clay mineral samples as schlieren within evaporite samples and, more often, as seams within the salt were analyzed. Due to the amounts of clay mineral needed for complete analysis, most samples were collected from the Duval No. 1 Potash Mine (3) and the remainder from drill core from the WIPP site (see Reference 3 for locations).

Interlayered chlorite-saponite is common in most samples of insoluble clay size ($< 2\mu$) materials ranging from 95% to 5% in the samples studied with a mean of 62%. Two samples contained saponite as the dominant clay mineral rather than the mixed layer saponite-chlorite. The illite content of the clay mineral assemblage ranges from 60% to less than 5% with a mean of 26%. Kaolinite was detected by electron probe study in twenty-two samples, but its abundance is unknown. The illite consists predominantly of the $2M_1$ polytype which is typical of detrital illite. In illite disseminated throughout the evaporite, Bodine (2) has reported other polytypes. Chlorite appears in significant quantities in only three samples; most is present in mixed layer saponite-chlorite. Talc was detected in seven samples. SEM studies reveal (3) that the kaolinite, too, is predominantly detrital in origin. While authigenic kaolinite is common in nature, it is usually restricted to relatively high alumina environments; very different from the Na-Mg-S-Cl evaporite environment. The lack of available aluminum for diagenetic reactions is also evidenced by the occurrence of saponite, a trioctahedral smectite with very little Al in exchangeable octahedral sites. We interpret the saponite-chlorite interlayered clay mineral as an incomplete alteration product formed from interaction of the Mg-rich brine with detrital, trioctahedral montmorillonite. This interpretation is supported by geochronologic data and SEM studies. Small euhedral crystals of halite embedded in Mg-rich clay were noted by SEM studies in several samples. No sulfate minerals were identified although all analyses showed some SO_3 in the total analysis.

TRACE ELEMENT DATA EXCLUSIVE OF THE LANTHANIDES

Fourteen clay mineral samples from water insoluble fractions of evaporite samples (n=5) and from seams in the Duval Potash Mine (n=9) were analyzed by NAA for eleven trace elements in addition to the lanthanides.

Some brief comments on sodium, iron and potassium are in order. The sodium content averages 0.8% with most of the sodium probably bonded in the mixed layer chlorite-saponite with minor amounts in chlorides. The relatively low sodium content of the clay minerals relative to magnesium is due to the ease of incorporation of the latter into octahedral sites. The potassium content of the samples

Table 1. Major Element Oxides, Average Trace Elements and Mineralogical Composition for Insoluble Clay Size Material Interbedded with Evaporite Salts

Element	Concentration	Element	Concentration	Oxide	Percent
Cs	3.0 ppm	La	22.8 ppm	SiO_2	32.5
Ba	540. ppm	Ce	39.2 ppm	TiO_2	0.4
Sc	9.1 ppm	Nd	44.6 ppm	Al_2O_3	11.5
Hf	2.5 ppm	Sm	4.6 ppm	Fe_2O_3	1.5
Ta	0.6 ppm	Eu	0.9 ppm	MgO	29
Cr	92.0 ppm	Tb	0.3 ppm	CaO	0.3
Rb	100.3 ppm	Tm	0.3 ppm	K_2O	1.0
Co	9.5 ppm	Yb	2.6 ppm	Na_2O	0.8
Ni	25.8 ppm	Lu	0.6 ppm	$H_2O(+)$	4.6
Sb	2.7 ppm	Total REE	118.9 ppm	$H_2O(-)$	13.3
Th	6.7 ppm	Clay mineral	abundance	Cl*	1
		chlorite-saponite	62%		
		illite	26%		
		kaolinite	11%		

*As element.

varies from 0.5 to 1.4% and is present in illite. The major oxides do not total to 100% in Table 1 due to uncertainties in the proportions of clay size material with evaporite material. The high MgO content and low Al_2O_3 contents are due to the abundance of the mixed layer saponite-chlorite, although the illite contributes significantly to the MgO content where abundant. The Fe_2O_3 content varies between 1 to 2%, giving the clay seams a characteristic reddish color. Of interest, the TiO_2 content averages 0.45 (3) which again attests to a detrital origin as the TiO_2 content of sea water is significantly lower than this. The clay mineralogical abundance data are given as part of Table 1. The cesium content of the samples ranges from 1.2 to 5.0 ppm. This concentration falls well within the range for Cs values from shales reported by Gaines (4). The cesium was most probably incorporated into the clay minerals during original formation and not due to clay mineral-brine exchange during incorporation into the evaporite. Barium is highly variable in the samples, ranging some three orders of magnitude from 25 to 2,500 ppm. The site(s) of barium are unknown. Some may be adsorbed on clay minerals while some may be incorporated into small amounts of barite (5). This problem should be explored further as radium will almost certainly be scavenged by any barite present. The scandium content of the samples ranges from about 7 to 11 ppm. This range is significantly higher than that for deep-sea carbonates (6) and reflective of scandium present in the clay mineral at time of incorporation into the evaporite sequence. Hafnium, tantalum and chromium contents reported in Table 1 are also consistent with typical values for marine shales (7, 8, 9). The rubidium present is camouflaged in potassium sites and ranges from 20 to 300 ppm. For samples with available potassium data, K/Rb ratios range from 30 to 250. Most authigenic marine clays yield ratios of K/Rb=250 which is close to the mean crustal ratio (9). Rubidium is known to be strongly adsorbed on clay minerals, especially illites and smectites, during weathering (10), and this preferential adsorption of Rb may yield low K/Rb ratios. Further, clay minerals preferentially remove Rb over K from sea water, thus lowering the K/Rb ratio observed in many marine sediments. This is consistent with the low K/Cs ratios noted above. Post-diagenetic Rb or Cs adsorption or ion exchange is not thought to be significant in view of the evidence for systemal closure of both primary evaporite minerals and detrital clays with respect to Rb and Sr (1). Cobalt varies in abundance from about 3 to 35 ppm (mean 9.5 ppm). Although cobalt is considered quite mobile in the overall geologic environment (11) and often concentrated in marine shales (12), our data do not show any pronounced enrichment but rather fall in the generally accepted range for clay minerals. Similarly, nickel varies in a normal manner from 14 to 40 ppm. Both the crustal and sea water abundance of antimony is quite low (13) (i.e., less than 1 ppm); hence, some of the higher antimony data in Table 1 may be due to analytical error. Alternately, antimony may be incorporated into iron sites. The thorium concentration varies from 4 to 14 ppm compared to the average shale value of 12 ppm (14). Because of the extreme insolubility of $Th(OH)_4$, most if it not all of the thorium must have been transported into the basin at the time of evaporite formation. Delayed neutron activation analysis of low-Na evaporites, including anhydrites and polyhalites (3), indicates the average evaporite thorium content to be less than 0.1 ppm.

LANTHANIDE DISTRIBUTION

Nine lanthanides were determined in the fourteen samples, and the data are presented in Table 1. For convenience, the average values of the lanthanides are reported. While the range in total lanthanides ranges from 454 to 22 ppm (mean 119 ppm), the distribution of the lanthanides within any particular sample is similar to that for any other sample. We have chosen to ratio the individual lanthanide average values for each element to that for the North American Shale (NAS) composite in conventional fashion; this is presented in Fig. 1. which also includes curves for pelagic sediments (15) and for sea water (multiplied by 10^7) (16). The curve for pelagic sediments is very nearly equal to that for most shales (15), and this is the typical curve which one would expect for detrital clay minerals for both marine and most non-marine shales. The curve for pelagic sediments is slightly depleted in total lanthanides relative to NAS. Further, the more pronounced depletion for lanthanum and cerium is due to the preferential fixation of the light lanthanides into calcareous sediments due to control by similar ionic radii for Ca^{2+} and

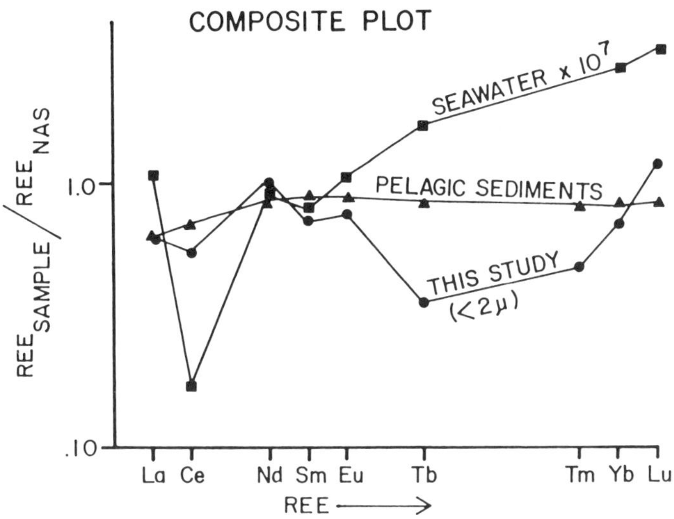

Fig. 1. Plots of the Average REE Concentration Relative to North American Shales of the Clay Minerals from the Salado Formation (This Study), Sea Water (Hogdahl, 1968) and Pelagic Sediments (Wildeman and Haskin, 1965)

La^{3+} and Ce^{3+}. The sea water lanthanide/NAS curve (x 10^7) shows a very pronounced cerium anomaly. It has been proposed that this depletion is due to formation of Ce^{4+} which is removed by precipitation into ferromanganese nodules (17).

For the results of the present study, most individual samples show curves very similar to the curve shown in Fig. 1. which is based on weighted averages of analyses for all samples. The data for terbium and thulium are less precise than for the other lanthanides (19) and considered as minimum values. Otherwise, a slight negative cerium anomaly is noted; the light lanthanides except for neodymium are slightly depleted relative to the curve for pelagic sediments, and the same increase in lanthanide/NAS for thulium through lutecium as for the sea water curve is noted. In the case of cerium in the clay minerals studied, the negative anomaly is slight and not as pronounced as that for sea water. A low cerium concentration relative to lanthanum and neodymium can easily be explained by some interaction between the clay minerals and the evaporite brine during diagenesis since the brine must certainly have been highly depleted in cerium. The apparent increase in the abundance of the heavier lanthanides relative to the lighter lanthanides is probably also a product of clay mineral-brine interaction as indicated by the disproportionately large amounts of Mg-rich clay minerals such as chlorite-saponite which formed from smectites during diagenesis. Since the ionic radii of the heavier lanthanides are closer to the ionic radius of magnesium (2+), this is a plausible explanation and one consistent with the observations of others (20, 21). Ion exchange between the clay minerals and evaporite minerals forming during authigenesis and diagenesis would favor incorporation of La^{3+}, Ce^{3+} into Ca^{2+}, Na^{+} sites of the evaporite minerals while the heavier lanthanides would be enriched in Mg^{2+} sites in the newly formed Mg-rich clay minerals. An alternate explanation would be one of mere dilution of the lanthanide distribution in the clay minerals by low-total lanthanide bearing evaporites. The ion exchange model is more likely, however, as the few evaporites analyzed (3) yield total lanthanide concentrations ranging from 0.05 ppm to 4.12 ppm (mean 1.34) with the two highest total lanthanide bearing samples (3) containing some insoluble oxide-clay mineral coatings which probably account for their relatively high values. If the others are considered, the range is 0.05 to 0.64 ppm with a mean of 0.24 ppm.

Collectively, these lanthanide data for the clay minerals and evaporites argue for relatively little exchange between the clay minerals and the brine during diagenesis. If exchange was a dominating process during diagenesis, then extremely depleted total lanthanide/NAS distribution and a pronounced negative cerium anomaly would be noted due to the extremely low abundance of the lanthanides in sea water. That this is not the case argues for very limited loss of some lanthanum-cerium from detrital clay minerals concomitant with an apparent enrichment in thulium-ytterbium-lutecium in the Mg-rich clay minerals. This enrichment is not due to incorporation of thulium-ytterbium-lutecium from the impoverished evaporite brine but rather is more likely due to a gradational depletion of the lanthanides from lanthanum through terbium which can substitute for calcium and/or sodium.

CONCLUSIONS

1. The trace element concentrations, exclusive of the lanthanides, for the clay minerals are significantly higher than for both anhydrites and polyhalites (3). This is a reflection of the detrital nature of the clay minerals and lack of significant loss of trace elements from clay minerals due to interactions with brine. Further, polyhalite trace element concentrations are essentially identical to those in anhydrites, suggesting early formation of polyhalite close to the end of diagenesis.

2. Where local concentrations of trace elements are noted in the evaporites, these are due to inclusions of very small amounts of clay minerals and/or oxide coatings on evaporite grains.

3. The lanthanide distribution patterns for the clay minerals are very close to the pattern for pelagic sediments but significantly different from the sea water pattern (Fig. 1.). This indicates

little exchange of the lanthanides of the clay minerals due to interaction with brine. The slight light rare earth depletion and heavy rare earth enrichment is due to the formation of Mg-rich clay minerals such as chlorite-saponite forming at the expense of original smectites (3). The source of the heavy lanthanides is from other clay minerals being altered as the overall abundance of the lanthanides in sea water is some six orders of magnitude lower than average clay mineral values and thus cannot account for the enrichment. Similarly, the slight loss of lanthanum and cerium is due to very limited incorporation of these elements into either the small amounts of evaporite minerals or oxide-hydroxide coatings present.

4. Despite the alteration of the original smectites during the formation of chlorite-saponite, the essential trace element chemistry of the clay minerals is preserved. In terms of the lanthanides, this presents a strong case for their retention in the detrital phase during diagenesis. Applied to possible overpack-brine interactions, the argument can be made for retention of both lanthanides and actinides in the overpack material closest to the canister in the presence of brine.

ACKNOWLEDGEMENTS

Financial support for this work was provided by the U.S. Department of Energy by Sandia Laboratories Contract 03-2065 with the University of New Mexico. The analyses were carried out at the Los Alamos Scientific Laboratory.

REFERENCES

1. D. G. Brookins, J. K. Register, M. E. Register and S. J. Lambert, "Long Term Stability of Evaporite Minerals: Geochronological Evidence," this symposium (1979).
2. M. W. Bodine, Jr., "Clay-Mineral Assemblages from Drill Core of Ochoan Evaporites, Eddy County," N. M. Bur. Mines Min. Res. Circ. 159, 21 (1978).
3. J. K. Register, "Rb-Sr and Related Studies of the Salado Formation, Southeastern New Mexico," M.S. Thesis, University of New Mexico (1979).
4. R. V. Gaines, "Cesium: Element and Geochemistry," Encyclopedia of Geochemistry and Environmental Sciences, 147 (R. W. Fairbridge, ed.; Van Norstrand Reinhold Company, NY, 1972).
5. O. Pilkey, "Barium: Element and Geochemistry," Encyclopedia of Geochemistry and Environmental Sciences, 62 (R. W. Fairbridge, ed.; Van Norstrand Reinhold Company, NY, 1972).
6. C. D. Curtis, "Scandium: Element and Geochemistry," Encyclopedia of Geochemistry and Environmental Sciences, 1060 (R. W. Fairbridge, ed.; Van Norstrand Reinhold Company, NY, 1972).

7. M. Fleischer, "Hafnium Content and Hafnium-Zirconium Ratio in Minerals and Rocks," U.S. Geol. Surv. Bull. 1021-A, 1 (1955).
8. K. Rankama and Th. G. Sahama, Geochemistry (Univ. Chicago Press, Chicago, 1950).
9. K. S. Heier and J. A. S. Adams, "The Geochemistry of the Alkali Metals," Phys. Chem. Earth 5, 253 (L. H. Ahrens, ed.; Macmillan Press, NY, 1964).
10. R. C. Reynolds, "Rubidium: Element and Geochemistry," Encyclopedia of Geochemistry and Environmental Sciences, 1050 (R. W. Fairbridge, ed.; Van Norstrand Reinhold Company, NY, 1972).
11. R. S. Young, "The Geochemistry of Cobalt," Geochim. Cosmochim. Acta 13, 28 (1955).
12. B. Mason, Principles of Geochemistry (John Wiley & Sons, NY, 1966).
13. V. M. Goldschmidt, Geochemistry (Clarendon Press, Oxford, 1954).
14. J. A. S. Adams, J. K. Osmond and J. J. W. Rodgers, "The Geochemistry of Uranium and Thorium," Phys. Chem. Earth 3, 321 (L. H. Ahrens, ed.; Macmillan Press, NY, 1959).
15. T. R. Wildeman and L. A. Haskin, "Rare Earth Elements in Ocean Sediments," Jour. Geophys. Res. 70, 2903 (1965).
16. O. T. Hogdahl, "Neutron Activation Analysis of Lanthanide Elements in Sea Water," Adv. Chem. 73, 308 (1968).
17. E. D. Goldberg, "Chemistry in the Oceans," Oceanography 67, 588 (M. Sears, ed.; Am. Assoc. Adv. Sci. Pub., 1961).
18. E. D. Goldberg, M. Koide, R. Schmitt and R. Smith, "Rare Earth Distributions in the Marine Environment," Jour. Geophys. Res. 68, 4209 (1968).
19. R. S. Della Valle, D. G. Brookins and J. P. Balagna, "Calibration Curves for Instrumental Activation Analysis for the Lanthanides," (in press).
20. D. Carroll and H. C. Starkey, "Effect of Sea Water on Clay Minerals," Clays and Clay Mins., 7th Ann. Conf., 80 (A. Swineford, ed.; 1960).
21. K. L. Russell, "Geochemistry and Halmyrolysis of Clay Minerals, Rio Ameca, Mexico," Geochim. Cosmochim. Acta 34, 893 (1970).

THERMAL GRADIENT MIGRATION OF FLUID INCLUSIONS IN SINGLE CRYSTALS OF SALT FROM THE WASTE ISOLATION PILOT PLANT SITE (WIPP)*

Edwin Roedder and H. E. Belkin

U. S. Geological Survey
959 National Center
Reston, VA 22092

ABSTRACT

Selected salt single crystals from core samples of bedded salt from the WIPP site have been tested to determine the rate of migration of their liquid-filled (fluid) inclusions in thermal gradients. Gradients of 1.5°C.cm^{-1}, maintained for periods of 3-10 days, at ambient temperatures of 108°-260°C, resulted in rates of movement of 1.2-5.4 cm.yr^{-1} for cubic inclusions 1 mm on an edge. Inclusions 0.1 mm on an edge moved only ∿30% as fast. Increase in ambient temperature and/or gradient increased the rate, in approximately direct proportion. The migration rate for inclusions in different parts of a given sample, however, may vary by a factor of 3, for unknown reasons, and some other interesting observed phenomena are still unexplained.

INTRODUCTION

Most salt beds contain at least some water in the form of liquid inclusions within salt single crystals, on grain boundaries, or in the larger intergranular cavities (i.e., pores). The storage of any heat-generating waste will establish thermal gradients in the salt. An aqueous liquid inclusion within or between crystals of a soluble salt, when placed in a thermal gradient, will continuously dissolve on the hot side and crystallize on the cold side of the cavity, thus causing the cavity, and its included liquid, to move up the thermal gradient toward the heat source. Inclusion movement is of conse-

*Work supported by U. S. Dept. of Energy, Albuquerque Oper. Office.

quence in the design of a waste repository, as the included brines are probably more corrosive to possible canister materials than even a saturated NaCl solution.

Several studies (1, 2, 3) have addressed the problem of the rate of migration of fluid inclusions in salt, experimentally or theoretically. As many variables may affect this rate, and as the interrelations of these variables are not fully understood, experimental determinations are desirable on material from each specific site to be considered. Hence, samples from segments of core from the ERDA no. 9 borehole within the WIPP site in SE New Mexico were subjected to thermal gradients to determine the rate of migration of the fluid inclusions that they are known to contain (4). Salt beds are polycrystalline. In this paper we report measurements only on the movement rates within single crystals. The route taken by the fluid after an inclusion contacts a grain boundary is a much more complex process that is difficult to quantify with meaningful experiments.

PREVIOUS WORK

Wilcox (5) summarized the extensive work (111 references) on the nature and rate of inclusion movement in various substances. Migration of inclusions also causes degradation of laser crystals grown at high temperatures by the Czochralski technique (6). Most liquid inclusions move up the thermal gradient, but if the vapor bubble in the liquid is large relative to the liquid, and particularly if boiling occurs in the inclusion in the gradient, the movement may be in the reverse direction, down the thermal gradient (7, 8, 9). The rate of movement is independent of inclusion size in many systems but is strongly (and directly) dependent on inclusion size in others (5); there may be a threshold size below which no movement occurs ($\sim$10 μm in KCl)(10). Large inclusions in several hosts break up during movement (5, 11). Even in a given host, many factors may affect the rate of migration, such as gravity, composition and surface tension of the liquid, inclusion size and shape, host-crystal anisotropy, strain and imperfections, external stress, volume percent of vapor bubble and presence of a foreign gas in it. In salt, the rate of movement in a given gradient can be expected to increase as ambient temperature increases, because the thermal coefficient of solubility, although small at room temperature, increases greatly as temperature increases. Higher ambient temperature also increases the migration rate by increasing the solubility, the diffusion coefficients and the interface kinetics in any host; the increase in rate was found to be particularly striking in NaCl (Ref. 5, p. 20). That the small inclusions in the WIPP site salt have not moved measurably during geological time in the geothermal gradient (4) is perhaps an indication of an exceedingly slow rate under the natural gradient.

It may also be a result of other weak factors, such as gravity, counteracting the thermal gradient effect.

SAMPLE MATERIAL USED

Segments of core from the ERDA no. 9 borehole, taken from the Salado Formation at depth intervals 2061.2-2061.6 ft (hereafter designated "2061") and 2618.5-2619.0 ft ("2618") were used in most of the study, but some scoping runs were made on material from 2665.0-2665.1 ft ("2665") and one sample from a core taken from the Vacherie salt dome, Webster Parish, LA, at a depth of 1432.5 ft (12). Only crystals of salt >1 cm were used in these studies, in order to provide test samples that were large enough for our procedure.

EXPERIMENTAL TECHNIQUE

Our test samples were rectangular blocks 1 cm square and $\sim$1-2 cm long, cleaved or cut from portions of single crystals of salt having suitable inclusions, with the final orientation of the block known relative to the crystallographic axes of the salt. In most runs, this orientation was parallel with the cubic (100) cleavage, but several other orientations were also used. To avoid damage to the salt and leakage of inclusions near the surface, cutting was done on a thin-bladed (0.012 in) wafering saw that was slow-speed, lubricated with cutting oil, and charged with very fine diamond grit. The sawn surfaces were smooth enough to permit optical examination and photography using only a cover glass and matching index liquid in lieu of a polish. The positions of the inclusions were photographed through the microscope against a series of fiducial marks (both nearby solid inclusions and fine lines scratched on the surface were used). The block was then placed in a 1x1 cm slot in the center of a cylindrical segmented Teflon (TFE) block (Fig. 1), in tandem with other similar salt blocks, with appropriately placed thermocouple junctions. This assembly was placed in a thermal-gradient furnace which had appropriate controls for both ambient temperature and superimposed gradients. No external stress or pressure was imposed on the sample. The gradient was established (in a horizontal direction) and maintained along the length of the cylinder, perpendicular to the section shown in Fig. 1. The samples were heated to the desired temperature gradient and ambient temperature slowly ($<30°C.hr^{-1}$) to avoid thermal stress and held constant under these conditions for several days. After slow cooling, the samples were rephotographed, and the movement of the inclusions relative to the fiducial marks was measured.

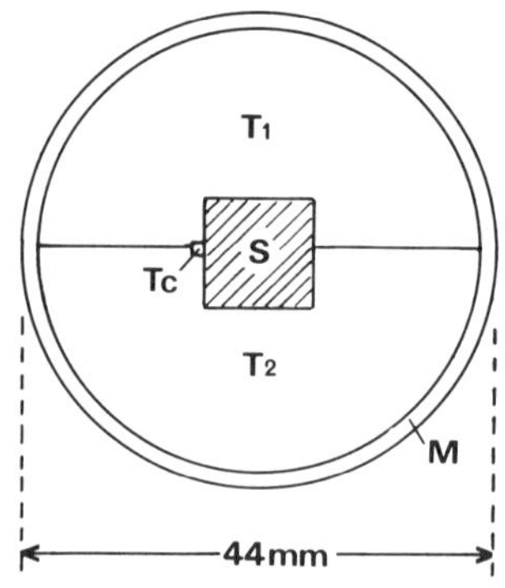

Fig. 1. Cross section of thermal-migration-experiment sample holder. The sample (S) consists of a 1x1 cm block of salt, in a closely fitting slot between two hemicylindrical teflon (TFE) blocks (T_1 and T_2), 41 mm in diameter, which fit closely into a closed-end inconel tube (M) of 1.5 mm wall thickness and 18 cm length. Thermocouples (Tc) are placed at known positions along the sample.

RESULTS

Sample 2061

The largest group of data was obtained on sample 2061 and is summarized in Fig. 2. The points on Fig. 2 were derived as follows. For a given run, the observed migration rates in $cm.yr^{-1}$ for each measured inclusion were plotted (linear scale) against the initial volume of the inclusion in μm^3 (logarithmic scale). A best-fit line was drawn through this data set on the basis of visual estimation. The intersections of this line with inclusion volumes were then plotted log-log as in Fig. 2. The original data points from individual inclusions in any given run showed some scatter on the plot of migration rate vs. log initial volume that differed from one sample to another. This scatter was usually $< \pm 50\%$ of the migration rate given by the best-fit line for any specific volume inclusion. (Examples are shown in Fig. 3, discussed later). Many reasons for scatter can be suggested, in addition to measurement uncertainties, such as differences in inclusion composition, shape or percent vapor, collisions en route (these were seen to have taken place), and sample strain. Most inclusions had no bubble at the start, but all had a bubble at the end of a volume percent proportional to the ambient temperature used. The inclusions expanded by deformation of the host salt during the slow heat-up period and were presumably liquid-filled during the runs.

Several features of Fig. 2 should be noted. First, a rather uniform decrease in migration rate takes place as inclusion volume decreases, so that the rates for inclusions of 10^9 μm^3 volume (i.e.,

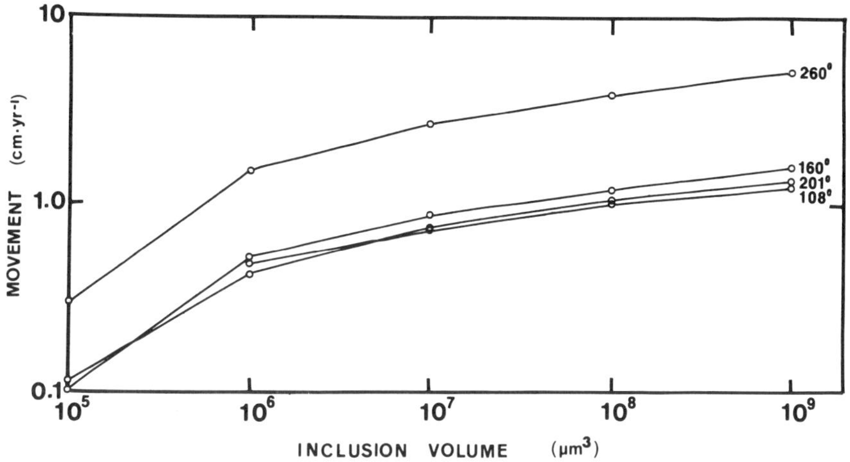

Fig. 2. Migration rates of various sizes of fluid inclusions in salt sample 2061, after 72-252 hours at 1.5°C cm^{-1} gradient and the ambient temperatures indicated. Each curve is based on smoothed data for a group (17 to 44) of inclusions.

cubes 1 mm on an edge) are about three times those for inclusions of 10^6 μm^3 volume (i.e., cubes 0.1 mm on an edge). The relatively few inclusions that were flattened perpendicular to, or elongated parallel to, the direction of motion were checked particularly for deviations in rate; most of these that did deviate appreciably did so in the direction of higher rates for the flattened and lower rates for the elongated inclusions, but the amount of deviation and the numbers of such inclusions were too small for any meaningful quantification of the differences. The rates for the run at 260°C ambient are ∿3.5 times those found for the other temperature runs for a given volume inclusion. The rate for runs at 108°, 160° and 260°C ambient show a consistent pattern, but that for the run at 201°C does not. In view of the differences we found between this sample and sample 2618 (see below), we suspect that the anomalously low rate for the 201°C run on sample 2061 may be a result of similar differences within parts of the same sample.

Effects of Crystallographic Orientation

Several runs were made to check the possible effects of the crystallographic orientation of the host relative to the direction of migration. Thirteen inclusions in a piece of sample 2061 cut to obtain movement perpendicular to the dodecahedron (110) yielded a rate identical with that of 27 inclusions cut to move perpendicular to the cube (100), within the precision of the measurements (Fig. 3).

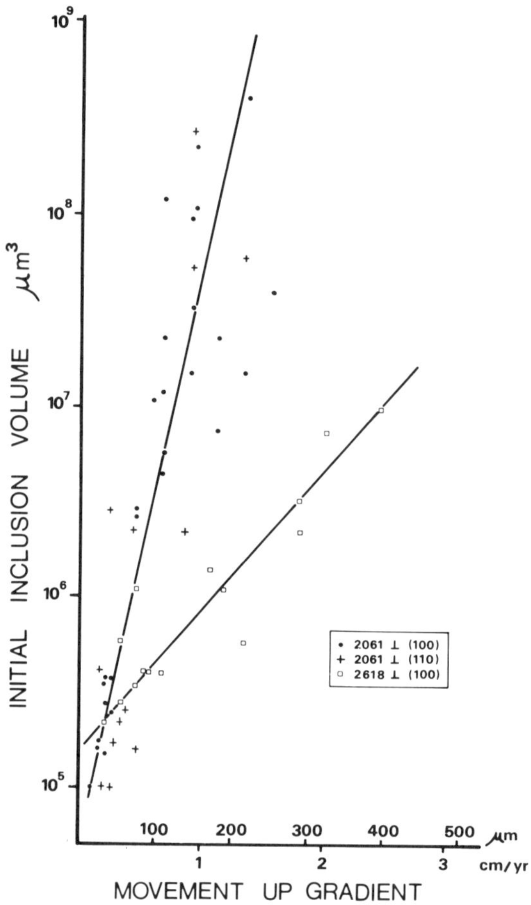

Fig. 3. Migration rates of several sizes and orientations of fluid inclusions in salt samples 2061 and 2618, run for 141 hours at a gradient of 1.5°C.cm^{-1} and an ambient temperature of 201°C.

The two pieces were run at 201°C ambient simultaneously, and the inclusions covered the same size range ($1x10^5$–$5x10^8$ μm^3). Another run, at 157°C ambient, again using two pieces of sample 2061, showed a rate for 44 inclusions moving perpendicular to (100) that was less than one third of that for 19 inclusions moving perpendicular to the octahedron (111). This difference may not be solely the result of the crystallographic orientation, however, as the rate for movement perpendicular to (100) was well under (and that for movement perpendicular to (111) was well over) the rates for other runs perpendicular to (100) on this same sample in this ambient temperature range, as given in Fig. 2.

Behavior of Small Inclusions

Small inclusions (10^4-10^5 μm^3) move so little that many are below the threshold for reliable detection of movement with our method (∿10 μm). With runs of the durations used here, this threshold corresponded to rates in the range 0.01-0.1 $cm.yr^{-1}$. Determinations on such small inclusions are of little practical interest, as usually very little of the total water content in any given salt is present as inclusions in this size range (4). Their movement, however, does have some theoretical interest. The best-fit line through the bulk of the data on a plot of movement vs. log initial inclusion volume, such as Fig. 3, was generally a straight line that appeared to intersect the ordinate at a threshold volume, below which no movement could be detected, and some inclusions of ∿10^5-10^6 μm^3 volume were found not to have moved measurably. However, on some of the runs, some inclusions as small as $2x10^4$ μm^3 and hence, well below the various apparent intercepts at "zero movement", were found to have migrated small but finite amounts. The sensitivity of the method used was too poor to permit quantification, but these points suggest that perhaps the best-fit line should not intersect the ordinate but should approach it asymptotically.

Intra- and Intersample Comparisons

Several runs were also made on sample 2618 to determine whether there were differences between samples, even though other studies of the fluid inclusions in these various samples had shown that intrasample variation was apparently as large as intersample variation (4). The results of one of these runs (Fig. 3) show that 15 inclusions in a sample of 2618 moved at rates more than three times those for sample 2061 run at the same time. Actually, the 2618 sample was run between the two pieces of 2061. No larger inclusions were present in the sample of 2618 shown in Fig. 3, so another run was made on 42 inclusions in sample 2618 at 202°C ambient and 1.5°$C.cm^{-1}$. Most of these were larger than 10^7 μm^3. This run yielded a line that had the same slope as that shown for 2618 on Fig. 3, but even greater movement rates. Thus, at 10^7 μm^3, the rate was 2.9 $cm.yr^{-1}$ rather than 2.4 $cm.yr^{-1}$ as shown in Fig. 3.

Another piece of sample 2618 was cut for movement perpendicular to (110) and run at 157°C ambient together with the two pieces of sample 2061, one perpendicular to (100) and the other to (111), referred to above. In this run, inclusions in sample 2618 migrated at a rate intermediate between those for the two pieces of 2061 (e.g., at 10^8 μm^3, the rates were 0.5 $cm.yr^{-1}$ [2061, (100)]; 1.3 $cm.yr^{-1}$ [2618, (100)]; and 1.7 $cm.yr^{-1}$ [2061, (111)].

From the above data, three-fold differences are readily apparent in the migration rates for inclusions in different pieces of a given WIPP sample and for different WIPP samples; differences due to crystal orientation alone may be of this same order of magnitude or may be negligible, as we have not eliminated the intra-sample variation. We have not recognized any differences between the parts of a given sample that yield different migration rates. Additional runs and tests are being planned in an attempt to isolate the several effects.

Comparison with a Sample from a Louisiana Salt Dome

Inclusions in some salt from cores in the Rayburn and Vacherie salt domes in Louisiana have been found to contain solutions that were close to simple $NaCl-H_2O$ brines, i.e., they contain relatively small amounts of other ions (12), rather than the bitterns present in the WIPP samples (4). As the presence of ions other than Na and Cl affects the solubility of NaCl, and as this solubility value enters into any theoretical calculation of the rate of migration (R. W. Potter II, pers. comm., 1979), we made a comparison run using a block of salt from the Vacherie dome, at 1432.5 ft depth, and WIPP samples 2061 and 2618, at 157°C ambient and $1.5°C.cm^{-1}$ gradient. The Vacherie sample was cut parallel to (100) and contained 16 small inclusions ($2.0x10^4$-$2.7x10^6$ μm^3). Only 3 of these were $>5x10^5$ μm^3, so the amount of movement was generally small, and some showed no detectable movement, but the apparent slope and migration rate on a plot like Fig. 3 were about the same as the WIPP sample 2061.

Migration Rates at Other Ambient Temperatures and Gradients

At the start of this work, some rough scoping experiments were set up to provide guidance for the experimental design of the migration runs. The temperature control and measurement were both much inferior to those used in the runs described above, but the values obtained (Table 1) provide at least some rough indications of the magnitude of the effects of these higher ambient temperatures resulting in faster migration.

Changes in Inclusion Shape with Movement

Most inclusions studied in this work were cubic or nearly cubic as found. At the end of the run, after the inclusions had moved tens or hundreds of micrometers, they were still nearly cubic if the original volume was $\lesssim\sim 2.7x10^6$ μm^3. Larger inclusions were quite different in shape at the end of the run regardless of the ambient temperature used, from 108° to 260°C. The advancing front surface

Table 1. Migration Rates Found in First Rough Scoping Experiments.

ERDA-9 Core Sample	Ambient temp (°C)	Gradient (°C.cm^{-1})	Time (hrs)	Inclusion size (μm^3)	Migration rate (cm.yr^{-1})
2665.0-2665.1'	250	4	4	5.5×10^9	109
2606.5-2609.9'	278	20	34-60	2.4×10^7	1.8
2606.5-2609.9'	340	20	31	1.0×10^6	10
2606.5-2609.9'	320	20	23	3.8×10^5	11
2606.5-2609.9'	320	20	23	6.4×10^4	8.8
2606.5-2609.9'	330	20	26	8.0×10^3	0.8
ERDA No. 7	248	1.0	84	4.3×10^7	6.7
ERDA No. 7	248	1.0	84	1.0×10^6	3.3*

*This value determined by the line-intercept method described in text, based on the largest inclusion (above) and 11 others.

became smaller though still flat and square in outline, and the sides became curved, forming a tapered, bullet shape. The rear face remained approximately the same size and shape as originally, but a thin sheetlike peripheral fringe of liquid had developed on all four edges, tapering off to the rear (Fig. 4). We do not know whether this shape developed early in the run and represents a dynamic equilibrium shape or whether it was still in the process of evolving at the end of the runs. Inclusions forced to migrate perpendicular to (110) and (111) formed similar fringes trailing off the outer edges. Somewhat similar trailing "veils" formed at the edges of migrating flat disk-like inclusions in synthetic KCl (10).

Reverse Motion (Down the Gradient) for Gas-Rich Inclusions

Only five gas-rich inclusions were found in the samples used in these runs. All five were in sample 2061 material and moved down the gradient, toward the cold end, at the following rates:

Ambient temp (°C)	Gradient (°C.cm^{-1})	Time (hrs)	Initial Size (μm^3)	Vol gas (%)	Migration rate (down)(cm.yr^{-1})
108	1.5	252	9.8×10^3	28	0.42
108	1.5	252	9.8×10^3	28	0.45
108	1.5	252	4.0×10^6	14	1.67
160	1.5	183	1.6×10^6	28	0.69
160	1.5	183	5.5×10^6	25	3.10

This movement was not only in the reverse direction from that for liquid inclusions, but was considerably faster than that for liquid inclusions of the same size (from 1.25 to 10 times greater).

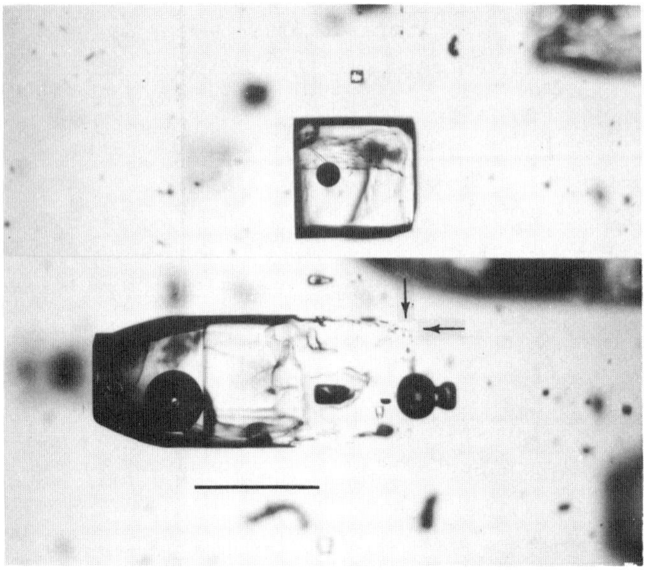

Fig. 4. Inclusions in salt sample 2618 before (above) and after (below) a 156-hour run at 202°C ambient and 1.5°C.cm^{-1} gradient. The large inclusion has split into gas-rich and liquid-rich parts that moved in opposite directions relative to the thermal gradient, which increased to the left. The original position of the inclusion can still be seen, outlined by minute specks of unidentified solids (see arrows). The fiduciary mark (a vertical scratch) is visible to the left of the inclusion in the upper photo; it is almost invisible in the lower photo, because of the illumination needed to see the (much larger) bubble, but a series of small specks to the right act as internal reference points. Scale bar = 500 μm.

Splitting of Inclusions into Vapor and Liquid

One run yielded a strange result that has yet to be duplicated. As far as we are aware, nothing was peculiar in the material used or the operation of the run. The material was sample 2618, cut to yield movement perpendicular to (100). The gradient was 1.5.cm^{-1} and the ambient temperature was 202°C, for 156 hours, a combination that had been used in other runs as well. At the end of the run, however, many of the measured inclusions were found to consist of two separate parts, a small, asymmetric, dumbbell-shaped, vapor-rich inclusion (∿5 vol % of original inclusion) that was now on the cold side of the original inclusion location, i.e., it had moved down the gradient, and a larger liquid inclusion, with a vapor bubble, that had moved up the gradient (Fig. 4). The actual migration distance of the two individual parts ranged widely from 15 to 725 μm; all inclusions >10^7 μm^3 in volume (12 total) had liquid moving farther than gas: liquid moved 437 to 725 μm up the gradient, and gas moved 15 to 475 μm down the gradient. Four of the seven inclusions having $<10^7$ μm^3 volume (down to $1.5x10^5$) had gas movements down the gradient that were greater than the corresponding liquid inclusions up the gradient. We cannot convert these movements to rates because we do not know at what stage during the run the splitting occurred; if it occurred at the start of the run, the liquid movements up the gradient yield rates in the same range as that found in other runs that did not exhibit splitting. A very few of the inclusions in some other runs showed similar splitting, but no other run yielded such a consistent splitting. We do not understand this phenomenon at present, but are investigating several possible explanations. We also do not understand the shape of the gas-rich part. Fig. 4 shows a typical example at the conclusion of the run; the vapor-rich inclusion consisted of two connected parts, a smaller, flattened, somewhat cubic mass of liquid at the front (in the direction of the movement), and a larger, almost spherical mass of gas (trailing), with a narrower neck between. A similar geometry of liquid and vapor parts of a vapor-rich inclusion moving down gradient has been seen in ADP crystals (5), but with quite different shapes.

DISCUSSION AND CONCLUSIONS

The rates of migration of fluid inclusions measured in these experiments are mostly well below those determined by Bradshaw and Sanchez (1) for inclusions in salt from Hutchinson, Kansas, but because the inclusions they used were much larger, the data would not be expected to be comparable. The volumes of the smallest inclusions they used ($8x10^9$ μm^3) were approximately an order of magnitude larger than our largest inclusions, and their largest were three orders of magnitude larger ($1x10^{12}$ μm^3). Although they do not specify the sizes of the individual inclusions they measured, a rough extrapolation of their data to the smaller size inclusions we ran (using Fig. 2), at $1.5°C.cm^{-1}$, yielded a migration rate only 34% higher than we found. Jenks (3) reviewed migration rate data for inclusions in KCl and in Kansas salt. He derived an empirical equation for the maximum migration velocity per unit temperature gradient for such salt as a function of temperature only that yields migration rates greater than we found for 1 mm^3 inclusions.

The application of these data to engineering design of the WIPP site is not simple. The bulk of the water in these beds is probably present in a relatively few large (>>1 mm) inclusions (4) whose rates of migration could not be measured by the procedures used. Even in the necessarily biased samples that were used, the bulk of the water was present in the relatively few inclusions ∿1 mm on an edge. Hence, the maximum migration rates found here (i.e., those for 1 mm inclusions) are the most nearly appropriate for calculating a <u>minimum</u> total rate of influx into a canister chamber.

Perhaps the most important aspect of the present work is the discovery that different samples from the same core, and even different parts from the same piece of core, can yield migration rates differing by a factor of 3. As differences of this magnitude exist between the otherwise apparently identical large crystals of salt selected for this study, it would be hazaradous to extrapolate these rates to the inclusions in the major part of the salt beds, which consists of smaller crystals than we used. Hence, it would be equally hazardous to extrapolate these rates to a calculation of expected brine inflow in a given storage scenario.

REFERENCES

1. R. L. Bradshaw and F. Sanchez, J. Geophys. Res. 74, 4209-4212 (1969).
2. K. A. Holdoway, Fourth Sym. on Salt, 1, (A. H. Coogan, ed., Northern Ohio Geol. Soc., Cleveland, 303-312, 1974).
3. G. H. Jenks, O.R.N.L. 5526, Dist. Cat. UC-70, 67 (1979).
4. Edwin Roedder and H. E. Belkin, Scientific Basis for Nuclear Waste Management, 1, (G. J. McCarthy, ed., Plenum Press, NY, 313-321, 1979).
5. W. R. Wilcox, Indust. Eng. Chem. 60, 13-23 (1968).
6. R. H. Hopkins, R. G. Seidensticker and A. M. Stewart, J. Crystal Growth 33, 223-231 (1976).
7. W. R. Wilcox, Indust. Eng. Chem. 61, 76-77 (1969).
8. T. R. Anthony and H. E. Cline, Acta Metallurg., 20, 247-255 (1972).
9. K. -H. Chen and W. R. Wilcox, Indus. Eng. Chem. Fund 11, 563-565 (1972).
10. T. R. Anthony and H. E. Cline, J. Appl. Phys. 42, 3380-3387 (1971).
11. T. R. Anthony and H. E. Cline, Acta. Metallurg., 21, 117-122 (1973).
12. E. Roedder and H. E. Belkin, U. S. Geol. Survey Open-File Report no. 79-1675, (1979).

THE APPLICATION OF THE PHYSICOCHEMICAL PROPERTIES OF BOILING BITTERNS TO THE INTERPRETATION OF BRINE MIGRATION EXPERIMENTS RELATED TO SALT REPOSITORIES

Michael A. Clynne, Robert W. Potter II*, and Valerie L. Thurmond**

United States Geological Survey
345 Middlefield Road
Menlo Park, CA 94025

An important unresolved question with respect to employing mined cavities in bedded rock salt as high level radioactive waste repositories concerns the behavior of small quantities of included bittern brines (Mg, Na, K, Ca//Cl, SO_4, Br-H_2O). These bittern brines have been shown to migrate at significant rates under the influence of a thermal gradient such as would be imposed by emplacement of high level waste (1). However, present estimates of the rate of migration of the bitterns range considerably because there is a lack of physicochemical data at the relevant compositions, concentrations, temperatures and pressures to solve the rate equations. This lack of appropriate physicochemical data for bitterns saturated with halite and the failure of theoretical models to address movement along grain boundaries have emphasized the need for field and large scale laboratory tests to measure the rate and quantities of brine which actually arrive at an individual waste canister (2, 3). In this report we present experimental data applicable to interpretation of the results of field and large-scale laboratory tests aimed at quantifying the brine migration phenomenon.

The amount of brine reaching the heat source in field and large-scale migration experiments can be determined by collecting steam and weighing the amount of condensate. The weight of condensate has been assumed equal to the amount of brine that has migrated to the heater (4). This assumption, however, is incorrect because the weight of condensate collected represents only a fraction of the actual amount

*Present Address: Occidental Petroleum, Irvine, CA.
**Present Address: University of Florida, Miami, FL.

of brine: the brine is composed of non-volatile salts and water whereas the steam condensate is essentially pure water. Two processes limit the amount of steam that can separate from the solid plus liquid assemblage at a given temperature. The boiling point elevation of the bittern is raised with increasing concentration, and water can be fixed by the precipitation of hydrated phases.

The total quantity of the bittern arriving at the canister at a given temperature can be related to the mass of collected steam condensate as follows:

$$\Sigma_b = \underline{c} \cdot N \qquad 1)$$

$$\underline{c} = \frac{1}{x \cdot f} \qquad 2)$$

where Σ_b is the total weight of bittern arriving at the heat source; at a given temperature $\underline{c}$ is a constant for each compositional variety of bittern, N is the weight of condensate collected at that temperature, f is the weight fraction of water in the bittern, and x is the weight fraction of water convertible to steam at a given temperature.

The value of $\underline{c}$ is unity only for pure water. In the case of a simple salt where x is approximately unity owing to an absence of hydrates and a low boiling point elevation for the salt-saturated brine, the only correction required is for the amount of salt in the brine. For example, sodium chloride at 200°C has a solubility in water of 31.898 weight percent (5); thus $\underline{c}$ using equation 2) is equal to 1.468, assuming $x \simeq 1$. However, for bitterns that form hydrated salts and have significant boiling point elevations, x is significantly less than unity, and therefore $\underline{c}$ would be correspondingly larger.

We have experimentally measured values of x for some bitterns in the system Na, Mg, K, Ca // Cl-H_2O with f values between 0.6012 and 0.7000 at temperatures from 114° to 200°C. Fig. 1. illustrates the apparatus employed to measure x which, with the f value for the brine, allows us to calculate $\underline{c}$ using equation 2). A 30 cc glass syringe was adapted by fitting the needle with a high pressure seal that could be attached to a valve block. The needle and a length of capillary were wrapped with water-soaked cloth strips to condense the steam to room temperature.

Experiments were conducted as follows:

(1) The autoclave was evacuated and loaded with a known mass of brine (±0.005 g).
(2) The autoclave was heated to the desired temperature and allowed to equilibrate.

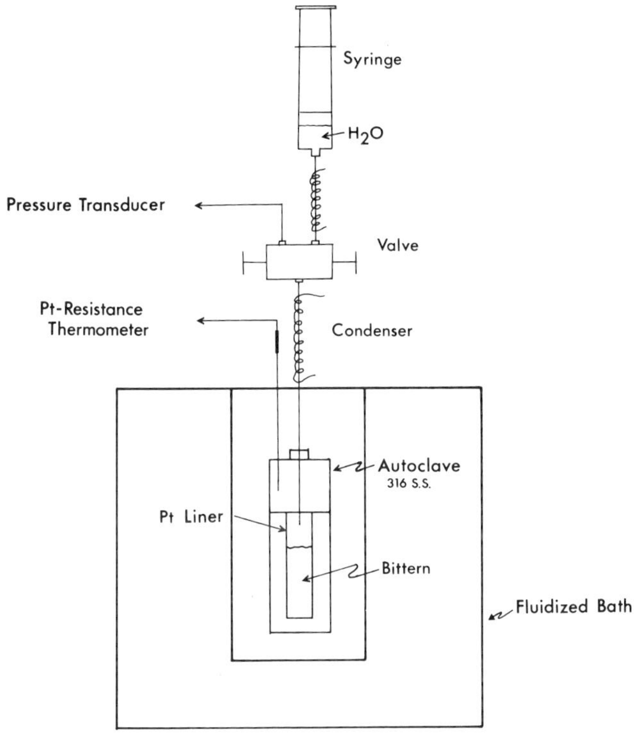

Fig. 1. Experimental apparatus used for measurement of x.

(3) Small quantities of steam were released into the sampling syringe.

(4) After the escape rate of steam had slowed so that it was indetectable visually, the valve was left open to allow the water in the syringe to equilibrate with the bittern in the autoclave for periods up to 6 hours. Throughout the whole experiment the cloth was kept wet and the autoclave maintained at a constant temperature.

(5) At the termination of the equilibration period, the plunger on the syringe was pulled upward to draw a partial vacuum ($\sim$25 mm Hg) and thus remove any condensate in the capillary. The valve was then closed, the syringe weighed and the amount of condensate determined by difference.

Note that the brine at T was allowed to equilibrate with the H_2O in the syringe (containing up to a few thousand ppm HCl transported in the steam) at 24°$\pm$2°C at a total P of $\sim$1 atm.

The experimental values of $\underline{c}$ range from 1.54 to 50.3; the higher values were obtained below 120°C. At 130°C, the weight of water extracted for periods of 2 hours, 4.5 hours and 6 hours was the same within ± 0.01 g. Between 120° and 200°C, the values of c range from 1.54 to 6.01. Fig. 2. illustrates the effect on c of the $CaCl_2$:$MgCl_2$ weight ratio at a constant f of 0.7. At temperatures above 150°C the value of c decreases continuously with increasing Ca:Mg. Below 150°C the value of c decreases with increasing Ca:Mg but then slowly increases at higher values of Ca:Mg. The value of c is also increased by saturating the bittern with NaCl as illustrated in Fig. 3.

Two important conclusions that can be drawn from these results are:

(1) The assumption that steam condensate equals the amount of brine can cause errors of 50 to 600% in estimating of the amount of brine; and

(2) At temperatures from 114° to 200°C, not all the water can be removed by boiling the bittern brines containing 30% total dissolved solids employed in this study.

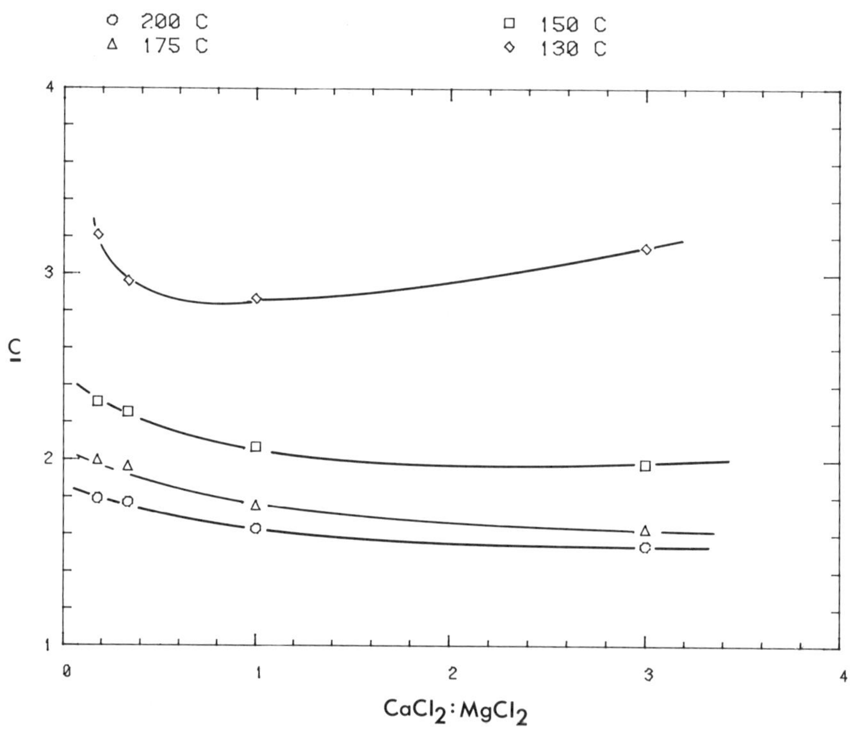

Fig. 2. The effect of the $CaCl_2$:$MgCl_2$ weight ratio (Ca:Mg) on the c value at constant f = 0.7 with NaCl = 5 wt% and KCl = 5 wt%.

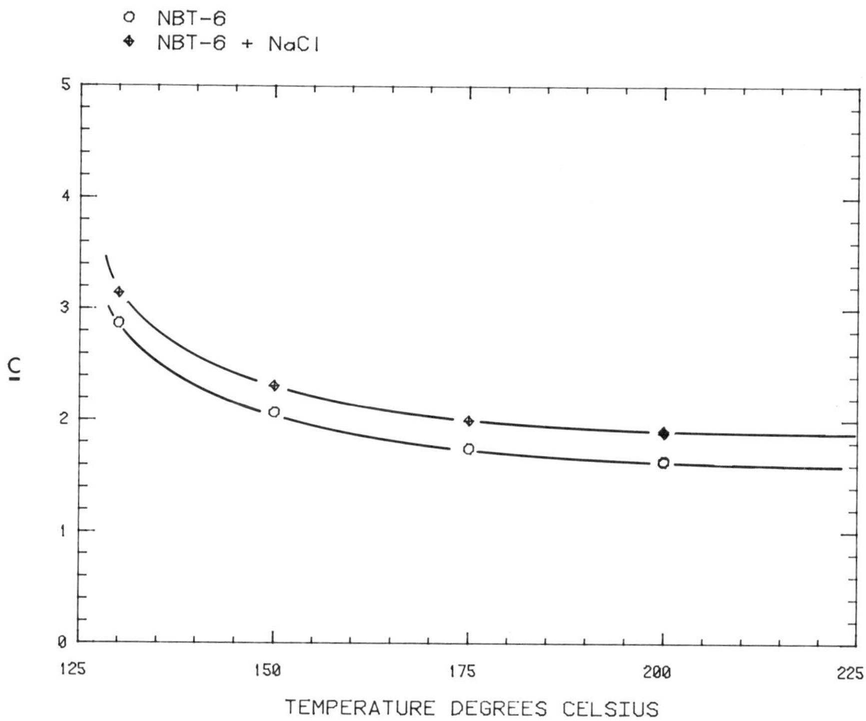

Fig. 3. The effect of saturating NBT-6 with NaCl is to raise the c value. NBT-6 is a 30% TDS bittern with a 1:1 $CaCl_2:MgCl_2$ ratio (10 wt% $CaCl_2$, 10 wt% $MgCl_2$, 5 wt% NaCl and 5 wt% KCl).

Our data can be used to indicate the approximate corrections to steam-condensate collections from field tests, provided the $CaCl_2:MgCl_2$ weight ratio in the bittern brine can be estimated and the temperature specified.

REFERENCES

1. E. Roedder and H. E. Belkin, "Application of Studies of Fluid Inclusions in Permian Salado Salt, New Mexico, To Problems of Siting the Waste Isolation Pilot Plant," Scientific Basis for Nuclear Waste Management 1, (G. J. McCarthy, ed; Plenum Press, NY, 313-322, 1979).
2. W. B. Krause and P. F. Gnirk, "Dome Salt Brine Migration Experiments at Avery Island, Louisiana," Proceedings of the National Waste Terminal Storage Program Information Meeting, October 1979. Office of Nuclear Waste Isolation 62, 97-100 (1979).

3. A. R. Sattler, H. C. Walker and T. O. Hunter, "In-situ Experiments to Support Development of the Waste Isolation Pilot Plant (WIPP)," Abstracts Symposium G, Annual Meeting of the Materials Research Society (1979).
4. R. L. Bradshaw and W. C. McClain, Project Salt Vault, Oak Ridge National Laboratory Report ORNL-4555, 356 (1971).
5. R. W. Potter, II, R. S. Babcock and D. L. Brown, "A New Method for Determining the Solubility of Salts in Aqueous Solutions at Elevated Temperature," Journal of Research of the U.S. Geological Survey, 5, 389-395 (1977).

MODELING OF BRINE MIGRATION IN HALITE

H. Cheung, M. E. Fuller * and E. G. Gaffney+

Lawrence Livermore Laboratory, Livermore, CA 94550;
*on leave from California State University Hayward, CA;
+Pacifica Technology, Del Mar, CA

INTRODUCTION

When canisters filled with radwastes are emplaced in a deep bedded salt repository the heat produced by the decaying waste will cause high temperatures and moderate thermal gradients to develop in the salt. The temperature and thermal gradient may be sufficient to cause the brine inclusions present to migrate towards the canister. The larger inclusions of SALT VAULT (1) were shown to migrate readily toward the emplacement hole while the very small ones were either unable to migrate or migrated very small distances (2). The rate of droplet migration can be limited either by ion diffusion through the liquid droplet or by the kinetics of ion removal from or deposition on the facets of the droplets. When the diffusivity and solubility of the solid in the liquid are large, the kinetic processes at the solid-liquid interface control the overall droplet velocity.

MODELS AVAILABLE

Four different models have been studied to determine their suitability as a working model for brine migration. All four models have several common factors, however, the remaining factors are present to account for several possible effects. The detail of each model leads to a certain degree of difficulty in applying it to the problem at hand.

This work was supported by the United States Nuclear Regulatory Commission under a Memorandum of Understanding with the U. S. Department of Energy.

Nernst Model

The first model considered was based upon the linear rate of growth of a crystal surface in a saturated solution (3). This model describes a plane crystal surface growing uniformly in a supersaturated solution. The growth rate is limited by the diffusion across a solute boundary layer at the surface. The rate of transport of solute is given by Fick's Law and the solute concentration at the interface is assumed to be the equilibrium value. The latter is true provided that the kinetic process is extremely rapid compared with the volume diffusion. The resulting equation is:

$$v = \frac{D}{\rho_s - C_\ell} \cdot \frac{\partial C_\ell}{\partial T} \cdot G_\ell \qquad 1)$$

where, v = linear rate of growth of crystal; D = diffusion coefficient; ρ_s = density of the crystal = C_s; C_ℓ = solute concentration at interface, $x = 0$; $\frac{\partial C_\ell}{\partial T}$ = slope of liquidus curve, at $x = 0$; and $\frac{\partial T}{\partial x} = G_\ell$ = thermal gradient in the liquid, at $x = 0$.

Equation 1) is dependent upon the values of the diffusion coefficient and the thermal gradient in the liquid. The diffusion coefficient can be estimated from viscosity data by using the Stokes-Einstein relationship (4):

$$D = \frac{kT}{6\pi r\eta} \qquad 2)$$

where, k = Boltzman constant; T = absolute temperature; η = viscosity;

r = average ionic radius, $\left[\frac{\left(r_{Na^+} + r_{Cl^-}\right)}{2}\right]$.

This method of estimation is subject to an error of less than an order of magnitude, but is considered to be the best available (5). Reliable data at elevated temperatures are badly needed. It is difficult to measure the thermal gradient in the liquid, G_ℓ, although it is very easy to measure the thermal gradient in the solid -- see Equation 4), below.

HOWS Model

The model developed by Hoekstra, Osterkamp and Weeks, as modified by Seidensticker (6), is based upon a study of brine pockets in glacial ice migrating under a thermal gradient. This model assumes that the shape of the inclusion does not change, therefore, the velocity of migration is independent of the aspect ratio of the inclusion and the direction of movement with respect to crystal orientation. Their consideration of thermal diffusion (Soret effect) indicated that it was less than 1% of the diffusion caused by concentration; subsequently, they neglected this effect. Thus, they calculated velocities from the equation:

$$v = D \frac{1}{C_\ell} \cdot \frac{\partial C_\ell}{\partial T} \cdot G_s \cdot \frac{3K_s}{2K_s + K_\ell} \cdot \frac{\rho_\ell}{\rho_s} \qquad 3)$$

where, K_s = thermal conductivity of solid; K_ℓ = thermal conductivity of brine solution; ρ_ℓ = density of the brine; G_s = thermal gradient in the solid and other symbols as previously defined, are within 1% of observed values for KCl and within 20% for NaCl.

The product of the thermal conductivity factor and G_s in Equation 3) approximates G_ℓ as indicated by the relationship (7):

$$G_\ell = \frac{K_s}{(1-F) \cdot K_s + F \cdot K_\ell} \cdot G_s \qquad 4)$$

where for salt K_s and K_ℓ change with temperature and are given by (T in °C):

$$\log K_s = 0.7618 - 1.37 \times 10^{-3} T \qquad 5a)$$

$$K_\ell = 0.5402 + 1.711 \times 10^{-3} T - 5.866 \times 10^{-6} T^2 \qquad 5b)$$

The factor F depends only on the aspect ratio of the droplet, and its value is 0.333 at X/L = 1.0 (cube and sphere) and increases to a value of 1.0 at X/L $\gg$ 1.0. L is the dimension in direction of migration and X is dimension perpendicular to direction of migration. Intermediate values of F are 0.53, 0.64 and 0.70 at X/L = 2.0, 3.0 and 4.0, respectively.

The parameters of Equation 3) are known or can be evaluated to some degree of reliability at the conditions expected in a repository. The thermal conductivity data are available for a wide range of temperatures, 0° – 340°C, as well as density data for a range of concentration and temperature (8,9).

Wilcox Model

Another model for impurity migration has been presented by Wilcox (10) to describe removing inclusions from crystals by gradient techniques (zone refining). He was especially concerned with inclusions formed during crystallization from solution. He assumed that diffusion through the zone is rate controlling, that the process is at steady state, there is no solubility of the solvent in solid, no convective mixing, constant properties, and no thermal diffusion (Soret effect). His derivation results in the equation:

$$v = D \cdot \frac{1}{m\,(1-w_o)} \cdot \frac{\rho_\ell}{\rho_s} \cdot G_\ell \qquad 6)$$

where v, D, G_ℓ, ρ_ℓ and ρ_s are as previously defined, and

w_o = weight fraction of crystalling material at the cold side of inclusion

$$\frac{\partial T}{\partial w} = m = \frac{\text{change in freezing point}}{\text{change in weight fraction}}$$

i.e., the slope of the liquidus curve on the crystal-solvent phase diagram.

A closer look at the factor 1/m allows it to be recognized as $\partial C_\ell/\partial T$ expressed in units of weight fraction rather than molarity. If $(1-w_o)$ were similarly expressed, we would have the molarity of the solvent.

Anthony-Cline Model

The most complete model has been presented by Anthony and Cline (11-18) and is expressed by:

$$v = D\,\frac{C_\ell}{C_s}\left[\left(\frac{1}{C_\ell}\cdot\frac{\partial C_\ell}{\partial T} + \sigma\right) G_\ell - \frac{1}{RT}\left(\frac{K}{L} + \frac{4\gamma\overline{V}_s}{XL}\right)\right] \qquad 7)$$

where, R = gas constant; σ = Soret coefficient; K = kinetics potential at interface; γ = surface tension of liquid; V_s= partial molar volume of solid; T = absolute temperature, °K and other symbols as previously defined. The four component terms can be represented as:

$$v = v_T + v_\sigma - v_K - v_\gamma \qquad 7a)$$

where, v_T = velocity due to ordinary diffusion; v_σ = velocity due to thermal diffusion (Soret effect); v_K = velocity due to kinetics at interface; v_γ = velocity due to surface tension at grain boundaries.

Because of the completeness of Equation 7), it is interesting to study the significance of each term. v_T is essentially the same as

Equations 1), 3) and 6). The effects of interface kinetics and grain boundaries are found in the third and fourth terms in Equation 7), i.e., v_K and v_γ. When a steady state for brine migration has been established within a crystal, there is no surface tension effect and v_γ is equal to zero. In this region, Anthony and Cline (18) quote values for K of 0.1 to 1.4 J/mole, depending upon approximate geometry of droplet.

When migration across grain boundaries occurs the surface tension must be considered and surface kinetics can become significant since most droplets change shape during migration; the surface tension is a function of the surface area exposed. When the droplet has reached a steady shape, Anthony and Cline (11) calculate a value of 0.13 J/mole for K for a particular size droplet, 38x38x10 microns. These are the only values for K available in the literature and they are for KCl. Anthony and Cline reported that during migration larger droplets tended to lose fragments, and thereby become an 'optimum' size for migration. Also, migrating droplets changed shape until their thickness (L) approached 8-10 microns, and attained an optimum aspect ratio, $X/L = <3$. Consequently, the following conclusions can be made:

1. As X approaches L, the surface tension effect goes to zero. The surface tension effect in only significant when crossing grain boundaries.

2. As L decreases, K increases but there is a limiting value for K as L approaches 10 microns.

3. The magnitude of the surface tension effect is about 10% of the magnitude of the kinetic interface effect.

4. The total volume of brine contained in droplets of $L<1$ mm, $X<1$ mm, would be rather small, so the contribution from surface effects toward total volume migration can be insignificant.

A COMPARISON

The comparison cases use the TRUMP computer code to generate the thermal fields around the buried canisters as functions of time and position. The power output of the canister (at the time of emplacement) is 3.5 kW for the HLW case and 0.55 kW for the spent fuel (10-yr old) case. The details of the TRUMP calculations are found in the work of Altenbach (19). The canister is assumed to be 1 ft. in diameter and 8 feet long for the HLW case and 16 feet long for spent fuel. Further, there is assumed to be no ventilation for heat removal from the repository itself and perfect thermal contact is assumed to exist between the waste canister and the salt.

A computer code, BRINE, was written (20) in FORTRAN and, with appropriate values in the input file, or the code itself, the output gives the brine accumulation as a function of time. With various output controls operating, such quantities as temperature, temperature gradient, diffusion coefficients, velocity, etc., can be printed for values of time and position desired. The thermal effects calculated for these two waste forms indicated that for the spent fuel the temperature remained below 60°C at the canister surface (maximum at 43 years) and the gradient in the salt was always less than 0.25°C/cm. With the HLW, however, canister surface temperatures were rising to above 500°C after 50 years (limit of TRUMP calculations), but the thermal gradient peaked at 32 years at 5.9°C/cm. The Anthony-Cline model predicts that droplet migration becomes effectively zero below some critical thermal gradient, and they observed this trend experimentally at about 1°C/cm (18). A comparison of the results of each model will be informative. The results of the BRINE calculations are given for each model in Table 1. Notice that all of the above models give results which are within a factor of two of each other.

CONCLUSIONS

Conclusion I

Equation 7) requires data which are not currently available in the literature. In particular, Equation 7) is the most complete model because it accounts for most known phenomena occurring in the migration process. However, two of the terms are droplet size dependent and one needs data on the distribution of these sizes and number of droplets of each size. Such data are not available in the literature. The smaller of these droplets make very little contribution to the total volume of brine present. Also, Soret coefficients are available only for temperatures below 45°C and almost no

Table 1: Results of Four Models Which Describe Brine Migration

	Volume of Brine Accumulated, at 0.5% water, liters		
	HLW (3.5 kW)	Spent Fuel (0.55 kW)	
Model	50 years	50 years	1000 years
Nernst	50.2	0.04	0.16
HOWS	89.4	0.17	0.78
Wilcox	63.8	0.06	0.23
Anthony-Cline (X=1, L=1 mm)	79.7	0.00	0.00

data are available for NaCl (21,22). Similarly, the value for the kinetic potential is available only for KCl. However, these surface effect terms make less than 10% contribution to the velocity. (See Conclusion II).

Conclusion II

The diffusion phenomenon is rate controlling, for all practical purposes. Equations 1), 3) and 6) had an original assumption that the migraion process was concentration diffusion controlled, i.e., kinetic effects and Soret effects were considered not significant. Consequently, it is surprising that these three models predict results in such close agreement with Equation 7).

Conclusion III

Neglecting the difference between the thermal gradients in the liquid and in the solid could affect the results by a factor of 2. All equations have the factor for thermal gradient in the liquid droplet. This is not easy to measure, but the thermal gradient in the solid can be measured quite readily.

Conclusion IV

The models may be ranked according to the difficulty of obtaining needed data: $1) < 6) < 3) \ll 7)$, or, the Nernst model is easier to apply than the Wilcox model, which is easier than the HOWS model, which is significantly easier than the Anthony-Cline model. Existing data are inadequate to determine the relative accuracy of the various models. Because Equation 3) has been successfully applied to the Project SALT VAULT experiment (23) it is tempting to use it as the model for brine migration, however, Equations 1) and 6) should also be considered. The Anthony-Cline model requires the greatest amount of estimation for the parameters needed, although the additional details may contribute little to the final result (see Conclusion II).

REFERENCES

1. R. L. Bradshaw and W. C. McClain, Project Salt Vault: A Demonstration of High Activity Waste in Underground Salt Mines, ORNL-4555 (1971).
2. K. A. Holdoway, "Behavior of Fluid Inclusions in Salt During Heating and Irradiation," Fourth Symposium on Salt, Vol. 1 (A. H. Coogan, ed., Northern Ohio Geol. Soc., Cleveland, 1974)
3. D. Elwell and H. J. Scheel, Crystal Growth from High-Temperature Solutions, (Academic Press, NY 1975).

4. W. J. Moore, Physical Chemistry, (4th Ed., Prentice-Hall, Inc., Englewood Cliffs, NJ, 1972).
5. D. G. Miller, Lawrence Livermore Laboratory, Livermore, CA (private communication).
6. P. Hoekstra, T. E. Osterkamp and W. F. Weeks, "The Migration of Liquid Inclusions in Single Ice Crystals," J. Geophys. Res. 70, 2180 (1966).
7. G. H. Jenks, Effects of Temperature, Temperature Gradients, Stress and Irradiation on Migration of Brine Inclusions in a Salt Repository, ORNL-5526, (1979).
8. W. R. Gambill, "Seven Answers from One Chart -- Physical Properties of Water," Chem. Eng., 139 (1959).
9. W. Unterberg, "Thermal Properties of Salt Solutions," Brit. Chem. Eng. 11, 494 (1966).
10. W. R. Wilcox, "Removing Inclusions from Crystals by Gradient Techniques," Ind. and Eng. Chem. 60, 13 (1968).
11. T. R. Anthony and H. E. Cline, "Thermal Migration of Liquid Droplets Through Solids," J. Appl. Phys. 42, 3380 (1971).
12. T. R. Anthony and H. E. Cline, "The Kinetics of Droplet Migration in Solids in an Accelerational Field," Phil. Mag. 22, 893 (1970).
13. T. R. Anthony and H. E. Cline, "The Interaction of Liquid Droplets with a Grain Boundary in Large Accelerational Fields," Phil. Mag. 24, 695 (1971).
14. H. E. Cline and T. R. Anthony, "The Thermomigration of Liquid Droplets Through Grain Boundaries in Solids," Acta Metall. 19, 491 (1971).
15. H. E. Cline and T. R. Anthony, "The Shape Relaxation of Liquid Droplets in Solids," Acta Metall. 19, 175 (1971).
16. T. R. Anthony and H. E. Cline, Acta Metall. 20, 247 (1972).
17. T. R. Anthony and H. E. Cline, "The Stability of Migrating Droplets in Solids," Acta Metall. 21, 117 (1973).
18. H. E. Cline and T. R. Anthony, "Effects of the Magnitude and Crystallographic Direction of a Thermal Gradient on Droplet Migration in Solids," J. Appl. Phys. 43, 10 (1972).
19. T. J. Altenbach, Interim Report on Nuclear Waste Depository Thermal Analysis, LLL, UCID-17865, 25 July (1978).
20. G. D. Duckworth and M. E. Fuller, BRINE--A Computer Program for Brine Migration, LLL (in press).
21. J. Schott, "Sur la Mesure des Coefficients de Soret par la Méthode Thermogravitationnelle," Compt. Rend. Acad. Sci. Paris, Series C 276, 459 (1972).
22. L. G. Longsworth, "The Temperature Dependence of the Soret Coefficient of Aqueous Potassium Chloride," J. Phys. Chem. 61, 1557 (1957).
23. R. L. Bradshaw and F. Sanchez, "Migration of Brine Cavities in Rock Salt," J. Geophys. Res. 74, 4209-4212 (1969).

LONG-TERM STABILITY OF EVAPORITE MINERALS: GEOCHRONOLOGICAL EVIDENCE

D. G. Brookins, J. K. Register, M. E. Register,
S. J. Lambert*

Department of Geology
University of New Mexico
Albuquerque, NM 87131

INTRODUCTION

The long term stability of evaporite minerals in areas being considered for the disposal of radioactive waste is critical in the overall assessment of such areas, especially when petrographic and other study indicate significant amounts of post-formational recrystallization. At the WIPP site in southeastern New Mexico, the age of the bedded evaporite minerals of the Castile and Salado Formations is known to be Late Permian, about 225 $\pm$ 15 MYBP (millions of years before the present). It is possible, however, that the major accumulations of potassium salts were formed by diagenetic-epigenetic processes in the early Triassic as the section was buried, local brines caused remobilization of some salts, and several stages of potassium-bearing layers resulted. These processes were probably complete by about 200 MYBP (1). Important potassium ore minerals (note: here referred to as primary even though they may not completely have formed until about 200 MYBP) include sylvite, KCl and langbeinite, $K_2Mg_2(SO_4)_3$; and minor amounts of carnallite, kainite, etc. are commonly interpreted to be primary. One of the most important secondary potassium-bearing minerals in polyhalite, $K_2Ca_2Mg(SO_4)_4 \cdot 2H_2O$, because of its water content. Petrographic study clearly indicates that the polyhalite is secondary to sylvite, yet just how much younger has not been quantitatively established.

In underground mine exposures some 70 kilometers to the southwest of the WIPP site, Tertiary lamprophyre dikes intrude the evaporite sequence (Kerr-McGee Potash Mine) and Pleistocene rubble

*Sandia Laboratories, Div. 4511, Albuquerque, NM 87185

chimneys are noted in the Mississippi Chemical Corporation Mine (2). In the former occurrence, K-Ar dates for the dike indicate emplacement at about 31 to 33 MYBP; while the rubble chimneys are interpreted (2) to have formed during the Pleistocene due to collapse of part of the evaporite sequence due to dissolution at depth. At the dike location, sylvite has been dissolved and a new generation of sylvite has been precipitated along with halite, polyhalite, and other minerals. In the rubble chimneys, the textural evidence suggests that many of the fragments of evaporite minerals are detrital; including halite, gypsum, anhydrite, polyhalite. Clay minerals and hydroxide-oxide coatings are common. If the formation of these rubble chimneys involved the presence of large volumes of water, then most of the evaporite minerals would be newly formed at the time of collapse and in-filling. Adjacent to the rubble chimneys, the bedded evaporites are highly distorted and textural relationships among the minerals difficult to establish, although many minerals appear to have flowed plastically as opposed to some process of dissolution-reprecipitation. It is important to distinguish between these possibilities, however, as the dissolution-reprecipitation model demands availability of water while the plastic flow model does not. Polyhalite also occurs in the distorted beds near the rubble chimney and, based on textural data, could be either Permian (i.e. relict grains, plastic flow) or Pleistocene (i.e. dissolution-reprecipitation accompanying rubble chimney formation).

Finally, it has been proposed by Bodine (3) and Tremba (4) that both continuous and episodic dissolution and recrystallization of the major evaporite salts has occurred throughout the Mesozoic and possibly into the Cenozoic, with a possible major period of recrystallization occurring in the mid-Cretaceous (4). Since these events would indicate availability of abundant fluids well after initial sedimentation, then, for radioactive waste site selection feasibility, it is critical to evaluate the absolute chronology of as many of these proposed events as possible. Many of the interpretations (3, 4) are based on radiometric age determinations, and this is the approach we have used to attempt to address some of the problems pointed out above.

RADIOMETRIC AGE DATING METHODS FOR EVAPORITES

Three major methods for the dating of evaporite minerals have been used at many locations (5). Of these, K-Ar dating has been most commonly applied because of the ease of the method, and some K-Ca dates have been attempted (5) although not from the WIPP site or nearby. Dating of the evaporite minerals by the Rb-Sr method has been attempted near the WIPP site (4) and elsewhere (5) although re-evaluation of many of these data is needed (6).

Potassium-Argon Dating

K-Ar dates have been reported (7) for langbeinite, sylvite, and langbeinite-sylvite mixtures from southeastern New Mexico. The K-Ar method has the advantage that, in most cases, minerals do not incorporate radiogenic ^{40}Ar into their structures at the time of formation, hence the amount of original daughter isotope (D_o = $^{40}Ar_o$ here) can usually be assumed equal to zero. Further, the potassium contents of minerals such as sylvite are high and easy to measure. A serious drawback to the K-Ar method for the dating of evaporite minerals is the ease with which radiogenic $^{*40}Ar$ is lost from the structures of some minerals. Minerals such as sylvite, or sylvite-halite mixtures, possess structures favorable to diffusion of $^{*40}Ar$ (5,8) resulting in K-Ar dates which are younger than their age of formation. Because ^{40}K decays both to ^{40}Ca (89%) and to ^{40}Ar (11%) some attempts at using the K-Ca method have been made for K-rich, Ca-poor minerals. Thus, successful K-Ca dates for sylvites have been reported (6) which is important, as it indicates that a solid species ($^{*40}Ca$) is more likely to be retained to the structure of a mineral such as sylvite than an inert gas like $^{*40}Ar$. Because of the many low K-Ar dates for sylvites, several authors (9,10) have doubted the application of this dating method to other evaporite minerals, including polyhalite. Polyhalite is triclinic and possesses more covalent bonding than isometric sylvite. Based on this information, a program to test the feasibility of using polyhalites for K-Ar radiometric age determinations from WIPP site evaporites has been undertaken.

The new data for polyhalites represent samples from several different sites in the WIPP-area and nearby environments. Two samples (MCC-M121 and 20/638.2) represent pure polyhalites from an inclusion in a rubble chimney and from highly distorted beds close

Table 1. Potassium-Argon dates

Sample No.	Mineral(s)	K-Ar Age (MYBP)	Reference
S-EPR2132A	Langbeinite	245 ± 10	7
S-EPR2132B	Sylvite	74 ± 8	7
S-EPR2132C	Langbeinite-sylvite	137 ± 8	7
S-EPR2132D	Langbeinite-sylvite	147 ± 10	7
S-EPR2132E	Sylvite	18 ± 8	7
DH ERDA 9,1784.2 ft.	Polyhalite	200 ± 7	*
DH ERDA 9,1759.5 ft.	Polyhalite	212 ± 7	*
DH ERDA 9,1215.2 ft.	Polyhalite	216 ± 7	*
MCC-M121	Polyhalite	198 ± 7	*
DH ERDA 9,1499.5 ft.	Polyhalite-sylvite	174 ± 6	*
DH AEC 8,1619 ft.	Polyhalite-sylvite-halite	154 ± 5	*
MB-76-22	Polyhalite	21.4 ± 0.8	*

to this chimney, both from the Mississippi Chemical Corporation Potash Mine. Another sample (MB76-22) is very pure polyhalite from the contact zone of the 32 ($\pm$ 1?) MYBP lamprophyre dike exposed in the Kerr-McGee Potash Mine. The remaining five polyhalites are from cores from various drill holes in the WIPP site proper and very closeby. Two of these, samples E9-99 and A8-18 are mixtures of polyhalite plus sylvite and polyhalite plus sylvite-halite as determined by X-ray diffractometry and chemical analysis; the remaining three samples (E9-15, E9-59, E9-84) are very pure polyhalites. The K-Ar dates for all five pure polyhalites (i.e., the two from the Mississippi Chemical Mine and the three from drill core) range from 198 $\pm$ 7 to 216 $\pm$ 7 MYBP with a mean of 208 $\pm$ 10 MYBP. This date is in excellent agreement with the Rb-Sr date for some of these same evaporite rocks (see below). More important, that the two samples from the rubble chimney and distorted beds nearby yield ages near 200 MYBP suggests that these samples were not affected by large amounts of water which has been postulated to accompany rubble chimney formation in the Pleistocene.

The two samples containing sylvite and sylvite-halite mixed with polyhalite yield K-Ar dates of 174 $\pm$ 6 and 154 $\pm$ 6 MYBP. These post-200 MYBP dates are interpreted to result from $^{*40}Ar$ loss due to the presence of sylvite.

The sample of pure polyhalite from the contact zone of the lamprophyre dike exposed in the Kerr-McGee Potash Mine yields a K-Ar date of 21.4 $\pm$ 0.8 MYBP. This polyhalite must be equal or younger in age than the dike. At present we cannot state with certainty if the polyhalite formed at about 21 MYBP or formed closer to 31 MYBP and then lost $^{*40}Ar$. Petrographic study of the contact zone samples indicates a new generation of sylvite as well as polyhalite. Our interpretation of the polyhalite date is that it represents a minimum date.

Rubidium-Strontium Dating

The Rb-Sr method has the advantage that ^{87}Rb-$^{*87}Sr$ decay occurs, the daughter $^{*87}Sr$ is a solid and retained in the original ^{87}Rb sites. Very few Rb-Sr dates have been attempted for evaporite minerals, however. One reason for this is the fact that Sr-Ca diadochy is more efficient than K-Sr diadochy and Ca in evaporite sequences is much more abundant than K, hence the initial $^{*87}Sr_o$ correction is commonly so large as to make determination of $^{*87}Sr$ (and thus age) virtually impossible. For our work we have chosen to use the Rb-Sr isochron method (5) in which growth of $^{*87}Sr$ with time is plotted as $^{87}Sr/^{86}Sr$ versus $^{87}Rb/^{86}Sr$. At t=0 the slope, $e^{\lambda t}-1$, of the isochron, is zero.

Table 2. Rubidium-strontium Data

Samples (no.)	Range in $^{87}Sr/^{86}Sr$	Range in Rb (ppm)	Range in Sr (ppm)
Whole rocks (18)	0.7068 - 0.7958	0.27 - 262.5	5.04 - 156.9
Evaporite minerals (41) (minus-polyhalite)	0.7071 - 1.6412	0.22 - 70.69	0.27 - 269.5
Polyhalites and anhydrites (6)	0.7078 - 0.7086	1 - 5	768 - 1560
Samples used for $(^{87}Sr/^{86}Sr)_o$ (27)	0.7071 - 0.7108	---	---

Notes: Rb and Sr determined by isotope dilution (10), samples from drill core at WIPP and nearby sites and from mine workings 30 - 40 km from WIPP site (x 10).

It is very difficult to separate very pure Rb-rich, Sr-free material to work with. Sylvites are usually well suited for age work, although even slight dissolution will result in loss of $^{*87}Sr$ because it is metastable relative to ^{87}Rb. Further, halite is often mixed with sylvite even in ore zones, and since Ca-Na-Sr diadochy is well documented, low Rb/Sr ratios commonly result. Polyhalites are not well suited for Rb/Sr age work because the Rb/Sr ratio is less than one as Sr is concentrated in late stage brines from which polyhalites form. In our work, we have attempted to determine the value for $(^{87}Sr/^{86}Sr)_o$ by isotopically analyzing Rb-poor-to-Rb-free phases such as anhydrite. These data, in Table 2 are from Register (10). The narrow range of $(^{87}Sr/^{86}Sr)_o$ from 0.7071 to 0.7108 for twenty seven data indicates a relatively homogeneous brine for the evaporites in terms of strontium isotopic composition. Further, the mean value, 0.7086 ± 0.0014 is within the range of values preferred for Permian sea water (5).

The Rb-Sr dating of evaporite minerals must be undertaken with care. The conventional whole rock method is virtually impossible to apply to evaporites due to a) coarse grain size of many evaporite minerals, b) inclusions of detrital mineral, and c) coatings of clay minerals and oxide-hydroxides which are in part detrital and in part authigenic. Register (10) attempted to date some whole rocks with a resulting apparent isochron are of 171 ± 36 MYBP with initial $(^{87}Sr/^{86}Sr)_o = 0.7113 \pm 0.0016$. No significance is attached to this date.

A more reliable method for dating evaporites is to separate the water soluble material from the samples. This has the advantage that most detrital material will be eliminated due to its highly insoluble nature. A possible disadvantage is that there may be different releases of Rb and Sr from the evaporite minerals due to their different solubilities. Register (10) has determined the age of the total forty one samples to be 199 ± 20 MYBP with initial $(^{87}Sr/^{86}Sr)_o$ = 0.7076 ± 0.0014. Five samples are clearly suspect, however. Elimination of these samples can be justified due to their anomalous Rb-Sr systematics. The resultant isochron age for the remaining thirty six samples is 214 ± 15 MYBP with initial $(^{87}Sr/^{86}Sr)_o$ = 0.7093 ± 0.0007. For all the isochron treatments, the least squares method of York (11) was employed.

While sylvite rich samples possess the highest $^{87}Rb/^{86}Sr$ ratios (10), it is significant that polyhalites fall within the 0.7086 ± 0.0014 initial $(^{87}Sr/^{86}Sr)_o$ range. If they formed much later than 200 MYBP, they would be expected to have lower total Sr contents due to lack of residual brines with which to interact and $^{87}Sr/^{86}Sr$ ratios well above 0.71 due to incorporation of $^{*87}Sr$ from the more readily soluble, radiogenic minerals such as sylvite, langbeinite, carnallite. That this is not the case supports the K-Ar interpretation that the polyhalites formed very close to the end of sylvite-langbeinite formation.

The Rb-Sr method has also been applied to detrital clay minerals to determine the extent of resetting of the Rb-Sr systematics by interaction with the evaporating brines at the time of sedimentation-diagenesis (10). If alteration is extremely severe, then the clay minerals would be expected to lose their $^{*87}Sr$ and exchange total Rb and Sr with the brines. This would result in removal of the pre-evaporite ages of the detrital clay minerals. We (6,10) have examined the Rb-Sr systematics of thirteen clay minerals presumed from SEM (scanning electron microscopy) to be detrital. The samples consist of various proportions of chlorite-saponite, illite, kaolinite, chlorite, saponite; and talc and/or serpentine are noted in others (10). Despite the abundance of Mg-rich phyllosilicates, the isochron age for these detrital clay minerals is 390 ± 77 MYBP with an apparent initial $(^{87}Sr/^{86}Sr)_o$ = 0.7122 ± 0.0049. This date indicates that the age(s) of source material(s) was not completely reset at the time of sedimentation during interaction with brines. Source areas capable of feeding detrital material to the evaporite basin could range from 1,000 MYBP to approximately 230-250 MYBP, but our data are too limited to identify any specific source areas.

DISCUSSION

Theories of substantial alkali-alkaline earth migration through the Salado Formation well after its' age of deposition and our proposed age of potash mineral formation (approximately 210 to 215 MYBP) have been advocated (3,4). Tremba (4) reported Rb-Sr dates of 129 ± 5 MYBP and 120 ± MYBP based on very few data (n=5) and tenuous arguments that a mid-Cretaceous tectonic event might account for two clusters of data through which the 129 and 120 MYBP isochrons could be constructed. Further, the initial $(^{87}Sr/^{86}Sr)_o$ intercepts for one isochron is well above 0.72, well beyond the range of $^{87}Sr/^{86}Sr$ initial ratios for the forty one data reported by us (10). In addition, data which did not fall on either isochron were discarded without explanation; these data, had they been included, would have increased the age. Still further, there is no geologic evidence for a mid-Cretaceous event affecting the Castile-Salado evaporite sequence. Finally, Tremba's (4) recommended model age based on composite samples, is 230-240 MYBP for the age of sedimentation for the Castile and Salado Formations.

The data reported here suggest an age of potash-ore mineral formation at 214 $\pm$ 5 MYBP which is within the lower error limit for the Permian-Triassic boundary of 230 $\pm$ 15 MYBP; thus refuting the earlier interpretations (3,4).

CONCLUSIONS

1) Our recommended Rb-Sr isochron date for potash-ore formation, and the minimum date of sedimentation for the Salado and Castile Formations of southeastern New Mexico is 214 $\pm$ 15 MYBP. This date indicates that these evaporite rocks have remained closed systems since pre-200 MYBP. 2) Our recommended date for major polyhalite formation is 208 $\pm$ 10 MYBP. This date includes a polyhalite inclusion from a rubble chimney and from highly distorted beds near the chimney from the Mississippi Chemical Corporation Mine which in turn indicate a lack of large amounts of water accompanying the formation of the rubble chimney. 3) Polyhalite-sylvite ($\pm$ halite) mixtures yield post-200 MYBP dates due to loss of $^{*40}Ar$; this is consistent with the work of many other investigators (6). 4) Polyhalite from the contact zone of a 32 $\pm$ 1MYBP amprophyre dike intrusive into the evaporites approximately 70 kilometers southeast of the WIPP site yields a K-Ar date of 21.4 $\pm$ 0.8 MYBP; this confirms the application of K-Ar dating of polyhalites to trace post-formational events in the evaporative sequence. 5) The forty one data with very low (i.e. less than 0.1) $^{87}Rb/^{86}Sr$ ratios yield initial $(^{87}Sr/^{86}Sr)_o$ = 0.7086 $\pm$ 0.0014. This value is within the limits of that for Permian sea water (5). 6) Rb-Sr dating of detrital clay minerals indicate an apparent age of 390 $\pm$ 77 MYBP which is interpreted to indicate incomplete clay mineral-brine interaction suring sedimentation and diagenesis. 7) Earlier suggestions for more-or-less continuous and significant alkali-alkaline earth migration post-200 MYBP (3,4) is refuted by our work. Instead we propose that the evaporites have remained closed to Rb and Sr since pre-200 MYBP as indicated by the Rb-Sr work and supported by the K-Ar polyhalite dates. This in turn is favorably interpreted as to the suitability of the WIPP site evaporites for retention of radioactive wastes.

ACKNOWLEDGMENT

This work was supported by Contract 05-4105 from Sandia Laboratories to the University of New Mexico from U.S. Department of Energy Funds, and by the Research Allocations Committee of the University of New Mexico.

REFERENCES

1. W. E. Dean and R. Y. Anderson, Salinity Cycles: Evidence for Subaqueous Deposition of Castile Formation and Lower Part of Salado Formation, Delaware Basin, Texas and New Mexico, N. M. Bur. of Mines. & Min. Res. Circ. 159,15 (1978).
2. R. Y. Anderson, K. K. Kietzke and D. J. Rhodes, Development of Dissolution Breccias, Northern Delaware Basin, New Mexico and Texas, N. M. Bur. of Mines & Min. Res. Circ. 159,47 (1978).
3. M. W. Bodine Jr., Clay-mineral Assemblages from Drill Core of Ochoan Evaporites, Eddy, County, N. M. Bur. of Mines & Min. Res. Circ. 159,21 (1978).
4. E. L. Tremba, Isotope Geochemistry of Strontium in Carbonate and Evaporite Rocks of Marine Origin, Ph.D. Disser., Ohio State Univ., 185 (1969).
5. G. Faure, "Principles of Isotope Geology", (John Wiley & Sons, Pubs., New York 1977).
6. D. G. Brookins, J. K. Register, M. E. Register and S. J. Lambert, "Burying High Level Waste," Nature 273,704 (1978).
7. J. H. Schilling, "K-Ar Dates on Permian Potash Minerals from Southeastern New Mexico," Isochron/West 6;37 (1973).
8. G. B. Dalrymple and M. A. Lanphere, "Potassium-argon Dating", (W. H. Freeman & Co., San Francisco, 1972).
9. G. Faure and J. L. Powell, "Strontium Isotope Geology", (Springer-Verlag, Heidelberg and New York, 1972).
10. J. K. Register, Rb-Sr and Related Studies of the Salado Formation, Southeastern New Mexico, M.S. Thesis, Univ. N. M. (1979).
11. D. York, "Least Squares Fitting of a Straight Line with Correlated Errors," Earth Plan. Sci. Lttrs. 5,320 (1969).

TRACER TESTS PERFORMED IN THE FIELD FOR WIPP IN SOUTHEASTERN NEW MEXICO*

Don Diego Gonzalez and Leslie R. Hill

Sandia Laboratories
Albuquerque, NM 87185

INTRODUCTION

A two-well recirculating tracer test began in October 1979 to define as accurately as possible the hydraulic character of a fractured carbonate aquifer, the Culebra dolomite member of the Rustler Formation at the WIPP site in Eddy and Lee Counties, New Mexico. The Culebra dolomite overlies a zone planned for isolation of transuranic contaminated waste generated by the United States defense programs. The storage zone of the proposed facility is in nearly pure halite and about 1400 feet (427 meters) below the Culebra dolomite, the most likely pathway for the migration of radionuclides to the biosphere in the event the repository is breached by groundwater.

Included in the definition of the hydraulic character of the Culebra aquifer are natural ground ater velocities, aquifer porosity and components of dispersivity.

This paper describes the proposed tracer test using sodium benzoate, a homologous series of chlorofluoromethanes and pentafluorobenzoic acid as tracers. Results of the test will be reported at a later date.

DESCRIPTION OF THE TRACER SITE

The tracer test will be conducted at a site 0.75 mi (1.2 km) southwest of the center of the proposed WIPP facility. The three

*This work supported by the U.S. Department of Energy.

major water bearing zones within the stratigraphic section above the proposed facility are two dolomite members in the Rustler Formation and the contact zone between the Rustler and Salado Formations (1). The site designated as Hydro-2 consists of three holes spaced in a right triangle array 50, 75 and 90 feet (15.2, 22.9 and 27.4 meters) apart. The holes have been designated H-2a, penetrating the upper dolomite section, the Magenta; H-2b, penetrating the lower dolomite section, the Culebra; and finally, H-2c, penetrating the Rustler Salado contact (1). Holes H-2b and H-2c spaced 75 feet (22.9 meters) apart have been completed to accommodate the two-well tracer test by penetrating the Culebra at a depth of 623 feet below land surface. The aquifer thickness has been determined as 19 feet (5.8 meters) (2), with hydraulic conductivities estimated as ranging from 0.01 feet/day to 0.05 feet/day (3).

DESCRIPTION OF THE PROPOSED TRACER TEST

Initially, the test will be designed on a conservative (low) estimate of hydraulic conductivity, using a withdrawal injection flow rate of 0.2 gallons/minute (0.76 liters/second). After steady state conditions have been reached and the results of a two-well pump test have been analyzed, better estimates for flow rates, injection concentrations, rates and sampling schedules can be made. The injection and recovery system will consist of a positive displacement pump suited to continuous use at low flow rates. The water and tracer flow will be monitored with a recording flow rate meter and a totalizing flow meter. Head measurements will be made throughout the test in packed-off horizons and in the wellbore above packers in both pumping and injection wells. The measurement of these parameters is essential to a two-well pumping test in that flow rates and heads represent boundary conditions in the flow equation and total volume pumped is necessary in the calculation of porosity. Tracer injection will be accomplished by a micro-metering pump and a pressure tank with diaphragm. The tracer injection system can simultaneously deliver tracers from three separate storage tanks, if necessary.

Water samples for tracer analysis will be collected both upstream and downstream from the injection point. Fluorocarbon analysis will be accomplished through use of a Varian 3700 gas-chromatograph equipped with a ^{63}Ni electron capture detector. The analysis for benzoate and pentafluorobenzoic will utilize a high performance liquid chromatograph with UV adsorption detection.

Quantitative interpretation of the results would utilize the analytical solution of Grove and Beetem (4) and a variety of recent numerical methods. Computer analysis will be performed as well as comparisons with existing models. Drs. G. Thompson, University of Arizona, and H. Bentley, Hydro Geo Chem Inc., will serve as principal consultants for the program in its entirety.

TRACERS

Sodium benzoate will be used as the principal tracer from which the aquifer parameters will be determined. Ground water tests with this compound show benzoate to exhibit little or no adsorption; however, its stability at low concentrations in water is questionable. Fluorocarbons will be injected as a backup to the benzoate tracer in the event the benzoate is degraded in the aquifer; if this occurs, the more sorbed but stable fluorocarbons can be used for the interpretive work.

The use of pentafluorobenzoic acid (PFB) as a tracer will provide an independent analysis for the determination of aquifer parameters that is less likely to be affected by degradation (sodium benzoate tracer runs a reasonable risk of biological degradation during the test). Further, the inclusion of two compounds as tracers with essentially the same sorptive properties (PFB and sodium benzoate) but significantly different molecular weights may enable interpretation of the test results in terms of matrix porosity. In principle, this is possible because the heavier anion PFM (molecular weight = 212) will exhibit a defusivity less than benzoate anion (molecular weight = 121) by an amount approximately proportional to the difference in molecular weights.

REFERENCES

1. S. J. Lambert and J. W. Mercer, Hydrologic Investigation of the Los Medanos Area, Southeastern New Mexico, 1977, SAND77-1401, Sandia Laboratories, Albuquerque, NM, Jan. (1978).
2. J. W. Mercer and B. R. Orr, "Geohydrology of the Proposed Waste Isolation Pilot Plant, Southeast New Mexico," U.S. Geological Survey, Water Resources Investigations in Review.
3. J. W. Mercer, U. S. Geological Survey, written communication.
4. D. B. Grove and W. A. Beetem, "Porosity and Dispersion Constant Calculations for a Fractured Carbonate Aquifer Using the Two-Well Tracer Method," Water Resources Research 7 (1), 128-134 (1971).

IN SITU EXPERIMENTS TO SUPPORT DEVELOPMENT OF THE WASTE ISOLATION PILOT PLANT (WIPP)*

A. R. Sattler, H. C. Walker and T. O. Hunter

Sandia Laboratories
Albuquerque, NM 87185

The main thrust of the WIPP in situ activities will be those associated with the WIPP early shaft complex to be planned and developed in the FY80-81 time frame. In this early shaft complex, there will be significant geological investigations. There will also be complexes for the following:

Large-scale rock mechanics design verification
Drifts with the heat load of anticipated waste
Experiments to address technical issues of: (1)

- TRU waste emplacement and retrieval (without radioactivity)
- Non-radioactive high level waste interactions
- Thermal structural experiments
- Stable nuclide migration
- Brine migration
- Permeability
- Borehole plugging
- Operation and design
- Instrumentation development and monitoring

Waste interaction investigations include corrosion, getter (multibarrier) development, nuclide transport (stable nuclides), i.e. those areas where preliminary information can be obtained without radioactivity. This project is (10/79) in the early planning and layout phase.

The borehole plugging program, largely independent of the early shaft, is continuing in southeastern New Mexico for the next few

*This work supported by the U. S. Department of Energy.

years. Initial activities centered about sealing a plug against the high pressure Bell Canyon aquifer (FY79). Activities in FY80 include potash core tests, a shallow wellbore test bank and media quantification experiments in southeastern New Mexico. In the short term, activities are currently underway at the Avery Island mine (in cooperation with ONWI) that include instrumentation development (hole closure device) and in situ performance verification of Sandia-developed stress gages for salt and heater development. Future activities at Avery Island will include scanning of intact salt with radar, a joint Sandia/ONWI investigation of brine migration (shutdown of ONWI heater at Site B) and possibly other activities. For some WIPP experiments, use of the tunnels at the Nevada Test Site are being considered. At present these include particulate and telemetry equipment experiments that are planned for FY80. Some limited access to southeastern New Mexico mines is also being investigated. Tentatively, our efforts at such mines would include a brine migration experiment, some hole closure, scanning of intact salt (radar and resistivity) and whatever other activities that could be arranged with the mine.

Finally, bench scale activities continue and new activities are planned centering around a joint Battelle PNL/Sandia bench scale experiment with real waste in salt. Most of the in situ activities now underway or in planning are summarized in Table 1. An asterisk denotes an activity where experimental work is underway. In these noted cases, results have been obtained from work at southeastern New Mexico mines, from the ONWI-run Avery Island facility in Louisiana and from large bench scale experiments at Sandia. The field work and results to date are briefly mentioned below.

THERMAL/STRUCTURAL

Bench Scale Experiments

A large block of salt, 1 m high and 1 m diameter, was instrumented with a heater in its center with surrounding thermocouples. Information was obtained on: (2)

Thermal conductivity -- isotropic for this block of bedded salt. Strict control must be obtained on external boundary conditions of experiments of this type in order to model adequately. More material data are needed for crushed salt for adequate modeling.

Instrumentation techniques -- Thermocouples with stainless steel sheaths or inconel sheaths gave the same reading side by side. Material used to fill thermocouple holes (suaerisen cement or crushed salt) did not affect the readings.

Corrosion coupons -- Fourteen corrosion candidate coupons, in a special hole in the block, were picked to represent a wide range

TABLE 1. In Situ Activities at the WIPP

	Relevance to WIPP			
	WIPP Oper	WIPP Design	WIPP Expts	WIPP Monit
TRU CHARACTERIZATION (Simulation)				
HLW INTERACTIONS				
Corrosion material and leaching coupons in heater fields			Yes	Yes
Bench scale experiments			Yes	
THERMAL/STRUCTURAL				
Bench scale experiments*			Yes	
Hole closure -- ambient and heated*		Yes	Yes	
Single heater experiments			Yes	
Mine deformation experiments*	Yes	Yes	Yes	Yes
Heated room			Yes	
Heated pillar			Yes	
NUCLIDE MIGRATION				
None (stable isotopes possible in heater fields)			Yes	
Getter experiments*			Yes	
PERMEABILITY				
In situ permeability in mine	Yes	Yes		
Nitrogen gas pocket analysis			Yes	
BRINE MIGRATION				
Bench scale experiments*			Yes	
In situ heater experiments			Yes	
BOREHOLE PLUGGING				
Salt plug tests at mine level		Yes		
Media quantification		Yes		
OPERATION AND DESIGN				
Drilling studies	Yes			
Simulated storage and retrieval/TRU	Yes			
Simulated TRU operation	Yes			
Simulated retrieval of HLW forms	Yes	Yes		
Radiation background*	Yes			
Mine ventilation studies	Yes	Yes	Yes	
Scanning of intact salt*	Yes		Yes	
Moisture exchange/mine vent system/outside air*		Yes	Yes	
Coring in hot salt			Yes	
INSTRUMENTATION DEVELOPMENT				
Thermal probe*		Yes		Yes
Stress and deformation sensors*		Yes		Yes
Borehole closure devices*		Yes		Yes
Heater development*		Yes		

of classes. Included were the following: (1) titanium C.P., (2) titanium-6V-4Al, (3) Hastelloy C-276, (4) Inconel 600, (5) zirconium, (6) Zircaloy-2, (7) Zircaloy-4, (8) SS-20Cb3, (9) SS-Nitronic 50, (10) SS-304, (11) 17-4 pH stainless steel, (12) 70-30 brass, (13) aluminum 6061, and (14) 1018 mild steel. The temperature at the bottom of the coupon hole was maintained between 100° and 150°C for about ninety days. Examination of all coupons confirmed that nothing corroded. The 70-30 brass, 1018 steel and 304 stainless steel coupons had a thin, nonadherent oxide film on the polished surface. The weight change was below experimental detection limits (3).

Deformation Experiments -- Pillar Collapse

The structural deformations following an ongoing mining operation in salt/potash were measured. Information was obtained on:

Structural deformation of rooms and pillars following sequential mining.

Instrumentation techniques -- It was necessary to develop instrumentation to measure large deformations and to develop an absolute level gage to distinguish floor heave from ceiling motion; the first field test of Sandia-developed stress gages for salt was made.

Hole Closure

Hole closure data for a given hole depth, as a function of time, are being taken in conjunction with testing a developmental closure gage.

NUCLIDE MIGRATION

Bench scale measurements of thermal conductivity and contact resistances for WIPP HLW backfill barrier material (clays, sands, salt) are in progress. Results obtained have confirmed that the thermal probe used in this study is accurate and repeatable. Calculated maximum probable errors for all values of thermal conductivity are about 5 percent. Dry backfill compositions and a compaction technique for the backfill mix have been identified. Corresponding thermal conductivities at room temperature were about 0.5 w/m °k, a value that is large enough to justify further consideration of these materials. (This will be presented as part of a separate paper.)

PERMEABILITY

Field data were taken from WIPP horizons from the AEC-7 borehole in support of laboratory data. In these tests compressed air was injected into the borehole and the resulting flow rates into the formation were measured. These flow rates were interpreted in terms of formation permeabilities. A series of tests were conducted using the Systems, Science and Software-designed guarded straddle packer system capable of measuring permeabilities down to 0.1 microdarcy. In these interval permeability tests, 100-foot lengths of borehole were isolated and the permeability of the surrounding formation examined. Tests were conducted at depths of 2270 feet and 1870 feet. Interval permeability test results indicate the 100-foot-thick salt layer centered at the 2264-foot depth has an in situ permeability between 12 and 21 microdarcies. Constant pressure flow measurements yield permeabilities ranging between 18 and 21 microdarcies. These tests were conducted for six- to eight-hour periods during which time the air flow should penetrate to a radial distance of approximately ten feet. Constant pressure flow measurements indicated the 100-foot-thick salt layer, centered at a depth of 1869 feet, had a permeability of three microdarcy (gas penetration approximately three feet). When this test interval was shut in, there was no observable pressure decay. This later response is not resolved at this time (4).

BRINE MIGRATION

A bench scale experiment (5) consisted of a large salt block, 1 m high and 1 m in diameter, with heater in the center. The heater core was flushed with nitrogen gas and the water content in the exhaust was instrumented with two independent techniques. Water release from the heated borehole was measured by continuously purging the sealed borehole with very dry nitrogen gas and measuring the water content in the exhaust gas with two independent techniques. Data were automatically recorded by a computer based data acquisition system. The heating of the salt block started on November 28, 1978, and terminated on April 16, 1979, after a series of steps at decreasing heater power. At the maximum heater power of 1.5 kW, the maximum temperature and temperature gradient at the surface of the heater hole were approximately 200°C and 12° c/cm, respectively. The total amount of water released from the heater hole during the course of the experiment was about 100 grams -- a small fraction of the estimated two to eight kilograms of water present in the salt block. Over 40 percent of the total water released occurred during the shutdown phase. The average and peak rates of moisture release during this phase were much higher than during either "power up" or the "steady state" phases. A similar surge of moisture was observed during a thirty-minute power outage. During the 69-day period when the heater power was held at 1.5 kilowatts, moisture was released at a monotonically decreasing rate. Complete data analysis and analytical modeling of Salt Block II have not been completed. Analyses using both fluid inclusion migration and vapor phase transport are being used. The former may be able to predict moisture release quantities during "power up" and "steady state" but cannot explain release rates during shutdown. The latter may be more appropriate to these transients. Mineralogical post-mortem analyses of Salt Block II are underway. In one examination, a slab of rock 1 cm thick was cut from a four-inch core taken adjacent to the heater hole. Transmitted light optical microscopy revealed the presence of elongate tracks of fluid inclusions, oriented radially with respect to the heater hole. Many of these tracks are terminated (at the "heated" end) with structures resembling nailheads, walking-stick heads and spindles. These tracks, together with the occurrence of carnallite encrusting the heater hole surface, indicate that a significant fraction of the 111 grams of water recovered in the Salt Block II experiment made its way to the heater hole in the liquid phase.

OPERATION AND DESIGN

Radiation Background

Background radiation measurements were taken in potash and salt mines. Data are being analyzed.

Scanning of Intact Salt (By Resistivity and Radar Techniques)

The resistivity techniques center on ideas taken from the Asse mine. The first resistivity measurements using four-point probe revealed the potential of the method (6). Resistivities vary from less than 100 OHM-meters, when "wet" salt or salt with clay impurities is present, to more than 60,000 OHM-meters when dry salt is encountered. Preliminary radar scanning measurements using reflected encountered. Preliminary radar scanning measurements using reflected radiation, under controlled conditions, could reveal voids, brine and metal in intact salt. In this first measurement, the ranges were short, approximately 50 feet or less, but they will be increased in subsequent measurements.

Moisture Exchange

Moisture exchange between the mine ventilation system and the outside air have been done under different conditions. In one instance, temperature pressure and relative humidity measurements have also been taken underground near the bottom of the inlet and exhaust shafts during the period from February through April, 1979. The humidity and temperature in the exhaust air showed little variation, essentially remaining constant around 35 percent and 23.5°C, respectively. The intake air showed some variation during the course of the experiments, and a range of 23 percent to 61 percent was observed on the relative humidity. The temperature at the mine level in the intake showed only a modest variation between February and April (11° to 23°C), and the daily variation was only 3° to 4°C. Because of the influence of salt particles in the air, proper experiment techniques for relative humidity were not easily developed, especially for the dewpoint hygrometers. These measurements and an earlier measurement show that certain mines in southeastern New Mexico can give up moisture to the outside, at least at certain times of the year. Various configurations of filters were investigated, ultimately establishing a reproducible, well calibrated system. In another instance, manual measurements were made at Avery Island for about a year for intake and exhaust air. The main variations are, as expected, on the inlet side. Data from these activities will ultimately yield heat transfer coefficients for typical shafts as well as flow coefficients when surface meteorological data are factored in. The data also suggest that conditioning of inlet air may be impractical.

Particulate Studies

The air particulate studies were performed during mining activities in a SENM mine and at a surface drill site during drilling operations (7). The underground data were collected about 2 km from the mining operations, while the surface measurements were made on the drill pad adjacent to the rig. Particle size distribution and total suspended material were measured. Data were taken using Anderson impactor samplers. The average particulate load on the surface (approximately 150 microgram/m^3) was less than that observed underground (approximately 340 microgram/m^3). Yet, during very dusty conditions, the surface level was observed to exceed 800 microgram/m^3. The surface site measurements could represent particulates ingested by the mine ventilation system under certain conditions.

INSTRUMENTATION DEVELOPMENT

Thermal Probe

A simple thermal probe to measure conductivities has been tested in the laboratory and in the Avery Island Mine. Data are being analyzed.

Stress and Deformation Sensors

A Sandia-modified IRAD gage and a Sandia strain gage stressmeter are undergoing in situ verification tests after earlier successful preliminary testing both in the laboratory and in the field (8).

Borehole Closure Devices

Two types of borehole closure devices have been developed: mechanical and capacitive. The early version of the mechanical gage, mainly for ambient closure measurements, is being tested at Avery Island and appears to be quite successful (see Hole Closure). A conceptual version of the capacitive gage, to be used for on-line closure measurements in a heater experiment (modeled after the one used at Asse), has been built, but field testing must await the design of a specific heater test.

Heater Development

A new heater concept using quartz lamps will be tested at Avery Island. Moisture collection apparatus may be included in that test as a target of opportunity.

REFERENCES

1. A. R. Sattler and T. O. Hunter, Pre-WIPP Experiments in Salt, Parts I and II, SAND79-0625, Sandia Laboratories, Albuquerque, NM, Aug. (1979).
2. T. A. Duffey for A. R. Sattler, "Final Report for the Salt Block I Test" (in preparation).
3. J. S. Braithwaite and M. A. Molecke, High Level Waste Canister Studies Pertinent to Geologic Isolation, SAND79-2111, Sandia Laboratories, Albuquerque, NM, Dec. (1978).
4. T. O. Hunter, Survey of Research and Development Activities in Support of a Waste Acceptance Criteria, SAND79-1035, Sandia Laboratories, Albuquerque, NM, Draft, June (1979).
5. J. J. Hohlfelder and H. C. Shefelbine, "Salt Block II Experiment Overview" (in preparation).
6. C. L. Elliot, "Preliminary Data and Comments, Underground Electrical Resistivity Experiments," Report to F. F. Dean, Feb. (1979).
7. L. W. Brewer, "Air Particulate Sampling at AEC-7 Site" (in preparation).
8. C. W. Cook and E. S. Ames, "Borehole Inclusion Stressmeter Measurements," Proceedings of 20th U. S. Symposium of Rock Mechanics, Austin, Texas, June (1979).

SHALE ROCKS AS NUCLEAR WASTE REPOSITORIES: HYDROTHERMAL REACTIONS WITH GLASS, CERAMIC AND SPENT FUEL WASTE FORMS*

W. Phelps Freeborn, Michael Zolensky, Barry E. Scheetz, Sridhar Komarneni, Gregory J. McCarthy** and William B. White

Materials Research Laboratory, The Pennsylvania State University, University Park, PA 16802

INTRODUCTION

The objectives of various contributions from this laboratory have been to simulate "worst case" situations, given a proposed choice of waste form, repository rock, and waste loading/waste age. The "worst case" seems to be the situation in which either young waste or high waste loadings produce a high temperature in a repository in which a fluid phase is present.

The pervasive assumption through most of this work is that the canister is somehow breached allowing chemical interaction between the waste form, the canister, and the repository rock all through the medium of some hot, highly mobile fluid. Most of our earlier studies have focused on the situation inside the canister and the reactions and extractions that might be expected if the waste form were to be in contact with a hydrothermal fluid. Chemical reactions with the repository rock also occur and the present report is devoted to this subject.

The present contribution deals with the reactions of four prototype waste forms which were mixed with deionized water and one of six reference shales. These mixtures were reacted at 100°C - 400°C at a pressure of 30 MPa. The main results reported concern the interaction

*Research Supported by DOE Through Battelle Office of Nuclear Waste Isolation, Contract E512-00500

**Present Address: Depts. of Chemistry and Geology, North Dakota State University, Fargo, ND 58105

Table 1. Chemical Composition* of Reference Shales

Shale	SiO_2	Al_2O_3	Fe_2O_3	FeO	S	H_2O^-	H_2O^+	C	CO_2
Antrim	59.8	12.9	3.27	2.87	2.88	1.25	6.72	6.28	0.19
Braillier	57.7	21.1	5.27	2.79	0.01	0.92	4.21	0.05	0.12
Catskill	67.3	13.5	4.01	2.63	0.01	0.64	2.46	0.10	0.49
Conasauga	54.7	21.3	4.15	4.59	0.63	0.96	4.97	0.31	0.26
Eleana	54.3	24.3	0.67	5.15	0.06	1.12	5.57	0.96	0.99
Salona	47.0	12.3	1.66	0.57	0.02	1.84	3.03	0.61	11.69

*Partial analyses. See (1). All compositons in weight percent

product phases, the role of alumino-silicate minerals in controlling the release of cesium in shale rocks, and the role of oxidation potential in controlling the migration of heavy metal radionuclides.

EXPERIMENTAL

No shale repositories have, as yet, been selected although the Eleana shale which underlies parts of the Nevada Test Site and the Conasauga shale which underlies parts of the Oak Ridge Reservation are considered of interest. Specimens of these shales and four others (Table 1) were selected to exhibit a range of chemical and mineralogical compositions. The waste forms were a simulated PW 7a calcine, PNL 76-68 borosilicate glass, an SPC-2 supercalcine ceramic, and two simulated spent fuels. One "spent fuel" consisted of 96.5 wt percent UO_2 with 3.5% simulated fission products; the other contained 60% UO_2 and 40% simulated fission products. Complete analyses of the prototype waste forms are given in Reference 1.

The waste forms and rock were ground and reacted in sealed gold tubes in the presence of water (typical weight ratio 1:1:1) for four weeks at temperatures of 100, 200, 300 and 400°C. At the end of the reaction period, the runs were quenched, the capsules opened and the contents analyzed. Observations of the residual solid products were made with the scanning electron microscope. Phases were identified by bulk X-ray diffraction and by determination of powder diffraction patterns of individual tiny reaction product crystals using the Gandolfi technique.

WASTE-ROCK INTERACTIONS

The evidence for waste-rock interactions is the appearance of small crystals of phases not present in the original waste-rock mix-

Table 2. Mineralogical Composition of Reference Shales

Shale	Description	Mineralogy
Antrim	Very fine-bedded dark grey shale with good foliation. Scattered masses of sulfides distrupting the bedding.	Quartz dominant, illite, chlorite, albite, and pyrite
Braillier	Olive-green shale with good foliation.	Quartz dominant, illite, chlorite and albite.
Catskill	Red, silty mudstone with indistinct bedding and foliation.	Quartz, albite, illite (= muscovite of microscopic description), chlorite and trace of calcite.
Conasauga	Finely bedded, heterogeneous siltstone, mixed red, green amd white beds.	Quartz dominant; with chlorite, albite, and illite and/or glauconite.
Eleana	Very dark grey massive siltstone with weak foliation.	Quartz, chlorite, illite, and pyrophyllite with lesser dolomite, potassium feldspar and albite.
Salona	Massive dark grey calcareous mudstone. No bedding or foliation.	Quartz and calcite dominant, albite and illite in lesser amounts.

ture. External morphologies of these products (Fig. 1.) were observed with the SEM; EDX and Gandolfi X-ray diffraction techniques permitted chemical and structural identifications. There are three sources for these reaction products: i) chemical reaction between the waste and the shale, ii) recrystallization of the waste form and iii) modification of the shale mineralogy due to higher temperatures.

The relative reactivity of the waste form in terms of new phase formation was calcine > spent fuel > glass > ceramic. At 100°C all reaction products had the appearance of wet mud. New crystalline interaction phases could be discerned easily in the 200°C runs. At 300° and 400°, extensive interactions were visible.

At 100°C the only apparent interaction is the enhancement of clay mineral diffraction peaks implying some improvement on the

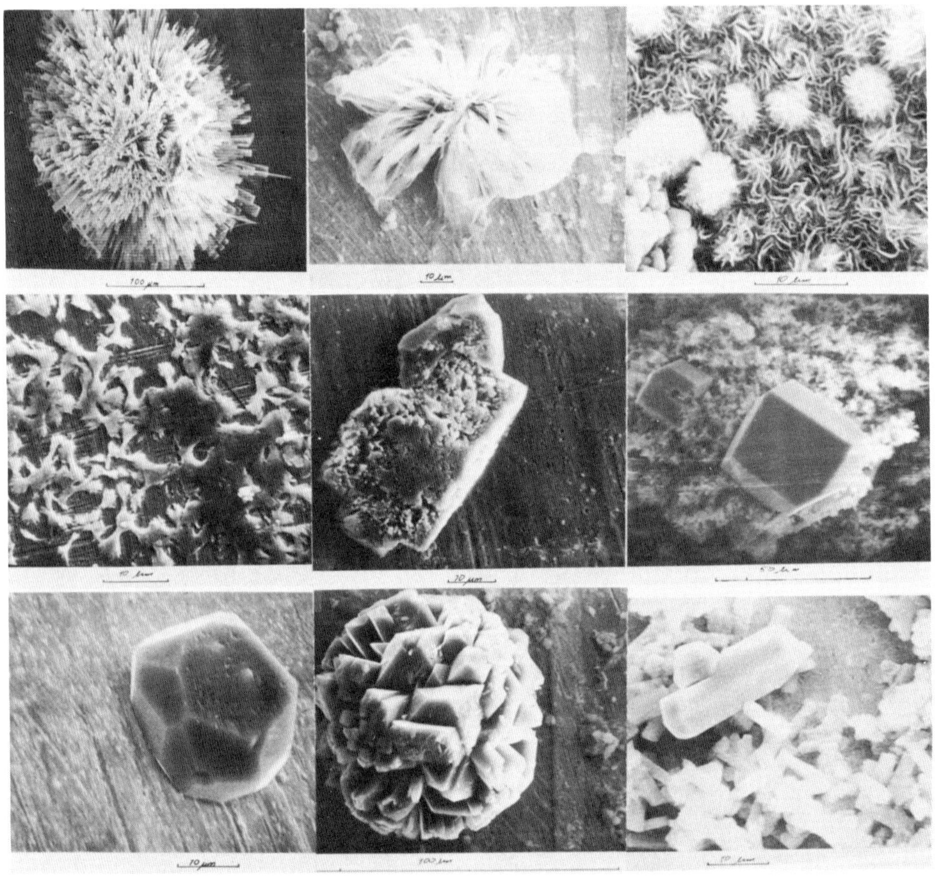

Fig. 1. Scanning Electron Microscope Images of Waste-Rock Interaction Products. Reading left to right and top to bottom: Weeksite, $(Na,K,Cs)_2(UO_2)_2(Si_2O_5)_3 \cdot 4H_2O$, Calcine + Salona, 300°C; Uranophane, $Ca(UO_2)_2Si_2O_7 \cdot 6H_2O$, Glass + Eleana, 200°C; Sauconite, $Na_{0.33}Zn_3(Al,Si)_4O_{10}(OH)_2 \cdot 4H_2O$, Glass + Brallier, 200°C; Xonotlite (?) $Ca_6Si_6O_{17}(OH)_2$, Glass + Antrim, 100°C, Powellite, $(Ca,Sr)MoO_4$, Spent fuel + Salona, 300°C; Willemite, Zn_2SiO_4, Glass + Catskill, 300°C; Pollucite, $CsAlSi_2O_6$, Calcine + Salona, 300°C; Analcime, $NaAlSi_2O_6 \cdot H_2O$ Glass + Eleana, 200°C; Albite, $NaAlSi_3O_8$, Glass + Eleana, 400°C.

crystallinity. At 200°C and above, new crystalline phases appear, the assemblage depending on the waste form, the accessory minerals in the shale and to a lesser extent the temperature. The principal shale minerals, quartz and clays, disappear at higher temperatures, partly by reaction with the waste materials and partly by recrystallization into feldspars and micas.

Weeksite, $(Na,K,Cs)_2(UO_2)_2(Si_2O_5)_3\ 4H_2O$, is the principal U-containing phase and appears as an interaction product between calcine, spent fuel, glass and to a lesser extent, supercalcine with all shales except the highly reducing Antrim. Other hydrated alkali uranyl silicate phases occur occasionally. Zinc silicates, mainly willemite, Zn_2SiO_4, occurs in many glass interaction products because of the reaction of zinc extracted from the glass with silica from the shale. Powellite, $CaMoO_4$, is a common phase formed by reaction of molybdenum from the waste with calcium present in the shales (or in the wastes). Pollucite is a common phase in all reaction products (it is a component of the ceramic waste form). Cesium is easily extracted and reacts readily with the aluminum and silicon-bearing minerals in the shales. In all, some 20 new mineral phases were recognized, some of them occurring infrequently.

THE ROLE OF ALUMINO-SILICATE MINERALS

Silica and alumina dominate the composition of shales as they also dominate the compositions of most igneous rocks. The ratio of alumina to alkali metals in shales is usually greater than 1:1 whereas in igneous rocks it is close to 1:1. Alumina occurs in shales as clay minerals, alumina-rich and alkali-depleted, with some feldspar whereas in igneous rocks alumina occurs almost exclusively in feldspars. Since shales are alkali-poor they should act as scavengers for alkali elements which would react with the clays to form framework silicates --zeolites or feldspars--with concurrent shift in the alumina:alkali ratio toward 1:1.

Reactions of alkali metals with shales occurs in the present experiments as the pervasive appearance of pollucite in many different combinations of shale, waste form, and temperature. There is no aluminum in any of the waste forms except for the SPC-2 supercalcine, so that the pollucite observed must have been formed by reaction between the shale and the waste.

Pollucite appears as a dominent reaction product in the 200°C runs but the pollucite lines become weaker in the diffraction patterns of the higher temperature reaction products. Does this mean that pollucite is again breaking down at higher temperatures and releasing cesium?

Combining experimental studies on analcime with what is known about the analcime-pollucite solid solutions, an interpretation of the experimental results can be given. Fig. 2. shows a series of isothermal sections through the system nepheline-$CsAlSiO_4$-quartz. The interpretation of the experimental results following this paragenetic scheme suggests that the first, abundant analcime-pollucite is quite silica-rich (a result in accord with semi-quantitative SEM/EDX analyses) and that with increasing temperature this phase shifts in composition toward an aluminum to silica ratio nearer to the nominal 1:2 for the pollucite phase. Continued increase in temperature results in the reaction of analcime with quartz producing feldspar and a new member of the analcime-pollucite solid solution, richer in the pollucite component than the original. Thus the cesium is not released; it is concentrated in the remaining pollucite, which is considerably reduced in abundance.

OXIDATION STATES AND OXYGEN BUFFERS

Many of the potential reactions in an underground nuclear waste repository involve a change in the oxidation state of one or more of

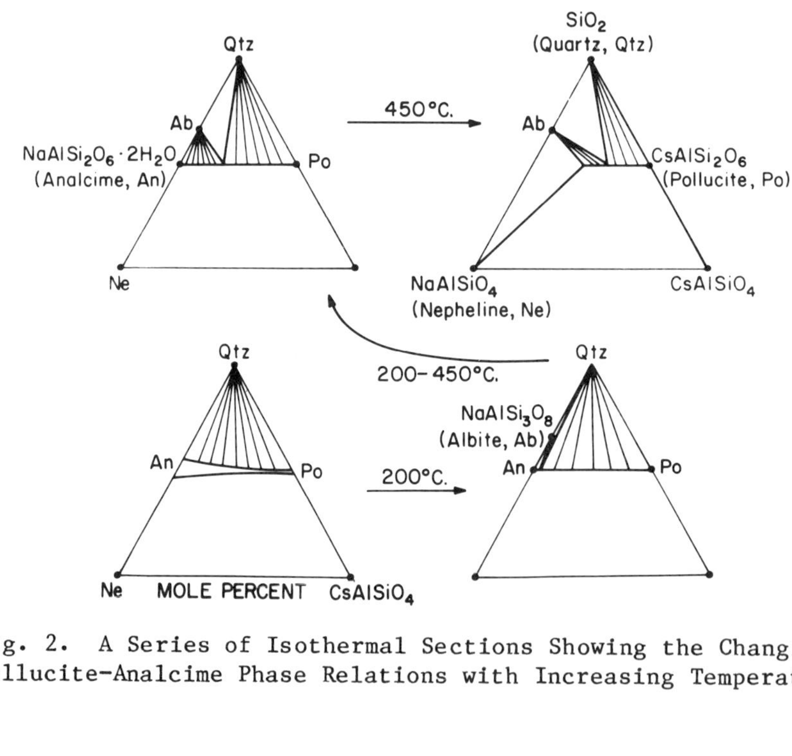

Fig. 2. A Series of Isothermal Sections Showing the Changing Pollucite-Analcime Phase Relations with Increasing Temperature

the elements involved. The mineral phases formed by these elements will be determined by the oxidizing or reducing potential of the rock and this in turn will be determined by the kind and abundance of oxidizer or reducer minerals in the rock. The principal constituents of shales, quartz, clay minerals, feldspars, are neutral in this respect.

The oxidation state is important because the solubility, and hence the mobility, of an element is different for different oxidation states. The solubility for uranium ranges from quite insoluble (less than 1 ppb) for uraninite with U^{4+} to weakly soluble (a few hundred ppb) for schoepite with U^{6+} (2). The ability of a shale to reduce uranium from 6+ to 4+ oxidation states would then also reduce the chance for escape of uranium and by implication the transuranic elements.

The evidence for the oxidation of uranium in the experiments reported here is the ubiquitous occurrence of weeksite. It occurs very commonly in reaction products between calcine or glass and the Salona, Catskill and Brallier shales which are poor in reducing accessory minerals. Weeksite does not occur or occurs very sparsely in the reaction products with the Antrim shale which is rich in or-

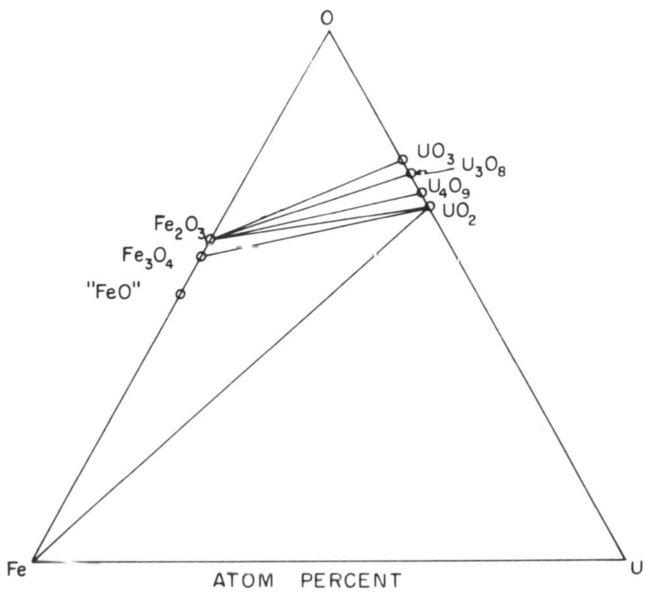

Fig. 3. Phase Relations for the System Fe-U-O Calculated from the Data of Naumov et al (3). Applicable to the Temperature Range 300-800°K.

ganic carbon and sulfides. Likewise, weeksite is generally absent from the reaction products of the Eleana and Conasauga shales.

It seems quite clear that highly reducing materials such as organic carbon and iron sulfides are sufficient to prevent the oxidation of uranium. The remaining question is whether hematite, which is the only redox oxide in many of the shales, can oxidize U^{4+}. Fig. 3. shows the phase relations. Uraninite is stable in the presence of Fe_2O_3 and indeed is found in association with hematite in geological environments. Thus iron oxide alone is not sufficient to oxidize uranium in the waste form although, as the phase diagram shows, it would also not reduce any uranium already present in the hexavalent state.

CONCLUSIONS

Prototype nuclear waste materials react with shale rocks in the presence of water to form at least 20 reaction product phases. These materials act to buffer or control the further release of radionuclides beyond the immediate reaction zone. In particular, aluminosilicate minerals act as scavengers for cesium and strontium which become immobilized as pollucite, powellite, and feldspar. Uranium and by implication transuranic chemistry is strongly influenced by the redox characteristics of the accessory minerals in the rock.

In essence, shales act as chemical barriers as well as physical barriers to the migration of radionuclides from the immediate vicinity of a waste canister.

REFERENCES

1. W. P. Freeborn, M. Zolensky, B. E. Scheetz, S. Komarneni, G. J. McCarthy and W. B. White, Hydrothermal Interactions Between Calcine, Glass, Spent Fuel, and Ceramic Waste Forms with Representative Shale Repository Rocks, ONWI/Sub/78/E-512-005, Topical Report 3 (in press).
2. D. Langmuir, "Uranium Solution-Mineral Equilibria at Low Temperatures with Applications to Sedimentary Ore Deposits," Geochim. Cosmochim. Acta 42,547-569 (1978).
3. G. B. Naumov, B. N. Ryzhenko and I. L. Khodakovsky, Handbook of Thermodynamic Data, (trans. Russian Rpt.) NTIS PB 226 722, 327 (1974).

HYDROGEOLOGIC CHARACTERISTICS OF FRACTURED ROCKS FOR WASTE ISOLATION - THE STRIPA EXPERIENCE *

J. E. Gale[1], O. Quinn[1], C. Wilson[2], C. Forster[1], P. A. Witherspoon[2], L. Jacobson[2]

[1]Department of Earth Sciences, University of Waterloo, Ontario N2L 3G1; [2]Lawrence Berkeley Laboratory, University of California at Berkeley, Berkeley, CA 94720

INTRODUCTION

Fractured crystalline and argillaceous rocks have been proposed as alternative host rocks for storage or disposal of high level radioactive waste. In a properly selected repository site the matrix porosity and permeability of these rocks should be such that no flow can occur through the rock blocks and hence the fracture system will provide the only pathway for radionuclides to migrate away from the repository. Our knowledge of the permeability of fractured rock masses has, until recently, been limited to borehole investigations in the upper 100 meters of the earth's crust. In general, these investigations have been conducted under the assumption that fractured rock masses can be treated as a slightly different form of porous media. This assumption has been carried forward into the more recent studies of the hydrology of fractured rocks conducted at depths of up to 1000 meters as part of nuclear waste disposal programs.

Nuclear waste disposal programs require an accurate description of the hydrology of fractured rocks. Thus one must begin to develop data bases that permit one to answer questions such as: 1)what is the role of fractures in determining the nature (isotropic or anisotropic) of fractured rock permeability? and 2) under what conditions if any, can fractured rock masses be treated as porous media or

*Work done under University of California Contract W-7405-ENG-48 for the U. S. Department of Energy.

equivalent porous media? The first question dictates that we must develop methods of characterizing a fracture system and its role in determining the hydrology of fracture systems in order to provide a framework within which to interpret local and large scale flow systems in fractured rock masses. Answers to the second question determine the type of borehole testing programs that will be undertaken in concept verification studies. Borehole testing programs must provide the data needed to develop hydraulic parameters that clearly describe how fluids move through fractured rocks. Testing programs based on porous media concepts will not provide sufficient data, if as is generally agreed, the permeability of fractured rocks is highly anisotropic.

Fig. 1 is a reasonable, though not necessarily complete, representation of the main pathways by which fluids move through fractured rock masses. In most cases it should be possible, once their location has been determined, to measure the hydraulic properties of the major shear and fracture zones by direct borehole tests. Determining the hydraulic properties of joints requires a different approach. This is a direct result of the relatively close spacing of joints compared to the major features and their distribution of orientations and spacings within a set. Also, in any rock mass the fracture or joint system consists of several sets all of which are involved in the flow of fluids. One approach to determining the hydraulic properties of the joint or fracture system is currently being pursued as part of a cooperative United States-Swedish research program on the fracture hydrology and thermo-mechanical characteristics of a fractured granite at the Stripa Mine in Sweden.

The main purposes of this paper are to describe the approach being taken to assess the directional permeabilities of the fractured granite at Stripa, to present typical results from the borehole

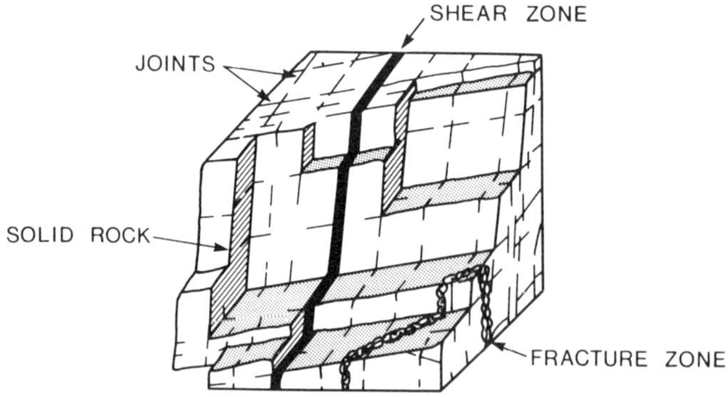

Fig. 1. Conceptual Model of Main Flow Paths in Fractured Rocks

testing program and to discuss the implications of these field test results for nuclear waste storage in fractured rocks.

General Geology and Borehole Locations

The Stripa mine, located in South Central Sweden, is an iron ore mine whose reserves have been exhausted. Excavations for the thermal and hydrology experiments have been mined in a medium grained granite under the north limb of an east-west plunging syncline of metamorphic rocks. The geology of the site has been described by Olkiewicz et al (1). Fig. 2 shows the distribution of outcrops in the immediate area of the test excavations and the location of the hydrology boreholes. The test excavations projected into plan view in Fig. 2 are located approximately 338 meters below the ground surface.

The surface boreholes consist of six water table wells, WT-1 through WT-6, a pump test well WT-7, and three long inclined boreholes SBH-1, SBH-2 and SBH-3 (Fig. 2). SBH-1 is an open, 76 mm diameter, diamond corehole, 385 meters in length that angles downward at 45° and passes over the top of the test excavations and terminates

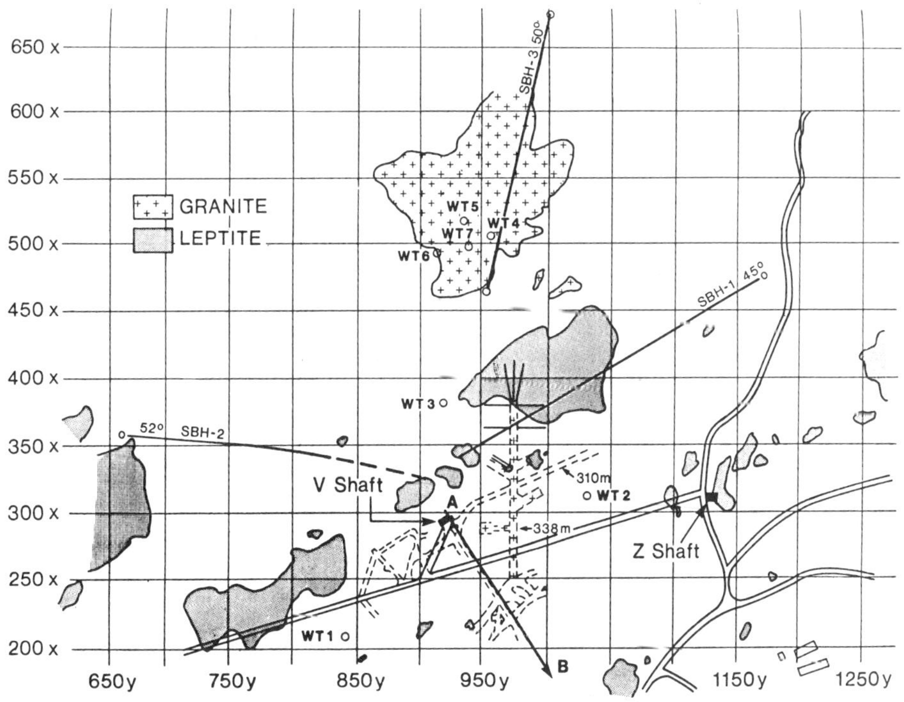

Fig. 2. General Geology and Hydrology Borehole Locations (Squares are 50 Meters on a Side)

at approximately the 290 meter level. SBH-2, also diamond cored, was drilled from the west towards the test excavations. This borehole is 365 meters in length, angles downward at 52 degrees and terminates in the position shown in Fig. 2 at approximately the 290 meter level. SBH-3, 315 meters in length, also diamond cored, is drilled from the north at an angle of approximately 50° south towards the test excavations, terminating in the position shown. All three inclined surface boreholes were oriented to optimize their intersection with the major fracture sets.

The subsurface hydrology boreholes (Fig. 3) are located at the north end of the test excavations. This group of boreholes are all diamond coreholes, 30 meters in length and 76 mm in diameter, except R-1 and R-6 which are 40 meters in length.

The geological log of the drill core, distribution of major fracture zones, RQD values and the bottom hole porewater pressures for SBH-1 are presented in Fig. 4. The porewater pressures, measured in a bottom hole, 3 to 4 meters long, cavity sealed off with an inflatable packer as the drilling advanced, show that at 100 meters or so below the surface the mine has reduced the water pressures in the rock mass.

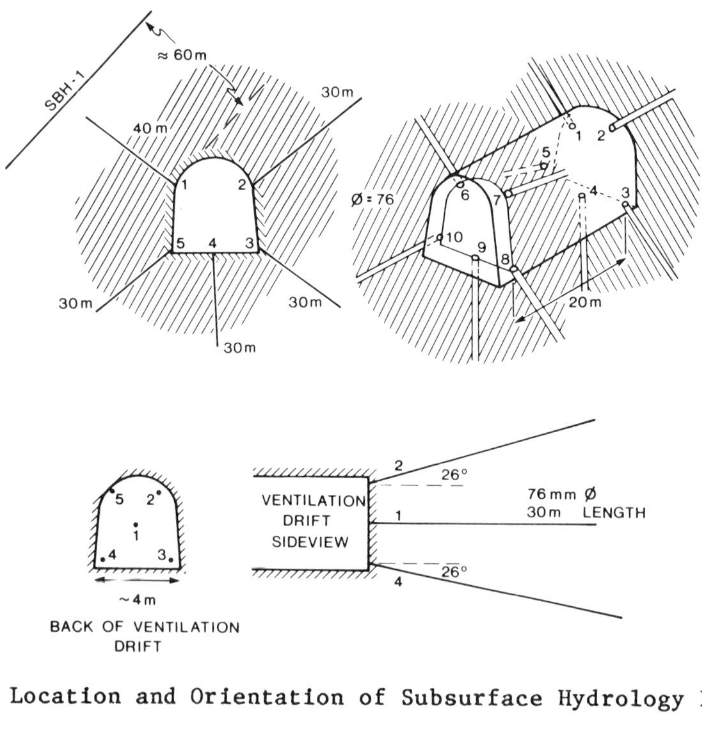

Fig. 3. Location and Orientation of Subsurface Hydrology Boreholes

APPROACH TO FRACTURE HYDROLOGY STUDIES AT STRIPA

The overall fracture hydrology program at Stripa (2) is summarized in Fig. 5 and consists of five basic field activities: 1) Assessing directional permeabilties; 2) geochemical and isotope studies; 3) tracer studies; 4) an underground ventilation experiment to measure rock-mass seepage and; 5) pump tests. The geochemical and isotope studies and the underground ventilation experiment are discussed in companion papers presented at this conference. The tracer studies have not been initiated at this time. Thus this paper deals primarily with the first part of the program, assessing directional permeabilities of the Stripa granite, which is the major activity of the fracture hydrology program.

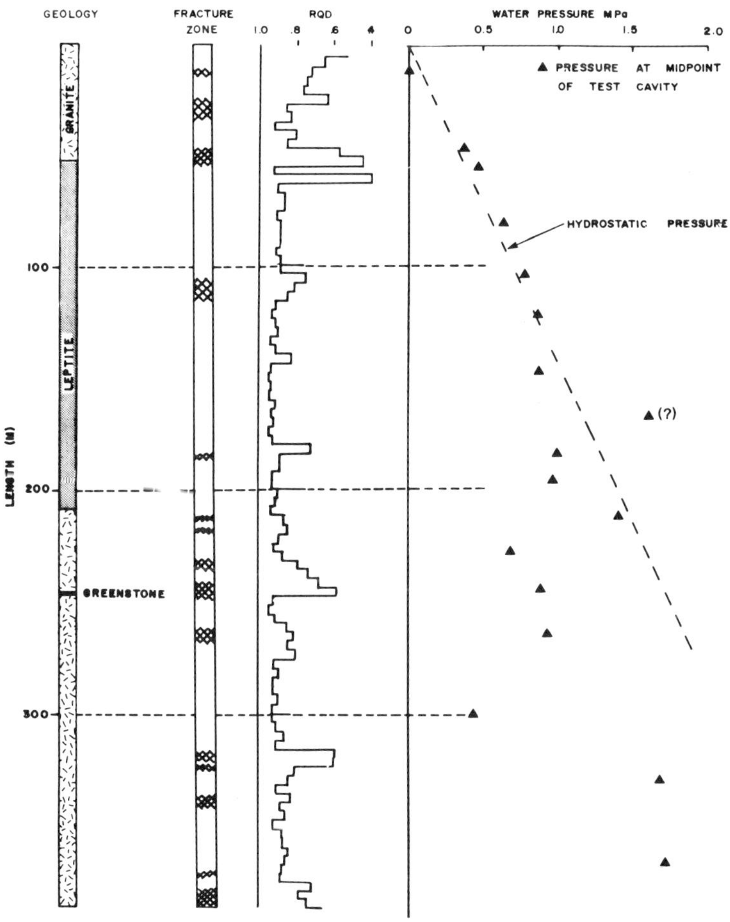

Fig. 4. General Geology, Fracture Zones, RQD Values and Bottom Hole Fluid Pressures, Measured During Drilling, for SBH-1

The mathematics of calculating directional permeabilities from fracture orientation and aperture data, using the parallel plate analogy of fracture flow, was first published by Romm and Pozinenko (3). Extensive work in this area has been performed by other workers (4,5,6,7). A related contribution on fracture orientation and density analysis has also been made (8). The basic approach consists of developing a permeability tensor from measured orientations and spacings of the fractures and assumed aperture distribution models. Principal permeabilities and their directions can be calculated from the eigenvalue and eigenvectors of the tensor following procedures outlined by Westergard (9).

The basic approach to assessing the directional permeabilities of the fractured granite at Stripa is based on this earlier work. In the work at Stripa, an attempt is being made to incorporate the effects of fracture continuity and fracture interconnection into the directional permeabilities calculations. Basic data on fracture orientations, spacings and continuity have been obtained by mapping the fractures in the surface outcrops and in the walls and floors of the subsurface excavations. Careful core drilling, core reconstruction and core orientation have been carried out in order to determine the variation in fracture geometry within the rock mass. The surface and subsurface data have been combined with the borehole data in an attempt to define the three-dimensional fracture system. The granite is highly fractured. In most parts of rock mass, four fracture sets are present. Borehole TV camera logs provide a check on fracture orientations obtained from oriented core studies.

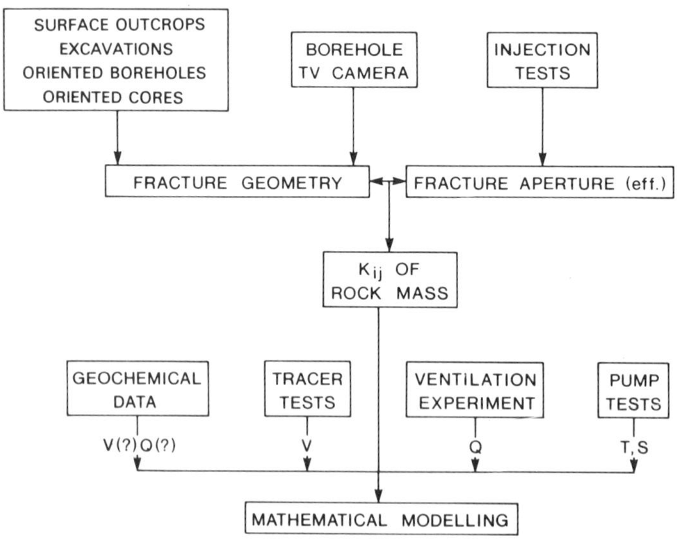

Fig. 5. Block Diagram of Fracture Hydrology Program at Stripa

The borehole injection test program is designed to provide information on the distribution of effective fracture apertures. The basic test equipment consists of a two packer assembly with downhole pressure and temperature probes. During each test the fluid pressures were measured both in the injection zone and above and below the injection zone. In all of the boreholes, packer tests were carried out using 1) a fixed packer spacing of two meters along the entire length of the boreholes; 2) packer spacings sufficient to isolate and test single fractures and; 3) packer spacing of approximately six to ten meters to test the main fracture-geophysical zones. This testing program enables one to develop fracture aperture distribution data for different parts of the rock mass from a statistical analysis of the fluid pressure, flowrate and fracture frequency data. In addition, apertures calculated from the single fracture tests provide a check on the fracture aperture distributions calculated from the systematic testing program. As shown in Fig. 5, the aperture data collected from this testing program can be combined with the measured fractured orientations and spacing to calculate the directional permeabilities of the rock mass.

Preliminary Results from Borehole Testing Program

The results of injection tests in 2 meter intervals in two sections of SBH-1 are shown in Fig. 6. Flowrates for these two meter packer spacings ranged from <0.01 ml/min. to approximately 3 ml/min. for both the 210 meter to 240 meter section and the 325 meter to 355 meter section.

In the subsurface boreholes, the excess pore pressures in the rock mass at the depth of the excavations resulted in flow out of the boreholes. Typical outflow rates, ranging from <0.001 ml/min. to about 6 ml/min., and in situ fluid pressures measured in two meter packer intervals over a 30 meter long borehole (R4) are given in Fig. 7. This borehole, R4, is drilled vertically downward from the floor of the ventilation drift.

The RQD values, as a rough measure of the degree of fracturing in the drill core, and a strip log showing the location of each fracture in the core are given in Figs. 6. and 7. for the two sections in SBH-1 and R4, respectively. Also the equivalent porous media hydraulic conductivity values for each 2 meter section have been calculated using the measured flow rates and head differentials at the wellbore. A radius of influence of 10 meters was assumed for these calculations.

Dewatering of the instrument and heater boreholes during the thermal experiments showed flowrates similar to those given for R4 and parts of SBH-1. Fig. 8 shows the volumes of water in terms of ml/min. pumped from two heater holes in the time-scaled heater experiment room. (The thermal experiments at Stripa are discussed in other papers presented at this conference). These boreholes are about 11 meters in length and the average rate of inflow, over several months, of 1 to 2 ml/min. represents the flow out of a fairly large volume of rock. Although injection rates up to 500 ml/min. have been measured in some test zones, the flowrates given in Figs. 6, 7 and 8 are generally representative of most of the granite rock mass at Stripa.

A detailed interpretation of the data presented herein in terms of the fracture hydrology of the Stripa granite, is beyond the scope of this paper. As a form of preliminary interpretation it is instructive to consider the measured flowrates in the context of the data presented in Table 1. In Table 1, we have calculated porous media hydraulic conductivities for a range of flowrates, subject to a given set of boundary conditions. If instead of a porous media, we have, as may very well be the situation, a single fracture intersecting the two meter test interval we can calculate the hydraulic conductivity of this single fracture. As shown in Table 1, there is a

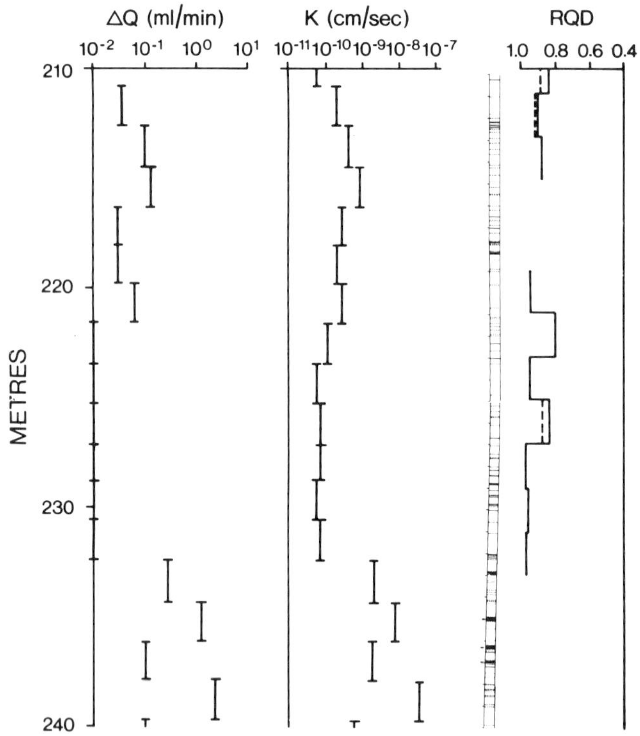

Fig. 6a. Injection Test and Fracture Data from SBH-1 210 m to 240 m Section

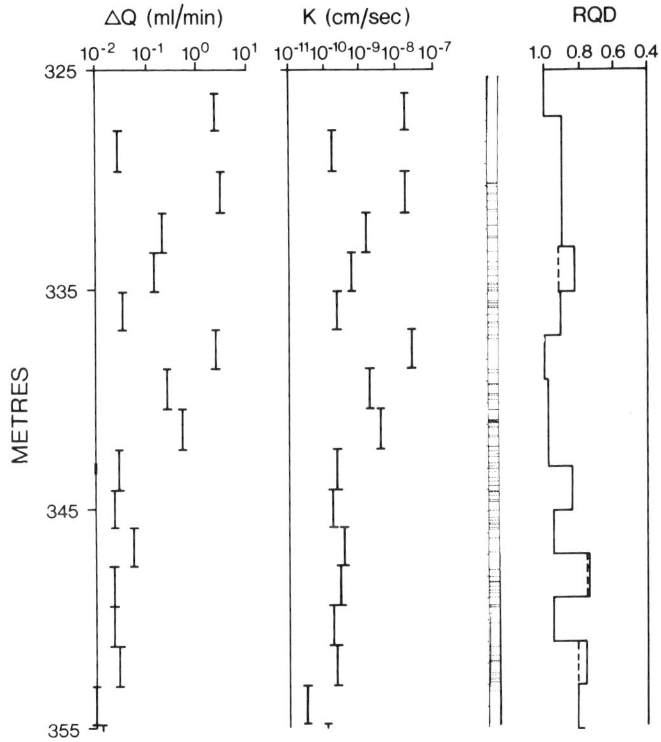

Fig. 6b. Injection Test and Fracture Data from SBH-1 325 m to 355 m Section

considerable difference in the impression created depending on whether we choose to express the results in terms of fracture or equivalent porous media hydraulic conductivities. One important point to remember is that the volume of fluid moving through the rock mass is generally independent of the model chosen to interpret the field data but that the rate of movement and the nature of the pathway followed is not.

SUMMARY

The results of borehole tests in fractured rocks are subject to widely different interpretations. Based on the results of the borehole testing program at Stripa the volumes of water moving through the fractured granite appears to be very small. Any attempt at calculating the rate at which the water is moving through the granite

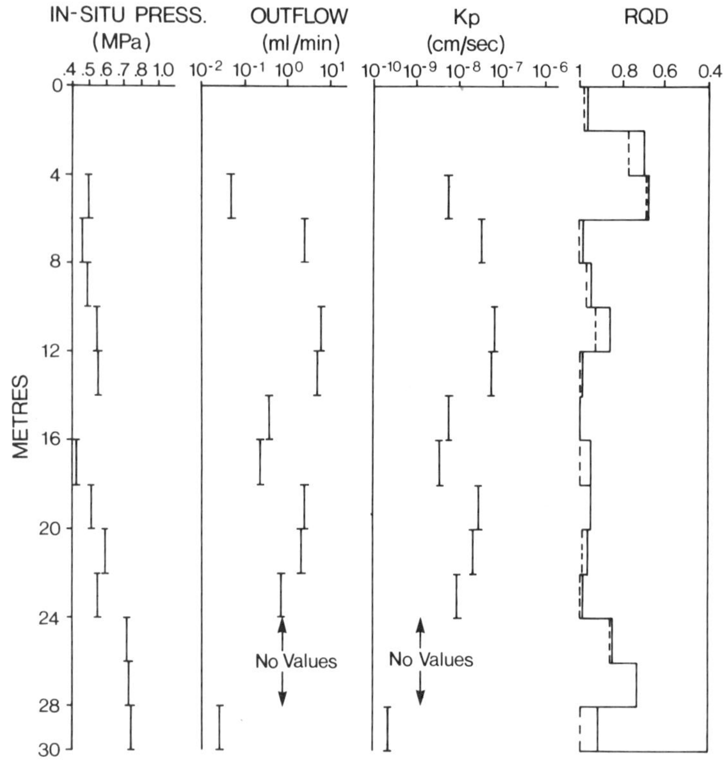

Fig. 7. Fluid Withdrawal and Fracture Data From R4 Borehole

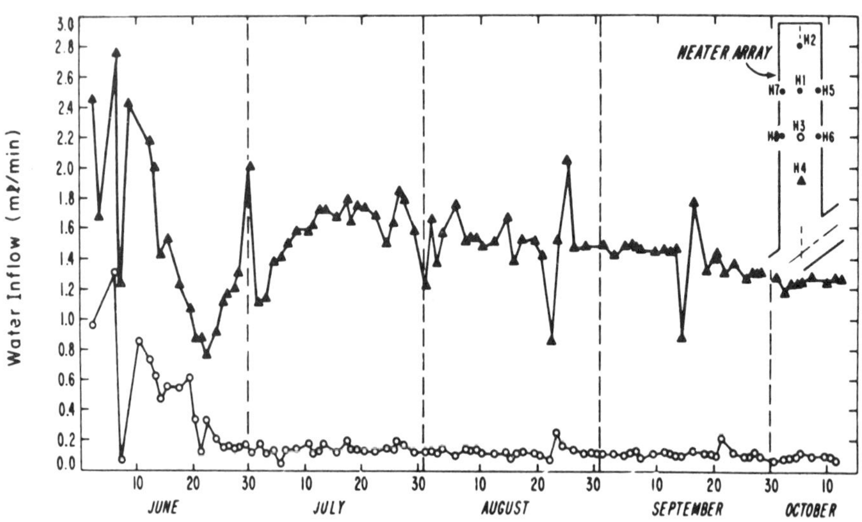

Fig. 8. Dewatering Data From Time-scaled Heater Boreholes

Table 1. Comparison of Equivalent Porous Medium Permeabilities and Fracture Permeabilities

k_p	Q		k_f	2b
cm/sec	ml/s	ml/min	cm/sec	M
4.5×10^{-5}	10	600	8.8×10^{-1}	1.04×10^{-4}
4.5×10^{-6}	1	60	1.9×10^{-1}	4.82×10^{-5}
4.5×10^{-7}	0.1	6	4.1×10^{-2}	2.24×10^{-5}
4.5×10^{-8}	0.01	.6	8.8×10^{-3}	1.04×10^{-5}
4.5×10^{-9}	0.001	.06	1.9×10^{-3}	4.82×10^{-6}
4.5×10^{-10}	0.0001	.006	4.1×10^{-4}	2.24×10^{-6}

Radial Flow Conditions, $r_w = .038$m,

$r_p = 10$m, $H_b = 0$, $H_w = 10$m

Fracture Spacing = 2m.

requires a systematic approach that involves a thorough analysis and integration of the basic fracture geometry and hydrology data. At Stripa this attempt will be supported by numerical modelling efforts designed to integrate the physical hydrology data with the results of the groundwater geochemistry and isotopic studies. Also the results of the large scale rock mass seepage study in the ventilation drift at Stripa will serve as a means of checking the accuracy and applicability of the approach outlined in this paper to the study of the hydrology of fractured rocks.

ACKNOWLEDGEMENTS

The fracture hydrology program at Stripa was initiated in August, 1977 and the field activities of the initial program will be completed in early 1980. During this period of time a large number of people have contributed to the program. It is impossible to identify all of the people involved, but we wish to give special thanks to D. McEdwards, J. Irving, M. Moebus, T. Doe, K. Billingshurst, R. Naden, D. Russel, P. Fransham, H. Norlander and P-A. Halen for their direct contributions to the program.

This work was supported by funds from the U. S. Department of Energy through the Lawrence Berkeley Laboratory under purchase order No. 478-3902 and through WRI contract 803-12.

REFERENCES

1. A. Olkiewicz, J. E. Gale, R. Thorpe, and B. Paulsson, Geology and Fracture System at Stripa, Lawrence Berkeley Lab. Report LBL 7079 (1979).
2. J. E. Gale and P. A. Witherspoon, An Approach to the Fracture Hydrology at Stripa - Preliminary Results, Lawrence Berkeley Lab. Rept. LBL 7079 (1979).
3. E. Romm and B. Pozinenko, "Investigations of Seepage in Fractured Rocks," Trudy VNIGRI (1979).
4. D. T. Snow, A Parallel Plate Model of Fractured Permeable Media, Ph. D. Thesis, Univ. of California, Berkeley, CA (1965).
5. M. A. Caldwell, The Theoretical Determination of the Fluid Potential Distribution in Jointed Rocks, M. Sc. thesis, Univ. of Witwatersrand (1971).
6. M. Parsons, "Determination of the Hydrogeological Properties of Fissured Rocks," Proc. 24th Geol. Cong. Montreal, Sec. II, (1972).
7. C. Louis and M. Pernot, "Three Dimensional Investigation of Conditions at Grand Maison Damsite," Proc. Symp. on Flow Through Fissured Rock, I. S. R. M., Stuttgart, Germany, Paper T4-F (1972).
8. L. Kiraly, "Statistical Analysis of Fractures: Orientation and Density," Geol. Rundsch. 59, 145-160 (1969).
9. H. M. Westergard, Theory of Elasticity and Plasticity, (Dover Publications Inc., New York 1969).

THE ROLE OF LARGE-SCALE PERMEABILITY MEASUREMENTS IN FRACTURED ROCK AND THEIR APPLICATION AT STRIPA*

P. A. Witherspoon[+], C. R. Wilson[+], J. C. S. Long[+], A. O. DuBois[+], J. E. Gale[++], M. McPherson[+++]

[+]Lawrence Berkeley Laboratory, [++]University of Waterloo, [+++]University of Nottingham

MOTIVATION FOR LARGE SCALE TESTING

The macropermeability experiment at the Stripa mine in Sweden is part of the Swedish-American Cooperative Program on Radioactive Waste Storage in Mined Caverns in Crystalline Rock (1,2). The experiment is an attempt to improve permeability characterization techniques for the analysis of regional groundwater flow through low permeability rock in the vicinity of a nuclear waste repository. This experiment consists of monitoring flow into, and pressure surrounding a 5 m x 5 m x 33 m long drift called the ventilation drift at the 335 m level of the Stripa mine (Fig. 1). This paper examines the theoretical problems associated with making such a measurement and describes the actual experiment now in progress.

Three problems arise in determining flow parameters for low permeability fractured rock. The first is to determine the minimum volume of rock for which the permeability tensor is representative of a larger rock mass and is amenable to a porous media method of analysis. The second problem is to determine this permeability tensor from field tests. The third problem is to assign permeabilities to the volumes of rock that are not directly examined in the field. The macropermeability experiment at Stripa is an attempt to increase our understanding of the first two of these problems. The third is not discussed here.

As the volume of the fractured rock sample increases from zero, the average permeability will oscillate as either fractured or solid

*Work done under University of California Contract W-7405-ENG-48 for the U. S. Dept. of Energy

rock is added to the sample. When the volume of rock becomes sufficiently large that permeability is no longer sensitive to the effects of individual fractures, the oscillations will cease. An average permeability can then be assigned to that volume of rock which is called the representative elementary volume (REV). Theoretically, volumes of rock the size of the REV can be treated as porous media for regional groundwater flow analyses. Increasing the volume further may ultimately cause additional oscillations if a different realm of fracturing is encountered. When a single permeability measurement is made on an arbitrary volume of rock there is no way of knowing a priori whether or not the measured permeability lies on the oscillating portion of the curve. A series of measurements on different scales must be made to determine if there is a REV smaller than the rock mass itself and the permeability associated with it.

In fractured rocks, where the discontinuities themselves may occupy areas on the order of 10^2 m^2, it is reasonable to expect REV's, if they exist, to be on the order of 10^4 or 10^5 m^3. The macropermeability experiment at Stripa will permit a measurement of the average permeability to 10^5 to 10^6 m^3 of rock. There is no assurance that this volume will be as large as the REV; however, the experiment taken along with other small-scale tests performed at the same site, should provide strong indications of the size and existence of the REV from this experiment.

The second problem is to determine the permeability of these large volumes from in situ tests. In high permeability soils or rocks, standard well tests can be run such that large volumes of the flow system are perturbed by the test in reasonable periods of time. However, in low permeability rocks, standard well tests may only affect the flow system within a few meters of the well. Determination of large scale permeability values can therefore be attempted

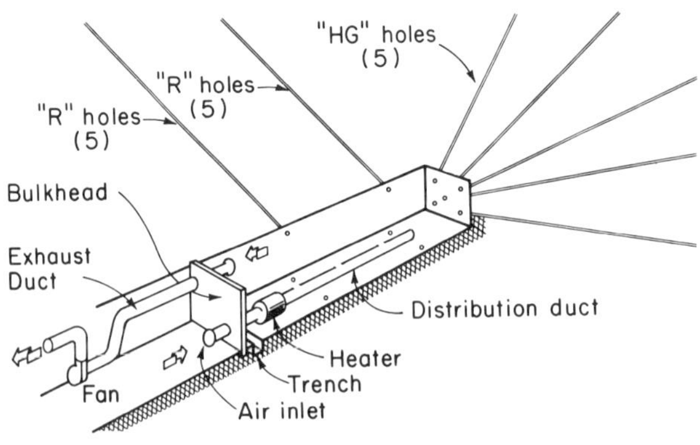

Fig. 1. Perspective Section Through the Ventilation Drift Showing Air Flow Pattern and Hydrology Instrumentation Boreholes.

in two ways. The first way is to synthesize large scale values from a series of small scale tests done in boreholes. The second way is to create a large scale sink (or source) which will perturb a large volume of the flow system, i.e., a macroscopic permeability test. Small scale borehole tests will probably remain the mainstay of hydrologic investigations since boreholes are the only practical means of extensively exploring deeply buried rocks. Therefore, reliable methods for predicting macroscopic permeability from borehole data are needed. In order to perfect such methods they will have to be checked by performing large scale in situ tests at the same site where the borehole data has been collected. The macroscopic permeability test at Stripa represents a large-scale measurement that will be used to check the analysis of an extensive series of small scale borehole tests (3).

MEASURING MACROSCOPIC PERMEABILITY

Any measurement of permeability must be made on a flow system with specified boundaries. The potential distribution on these boundaries as a function of time and the flux through each boundary as a function of time must be inferred or measured. Then the effective permeability can be calculated using Darcy's law. Most in situ tests cannot be constructed such that all these conditions are exactly met. Some assumptions about boundary conditions are usually made in the analysis. Theoretically, flow into the ventilation drift will approach steady conditions, but will never actually reach equilibrium. In order to use steady state analysis the experiment will continue for 3 to 6 months or until there is no discernible change in pressure and temperature. However, both the definition of boundary conditions and the measurement of flow into the drift are problems of some magnitude.

Theoretical Considerations in Pressure Measurement

The ventilation drift can be idealized as a long, but finite cylindrical sink. The inner boundaries of this flow system are well defined. The location of the boundary is the wall of the drift. The pressure on that boundary is essentially zero since the drift does not fill with water. The outer boundaries of the flow system are more difficult to characterize. Several possible approaches can be taken. One is to assume that at some distance from the drift the flow field is undisturbed and the hydraulic potential is constant. However, the hydrology of the region surrounding the drift is to a large extent dominated by the extensive Stripa mine workings. Piezometric profiles in the vicinity of the ventilation drift are irregular. An assumption of undisturbed conditions would not be valid.

Another method is to define a convenient boundary surface and measure the pressure distribution on this surface. Alternatively, pressure could be measured at an array of points in space such that an isopotential surface could be located in space. In an ideal, homogenous porous medium this method would present little problem. The open interval of the piezometer would intersect a large number of pores and grains and the pressure measured would be a physically averaged pressure. In a fractured system, however the piezometer interval may only intersect a few fractures. Since the direction of flow in these particular fractures is not necessarily in the direction of the average gradient as shown in Fig. 2, the fractures sampled will not necessarily yield the average gradient. Without knowledge of the average gradient, the average or equivalent permeability cannot be calculated. The problem is equivalent to predicting the permeability of a porous medium from measurements of pressure in only a few pores.

As a partial solution to the above problem, piezometers can be installed parallel to the theoretical isopotential surfaces. These peizometers can then be open the whole width of the flow zone. In this case many fractures will intersect the piezometer and a physically averaged pressure will be recorded. Two problems are presented by this arrangement. The first is that long sections of open borehole may create a significant leakage path and increase the flow into the drift. This will in turn increase the measured permeability. The closer the open borehole is to the drift the worse this problem is likely to be. Another problem is that this method of pressure measurement allows axial averaging but no circumferential averaging. The variability of pressure in the circumferential direction can be just as erratic as in the longitudinal direction. The only way to ultimately insure a given set of boundary conditions is to create those conditions artificially. This can be accomplished by drilling a ring of closely spaced boreholes around the drift and

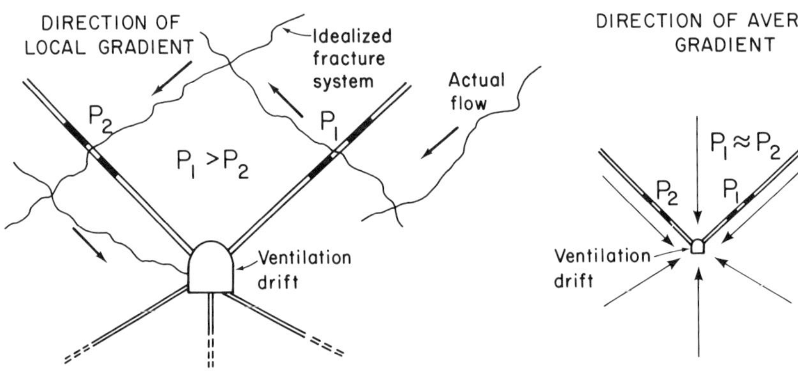

Fig. 2. Comparison of Local and Average Gradients, p_i = Pressure at Point i

maintaining them at constant hydraulic head. This method may be difficult and costly to construct, and it will not provide any information about anisotropy.

Pressure Measurement at Stripa

At Stripa 15 boreholes 30 to 40 m long have been drilled from the ventilation drift into the surrounding rock (Fig. 1). These boreholes are divided into 3 groups of 5 each. Two groups (the R-holes) are radial holes and one group (the HG holes) extend from the face of the drift. Each of these holes has been instrumented to monitor pressure. A total of 94 packers have been installed at approximately 5 m intervals. This prevents the boreholes from acting as drains into the drift and creates a 3-dimensional array of 94 zones, 90 of which are individually connected by tubing to pressure gauges. During the experiment air in the drift will be kept at constant temperature. Pressures in the boreholes and water flow into the drift will be monitored until they are steady. Then the air temperature in the drift will be raised and the experiment repeated.

Between constant temperature experiments two further experiments will be conducted. The first will be small scale permeability tests in which each zone in one or two boreholes will be drained. The flow rate from each zone will be monitored separately until approximately steady conditions are reached. By repeating this test after each constant temperature experiment it may be possible to detect local changes in permeability due to changes in temperature. These changes can then be correlated to changes in permeability detected for the whole macroscopic test.

At Stripa we will not be able to examine average pressures parallel to the theoretical average isopotentials. However, since the average gradient is not expected to be very large at the outer ends of the boreholes, we will be able to approximately examine the effect of zone length on pressure by hydraulically interconnecting adjacent zones through the pressure tubing and allowing the system to come to equilibrium. Head losses in the tubing will be known and the pressure in each of the interconnected zones can be calculated. Since flow rates between zones are in most cases expected to be quite low, the pressure in the zones should be nearly equal. The average of the pressures will be taken as an estimate of the pressure the zones would have if the packer between them were removed.

Theoretical Considerations in Flow Measurement

In most permeability tests, the water can be collected in a tube or pipe and the flow rate measured in a straight forward manner. However, in the low permeability rock of this experiment, the flow rate into the drift is so low and the surface area is so great that a significant proportion of the inflow would be lost to evaporation and the remainder would collect on the floor of the drift too slowly to measure in a reasonable amount of time. For example, inflow from the fifteen open boreholes at Stripa is about 0.8 liters per minute which is equivalent to collecting 0.8 mm per day over the floor area of the drift. This inflow is expected to decrease after these boreholes are sealed.

One of the purposes of this experiment is to examine the utility of measuring these low flow rates by evaporating all the water into the ventilation air while measuring the change in water vapor content between the incoming and exhaust air streams. To operate successfully, the ventilation system should provide an air flow rate and temperature capable of vaporizing and carrying away all of the influx water (4). In addition, all parameters must be steady enough to allow identification of the equilibrium conditions. Also, the differential relative humidity between the air entering and leaving the experiment area must be great enough to provide acceptable water vapor measurement accuracy. As a point of reference, 1.0 m^3/s (2,100 cfm) of air flowing at normal atmospheric pressure, 20°C, and 100% relative humidity will transport about one liter of water vapor per minute.

For the macropermeability experiment it is necessary to measure: 1) the barometric pressure; 2) the air velocity in the exhaust duct; and 3) the wet and dry bulb temperatures of the incoming and exhaust air streams. These measurements allow us to calculate the incoming and exhaust mass flow rates of both the air and the water vapor associated with it. The difference between the incoming and exhaust water vapor flow rates is the amount of water which has evaporated from the surfaces of the drift. The floor of the drift and the surfaces of the walls can be observed to assure that all of the water is being evaporated as it arrives, without significant puddling. To maximize the accuracy of water vapor transport calculations, one wishes to operate with the exit air stream as near to saturation as possible. Nomograms have been prepared for a variety of exit air temperatures to assist in the selection of air flow rates and heater power levels that will maintain an acceptable (±20%) level of accuracy. One of these is shown in Fig. 3. It is anticipated that several months may be required to establish near-equilibrium conditions in the rock temperature and water vapor flow regimes. The experiment will be initiated with a room temperature control setting as near as possible to the preexisting ambient air temperature. Only enough heat will be added to assure that the exhaust air is below saturation and that all available water is evaporated. Subsequently warmer equilibrium room temperatures will be established to investigate the impact of that temperature upon the water influx.

Flow Measurement at Stripa

A sketch of the experimental setup was shown in Fig. 1. An air- and vapor-tight bulkhead seals off a 33 m length of the ventilation drift. This wall consists of a structural wooden frame covered with a 0.15mm thick sheet of PVC, as a vapor barrier, and fiberglass insulation to prevent condensation. The vapor barrier and the structure are keyed into a 20mm deep notch on all four sides, with foamed plastic sealing in the notch to prevent the loss of water through the blast damaged surface rock. A frozen-food-locker type door provides access through the bulkhead.

The general mine ventilation system delivers about 1.4 m^3/s of fresh air to a point just outside the bulkhead. The experimental ventilation system admits a portion of this fresh air into the sealed room to pick up the water (as vapor) which flows into the sealed room through the surrounding rock. This secondary ventilation system is driven by a fan located in the exhaust duct so that a slightly negative pressure will exist within the sealed room. Any air leaking through the bulkhead will be into the room and it will be the same air as that entering through the inlet duct. In this way, the inlet wet and dry bulb measurements will be representative of both the intentionally admitted air and the leakage air. The flow measuring station as well as the exhaust wet and dry bulb sensors are in the exhaust duct. These sensors sense the total air flow including leakage. The air entering the sealed room passes through an electric duct heater. It is then distributed through an insulated duct with multiple openings along the length of the room. The air issuing from these openings will be directly preferentially onto any damp spots on the rock surface (most of the water arrives by seeping along discrete fractures). We expect to maintain the

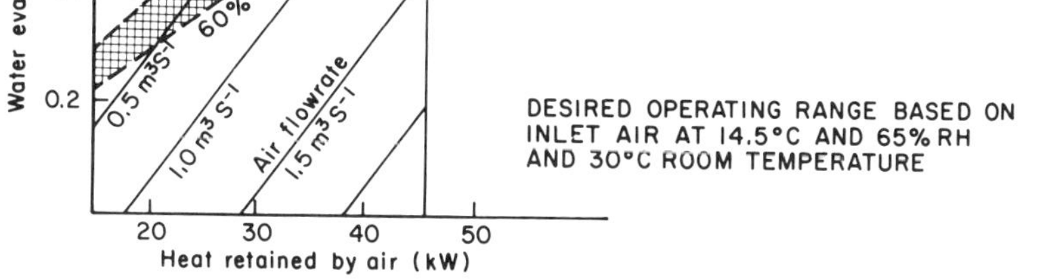

Fig. 3. Plot of Heat Retained Versus Evaporation Rate.

walls of the drift in an almost dry condition, with fans used where necessary to provide a high air velocity over particularly damp spots.

The climatic conditions within the room will be controlled by manually selecting a suitable air flow rate (by adjusting a damper on the fan exhaust) and then adjusting the heater power as required to maintain the desired air temperature in the room. The energy input from the heater is utilized in three ways: 1) the sensible heat of the exhaust air will be increased; 2) the heat of vaporization for the water will be provided; and 3) the rock will be heated. As equilibrium conditions are approached the third contribution would become negligible. The 45kW heater is divided into six sections which are separately controlled. Five sections are under manual control and one is under automatic on-off control with feedback from a room-temperature sensor. The heater voltage is regulated to isolate it from the large voltage fluctuations observed in the mine power system. A 75 mm deep trench has been cut across the floor of the drift just inboard of the bulkhead. If the trench and floor of drift remain dry, we shall conclude that all available water is being evaporated at its arrival rate.

SUMMARY

Completion of the macropermeability experiment will provide: (i) a direct, in situ measurement of the permeability of 10^5 to 10^6 m^3 of rock; (ii) a potential method for confirming the analysis of a series of small scale permeability tests performed in surface and underground boreholes; (iii) a better understanding of the effect of open borehole zone length on pressure measurement; (iv) increased knowledge of the size and existence of a representative elementary volume in fractured rock; (v) a basis for evaluating the ventilation technique for flow measurements in large scale testing of low permeability rocks.

REFERENCES

1. P. A. Witherspoon, and O. Degerman, Swedish-American Cooperative Program on Radioactive Waste Storage in Mined Caverns, Program Summary LBL-7049, SAC-01, Lawrence Berkeley Laboratory, Berkeley, CA (1978).
2. P. A. Witherspoon, Rock Mass Characterization for Storage of Nuclear Waste in Granite, LBL-8570, SAC-18, Lawrence Berkeley Laboratory, Berkeley, CA (1979).
3. J. E. Gale, and P. A. Witherspoon, An Approach to the Fracture Hydrology at Stripa: Preliminary Results, LBL-7079, SAC-15, Lawrence Berkeley Laboratory, Berkeley, CA (1979).
4. M. J. McPherson, "Psychrometry: The Measurement and Study of Moisture in Air," Nottingham University Mining Magazine, University of Nottingham, United Kingdom.

CHARACTERIZATION OF DISCONTINUITIES IN THE STRIPA GRANITE*

Richard K. Thorpe

University of California
Lawrence Berkeley Laboratory
Berkeley, CA

INTRODUCTION

An important problem in the storage of radioactive waste in fractured crystalline rock is the characterization of natural fractures for both thermomechanical and hydrologic purposes. From a thermomechanical standpoint fractures can introduce nonlinear behavior in the rock mass response to heating. Hydrologically, fractures represent the principal flow paths for radionuclide transport away from a repository. In the Swedish-American Radioactive Waste Storage Program at Stripa, Sweden, the fracture system is being characterized over either a local or large scale according to the needs of the experiment (1). The large scale studies are hydrologically oriented (2), whereas this paper deals with local fracturing around one of the underground heater tests at Stripa. The experiment, termed the "time-scaled" test, is located in a cavern 335 m below surface in massive granitic rock. While the site geology and experiment are described elsewhere (3), the purpose here is to present observational fracture data that can serve as input for thermomechanical interpretation of the test.

METHODOLOGY AND RESULTS

Two approaches are used here to characterize the local fracture system. First, major discontinuities are identified in the test area

*Work done under University of California Contract W-7405-ENG-48 for the U. S. Dept. of Energy.

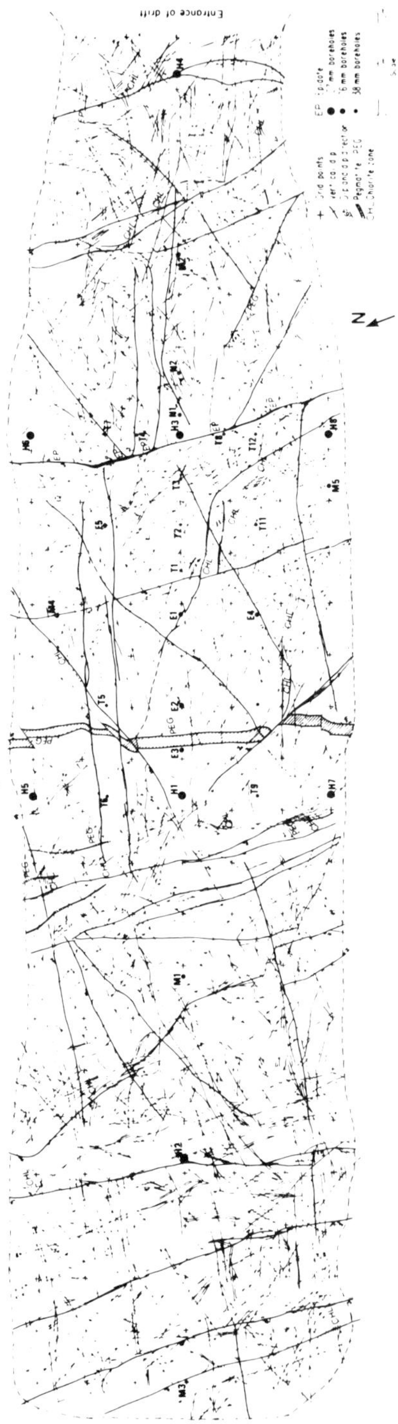

Fig. 1. Detailed Fracture Map of the Floor of the Time-Scaled Experiment Drift

so that they can be modeled as discrete elements of weakness (4). While these features probably have a major role in the rock mass behavior, they comprise only a small percentage of the fracturing. Most of the other fractures are discontinuous in their own planes, hence the second approach involves defining fracturing statistically in terms of orientation, spacing and length of joints. Compatibility of the two sets of results will be demonstrated.

Characterization of Major Discontinuities

The success to which major discontinuities can be delineated within a rock mass depends on their continuity and the quality of subsurface information from which they can be identified. Surficial mapping is an aid in assessing the continuity of fractures, and at Stripa, the walls and floors of the heater experiment drifts have been mapped in detail to show all fractures longer than about 0.3 m. The outcrops of features were mapped at a scale of 1:20, using a 1-by-1 m reference grid painted on the rock surface. Information such as rock type variations, fracture fillings, or clear signs of faulting was noted during the mapping. The detailed fracture map developed in this manner for the time-scale drift floor is shown in Fig. 1.

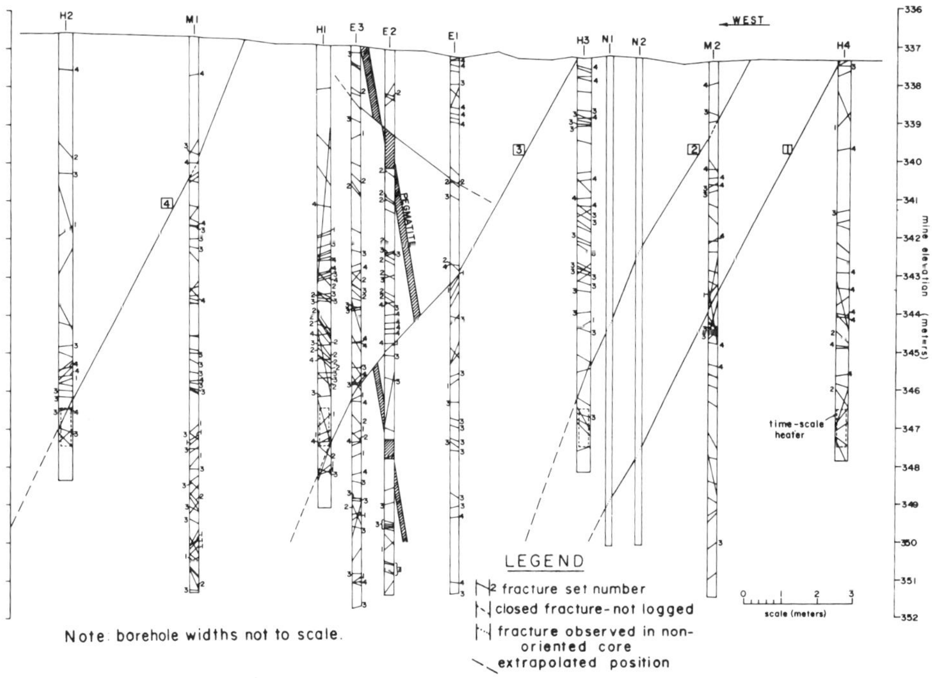

Fig. 2. Profile of Fractures Along Centerline of Time-Scale Drift

Heaters for the experiment are placed 10 m below the drift floor therefore only the most prominent and continuous features in the map are likely to extend through the heated region and affect the rock mass behavior. Accordingly, only the prominent faults striking transverse to the drift were extrapolated downward and correlated with features in the borehole fracture logs. The correlation of the features was based on observational evidence; i.e., similarity of orientations, coatings, and surface characteristics and proximity to the extrapolated position. Other significant discontinuities may exist near the heaters and not intersect the drift; however, none could be identified with confidence using the observational technique. Results of the discrete characterization are illustrated by Fig. 2., which shows the inferred profile of four shear surfaces that pass through the heater array. These features offset or truncate other discontinuities, and their filling minerals of chlorite, calcite, epidote and clay are several times thicker than the fillings of other fractures. Fault number 3, which is the most prominent and well-defined of the set, apparently offsets a 20 cm-wide pegmatite dike as shown in the figure.

Characterization of Jointing

In order to describe the abundant jointing between the major features, it is preferable to adopt a statistical approach that incorporates both borehole and surficial data. The important parameters are the dominant joint orientations, i.e., joint sets and the distributions of joint spacings and trace lengths. Joint orientations and spacings have been derived from borehole fracture logs in the experimental area, and trace lengths have been compiled from Fig. 1. The jointing can be separated into four distinct sets according to Fig. 3. Assigning these set designations to fractures in Fig. 1. allows a compilation of trace lengths to be made. A typical histogram of the data for one joint set is shown in Fig. 4. When these values are plotted in lognormal probability form, the cumulative frequency distributions in Fig. 5. are obtained.

Joint spacing distributions are described in a similar manner by assigning appropriate set designations to fractures logged in the boreholes. Vertical spacings are then computed as the distance between consecutive fractures of the same set. Fig. 6. is a lognormal probability plot of the data with the respective distributions shown. Several errors are inherent in the above approach. First, inclined fractures are under-represented by vertical boreholes. This bias could be reduced by incorporating the floor map in the spacing analysis. For the data presented here, the spacing normal to joints is found by multiplying the vertical spacing by the cosine of the mean dip angle of the set. Similarly, horizontal features are missing from the floor map, hence their length distributions are not given. Also, there are biases in the trace length data caused by

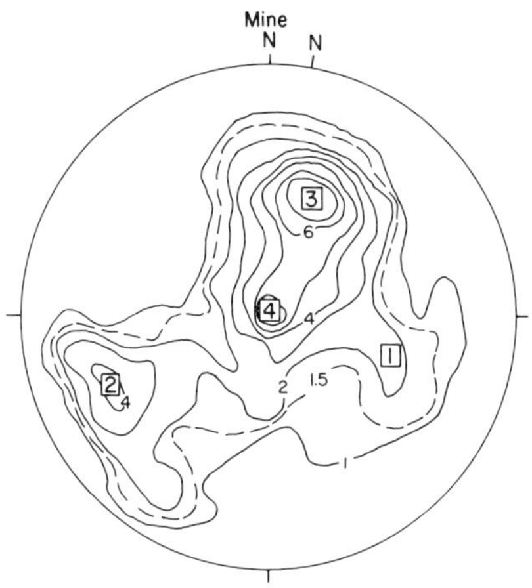

Fig. 3. Stereonet Plot of Fracture Poles for Time-Scaled Experiment

a) the minimum length of fractures that were mapped (0.3 m) and b) the dimensions of the drift. These biases prevent the distributions from being used to predict extreme values outside the range of data upon which they are based.

DISCUSSION

The virgin state of stress has been measured in the underground test site (5) and this information helps to validate the above findings. Fig. 7. is a stereographic plot of the principal stresses and the mean pole directions to the four fracture sets in the time-scale experiment. The pole of joint set 1, $\hat{n}_1$, corresponds to that of the four faults traced through the rock mass. Resolving the principal stresses into shear and normal components on the mean fault plane as shown in Fig. 8. yields a theoretical shearing azimuth of 242°. The azimuth of slickensiding on the faults was measured as 240° by field observation and inspection of the fracture intercepts in the core. Reconstruction of the inferred configuration of the fractures by orthographic projection and physical modeling shows the intercept points to be nearly colinear in this direction (3). The origins of the jointing are still uncertain, but it appears likely that set 3 is the youngest of the four sets and may be associated with continuing post-glacial uplift. Three observations in support of this claim are offered: 1) The set lies perpendicular to the existing minimum principal stress (Fig. 7.). 2) Joints of this set are predominently filled with calcite rather than the more common chlorite found in other joints, which suggests relatively recent transport and deposition. 3) The joints tend to cross or offset nearly all other joints when observed in the core samples (5).

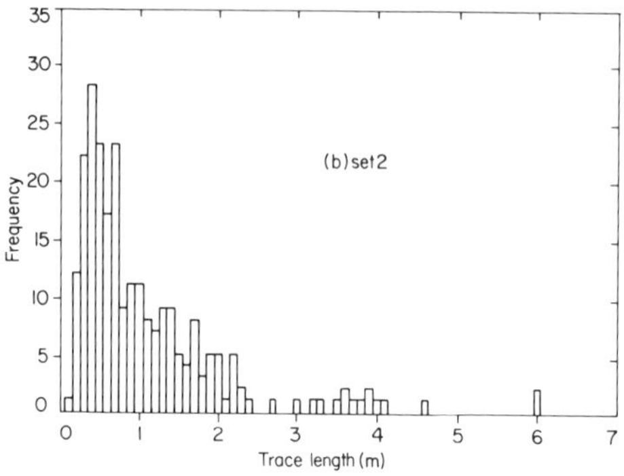

Fig. 4. Histogram of Joint Lengths

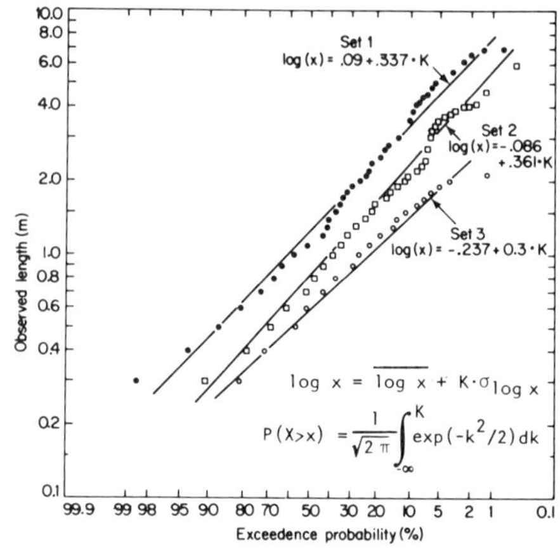

Fig. 5. Lognormal Joint Length Distributions

Statistically, joints of set 3 are shorter and more closely spaced than are the others (Figs. 4. and 5.). A shear-to-normal stress ratio of 4.70 has been calculated for the mean plane of the set, which is over twice that for the other joint sets (3). Thus, if thermal loading or some other perturbation were to propagate or coalesce these fractures, local stability problems might arise. Fractures in Set 2 tend to be rough and irregular, suggesting an extension mechanism. Clearly, these could not result from the state of stress reported here, but more likely from a condition where the major principal stress was perpendicular to its present

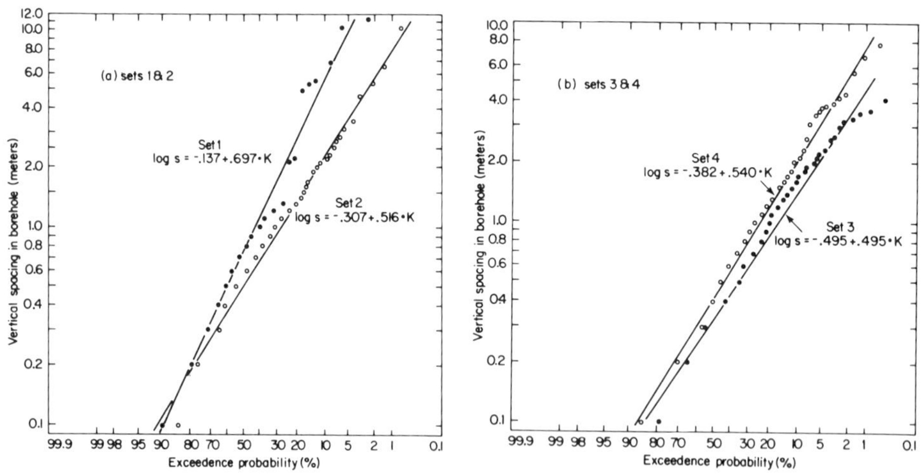

Fig. 6. Lognormal Plots of Joint Spacings

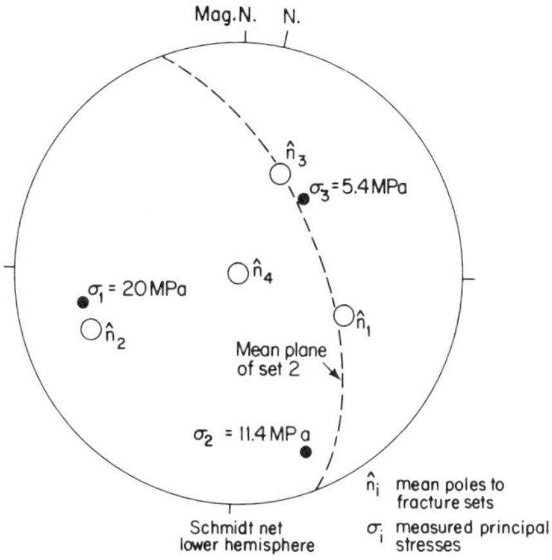

Fig. 7. Plot of Principal Stresses and Joint Sets

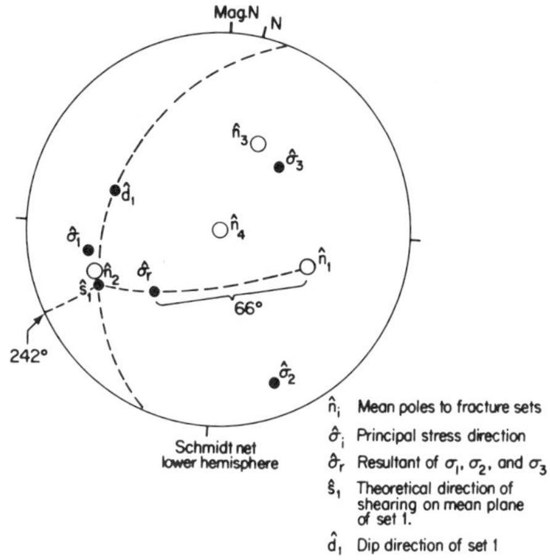

Fig. 8. Determination of Stress Components on Mean Plane of Joint Set 1.

orientation. Such conditions could occur by a) glacial loading, in which the major principal stress would be closer to vertical, b) stress changes during cooling of the pluton, or c) viscous drag during the intrusion process (6). The othogonality of the joint sets indicates that there is probably a reasonable explanation for their occurrence, yet it is tenuous to speculate on this with limited knowledge of the regional geology.

Certain observations can be made regarding the engineering significance of joint sets 1, 2, and 4. First, with set 2 lying perpendicular to the major principal stress it would be unlikely for these joints to deform appreciably or otherwise affect the stability of the rock mass under moderate stress perturbations. Secondly, radial stress relief around the opening has reduced the compression across the horizontal joints, which should increase the nonlinearity of their deformation behavior. Finally, joint set 1 is effectively represented by the four faults in Fig. 2., since most instability along planes at this orientation would be accomodated by these major weaknesses. The fault strengths are anisotropic, however, due to their lack of planarity transverse to the direction of slickensiding (3). This will be properly represented if numerical models utilize the irregular surface configurations that have been developed (3), or alternatively, if anisotropic shear strength parameters are assigned.

CONCLUSIONS

The above results pertain to nearfield behavior of a rock mass, and as such they represent a practical limit to the degree to which discrete subsurface discontinuities can be defined by surface mapping and cross-correlation of observational borehole data. Discontinuities on the scale of the rock mass being studied can dominate its response, and should therefore be described deterministically. Alternatively, if features are of a scale much less than the rock mass in question, a statistical approach is appropriate.

It has been demonstrated that in addition to facilitating a discrete characterization of major features, detailed mapping of fractures and thorough logging of core samples can yield valuable statistical information. Efficient collection of such data requires prior knowledge of the engineering significance of parameters to be measured. The experience gained at Stripa in characterizing fractures and determining their behavior will contribute to this knowledge.

REFERENCES

1. P. Witherspoon, J. Gale, P. Nelson, T. Doe, R. Thorpe, C. Forster, and B. Paulsson, "Rock Mass Characterization for Storage of Nuclear Waste in Grantite," Proc. 4th ISRM Congress, Montreaux, Switzerland (1979).
2. J. Gale, O. Quinn, D. McEdwards, P. Witherspoon and T. Doe, "Hydrogeologic Characterizations of Fractured Rocks for Waste Isolation - Results from Stripa," Proceedings of This Symposium (1979).
3. R. Thorpe, Characterization of Discontinuities in the Stripa Granite - Time-scale Experiment, Lawrence Berkeley Laboratory Report LBL-7083, SAC-20, Berkeley, CA (1979).
4. R. E. Goodman, Methods in Geological Engineering in Discontinuous Rocks (West Publishing Co., St. Paul, MN 1976).
5. H. Carlsson, Stress Measurements in the Stripa Granite, Lawrence Berkeley Laboratory Report LBL-7078, SAC-04, Berkeley, CA (1978).
6. R. Balk, Structural Behavior of Igneous Rocks, (Geological Society of America, Memoir 5, 1937).

ROCK INSTRUMENTATION PROBLEMS EXPERIENCED

DURING IN SITU HEATER TESTS*

E. Binnall, A. DuBois and R. Lingle**

Lawrence Berkeley Laboratory, Berkeley, CA
**Terra Tek, Inc., Salt Lake City, UT

INTRODUCTION

A cooperative Swedish-American program is underway to investigate the suitability of granite as a deep underground storage medium for nuclear waste isolation. The experimental program, located in an inactive iron mine at Stripa, Sweden, includes three heater experiments to evaluate the thermomechanical response of a granite mass when subjected to the heat load of simulated canisters of high level nuclear waste. These experiments require the measurement of temperature, rock displacement and stress during one and a half years of heating and cool-down. The first heaters were turned on June 1, 1978 and the last heater turned off September 26, 1979. Rock temperatures over 300°C have been measured and the mechanical response of the rock has been measured at temperatures to 150°C. Measured heater skin temperatures have approached 500°C. Few of the available instruments are designed to meet the accuracy and reliability requirements at the elevated temperatures encountered in this type of experiment. Therefore, available instruments have been modified and tested; and calibration techniques have been developed in an attempt to meet more strigent requirements. The modified instruments have produced a wealth of valuable data at Stripa in spite of some specific problems discussed here (1).

INSTRUMENTATION AND DATA ACQUISITION SYSTEM

Four types of instruments were used at Stripa; 389 thermocouples for temperature, 35 rod extensometers for displacement,

*Work done under University of California Contract W-7405-ENG-48 for the U. S. Dept. of Energy.

30 U. S. Bureau of Mines (USBM) borehole deformation gages and 26 IRAD (Crearé) vibrating wire stress gages for stress determination. These sensors are installed in vertical and horizontal instrumentation boreholes strategically located around vertical heater boreholes (1). The heater experiments are located in two drifts 340 meters below the surface. A time-scale experiment is located in one of the drifts and two full-scale experiments in opposite ends of the other drift. A third drift provides access for horizontal instrumentation. All signal conditioning circuits, data loggers, computer scanners, analog-to-digital converters and interfacing hardware are housed in sheds located in the two heater drifts. More than 750 sensor signals are digitized and transmitted to a ModComp IV computer (2) located a few tens of meters from the heater experiments. Selected samples of IRAD gages and of medium and high temperature thermocouples were laboratory calibrated. All USBM gages, extensometers and low temperature thermocouples were calibrated in the field. The field calibrations were performed on-line with the instruments integrated into the computerized data acquisition system.

THERMOCOUPLES

Rock, instrument and heater canister temperatures are measured using Chromel-Alumel (ANSI type K) thermocouples connected to ice-point temperature references manufactured by Kaye Instruments. Thermocouples were purchased from Omega Engineering and, where possible, were obtained from the same wire lot to minimize variability due to differences in metallurgy. Three types of thermocouple wire covering were used:

- TFE Teflon capable of continuous temperatures to 260°C
- 304 stainless steel sheath for temperatures to 400°C
- Inconel 600 sheath capable of temperatures to 1150°C

Teflon insulated thermocouples were manufactured with extruded Teflon insulation over the individual wires and a fused Teflon tape wrap over the wire pairs. Junctions were coated with a two component RTV 60 silicone rubber compound. A desired precision of ±0.5°C over a temperature range of 10° to 220°C for these "low temperature" thermocouples necessitated individual calibration. A fourth order polymonial curve was fitted to the calibration data, resulting in a set of five coefficients for each thermocouple. These were stored by the computer for on-line conversion from thermocouple voltage to temperature. After installation, some thermocouples developed water leakage at the RTV coated junctions causing erroneous readings due to ground currents. This problem was solved by removing all electrical ground points other than those that could occur at the thermocouple junctions. In some of the grouted boreholes, head pressure forced the water along the thermocouple wire between the two layers of Teflon insulation and into the electronics enclosures. This was solved by cutting a short slit in the outer layer of insulation just after the thermocouple exited the borehole.

Sheathed thermocouples were used where higher temperatures were expected. Inconel 600 sheath was used for thermocouples attached to heater canisters and 304 stainless steel sheath was used where temperatures might exceed 200°C but remained well below 400°C. All sheathed thermocouples were 1/16" ouside diameter with magnesium oxide (MgO) electrical insulation and with junctions sheathed but ungrounded. Lengths were selected to allow sheathed wire to exit the boreholes into an ambient temperature environment (10-20°C) before entering a transition joint to Teflon insulated type K extension wire. The sheathed thermocouples were heat treated at just below 600°C for 50 hours to stabilize an upward shift in output voltage that occurs with thermocouples using Chromel at elevated temperatures. This forced an upward but nonrecurring voltage shift equivalent to approximately 0.5°C per 100°C above 0°C. Unfortunately, the heat treatment also sensitized the stainless steel sheaths to corrosion, subsequently requiring the replacement of over 50 thermocouples. The Inconel sheaths have been trouble free. Inconel sheathed thermocouples were inserted down long tubular thermocouple wells attached to the sides of the time-scale-heater borehole casings. Closed at the bottom near the heater elements, these wells trapped moisture causing a boiling and condensing cycle within the tubing. The result was erratic temperature readings until the moisture could be removed. Holes in the well bottoms would have avoided this problem.

Recommendations for future experiments include:

- Use Inconel sheathed thermocouples exclusively. At least avoid stainless steel sheath.
- Use grounded junctions in enclosed sheaths.
- Obtain all thermocouple wire from single material melts.
- Retain control samples to check long term stability.
- Install tubing for traveling thermocouples.
- Thermocouples removable for recalibration or replacement.
- Avoid closed thermocouple wells that can capture water.

ROD EXTENSOMETERS

A rod extensometer is a common device for measuring changes in the axial length of a borehole (3). Terrametrics Inc. (Golden, CO) model 4CSLT extensometers were used. Basically these have performed well with interim maintenance and minor field modifications. Each extensometer measures the displacement of four downhole anchor points with respect to the borehole collar over a range of ±13mm. One spring tensioned rod up to 13 meters long connects each anchor point to one of the four linear variable differential transformer (LVDT) type sensors located at the borehole collar. The major elements of these extensometers (Fig. 1.) are: 1) the anchor system; 2) the anchor-to-collar rod connection mounted inside of a waterproof flexible conduit; 3) a head assembly which includes the rod tensioning system (100 lbs/rod) and the displacement sensors; and 4) several thermocouples for sensing the temperature profile along the connecting rods.

On the basis of anchor system tests: 1) Dow Corning X2-1162 silicone fluid was selected as the anchor pressurizing fluid capable of withstanding long time service to 200°C; 2) the hydraulic system was demonstrated to allow control of anchor pressure with changing temperature; 3) anchor creep under a 100 pound axial load at 190°C for fifteen days was shown to be less than 2 microns even for an anchor which had been depressurized and where the grout had deteriorated; and 4) common portland cement grout was demonstrated to maintain adequate strength at temperatures to 200°C.

Thermal expansion measurements were made on several superinvar rods, selected at random from the available stock (a single melt). These measurements were found to be reproducible after the stock had been heat treated at 225°C for 5 hours. The averaged expansion values were utilized in the computer data reduction program to calculate a thermal expansion correction. At Stripa this correction was as much as 20% of the gross displacement. The temperature profile along the rods is estimated from the four to six measured values by using a cubic spline curve fitting procedure. However, this involves some assumptions, based on the symmetry of the thermal field, which have been found not to be entirely valid. We would, in the future, expect to install a tube down the length of the extensometer to enable a traveling thermocouple to survey the actual thermal profile several times during the experiment.

Only one of the displacement sensors has failed during operation; however, internal rod friction has caused the displacement output to move in a stepwise fashion. We conclude that most of this friction effect resides in the spring tensioning and guiding elements of the head since up to 80 microns of stored rod displacement was released when the heads of the extensometers were rapped with a screw driver handle. A routine has been instituted to rap each instrument head in this fashion several times per week to continuously release the friction. No technique has been devised for evaluating any possible downhole friction effects, other than to observe that the indicated displacements no longer include significant stepwise movements. Some design changes, such as modifying the rod spacings and the bushing materials, can reduce the friction, and a vibrator can be added to the head to periodically release the stored energy.

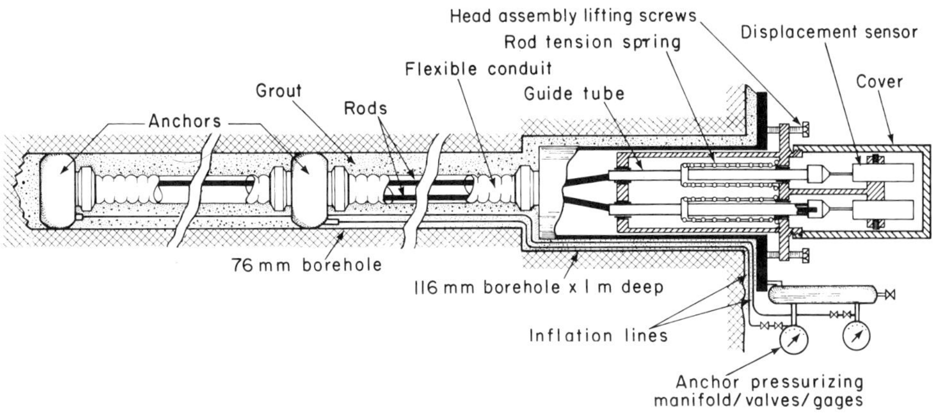

Fig. 1. Sectional View of Extensometer, Two Anchor Version

In spite of the existence of a flexible metal conduit with a silicone rubber outer sleeve to protect the rods and thermocouples, water has accumulated in most of the horizontal extensometers and is probably also present in the vertical units. Water may corrode the rods and thermocouples, and create deposits inside of the rod guide bushings. Electrical connections at screw-type terminal strips within the extensometer heads have caused trouble due to oxidation and corrosion. These have been replaced with solder connections and were waterproofed where necessary. The head covers have been drilled to allow any leakage water to drain. We will investigate the source of these water leaks at the conclusion of the heater experiments.

VIBRATING WIRE STRESSMETERS

Vibrating wire stressmeters (4) manufactured by IRAD GAGE Inc. (Lebanon, NH) have been installed in 38 mm diameter boreholes surrounding the full scale electrical heaters at Stripa. Two stress meters oriented at 90° to each other are wedged into each of eight horizontal holes and five vertical holes to measure the principal stress changes in the rock resulting from the thermal gradients.

The vibrating wire stressmeter (Fig. 3.) operates on the principal that change in the tension on a wire causes a change in its fundamental period of vibration. The IRAD stress meter consists of a hollow steel cylinder with a highly stressed steel wire stretched across a diameter of that cylinder. When installed in a borehole, the cylinder is preloaded across the borehole, in the direction of the wire axis, by means of a sliding wedge and platen assembly. In this way the gage becomes a "hard" inclusion and the wire undergoes a change in tension as the steel cylinder deforms from stress changes in the surrounding rock. This change in wire stress is

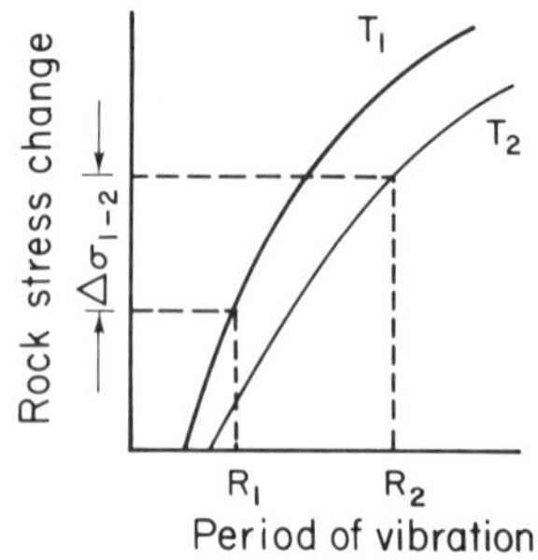

A plot of $\Delta\sigma = C_1(T) \times (R+\Delta R)^{-2} + C_2(T)$ for two different temperatures. $C_1(T)$, $C_2(T)$ and ΔR are functions of temperature (T) and of the room temperature calibration values.

$\Delta\sigma_{1-2}$ = the change in stress between the time and temperature at installation (subscript 1) and some later time (subscript 2).

Fig. 2. IRAD Gage Data Reduction Method

measured by exciting the wire and measuring its period of vibration (160 to 400 microseconds).

Preliminary tests demonstrated that the IRAD gages require individual calibration and that these calibrations are temperature dependent. Laboratory calibration in a sample of Stripa granite at ambient and at elevated temperatures provided a method for calculating (Fig. 2.) the rock stress changes as a function of measured gage temperature (T) and period of vibration (R). Note that R_1 (Fig. 2.) is the gage reading after the gage has been preloaded, by wedging it into the hole, and allowed to reach thermal equilibrium. The gage cannot measure the state of stress then existing in the rock unless that stress is released by overcoring. At Stripa the IRAD gage senses only stress changes subsequent to its installation.

Most of the IRAD gages have operated reliably for approximately one and a half years. Three gages have failed and one of these was sectioned, revealing a severely rusted wire. Other gage failures will be investigated at the conclusion of the experiment. The principal problem with these gages is the uncertainties of the data reduction. The gage calibration "constants" are sensitive to rock

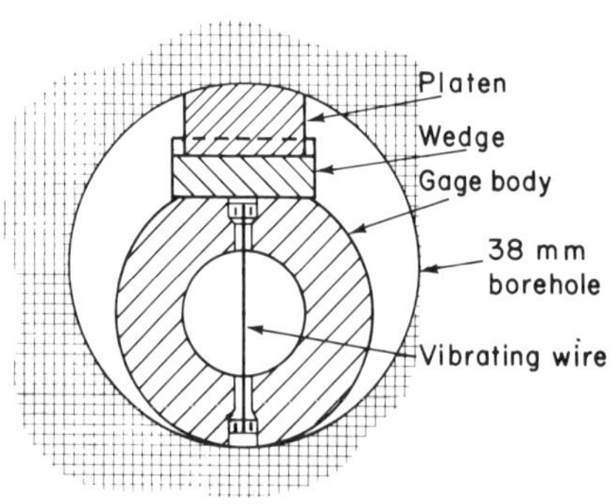

Fig. 3. IRAD Gage Cross-Section

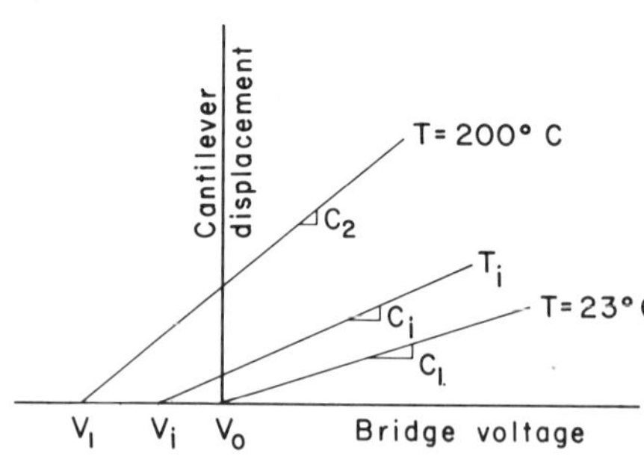

Fig. 4. USBM Gage Calibration Curve

modulus and coefficient of expansion as well as to the exact seating of the preloading wedge system. In spite of calibration in a uniaxially loaded sample of the host rock, considerable uncertainties remain in the calculated stress change, particularly under conditions of varying temperature and cyclic loading.

USBM GAGES

The US Bureau of Mines "soft" borehole-inclusion gage utilizes three pairs of opposed cantilever beams to sense changes in the dimensions of a borehole. These measurements across three diameters at 60° separation allow the principal stresses in the rock to be calculated. Strain gages are mounted on the top and bottom of each cantilever. The four strain gages on opposing cantilevers make up one bridge circuit. The gage was developed to measure in situ stresses by the overcoring technique at ambient temperature (5).

For the Stripa application the gage data reduction scheme was adapted for operation over a long time period at elevated temperatures, and several modifications have been made in the gage (manufactured by Rogers Arms and Machine Co., Grand Junction, CO):

1. The cantilever arm was changed to 17-4 pH stainless steel;
2. High temperature strain gages and solder, and Teflon-insulated wire were used;
3. Individual excitation was provided for each bridge;
4. A thermocouple was mounted on each USBM gage.

The USBM gage responses to temperature and strain were determined during laboratory tests at temperatures up to 200°C. The gage output was found to include a voltage offset at constant displacement which was proportional to temperature and different for each cantilever pair. There was also a linear change in slope (voltage versus displacement) of 6% from 20° to 200°C as indicated in Fig. 4.

The USBM gages have experienced numerous failures in service at Stripa. Sixteen of the twenty gages installed in vertical holes and two of the ten installed in horizontal holes have failed. The failures have been caused by water entering the gage housing--probably through or around the electrical cable. These failures occurred as either a short circuit at a strain gage or an open circuit due to extensive corrosion of the pins in the connector joining the cable to the strain gage. The leaks occurred in spite of a regular dewatering operation which assured that the water level in any borehole was always below the gage.

To correct these problems, the gages were removed, rebuilt, recalibrated and leak tested before reinstallation. The rebuilt gages incorporated several design changes to improve the integrity of the hermetic seal at the cable connection and to replace the pin connections with soldered connections. The gages passed a leak test procedure which subjected them to several days of immersion in water at 20 psig and 100°C. Those gages reinstalled in vertical boreholes were filled with a silicone fluid to prevent moisture from collecting in critical areas. It is still too early to assess their performance.

CONCLUSION

A large amount of valid data has been obtained from the Stripa experiments by the use of modified and calibrated off-the-shelf rock instrumentation. The elevated temperatures and the need for long time integrity in a wet environment revealed unsolved problems. The identification and solution of these problems would greatly benefit future experiments. Obviously more effort should be applied to the development of instruments to meet the environmental, reliability, and accuracy requirements for test site observations.

ACKNOWLEDGEMENT

This work was supported by the Office of Nuclear Waste Isolation of the U. S. Department of Energy under Contract No. W-7405-ENG-48.

REFERENCES

1. T. Schrauf, H. Pratt, E. Simonson, W. Hustrulid, P. Nelson, A. DuBois, E. Binnall, and R. Haught, Instrumentation Evaluation Calibration and Installation for the Heater Experiments at Stripa, Lawrence Berkeley Laboratory, LBL-8313, to be released (1980).
2. M. B. McEvoy, Data Acquisition, Handling and Display for the Heater Experiments at Stripa, Lawrence Berkeley Laboratory, LBL-7062, SAC-14 (1979).
3. Committee on Field Tests, "Suggested Methods for Monitoring Rock Movements Using Borehole Extensometers," International Society for Rock Mechanics Commission on Standardization of Laboratory and Field Tests, Document No. 5 (Pergamon Press Ltd. Great Britain 1978).
4. I. Hawkes and W. V. Bailey, Low Cost Cylindrical Stress Gage, U. S. Dept. of Commerce, NTIS PB243-374/A5, Springfield, VA (1973).
5. V. E. Hooker, J. R. Aggson, and D. L. Bickel, Improvements in the Three-Component Borehole Deformation Gage and Overcoring Techniques, U. S. Bureau of Mines, Report of Investigations No. 7894 (1974).

GEOCHEMISTRY, ORIGIN AND AGE OF GROUNDWATERS IN THE STRIPA (SWEDEN) TEST MINE*

P. Fritz, J. F. Barker and J. E. Gale

Department of Earth Sciences
University of Waterloo
Waterloo, ON, N2L 3G1

INTRODUCTION

The evaluation of a potential repository site for nuclear waste must include premining assessment of the hydrology and geochemistry of groundwaters which eventually will fill the repository in the post-emplacement phase. Neither can be predicted a priori, and both will strongly depend on the regional geology of the area. This paper presents aspects of geochemical observations made on groundwater from the Stripa test site in Sweden, where a granitic pluton adjacent to a 400 year old iron ore mine was mined for a variety of tests relevant to rock stability, fracture hydrology and geochemistry within the context of nuclear waste disposal.

Groundwater in crystalline rocks is usually thought to have a low total dissolved solids (TDS) content, with its constituents reflecting the instability of the specific mineral assemblage of the rock in which the water is found. Within certain limits this is true for many shallow groundwater systems in crystalline rocks. Deep groundwaters in crystalline rocks, however, may have circulated over large distances and so have encountered very different mineralogies than those of their present host. Some highly saline groundwaters found in granitic rocks of the Fennoscandian and Canadian Shields may have formed by dissolution of evaporites present in metasedimentary sequences associated with igneous intrusions. Thus, the occurrence of saline waters in many areas on both Shields is not necessarily linked to the presence of fossil sea water, as is thought to be the

*Work done under University of California Contract W-7405-ENG-48 for the U.S. Dept. of Energy.

case for occurrences close to the Baltic Sea on the Fennoscandian Shield (1). Another important factor is time: deep groundwaters in crystalline rocks may be very old, and slow reactions between groundwater and high temperature silicates could gradually lead to higher and higher salinities because thermodynamic equilibria between such silicates and low temperature groundwaters cannot be established.

In the Stripa study, geochemical and environmental isotope (^{18}O, ^{2}H, ^{13}C, etc.) analyses are being used to define the origin of groundwaters and to describe their geochemical evolution. Naturally occurring radioisotopes, particularly ^{3}T, ^{14}C and members of the uranium decay series, are employed to estimate the ages of groundwaters encountered. Groundwater was obtained from boreholes flowing into the test site at about 370 m depth and from isolated intervals of a flowing (10L/h) borehole drilled by the Swedish Geological Survey (SGU) from a 410 m mine level to about 940 m below ground level. Shallow water table wells and private water supply wells in the area were also sampled. Sampling and analytical techniques have been described by Fritz et al. (2).

GROUNDWATER CHEMISTRY

The available data are summarized in Table 1. They show an increase in total dissolved load with depth. This increase is due primarily to Ca^{++}, Na^{+}, Cl^{-} and some sulphate. Bicarbonate (or total inorganic carbon) decreases dramatically below $\sim$100 m depth and both Mg^{++} and K^{+} contents drop below 1 mg/L in mine waters from much higher levels in the shallow groundwaters.

Especially remarkable, however, is the rise in pH with values as high as 9.8 in the deepest groundwaters discharging from the fracture systems. This rise in pH is probably linked to the dissolution of primary silicates such as feldspars and the formation of clay minerals. These processes release calcium which causes continuous saturation of the mine waters with respect to calcite. Thus, as Ca^{++} is freed from feldspars, calcite will precipitate which explains its ubiquitous presence as a fracture mineral as well as the decrease of total inorganic carbon (TIC) with depth.

Other common fracture minerals are quartz, chlorite and sericite (muscovite). Their formation is also linked to the dissolution of primary silicates. Silica contents in these waters never exceed 16 mg/L, and quartz saturation is reached even in shallow groundwaters. Chlorite and sericite formations are probably responsible for the loss of Mg^{++} and K^{+}, respectively.

The increased Na^{+} concentration at depth could be explained by plagioclase dissolution, but it is difficult to determine the origin of Cl^{-}, whose concentration approaches 500 mg/L in the deep waters.

Table 1. Geochemical Data on Groundwater from the Stripa Granite Area

Sample locality	depth(m)	t°C'	pH	HCO_3^-	Ca^{++}	Mg^{++}	Na^+	K^+	Cl^-	$SO_4^=$	SiO_2
		FIELD DATA					LABORATORY ANALYSES				
SHALLOW GROUNDWATERS											
Water table well #2	< 10	4.2	5.15	2.9	2.0	1.2	2.1	0.25	2.5	11.8	8.2
Water table well #3	< 10	6.0	6.62	34.1	8.9	1.6	4.7	1.5	2.3	7.4	9.4
Private well #5	42	5.8	6.2	17.3	5.7	1.5	2.7	1.1	2.3	13.1	10.0
Private well #4	40	10.7	8.05	176.8	31.8	10.5	10.9	4.3	2.4	10.3	16.4
Private well #2	60	7.7	7.35	83.6	19	4.5	4.5	1.4	5.2	13	8.8
SBH-3	89-104	6.3	7.89	142.5	33.9	4.5	12.5	1.7	3.7	8.8	12.5
330 m LEVEL GROUNDWATER											
R1, ventilation drift	~330	11.7	8.89	97.4	15.4	0.27	44.7	0.2	35.4	3.0	11.2
R9, ventilation drift	~320	n.d.	8.55	68.4	18.0	0.25	40.6	0.13	49.6	1.3	12.1
M3 timescale room	340	14.5	9.10	84.4	16.3	0.25	51.5	0.2	56.0	3.1	11.1
410 m LEVEL BOREHOLE											
	418-449	8.2	9.16	50.3	19.6	0.29	64.0	0.2	90.5	5.4	11.3
	801-838	8.0	9.60	9.4	110	0.18	177	0.55	449	55.4	10.2
	801-838	8	9.44	8.3	109	0.2	176	0.5	439	42.9	10.3

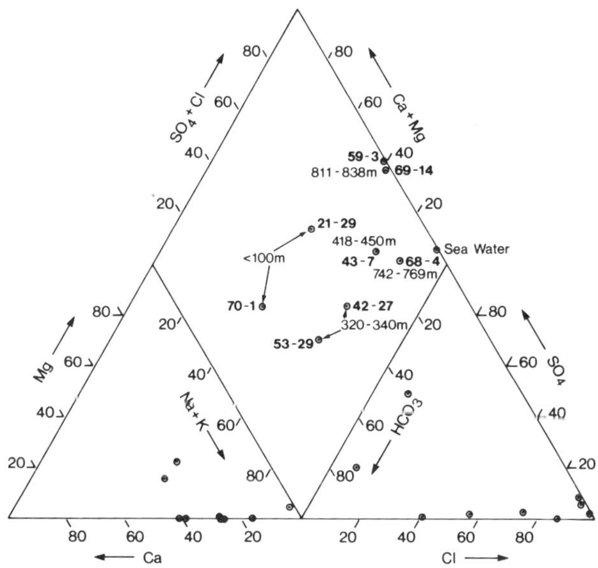

Fig. 1. Piper Diagram Showing Chemical Analyses of Principle Water Types Found in the Granite at Stripa (Sweden)

If it is attributed to fossil sea water, then very significant geochemical changes must have modified the chemistry of this water, and simple mixing between fresh water and this sea water cannot explain the observed chemistry. This is shown in Fig. 1. where the chemical compositions of the different waters are composed on a Piper triangular diagram.

It is noteworthy that elsewhere on the Fennoscandian Shield, evidence for the presence of fossil sea water is much stronger. For example, Mg^{++} is not a trace constituent and increases with increasing salinity (1). This is not the situation in Stripa groundwaters. Thus, the geochemical history of the deep Stripa groundwaters is more complex than that of saline groundwaters from other localities in similar rocks, and the deep groundwaters at Stripa are apparently not related to fossil sea water which would have infiltrated less than 10,000 years ago.

An interesting observation is that the chemistry of water discharging in fractures between ∿300 and ∿770 m below ground surface are very similar (Table 1), but below 770 m a sharp break occurs and the saline waters appear. Thus, different types of water discharge from different fracture systems and little mixing appears to exist between them. To substantiate this observation, an extension of the 410 m level borehole to greater depth would be needed. At present, only one sampling point for the saline water exists and does not suffice to recognize concentration gradients which could indicate mixing.

STABLE ISOTOPE ABUNDANCES

At all sampling points, ^{18}O, ^{2}H and ^{13}C contents were determined to obtain some information on the origin of these waters. The results of the ^{18}O and ^{2}H analyses* are shown in Fig. 2. In Fig. 3., Cl^- contents are compared with ^{18}O concentrations.

The comparison of ^{18}O and ^{2}H data shows that all groundwaters sampled fall close to the global meteoric waterline. This signifies that evaporation and/or isotope exchange with oxygen bearing minerals at elevated temperatures are insignificant. They are thus "normal" groundwaters for which ^{18}O and ^{2}H contents reflect climatic conditions in the recharge area.

As a general rule, lower heavy isotope concentrations signify lower average annual temperatures at the recharge area. Therefore, the deep saline groundwaters which have the lowest ^{18}O and ^{2}H contents must have recharged at lower average annual temperatures than the shallower groundwaters. This has been confirmed by rare gas analyses done on samples collected at all levels (2, 3), and there-

*The ^{18}O and ^{2}H analyses are reported as δ $^{o}/oo$ values with reference to SMOW. A δ ^{18}O = -10 $^{o}/oo$ signifies that the sample has 10 $^{o}/oo$ (permil) less ^{18}O than the reference standard which closely reflects average ocean water. The ^{13}C contents are reported as δ $^{o}/oo$ values with respect to the standard PDB, a calcitic Belemnoid rostrum from the Pee Dee Formation in the Carolinas.

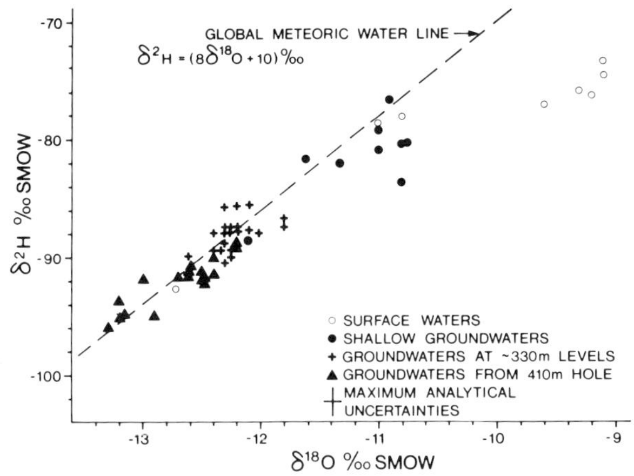

Fig. 2. Comparison of $\delta^{18}O$ and $\delta^{2}H$ values of water samples collected from surface waters, shallow groundwaters and water from the granite at Stripa. Note the closeness to the meteoric water line of all but the surface water samples.

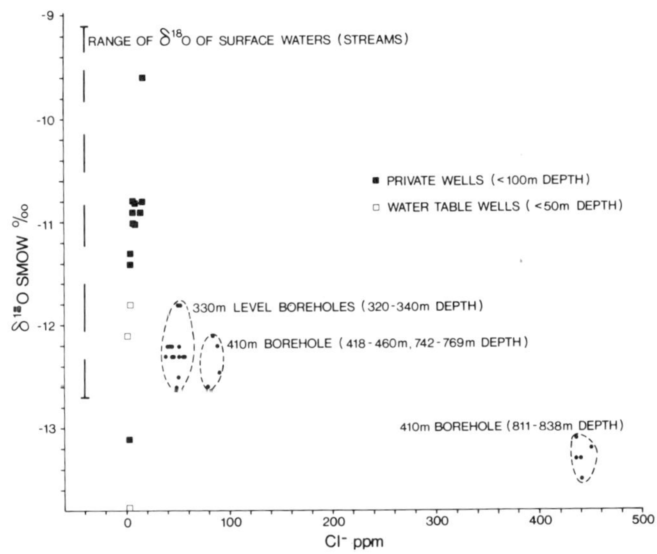

Fig. 3. This comparison of Cl^{-} with $\delta^{18}O$ values shows that in the Stripa granite, different fracture systems carry different types of water.

fore, one must conclude that the deep groundwaters have a different origin than the shallower ones.

The comparison of Cl^- concentrations and $\delta^{18}O$ values in Fig. 3. further substantiates that fracture systems at different depths carry different types of water. At Stripa the regional picture is missing, and therefore, we cannot discuss the movements and origins of the different groundwater masses beyond the observation that average annual temperatures in the recharge areas of the deeper mine water system must have been slightly lower than in the shallow ones. This probably signifies that the deeper groundwaters are much older than the shallow ones.

Because of the lower ^{18}O isotope contents in the deeper groundwater, one could argue that they might represent subglacial recharge. This is disputed by the ^{13}C analyses: all waters from the mine levels have $\delta^{13}C$ values close to or below -15 ‰ which indicates that biogenic carbon is present in the dissolved organic carbon. Barring significant bacterial activities in the subsurface, this would signify that these waters infiltrated through soil horizons, i.e., were generated during an interglacial period. Therefore, they must be younger than about 8,000 to 10,000 years or older than about 25,000 years.

The question of the presence of fossil sea water can also be addressed. The $\delta^{18}O$ values of modern oceans, including the Baltic Sea, are much higher than the ^{18}O contents of the deep waters. As decreasing ^{18}O and ^{2}H contents with increasing depth and salinity are observed, no evidence for the presence of isotopically enriched seawater is found, unless the fresh water component of the deep groundwater had much lower heavy isotope contents than those measured. Mixing of the two components could then have produced the waters sampled. Therefore, stable isotope data do not indicate that fossil seawater is present, yet do not conclusively exclude it, either. As already mentioned, further drilling and sampling at greater depths is needed for more conclusive discussions.

GROUNDWATER AGE DETERMINATIONS

The most difficult and inconclusive part of this project was the attempt to date the groundwaters which discharged at the various mine levels. All techniques employed, with the exception of tritium analysis, rely on an understanding of the geochemical evolution of these waters, which is still rather incomplete because of the inability to assign a clear origin to the saline groundwaters.

Tritium was not encountered (< 0.5 T.U.)* in any of the mine levels, which signifies that the waters in the different fracture systems do not have a component which is younger than 30-40 years.

*One T.U. = 1 ^{3}H atom per 10^{18} 'H' atoms.

However, tritium levels approaching 100 T.U. were measured in all shallow groundwaters and--interestingly enough--also in waters seeping into the old mine workings. This was surprising because the sink generated by the old iron ore mine could have drained the fracture systems in the adjacent granite. However, drilling in the old mine workings and the proximity of open pits might have generated preferential paths for modern surface waters to penetrate the old mine.

The most commonly used tool to date groundwaters is ^{14}C, and an extensive sampling program for such age-determinations was carried out. However, very major problems were encountered because of the very low dissolved inorganic carbon content in the deep waters. This required the treatment of 2,000-3,000 liters of water in order to obtain sufficient carbon for a conventional ^{14}C analysis (3). The results obtained ($<$ 3 pmC* at the 330 m levels during the initial sampling) indicate that this water and probably the waters discharging from the 410 m level holes exceeded 20,000 years in age (Table 2). Unfortunately, because of possible contamination problems, the ages for samples from the 410 m level can at present not be better defined.

The uranium decay series offers a number of potential dating tools: a) changing uranium activity ratios, b) helium accumulations with time in closed system and c) radium-radon relationships. All were evaluated.

The $^{234}U/^{238}U$ activity ratios in the mine waters change from 10.4 at the 330 m levels to about 6 at the top of the 410 m level hole ($\sim$ 440 m below ground surface) and to almost 4 at the bottom of the 410 m hole in the "saline" waters. For model considerations, it must be assumed that these waters underwent similar geochemical evolutions (although there is no direct evidence for it) and therefore, that all deeper waters once had a uranium isotopic composition similar to the one found today in the water at the 330 m levels. Furthermore, no exchange with fracture minerals can be permitted. If these conditions were met, then the "closed system" evolution of the activity ratios, i.e., their movement from 10 to 4 towards a secular equilibrium, is a function of time. Water ages exceeding 100,000 years are then calculated for waters from fractures below the 410 m level.

Somewhat lower ages are determined from the He concentrations in the mine waters. The He content of mine waters is several orders of magnitude above atmospheric saturated levels (4.9×10^{-8} cm^3He/cm^3H_2O). Waters from the M-3 flowing borehole at the 330 m level contained 3.6 - 3.0×10^{-4} cm^3He/cm^3H_2O, and the 801-838 m depth interval of the

---

*pmC = percent modern carbon as defined by 95% of the activity of the oxalid acid standard distributed by the National Bureau of Standards.

Table 2. Estimated Groundwater Ages (Years) at the Stripa Test Site

	$\delta^{18}O$ $^{o}/oo$ SMOW	$\delta^{13}C$	^{14}C	$^{234}U/^{238}U$ Activity Ratios	$^{234}U/^{238}U$ + Radon	$^{226}Ra-^{222}Rn$	^{4}He	$^{40}Ar/^{36}Ar$
SHALLOW <100 m	-10.9	<10,000 or >25,000	MODERN					INCREASING AGE ↓
330 m LEVELS	-12.2		>20.000		12,000 - 35,000	>8000	n x 10^4	
410 m BOREHOLE TOP	-12.3		>25.000	>2.5 x 10^5	10,000 - 30,000		n x 10^4	
410 m BOREHOLE BOTTOM	-13.2		?	>4 x 10^5		>8000	n x 10^5	VERY OLD?
	(1)		(2)	(3)			(4)	(5)

REMARKS:
(1) Lower ^{18}O contents signify lower recharge temperatures provided global atmosphere circulation patterns did not shift.
(2) These age brackets assume insignificant bacterial activities such as methane oxidation.
(3) Closed system evolution without rock-water interaction is assumed.
(4) The estimated values are dependent on uranium/thorium contents of the rocks. Values typical for the Stripa granite were assumed (30-160 ppm).
(5) No age estimates were attempted because no information on argon release rates from the rocks was available.

SGU borehole contained 1.4 - 0.86×10^{-3} $cm^3 He/cm^3 H_2O$. These elevated He contents result from alpha decay in the uranium-thorium decay series. Assuming closed system conditions without loss or gain of He at the sampling point and knowing U-Th concentrations of the rock and He release rates, then He ages for the groundwaters can be calculated. These exceed several tens of thousands of years (Table 2).

If ^{222}Rn accumulates as a recoil product and is in equilibrium with ^{226}Ra, then ^{234}U excess and ^{222}Rn activities will eventually reach equilibrium. Ages calculated by this model indicate 10,000-35,000 years for the different mine waters. An extension of this approach considers the ^{226}Ra concentrations in the rock minerals and the ^{226}Ra in the water. If equilibrium between the two exists, i.e., if the recoil rate from the rock equals decay in solution, then the waters must be at least 8,000 years old (5 half lives of ^{226}Ra). There is evidence that this is the case supporting again the assumption that the waters presently discharging at the different mine levels from fracture systems in the granite are indeed many thousands of years old. The atmospheric $^{40}Ar/^{36}Ar$ ratio is 295.5. In a closed system it will increase if ^{40}K decay occurs. This is the case in Stripa where values increase to 302.8 at the 330 m level and to 315.6 at the bottom of the 410 m hole. This enhancement is almost certainly a function of age and would indicate that the mine waters are indeed very old.

CONCLUSIONS AND SUMMARY

Environmental isotope and chemical analyses on waters collected at various depths from water-bearing fracture systems in the granitic pluton show that different types of groundwater are present. Shallow groundwaters (< 100 m depth) are modern and infiltrated locally. They are characterized by a low TDS content but by a variable composition which depends on the local geology.

Deeper groundwaters show higher salinities. Expressed through Cl^- analyses, values rise from < 10 mg/L Cl^- in shallow groundwaters to ∿500 mg/L at depth. Elsewhere, it has been postulated that such increases were due to the presence of small amounts of fossil seawater, and a similar explanation cannot be totally excluded for the Stripa groundwaters. However, if fossil seawater were present, it must have undergone very significant changes through rock-water interactions. This differs from other occurrences of fossil seawater and could suggest that the Stripa waters (including the possible seawater component) are older than these. Any alternate explanation which does not invoke the presence of seawater must also rely on geochemical evolutions over long timespans. Environmental isotope analyses support this argument. ^{18}O and ^{2}H contents show that the deep waters infiltrated during a colder climatic period and $\delta^{13}C$ values indicate that this occurred through a soil cover and not during glaciation.

Groundwater age determinations were attempted using ^{14}C and elements of the uranium decay series. Although it was not possible to obtain absolute ages, all techniques indicate that these waters are many thousands of years old. This implies that the hydraulic sink generated through the mining activities in preceding centuries was not able to drain these waters either because the reservoirs are enormously big or because the horizontal conductivity of the fracture systems is limited. Studies are in progress to test these hypotheses.

ACKNOWLEDGEMENTS

The geochemical program of the Stripa project could not have been carried out without the help of the staff of the Lawrence Berkeley Laboratory (LBL), University of California, U.S.A.; Kärnbränslesäkerhet (KBS) and Sveriges Geologiska Undersökning (SGU), both of Stockholm, Sweden; Department of Earth Sciences, University of Waterloo, Canada; and the International Atomic Energy Agency, Section of Isotope Hydrology, Vienna, Austria. The assistance of Stalllbergsbologen personnel, particularly P-A. Halén and O. Hagström, is also gratefully acknowledged. The sampling program at Stripa was carried out with the particular assistance of T. Doe (LBL), C. Forster and

O. Quinn (University of Waterloo), and K-E. Almen, P. Hammargren, K. Hansson and L. Ekman, all of SGU.

This work was supported by funds from Contract #W-7405-ENG-48 to the Lawrence Berkeley Laboratory under Purchase Order 478-3902 and through WRI Contract 803-12 as well as funds from the National Research Council of Canada given to P. Fritz (Grant #A7954).

REFERENCES

1. G. Jacks, "Chemistry of Some Groundwaters in Igneous Rocks," Nordic Hydrology 4 (4), 207 (1973).
2. P. Fritz, J. F. Barker and J. E. Gale, "Geochemistry and Isotope Hydrology of Groundwaters in the Stripa Granite," Univ. of California, Lawrence Berkeley Laboratories, Berkeley, CA, Rep. LBL-8285, 107 (1979).
3. P. Fritz, J. F. Barker, J. E. Gale, J. N. Andrews, R. L. F. Kay, D. J. Lee, J. B. Cowart, J. K. Osmond, B. R. Payne and P. K. Witherspoon, "Geochemical and Isotopic Investigation at the Stripa Test Site (Sweden)," Proc. Symp. Underground Disposal of Radioactive Wastes, IAEA, Otaniemi, Finland, July 1979, in press.

RECENT STUDIES ON RADIATION DAMAGE FORMATION IN SYNTHETIC NaCl AND NATURAL ROCK SALT FOR RADIOACTIVE WASTE DISPOSAL APPLICATIONS*

K. J. Swyler, R. W. Klaffky** and P. W. Levy

Brookhaven National Laboratory
Upton, NY 11973

ABSTRACT

Radiation damage formation in natural rock salt is described as a function of irradiation temperature and plastic deformation. F-center formation decreases with increasing temperature while significant colloidal sodium formation occurs over a restricted temperature range around 150°C. Plastic deformation increases colloid formation; it is estimated that colloid concentrations may be increased by a factor of 3 if the rock salt near radioactive waste disposal canisters is heavily deformed. Optical bandshape analysis indicates systematic differences between the colloids formed in synthetic and natural rock salts.

INTRODUCTION

A continuing study (1, 2) of radiation damage in synthetic and natural rock salts is being carried out in support of the radioactive waste isolation program. The most recent measurements have been concentrated on natural rock salts from different localities; results obtained with natural samples from the AEC-8 drill hole in Southeastern New Mexico will be emphasized in this paper.

---

*Research supported by the U.S. Department of Energy Office of Waste Isolation at Battelle-Columbus and by the U.S. Department of Energy Division of Basic Energy Sciences.

**Now at the Los Alamos Scientific Laboratory, Los Alamos, NM.

Earlier investigations (1-5) have established the importance of radiation-induced colloidal metal formation in radiation damage considerations. Colloidal sodium formation in NaCl is believed to occur by virtue of the diffusion and aggregation of radiation-induced F-centers (negative ion vacancies containing trapped electrons). Under suitable conditions this aggregation stabilizes the defects, and substantial concentrations of colloidal sodium (and complimentary defects) may be achieved. This process is particularly important in the present context since, in synthetic NaCl, maximum colloid formation occurs at temperatures ($\sim$150°C) within the range anticipated on radioactive waste repositories.

A theory of radiation-induced colloid formation has been advanced by Jain and Lidiard (6). This model stresses the importance of temperature, dose rate and dislocation density as controlling factors in colloid growth. Previously (1, 2), the effects of these and other parameters on radiation damage formation have been systematically determined in synthetic NaCl and initially examined in natural salt. These measurements have demonstrated the importance of dislocation related effects in colloid formation and uncovered systematic differences between the colloid formation kinetics in synthetic and natural rock salts. The present paper describes the detailed effects of temperature and strain on radiation-induced F-center and sodium colloid formation in natural rock salt (AEC-8). In addition, the equivalence of colloid formation in synthetic and natural samples is investigated by comparison of optical bandshapes.

EXPERIMENTAL

The experimental techniques have been described previously (1, 2) and will be mentioned only briefly here. Optical absorption bands associated with particular defects are monitored as the samples are continuously electron irradiated at controlled temperatures. A large number of absorption spectra are recorded at successive times during the irradation. These can be used to construct "growth curves" of absorption (or defect concentration) vs. time (or irradiation dose). The composite spectra can be precisely resolved into F-center and sodium colloid bands by a numerical procedure discussed earlier, yielding growth curves which are accurately corrected by absorption band overlap. Moreover, the resolved colloid bandshape (which generally evolves with time) contains information about the colloid particle size distribution.

TEMPERATURE DEPENDENCE OF RADIATION DAMAGE FORMATION IN ROCK SALT

The growth of F-center and colloid band absorption in AEC-8 rock salt during irradiation at different temperatures is shown in Fig. 1. The behavior is generally similar to that reported previously for

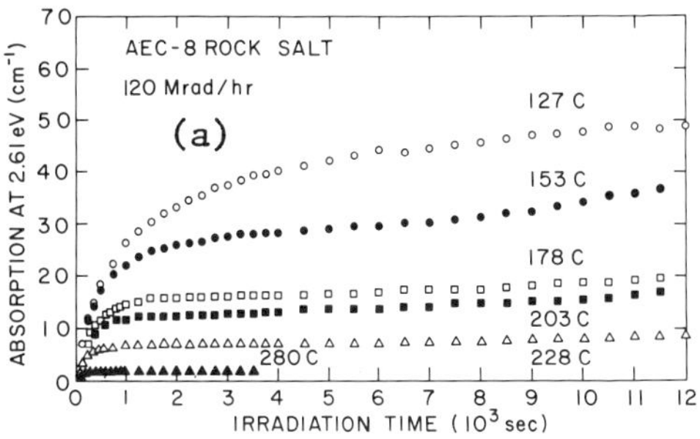

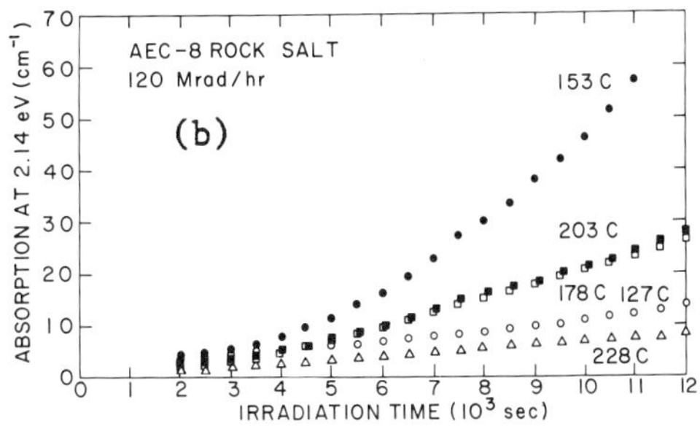

Fig. 1. Growth of F-Center (a) and Sodium Colloid (b) Absorption in AEC-8 Rock Salt During Irradiation at Various Temperatures. The data are not corrected for overlap of the F-center and colloid bands.

synthetic NaCl (1, 3, 4): F-center growth curves approach saturation levels which decrease with increasing temperature, while significant colloid formation occurs only after the F-center curves are near or at saturation. This colloid formation occurs only over a restricted temperature range (Table 1).

In the Jain-Lidiard theory (6), F-center diffusion is taken to be a rate limiting process in colloid formation. The theory treats F-center diffusion as a thermally activated process with a single activation energy and predicts that F-center saturation should be achieved at successively lower radiation doses as the temperature is

Table 1. Synthetic and Natural Rock Salt Colloid Absorption Band Growth Rates Measured During Irradiation at 120 Mrad/hour

AEC-8 Rock Salt		H-2 Synthetic NaCl	
Irradiation Temp (°C)	Growth Rate* (cm^{-1}/hr)	Irradiation Temp (°C)	Growth Rate* (cm^{-1}/hr)
		100	37.8**
127	4.28	127	21.6
153	34.0	150	79.0
		163	72.0
178	11.0	171	19.8
		187	43.0
203	12.0	200	1.31
227	2.5	226	4.31
		250	nil
280	nil		

*Determined from derivatives of growth curves, such as Fig. 1B., at 10,500 sec irradiation time. Such values are indicative of the maximum observed growth rates.
**Possibly an overestimate.

increased. This trend is clearly evident in the data of Fig. 1A. The theory also indicates that the activation energy for F-center diffusion may be obtained from the slope of a semilogrithmic plot of F-center saturation levels vs. reciprocal temperature. Such a plot is shown in Fig. 2. As in synthetic NaCl (2), the data do not appear to represent a single thermally activated process over the entire temperature range. Over the temperature region where colloid formation is observed (roughly 125° to 225°C), an activation energy of 0.6 eV is obtained. This value is substantially greater than the 0.4 eV obtained over the same region in synthetic NaCl and fairly close to the value of 0.8 eV assumed by Jain and Lidiard.

The maximum observed colloid absorption band growth rates at different temperatures, as determined from the data in Fig. 1B., are given in Table 1. For comparison, similar data for synthetic NaCl (1) are also included. In both materials, the maximum colloid forma-

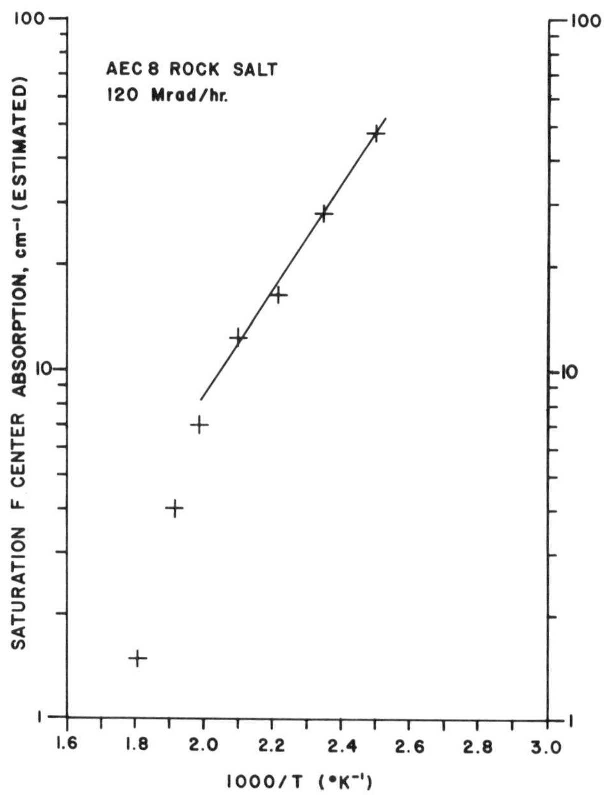

Fig. 2. Plot of F-Center Absorption at Saturation vs. Reciprocal Temperature for AEC-8 Rock Salt. The absorption values were estimated from the data in Fig. 1A. after making an approximate correction for colloid band overlap.

mation rate occurs around 150°C at the dose rates used here. However, in the synthetic salt, colloid formation rates are greater than in natural salt.

EFFECTS OF STRAIN ON RADIATION DAMAGE FORMATION IN ROCK SALT

Irradiation-induced colloid formation in AEC-8 rock salt is increased by uniaxial compressive strain applied prior to irradiation. This effect is illustrated by Fig. 3. In samples deformed more than 2 or 3%, rapid colloid formation begins nearly at the onset of irradiation. Thereafter, colloid absorption increases almost linearly with irradiation dose. In undeformed samples, an appreciable radiation dose is necessary to initiate colloid formation, and the colloid growth curves are more nearly parabolic over the initial rapid growth region.

Qualitatively, the effects described above are similar to those encountered previously in synthetic NaCl (1). There are, however, systematic differences in the detailed behavior. For example, quasilinear growth of colloid absorption at low total doses was not observed in synthetic samples deformed as much as 6%. The Jain-Lidiard theory predicts that, once nucleation is completed, the concentration of colloidal sodium should increase linearly with irradiation dose at sufficiently high dislocation densities. It appears that these con-

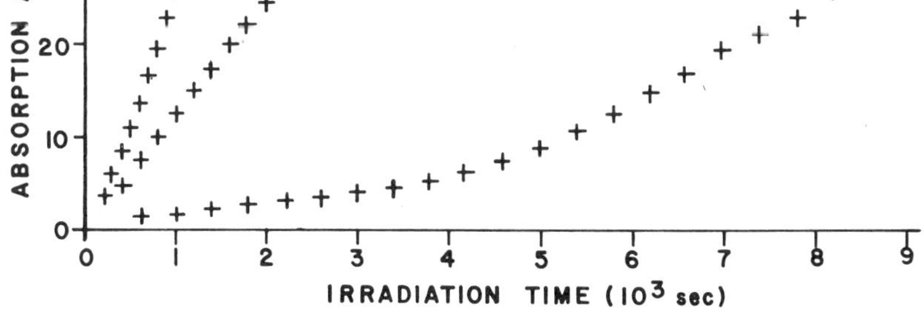

Fig. 3. Growth of Colloid Absorption During Irradiation of Plastically Deformed AEC-8 Rock Salt. The data are not corrected for F-band overlap.

ditions may be achieved in plastically deformed natural rock salt at low radiation doses.

The effect of plastic deformation on defect concentrations in natural rock salt at repository doses ($\sim 10^{10}$ rad) may be estimated by extrapolating the colloid growth curves. The F-center concentrations are not significant since the F-center saturation levels are not greatly modified by plastic deformation. Growth rates for colloid absorption in strained and unstrained AEC-8 samples are given in Table 2. A linear extrapolation based on these data indicates that at repository doses colloid concentrations will increase only moderately ($\sim$30%) in rock salt deformed 4%. However, an 11% deformation would increase the colloid concentration by a factor of 3.

ANALYSIS OF COLLOID BANDSHAPES IN SYNTHETIC AND NATURAL ROCK SALTS

There are three prominent differences between the radiation-induced colloid absorption spectra of sythetic and natural rock salts:

1. In synthetic samples at 150°C, the bands are comparatively broad and definitely sharpen as they increase in intensity. In natural rock salt, the bands are narrow and do not sharpen appreciably during irradiation.

2. In natural samples, the peak positions shift to lower energies as the colloid band grows. This effect is particularly evident in samples irradiated at temperatures above 150°C. This shift is barely discernible in synthetic NaCl.

3. In natural samples, an absorption band near 390 nm (3) accompanies the colloid band. This band is most prominent in samples

Table 2. Colloid Absorption Band Growth Rates Measured During Irradiation at 150°C in Strained and Unstrained AEC-8 Rock Salt

Treatment (% strain)	Growth Rate* (cm^{-1}/hr)
0	33
0	28
0	33
3.9	44
11	102

*Determined from derivatives of growth curves, such as Fig. 3., at the upper measurement limit. Linear extrapolation to repository doses predicts a 300% increase in colloid sodium at 11% strain.

irradiated at temperatures above 150°C. It is only faintly evident or wholly absent in synthetic NaCl.

Evidently, the colloids in synthetic samples are not entirely equivalent -- either different stages of colloid growth are being observed or different defects occur in the two types of crystals. Carefully resolved colloid bands measured in synthetic and natural crystals at 150°C have been fitted to a form of the Mie theory which includes particle size effects (7). The analysis indicates that, for comparable concentration of metallic sodium, the colloid particles are smaller (1.6 vs. 2.0 nm radius) and more numerous (7×10^{14} vs. 3×10^{14} colloids/cm^3) in synthetic NaCl than in natural rock salt. This result suggests differences in the colloid nucleation. Interestingly, the analysis finds that the colloid radii increase with radiation dose in synthetic NaCl but remain practically constant in natural rock salts. It is not apparent why these differences in colloid particle growth characteristics occur and further measurements are planned.

DISCUSSION AND SUMMARY

The temperature dependence of F-center and colloidal sodium particle formation in AEC-8 rock salt is in good agreement with previous work. At the dose rates employed in these measurements, detectable colloid formation occurs over the temperature range extending from about 125° to 225°C with a maximum near 150°C. Near this maximum, substantial metallic sodium concentrations may be anticipated at repository doses. Some aspects of the F-center temperature dependence are in semi-quantitative agreement with theory. It does not appear at this point, however, that F-center diffusion during irradiation can be completely described by a single thermally activated process.

Irradiation-induced defect formation in natural rock salt is enhanced by plastic deformation. Near the maximum colloid formation temperature, certain features of this strain enhancement support the Jain-Lidiard theory if it is assumed that the primary effect of strain is to generate dislocations. When extrapolated to repository doses, these results indicate that an 11% strain would increase the colloid content by a factor of 3.

An analysis of colloid band shapes indicates that the previously reported differences between the colloid formation kinetics in synthetic and natural samples are accompanied by differences in the colloids themselves. The origin of these differences is unclear at present; possibly it lies in the nucleation process. Also, there is some evidence suggesting that F-center diffusion during irradiation occurs more readily in synthetic samples: the apparent activation energy for F-center diffusion is lower, colloid formation rates are greater, and colloid formation occurs at lower temperatures in synthetic NaCl than in natural rock salt samples.

REFERENCES

1. K. J. Swyler, R. W. Klaffky and P. W. Levy, "Radiation Damage Studies on Synthetic NaCl Crystals and Natural Rock Salt for Waste Disposal Applications," Scientific Basis for Nuclear Waste Management, 1, 349 (G. J. McCarthy, ed., Plenum, NY, 1979).
2. R. W. Klaffky, K. J. Swyler and P. W. Levy, "Radiation Damage Studies on Synthetic NaCl Crystals and Natural Rock Salt for Waste Disposal Applications, II," Ceramics in Nuclear Waste Management, Conf. 790420, Proc. of an Int. Symp., Cincinnati, OH, April 30-May 2, 1979, 310 (T. D. Chikalla and J. E. Mendel, eds. Tech. Info. Center, U.S. Dept. of Energy, Washington, 1979).
3. L. W. Hobbs, M. Saidoh and V. M. Orera, "Void Analogues in Irradiated Alkali Halides," Int. Conf. on Defects in Insulating Crystals, Gatlinburg, TN, 172 (1977).
4. L. W. Hobbs, "Point Defect Stabilization in Ionic Crystals at High Defect Concentrations," J. Physique Colloq. 37, C7-3 (1976).
5. G. H. Jenks and C. D. Bopp, Storage and Release of Radiation Energy in Salt in Radioactive Waste Repositories, ORNL-5058, Oct. (1977).
6. U. Jain and A. B. Lidiard, "The Growth of Colloidal Centers in Irradiated Alkali Halides," Phil. Mag. 35, 245 (1977).
7. V. Kreibig and C. V. Fragstein, "The Limitation of Electron Mean Free Path in Small Silver Particles," Z. Phys. 224, 307 (1969).

PHYSICO-CHEMICAL PROCESSES IN ROCK SALT UNDER IRRADIATION

V. I. Spitsyn, S. A. Kabakchy, L. I. Barsova

USSR State Committee of Atomic Energy

INTRODUCTION

We have investigated both color center formation and thermoluminescence techniques as possible tools for determining radiation doses absorbed by γ-irradiated rock salt. Experiments were performed with rock salt samples taken from two different deposits in the USSR Caspian region. Experimental conditions simulated the high dose and temperatures typical (1) of those expected for storage of high level nuclear wastes in salt strata.

We were also interested in radiation-induced chemical transformations which occur in liquid brines. Rock salt formations usually contain both micro- and macro-inclusions of liquid brines. We studied the radiolysis of saturated aqueous sodium chloride samples to investigate effects of radiation upon brine inclusions.

THERMOLUMINESCENCE MEASUREMENTS

Luminescence can be observed (Fig. 1) in thermally annealed (at 300 - 600°K) samples of rock salt irradiated at 300°K. Table 1 summarizes the temperatures where maximum luminescence is observed.

The luminescent maxima temperature for Peak I for these two irradiated salt samples was independent of dose rate and absorbed dose (0.03 - 20 Mrad). Activation energies of luminescence were calculated using the procedure given in Reference 2. For Sample 1 in the region of Peak I, the activation energy of luminescence is 18 kcal/mol, independent of absorbed dose (Table 2). In the region of Peak III, however, the activation energy of luminescence decreases

Table 1. Temperatures for Maximum Luminescence in Irradiated Rock Salt

Rock Salt Sample Number	Peak Luminescence I	II	III
1	333 ± 3°K	Absent	543 ± 5 K
2	333 ± 3°K	356 ± 2°K	543 ± 5 K

from 40 kcal/mol to 24 kcal/mol as the absorbed dose increases from 0.1 to 10 Mrad. The activation energy is equal to 23 kcal/mol, in the region of Peak III for a rock salt sample irradiated to dose of 230 Mrad.

The activation energy of luminescence Peak I was not altered by heating rock salt samples to 579°K before irradiation. On the other hand, at doses in the range 2-10 Mrad, the activation energy of luminescence Peak III of samples heated to 570°K before irradiation is 40 kcal/mol compared to 24.4 - 40.8 kcal/mol for unheated samples. Thus, the dependence of activation energy on absorbed dose is not apparent for samples heated prior to irradiation.

The activation energy for luminescence for irradiated rock salt samples may depend upon the particular geographical location from which the salt samples are taken, as evidenced by the results for Sample 2 (Table 1). Luminescence maxima have been observed at the same temperatures for different salt samples. But, activation energies of luminescence Peak III differ, under the same experi-

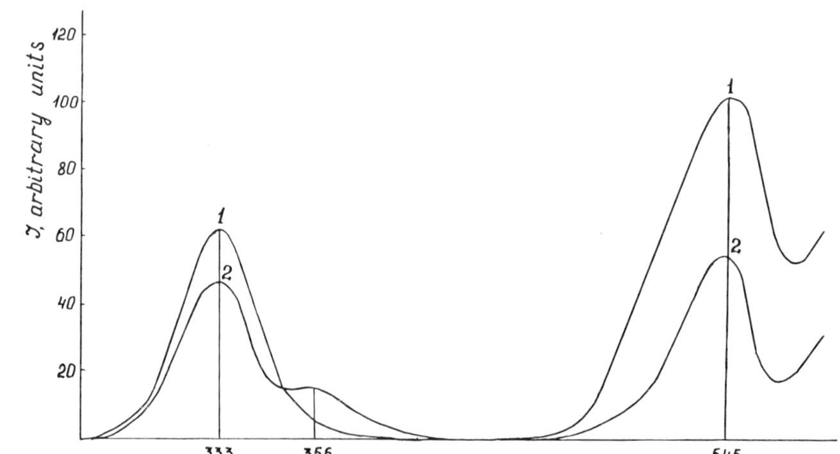

Fig. 1. Luminescence of Thermally Annealed Rock Salt (Samples 1 & 2)

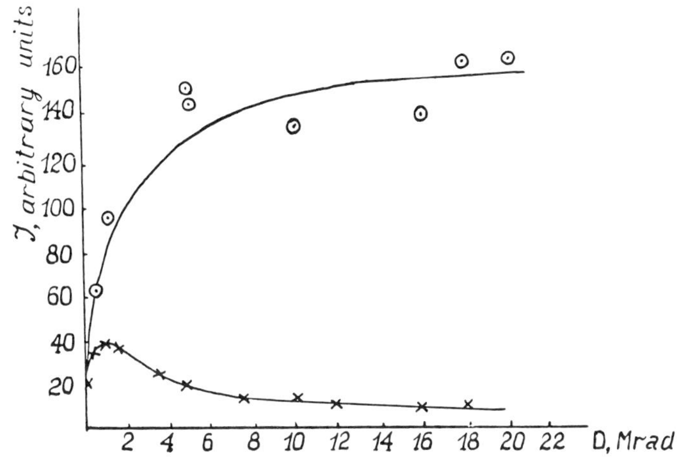

Fig. 2. Dose Dependence of Luminescent Intensity for Peak I (·) and Peak 2 (x)

mental conditions, by as much as 10-15 kcal/mol. These energy differences arise because of the difference in amount and type of impurities in the rock salt samples.

We also evaluated the dependence of luminescence peak intensity, at 333°K and 545°K, upon doses absorbed by rock salt (Sample 1). As indicated in Fig. 2, the intensity of luminescence (Peak I) increases with increased dose. An intensity maximum occurs at a dose of about 1 Mrad and the intensity falls by a factor of eight as the dose is

Table 2. Activation Energy of Luminescence as a Function of Absorbed Radiation Dose

Radiation Dose (Mrad)	Activation Energy (kcal/mol) of Luminescence Peak I	III
0.10	18.1	38.4
0.23	19.7	40.8
1.36	18.7	25.8
5.00	14.1	27.6
10.00	19.4	24.4

increased to 20 Mrad. At doses higher than 20 Mrad, Peak I luminescence is not observed. When the sample temperature is maintained at 543°K during irradiation, however, the intensity of luminescence Peak III continues to increase even at doses as high as 200 Mrad. We observed that, at similar absorbed doses, the intensity of luminescence peaks for rock salt Sample 2 were lower than that of peaks for Sample 1.

We have also investigated the luminescence of salt samples irradiated at temperatures as high as 600°K. Peak I of such luminescence is observed only for samples irradiated at temperatures below 370°K. The threshold temperature, i. e., the temperature above which luminescence Peak I is not detected, is a function of the location from which the rock salt sample is taken; luminescence Peak III occurs for all samples irradiated at temperatures below 600°K. The intensity of Peak III, however, decreases (Fig. 3) if the temperature is increased during irradiation.

The intensity of luminescence peaks of irradiated rock salt samples examined in this work was also a function of the time between irradiation and the time luminescence spectra are recorded. For example, the intensity of luminescence Peak I diminished rapidly

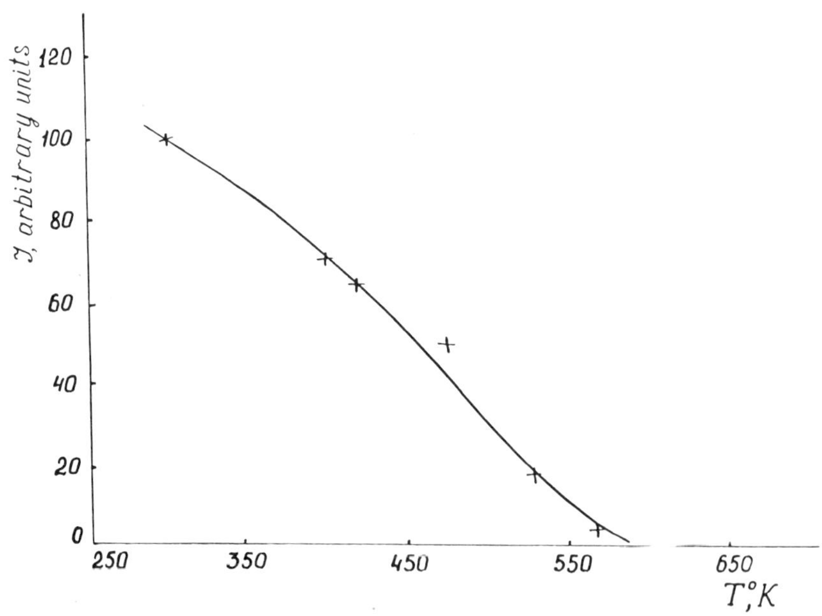

Fig. 3. Decrease in Luminescent Intensity (Peak III) as a Function of Sample Temperature during γ-Irradiation

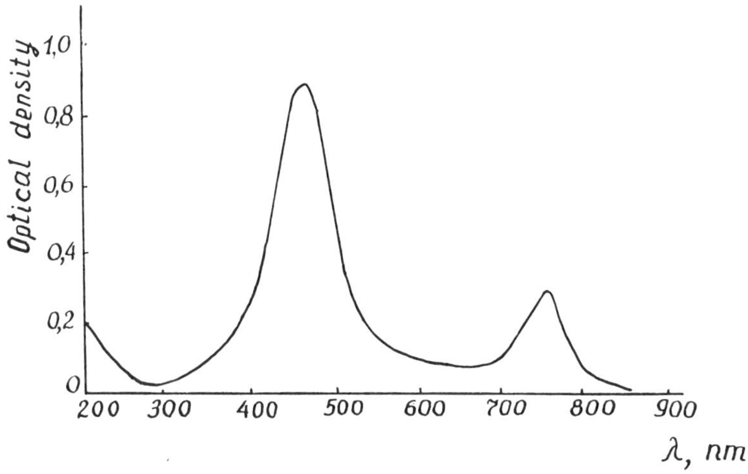

Fig. 4. Room Temperature Optical Absorbance Spectrum of γ-Irradiated Rock Salt

after irradiation, and Peak I disappeared completely only four hours after irradiation. The intensity of luminescence Peak III, however, did not change even one month after irradiation.

In general, our results showed that the thermoluminescence of irradiated rock salt samples taken from one particular geographical location had certain specific regularities such as the number of luminescence peaks, the temperatures at which peaks occur, the activation energy of luminescence, and the dependence of luminescence intensity on absorbed dose.

COLOR CENTER FORMATION

The room temperature optical absorbance spectrum of γ-irradiated rock salt samples is shown in Fig. 4. Two well resolved bands with maxima at 460 and 750 nm are attributed to formation of F and M color centers, respectively (3). Fig. 5 illustrates the variation in number of M-color centers with irradiation temperature for rock salt Sample 1. The number of color centers correlates linearly (Fig. 6) with luminescence intensity for salt samples receiving equal radiation doses at different temperatures. Such correlation indicates that the rate of color center formation is a strong function of the particular geographical location and geologic strata from which the rock salt sample was taken. A similar dependence of thermoluminescence properties of irradiated rock salt samples was also noted, as pointed out earlier.

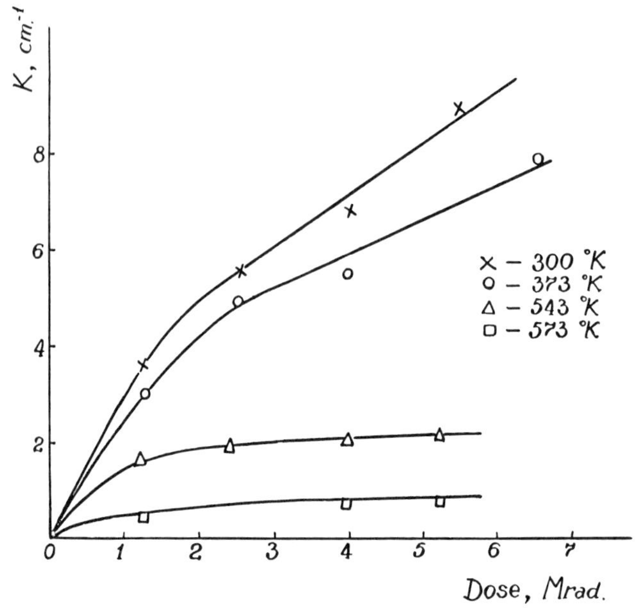

Fig. 5. Dose Dependence of M-center Density for Various Sample Temperatures

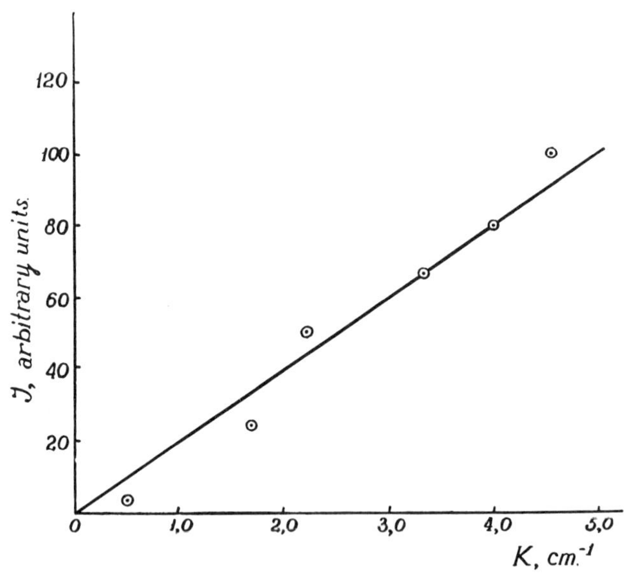

Fig. 6. Correlation between the Density of Color Centers and Thermoluminescent Intensity for γ-Irradiated Rock Salt

RADIOLYTIC-GAS FORMATION

Gaseous hydrogen and oxygen form during γ-irradiation of saturated aqueous solution of NaCl. The radiolysis yield, GH_2, is 0.42 molecules/100 ev independent of temperature over the range 0-85°C or the presence of oxygen. Oxygen yields were determined by irradiation of deaerated saturated solutions of NaCl. The GO_2 value is 0.08 molecules/100 ev at room temperature. Radiolytic H_2 and O_2 yields from saturated NaCl solutions are not significantly affected by the presence of various ions characteristically present in actual brine solutions (4).

SUMMARY AND CONCLUSIONS

Results obtained in this study demonstrate that color center formation and thermoluminescent properties of irradiated rock salt can be used to:

1. Identify salt formations suitable for storage of nuclear wastes; and
2. Determine radiation doses absorbed by salt deposits.

Furthermore, our data for radiolytically generated H_2 and O_2 (from saturated NaCl solutions) can be used to estimate the amount of gas which may accumulate in brine lenses in rock salt strata used to store radioactive wastes. These data may also be used to calculate gas pressures in such brines.

REFERENCES

1. W. C. McClain and A. L. Boch, Nuclear Tech. 24, 3 (1974).
2. Ch. B. Lushchik, Doklady An SSSR 101, 641 (1955).
3. J. J. Markham, F-Centers in Alkali Halides (Academic Press, NY 1966)
4. A. Swollow, "Radiation Chemistry," Atomizdat, M, (1978).

GAS GENERATION FROM TRANSURANIC WASTE DEGRADATION*

Martin A. Molecke

Sandia Laboratories
Albuquerque, NM 87185

INTRODUCTION

The degradation of existing forms of defense-related transuranic (TRU) contaminated waste, particularly with regard to gas generation potential, has been extensively studied over the past several years. This area of study, particularly under environmental conditions postulated for deep geologic terminal isolation in a bedded salt repository, is a major segment of the Sandia Laboratories' Waste Isolation Pilot Plant (WIPP) Transuranic Waste Characterization Program (1, 2). A primary purpose of this program is to support the Department of Energy in establishing waste acceptance criteria for TRU waste forms to be disposed of in the WIPP.

This review includes data on gas generation rates from several common TRU waste forms, various degradation mechanisms, resultant gas compositions and comparisons between generation modes. Many of the experimental investigations to be described are still in progress (2); some data are preliminary and may be subsequently refined.

The TRU-contaminated waste forms considered in this study include:

1. CELLULOSICS (paper, wood, cotton)
2. PLASTICS (polyethylene, polyvinyl chloride)
3. RUBBERS (neoprene, Hypalon)
4. ORGANIC COMPOSITE (35 wt% mixed cellulosics, 23% polyethylene, 12% PVC, 15% neoprene and 15% Hypalon)

---

*This work is supported by the U.S. Department of Energy.

5. PROCESS SLUDGES (inorganic oxides plus 60-70 wt% water)
6. CONCRETE-TRU ASH (developmental waste form)
7. MILD STEEL (waste container or scrap metal)
8. ASPHALT (European low level waste encapsulant)

The organic matrix TRU wastes are usually packaged in 210-liter steel drums which have a high density polyethylene liner. Based on an inventory of TRU wastes temporarily stored at the Idaho National Engineering Laboratory (INEL) (3), the calculated "average" weight of a drum of organic composite waste is about 51 kg (net); such a drum contains approximately 0.5 g of weapons grade plutonium (primarily ^{239}Pu). This corresponds to 0.04 Ci of alpha activity per drum.

The degradation mechanisms investigated in the Sandia-sponsored studies are:

1. RADIOLYSIS
2. BACTERIAL
3. THERMAL (decomposition and dewatering)
4. CHEMICAL CORROSION
5. ALPHA DECAY (He generation)

The major gases produced by waste degradation, in approximate order of abundance and significance, are:

H_2, CO_2, CO, H_2O, CH_4, O_2, NO_x, He.

Concerns about TRU waste degradation and gas generation can be divided into two distinct phases: 1. the short-term or operational phase of the repository encompassing waste emplacement, routine operations and decommissioning; and 2. the long-term passive phase of terminal isolation following decommissioning. Short-term concerns include formation of flammable or explosive concentrations of gases in the waste containers and in the waste storage rooms, providing fuel for a fire with a credible initiation event, and subsequent dispersal of particulate contamination. The most significant long-term concern involves the accumulation of large quantities of gases in the sealed repository with possible pressurization and fracture of the geologic formation surrounding the repository.

TRU waste degradation studies have been conducted under environmental conditions representative of those expected in a bedded salt repository and under credible overtest conditions. This is for the purpose of formulating, then validating, analytical predictive models for such waste repository interactions. The models and the data are necessary for consequence and safety assessment studies.

The acquisition and interpretation of data in this study include major contributions from the Los Alamos Scientific Laboratory, Savannah River Laboratory, the University of New Mexico Biology Department and Sandia Laboratories, all in support of the WIPP TRU Waste Characterization Program (1,2,4,5)

RESULTS AND DISCUSSION

Gas generation rates measured in these studies are presented in Table 1 in terms of moles of gas per year per drum of TRU waste. These rates, the observed ranges of initial values and estimated uncertainties, do not take into account factors which may tend to decrease rates over a period of years or more. These factors include pressure, localized matrix depletion due to radiolysis, unfavorable geochemical or gaseous environment, etc. Gas generation rates due to different mechanisms are not additive; the resultant (non-gas) degradation products from one mechanism, e.g., radiolysis or thermal, may not be optimal starting materials for another mechanism, e.g., bacterial. However, multiple modes of degradation can proceed simultaneously.

Radiolysis

Radiolytic degradation was initially assumed to have the greatest potential for generating significant quantities of gas. It has received the greatest amount of experimental attention over the past several years and, consequently, is the best characterized of the various degradation mechanisms.

Organic matrix materials have been doped with from 1.5 to 2500 alpha curies/drum of waste in order to generate gas in an experimentally convenient period of time. Gas production was measured by monitoring pressure increase in a sealed system (4). A dose rate effect on gas production has been tentatively identified: for high dose rates (high plutonium loading), gas generation decreases exponentially with time, primarily due to localized matrix depletion near individual alpha-emitting particles. For existing organic combustible TRU-contaminated wastes, with an average Pu loading of 0.5 g/drum, the gas generation rate will be proportionately lower but may remain fairly constant for hundreds of years.

A pressure effect on gas generation rates for cellulosics and plastics was observed at pressures as low as 700 kPa (100 psi). Increasing the pressure resulted in a decreased gas generation rate. At a pressure of 1×10^4 kPa (1500 psi), radiolytic gas generation essentially ceased (4). This pressure effect could have a significant impact on total quantities of gas that can be generated before assumed back reactions become effective. No pressure effect has been noted for concrete-TRU ash waste systems up to 1400 kPa (200 psi).

Table 1. Comparative Gas Generation Rates

Mechanism/Matrix	Comments	Gas Generation Rate (moles/year/drum*)
BACTERIAL (5)		
Composite, Aerobic		0-(0.9-5.5)-12**
Composite, Anaerobic		0-(1.2-4.2)-32
Plywood Box*, Aerobic		0-(0.44-2.2)-3.0
Plywood Box*, Anaerobic		0-(1.1-3.7)-4.1
Asphalt, Aerobic		0-(0.1-2.6)-8.4
Asphalt, Anaerobic		0-(0-1.9)-4.8
THERMAL (4)		
Paper	(70°C)	0.5-(1.3)-2
Composite	(40°C, calc.)	0-(0.02-0.2)-0.4
RADIOLYSIS (4, 6)		
Cellulosics	(0.04 Ci)	0.002-(0.005-0.011)-0.012
Polyethylene	(0.04 Ci)	0.003-(0.007)-0.008
Polyvinyl Chloride	(0.04 Ci)	0.01-(0.03-0.042)-0.08
Composite	(0.04 Ci)	0.002-(0.005)-0.006
Asphalt	(7.7 Ci)	0.1-(0.15-0.76)-1.0
Concrete-TRU Ash	(15 Ci, poured)	0.03-(0.045-0.93)-1.0
Concrete-TRU Ash	(15 Ci, heated)	0.0002-(.0005-.035)-0.05
CORROSION (5)	(Mild Steel)	0-(0)-2.0
ALPHA DECAY	(He, 15 Ci)	0.00002
OVERALL AVERAGE		
Existing INEL TRU Wastes (3) (Averaged Over Volume)		0.0005-(0.3-1.4)-2.8

* volume = 0.21 m (3).
** lower limit - (most probable range) - upper limits; with estimated uncertainties.

Temperature was found to have a synergistic effect on radiolytic gas generation. A 43% increase in gas generation rate was noted for a temperature increase of 30°C (from 20° to 50°C), and a 50°C increase (from 20° to 70°C) produced a 70% increase. This effect decreases in importance with time. At 70°C this increase has, after several months, apparently leveled off at +30% (4).

The gas generation rate of a developmental concrete-TRU ash waste matrix was measured, as a function of many variables, at

Savannah River Laboratory. This matrix consists of approximately 30 wt% incinerator ash integrally mixed in concrete. The gas generation potential of this waste form was significantly decreased either by drying the cured concrete at 100°-200°C or by adding nitrate or nitrate scavenging agents. Some data for this are presented in Table 1; details are discussed elsewhere (2, 6).

Explosive concentrations of radiolytically generated gases have been found in temporarily stored, ^{238}Pu highly-contaminated drums of organic matrix wastes. Routine gas sampling and analysis, venting or replacement of in-drum gases with an inert gas for such wastes has been suggested.

Radiolysis yields primarily H_2 (60% of total gases from cellulosics, 95% from polyethylene, 90% from PVC and almost 100% from concrete-ash); the remainder is CO_2 and CO, predominantly. Water decomposes to H_2 and O_2. Oxygen becomes almost totally depleted in these systems due to the formation of carbon oxides.

Bacterial Degradation

Transuranic-contaminated organic matrix wastes can serve as the energy source or nutrient for microbial growth. Experiments at the University of New Mexico are currently quantifying the gas production rates of organic composite waste, asphalt and sawdust (representative of the plywood in FRP-coated plywood boxes used for packaging non-combustible TRU wastes). Tests are being conducted in both dry or wet (water, brine and nutrient solution) systems. Compared with data from other mechanisms in Table 1, it is obvious that bacterial degradation has the greatest potential to generate significant quantities of gas. Some bacteria or fungi will definitely be present in the existing matrices or on their containers following temporary storage. Gas generation rates will depend on how well the bacteria flourish and may vary widely.

Increases in temperatures from ambient up to 40° or 50°C generally yield increased gas production in both aerobic and anaerobic environments. Rates generally decrease at 70°C. Degradation of the cellulosic component of the waste appears to be the source of most of the gas. Rubbers are attacked at a much slower rate; it is not likely that polyethylene will be bacterially attacked. Essentially all gas produced is carbon dioxide. Methane evolution is expected but not observed; this is probably due to lack of an adequate methanogenetic inoculum and/or lack of strict anaerobic conditions. This experimental effort is continuing in an effort to quantify bacterial gas generation rates and total gas quantities as a function of time, system pressure, and actinide contamination.

Thermal Degradation

The thermal degradation rates of paper, polyethylene and composite at 70° and 100°C have been calculated on the basis of measured (4) gas pressure increases. For paper and composite, the generated gases consisted of 80% CO_2 with a smaller amount of CO. The gas generation rate for composite at 40°C has been calculated but not as yet measured; surveillance-type long-term measurements are continuing in order to minimize experimental uncertainties due to low generation rates and other factors. Qualitatively, NaCl and $FeCl_2$ appear to catalyze thermal degradation at 40°C. Both are likely to be found in a salt mine repository. Water vapor release from process sludges was measured at 25° to 100°C. This thermal dewatering can be quite significant, even at 25°C.

Chemical Corrosion

The corrosion and gas generation rates for mild steel (TRU waste containers, contaminated metal scrap), were measured at 25°C under dry, moist and brine-inundated conditions. Hydrogen will be generated by corrosion only in an anaerobic and wet or inundated storage environment. In an air atmosphere with moisture present, oxygen will be consumed. Under expected repository environmental conditions, the corrosion of steel is not expected to yield significant quantities of gas.

Alpha Decay

The maximum amount of weapons grade plutonium that can be shipped (Department of Transportation regulations) in a standard 210-liter (DOT 17C) drum is 200 g or 15 Ci of alpha emitters. This is significantly more plutonium than is found in existing TRU wastes. The helium released from alpha decay for 200 g of weapons grade plutonium (in Table 1) is insignificant compared to that released by other mentioned mechanisms.

CONCLUSIONS

Based on the experimental data presented, we can rank order the relative quantities of gas generated by individual mechanism for existing forms of TRU-contaminated wastes under expected WIPP environmental conditions (with appropriate qualifying comments):

1. BACTERIAL (greatest potential, but large variability)
2. THERMAL (at 40°C; probably insignificant at 25°C)
3. RADIOLYSIS (most predictable; potential short-term risks)

4. CORROSION (under the dry conditions expected in WIPP)
5. ALPHA DECAY (insignificant)

Calculations of total gas accumulation in a sealed repository, laboratory and in situ measurements of gas dissipation through rock salt permeability, and rock mechanics data have been coupled to predict the mechanical response of the mine to gas pressurization; these results are presented elsewhere (5). Long-term concerns on gas pressurization have thus been analyzed; results indicate that waste processing to minimize such gas generation is not technically required.

Short-term considerations of gas generation must also be put into perspective. Radiolysis can and does produce combustible and, in certain cases, explosive gas mixtures in drums of temporarily stored TRU wastes and in a relatively short period of time. These potentially hazardous conditions should be considered for handling, transportation and mine emplacement operations -- and for accident consequence assessments related to these operations. The sum of all these analyses is a major consideration in determining whether existing forms of defenserelated TRU wastes are acceptable for safe isolation in the WIPP or whether they must be processed to alternate, more degradation resistant forms. Administrative and technical assessments concerning the disposition of these wastes remain to be made.

REFERENCES

1. M. A. Molecke, Waste Isolation Pilot Plant Transuranic Wastes Experimental Characterization Program: Executive Summary, SAND78-1356, Sandia Laboratories, Albuquerque, NM, Nov. (1978).
2. M. A. Molecke, Gas Generation from Transuranic Waste Degradation: Data Summary and Interpretation, SAND79-1245, Sandia Laboratories, Albuquerque, NM, Nov. (1979).
3. H. C. Shefelbine, Preliminary Evaluation of the Characteristics of Defense Transuranic Wastes, SAND78-1850, Sandia Laboratories, Albuquerque, NM, Nov. (1978).
4. S. T. Kosiewicz, A. Zerwekh and B. Barraclough, Studies of Transuranic Wastes Under Conditions Expected in the Waste Isolation Pilot Plant, Interim Summary Report, Oct. 1, 1977 to June 15, 1979, LA-7931-PR, Los Alamos Scientific Laboratory, Jan. (1980).
5. Summary of Research and Developmental Activities in Support of Waste Acceptance Criteria for WIPP, Sandia Laboratories, SAND79-1305, Sandia Laboratories, Albuquerque, NM, Nov. (1979).
6. N. E. Bibler, "Radiolytic Gas Generation in Concrete Made With Incinerator Ash Containing Transuranium Nuclides," Materials Research Society Symposium, Boston, MA, Nov. (1979).

THE REACTION OF HYDROGEN AND OXYGEN IN THE PRESENCE OF CONCRETES INCORPORATING SIMULATED RADIOACTIVE WASTE*

S. Katz

Oak Ridge National Laboratory
P.O. Box X
Oak Ridge, TN 37830

INTRODUCTION

The radiolysis of contained water, with the potential of pressurization due to the accumulation of the radiolytically produced gases, has been identified as a major concern for radioactive wastes fixed in cementitious matrices (1). One estimate (1) of the pressure buildup is that about 100 atm could accumulate in a standard canister containing alpha waste after 100,000 years. However, even for this pressure, the rate of gas production would be small. For example, a standard 200-liter canister containing 10 g of ^{239}Pu dispersed in concrete containing normal concentrations of unbound water would generate a maximum of about 1 ml of H_2 per day. Since the radiolysis of concrete is dependent on the amount of unbound or free water present in the system, removal of all or most of this water would greatly reduce this rate (2). Regardless of the production rate, however, pressurization from the accumulation of hydrogen and oxygen would be avoided if there were sufficient recombination of these gases during or after their generation. These studies were undertaken to determine the recombination rates obtained with special cementitious matrices.

Concrete as a waste host is being investigated at Oak Ridge National Laboratory in a program directed at the investigation of FUETAP (Formed Under Elevated Temperature And Pressure) concrete (3). This material, because of the composition and method of preparation,

*Research sponsored by the Office of Nuclear Waste Management, U.S. Department of Energy under contract W-7405-eng-26 with the Union Carbide Corporation.

offers some distinct advantages over standard concretes as a waste fixation medium. FUETAP is solidified and cured for 24 hr at 100 to 250°C and ≤4.1 MPa (600 psi). The properties that are demonstrated by specimens containing simulated waste and prepared under these conditions include good physical strength, excellent resistance to the leaching of the contained nuclides, a tolerance for some nitrate in the waste, and a very low free-water concentration. The FUETAP product can be classified as a low-temperature ceramic. This paper presents the results of an investigation of the reaction of hydrogen and oxygen in the presence of FUETAP concrete.

The reaction of hydrogen and oxygen has been studied exhaustively (4-6). These elements do not combine in the gaseous state at ambient temperatures; however, they do interact when the gases are in contact with a catalytic surface with a recombination rate that is directly related to the surface area and the temperature. Thus the basis for this study was the realization that FUETAP concrete containing radioactive waste or perhaps some other added constituent could provide a surface to catalyze the recombination of any radiolytically produced hydrogen and oxygen.

EXPERIMENTAL PROCEDURE

In this investigation, the rate of reaction between hydrogen and oxygen was measured by determining the yield of water, or the depletion of oxygen, as mixtures of these gases flowed through a bed of the test material. Three cementitious mixtures were examined - a neat cement paste and two FUETAP concretes. The FUETAP concretes (Table 1) were prepared with and without 2.8 wt % $NaNO_3$ present. Alumina and ferric oxide were used as reference materials (Table 2). The concrete specimens were crushed, and 300-g samples of the fragments were placed in a stainless steel flow-through reactor. The alumina reference material consisted of 300 g of 15- to 20-mesh particles and the ferric oxide of 2 g of 10^2-m^2/g surface area material diluted with 18 g of alumina. The alumina served as a nonreactive substrate, and the ferric oxide served as a reactive substrate for comparison with the two FUETAP concretes containing waste.

The three dynamic-flow test systems that were used to determine the reaction rate between hydrogen and oxygen differed principally in the type and range of the analyzer. One analysis was a magnetic susceptibility instrument capable of measuring the oxygen content in the 0 to 5% range. The other two analyses were electrolytic water sensor units, one with a range of 0 to 20,000 ppm and the other with a range of 0 to 1000 ppm.

Table 1. FUETAP Concrete Formulation

Component	wt %
Cement	38.9
Fly ash	9.3
Clay	9.3
Quick-gel	0.4
Water	27.3
Simulated waste[a]	14.8

[a]Composed of 75.2% Fe_2O_3, 14.9% Al_2O_3, 4.9% Cr_2O_3, 4.8% NiO, 0.14% SrO, and 0.04% CeO_2.

Table 2. Substrates Selected for Study of the Hydrogen-Oxygen Reaction Rate

Substrate	Waste solids added (%)	Test quantity (g)	Substrate form
Neat paste	0	300	1- to 10-cm fragments
FUETAP concrete			
Simulated waste with 2.8% nitrate	14.3	300	1- to 10-cm fragments
Simulated waste without nitrate	14.8	300	1- to 10-cm fragments
Alumina	0	300	15- to 20-mesh crystals
Ferric oxide	0	2[a]	10 m^2/g powder

[a]The 2 g of ferric oxide was mixed with 18 g of 15- to 20-mesh alumina.

Oxygen Depletion Tests

A schematic diagram of the experimental layout used to measure the difference in oxygen concentration after passing gas over the various substrates is shown in Fig. 1. The measurements were made by sampling the gas before and after it had passed over the substrate or bypassed the sample. The feed gas, which was flowing at a rate of 75 ml/min, was 75% hydrogen and 25% air. Steady-state conditions were achieved by flowing the gas mixture through the substrate for 24 hr. The oxygen depletion, which was calculated from the difference in oxygen content of the gas stream alternately directed through the substrate and the bypass, was then plotted against the system temperature.

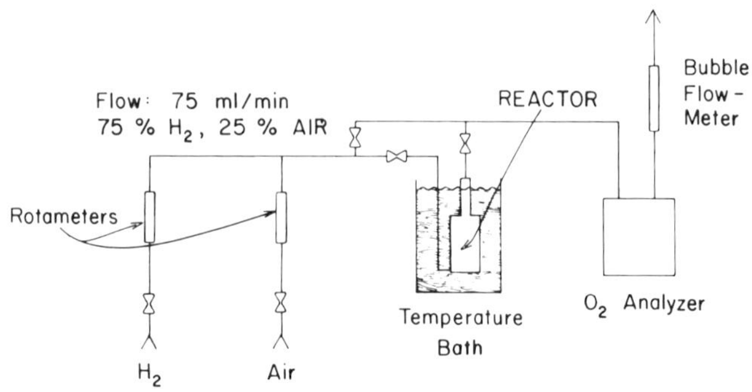

Fig. 1. Oxygen Depletion Test System

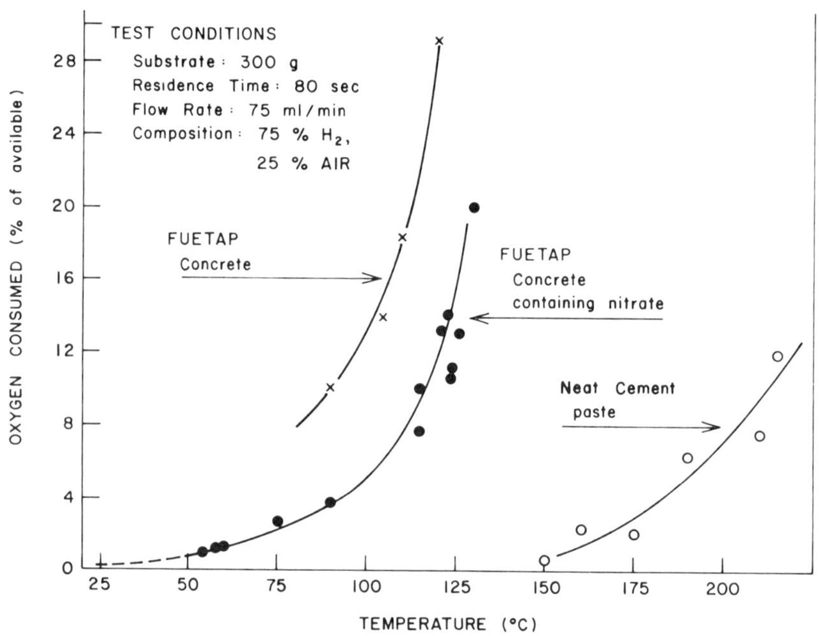

Fig. 2. Data Using Oxygen Depletion Test System

The oxygen depletion for the FUETAP samples containing simulated waste was significant and exceeded 20% of the available oxygen at temperatures about 130°C. Extrapolation of the data to ambient temperatures indicates that about 0.1% of the available oxygen would have been depleted when it contacted the gas and the waste-containing substrates for 80 sec. Oxygen depletion with the neat cement paste substrate were very low and essentially undetectable at temperatures below 140°C (Fig. 2.).

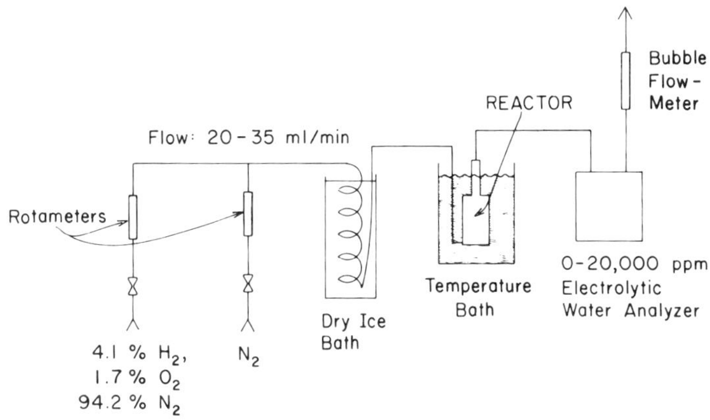

Fig. 3. Test System for High Range Water Yield

Table 3. Water Yield (ppm) Obtained with 0- to 20,000-ppm Water Yield Test System

	Substrate					
	Neat paste		FUETAP concrete with waste, without nitrate		FUETAP concrete with waste and nitrate	
Feed gas[a]	Test	N_2	Test	N_2	Test	N_2
Test temperature, °C						
27	23	20	7700	20	6400	20
28			5600	30	6700	20
100-110			5200	100	6700	30
120-130	900	35	7700	100	7700	100

[a]Test feed gas, 4.1% hydrogen, 1.7% oxygen, and 94.2% nitrogen; background feed gas, 100% nitrogen; residence time, 3 ± 0.5 min.

Water Yield Test - High Range

The schematic diagram of the experimental arrangement used to measure the yield of water in the range 0 to 20,000 ppm is shown in Fig. 3. Operation with this detector required that the gas flow be maintained at 20 mL/min, and a relatively long time was required to achieve steady-state conditions. In some cases corrections were necessary to compensate for long-term flow rate drift. The measurements at ambient temperatures were carried out on substrates that had been conditioned at 200°C in a stream of dry gas so that the background was reduced to less than 5 ppm. These results (Table 3) demonstrated that a significant catalytic reaction occurs at ambient as well as at higher temperatures in the presence of FUETAP concrete specimens containing waste. Neat cement paste yielded a significantly lower recombination rate.

Water Yield Test - Low Range

The use of a lower range water analyzer (0 to 1000 ppm) permitted a flow rate of 100 mL/min, and thus, steady-state conditions were obtained more rapidly than in the higher range test (Fig. 4). As before, the substrates used were pretreated at 200°C with a flow of dry gas until a background of less than 1 ppm was achieved at ambient temperatures. Three substrates were used with this system: alumina, ferric oxide, and FUETAP concrete with simulated waste but no nitrate. The feed gas composition was ≥95% hydrogen, 0.1 to 1% oxygen, and the remainder nitrogen except in the alumina test; in the latter test the gas mixture was maintained at 95% hydrogen, 1.2% oxygen, and 3.8% nitrogen.

Although the results from the system show high rates of recombination over the FUETAP concretes and the ferric oxide (Table 4), in both cases the water yield did not reach a steady state. At temperatures below 60°C, the water yield slowly increased and then exceeded 1000 ppm. In the 60 to 100°C range, the yield of water varied in a cyclic manner as shown in Fig. 5. The results from the inert alumina substrate were consistently low and showed only a slight increase with temperature. As for the cyclic behavior, the periods ranged from 20 min to greater than 40 hr depending on the hydrogen to air (i.e., oxygen) ratio. Each cycle was characterized by a gradual rise to 300 ppm and then a rapid rise to over 1000 ppm. After a short time at this level, the yield dropped rapidly to

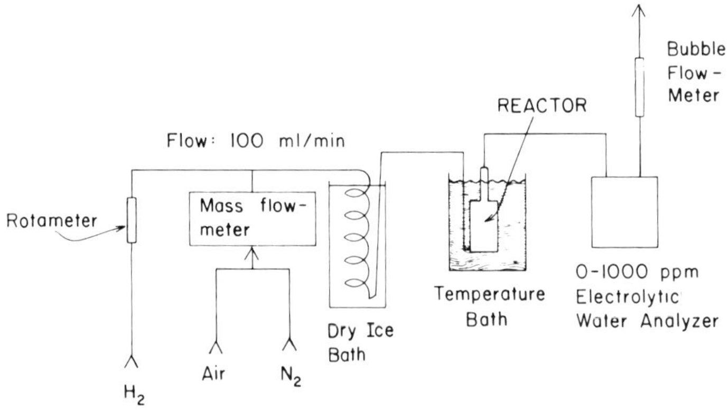

Fig. 4. Test System for Low Range Water Yield

Table 4. Water Yield (ppm) Obtained with 0- to 1000 ppm Water Yield Test System[a]

Temperature (°C)	Substrate: FUETAP concrete with waste, without nitrate	Alumina, 15 to 20 mesh	Ferric oxide, 10 m^2/g
25-60	100 to >1000		100 to >1000
60-100	100 to 1000 (cyclic)		100 to >1000 (cyclic)
170	>1000	0.2	>1000
190	>1000	0.4	>1000
210	>1000	3	>1000

[a]Test feed gas for concrete and ferric oxide, >95% hydrogen, 0.1 to 1% oxygen, remainder nitrogen; test feed gas for alumina, 95% hydrogen, 1.2% oxygen, 3.8% nitrogen; background feed gas, 100% hydrogen; residence time, 1 min for concrete and alumina, 0.1 min for ferric oxide.

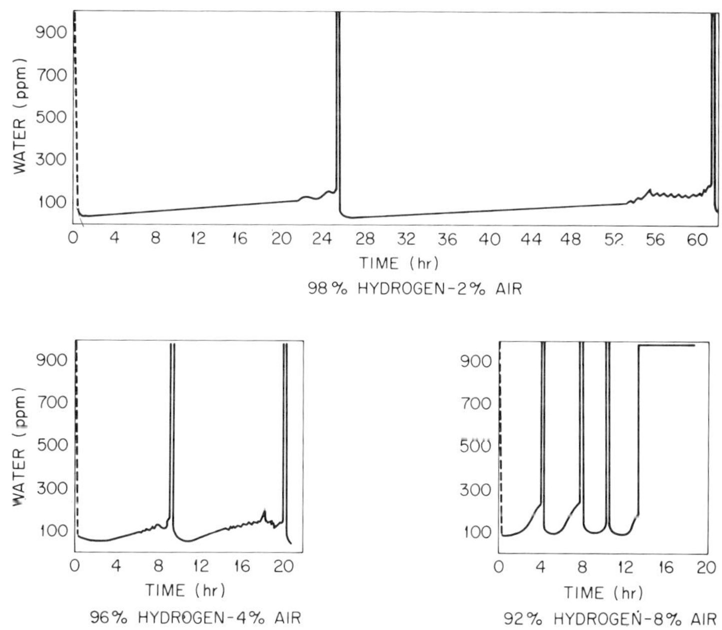

Fig. 5. Cyclic Reaction of Hydrogen and Oxygen

approximately 100 ppm and the cycle resumed. Such cyclic behavior has been observed in other catalytic studies (7) and was accounted for by postulating the existence of two or more possible reaction routes which proceed at different rates and for which a triggering mechanism exists.

Table 5. Combination of Hydrogen and Oxygen in Contact with FUETAP Concrete (No Nitrate) at Ambient Temperatures

	Test		
	Oxygen depletion	High range	Low range
Gas flow, ml/min	75	20-35	100
Gas composition			
Hydrogen, %	75	4.1	>95
Oxygen, %	5	1.7	0.1 to 0.5
Residence time, min	1.25	3.3	1
Yield	250 ppm	7000 ppm H_2O	100 to >1000 ppm H_2O
Yield, milliliters of H_2 per day	>8	250	14 to 140

CONCLUSIONS

The data obtained with each of the three experimental systems show that there is a high hydrogen-oxygen recombination rate in presence of FUETAP concretes containing simulated waste (Table 5). A 300-g sample will catalyze the recombination of greater than 8 ml of H_2 per day (Table 5). Because the radiolytic gas produced by 10 g of ^{239}Pu dispersed in 200 liters of ordinary concrete is only about 1 ml of hydrogen per day, the data suggest that radiolytic gas pressurization may not be a problem with FUETAP concretes containing waste provided that adequate contact is maintained between the gases and the available solid surfaces.

REFERENCES

1. N. E. Bibler, SRL DP-1464, Jan. (1978).
2. N. E. Bibler, Savannah River Laboratory, personal communication, Sept. (1979).
3. J. G. Moore, E. Newman and G. C. Rogers, ORNL/TM-6573, March (1979).
4. P. H. Emmett, ed., Catalysis - Fundamental Principles (Part 1), Vol. I (Reinhold Publishing Co., 160, 1954).
5. P. C. L. Thorne and E. R. Roberts, Fritz Ephraim Inorganic Chemistry, (Interscience Publishers, Inc., 401-406, 1948).
6. V. Ponec, Z. Knor and S. Černý, "Chemisorption and Interaction of Hydrogen and Oxygen on Transition Metals," Proceeding of the Third International Congress on Catalysis (W. M. H. Sachtler, G. C. A. Schuit and P. Zwietering, eds; North-Holland Publishing Co., Amsterdam, 363, 1965).
7. V. Hlaváček and J. Votruba, "Hysteresis and Periodic Activity Behavior in Catalytic Chemical Reaction Systems," Adv. Catal. 27, 59 (1978).

RADIOLYTIC GAS GENERATION IN CONCRETE MADE WITH INCINERATOR ASH CONTAINING TRANSURANIUM NUCLIDES*

Ned. E. Bibler

Savannah River Laboratory
E. I. du Pont de Nemours and Company
Aiken, South Carolina 29801

ABSTRACT

The effects of various factors on H_2 generation by alpha radiolysis of concrete containing TRU incinerator ash were studied. Methods for reducing H_2 generation were investigated. Samples of portland and high-alumina cement containing up to 30% calcined ash (dry basis) were doped with $^{238}PuO_2$. Gas pressures were measured as a function of radiation dose; gas compositions were determined. Gas yields were calculated in terms of G values (molecules produced per 100 eV of alpha energy absorbed). These yields were used to estimate pressures in containers of radioactive concrete waste during storage.

INTRODUCTION

The action of α, β, or γ radiation of water produces hydrogen gas(1). Thus, nuclear waste forms, such as concrete, that contain H_2O can produce H_2. Potentially, this can lead to pressurization of storage containers, or to the production of flammable or explosive gaseous mixtures.

This paper summarizes results of an extensive investigation (2) to estimate the rate of radiolytically generated H_2 evolution from concrete containing transuranium (TRU) waste. Other studies showed that ^{60}Co-γ and ^{3}H-β radiolysis of concrete also produce H_2 (3-5).

*Work done under USDOE Contract No. AT(07-2)-1.

The gas pressures generated during γ radiolysis of concrete reaches a low equilibrium value because of a back reaction between OH radicals and H_2. Earlier studies (3,5) of alpha radiolysis of concrete showed that an equilibrium pressure was not attained at H_2 pressures up to 1.1 MPa indicating considerable H_2 accumulation in the container. The present study was initiated to estimate the radiolytic rate of gas production from concrete TRU incinerator ash during long term storage in the Waste Isolation Pilot Plant in New Mexico (6). The effects of various factors on this rate, along with methods to reduce it, were investigated.

EXPERIMENTAL PROCEDURE

Simulated TRU incinerator ash was prepared by calcining at 800°C the ash from impartial incineration of typical combustible laboratory wastes such as tissues or polyethylene tubing. Principal components of the ash were Ca (added as heat retardant to the plastics) and Ti (added as whitener). Carbon content of the calcined ash was nominally $<1\%$.

Mixtures of 70 wt % cement [Portland Type I (P-1) or high-alumina (HAC)] and 30 wt % ash (dry basis) were prepared with a known amount of $^{238}PuO_2$. After thorough mixing, sufficient water was added to give a workable paste (initial water/cement = 0.8) which was then poured into a cylindrical steel mold (10 cm x 2 cm O.D.). After curing 5 to 30 days at ambient conditions, the final composition of the specimens (total weight = 60 g) was nominally 50 wt % cement, 20 wt % ash, and 30 wt % water containing 0.1 to 1.0 Ci of ^{238}Pu. In the presence of air, the specimen was sealed to a pressure transducer and the pressure increase recorded. After a known time, the gas was sampled and its composition determined by gas chromatography. G values (molecules produced/100 eV of α energy absorbed) were calculated from both the final gas composition and from the rate of pressure increase using the ideal gas law, the void volume in the container, and the amount of ^{238}Pu present.

GAS YIELDS

Typical pressure increases for three concretes are shown in Fig. 1. Sample P-1 and HAC-1 contained simulated ash doped with $^{238}PuO_2$, and sample HAC-2 contained ash prepared from incineration of waste contaminated with a ^{238}Pu solution. Thus, HAC-2 resembled more closely an actual waste form. These three samples were cured unsealed at ambient conditions for approximately a month. When non-radioactive concretes were sealed to the transducer, no pressure was produced, confirming that radiolysis caused the pressure increases.

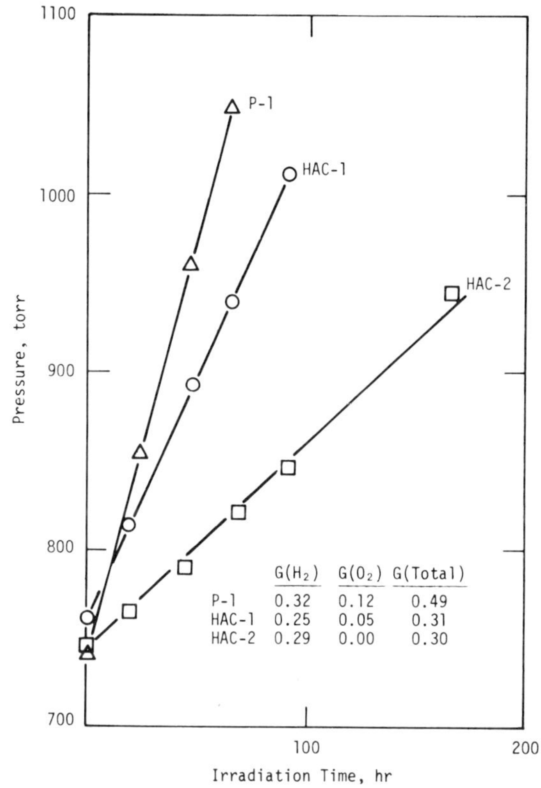

Fig. 1. Pressure Increase from Radiolysis of Portland or High Alumina Concrete Containing ^{238}Pu

The slopes of the three lines in Fig. 1. differ because of differences in the G values and dose rates for the three concretes. $C(H_2)$ and $G(O_2)$ values were calculated from the changes in the partial pressures of H_2 and O_2 during the test. The partial pressure of N_2 was unchanged by radiolysis. G (total) is the 100-eV yield of all gaseous molecules and was calculated from the slopes of the lines. Within experimental error, $G(H_2) + G(O_2) = G$ (total) as expected if if the N_2 pressure is unchanged. Less than a stoichiometric amount of O_2 was formed, an indicated by the ratio $G(O_2)/G(H_2)$ being less than 0.5 (Fig. 1). This agrees with the α radiolysis of liquid water where $G(O_2)/G(H_2)$ is only 0.13 (7). The reason for O_2 not being produced with HAC-2 at this time is unknown. Less than a stoichiometric amount of O_2 being formed indicates that some other oxidation product of water is being formed. Perhaps this is H_2O_2 as it is in liquid water. Another study showed evidence of this species in γ irradiated concrete (3).

Dose rates for the tests shown in Fig. 1. were nominally 10^{17} eV/ (min) (gram of concrete), with HAC-2 having the lowest rate. This dose rate is about 1000 times larger than that expected in actual waste. For example, the dose rate in a 210-liter drum 90% full of concrete containing 200 g ^{239}Pu is only $4x10^{14}$ eV/(min) (gram of concrete). If H_2 is formed as it is in liquid water (recombination of H atoms or hydrated electrons), no change in $G(H_2)$ at lower dose rates is predicted and none was observed over dose rates of 10^{16} to 10^{17} eV/ (min) (gram of concrete). Over the range 23 to 100°C, $G(H_2)$ is independent of temperature for concretes containing the same amount of water. This agrees with the radiolysis of liquid H_2O where $G(H_2)$ is not significantly affected by temperature (1).

REDUCTION OF $G(H_2)$

Lower rates of H_2 production would be desirable both for short and long term considerations. Two methods were found to reduce $G(H_2)$: 1) addition of H atom scavengers to concrete, 2) removal of H_2O by heating concrete or using pressed concrete. The latter method would be effective over the long term only if water were excluded from the concrete during storage. The best H atom scavengers for reducing $G(H_2)$ were NO_3^- and NO_2^-. These were supplied by adding $NaNO_3$ and $NaNO_2$ to the water used to make the concrete. Maximum concentrations of these ions in the solutions were 6M for NO_3^- and 5M for NO_2^-. $G(H_2)$ with NO_3^- was reduced to 0.03. $G(H_2)$ with NO_2^- was reduced to 0.053.

Radiolysis caused O_2 production in the presence of NO_3^- (G=0.37) and O_2 consumption in the presence of NO_2^- (G = -0.08). Also, when NO_2^- was present, a slight amount of N_2O was produced [$G(N_2O)$ = 0.04] indicating reduction of the NO_2^-. Relative decreases in $G(H_2)$ upon addition of NO_3^- or NO_2^- ions are shown in Fig. 2. The data for NO_3^- ions follow the trend shown by the solid line which depicts the effect of these ions in the γ radiolysis of liquid water (7). This similarity indicates that these ions are reacting as they do in liquid water. Further, this suggests that the reaction for H_2 formation from α radiolysis of concrete may be identical to that for liquid water. The effect of NO_2^- ions on $G(H_2)$ from α radiolysis of liquid water has not been determined.

One method of obtaining a concrete with low water content is to prepare a pressed concrete. This concrete is prepared by pressing a cement-ash pellet containing very little water (about 2 wt %), at 173 MPa (8). Gas production from three pressed concretes containing portland Type I cement doped with $^{238}PuO_2$ was measured. Two concretes contained simulated TRU incinerator ash. The third contained bone char (a material for absorbing Pu from waste streams). $G(H_2)$ values measured from the amount of H_2 produced were 0.009 and 0.005 for the concretes containing ash and 0.003 for the one containing

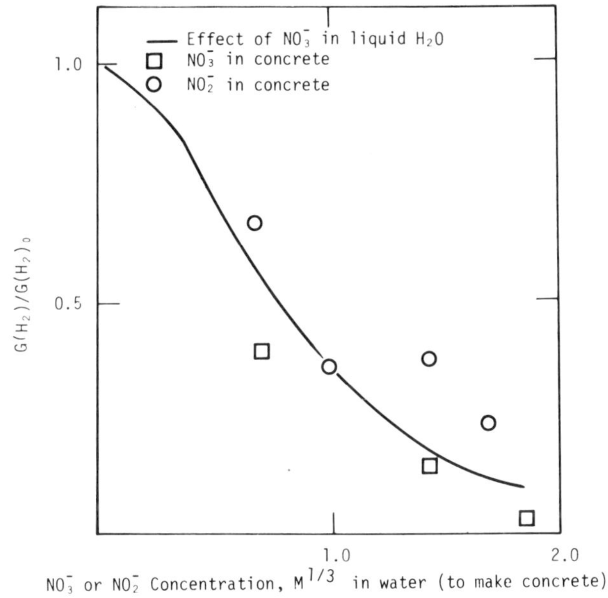

Fig. 2. Lowering of $G(H_2)$ in α Radiolysis of Concrete by Addition of NO_3^- or NO_2^- Ions

bone char. Pressure increases were insufficient to calculate values for G(total) because of the relative large void volumes of the containers. To determine the consequence of this type of concrete becoming wet, pressure measurements were made while a pressed concrete containing ash was submerged in water. $G(H_2)$ increased to 0.13. O_2 was produced [$G(O_2)$ = 0.03] and G(total) was 0.16. Weight measurements indicated that the sample had sorbed 16 wt % H_2O. Clearly, to retain the low rate of H_2 production, water must be excluded from the concrete during storage.

Another method of reducing the water content is to heat the concrete to remove free water after the concrete has cured for a sufficient time. Most of the H_2O in concrete is added to achieve a workable paste that can be poured or cast rather than to cause the hydration reactions that give concrete its strength (9). The drastic reduction in the pressurization rate caused by prior heating at 200°C to remove H_2O is shown in Fig. 3. Heating at 90°C did reduce the G values, but not as significantly as heating at 200°C. The concrete used to obtain the results in Fig. 3. has been cured at room temperature in a moist environment for 36 days before it was heated and thus had attained at least 70% of its final strength. It was dried by heating in an oven in the steel cylinder into which it was cast. At 90°C, the sample showed a continuous weight loss as long as it was heated (34 hr). At 200°C weight loss had ceased after 16 hours of

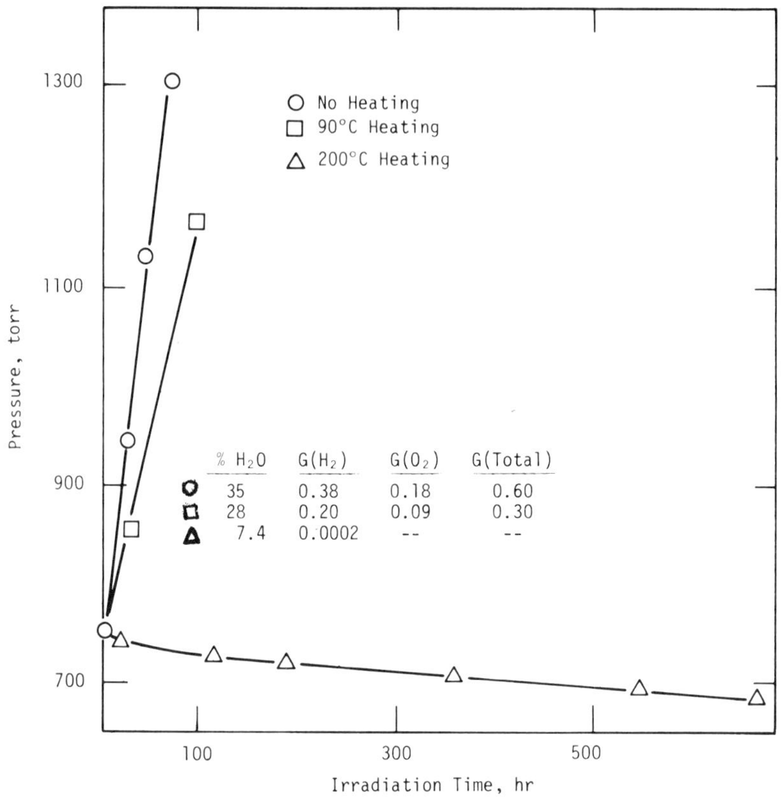

Fig. 3. Pressure from Alpha Radiolysis of a Previously Heated Portland Concrete Waste Form

heating. In both tests, the temperature of the concrete had returned to ambient before the sample was sealed to the transducer.

A very low value for $G(H_2)$ was obtained for the concrete that has been heated to 200°C. Low $G(H_2)$ values were also found for other concretes after they had been heated at 200°C. For HAC-2 (Fig. 1.) and for a pressed concrete, $G(H_2)$ values of $5x10^{-4}$ and $3x10^{-4}$, respectively, were obtained. The pressure above these samples also decreased when they were sealed to pressure transducers. A pressure decrease was also observed after a nonradioactive concrete after it has been heated to 200°C.

The results in Fig. 3. indicate that the water driven off concrete at temperatures below 200°C is radiolytically reduced more easily than the remaining water. The remaining water is probably involved in the hydration reactions that give the concrete its strength. When water was again added to the concrete used to obtain the data in Fig. 3., $G(H_2)$ and $G(O_2)$ increased to nearly their original values, 0.34 and 0.18, respectively.

RADIOLYTIC PRESSURES IN STORAGE CONTAINERS

The above G values can be used to calculate pressures generated by radiolysis in a storage container of concrete containing TRU waste. The equation for a concrete containing several TRU iostopes is:

$$P = \frac{[1.2 \times 10^{18}\ \alpha/(\text{year})(\text{Ci})]\ [G(\text{total})]\ RT}{100\ NV} \sum_i \frac{Ci_i E_i (1-e^{-\lambda_i t})}{\lambda_i}$$

where Ci_i, E_i, and λ_i are the number of curies, the α energy, and the α decay constant, respectively of the i-th isotope. Pressures for H_2 and O_2 can also be calculated using $G(H_2)$ or $G(O_2)$. Results for H_2 generation in a 210-liter container 90% full of concrete (10% porosity) containing 200 g of ^{239}Pu are shown for various values of $G(H_2)$ in Fig. 4. The lines were calculated by assuming the $G(H_2)$ was not affected by the H_2 pressure and that a slow thermal recombination reaction of H_2 and O_2 was not occurring over the long storage times. A previous α-radiolysis test has shown that $G(H_2)$ is not significantly affected by H_2 pressures up to 1.1 MPa (3). Also, no evidence for a thermally induced recombination reaction has been observed at the α dose rates used thus far.

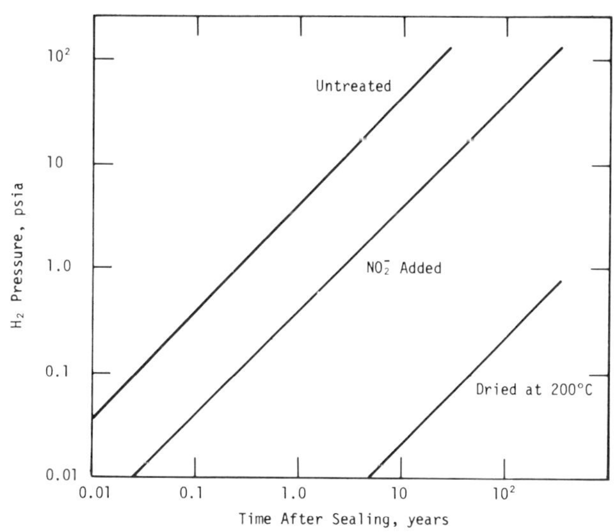

Fig. 4. Calculated Radiolytic H_2 Pressure in a 210-liter Container 90% full of Portland Concrete and 200 g ^{239}Pu

The effect of adding NO_2^- to the concrete or of drying the concrete is clearly evident in Fig. 4. With untreated concrete, considerable H_2 pressures can be generated after intermediate storage times. Also, with the untreated concrete, a flammable mixture (4% H_2 in air) can be present after only about 40 days. With concrete dried at 200°C more than 10^2 years would be necessary to generate a flammable mixture. If less ^{239}Pu is present, longer times would be necessary.

CONCLUSIONS

H_2 and O_2 will be produced by α radiolysis during storage of canisters of portland or high alumina concrete containing TRU waste. The rate of H_2 production can be decreased by adding NO_3 or NO_2 ions to the concrete or by drying the concrete. NO_3^- addition increases O_2 production while NO_2^- addition decreases O_2 production. Drying at 200°C reduces the rate of H_2 production by a factor of approximately 2000 while reducing the water content by a factor of only 5. If H_2O is again added to the concrete, the gas production rates return to nearly their original values.

REFERENCES

1. I. G. Draganic' and Z. D. Draganic', The Radiation Chemistry of Water (Academic Press, New York, 1971).
2. N. E. Bibler, USDOE Report DPST-78-150-2, Du Pont Company (1979).
3. N. E. Bibler, USDOE Report DP-1464, Du Pont Company (1978).
4. W. E. Tadlock and G. C. Asbell, USDOE Report MLM-2620, Tritium Waste Controls, October-March 1979 (1979).
5. N. E. Bibler and E. G. Orebaugh, USDOE Report DP-1459, Du Pont Company (1977).
6. M. A. Molecke, Trans. Amer. Nucl. Soc. 27, 440 (1978).
7. N. E. Bibler, J. Phys. Chem. 78, 211 (1974).
8. L. M. Klinger, D. H. Batchelder, and E. L. Lewis, USDOE Report MLM-2575 (1978).
9. F. M. Lea, The Chemistry of Cement and Concrete, 3rd Ed. (Chemical Publishing Company, New York, 1971).

INFLUENCE OF IONIZING IRRADIATION ON THE PROPERTIES OF ALUMOSILICATE MINERAL ROCK REPOSITORIES OF RADIOACTIVE WASTES

V. I. Spitsyn, V. D. Balukova, I. M. Kosareva and E. I. Evko

Institute of Physical Chemistry of the Academy of Sciences of the U.S.S.R., Moscow

INTRODUCTION

The study of changes occurring in minerals which may occur in the rocks of geological repositories for radioactive wastes is of current concern in many countries; such changes may be caused by ionizing radiation fields produced by the radioactive wastes. One way to study these changes is to subject the minerals to controlled exposures of γ-rays and electron beams.

Alumosilicate minerals have a low electrical conductivity. Therefore, when exposed to external irradiation such as an electron beam, a volume charge is accumulated. This charge creates a field and reduces the electron's path. Hence, major changes in alumosilicate minerals occur in surface and near surface layers. It is well known that bonds among lattice elements in such minerals are rather strong, and the possibility of the shift of atoms from their regular positions would be expected to be small. The transfer of electrons from the anion to the cation may be caused by secondary irradiation effects, although the possibility of the rupture of the chemical bond inside the ion may not be excluded.

Therefore, external irradiation may cause changes in energy characteristics of the irradiating electron beam which in turn affect the surface microstructure of the minerals. This, then, may influence the sorption parameters of minerals during the sorption of components from the liquid phase.

In repositories, rocks are represented by mineral assemblages with one or more dominant types of minerals. Rocks always contain small quantities of bonded water. Irradiation of minerals is accom-

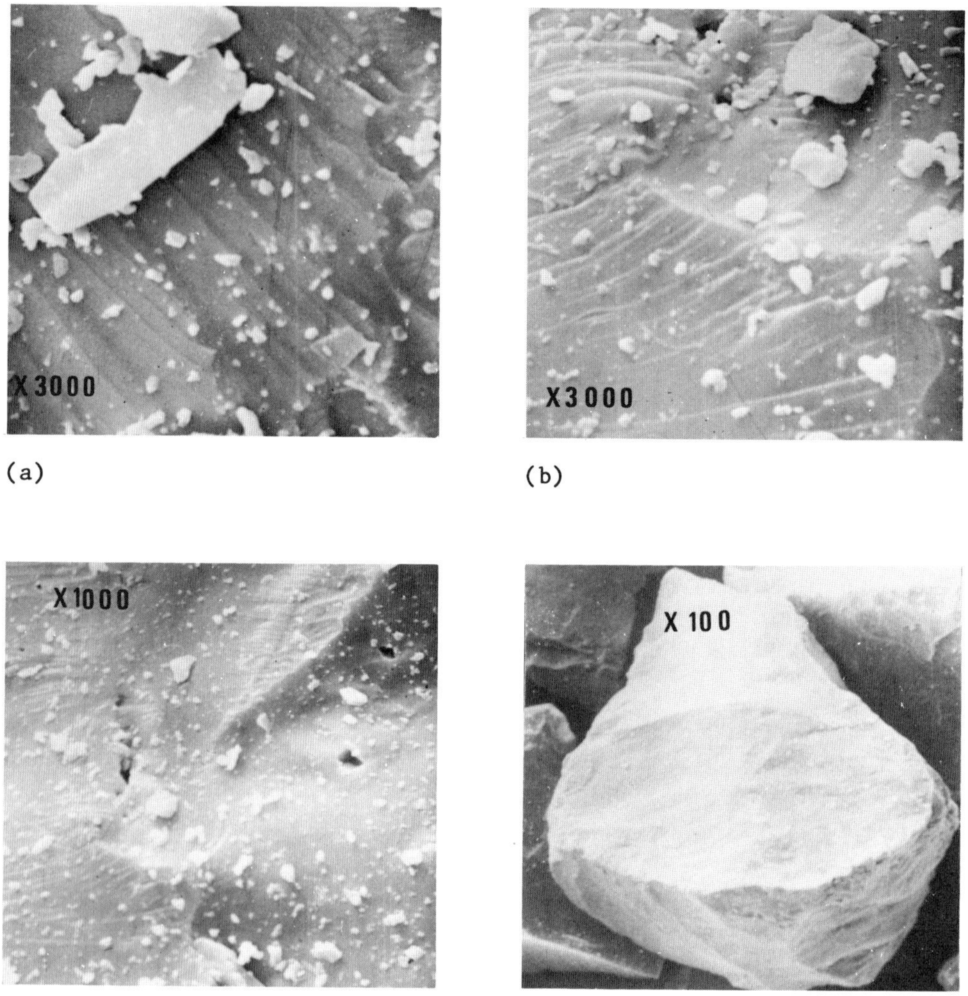

(a) (b)

(c) (d)

Fig. 1. Scanning Electron Microscope Pictures of the Initial (Non-Irradiated) Quartz

panied by water radiolysis, gas formation and mechanical stress effects. All of these processes may bring about changes in the stability of the rock (1-4).

SIMULATION EXPERIMENTS

To follow the changes in the surface of minerals under the influence of electron and γ-irradiation, electron-microscopic studies

Fig. 2. Scanning Electron Microscope Picture of γ-Irradiated Quartz (γ-Source ^{60}Co, Dose Rate-2.63 M Rad/Hour, Absorbed Dose-6.2x10^{9} Rad)

of two types of minerals were done. These minerals were quartz and kaolinite. Fig. 1. and 2. show electron-microscopic pictures of non-irradiated and irradiated quartz (Co^{60}, dose rate 2.63 M rad/hour, absorbed dose 6.2x10^{9} rad). Irradiation was carried out with dry samples. Structural and textural changes and other peculiarities can be seen in some of the details of the microrelief of the specimens. The photomicrograph (Fig. 1a.) shows that the microrelief of non-irradiated samples has a terraced shape. Horizontal platforms can be seen on the surface, and their extension is considerable in comparison with the height of the terraces (0.1-0.4 mkm). The steep slopes of the terraces correspond to the steps of the microrelief which form grooves on the quartz chips (Fig. 1b.). The surface of the natural dissolution of quartz grains is typically represented by a surface pocked with pickling holes (Fig. 1c.). In Fig. 1d., a general view of a non-irradiated quartz polycrystal is shown.

As it is seen from Fig. 2., irradiation loosens and increases the sample surface. The depth of this phenomenon is 1 mkm. A great number of isometric crystals appear on the surface. Large polycrystals have microrelief similar to that of non-irradiated crystals (pickling holes, chips along non-crystallographic axes). Quartz is considered to be radiation-stable and the effects appear to be variable. The initial sample must have contained either water or one-dimension dislocations typical of natural conditions which, due to irradiation influence, form canals, microcracks and cracks. To determine the heterogeneity of the initial material, chemical pickling of the mineral by hydrofluoric acid was performed. The formation of microcrystals on the chips of the initial (non-irradiated) quartz was observed. Samples of irradiated quartz with a large number of missing crystals look like quartz samples after pickling. Thus, the main reasons for the changes in the surface during irradiation of natural quartz may be the presence of liquid elements.

Kaolinite is one of the most wide spread flaky minerals. IR-spectra and X-ray diffractograms of the initial kaolinite showed that the sample has a perfect structure. A thermal analysis showed an endo-thermal event at 530°C which is caused by the loss of bound water and mineral amorphization. An exo-thermal event occurs at 950°C due to the crystallization of amorphous products of kaolinite and formation of mullite. The quantity of bound water amounted to 1% based upon calculations of the endo-thermal event. The comparison of electron-microscopic pictures of non-irradiated and irradiated kaolinite (Fig. 3. and 4., respectively) showed an increase in the number of large aggregates (2.5-3 mkm) and a corresponding decrease in the galluasite content with microcrystals appearing on the flat surfaces of the crystals. Photographs enlarged 10^4 times testify to the sticking of small (0.5-0.05 mkm) aggregates.

At the same time the γ-irradiation studies were being conducted, we investigated the influence of an electron beam on these minerals. γ-irradiation penetrates throughout the particle which results in uniform ionization and excitation. The electron beam, on the other hand, produces very localized effects. As a rule, the electron

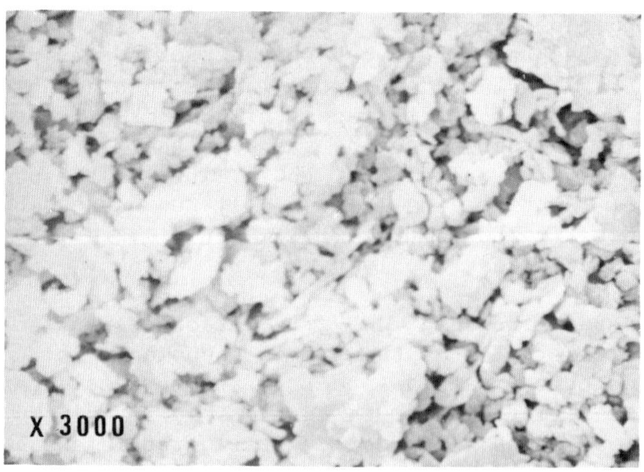

Fig. 3. Scanning Electron Microscope Picture of the Initial (Non-Irradiated) Kaolinite

Fig. 4. Scanning Electron Microscope Picture of γ-Irradiated Kaolinite

bombardment of minerals leads to the formation of structural defects and accumulation of the volume charge. Fig. 5. illustrates the changes in the kaolinite surface which are caused by the electron beam influence. The photos show the appearance of "pores" (accelerating voltage E_0=100 keV). Vacuum thermal treatment at 273°C and 694°C does not result in the formation of such "pores". Only the combination of temperature and electron beam action produces the above mentioned effect both for non-irradiated and irradiated samples.

Continuous action of the electron beam also leads to the amorphization of the surface of the grains. The microdiffraction analysis showed that, on the whole, no changes occur in the kaolinite structure. All of these changes are produced at the surface. The limiting accelerating voltage at which "pores" begin to appear for kaolinite ranges from 25-100 keV. At 25 keV, no effect is observed even during a continuous irradiation. Irradiation of quartz with the electron beam under the same conditions does not cause any changes in the surface.

Table 1. The Intensity of Characteristic Lines of Kaolinite Components

	Initial Sample				Irradiated Sample (Co^{60})			
Elements	Edge	Center	Edge	Average	Edge	Center	Edge	Average
Si	52	50	54	52	45	46	43	44.7
Al	40	46	44	46	41	41	37	39.6
Ti	8	8	13	9.7	6	7	7	6.7

X-ray microanalysis was used to determine the elemental distribution in kaolinite in the near surface layer (1 mkm). The results are listed in Table 1. If the intensity of the components of non-irradiated samples was taken to be 100%, then the intensity of irradiated components in the near surface layer was found to be reduced by 14% for Si, by 14% for Al, and for Ti by 22%. In the near surface layer, the concentration of basic and admixture components is smaller than in the bulk. The proportional changes in the content of Si and

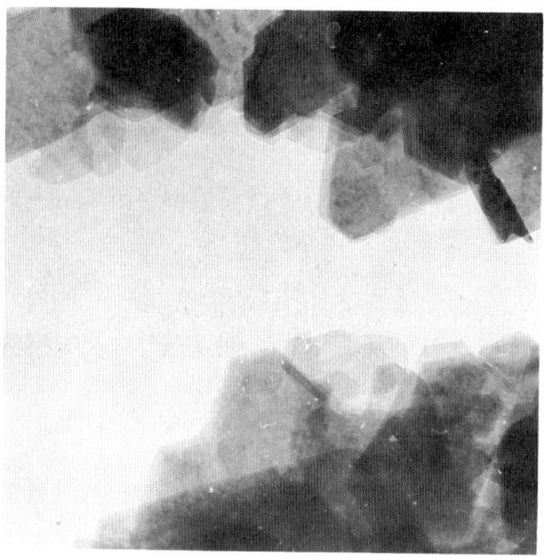

(a)

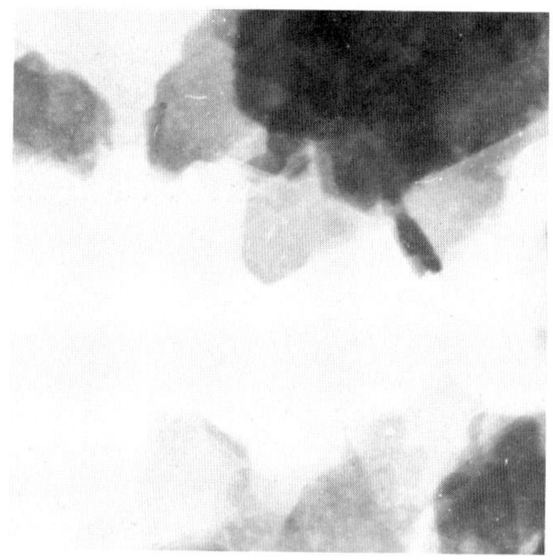

(b)

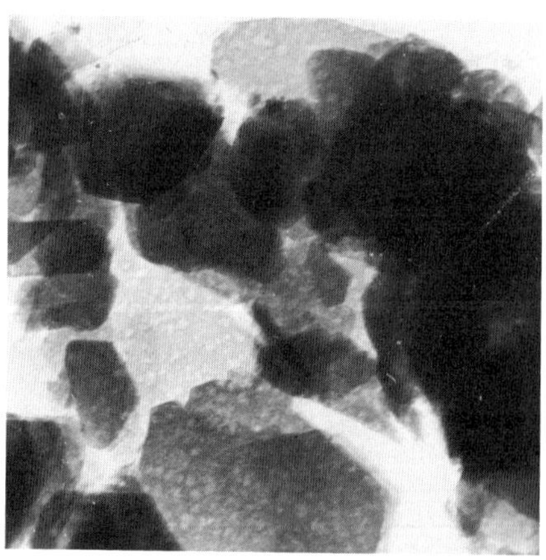

(c)

Fig. 5. Changes in Kaolinite Due to Irradiation by Electron Beams

Al observed after irradiation agree with the interpretation of an increase in the volume of the near surface layer.

CONCLUSIONS

1. The influence of γ-irradiation and electron beams on the surface structure of natural samples of the minerals, quartz and kaolinite, have been studied by the electron-microscopic method.

2. The increase in volume after irradiation of the near surface layers of natural minerals was observed in those specimens containing water.

3. The electron beam influence leads to the formation of "pores" in the near surface layers of natural kaolinite. γ-irradiation may result in an increase in the number of large aggregates of defects.

REFERENCES

1. A. A. Vorobiov and B. A. Kononov, Prokhozdenije Electronov Cherez Veshestvo (Tomsk, izd. Tomskogo Gos. Universiteta, 1966).
2. V. I. Vasiljev, Metodi Electronnomikroskopicheskogo Izuchenija Skolov Kvartsa (Moskva, 1969).
3. V. V. Gromov, Electricheski Zarjad v Radioactivhikh Materialah (Vestnik AN SSSR N°2, 1979).
4. L. A. Shternberg, L. A. Gordienko, "Decorirovanije Defectov Qvartsa," J. Kristallografija 14, (5), 845-851 (1961).

^{99}Tc, Pb and Ru MIGRATION AROUND THE OKLO NATURAL FISSION REACTORS

A. Gancarz, G. Cowan, D. Curtis and W. Maeck*

Los Alamos Scientific Laboratory, Los Alamos, NM 87545
*Idaho National Engineering Laboratory, Idaho Falls, ID 83401

INTRODUCTION

The generally favored method for disposal of commercially generated radioactive wastes is deep geologic burial (1). Existing information supports the belief that appropriate geologic media in conjunction with engineered barriers will be adequate to isolate the waste from the biosphere (1). Because of the long half-lives of many of the radionuclides, the required periods of isolation are much longer than recorded human history, therefore evaluation of the reliability of particular geologic media to isolate the waste must rely heavily upon analytic models of radionuclide transport in geologic environments. Studies of element migration in natural systems can yield important constraints for such analytic transport models. Of particular value are the remains of naturally occurring fission reactors found at the Oklo uranium mine in Gabon, Africa. The uranium rich zones began to undergo sustained nuclear fission 2×10^9 years ago and remained critical for several hundred thousand years (2). Approximately 10 tons of fission products and 4 tons of ^{239}Pu were generated. Hence, Oklo provides a unique opportunity to study movement of fission-produced materials which have resided in a geologic media for times greater than those of concern in the storage of radioactive wastes (3,4). Such studies will be valuable in evaluating transport models and can provide information regarding chemical and physical conditions, time scales, and processes operative in the movement of elements from a well-characterized reactor waste source.

The Oklo deposit is in a Precambrian basin filled with sedimentary and volcanic rocks, the Francevillian Series. The series is several kilometers thick. The uranium deposits occur in a thin,

4 to 10 meter thick, sandstone unit near the base of the series. Within the ore zones, nine natural fission reactors have been identified. Fission, neutron capture and radioactive decay processes have significantly altered the isotopic composition of the materials in the reactors and the immediate vicinity. Consequently, elements transported from the reactor zones into the surrounding environment can, in principle, be identified unambiguously. Applying this principle, we have attempted to determine transport paths, define rates of movement and identify processes controlling the movement of these isotopically anomalous elements. Here we report the results of a study of the migration of fission product ruthenium and ^{99}Tc and of lead produced by uranium decay.

RUTHENIUM AND ^{99}Tc MIGRATION

Relative proportions of ^{235}U and ^{238}U observed in unirradiated natural materials are effectively constant at any time (5). As a result of sustained nuclear fission, ^{235}U is strongly depleted while the effects of neutron irradiation on the abundance of ^{238}U are small. Many rocks from Oklo show shifts in ^{235}U/^{238}U from the normal ratio which are essentially a direct measure of the extent of ^{235}U fission. In conjunction with uranium element abundances, these shifts can be used to calculate the amount of fission products that a sample would contain had it been in a chemically closed system since the nuclear transformations began. Ruthenium is a rare element, constituting <1 ppb of the earth's crust. It is particularly suitable for identifying the occurrence of fission in natural systems because, in contrast to its low natural abundance, ruthenium isotopes constitute $\sim$8.5% of the thermal neutron induced fission products of ^{235}U. Three of the isotopes, ^{101}Ru, ^{102}Ru, and ^{104}Ru, are produced promptly by the decay of short lived radioactive fission products. The fourth fission produced isotope, ^{99}Ru, results from the decay of ^{99}Tc, which as a relatively long half-life of 2.13×10^5 years. Uranium ores contain measurable quantities of ruthenium produced by thermal neutron fission of ^{235}U and by spontaneous fission of ^{238}U in addition to the native element. The proportion of each of these three components can be determined by measuring the relative ruthenium isotopic composition. An imbalance of fissiogenic ruthenium relative to the quantity calculated from ^{235}U depletion is indicative of an open chemical system with the gain or loss of ruthenium and/or uranium.

Table 1 contains uranium element and isotope abundances for ten samples from Oklo. As shown on Fig. 1., the samples are peripheral to the reactor zones. Assuming that normal uranium contains 0.7205% ^{235}U, all samples but one are deficient in this isotope relative to ^{238}U. From these data and the ^{235}U thermal neutron fission yields (6), the amounts of ^{99}Ru and ^{101}Ru have been calculated. These calculated quantities are identified as δ^{99}Ru and δ^{101}Ru in Table 1. Table 1 also contains ruthenium element and isotopic abundances with the native ruthenium component subtracted. This modification to the measured data was done using ^{100}Ru, a naturally occurring isotope not produced by ^{235}U thermal fission or ^{238}U spontaneous fission. From these data the amount of fissiogenic ^{99}Ru has been calculated and labeled ^{99}Ru in the table. The ratios of ^{99}Ru to $\delta^{99}Ru$ vary by a factor ~ 200, indicating significant separation of ^{99}Ru and/or its precursor ^{99}Tc from uranium. Chemical fractionation of ^{99}Tc from ruthenium alters the abundance of ^{99}Ru relative to the other fissiogenic ruthenium isotopes. Therefore, the ratios of ^{99}Ru to these other isotopes are not necessarily proportional to the relative abundances of ruthenium produced by neutron induced ^{235}U fission and spontaneous ^{238}U fission. In contrast, ^{101}Ru and ^{102}Ru are produced promptly by fission, and chemical fractionation cannot produce isotopic shifts. Consequently, $^{101}Ru/^{102}Ru$ is a direct measure of the proportions of the two fissiogenic components. With one exception, ruthenium from ^{235}U fission constitutes 78% to 99% of the fissiongenic Ru. Sample KN-236 has lost all of this component and contains only the products of ^{238}U spontaneous fission. Calculated amounts of ^{101}Ru produced by ^{235}U thermal neutron induced fission, $^{101}Ru(th)$, compared to $\delta^{101}Ru$ range over a factor of 65. These ratios indicate separation of ruthenium from uranium. Both $^{99}Ru/\delta^{99}Ru$ and $^{101}Ru(th)/\delta^{101}Ru$ have values less than one, showing that some samples have lost ruthenium and/or technetium relative to

Table 1. Uranium and Fission Product Ruthenium Data from Oklo Ores

Sample	Wt.% U	^{235}U[a] (atom %)	picog f.p.Ru[b] / g rock	^{99}Ru[b] / ^{101}Ru	^{102}Ru[b] / ^{101}Ru	^{99}Ru[c] / $\delta^{99}Ru$	$^{101}Ru(th)$[d] / $\delta^{101}Ru$	$\Delta*$[e]	T_f 10^5 yr.
KN 224	3.06	0.7190	666	3.11	0.843	4.34	1.58	2.62	-
KN 236	1.52	0.6475	0.7	0.881	1.12	0	0	-	-
KN 241	8.56	0.7200	39.6	1.05	0.829	0.15	0.17	0.87	6.3
KN 244	3.0	0.7205	-	-	-	-	-	-	-
KN 255	5.34	0.7197	849	3.79	0.841	6.44	1.95	3.16	-
KN 258	1.74	0.7196	19.4	1.17	0.885	0.21	0.17	1.01	-
KN 266	8.37	0.7204	34.9	0.762	0.856	0.54	0.76	0.64	3.2
KN 268	0.993	0.7193	1.7	1.19	0.822	0.03	0.03	0.98	12.1
KN 270	12.2	0.7189	1420	0.742	0.855	0.92	1.34	0.63	3.1
SCO 2252	3.0	0.7201	7.4	2.80	-	-	-	(2.31)[f]	
^{235}U(thermal fission)[h]				1.201	0.833				
^{238}U(spontaneous fission)[h]				0.828	1.105				

[a] Naudet and Renson.[7]

[b] Fission product Ru. Natural Ru component subtracted on the basis of ^{100}Ru.

[c] $\delta^{99}Ru$ is the ^{99}Ru produced from $^{235}U(n_{th},f)$. The $\delta^{99}Ru$ was calculated from the uranium data, assuming ^{235}U = 0.7205 atom % and the fission yield data of Maeck *et al.*[6]

[d] $^{101}Ru(th)$ is the ^{101}Ru in the sample produced by $^{235}U(n_{th},f)$. $^{101}Ru(th)$ is determined from the ratio of $^{102}Ru/^{101}Ru$. $\delta^{101}Ru$ is calculated in the same manner as $\delta^{99}Ru$.

[e] $\Delta*$ is the $^{99}Ru/^{101}Ru$ ratio (with both isotopes corrected for a spontaneous ^{238}U fission component) divided by the $^{99}Ru/^{101}Ru$ ratio for $^{235}U(n_{th},f)$.

[f] The value given is derived from the measured $^{99}Ru/^{101}Ru$ ratio without corrections to the data.

[h] Maeck *et al.*[6]

uranium, and values greater than one, indicating the enrichment of these fission products relative to uranium. The extreme cases are sample KN-268 which has lost 97% of its ^{235}U fission products and sample KN-255 which is enriched in fissiogenic ^{101}Ru by 95%. The quantity Δ^* measures fractionation of ruthenium from ^{99}Tc. It symbolizes the measured $^{99}Ru/^{101}Ru$ with both the native and ^{238}U spontaneous fission components subtracted, divided by $^{99}Ru/^{101}Ru$ produced by ^{235}U thermal neutron induced fission. Values of Δ^* greater than one indicate enrichment of ^{99}Tc relative to ruthenium and values less than one indicate ^{99}Tc depletion. The values shown in Table 1 range from 0.63 to 3.16.

Because ^{99}Tc decays to ^{99}Ru, limits can be established on the time that elapsed between the end of ruthenium and technetium production by nuclear fission and the loss of ^{99}Tc by chemical fractionation. If it is assumed that the fission products were produced instantaneously and that ^{99}Tc was removed completely from ruthenium at some later time, then the time interval, T_f, is

$$T_f = \frac{1}{\lambda} \ln \frac{1}{1-\Delta^*}$$

where λ is the ^{99}Tc decay constant. For the samples analyzed, T_f ranges from 0.3 to 1.2 million years. Our analyses show that ruthenium, ^{99}Tc, and/or uranium have been relocated relative to each other since the reactors operated. Samples are both enriched and depleted in ruthenium and ^{99}Tc relative to uranium. Similarly, ^{99}Tc is enriched relative to ruthenium in some samples and depleted in others. The stratigraphic unit from which the samples were collected, and which contains the natural reactors, dips to the east. Therefore, samples collected east of the reactors (see Fig. 1.) lay below the reactors, and samples to the west lay above them. Samples to the east have preferentially lost ruthenium and ^{99}Tc relative to uranium and have

Table 2. Uranium-Lead Data from Oklo Ores and Rocks

Sample		Wt.% U	Wt.% Pb	$^{208}Pb/^{206}Pb$	$^{207}Pb/^{206}Pb$	$^{204}Pb/^{206}Pb$
F_B	PYRITE	0.00005	0.0087	2.3507 ± 4	1.0075 ± 1	0.0662 ± 1
SC50-2273	BASAL CONGLOMERATE	0.011	0.28	0.07982 ± 2	0.14277 ± 4	0.000945 ± 1
SCO-2261		0.0014	0.068	0.08784 ± 1	0.15413 ± 2	0.000886 ± 1
KN230-2460b		0.039	0.16	0.01671 ± 3	0.13829 ± 11	0.000371 ± 1
SC48bis/2228		0.020	0.095	0.11329 ± 2	0.16210 ± 2	0.001973 ± 4
KN-5	ORE	0.30	0.13	0.02148 ± 1	0.13566 ± 7	0.000462 ± 1
KN-6		1.13	0.33	0.01353 ± 1	0.12326 ± 4	0.000238 ± 2
KN-56		0.59	0.24	0.02119 ± 1	0.13138 ± 4	0.000425 ± 2
KN-61		0.28	0.13	0.05772 ± 1	0.14594 ± 2	0.001160 ± 2
KN-76		0.59	0.20	0.01187 ± 1	0.11584 ± 1	0.000186 ± 1
KN-187		3.5	0.64	0.008928 ± 2	0.10718 ± 1	0.000178 ± 1

either lost ^{99}Tc relative to ruthenium or show no fractionation of the two elements. Samples to the west are enriched in both ruthenium and ^{99}Tc relative to uranium and ^{99}Tc relative to ruthenium. Assuming that uranium has been quantitatively retained, then ruthenium and ^{99}Tc have been removed from rocks below the reactors and redeposited above the reactors roughly 10 meters away. The hot aqueous fluids moderating the reactors may have been responsible for transporting these elements.

LEAD MIGRATION

As discussed previously, ruthenium and technetium isotopic systematics permit the evaluation of element migration for times on the order of 10^5 to 10^6 years. Because of the long half-lives of the uranium isotopes, uranium-lead systematics can be used to study element migration over much longer times, e.g., the 2 billion years since the Oklo reactors operated. Early studies of Oklo ores from the reactor zones and the area around them showed depletion of Pb relative to the amounts calculated from uranium decay (8,9,10,11). Losses range from 20% to 80%; the average being $\sim$50% (11). Gancarz (11) attributed these deficiencies to lead loss from the host mineral, uraninite, by continuous diffusion for the past 2.05×10^9 years. These previous studies suggest additional work to delineate transport paths from the ores by searching for the missing lead and to clarify the mechanism by which lead escapes from uraninite. Table 2 contains uranium and lead element abundances and lead isotope ratios for samples which were chosen from a large number of core samples containing materials from above and below the uranium rich regions. "Non-ore" samples are from a conglomeratic layer 3 to 5 meters underneath the ore and reactor zones. Rocks from this basal conglomerate were the only ones to contain either micro or macroscopic crystals of pyrite (FeS_2) and/or galena (PbS).

Lead from uranium ores (radiogenic lead) is rich in ^{207}Pb and ^{206}Pb, the stable end products of the ^{235}U and ^{238}U radioactive decay chains, respectively, and poor in ^{208}Pb and ^{204}Pb. Ratios of $^{204}Pb/^{206}Pb$ in the Oklo ores shown in Table 2 range from 0.00018 to 0.0012, reflecting the enrichment of radiogenic lead. In contrast, lead from typical rock (common lead) contains relatively less ^{207}Pb and ^{206}Pb and typical $^{204}Pb/^{206}Pb$ values range from 0.05 and 0.10. Authigenic pyrite from rock stratigraphically above the ore bearing horizon, sample F_B in Table 2, has a $^{204}Pb/^{206}Pb$ ratio of 0.0662. This isotopic composition of lead is that for normal rock formed $2.05 \pm 0.03 \times 10^9$ years ago (12). This time is in good agreement with the age of uranium mineralization measured by the uranium-lead systematics (11) and the time of nuclear criticality determined by neutronic considerations of neodymium and uranium isotopic abundances (2). We consider the authigenic pyrite lead to be representative of lead incorporated into rocks at Oklo when they formed. This lead is referred to as

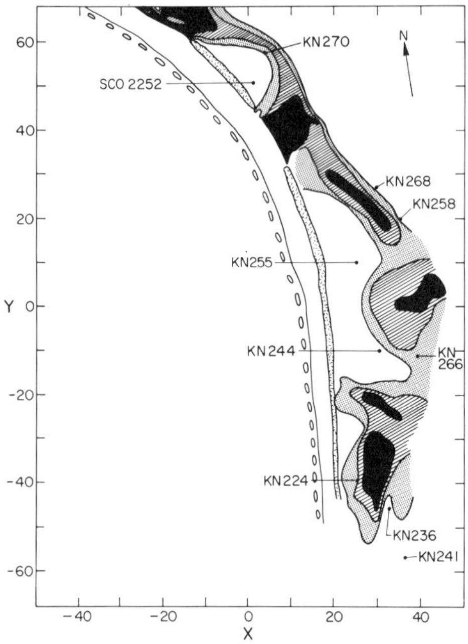

Fig. 1. Locations of samples that were analyzed for ruthenium isotopic composition to investigate ^{99}Tc and fissiogenic Ru migration. The Oklo reactor locations are shown in black. Concentrations of uranium in the ores decrease away from the reactor zones, as indicated by the contours. The abscissa and ordinate units are in meters.

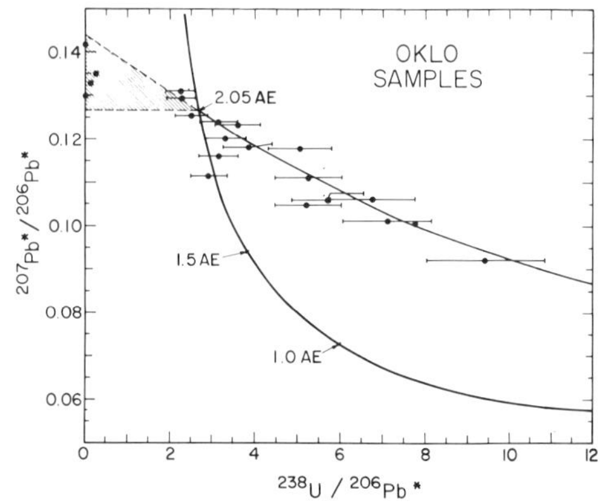

Fig. 2. A lead-uranium evolution diagram. Lead deficient samples to right of the concordia curve are all uranium ores. Rocks from the basal conglomerate plot near the abscissa in the dashed region. This is the region for materials containing excess lead lost by continuous diffusion from ores

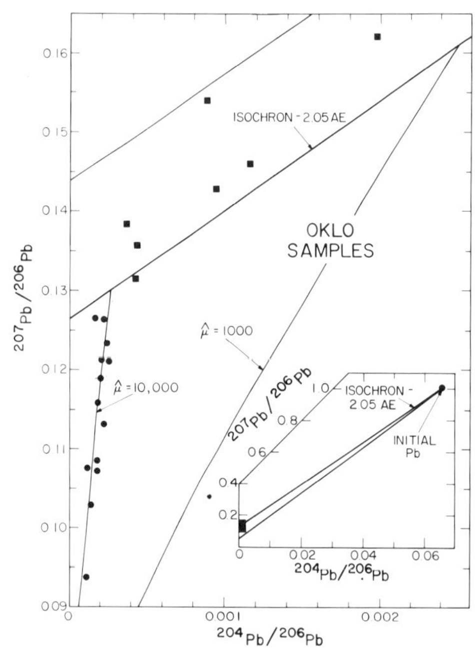

Fig. 3. A plot of $^{207}Pb/^{206}Pb$ versus $^{204}Pb/^{206}Pb$. Samples that have lost lead by diffusive processes are enclosed in the triangular region shown in the inset. An enlargement of the dark area of the inset includes all the data from Oklo rocks. Circles represent lead deficient ores as determined from Fig. 2. Lead rich rocks including those from the basal conglomerate are plotted as squares.

initial lead. Rocks at Oklo contained common initial lead when they formed. If rocks now contain radiogenic lead, then they are either enriched in uranium or have inherited lead from a uranium rich source. In the latter case, we would conclude that the rocks are parts of lead transport paths. Relative proportions of radiogenic lead and uranium in the samples are shown on Fig. 2. The lower curve is the concordia curve, i.e., the locus of points for closed system lead evolution from pure uranium between the time indicated on the curve to the present. Data to the right of the curve indicate lead loss relative to uranium and data to the left reflect relative lead gain. Asterisks, as in Fig. 2, indicate lead isotopic data with initial lead contribution subtracted. Data previously reported by Gancarz (11) are included in the figure. All four samples from the basal conglomerate have $^{238}U/^{206}Pb^*$ values less than 0.25 and plot to the left of the concordia curve showing them to be enriched in radiogenic lead. Ratios of $^{204}Pb/^{206}Pb$ in these samples range from 0.00037 to 0.00197 one to two orders of magnitude smaller than the initial lead. There is no doubt that rocks in the basal conglomerate contain lead which originated in the overlying uranium ores.

In contrast to these basal conglomerate samples, which plot to the left of the concordia curve, the ore samples in general plot to the right of the curve reflecting relative lead loss. The curve above the concordia curve is the locus of points assuming loss of lead by continuous diffusion from 2.05×10^9 years to the present. Rates of diffusion are governed by the diffusion coefficient, the radius of the host mineral grains and the lead concentration gradient (13). These new data are consistent with lead loss by diffusion from uraninite. The dashed region to the left of concordia represents samples that have lost no lead and that have gained the diffused lead. Samples of basal conglomerate all plot in this portion of the diagram. These data confirm our hypothesis that lead is lost from host uraninite by a diffusion mechanism. The conglomerate underlying the uranium deposit contains this diffused lead and is, therefore, both a transport path and repository for lead from the ores. The significance of the variation of $^{207}Pb/^{206}Pb$ in the conglomerate and of the data which plot below the diffusion trajectory remains unclear. Fig. 3. is a diagram that shows lead isotope data, uncorrected for initial lead. The curves below the isochron line are model curves for the loss of both initial lead and radiogenic lead by diffusion from uraninite. Circles represent data for rocks depleted in radiogenic lead as deduced from Fig. 2. Agreement between these data and the model curve where the parameter $^{238}U(today)/^{204}Pb$ (initial) $\equiv (\hat{\mu})$ = 10,000 indicates that both radiogenic and initial lead were lost from uraninite by diffusion. The two parallel lines, one labeled "isochron" in Fig. 3., define the region for samples that have retained lead and gained the lead lost by diffusion from uraninite. All samples enriched in radiogenic lead relative to uranium, as identified by squares in the figure, fall within this area. As with the uranium-lead systematics, lead isotope data alone show that lead was lost from uraninite by diffusion. Subsequently this lead was transported 3 to 5 meters and deposited in the underlying conglomerate.

CONCLUSIONS

This work demonstrates the utility of the Oklo uranium ore deposit and natural fission reactors as a long time scale analogue for man-made radioactive waste repositories. It has been shown that the ores and nearby rocks were open to the loss and gain of ^{99}Tc, ruthenium, and lead relative to uranium. Identified regions of element deficiencies and those which are correspondingly enriched are separated by less than 10 meters. However, more extensive sampling is required to define the overall extent of the element migration. Element fractionation took place on at least two vastly different time scales; ^{99}Tc was fractionated from ruthenium within one million years of the end of reactor criticality. Lead-uranium fractionation has been ongoing for most of the two billion years since the ores were formed. Diffusive loss of lead from host uraninite appears to be an important process in the fractionation of lead from uranium.

REFERENCES

1. Management of Commercially Generated Radioactive Waste, U. S. Department of Energy, DOE/EIS-00460 Vol. 1 (1979).
2. Cowan, Bryant, Daniels and Maeck, The Oklo Phenomenon, IAEA Vienna, 341 (1975).
3. Walton and Cowan, The Oklo Phenomenon, IAEA Vienna, 499 (1975).
4. Bryant, Cowan, Daniels and Maeck, Actinides in the Environment, ACS Symp. #35, 89 (1975).
5. Cowan and Adler, Geochim. Coshim. Acta. 40, 1487 (1976).
6. Maeck, Delmore, Eggleston and Spraktes, Natural Fission Reactors, IAEA Vienna, 521 (1978).
7. Naudet and Renson, Natural Fission Reactors, IAEA Vienna, 131 (1978).
8. Lancelot, Vitrac and Allegre, Earth and Plant. Sci. Lett. 25, 189 (1975).
9. Hagemann, Lucas, Nief and Roth, Earth and Plant. Sci. Lett. 23, 170 (1974).
10. Devillers, The Oklo Phenomenon, IAEA Vienna, 293 (1975).
11. Gancarz, Natural Fission Reactors, IAEA Vienna, 513 (1978).
12. Doe and Stacey, Economic Geology 69, 757 (1974).
13. Tilton, J. Geophys. Res. 65, 2933 (1960).

PARAMETERS AFFECTING RADIONUCLIDE MIGRATION IN GEOLOGIC MEDIA*

B. R. Erdal, B. P. Bayhurst, W. R. Daniels, S. J. DeVilliers, G. H. Heiken, F. O. Lawrence, M. L. Sykes, J. R. Smyth, J. L. Thompson, E. N. Vine, and K. Wolfsberg

Los Alamos Scientific Laboratory
Los Alamos, NM 87545

INTRODUCTION

If geologic repositories are used for the isolation of nuclear waste, intruding groundwater could lead to dissolution of radionuclides and their subsequent transport to the biosphere. Interaction with the geologic media is one of the most important factors retarding such transport. It is therefore necessary to understand the many interrelated factors and processes that govern radionuclide transport. The effects of some of the parameters that may influence sorption and transport properties are being systematically investigated (1-6) at the Los Alamos Scientific Laboratory. Earlier studies were performed on three rock types specific to the Nevada Test Site (NTS). These are a quartz monzonite porphyry (Climax Stock), an argillite (Eleana Formation), and several varieties of rhyolitic tuff (Jackass Flats), described in Refs. 1-4. Several additional tuff samples have now been investigated (5, 6).

The equilibrium distribution coefficient, K_d, for the distribution of radioisotopes between aqueous and solid phases is conventionally defined as the concentration in the solid phase divided by the concentration in solution. Since it is not known whether equilibrium is achieved for the interactions of radionuclides with geologic media, in this report the value is called the sorption ratio which is otherwise identical to the distribution coefficient but does not imply equilibrium.

*This work was supported by the U. S. Department of Energy.

STATIC (BATCH) MEASUREMENTS

The experimental methodology used to perform laboratory studies of some of the variables that may influence radionuclide transport was presented in Refs. 1-4. Initial results on various parameters were also given for the quartz monzonite, argillite and tuff from Jackass Flats.

Representative values for the sorption ratios for the additional tuff samples from Yucca Mountain are given in Table 1. Generally, the conclusions reached earlier (1-4) for other rock types are also valid for these samples (6). However, in contrast to the earlier studies, there were no consistent correlations between the sorption ratio and particle size for either the sorption or desorption measurements. For example, the sorption ratio values for strontium, cesium and barium on YM-38 decreased as the sieve fraction size decreased, while the values for the other samples tended to increase or remain approximately constant with decreasing size. X-ray diffraction analyses indicate no difference in X-ray patterns or intensities for the different size fractions. Therefore, unless the difference in sorption behavior is due to a relatively minor component which is not resolved by X-ray diffraction, it is unlikely that these results are due to mineral fractionation. A possible explanation is that the smaller fraction ($<106\ \mu m$) had been ground to particles smaller than natural grain sizes, thus exposing different surfaces and perhaps destroying others.

Several experiments have been performed to assess the dependence of sorption on groundwater composition. The initial composition of the two waters were (in mg/ℓ) Na(10), K(5), Ca(10), Mg(2), SO_4(5), and Cl(4.5) for the "dilute" water (called Water I) and Na(50), K(5), Ca(50), Mg(20), SO_4(70), and Cl(15) for the "concentrated " water (called Water II). The 250-350 μm sieve fractions from the various materials were used with the same pretreatment and experimental techniques described earlier. The sorption of strontium, cesium, barium, cerium and europium was studied at about the same concentrations as used earlier. Contact times of 28 and 60 days were used.

If other uncontrolled variables are unimportant, the composition of the water is a major factor governing sorption (Table 2). As expected, the sorption ratios for strontium, cesium and barium decreased when Water II was used, due to the increased competition for sorption sites. For cerium and europium, the sorption ratios tended to increase with increasing ionic strength. This may be due to the enhanced tendency to form radiocolloids as the sulfate concentration increases. The sorption behavior of cerium and europium is evidently quite different than that for cesium, strontium and barium. One should note that Water I and Water II are somewhat similar to the waters used for the earlier granite (2) and argillite (3) studies, respectively. The sorption ratio values obtained using

Table 1. Representative Sorption Ratios (mℓ/g) for Tuff from Yucca Mountain[a]

Element	YM-22	YM-38	YM-54
Sr	53, 63[b]	12 000, 20 000	90, 97
Tc(VII)	0.3, 1.2	0.2[c], 2.0[c]	
Cs	340, 400	8 600, 13 000	250, 310
Ba	980, 1 000	66 000, 190 000	620, 660
Ce(III)	1 400, 6 100	830, 3 800	140, 990
Eu(III)	1 400, 3 600	2 300, 8 700	500, 1 800

[a]Atmospheric conditions.
[b]The second value listed for each element is that obtained from the desorption measurements.
[c]Obtained using a sample (YM-49) from the 678 m depth which is mineralogically similar to YM-38.

these new waters are very consistent (within 20%) with the previous values, showing that our experimental method gives consistent results.

Under the generally reducing conditions expected from groundwater not in contact with the atmosphere, one can anticipate that the sorption behavior of some elements would be different from that under atmospheric conditions. One could also expect some differences in the sorptive capacity of a rock if its surface had been altered by exposure to air. These effects were investigated by performing studies in a pure nitrogen atmosphere having ≦0.2 ppm oxygen and ≦20 ppm carbon dioxide present. It should be emphasized that these studies are perhaps not truly representative of the conditions to be found in deep geologic systems since no carbon dioxide or soluble carbonate was intentionally present. This could be particularly important for uranium(VI), which is strongly complexed by carbonate ions. However, since the pH values observed when the rock was present were not appreciably different than those observed under atmospheric conditions, the rocks themselves may have supplied some carbonate ion. All of the operations (including the grinding and sieving) needed to make the batch measurements were performed under these controlled atmosphere conditions using the same granite, argillite and tuff from Yucca Mountain as described earlier.

Sorption ratios for cesium, strontium and barium were generally somewhat less than those observed under atmospheric conditions for the argillite and tuff samples, but they were about the same for the granite samples. In contrast, the sorption ratio values for cerium and europium were appreciably higher for granite and argillite under the controlled atmosphere conditions, while the values for tuff were unchanged. The sorption ratio values for technetium under the reduced oxygen and carbon dioxide conditions were consistently higher (by more than a factor of 40) than under atmospheric

Table 2. Effect of Water Composition on the Sorption Ratio (mℓ/g)[a]

Sample[b]	Water	Cs	Sr	Ba	Ce	Eu
CS-5	I	180	8.2	63	180	200
	II	230	5.0	28	>10 000	4 800
CN-2	I	2 400	700	2 300	5 810	8 340
	II	1 600	170	700	12 400	10 700
JA-18	I	10 900	23 600	17 200	945	1 130
	II	740	5 030	46 400	5 200	1 700
JA-32	I	130	84	587	2 870	1 690
	II	85	18	656	>3 000	23 500
JA-37	I	1 330	415	840	27 100	12 000
	II	844	174	375	>36 000	12 700

[a]Only the data for the 28-d contact time is shown; see Refs. 5 and 6 for the 60-d results.

[b]CS means Climax Stock quartz monzonite (Ref. 2), CN means an Eleana argillite (Ref. 3), and JA means a tuff from Jackass Flats (Ref. 4).

conditions for all three rock types, which supports the conclusion (7) that pertechnetate ion can be chemically reduced to a less soluble form. Similar measurements for uranium on the granite and argillite indicated that the sorption ratios were higher (by about a factor of 10) for the < 75-μm sieve fractions under atmospheric conditions than in the absence of oxygen and carbon dioxide, and essentially equal for the remainder of the material (75-500 μm). Elemental analyses of these fractions did not indicate any difference between the samples, particularly in the total iron content. Perhaps mineral fractionation occurred in the sieving process. However, the rather striking difference in the measured sorption ratio values may be due more to the absence of carbon dioxide than to the absence of oxygen.

In our usual experimental procedure for plutonium and americium batch sorption experiments, the groundwater mixtures were separated after contact by centrifuging the aqueous phase three times for at least 1 h at about 12 000 rpm (32 000 g) with careful transfer of the water following each centrifuging run. This process produces plutonium solutions which generally can be passed through a polycarbonate filter membrane with pores as small as 0.05 μm without loss of radioactivity. Some americium is normally removed when such solutions are centrifuged. A microautoradiographic technique has been used to examine the filter membranes. For solutions from sorption experiments, α-ray "stars" or large radioactive clusters were observed, indicating the presence of large agglomerates containing many americium atoms even after the rigorous centrifuging process. For solutions from desorption experiments, the filter membranes showed americium only as single α-tracks, indicating that agglomerates of the type observed in the centrifuged solution from the sorption experiments were not present in the desorption solutions and that americium may sorb on the membrane itself. Apparently, such clusters were neither removed from the solid phase once they had been "sorbed" nor formed in the desorption water.

Considering the above results, we have decided to take the conservative approach in calculating americium sorption ratios and use the results from the solutions after they have been centrifuged but not filtered. This is reflected in the values given in Refs. 1, 5 and 6.

MICROAUTORADIOGRAPHY

Microautoradiography has been employed (8) at LASL as another technique for studying the sorption properties of certain elements on selected types of rocks. Microautoradiography can indicate the specific mineral phase responsible for the sorption and can give insight into the state of aggregation of the sorbed material. Since many of the transuranic elements tend to form polymeric or colloidal species in solution at groundwater pH values, observation of the degree of aggregation in adhering species is very important. This technique has been used for studies of the behavior of ^{233}U, ^{237}Np, ^{239}Pu and ^{241}Am.

The results indicate that the majority (>90%) of the ^{233}U activity sorbed onto the quartz monzonite thin-sections was confined to the clay-rich alteration bands in the plagioclase which comprises less than one percent of granites. Most of the ^{233}U is sorbed on the clay matrix in the argillite while very little sorption occurred on the detrital quartz and secondary calcite. For the tuff, most of the ^{233}U was localized in the secondary zeolite minerals (clinoptilolite) as compared to the phenocrysts of quartz and feldspar. In general, ^{237}Np, ^{239}Pu and ^{241}Am showed the same trends as the uranium when sorption was definitely observed; however, stellate patterms of α-tracks frequently appeared on all rock types and did not correlate in any obvious way to the specific minerals in the thin-sections.

A number of autoradiograms were prepared using ^{85}Sr, ^{63}Ni, ^{233}Pa and ^{90}Sr. Strippable emulsion, liquid emulsions, and a film-backed emulsion were utilized. The beta tracks (generally spots) were readily visible at moderate densities against a clear background, but were difficult or impossible to identify except at very high densities against the mottled backgrounds typical of rock thin-sections.

DYNAMIC (COLUMN) STUDIES

The logical next step in extending our knowledge of the behavior of radionuclides in geologic systems was to measure migration rates. For crushed rock columns, assuming equilibrium between water and rock, the ratio of the velocity of a radionuclide to the groundwater velocity, the retardation factor, is given by $K_d(\rho/\varepsilon)+1$, where ρ is the bulk density and ε is the porosity (free column volume/bed volume). In fact the nonequilibrium "sorption ratios" are more appropriate in this expression for short term experiments than are the equilibrium K_d values.

Initial experiments have been performed (5, 6) using the same materials and pretreated waters as used in the batch measurements, in order to compare the results with those from the "batch" tests. A simple, upflow, 0.4-0.5 cm inside diameter by 2-3 cm high column, lined with Teflon, with polyethlene bed supports on each end, was developed. Water delivery rates, as controlled by syringe pumps, were usually below 0.05 mℓ/h ($\sim$30m/y). Actual delivery rates were determined by weighing the eluate collected during a known time period. The free-column-volumes were usually determined by use of both triatiated water and $^{131}I^-$; no difference was observed between the two methods. The tracers, ^{85}Sr, ^{137}Cs and ^{133}Ba (all about 10^{-9} m), used singly or as a mixture, were usually added in a minimal (5-15 $\mu\ell$) volume. Since the elution peaks were reasonably symmetrical, the volume were 50% of the activity had been eluted corresponded closely with the peak concentration in the elution curve. This volume was used to calculate the retardation factor.

The sorption ratios obtained (Table 3) from the column data were always less than those obtained using the batch technique. In most cases, the agreement was within a factor of two to three; the largest differences were about a factor of five for ^{133}Ba. This may be due to the rate of sorption on the rock being slow compared to the water flow rate. In comparing these results with batch data, we note that diffusion into the solid may contribute in batch experiments that last weeks or months but could be negligible in a column having a comparatively short, one day, residence time.

ACKNOWLEDGEMENTS

The assistance of the following Los Alamos Scientific Laboratory personnel is acknowledged: R. D. Aguilar, S. Maestas, P. Q. Oliver, P. A. Elder, M. E. Lark and L. M. Mitchell.

This work was supported in part by the Waste Isolation Safety Assessment Program being conducted by the Office of Nuclear Waste Isolation which is managed by Battelle Memorial Institute under contract with the Department of Energy (DOE), and by the Nevada Nuclear

Table 3. Comparison of Column and Batch Sorption Ratios (mℓ/g)[a]

Column Material	Sr	Cs	Ba
YM-22	30 (50)	122 (287)	355 (652)
YM-54	44 (84)	92 (247)	126 (652)
		80 (247)	
	44 (84)	113 (247)	122 (652)
JA-32	52 (56)		
	37 (56)		
	35 (56)		
CN-1	43 (138)	b (1 830)	
CN-2	52 (56)	b (1 580)	
CS-5	9 (14)	b (328)	32 (154)
CS-7	15 (18)	b (347)	44 (175)

[a]The sorption ratios obtained on the same materials using the batch technique are in parentheses.
[b]The elution is not yet complete.

Waste Storage Investigations project and the Radionuclide Migration project, both managed by the Nevada Operations Office of the DOE.

REFERENCES

1. B. R. Erdal, B. P. Bayhurst, B. M. Crowe, W. R. Daniels, D. C. Hoffman, F. O. Lawrence, J. R. Smyth, J. L. Thompson, and K. Wolfsberg, "Laboratory Studies of Radionuclide Transport in Geologic Media," IAEA/NEA Symposium on the Underground Disposal of Radioactive Wastes, Otaniemi, Finland, 2-6 July 1979 (in press).
2. B. R. Erdal, R. D. Aguilar, B. P. Bayhurst, W. R. Daniels, C. J. Duffy, F. O. Lawrence, S. Maestas, P. Q. Oliver, and K. Wolfsberg, Sorption-Desorption Studies on Granite, LA-7456-MS (1979).
3. B. R. Erdal, R. D. Aguilar, B. P. Bayhurst, P. Q. Oliver, and K. Wolfsberg, Sorption-Desorption Studies on Argillite, LA-7455-MS (1979).
4. K. Wolfsberg, B. P. Bayhurst, B. M. Crowe, W. R. Daniels, B. R. Erdal, F. O. Lawrence, A. E. Norris, and J. R. Smyth, Sorption-Desorption Studies on Tuff. I. Initial Studies with Samples from the J-13 Drill Site, Jackass Flats, Nevada, LA-7480-MS (1979).
5. B. R. Erdal, Ed., Laboratory Studies on Radionuclide Distributions Between Selected Groundwaters and Geologic Media, October 1, 1978 - September 30, 1979, Annual Report, LA-8088-PR (1979).

6. E. N. Vine, R. D. Aguilar, B. P. Bayhurst, W. R. Daniels, S. J. DeVilliers, B. R. Erdal, F. O. Lawrence, S. Maestas, P. Q. Oliver, J. L. Thompson, and K. Wolfsberg, Sorption-Desorption Studies on Tuff. II. A Continuation of Studies with Samples from Jackass Flats, Nevada and Initial Studies with Samples from Yucca Mountain, Nevada, LA-8110-MS (1980).
7. E. A. Bondietti and C. W. Francis, "Geologic Migration Potentials of Technetium-99 and Neptunium-237," Science 203, 1337 (1979).
8. J. L. Thompson and K. Wolfsberg, Applicability of Micro-Autoradiography to Sorption Studies, LA-7609-MS (1979).

RADIONUCLIDE TRANSPORT IN A DOLOMITE AQUIFER*

R. G. Dosch and A. W. Lynch

Sandia Laboratories**
Albuquerque, NM 87185

ABSTRACT

The most probable paths for radionuclide transport between the biosphere and the proposed bedded salt repository (WIPP) in southeastern New Mexico are through dolomite formations which are the only currently water-bearing units overlying the repository horizon. This paper describes some of the experimental work performed to determine the potential for radionuclide migration in these formations. Included are the results of radionuclide sorption studies on natural dolomites (Magenta and Culebra) and on synthetic dolomite (proto-dolomite). Sorption on natural dolomites was evaluated with respect to core sample location, ground water composition, nuclide concentration and water volume to rock mass ratios used in making the measurements. The results show that a discrete value such as a distribution coefficient (Kd) assigned to a given radionuclide does not accurately describe its interactions with site specific geological formation.

INTRODUCTION

The final links in the most probable path for radionuclide transport between the biosphere and the proposed bedded salt repository (WIPP) are dolomite formations which represent the only currently water-bearing units overlying the repository horizon.

*This work supported by the United States Department of Energy (DOE) under Contract DE-AC04-76-DP00789.
**A DOE facility.

Batch distribution coefficient (Kd) experiments have shown that radionuclides are efficiently absorbed by dolomites under some conditions (1). Based in part on some of this data, mathematical modeling associated with safety assessment efforts predicts significant retardation of radionuclides in ground water relative to ground water movement. This study was initiated to obtain a better understanding of the sorption process on dolomite and thus contribute to the reliability of mathematical extrapolation of observed phenomena. The results show that a discrete value such as a Kd assigned to a given radionuclide does not accurately describe its interactions with a site specific geological formation.

RADIONUCLIDE SORPTION ON NATURAL DOLOMITE

Two water-bearing dolomite beds are recognized in the WIPP site area and range between 20 and 30 feet thick. The upper dolomite member, Magenta, is a finely crystalline, dense dolomite; and the lower, Culebra, is vuggy and commonly associated with anhydrite (2). Since the formations are not homogeneous, measured sorption properties of radionuclides on these formations are dependent on both the borehole location and depth from which core samples used in a particular measurement were obtained. Table 1 gives some sorption coefficients (Kd) for Cs, Sr and Pu on Magenta and Culebra dolomite obtained from different boreholes near the WIPP site. Techniques used in determining and expressing the Kd values are described elsewhere (1). The water used for the aqueous phase was a simulant containing approximately 10,000 ppm of dissolved solids with a composition similar to that of water obtained from boreholes H-2b and H-2c in the Culebra formation (3). The results in Table 1 were extracted from a larger body of data describing the variability of sorption properties in Magenta and Culebra dolomite (4) and represent the range of sorption behavior observed for these particular nuclides.

Table 1. Nuclide Sorption on Magenta and Culebra Samples From Various Locations at the WIPP Site

Borehole/Depth		Formation	Kd, ml/g		
			Cs	Sr	Eu
H5A	800'	Magenta	215	2	280
H5A	811'	Magenta	10	9	126
H5A	818'	Magenta	0.5	22	24
H4B	517'	Culebra	9	0.6	1100
H4B	526'	Culebra	1.2	32	530
H5B	906'	Culebra	68	0.6	4460

RADIONUCLIDE SORPTION ON SYNTHETIC DOLOMITE (PROTO-DOLOMITE)

The variability of the sorption results on natural dolomite suggests that sorption is associated in part with minor or trace mineral phases in the dolomite. This heterogeneity in sorption properties was demonstrated in electron microprobe examinations of Magenta which had been equilibrated with concentrated solutions of stable isotopes of some of the fission product elements (5). A synthetic dolomite, proto-dolomite, prepared by the method of Gaines (6) with analytical grade reagents, was used in an attempt to determine the sorption properties of the matrix of Magenta and Culebra. Two materials were prepared and were shown to be a mixture of proto-dolomite and brucite (Prep 1) and pure proto-dolomite (Prep 2), respectively, by comparison of their X-ray diffraction patterns with previously published data (7, 8). Sorption coefficients for several radionuclides on the synthesized materials are given in Table 2. Sorption coefficients for natural dolomite measured under the same conditions are included for comparison. The data show that the sorption properties of the natural and synthetic materials are very similar, suggesting that the dolomite matrix of Culebra and Magenta can contribute to radionuclide sorption.

There are several probable explanations for the apparent discrepancy between the microprobe study showing heterogeneous sorption and the proto-dolomite work indicating sorption on the dolomite matrix. High solution concentrations ($\geq$50 ppm to 1 $\underline{M}$) of stable fission product isotopes used in the microprobe work probably led to precipitation in localized areas where solubility products were exceeded. Strontium sorption, for example, was associated with gypsum inclusions in the natural dolomite. Impurities, such as clay minerals, with higher exchange capacities per unit volume than the matrix material would also contribute to apparent heterogeneous

Table 2. Comparison of the Radionuclide Sorption Properties of Natural Dolomite and Synthetic Proto-dolomite

	Proto-dolomite Kd, ml/g		Natural dolomite Kd, ml/g	
	Prep #1*	Prep #2**	Culebra	Magenta
Cs	25	76	9	4
Sr	16	9	5	5
Eu	5×10^4	4×10^5	1×10^4	1×10^4
Pu	5×10^3	2×10^3	7×10^3	2×10^3
Am	7×10^3	2×10^4	2×10^4	2×10^3
Tc	0	0	1	1.5

*XRD patterns showed mixture of proto-dolomite and brucite.
**XRD patterns showed only proto-dolomite.

sorption, particularly if the concentrations of sorbed species on the dolomite matrix were below microprobe detection limits.

THE EFFECT OF GROUND WATER COMPOSITION AND NUCLIDE CONCENTRATION ON RADIONUCLIDE SORPTION

Both groundwater composition and nuclide concentration are important parameters in sorption studies associated with radionuclide migration since most natural materials exhibit low cation exchange capacities and non-linear absorption isotherms. Shiao, Rafferty, and Meyer (9) found in studies with montmorillonite clays that the Kd values for Cs, Sr and Eu decreased gradually with increasing concentrations until approximately 10% of the total exchange capacity of the clay was achieved. This was followed by a more precipitous decrease at higher loadings. Similar behavior was noted in this work in the case of Cs sorption on Culebra dolomite and is evidenced by the results given in Table 3.

Table 3. Dependence of Cs Distribution Coefficients on Culebra Dolomite on Cs Concentrations

1. Initial Cs Concentration Range: 8.4×10^{-11} to 8.4×10^{-8}M

[Cs] initial, mole/l	Kd, ml/g	K', ml/g
8.4×10^{-11}	34	1.4
8.4×10^{-10}	24	1.4
8.4×10^{-9}	21	1.7
8.4×10^{-8}	12	1.3

$$K' = [Cs]_{rock} / [Cs]_{liquid}^{.87}$$

2. Initial Cs Concentration Range: 8.4×10^{-8} to 7.5×10^{-4}M

[Cs] initial, mole/l	Kd, ml/g	K'', ml/g
8.4×10^{-8}	12	0.19
7.6×10^{-7}	6.4	0.18
7.5×10^{-6}	3.2	0.16
3.8×10^{-5}	2.3	0.18
7.5×10^{-5}	1.8	0.17
7.5×10^{-4}	1.2	0.20

$$K'' = [Cs]_{rock} / [Cs]_{liquid}^{.75}$$

The problem presented by this behavior in terms of current transport modeling lies in choosing a discrete value of Kd to describe the sorption of Cs on Culebra. This purpose would be better served using empirical constants, K' and K'', where $K' = [Cs]_{rock}/[Cs]_{liquid}^{.87}$ and $K'' = [Cs]_{rock}/[Cs]_{liquid}^{.75}$ over initial Cs concentration ranges of 8.4×10^{-11} to 8.4×10^{-8}M and 8.4×10^{-8} to 7.5×10^{-4}M, respectively. Average values of 1.5 ± 0.1 ml/g and 0.18 ± 0.01 ml/g were found for K' and K'', respectively.

The effect of ground water composition on Kd is of particular importance in the case of repositories planned for halite formations where any water escaping from the repository would most likely be a saturated brine. Subsequent dilution in other ground waters could result in widely varying compositions, particularly with respect to sodium concentrations. Values of distribution coefficients of greater than 600 are observed for Cs on Magenta dolomite in deionized water. The effect of sodium ion concentration on Cs sorption on Magenta is shown in Table 4 over a sodium concentration range of 0.25 to 5.0 m. This range was produced by adding varying amounts of brine representing water in contact with a halite zone to a synthetic ground water typical of the WIPP site area. Again, an empirical relationship, $K''' = \left([Cs]_{rock}/[Cs]_{liquid}\right) \times [Na]^{0.52}$, can be used to describe sorption as a function of sodium concentration, where an empirical constant, K''', was found to be 57 ± 4 for the concentration range studied.

Table 4. The Effect of the Presence of Sodium on Sorption of Cesium on Magenta Dolomite

[Na], mole/liter	Kd, ml/g	$K''', \frac{ml}{g} \times \left(\frac{mole}{liter}\right)^{.52}$
0.25	101	50
0.50	80	56
0.75	66	57
1.0	59	59
1.25	55	62
2.0	43	61
2.5	38	61
3.75	27	54
5.0	22	50

$$K''' = Kd \times [Na]^{0.52}$$

THE EFFECT OF LIQUID-TO-SOLID RATIOS USED IN LABORATORY SORPTION MEASUREMENTS

The liquid-to-solid ratio used in measuring distribution coefficients by batch contacting of a ground water containing dissolved radionuclides with a solid has been shown to affect experimental results. Due to practical considerations such as the need for liquid-solid phase separation prior to analyses and limited availability of core sample, the tendency is to use liquid-to-solid ratios (>10 ml/g) which are much higher than those found in typical geological formations (<<1 ml/g). An example of the influence of this ratio for the sorption of U(VI) on Magenta and Culebra dolomites is shown in Table 5. The liquid phase used in these measurements was Solution C (1), a ground water simulant typical of the WIPP area which was doped with a mixture of ^{236}U and ^{237}U to a concentration of 8×10^{-10}M U(VI). The ^{237}U to ^{236}U isotopic ratio was 1.8×10^{-4}. The solid phases were -200 mesh Magenta and Culebra dolomite from the 749' depth of the AEC#8 borehole and the 900' depth of the H5B borehole, respectively. Two sets of experiments were done using as-received dry core material and the same material which had been pre-washed with the ground water simulant prior to contact with the doped simulant. Liquid phase U(VI) concentrations were determined by gamma spectroscopy using the 208 Kev peak of ^{237}U.

A significant trend of decreasing Kd with decreasing liquid-to-solid ratios was observed in all cases. The generally higher values observed for the pre-washed core sample may reflect a lower solution ionic strength due to removal of highly soluble components of the

Table 5. The Effect of Liquid-to-Solid Ratios on Distribution Coefficients for U(VI) on Dolomites

Material	Borehole/Depth	Liquid-to-Solid Ratio, ml/g 5	15	25	50	100
Culebra As-received	H5B 900'	2.1* (8.0)**	4.4 (8.0)	5.6 (7.7)	13.6 (7.7)	38 (8.2)
Culebra Pre-washed	H5B 900'	7.8 (7.6)	13.6 (7.5)	18 (7.4)	34 (7.4)	80 (7.2)
Magenta As-received	AEC#8 749'	2.3 (8.1)	5.1 (7.9)	8.5 (7.9)	17.4 (7.7)	43 (8.0)
Magenta Pre-washed	AEC#8 749'	13 (7.7)	20 (7.5)	29 (7.4)	52 (7.4)	132 (7.3)

*The Kd values are given in units of ml/g.
**The values under the Kd in parentheses are the final pH values.

core sample which occurs during the pre-wash. There is no apparent correlation of Kd with pH as these values either remain nearly constant or increase slightly as the liquid-to-solid ratio decreases. Increasing pH is generally associated with increasing rather than decreasing sorption coefficients for cations on inorganic materials. The data of Table 5 suggests that interaction of U(VI) with dolomite involves precipitation (described by a solubility product and independent of the liquid-to-solid ratio) rather than an ion exchange sorption equilibrium.

This trend raises questions about the validity of extrapolating laboratory results to field conditions since U(VI) sorption appears to decrease as field conditions are approached. However, Cs sorption does not appear to be dependent on liquid-to-solid ratios. Cs distribution coefficients determined in laboratory experiments using liquid-to-solid ratios in the range of 50 to 100 have been successfully used to predict transport velocity through one-inch thicknesses of monolithic Magenta core sample (10). The liquid-to-solid ratio in saturated Magenta core is approximately 0.15.

SUMMARY

The results presented in this paper show that a discrete value for the distribution coefficient assigned to a given radionuclide does not accurately describe its interaction with a site specific geological formation. Empirical relationships such as those given above can account for changes in solution chemical parameters such as nuclide concentration and ground water composition. If variations in parameters such as liquid-to-solid ratios which are unique to laboratory measurements are determined to be important for a given set of conditions, their effects can probably be minimized by designing experiments which approximate field conditions as closely as practicable. Variation in sorption properties with changes in the geological formation being studied and the local hydrology is a more difficult problem, and additional information is needed in this area to provide a sound basis for evaluating potential radionuclide migration associated with a repository site.

REFERENCES

1. R. G. Dosch and A. W. Lynch, Interaction of Radionuclides with Geomedia Associated with the WIPP Site in New Mexico, SAND78-0297, June (1978).
2. S. J. Lambert and J. W. Mercer, Hydrologic Investigations of the Los Medanos Area, Southeastern New Mexico, 1977, SAND77-1401, Jan. (1978).

3. D. W. Powers, S. J. Lambert, S. E. Shaffer, L. R. Hill and D. W. Weart, Geologic Characterization Report, WIPP Site, Southeastern New Mexico, SAND78-1596 (1978).
4. R. G. Dosch and A. W. Lynch (to be published in a SAND report).
5. T. E. Hinkebein and P. F. Hlava, Microstructural Interactions of Geologic Media with Waste Radionuclides, SAND78-0108 March (1978).
6. A. M. Gaines, "Proto-dolomite Synthesis at 100°C and Atmospheric Pressure," Science, 183, 518 (1974).
7. G. Baron, "The Synthesis of Dolomite; Application to the Phenomenon of Dolomitization," Rev. Inst. Fr. Petrole, 15, 3-68 (1960).
8. F. R. Siegel, "Factors Influencing the Precipitation of Dolomitic Carbonates," State Geological Survey of Kansas Bulletin 152, 1961 Reports of Studies, Part 5, 127-158, December (1961).
9. S. Y. Shaio, P. Rafferty and R. E. Meyer, "Ion Exchange from the Los Medanos Area in Southeastern New Mexico," 1979 Annual Meeting of the American Nuclear Society, Atlanta, Georgia, June 3-8 (1979).
10. A. W. Lynch and R. G. Dosch, "Migration of Cesium in Dolomite from the Los Medanos Area in Southeastern New Mexico," 1979 Annual Meeting of the American Nuclear Society, Atlanta, Georgia, June 3-8 (1979).

CHEMICAL REACTIONS IN THE BEDROCK-GROUNDWATER SYSTEM OF IMPORTANCE FOR THE SORPTION OF ACTINIDES

G. W. Beall, B. Allard*, T. Krajewski** and G. D. O'Kelley

Transuranium Research Laboratory
Oak Ridge National Laboratory
Oak Ridge, TN 37830

INTRODUCTION

Most suggested alternatives for deep underground disposal of high level radioactive wastes rely upon several independent barriers like resistant canister materials, waste forms of low solubility in groundwaters and the use of back-fill material of low permeability and with nuclide retaining capacity. These barriers would retard the eventual release of radionuclides from the repository into the groundwater/bedrock system. The final and the only non-engineered barrier would be the host rock itself. It would be desirable if the rock alone would be able to retain the long-lived radionuclides coming from the waste for long enough times to allow decay to harmless activity levels before they might reach the biosphere.

The biological hazards from high level reprocessing waste as well as from unreprocessed spent uranium fuel are dominated by the actinides and their daughter products (americium, plutonium, neptunium, thorium, radium) from about three hundred years after discharge from the reactor up to millions of years. In order to allow predictions of the migration of the actinides in the ground, the chemical behavior of these elements in groundwater under environmental conditions and their interactions with geologic media must be studied in detail. Under the anoxic conditions to be expected in a deep underground repository in igneous rock (1), the actinides would be expected to exist predominantly in the tri- and tetravalent states (2).

*On leave from Dept. of Nuclear Chemistry, Chalmers University of Technology, S-41296 Göteborg, Sweden.
**Dept. of Chemistry, University of Tennessee, Knoxville, TN 37916.

The aqueous chemistry of the actinides in these valence states and at environmental pH-levels is largely dominated by hydrolysis reactions (2). In oxic systems like surface waters, the penta- and hexavalent states would predominate (for uranium, neptunium, and possibly also plutonium). For these higher valence states, considering AnO_2^+ and AnO_2^{2+} species, the aqueous chemistry would be entirely different at environmental pH levels, with a much lower degree of hydrolysis and higher overall solubility (2, 3). In this paper, studies of the sorption of americium (trivalent) and neptunium (pentavalent under oxic conditions) on some major rock-forming minerals of igneous rocks and accessory minerals are reported.

MINERALS AND GROUNDWATER IN IGNEOUS ROCKS

The igneous rocks, which are usually classified according to mineralogic composition, grain size and origin, are composed of a small number of major rock-forming minerals. The main constituents of silica-rich rock types like granite are quartz, feldspars, micas and amphiboles, while silica-poor rocks like basalt are dominated by feldspars, pyroxenes and olivine. Moreover, there are some common accessory minerals which would be of importance in a natural system, either for controlling the redox potential or as sources of complexing anions, such as pyrites, apatite and marble. The minerals and composition of the two granites employed in these studies have been reported previously (4, 5).

Based on prevailing mineral/groundwater equilibria and available groundwater analysis data (1, 5), an artificial groundwater was prepared for use in these studies whose composition has been reported previously (4). The composition of this water was not significantly altered by contact with the minerals for several months, except for quartz (giving an increased concentration of total SiO_2 by a factor of 4-5) and hornblende (giving an increased concentration of Ca by a factor of 3).

EXPERIMENTAL

The sorption of americium and neptunium was measured as a function of pH for the minerals and rocks using a batch technique described in detail previously (4). All minerals were obtained from Ward's Natural Science Establishment and the rocks from Pacific Northwest Laboratory (6). The purity of the minerals was checked by X-ray diffractometry only. Existing impurities were reasonably low (within 5 percent).

The autoradiographic technique employed in these studies used a Polaroid 545 film holder and Polaroid type 57 film. The granite slabs were mounted in a light-tight box and then covered with a ZnS

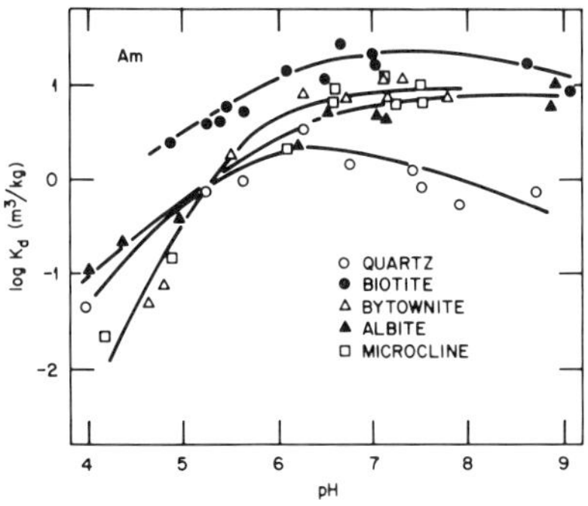

Fig. 1. Distribution Coefficients vs. pH for ^{241}Am on Selected Minerals

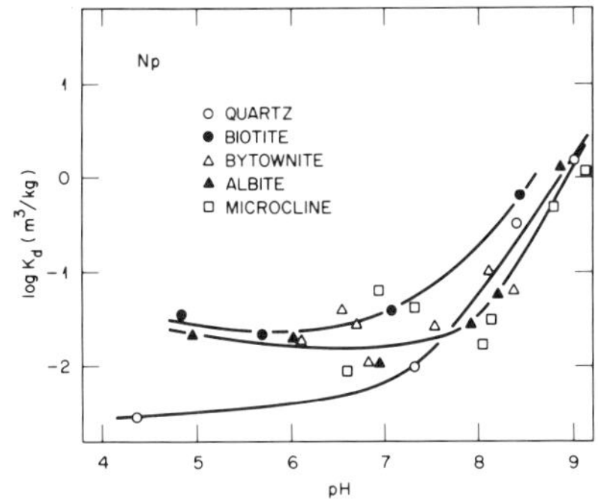

Fig. 2. Distribution Coefficients vs. pH for ^{235}Np on Selected Minerals

coated mylar scintillator. This assemblage was placed in direct contact with the Polaroid film by a lever actuated from outside the light-tight box.

RESULTS AND DISCUSSION

The sorption in terms of distribution coefficients (K_d; moles/kg solid per moles/m^3 water) for americium and neptunium is given in Fig. 1. and 2. for the various minerals.

For americium, similar sorption isotherms were obtained for all the minerals studied. The distribution coefficients increase from low values at pH 4 to their maximum values at pH 6.5-8 and then tend to level out or slightly decrease at higher pH. Among the common minerals of igneous rocks, the order of increasing sorption is quartz, the feldspars with augite and olivine, kaolinite and biotite. Hornblende deviates from the general behavior of the other minerals, probably due to the presence of calcite as an impurity which strongly buffers pH to 7-8.5. The difference between the low-sorbing quartz and the high-sorbing biotite is, however, only about one order of magnitude in terms of distribution coefficients. The cation exchange capacities of the minerals seem to be correlated to their specific surfaces, although there is no quantitative correlation with the distribution coefficients. In most of the systems, however, higher distribution coefficients were obtained for minerals with high surface area or cation exchange capacity. Sorption studies in 4 M NaCl show no significant salt dependence on the sorption in the pH range 4-9 with the exception of biotite and kaolinite (4). This result excludes ion exchange as the predominant sorption mechanism. In the case of biotite and kaolinite, the salt dependence only appears at pH below 5. Hydrolysis of americium in aqueous solution would be noticeable above pH 5 (ca. 1% hydrolyzed at pH 5, 50% at pH 6.5 and more than 99% at pH ≥ 8) considering the hydrolysis behavior of a typical trivalent actinide (2). Thus, the hydrolysis in the aqueous phase seems to be one parameter that can account for the shape of the K_d vs. pH curves. It should be pointed out that the concentration level (2×10^{-9} M) is below the expected total solubility in the pH range that was studied (2, 3); therefore, these sorption reactions cannot be considered as precipitation phenomena.

Also for neptunium, similar sorption isotherms were obtained for the various minerals, with low distribution coefficients at pH below 7-7.5 and then a sharp increase with increasing pH. The order of increasing sorption seems to be the same as for americium, and with a difference between quartz and biotite about one order of magnitude for the distribution coefficients. Sorption studies in 4 M NaCl indicate a slight salt dependence that for most of the minerals is opposite to what would be expected for an ion exchange process, since slightly higher distribution coefficients were obtained in 4 M NaCl than in the groundwater (7). This may, however, be due to the presence of carbonate in the groundwater, leading to the formation of anionic carbonate complexes. (For biotite and kaolinite, lower distribution coefficients were obtained in 4 M NaCl than in groundwater). The hydrolysis of pentavalent neptunium in aqueous solutions would be noticeable at pH abouve 7 (ca. 1% hydrolyzed at pH 7, 50% at pH 8.5 and more than 99% at pH ≥ 10) (2). Thus, just for americium, there seems to be a relationship between sorption and hydrolysis in the aqueous phase. It can be concluded that the pH of the aqueous phase seems to be one of the most important parameters that determine the sorption on a solid surface.

The sorption on the two granite types is given in Fig. 3. The sorption isotherms fall within the expected range covered by the curves for the individual minerals indicated by dashed lines. The distribution coefficients of the rocks are, however, not equal to the weighted average values of the distribution coefficients of the individual discrete minerals of the rock. The only apparent difference in composition between the Climax Stock and Westerly granites that could explain the differences in observed distribution coefficients is the higher biotite concentration in the Climax Stock granite (5). Probably different degrees of weathering and decomposition would largely influence specific surface area and sorption characteristics. An error due to fractioning of the rock minerals in the crushing/sieving process cannot be excluded. Regardless of these uncertainties, a fair prediction of the distribution coefficient for the rock at a certain pH using data for the individual minerals seems to be feasible. Since the sorption values on the various minerals are similar, an estimated value for rock would always be within a factor of three from the correct value. Such an estimate would be fully adequate for the use in a migration or safety analysis and may not be worse than many values based on single measurements reported in the literature.

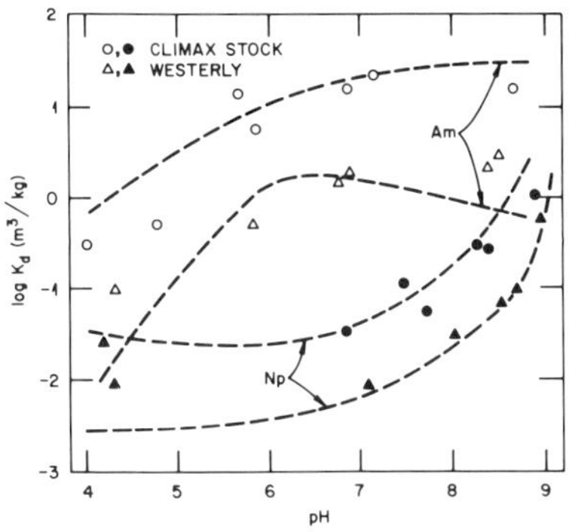

Fig. 3. Distribution Coefficients for ^{241}Am and ^{235}Np vs. pH on Granites

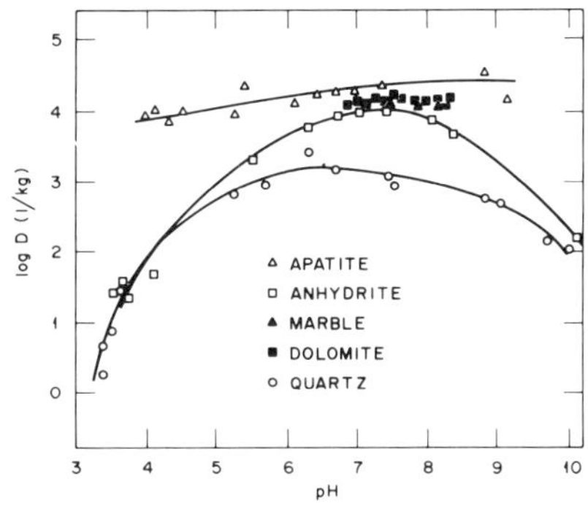

Fig. 4. Distribution Coefficients for ^{241}Am on Accessory Minerals vs. pH

A second mechanism of sorption that has been observed for both Am and pentavalent Np is one involving minerals that contain strong complexing anions. Fig. 4. gives sorption of Am on four such minerals. It can be seen in the case of apatite, marble and dolomite that the sorption is quite strong and pH independent in contrast to the quartz sorption. This strong sorption is most likely due to surface complex formation reaction with PO_4^{3-} and CO_3^{2-} groups. A similar behavior toward Np has been observed for apatite but not for the carbonate containing minerals. This type of sorption reaction could have strong implications for waste isolation since apatite is fairly commonly dispersed in igneous systems and the carbonate minerals are quite common in various sedimentary rock formations.

A third and final mechanism of sorption involving pentavalent Np reduction by Fe(II)-containing minerals is illustrated in Fig. 5. The sorption in these experiments was conducted on Climax Stock granite slabs. The slab picture in Fig. 5. shows pyrite grains as white, quartz as light gray, the feldspars as dark gray, and biotite as black. It can be seen that on the granite with oxidized surface, the sorption is not specific for any mineral. In contrast, the sorption on a fresh surface in air exhibits strong correlation to the Fe(II)-containing minerals (pyrite and biotite). This sorption reaction is enhanced further by purging the solution with nitrogen to get rid of excess oxygen. This reduction of Np by Fe(II)-containing minerals will be of great significance in evaluating migration of radioactive waste through geologic media, not only for Np but for Tc, Pu and U.

CONCLUSIONS

There appears, based on data presented in this paper, that there are three major mechanisms operating to immobilize actinides that might leach out of a nuclear waste repository. The first mechanism

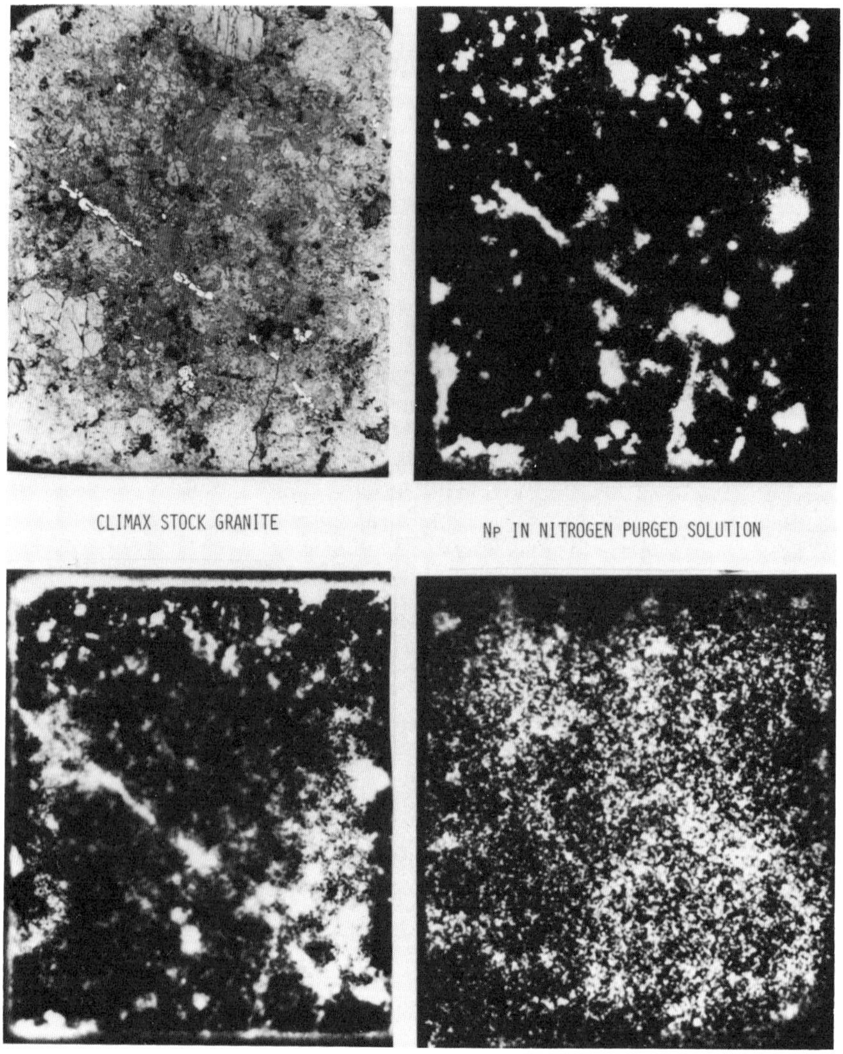

Fig. 5. Autoradiograph of ^{237}Np on Climax Stock Granite

appears to involve sorption that correlates with hydrolysis or carbonate complex formation reactions. These reactions have their strongest sorption for Am in the pH range 7-9 and for Np 8-10. These are both reasonable pH's one could expect in a deep geologic repository. The second mechanism seems to involve complex formation on the surface of certain minerals that contain such anions as phosphate and carbonate. Energetically, this reaction appears to be stronger than the first surface sorption reaction and is very pH independent. The third and last mechanism involves reduction of Np by Fe(II)-containing minerals. This is possibly the most important of the sorption mechanisms for nuclear waste isolation since the higher valent actinides tend to be the most mobile. This could mean that NpO_2^+, UO_2^{2+}, TcO_4^- and PuO_2^{2+} could be reduced to their less mobile NpO_2, UO_2, TcO_2 and PuO_2 states, respectively.

ACKNOWLEDGEMENTS

Fruitful discussions and advice by Drs. R. L. Hahn and R. E. Meyer are gratefully acknowledged. This work was sponsored by the Division of Nuclear Sciences, Office of Basic Energy Sciences, U. S. Department of Energy under contract (W-7405-eng-26) with the Union Carbide Corporation and EY-76-S-05-4447 with the University of Tennessee. B. Allard was partially supported by the Office of Nuclear Waste Isolation under contract with Battelle Memorial Institute's Waste Isolation Safety Assessment Program and partially by the Swedish National Council for Radioactive Waste Management.

REFERENCES

1. G. Jacks, Groundwater Chemistry at Depth in Granites and Gneisses, KBS Technical Report 88, Karnbranslesakerhet, Stockholm (1978).
2. B. Allard, H. Kipatsi and J. O. Liljenzin, "Expected Species of Uranium, Neptunium and Plutonium in Neutral Aqueous Solution," submitted to J. Inorg. Nucl. Chem. (1979).
3. C. F. Baes and R. E. Mesmer, The Hydrolysis of Cations (John Wiley & Sons, Toronto, 1976).
4. B. Allard and G. W. Beall, "Sorption of Americium on Geologic Media," J. Envir. Sci. Health 14A (6), (1979).
5. L. L. Ames, Waste Isolation Assessment Program Controlled Sample Program, Publ. No. 1. Characterization of Rock Samples, PNL-2797, UC-70, Pacific Northwest Laboratory, Richland, WA (1978).
6. L. G. M. Baas Becking, I. R. Kaplan and D. Moore, J. Geol. 68, 243 (1960).
7. G. W. Beall, B. Allard and T. Krajewski, to be published.

INTERACTION OF RADIONUCLIDES WITH GEOMEDIA FROM THE NEVADA TEST SITE*

A. W. Lynch and R. G. Dosch

Sandia Laboratories**
Albuquerque, NM 87185

ABSTRACT

Aqueous/solid distribution coefficients have been measured for ^{137}Cs, ^{85}Sr, ^{99}Tc, ^{152}Eu, ^{144}Ce, ^{238}Pu, ^{243}Am and ^{244}Cm on powdered argillite and hornfels obtained from the Nevada Test Site. The fluid phases included a groundwater simulant and deionized water preequilibrated with powdered argillite. Distribution coefficients for lanthanides and actinides are generally high, and Eleana shale was found to effectively sorb technetium.

INTRODUCTION

Among the sites being investigated as a possible repository for radioactive waste are geologic formations at the Nevada Test Site (NTS). One of the studies included in the evaluation is the measurement of distribution coefficients (Kd's) of radionuclides in various aqueous media on the geomedia in the area. Argillite from two areas, Syncline Ridge and Calico Hills, and hornfels from Calico Hills were among the materials initially chosen for study. Both argillites are from the Eleana formation and were chosen to test the variability of the distribution coefficients.

One concern of the safety assessment of any proposed nuclear waste repository is the potential for groundwater flowing to the

*This work supported by the U. S. Department of Energy (DOE) under Contract DE-AC04-76-DP00789.
**A DOE facility.

waste to leach the waste form and transport radionuclides to the biosphere. Sorption of the migrating radioisotopes by the surrounding rock could have a significant impact on the rate with which the radionuclides migrate. Distribution coefficients for ^{137}Cs, ^{85}Sr, ^{99}Tc, ^{152}Eu, ^{243}Am and ^{238}Pu were measured for a simulated groundwater and both powdered hornfels and argillite. Additional measurements were made with argillite from Syncline Ridge in contact with several other groundwater simulants containing the above isotopes plus ^{144}Ce and ^{244}Cm.

The Kd values obtained in the measurements varied over a wide range depending on the rock, the radionuclide and its concentration in solution, and the concentration of other ions in solution. Technetium, for example, did not sorb on hornfels but was strongly sorbed by argillite. The general trends observed were an increase in Kd with decreasing nuclide concentration and lower Kd's in the preequilibrated deionized water as compared to both preequilibrated and as-prepared NTS simulants. There were, however, a number of exceptions to these trends.

GEOMEDIA

The rock samples used were obtained from the following core samples drilled at the Nevada Test Site:

1) Syncline Ridge argillite - borehole UE17E - 1805' horizon
2) Calico Hills argillite - borehole UE25H#3 - 1081' horizon
3) Calico Hills hornfels - borehole UE25H#3 - 1678' horizon

X-ray analysis (1) of the argillite from borehole UE17E-1805' horizon showed the following constituents: quartz-SiO_2 and illite, a variable composition K-Al silicate, along with lesser or trace quantities of chlorite - $(Mg, Fe, Al)_6(Al, Si)_4O_{10}(OH)_8$; ferroan dolomite - $Ca(Mg_xFe_{1-x})(CO_3)_2$; siderite - $FeCO_3$; microcline - $KAlSi_3O_8$; pyrophyllite - $Al_2Si_4O_{10}(OH)_2$; and sodic plagioclase - (Na,Ca)(Al,Si) $AlSi_2O_8$. A quantitative analysis is given elsewhere (2).

Analyses by emission spectroscopy (3) which compared the argillite with hornfels from Calico Hills showed no significant difference in qualitative elemental composition. A quantitive analysis (4) for carbon, sulfur and nitrogen gave the following results:

Rock	C	S	N
Hornfels	<0.05%	0.028%	<0.05%
Argillite	1.35%	0.032%	<0.05%

All rock samples used in this work were -200 mesh (< 74 micron) powder. A piece of core (50 to 100 grams) was crushed in a tungsten carbide ball mill and/or an alumina mortar and pestle until it completely passed through a 200 mesh sieve.

LIQUID PHASE

The aqueous phase used in the majority of the measurements was a simulated groundwater with a composition based on an analysis from the Tippipah limestone which overlies the Eleana and uppermost quartzite within the Eleana formation (5). This simulant composition represented a solution which may enter the repository through a large fracture. Other solutions include the same simulant preequilibrated with the sample rock to simulate a natural groundwater solution that is slowly migrating through the area surrounding the repository and a preequilibrated deionized water solution. Preequilibration of the aqueous solutions with rock were accomplished by continuously agitating a mixture of 20 grams of crushed argillite (-200 mesh) with one liter of solution for 3 weeks, then filtering (medium glass frit). All measurements were made in solutions containing a single radionuclide. The distribution coefficients were determined at two or more concentrations of each radionuclide. All the solutions were analyzed prior to use and the results are given elsewhere (2).

EXPERIMENTAL PROCEDURE

Samples of the argillite and hornfels (1 gram or less) were weighed directly into 30 ml capacity polyethylene dropping bottles. The solutions were doped with a single radioisotope in one of two ways:

1) Addition of stock isotope solution directly to the simulant groundwaters, or
2) The stock isotope was dried at ambient temperature and then redissolved in the simulant groundwater.

Method 2) was used in an attempt to prevent precipitation in the control solutions. All controls were filtered through 0.8μ (Gelman Acropore) filters before adding them to the rock samples.

The measurements made on the argillite from Syncline Ridge used method 1) exclusively. The liquid to solid ratio was 50 ml/g. The subsequent measurements on the Calico Hills materials used doping method 1) for the Cs, Sr, Eu and Tc; method 2) for the Pu and Am, and a liquid/solid ratio of 30 ml/g. Contact times of one month were used for all samples.

Sampling involved placing an 0.8 micron filter over the bottle opening, replacing the cap and collapsing the bottle to force the liquid through the filter. Known volumes of the clarified solutions were used for analyses. Problems associated with multiple sampling and competing ion effects were circumvented by using a different simulant-argillite mixture containing a single radionuclide for each measurement. Control volumes of each doped simulant solution (containing no rock sample) were carried through an experimental and sampling procedure identical to that used for the mixtures. A more detailed description of the sampling procedure may be found elsewhere (6).

The fission product activities were determined by gamma ray spectroscopy using a NaI well crystal. Actinide activities were determined by liquid scintillation counting.

DISTRIBUTION COEFFICIENTS

The Kd's presented in Table 1 were calculated using Equation 1):

$$Kd = \frac{\text{Control Activity} - \text{Final Activity}}{\text{Final Activity}} \times \frac{\text{Simulant Volume, ml}}{\text{Weight Solid, G}} \qquad 1)$$

where activities refer to solution phase activity and the control and final activities refer to those in the control simulant and the filtered simulant-rock mixture, respectively. The Kd values calculated represent total sorption coefficients and are valid only for the conditions set forth in each individual experiment. Parameters such as hydrolysis and absorption on container walls may also contribute to the calculated Kd; however, control samples were used to minimize the latter. The results in most cases are given as both apparent Kd's and overall Kd's. The apparent Kd is the first value given in Table 1 and was calculated using the activity in the control simulant after the contact period. The values in parentheses were calculated using the initial activity as the control activity in Equation 1). The Kd thus calculated includes all sorption mechanisms and is referred to as overall Kd's.

^{137}Cesium

Little if any of the original activity added to the controls is lost; therefore, the overall values compare closely with the apparent Kd's. All values are quite high (greater than 10^3), with the lowest value being obtained with argillite in equilibrated deionized water (1.3×10^3 ml/g) and the highest value coming from the argillite from Calico Hills in groundwater simulant. All values agree well within a factor of 10. The results indicate that no Cs concentration dependence is evident at the low levels of total Cs (1.0 to 0.01 μCi/ml, 0.01 to .0001 ppm) added to the solutions.

Table 1. Distribution Coefficients (1)

Isotope/Rock	Non Equilibrated NTS Simulant 1.0$^{(2)}$	Non Equilibrated NTS Simulant 0.1$^{(2)}$	Non Equilibrated NTS Simulant 0.01$^{(2)}$	Argillite Equilibrated Solutions NTS Simulant 0.1$^{(2)}$	Argillite Equilibrated Solutions D.I. Water 0.1$^{(2)}$
^{137}Cs					
Argillite (S.R.)$^{(3)}$	3.1×10^{3}	3.3×10^{3}	4.0×10^{3}	2.2×10^{3}	1.3×10^{3}
	(3.6×10^{3})	(3.6×10^{3})	(4.2×10^{3})	(2.2×10^{3})	(1.3×10^{3})
Argillite (C.H.)$^{(3)}$	--	5.1×10^{3}	5.1×10^{3}	--	--
Hornfels (C.H.)$^{(3)}$	--	2.4×10^{3}	2.4×10^{3}	--	--
^{85}Sr					
Argillite (S.R.)	41	50	43	$<0^{(4)}$	4.1×10^{2}
	(58)	(58)	(57)	(287)	(2.1×10^{3})
Argillite (C.H.)	--	65	69	--	--
Hornfels (C.H.)	--	22	23	--	--
^{152}Eu					
Argillite (S.R.)	--	1.4×10^{4}	4.5×10^{3}	37	18
		(4.2×10^{4})	(7.8×10^{4})	(1.7×10^{5})	(3.3×10^{3})
Argillite (C.H.)	--	9.2×10^{4}	5.8×10^{4}	--	--
		(9.7×10^{4})	(6.6×10^{4})		
Hornfels (C.H.)		8.9×10^{4}	9.7×10^{4}	--	--
	--	(9.4×10^{4}	(1.1×10^{5})	--	--
^{144}Ce					
Argillite (S.R.)	--	4.4×10^{4}	∞	2.9×10^{4}	26
		(9.0×10^{4})	(∞)	(1.5×10^{5})	(5.8×10^{3})
^{238}Pu					
Argillite (S.R.)	3.3×10^{4}	2.0×10^{3}	1.1×10^{3}	1.4×10^{2}	9.3×10^{2}
	(1.1×10^{5})	(1.2×10^{5})	(1.1×10^{5})	(5.5×10^{4})	(1.8×10^{4})
Argillite (C.H.)	--	1.6×10^{4}	6.8×10^{3}	--	--
		(2.0×10^{4})	(8.3×10^{3})		
Hornfels (C.H.)	--	3.0×10^{4}	7.4×10^{3}	--	--
		(3.8×10^{4})	(9.2×10^{3})		
^{243}Am					
Argillite (S.R.)	5.9×10^{3}	2.3×10^{4}	77	3.2×10^{4}	78
	(1.2×10^{5})	(2.6×10^{5})	(8.8×10^{5})	8.5×10^{4})	(7.3×10^{3})
Argillite (C.H.)	--	1.9×10^{4}	2.2×10^{4}	--	--
Hornfels (C.H.)	--	2.9×10^{4}	2.6×10^{4}	--	--

Table I. Distribution Coefficients(1) (cont.)

Isotope/Rock	Non Equilibrated NTS Simulant $1.0^{(2)}$	$0.1^{(2)}$	$0.01^{(2)}$	Argillite Equilibrated Solutions NTS Simulant $0.1^{(2)}$	D.I. Water $0.1^{(2)}$
^{244}Cm					
Argillite (S.R.)	$4.8x10^4$	--	$2.6x10^2$	$9.0x10^2$	$1.2x10^2$
	$(2.8x10^5)$	$(8.0x10^3)$	$(1.1x10^5)$	$(2.2x10^5)$	$(1.5x10^4)$
^{99}Tc		$0.01^{(2)}$	$0.001^{(2)}$	$0.001^{(2)}$	$0.001^{(2)}$
Argillite (S.R.)	--	16 (30)	196 (244)	72 (80)	38 (36)
Argillite (C.H.)	--	155	$6.5x10^3$	--	--
Hornfels (C.H.)	--	0	0	--	--

(1) Distribution coefficients in ml/gm.
(2) Activity added in μCi/ml. (See Ref. 5 for element concentration)
(3) S.R. - Syncline Ridge, Hole UE17E, 1805 feet deep.
C.H. - Calico Hills argillite, UE25A #3, 1081 feet deep.
Calico Hills hornfels, UE25A #3, 1678 feet deep.
(4) Final activity in control lower than final activity in sample.

Note: Final pH's ranged from a low of 7.6 in the non equilibrated simulant samples to a high of 8.8 in the equilibrated D.I. water samples.

Strontium

The results of the strontium Kd measurements vary widely and appear to be strongly dependent on the solution used in the experiment. The reason for the negative value given for the Sr in the pre-equilibrated NTS simulant is that more ^{85}Sr activity was found in the sample solution than in the control after the one-month equilibration. This suggests further solution changes may have occurred after contact with the rock. The Kd values ranged from 22 to 69 for all rocks in the simulant while an apparent Kd of $4.1x10^2$ ml/gm was obtained with argillite in equilibrated DI water. Stable strontium is included in the simulants (∿1 ppm) but not in the equilibrated deionized water. The low ionic strength, as well as the lack of strontium in the equilibrated deionized water, is probably the reason for the high Kd measured.

Technetium

One of the most significant results of the Kd survey was that technetium is removed from solution by the argillite under oxidizing conditions. This phenomenon appears to be fairly unique among the geomedia from the various potential repository sites around the country and is probably associated with the carbon found in the Eleana argillite.

While the hornfels is chemically very similar to the argillite lying above it in Calico Hills, it appears to have no affinity for technetium. A study to determine the mechanism for TcO_4^- removal by argillite is in progress.

Europium and Cerium

Mechanisms other than sorption on the rock usually are a significant factor when making Kd measurments with europium. This is shown by the large difference between the apparent and overall Kd values for ^{152}Eu for argillite in the equilibrated solutions (37 vs 1.7×10^5 ml/gm). However, the problem was almost negligible in the experiments with as-prepared simulant.

Similar comments may be made about the cerium results. The results do indicate, however, that argillite is a very efficient getter for both cerium and europium (Kd $> 10^3$ ml/gm). The low Kd values given for the equilibrated solutions are a result of the loss of > 99% of the activity from the filtered control.

Americium

A wide range of Kd values was obtained for americium. The apparent Kd ranged from a low of 77 to a high 3.2×10^4. The low values (77, 78 ml/cm) are due to loss of control activity. Americium is efficiently absorbed on both argillite and hornfels. No ^{243}Am concentration dependence is evident.

Plutonium

Large Kd values were measured for all the samples in contact with the ^{238}Pu. Contrary to the results that might be expected from classical ion exchange processes, the Kd's for argillite and hornfels with ^{238}Pu appear to increase with increasing ^{238}Pu concentration over the range of 0.01 to 1.0 μ Ci/ml.

Curium

^{244}Cm was not contacted with the Calico Hills samples; however, the range of Kd values for ^{244}Cm on the argillite from Syncline Ridge appears to be similar to that of the Am and Pu. Loss of control activity is also a problem with curium.

SUMMARY

The major factors controlling trace radionuclide distribution coefficients associated with the NTS samples are the same as those found with materials and solutions from the proposed WIPP site in New Mexico (6). These include solution composition, radionuclide concentration and mechanisms which cause the loss of activity in control solutions. For nuclides which do not undergo hydrolysis in the pH range of interest (Cs, Sr and Tc), the overall Kd's agree generally well with the apparent Kd's. For the other isotopes studied, solution composition was very important. Distribution coefficients for the lanthanides and actinides were generally found to be high. However, differences between the apparent and overall Kd's of up to five orders of magnitude were not uncommon.

The mechanism for the loss of activity in the control solutions may be hydrolysis, precipitation due to high pH, etc. However, neither the loss of activity from control solutions nor the final activity in the argillite solution mixtures can be directly correlated with pH or pH changes during equilibration. The overall Kd's appear to provide the most useful empirical data because they include the effects of all mechanisms which result in the removal of radionuclides from the liquid phase.

Based on two samples taken a considerable distance from each other (Calico Hills and Syncline Ridge), Eleana argillite appears to react similarly with the radionuclides tested and is not sensitive to the area from which it was taken. Argillite and hornfels appear to have similar affinities for radionuclides tested with the exception of technetium. Technetium was effectively sorbed on Eleana argillite.

REFERENCES

1. G. T. Gay, Sandia Laboratories, Unpublished Data.
2. R. G. Dosch and A. W. Lynch, Interaction of Radionuclides with Argillite from the Eleana Formation on the Nevada Test Site, SAND78-0893, February (1979).
3. J. F. Wolcott, Sandia Laboratories, Unpublished Data.
4. S. L. Erickson, Sandia Laboratories, Unpublished Data.
5. G. Dinwiddie, U.S. Geological Survey, Preliminary Analytical Results.
6. R. G. Dosch and A. W. Lynch, Interaction of Radionuclides with Geomedia Associated with the Waste Isolation Pilot Plant Site in New Mexico, SAND78-0297, June (1978).

RADIONUCLIDE SORPTION STUDIES ON ABYSSAL RED CLAYS*

Kenneth L. Erickson

Sandia Laboratories**
Albuquerque,NM 87185

INTRODUCTION

The equilibria for the sorption of radionuclides by a widely distributed abyssal red clay are being experimentally investigated. An earlier paper (1) summarized a portion of the investigation concerned with sorption of Rb, Cs, Sr, Ba, Ag, Cd, Ce, Pm, Eu and Gd from aqueous 0.68 N NaCl solutions containing a single nuclide of interest. This paper summarizes more recent work concerned with sorption of Tc, U, Pu, Am or Cm from 0.68 N NaCl solutions and with determining the clay's relative sorption selectivities for cesium, barium and cerium.

EXPERIMENTAL

Acquisition of and chemical and mineralogical analyses of the red clay samples were discussed in the earlier paper and by Heath (1, 2). The principal results of the analyses were that the clay was dominated by iron-rich smectite and contained lesser, varying amounts of phillipsite and five to six weight percent hydrous iron and manganese oxides. Sorption equilibrium data for Tc, U, Pu, Am and Cm in 0.68 N NaCl solutions and for determining relative sorption selectivities were obtained using batch equilibration techniques similar to those discussed in the earlier paper. Typical equilibration times were on the order of two to four days.

---

*This work supported by the United States Department of Energy (DOE), under Contract DE-AC04-76DP00789.
**A DOE Facility

RESULTS

Summary of Previous Results

The distribution coefficient K_{D_i} for nuclide i is defined by

$$K_{D_i} = \frac{\text{solid-phase equilibrium concentration}}{\text{solution-phase equilibrium concentration}}$$

For the nuclides studied, the distribution coefficients varied non-linearly from about 10 to 10^5 ml/gm as the solution-phase nuclide concentration varied from about 10^{-2} to 10^{-11} M, respectively. The distribution coefficients were relatively insensitive to pH at values between 6 and 8, but with the exception of cesium were greatly diminished at pH 2.7.

At least two mechanisms appeared to be involved with the sorption of each nuclide. One mechanism is believed to be an ion-exchange phenomena associated with the silicates and dominates sorption at higher solution-phase concentrations (greater than 10^{-3} M). The second mechanism dominates at lower solution-phase concentrations (less than about 10^{-4} M) and for cesium is apparently associated with the silicates. However, for the other nuclides, except rubidium, the second mechanism is apparently associated with the hydrous metallic oxides.

Results for Technetium in 0.68 N NaCl Solutions

During preparation of samples for the batch equilibration experiments, both solutions and sediment came in contact with atmospheric oxygen. Furthermore, the polyethylene bottles used for sample containers were permeable to oxygen. Therefore, the batch equilibration experiments were essentially done under mildly oxidizing conditions. The technetium isotope used was ^{99}Tc, which was added as the pertechnitate anion at initial technetium concentrations of about 2.5 x 10^{-5} mg-atom/ml. The resulting distribution coefficients for technetium were essentially zero at solution-phase pH values of both 8.2 and 2.6.

Results for Actinide Elements in 0.68 N NaCl Solutions

Preliminary sorption experiments were conducted using U, Pu, Am and Cm. For these experiments, the initial oxidation states of the nuclides were not determined. However, as supplied by the manufacturer the oxidation state of the uranium should have been (VI); the oxidation state of the plutonium was probably predominately (IV), and that of americium and curium was probably (III). In particular the solution in the shipping vial for the americium had a distinctly pink color, characteristic of Am(III).

Uranium. Distribution coefficients for uranium in solutions having pH values of 2.8 and 7.1 were determined using an initial solution-phase concentration of about 3.1×10^{-8} mg-atom/ml. The resulting data are shown in Fig. 1. The magnitude and the strong pH dependence of the distribution coefficients are similar to those observed with the lanthanides (Fig. 2). Unlike experiments with the lanthanides or with plutonium, americium and curium, the uranium experiments were apparently not complicated by decreases in solution-phase tracer concentrations in "blank" samples. Furthermore, additional sorption experiments with uranium were conducted using clay from which the hydrous iron and manganese oxides had been leached. The solution-phase pH for these experiments varied from 7.4 to 8.2, and the resulting distribution coefficients were essentially zero. Furthermore, it was found that the uranium could subsequently be quantitatively removed from the solution phase by coprecipitation with $Fe(OH)_3$. Such results indicate that the same mechanisms which dominated sorption of the lanthanides also probably dominate sorption of uranium, particularly the mechanisms thought to be associated with the hydrous oxide phases.

Americium and Curium. Distribution coefficients for americium and curium in solutions having pH values of 2.7 and 6.9 were determined using initial solution-phase concentrations of 3.6×10^{-6} and 6.7×10^{-9} mg-atom/ml, respectively. The resulting data are shown in Fig. 1 and are compared with similar data for some lanthanide elements in Fig. 2. The magnitude and strong pH dependence of the distribution coefficients for americium and curium are similar to those observed for the lanthanides and again appear to indicate that the same mechanisms dominate sorption. (Note: The solution phase concentrations for promethium which were given in the earlier paper (1) were in error; the correct values are given in Fig. 2). Also as observed with the lanthanides, significant decreases in solution-phase tracer concentrations occurred in "blank" samples. Some typical data have been included in Fig. 1.

The method discussed in the earlier paper was used to calculate distribution coefficients using data from the sorption and "blank" experiments. Since the competing processes are not yet understood, the calculation of sorption distribution coefficients from experimental data involves substantial uncertainties, which in Fig. 1. and 2. may account for the scatter in the lanthanide and actinide data for low solution-phase nuclide concentrations and high pH values.

Plutonium. Distribution coefficients for plutonium in solutions having pH values of 2.7 and 6.9 were determined using an initial solution-phase concentration of 6.4×10^{-8} mg-atom/ml. The resulting data are shown in Fig. 1. At comparable solution-phase concentra-

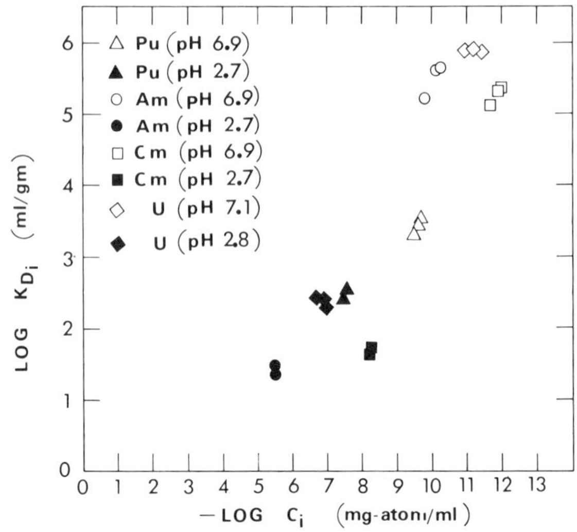

Fig. 1. Distribution Coefficients for U, Pu, Am and Cm in 0.68 N NaCl Solutions and Data from Actinide "Blank" Samples.

tions, the magnitude of the distribution coefficients for plutonium appear somewhat less than for americium, curium and the lanthanides. Furthermore, the pH dependence of the distribution coefficients for plutonium does not appear to be as pronounced. Significant decreases in solution-phase plutonium concentrations occurred in "blank" samples, and some typical data have been included in Fig. 1. Distribution coefficients for plutonium were calculated in a manner similar to those for americium, curium and the lanthanides. At higher pH values, decreases in the plutonium concentrations were much less than the decreases observed for the americium and curium "blanks". Assuming the oxidation state of the plutonium was predominantly (IV), then the "blank" samples imply that at pH 8.3 considerably more plutonium apparently remained in solution than would have been expected based on the solubility product for $Pu(OH)_4$(3). Therefore, the data for plutonium appears to indicate that the dominant sorption mechanisms are probably the same as for americium, curium and the lanthanides; however, the competing solution-phase reactions involving plutonium are more complicated.

Relative Sorption Selectivities

As an initial step toward obtaining multicomponent sorption equilibrium data, two-component experiments in which a nuclide of interest is sorbed from a 0.68 N chloride solution of a competing nuclide other than sodium have been conducted. For such experiments,

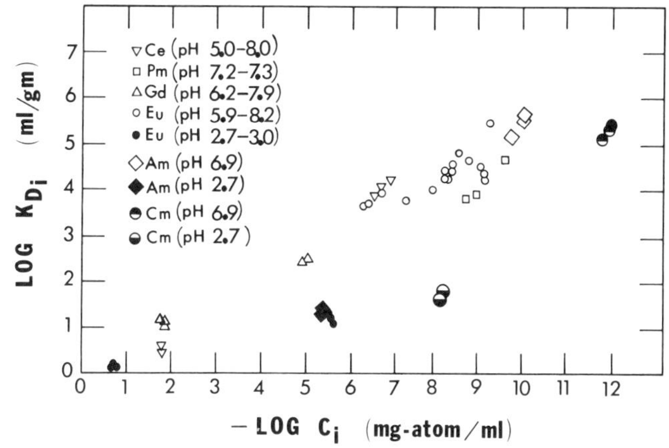

Fig. 2. Distribution Coefficients for Ce, Pm, Gd, Eu, Am and Cm in 0.68 N NaCl Solutions.

the preference of the clay for species i rather than species j can be represented by the selectivity coefficient K^i_j which is defined by

$$K^i_j = (\bar{C}_i/C_i)/(\bar{C}_j/C_j) = K_{D_i}/K_{D_j}$$

where $\bar{C}_i$ and $\bar{C}_j$ denote, respectively, the concentrations of species i and j associated with the clay phase. The coefficient K^i_j includes selectivity effects resulting from interactions associated with both solution and clay phases (4, 5). For the experimental conditions used, $K_{D_j} \approx 1$, and $K^i_j \approx K_{D_i}$. If $K_{D_i} > 1$, the clay is selective for species i. Since at higher solution-phase concentrations, the silicate phases appear to dominate sorption, it was felt that values of K^i_j corresponding to nuclide concentrations greater than 10^{-3} M would reflect relative selectivities associated with the silicates. Similarly, it was felt that values of K^i_j corresponding to solution-phase concentrations on the order of 10^{-6} M and less would reflect selectivities associated with the hydrous oxides.

The two-component batch equilibration experiments were conducted at 4°C and atmospheric pressure and involved the following pairs of nuclides; cesium and barium, cesium and cerium, and barium and cerium. Values of K^i_j were determined for each nuclide at two different concentrations. The higher was about 10^{-3} M and the lower was on the order of 10^{-6} M or less. Solution pH values for the various pairs of nuclides ranged between 5.2 and 7.2. From the experimental data (6), relative sorption selectivities associated with both silicate and hydrous metallic oxide phases were determined. The silicate phases appear to be selective for cesium over barium and cerium; however, the precision in the experimental data is insufficient to resolve the selectivity of the silicates for barium relative to cerium. Analysis of data from the previously reported experiments indicated that the silicates should probably be selective for barium over cerium (1). The hydrous metallic oxide phases appeared to be selective for cerium over barium and for barium over cesium.

CONCLUSIONS

The same mechanisms which dominate sorption of the lanthanides also apparently dominate sorption of the actinides. However, in the case of plutonium the effects of competing solution-phase reactions are more pronounced.

The relative sorption selectivities determined from the two-component experiments with cesium, barium and cerium generally appear consistent with results anticipated from earlier experiments involving cesium, barium or cerium in 0.68 N NaCl solutions. However, more refined and precise experiments are required in order to definitely resolve selectivities between cerium and barium in the silicate phases.

REFERENCES

1. K. L. Erickson, "Radionuclide Sorption Studies on Abyssal Red Clays", Radioactive Waste in Geologic Storage, (S. Fried, ed., American Chemical Society, Washington, D.C. 1979).
2. G. R. Heath, "Seabed Disposal Program: Geochemical and Sedimentological Studies of North Pacific Sediments", Seabed Disposal Program Annual Report January-December 1975, (D. M. Talbert, ed., Sandia Laboratories, Albuquerque, NM, Report No. SAND76-0257, May 1976).
3. John A. Dean, Large's Handbook of Chemistry, (McGraw-Hill, New York, Eleventh edit., 1973).
4. F. Helfferich, Ion Exchange (McGraw-Hill, New York, 1962).
5. D. Reichenberg, "Ion-Exchange Selectivity", Ion Exchange, Vol. 1, (J. A. Marisnsky, ed., Marcel Dekker, Inc., New York 1966).
6. K. L. Erickson, to be published.

MEASUREMENT OF PENETRATION DEPTHS OF PLUTONIUM AND AMERICIUM IN SEDIMENT FROM THE OCEAN FLOOR*

S. Fried, A. Friedman, J. Hines, R. Sjoblom, G. Schmitz and F. Schreiner

Chemistry Division, Argonne National Laboratory
Argonne, IL 60439

INTRODUCTION

The clay-like sediment covering the ocean floor constitutes the last barrier that shields the biosphere from contamination by radionuclides stemming from the nuclear wastes of a subseabed repository. In the event of a failure of the engineered barriers, the mobility of the released radionuclides in the sediment determines the rate and the extent of entry into the water of the ocean. In this article, we present the initial results of measurements designed to determine the mobility of transuranium elements in sediment from the ocean floor. Our data indicate very low migration rates and imply strong chemisorptive interaction with the sediment.

EXPERIMENTAL PROCEDURE

The method chosen to explore the diffusion behavior of selected radionuclides in the ocean floor sediment consists of the direct measurement of the penetration depth into samples of the sediment from a supernatant solution. Fig. 1. shows the makeup of a sediment sample for the diffusion experiment. The sediment is contained in the center hole of a paraffin collar of 22 mm outer diameter and forms a cylinder of 13 mm diameter and approximately 15 mm height. Care is taken to pack the sediment into the collar in such a way that no voids or bubbles exist in the sample. The paraffin collar is supported by a brass holder to which it has been attached by melting of

*Funded under the auspices of the SUBSEABED Nuclear Waste Disposal Program, Contract No. 13-0607, Sandia Laboratories, Albuquerque, NM.

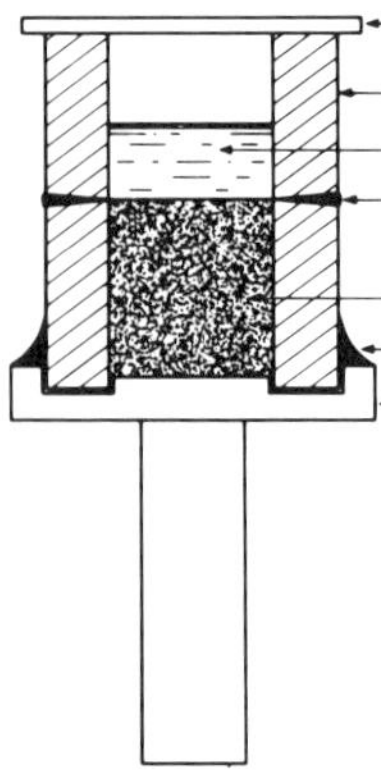

Fig. 1. Sediment Sample for Migration Measurements

the paraffin. A short rod extends from the bottom of the brass holder by which the sample can be mounted on the stage of a microtome.

After the sediment has been packed into the collar, the surface is smoothed and made flush with the paraffin rim. A short, approximately 8 mm high, paraffin extension ring is then fused onto the wax collar so that the sediment can be covered with solution. The samples were left in contact with 0.2 to 0.3 cm^3 of artificial sea water (1) for an extended period to saturate the sediment with liquid and equilibrate it with the sea water. For a diffusion experiment, the sea water covering the selected sample was replaced by a solution containing one or more of the radionuclides ^{237}Pu, ^{241}Am, ^{22}Na and ^{125}I in sea water. The total amount of an individual tracer element applied to a sediment sample was of the order of 0.3 microcurie. The tracer solutions were left in contact with the samples for time periods varying from 1 hour to 100+ days. Evaporation of the solution was inhibited by closing the top of the samples with cover glasses fused onto the paraffin extension ring. During the time allowed for contact of the sediment with tracer solutions, the samples were stored at ambient temperature varying no more than $\pm$ 1 K.

After an appropriate contact time, the diffusion experiments were terminated by removal of the solution without disturbance of the surface of the sediment. The extent to which the tracer elements had penetrated into the sediment cylinders was determined by sectioning of the cylinders with a refrigerated microtome. With few exceptions, the samples were sectioned after freezing, a procedure which generally yielded dry, crumbling material. Each section of approximately 0.050 cm thickness was collected on a small piece of tissue paper and transferred to a plastic counting vial. In the case of sections obtained from non-frozen sediment, the material had to be collected by wiping it off the blade of the microtome. The concentration of the radioactive tracers in each section was measured by placing the counting vials in the well of a sodium iodide crystal scintillation counter and by determining the pulse rate of γ- or X-radiation characteristic for the individual tracers by pulse height analysis.

RESULTS AND DISCUSSION

Of 15 sediment samples that have been sectioned to date, 12 have yielded usable results for the penetration behavior of plutonium, americium, sodium and iodide ions. A listing of the types of sediment used, the type of radioactive tracers and the time span for which each sample was left covered with solution is presented in Table 1. The sediment used to make up the samples stemmed from two sources: the lower end of a 10-foot-long core of 4.5-inch diameter taken in the mid-Pacific ocean (30°47.0'N, 157°42.9'W) from a depth of 5821 meters and a sample labeled "TT-141-II #7G," presumably from a deeper horizon of the sediment*. The former core contains illite clay minerals and is therefore designated in Table 1 as "illitic, core 11"; the latter material contains smectites and is referred to as "smectitic, TT-141-II." The tracer nuclides were dissolved in artificial sea water that had been prepared according to a formula reported by Kester, et al. (1). Concentration levels of the radionuclides in the original solutions ranged from approximately 10^{-6} mol/ℓ for ^{241}Am to 10^{-8} mol/ℓ for ^{237}Pu. The choice of the particular isotope of each radionuclide that was applied as tracer was made by type of radiation. The nuclides listed in Table 1 are all distinguished by emission of characteristic X- or γ-radiation. The electromagnetic radiation permits identification and assay of samples by scintillation counting without prior chemical purification. The contact times during which the samples were covered with the solution of tracers ranged from approximately 1 hour for the iodide sample No. 200605 to nearly 120 days for sample No. 210503. By sectioning samples after varying time periods, it was hoped that coefficients for the effective diffusivity could be determined. The activity of a given radionuclide observed in each section of a sample was converted to specific activity by referring it to a constant section thickness of usually 1 mm and was corrected to 100% cross-sectional area in cases in which only part of the cross section was removed by the microtome. The specific activities were used to generate histograms representing the relative variation of the tracer concentration with depth and were scaled in such a manner as to obtain unity relative concentration near the top of the sample.

*The smectitic sediment was supplied by R. Heath of Oregon State University, Eugene, OR.

Table 1. Summary of Sectioned Samples of Ocean Floor Sediment

Sample Identification Number	Type of Sediment	Type of Tracer	Contact Time in Seconds
090301	illitic, core II	$^{22}Na+^{237}Pu$	$3.89x10^6$
100701	smectic, TT-141-II	$^{125}I+^{241}Am+^{237}Pu$	$1.533x10^6$
100702	smectic, TT-141-II	$^{125}I+^{241}Am+^{237}Pu$	$4.640x10^6$
200605	illitic, core II	^{125}I	$0.00426x10^6$
210501	illitic, core II	^{241}Am	$0.0756x10^6$
210502	illitic, core II	^{241}Am	$2.412x10^6$
210503	illitic, core II	$^{125}I+^{241}Am+^{237}Pu$	$10.356x10^6$
210504	illitic, core II	^{241}Am	$1.193x10^6$
210606	illitic, core II	$^{125}I+^{241}Am+^{237}Pu$	$2.401x10^6$
300301	illitic, core II	$^{22}Na+^{241}Am+^{237}Pu$	$4.15x10^6$
300303	illitic, core II	$^{125}I+^{241}Am+^{237}Pu$	$0.52x10^6$
300304	illitic, core II	$^{125}I+^{241}Am+^{237}Pu$	$0.69x10^6$

A histogram showing the relative concentration of iodide ion as a function of distance from the top surface of the sediment is shown in Fig. 2. The data are those obtained from the sectioning of sample 200605, and the width of the individual bars indicates the width of each section. Also plotted in Fig. 2. are the relative concentrations of iodide observed for sample No. 210606. Even though there is considerable scatter of the data points, it is obvious that after $2.4x10^6$ sec, the iodide has permeated the sample 210606 and is at a nearly uniform concentration. This is not the case for the sample 200605 which was sectioned after only 4260 s of contact with the tracer. The variation of the iodide concentration can best be approximated by the curve calculated for the diffusion coefficient of $D_{I^-} = 2.5x10^{-4}\ cm^2 \cdot s^{-1}$. By comparison, the coefficient of self-diffusion of iodide ion has been measured by Friedman and Kennedy (1) who found a value of $2x10^{-5}\ cm^2 \cdot s^{-1}$ for D_{I^-} in 0.5 molar solutions. The aggregate information on the mobility behavior of iodide ion gathered through the sectioning of seven iodide-containing samples shows clearly that the iodide ion diffuses in the sediment without retardation and at approximately the same rate as in sea water alone.

The sectioning data of americium and plutonium do not yet permit quantitative evaluation of the mobility in terms of an effective diffusion coefficient. An example for the variation of the tracer element activity with depth is shown in Fig. 3. Plotted is the observed specific counting rate, i.e., the counting rate per unit thickness of each section, for sample No. 210503 which had been in contact with a sea water solution containing ^{125}I, ^{241}Am and ^{237}Pu in small concentrations.

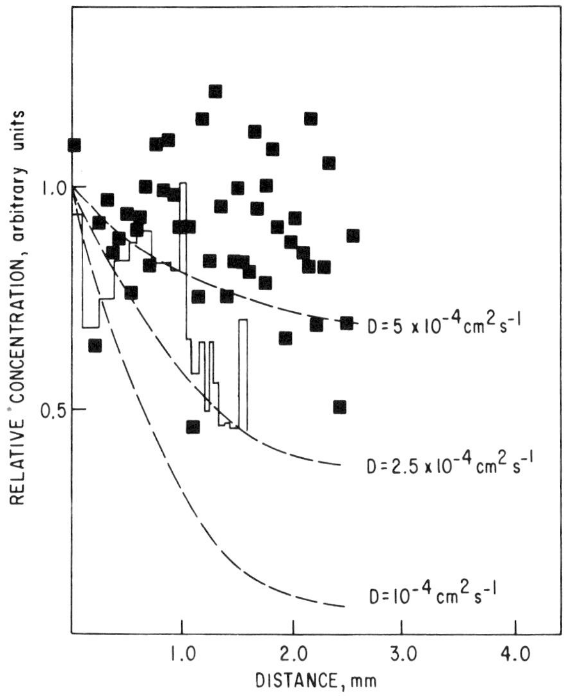

Fig. 2. Diffusion of Iodide-Ion in Illitic Sediment. Solid line = Iodide concentration after 4260 seconds. Black squares = iodide concentration after 2.4×10^6 seconds. Dashed lines = concentration profiles calculated for selected values of the diffusion coefficient.

In order to facilitate comparison of the sectioning data obtained on different samples, the specific activity was converted into a relative concentration by multiplication of each section's specific activity with a constant factor that differed from sample to sample. This factor was chosen in such a manner as to obtain values near unity for the relative concentration near the top surface of the sediment cores. Since the first few sections are likely to be more affected by uncertainties than the deeper ones, the reduction factor was estimated on the basis of sections taken at 0.2 and 0.4 mm depth. It is evident from Fig. 3. and from the scatter of the relative concentration data that the precision of the observed activities suffers from experimental errors. Sources for these errors, among others, are the difficulty in determining the exact thickness of a section, the distortion of the original shape of the sediment cylinder during freezing, and the statistical uncertainty inherent in low counting rates.

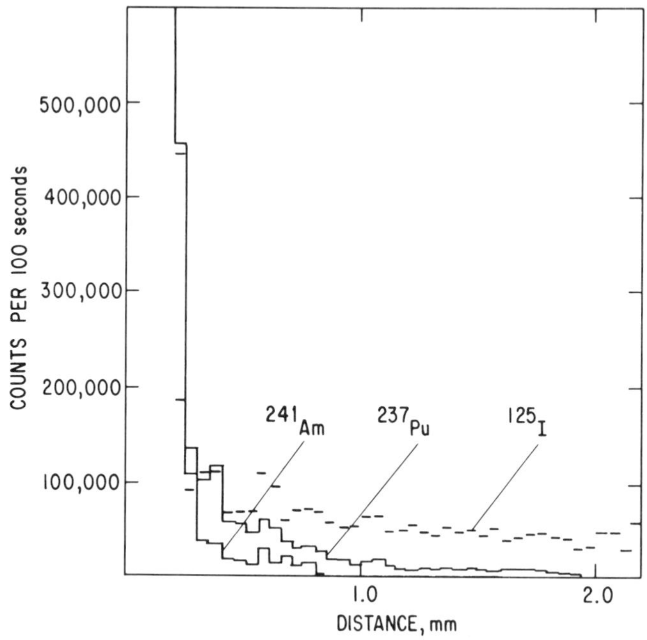

Fig. 3. Tracer Concentrations Measured in a Sample of Illitic Sediment After 119.85 Days ($10.36x10^6$ Seconds). For ^{237}Pu, the ordinate scale has to be divided by 5.

The data obtained by counting the sectioned samples can be approximated by curves that are calculated as solutions to the linear diffusion equation

$$\frac{\partial c}{\partial t} = D \cdot \frac{\partial^2 c}{\partial x^2} \qquad 1)$$

with the following boundary conditions: The diffusion takes place in a one-dimensional bounded region of 30 mm total length. The initial distribution of the diffusing material is such that the concentration in arbitrary relative units has the value 2 in the left half of the region and zero in the right half. The concentration curves as functions of the time for the right half compartment will be representative of the variation of the relative concentration with distance and time in a homogeneous sample of sediment of uniform cross section and 15 mm length, in close approximation of the length of the actual samples. The characteristic shape of the concentration curves over the 15 mm distance is a function of the product of diffusion coefficient and time as a parameter.

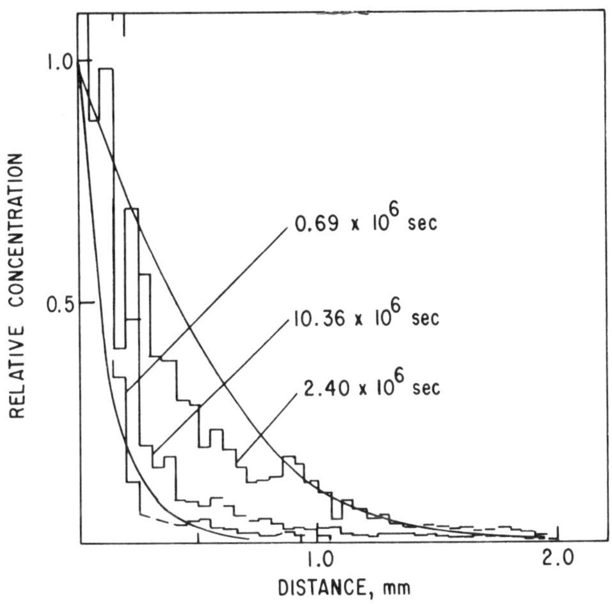

Fig. 4. Penetration of Illitic Ocean Floor Sediment by Plutonium

In Fig. 4., the data for the samples 300304, 210606 and 210503 are plotted as histograms for the first two millimeters, and two curves calculated for values of the product D·t of $5x10^{-8}$ and $2x10^{-7}$ m^2, respectively, are also shown. These two curves seem to bracket the experimental data. The first feature to be noted is the absence of any significant difference regarding the concentration variation of the experimental data. The three histograms refer to the observed activity of ^{237}Pu in illitic ocean floor sediment after having been in contact with the tracer for $0.69x10^6$, $2.40x10^6$ and $10.36x10^6$ seconds. There is no clear evidence for diffusion of the plutonium, since the shape of the concentration curves is similar in all three cases. By the time the sectioning has progressed beyond a 2 mm distance, the activities observed are too close to background to yield reliable information. It is important to note that the data for $10.36x10^6$ sec show a curve shape that is no shallower than those for $2.40x10^6$ seconds.

It follows that these data permit an estimate for the upper limit of the effective diffusion coefficient only. The numerical value for the effective diffusion coefficient of plutonium in illitic ocean floor sediment consistent with the two curves in Fig. 4. is:

$$D_{eff}(Pu) = (1.2 \pm 0.7) \times 10^{-10}\ cm^2\ s^{-1}.$$

In a similar fashion as described for plutonium, the upper limit for the effective diffusion coefficient of americium can be estimated. The result of the estimate is:

$$D_{eff}(Am) = (1.1 \pm 0.8) \times 10^{-10}\ cm^2\ s^{-1}.$$

For the corresponding upper limits of the effective diffusion coefficients in smectitic sediments, the data of Fig. 4. yield the values:

$$D_{eff}(Pu) = (2.7 \pm 1.6) \times 10^{-10}\ cm^2 \cdot sec^{-1} \text{ and}$$

$$D_{eff}(Am) = (1.2 \pm 1.0) \times 10^{-10}\ cm^2 \cdot s^{-1}.$$

As judged by these numbers, the diffusivity of both americium and plutonium is less than that of the relatively unaffected iodide ion by a factor of one million. This dramatic reduction over the mobility in solution cannot be accounted for by the larger mass or the higher charge of the ions. It must be caused by either a chemical transformation of the simple cationic species in solution, such as a hydrolysis and the attendant formation of an insoluble hydroxide, or by a very strong interaction with the constituent minerals of the sediment. Whether it is one or the other or perhaps a combination of these effects is currently being investigated in the laboratory.

REFERENCES

1. D. R. Kester, I. W. Duedall, D. N. Conners and R. M. Pytkowicz, Limnology and Oceanography 12, 176 (1967).
2. G. R. Heath, SAND76-0256, Sandia Laboratories, Albuquerque, NM.
3. A. M. Friedman and J. W. Kennedy, Jour. Amer. Chem. Soc. 77, 4499 (1955).
4. F. Schreiner, Nuclear Technology (to be published).

THE RADIOLYTIC OXIDATION AND REDUCTION OF PLUTONIUM[+]

Sherman Fried, Arnold Friedman, J. C. Sullivan, Kenneth Nash, Donald Cohen and Ruth Sjoblom

Chemistry Division, Argonne National Laboratory
9700 South Cass Avenue, Argonne, Ill 60439

INTRODUCTION

It has been shown on several occasions that all other things being equal, the rate of migration of plutonium through an aquifer is greatly influenced by its oxidation state. It is a reasonable expectation that other actinides exhibiting multiple oxidation states will behave similarly. By and large the more highly charged ions in solution move through rock strata more slowly than those of lesser charge. Thus from comparative experiments on the relative migration rates of Pu(IV) and Pu(III) we would expect Pu(III) to move about 10 times faster than Pu(IV)* and it has been shown PuO_2^{2+} moves about 250 times more rapidly than Pu(IV) (1).

Since one of the more important parameters of safety assessment is radionuclide migration rate, it can be seen that it becomes important to know the oxidation states of radionuclides (actinides for the purpose of this report) existing in possible leachates from the material originally emplaced in a repository. It was with this in mind that a study of the "intrinsic stability" of plutonium oxidation states in groundwater solutions was undertaken. One of the

*Americium(III) was used instead of Pu(III) for the comparative experiments because of the difficulty of maintaining Pu(III) in that state in air at a pH of about 5-6. It was felt that the behavior of Am(III) would be sufficiently close to that of Pu(III).

[+]This work was sponsored by the Sub-Seabed program and the Waste Isolation Pilot Plant (WIPP) under the auspices of Sandia National Laboratories.

most prominent proposals for a waste repository is the WIPP (Waste Isolation Pilot Project) which intends to utilize bedded salt strata as the isolating medium. Groundwater in that context is saturated brine and is made up to specifications issued by Sandia Laboratories and designated as "Solution A" (2).

Concomitant with the studies of plutonium oxidation states in "Solution A", parallel studies were made in dilute brines (Copenhagen Seawater) since emplacement of radwastes in subseabed sediments is also under consideration. Subseabed studies were also carried out in distilled water to examine the stability of the oxidation states of plutonium unperturbed by the presence of large concentrations of extraneous ions. Some of these studies have been made under anoxic conditions (to avoid the perturbing effects of oxygen) and others were made in the presence of air since it was not known whether groundwater in a WIPP-type repository will contain oxygen or not. In the case of subseabed, bottom waters in some areas of the oceans may contain amounts of oxygen in solution corresponding to 10% of that of surface waters and in other areas, e.g., the Black Sea, the water contains no free oxygen. In some cases the pH was controlled and in others it was allowed to vary as plutonium was converted from one oxidation state to another. It should be made clear that the latter case (of uncontrolled pH) was not one of choice but rather was imposed upon the experiment by virtue of the fact that some of the reaction vessels were sealed _in vacuo_ and it was not possible to monitor the pH without losing the integrity of the experiment.

EXPERIMENTAL METHOD

The plutonium solutions employed were relatively concentrated, approximately 0.02 M. This high concentration enabled the determination of the oxidation state by spectrophotometric observation of optical adsorption peaks corresponding to a specific oxidation state. Monitoring the progress of the change in oxidation state was accomplished by repeated spectrophotometric observations. Plutonium was prepared in single oxidation states as the chloride. Whenever possible the use of oxidizing or reducing reagents that would leave residues was avoided since the presence of extraneous cations or anions would only complicate the interpretation of the observations. Thus, Pu(III) was prepared by electrolytic reduction of Pu(IV) and Pu(VI) was prepared by oxidation of Pu(IV) by ozone. The plutonium isotope used in these experiments was ^{242}Pu.

In the case of anoxic preparations, a quantity of pure plutonium compound (as the chloride) was introduced into a glass ampoule. To this was added a known volume of "Solution A". The combined solutions were evaporated in a stream of nitrogen at room temperature in order to avoid possible decomposition or oxidation. The ampoule containing the dried mixture of salts was connected to a vacuum line and the system pumped until the pressure was of the order of 10^{-5} mm Hg.

At this point water was distilled onto the dried salt mixture (in the same high vacuum) until enough had been added to reach the original volume of the solution (Pu + Solution A). Thus, the "Solution A" was reconstituted with the Pu in such a way that it was oxygen free. At this point the glass ampoule was sealed off and was ready for continued spectrophotometric observation. The foregoing description applies to the preparation of all oxidation states of plutonium in the various solvents when it was desired to study them under highly anoxic conditions.

Another method for the preparation of anoxic solutions of plutonium was the use of a controlled atmosphere box. It was possible to lower the oxygen content of its atmosphere to 1-2 parts per million. The use of this box permitted the study of reactions in openable containers (in the box) so that changes in pH and E_h could be monitored by a pH and E_h electrode in the box or if desired, the pH could be controlled by addition of small amounts of the appropriate reagents. Withdrawal of aliquots from the main batch enabled the monitoring of samples by the spectrophotometer. Other solutions of plutonium were prepared in air using stoppered spectrophotometer cells as reaction vessels. These preparations could be monitored for pH from time to time by removing the stopper and insertion of a glass electrode.

EXPERIMENTAL RESULTS

To anticipate, it should be said at this point that all evidence indicates that the oxidation state of plutonium in solution is affected by radiolysis of the water. In particular, Pu(III) and Pu(IV) seem to be oxidized. This appears to be true whether air is present or the solution is oxygen-free and whether the solution is acidic or basic or in concentrated brines or pure water. This result is surprising in view of the fact that previous work states that the most common effect of radiation is to <u>decrease</u> the oxidation number (3).

The extent or the rate of the reactions may vary from solution to solution but in the main the same effects are observed. Indeed, the observations seem to indicate that water radiolysis will be a major factor in determining the oxidation state of plutonium solutions in the near vicinity of the radiation field of a waste repository. In fact, it can be postulated that even small quantities of plutonium carried away from the repository by some leaching and transport process and subsequently adsorbed on some mineral in a rock stratum will also be subjected to the oxidation effects of radiolysis from its own alpha activity.

The reasons for making the statement given above is as follows: an anoxic solution of Pu^{+3} in water made some two years ago was found to oxidize slowly to Pu(IV) and Pu(VI). This was demonstrated by observing the decrease in concentration of Pu(III) in solution spectrophotometrically and then isolating the Pu(IV) and Pu(VI) components by chemical separation procedures. Examination of the kinetics showed that the reaction could be expressed as first order with a half time of Pu^{+3} concentration of about 1.3 x 10^3 days.

It was considered possible that this oxidation was due to radiolysis. If radiolysis is a factor in these reactions, then the rate ought to correspond to the intensity of the radiation field. Accordingly very small amounts of ^{244}Cm were added, increasing the level of alpha activity from ten to one hundred times above that of the plutonium alone. The amount of ^{244}Cm required was only a few micrograms and it is unlikely that any chemical effect of curium would be observed since the predominant if not sole oxidation state of curium is plus three. It was found that the oxidation of the Pu^{+3} was speeded up in amounts roughly proportional to the level of added alpha activity. Thus, the question of <u>intrinsic</u> stability of plutonium oxidation states is probably moot since all plutonium is radioactive and indeed in a real repository where substantial radiation fields exist radiolysis effects on oxidation states will be of paramount importance.

In view of the foregoing it was decided to carry out all experiments in the presence of enough ^{244}Cm to raise the level of alpha activity one hundred fold over the original alpha activity of solution due to ^{242}Pu alone. In this way changes could be observed in reasonable time spans and the prior demonstration of proportionality of redox rate to alpha activity justified these experiments. It is proper to point out at this time that the effect of radiolysis is obviously mainly on the water rather than the plutonium directly since the concentration of the water is so much greater than the plutonium. It is observed that the overall effect of radiolysis is to oxidize Pu(III) to Pu(IV) and thence to Pu(VI). After a time, Pu(VI) appears to be reduced and a cycle of oxidation-reduction can be initiated.

CONCLUSIONS

The fact that the radiolytic oxidation of Pu exhibits pseudo first order reaction kinetics is not surprising. The reaction is probably between the Pu ion and OH radical. The OH radical is maintained at a constant concentration by its replenishment from the radiolysis of the water. The only other reactant is apparently the Pu. The rate will then depend only on the concentration of the Pu and hence the reaction appears to be first order. As the Pu(III) is depleted, the OH radicals can accumulate more rapidly than they are

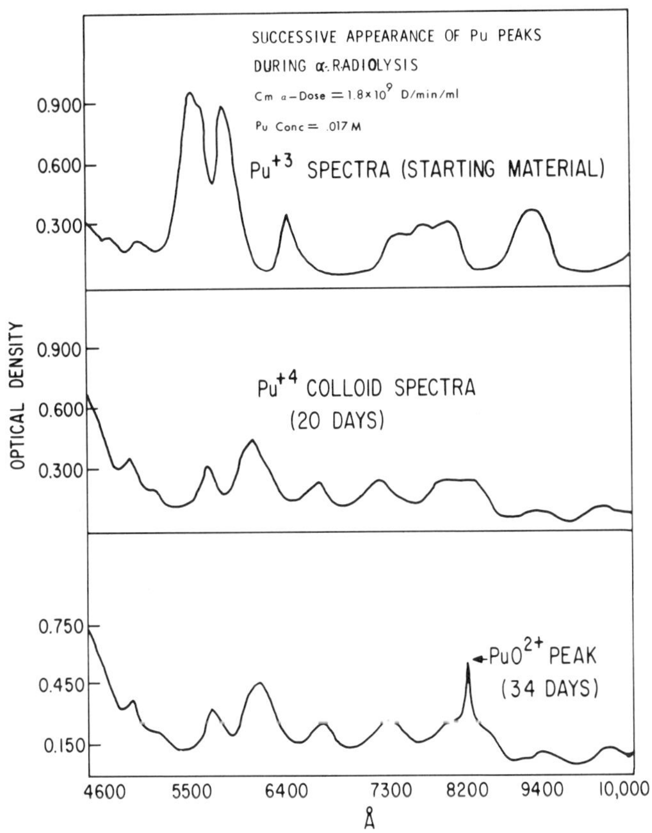

Fig. 1. Cyclic Behavior of Pu^{+3}, Pu^{+4} and Pu^{+6}

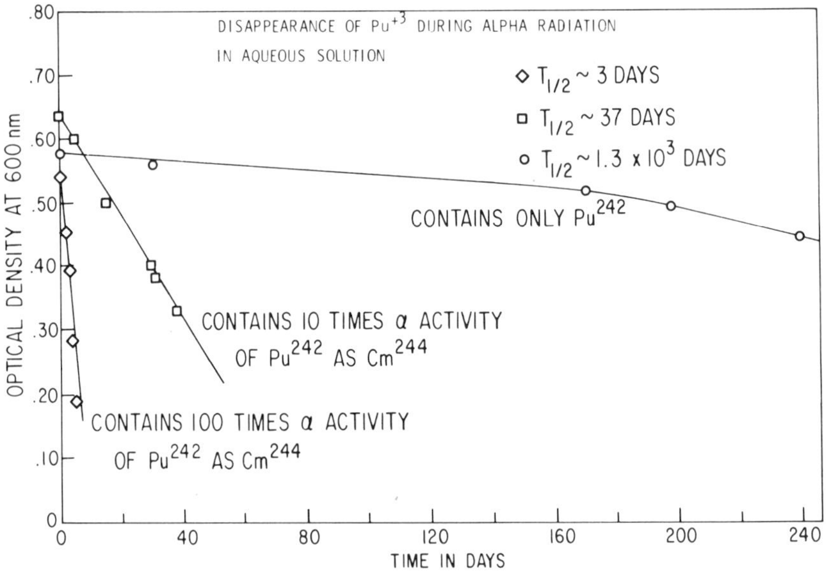

Fig. 2. Loss of Pu^{+3} as a Function of Time for Various Radiolysis Rates

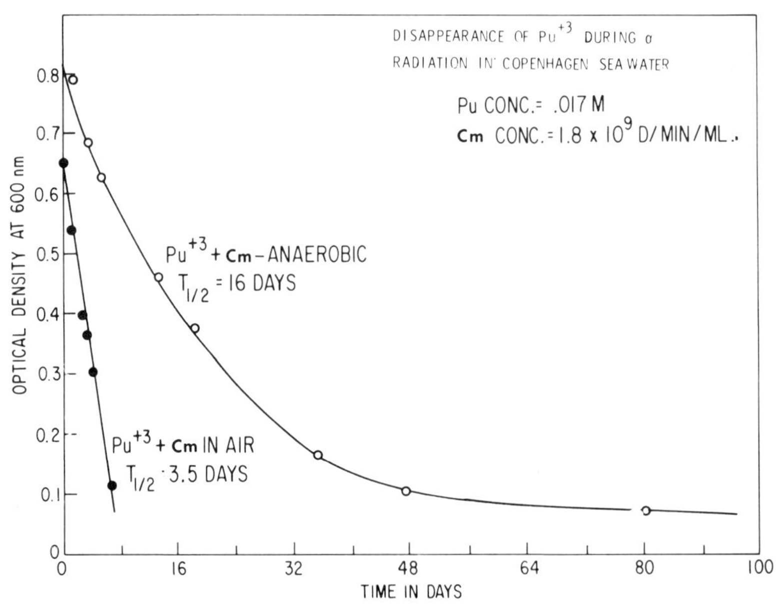

Fig. 3. Loss of Pu^{+3} as a Function of Time for Copenhagen Seawater

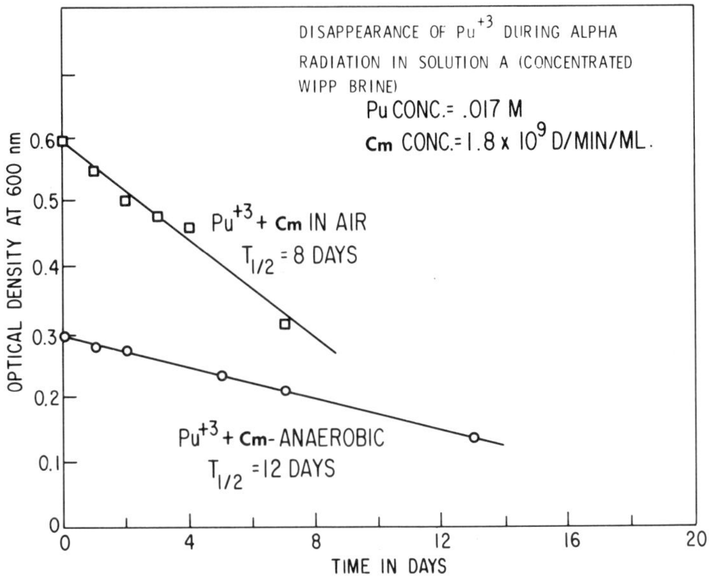

Fig. 4. Loss of Pu^{+3} as a Function of Time for WIPP Brine A.

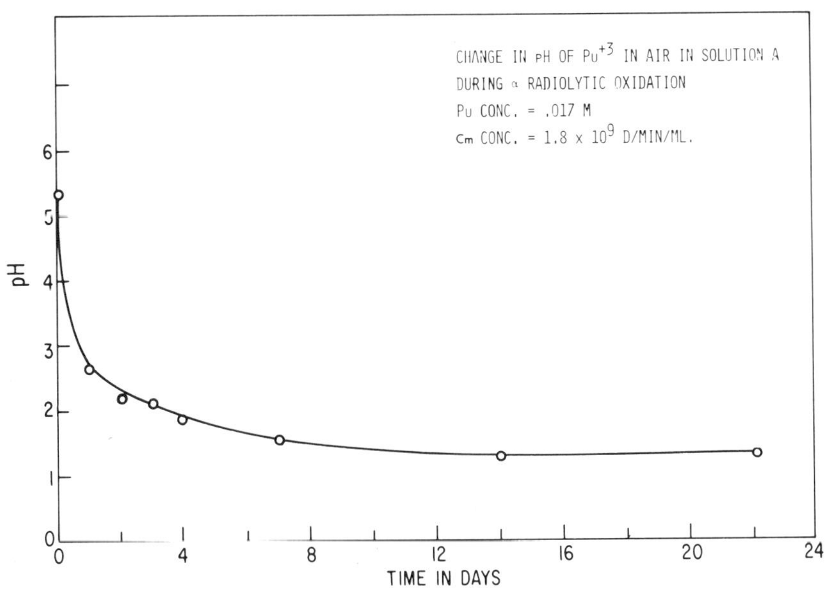

Fig. 5. Change in pH of Solution A Containing Pu^{+3} in Air.

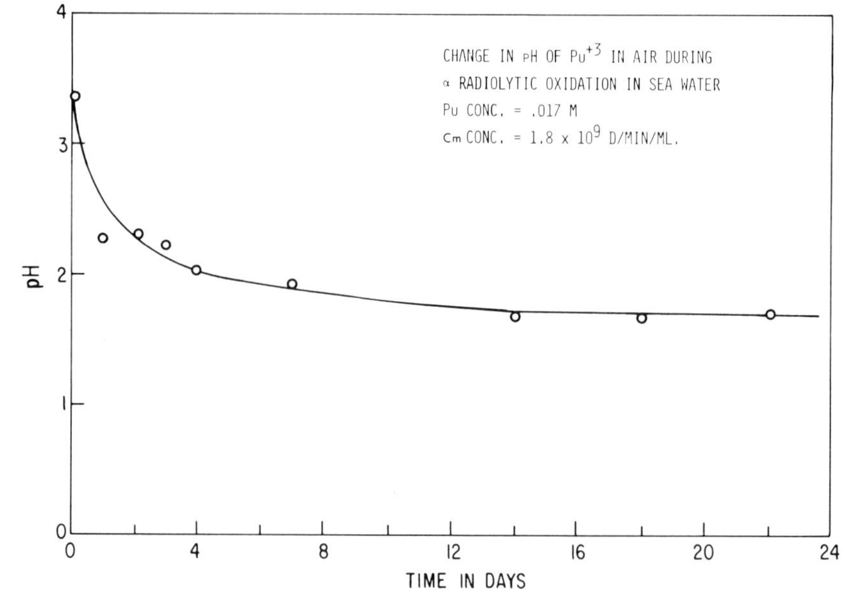

Fig. 6. Change in pH of Seawater Containing Pu^{+3} in Air

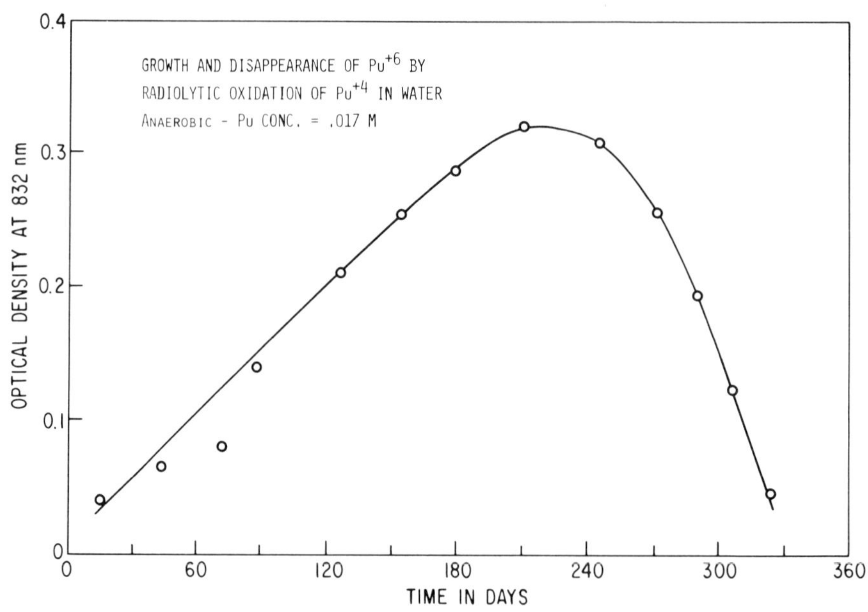

Fig. 7. Variations in Pu^{+6} in Water Under Anaerobic Conditions

consumed and they may combine to form H_2O_2. In any event the concentration of OH radicals will now change as the reaction proceeds and the rate will no longer be first order.

The apparent paradox of radiolysis acting both to oxidize and reduce plutonium can be explained if certain reactions are postulated. They are:

1) Water is decomposed by radiation into OH radicals and hydrogen atoms or hydrogen atoms plus hydrogen atoms plus hydrated electrons.

2) The hydroxyl radical is an oxidizing agent oxidizing Pu(III) and its products successively to Pu(VI).

3) The Pu(III)-hydroxyl radical oxidation reaction is fast (the second order rate constant is of the order of 10^8). The Pu(IV) to Pu(V) is slow, while the Pu(V) to Pu(VI) is fast.

4) When all the Pu(III) is oxidized, the OH radicals accumulate (for lack of substrate) and begin to react with each other to form hydrogen peroxide.

5) Hydrogen peroxide reacts with PuO_2^{++} [Pu(VI)] to reduce it according to the equation

$$2PuO_2(OH)_2 + H_2O_2 \rightarrow 2PuO_2^+ + O_2 + 2H_2O + 2OH^- \qquad 1)$$

The PuO_2 may then disproportionate to $Pu(OH)_4$ and PuO_2^{++} Thus, the Pu(VI) is reduced after all the Pu(III) is depleted and H_2O_2 is formed.

The reaction is made plausible by observing that the addition of H_2O_2 to PuO_2^{++} results in a precipitate and an increase in pH as required by equation 1) and in other experiments, Pu(V) has been spectrophotometrically observed during the reaction of Pu(VI) with H_2O_2 at pH 8-8.5. Furthermore, on standing for a time some of the Pu(VI) reappears. This is consistent with the known disproportionation of Pu(V) into Pu(VI) and Pu(IV) according to the equation

$$2Pu(V) \rightarrow Pu(IV) + Pu(VI) \qquad 2)$$

The results of our experimental observations are summarized in the following set of graphs. These graphs show the changes of pH, the oxidation of the various plutonium species in various solutions as well as the final reduction of Pu(VI). They display the reactions under anoxic conditions and in the presence of air as well as in various media. The data shown in the various graphs are only for comparison and cannot in themselves be used to predict the concentrations of various plutonium species in the indicated solutions in a real repository. These calculations can only be made when the rate laws governing the reactions are completely elucidated. Likewise, the changes observed in E_h and pH in the laboratory sized samples cannot be directly translated into the corresponding changes in very dilute solutions such as would result from the leaching of the radioactive source emplaced in a repository. Clearly the reaction rates will change over many orders of magnitude from laboratory conditions to field conditions and possibly even the mechanism will change. It is to be expected that the rock strata themselves through which the radionuclide migrate will modify and "buffer" the pH and E_h of the solutions.

The net result of these series of reactions is the plutonium may undergo a cycle of oxidation and reduction. This is a composite effect, since it is the sum of radiolytic oxidation and reduction and disproportionation and reproportionation reactions. The period of the cycle must depend on concentration of Pu, radiation field intensity, pH of solution, temperature and formation of complexes. Thus, the migration characteristics of all of the oxidation states of Pu [with the possible exception of Pu(VIII)] will have to be considered. It should also be pointed out that these radiolysis effects are probably not confined to plutonium. Neptunium, uranium, technetium and iodine are probably among the multivalent ions that will be affected and their various migration characteristics will have to be determined and will also have to be factored into the safety assessment of a repository.

REFERENCES

1. S. Fried and A. Friedman, Retention of Plutonium(VI) on Los Alamos Tuff, Waste Management 76, 206, CONF-761020, Proceedings of the Symposium on Waste Management, Tucson, Arizona, Oct. 3-6, 1976. National Technical Information Service, U.S. Dept. of Commerce, Springfield, Virginia 22161.
2. R. G. Dosch and A. W. Lynch, Interaction of Radionuclides with Geomedia Associated with the Waste Isolation Pilot Plant Site in New Mexico, SAND78-0297, June 1978.
3. J. M. Cleveland, The Chemistry of Plutonium, pp. 38-46, published by the American Nuclear Society, LaGrange Park, Ill 60520 (1979).

TRANSPORT THROUGH DEEP AQUIFERS OF TRANSURANIC NUCLIDES LEACHED FROM VITRIFIED HIGH-LEVEL WASTES

A. Avogadro, C. N. Murray and A. De Plano

Commission of the European Communities
Joint Research Centre - Ispra Establishment
21020 Ispra (Varese), ITALY

INTRODUCTION

Methods for analysis of the safety of long term storage of waste from nuclear power production in the European Community are under study at the Joint Research Centre of Ispra (Italy) (1). The methodology developed for risk assessment of radioactive waste storage in geological formations is based on the quantitative evaluation of the barriers which are present or may be interposed between the waste and mankind.

The retention by deep soils of released radioelements during transport by ground water constitutes the so-called geochemical barrier. This barrier takes account of the sorption phenomena which accompany the radioisotope transport by water through porous underground media and which can cause delays in radioactivity appearance at the land surface following loss from geological segregation. The modelling of this barrier relies upon a mathematical treatment of the physico-chemical phenomena which occur during the migration through the soil column. It must be stressed, however, that, especially in the case of transuranic nuclides which are important isotopes for the long term risk assessment, data presently available are only rough estimates based on scarce experimental evidence. An example of this is our present knowledge of plutonium chemical forms in aqueous media (2); complex stability, redox reactions, precipitation and filtration processes, colloid aging etc. are all phenomena not yet known in quantitiative terms. A better understanding of all these aspects is clearly needed to allow reliable quantitative modelling of the sorption barrier.

Experimental studies to investigate the physico-chemical behavior of transuranic elements leached form a glass matrix, under conditions similar to those of natural groundwaters, are being developed in our laboratories (3). We assume that as a consequence of a postulated accidental event water penetrates the repository and partially dissolves the actinide-bearing glass, moving afterwards through a surrounding porous formation.

METHODS AND MATERIALS

In order to simulate the expected conditions of glass leaching and underground transport in the laboratory, a water pathway is established which flows over the actinide-bearing glass, through a filter and then up through columns containing soil samples. Water fractions are continuously collected and analyzed at the column outlet; the water velocity through the columns corresponds to a groundwater movement of about 25 m/yr. The experiments are carried out over periods of 1 to 6 months, and at the end the columns are cut into thin sections and the distribution profile of transuranic isotopes is measured. The setup is schematized in Fig. 1. A special glove box has been installed for these experiments with several columns filled with typical subsoils being studied simultaneously. The column dimensions are 200 mm in length and 26 mm in diameter.

Borosilicate glasses containing separately ^{238}Pu, ^{241}Am and ^{237}Np, have been produced in the laboratory; the production procedure consists of three successive fusions of the glass matrix powder (60% SiO_2, 18.8% B_2O_3, 6.2% Al_2O_3, 15% Na_2O) containing 20% of simulated waste oxides. The concentration of the actinides corresponds to the composition of the vitrified high level wastes, after a cooling time of 1000 years. The glasses are ground and sieved, and the resulting material (1 - 2 and 0.5 - 1 mm nominal size range) is used as a contamination source in the leaching and migration experiments. Radiochemical and metallographical analysis of the glass fragments showed homogeneous distribution of transuranic nuclides and of metallic inclusions. The leaching rates for both size fractions at room temperature were determined to be 1.2×10^{-6} for neptunium, 1.12×10^{-6} for plutonium and 6×10^{-7} $g/cm^2/day$ for americium respectively.

The typical formation we chose for the initial set of experiments is the porous bed which overlies the Boom clay formation in Belgium. This is a sand layer with porosity of 34% and permeability of the order of 10^{-4} cm/sec (0.1 darcy) containing 20.8 wt % of glauconite (4). This clay material is a magnesium-iron alumino-silicate of the illite type, having a cation exchange capacity of 28 - 30 meq/100 g. The bulk density of the experimental columns is 2 $g.cm^{-3}$.

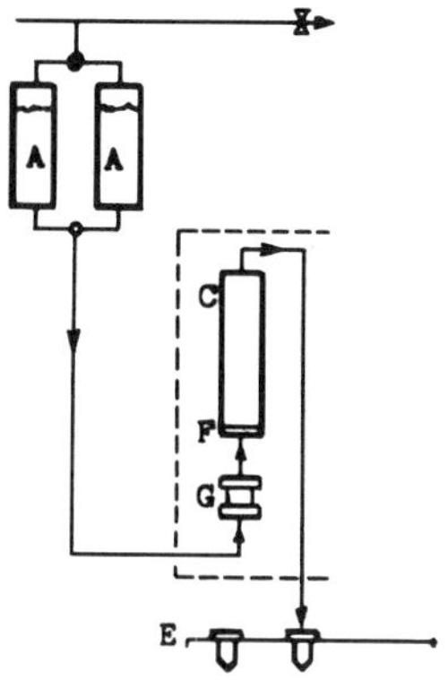

Fig. 1. Experimental Setup for Studying the Interaction of Glass Leachant with Deep Soils

The synthetic water used has a composition typical of the aquifer where the glauconite samples were collected (Ca^{++} 3.0, Mg^{++} 3.2, Fe(total) 0.1, Na^{+} 55, K^{+} 7.7, Cl^{-} 6.4, $SO_4^{=}$ 0.5, ΣCO_2 201mg/l, pH 8.35, Eh 150 mv).

During the experimental period, the water reservoirs were always protected from contact with the atmosphere in order to avoid oxygen dissolution and to control the redox potential. The soil columns were equilibrated for several days by percolation before the introduction of the radioactive glass in the water pathway. The pH of the water before and after the soil column remained constant during the entire experimental period.

The ^{241}Am in water and soil samples was measured directly by gamma-counting; ^{238}Pu was determined by total alpha liquid scintillation; soil samples were treated with HNO_3-HF followed by solvent extraction before counting by liquid scintillation. ^{237}Np was determined by alpha-spectrometry (Si-barrier detector, evaporated source) or by gamma counting after 4 months equilibration of the ^{237}Np daughter ^{233}Pa (300 KeV).

RESULTS AND DISCUSSION

In Table 1 results are given on the percentage distribution of actinides leached from contaminated glass. The partitioning was obtained in an experiment over a six month period using the set-up illustrated in Fig. 1. As the operating conditions were identical for all three isotopes, the quantity retained by the filters inserted before the soil columns does not appear to depend upon the glass but on the characteristics of the leached radioelements. On the basis of the 10 μm filter americium turns out to be the most insoluble

Table 1. Percentage Distribution of Actinides After Six Months of Percolation

	10 μm filter	soil column	column outlet
^{237}Np	1.3	96.4	2.3
^{238}Pu	25.3	74.67	0.03
^{241}Am	60	39.96	0.04

element followed by plutonium. Neptunium on the other hand appears to be almost completely soluble.

The concentrations in the water directly after contact with the alpha-bearing glasses for americium and plutonium are 10^{-9}M and for neptunium 3.7 x 10^{-8}M. At the outlet of the soil columns the concentrations continually maintained a constant value of $2 \cdot 10^{-13}$M for americium and plutonium and $8 \cdot 10^{-10}$M for neptunium. Two ion exchange resin columns (BioRad AG 50Wx8 and AG 1, 200-400 mesh) were placed in series with the soil column. It was found that all the plutonium and americium released from the soil columns were absorbed on the anion resins. This confirms the existence of negatively charged soluble species, probably being of a carbonate or hydroxycarbonate complex form (2). Results for neptunium behavior are not yet available.

The percentage distribution of the activity fixed to the soil columns are shown in Fig. 2 for the three radionuclides.

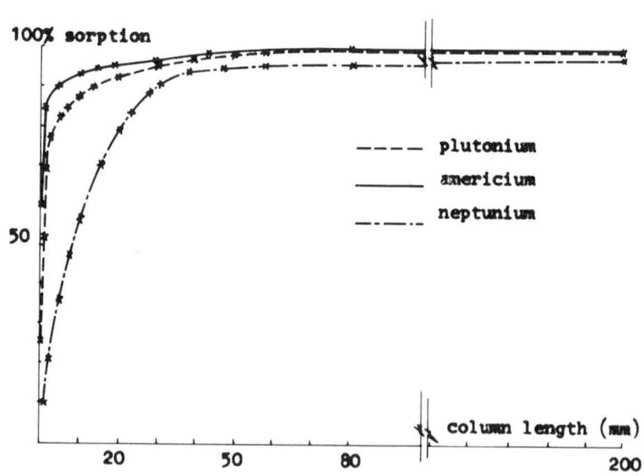

Fig. 2. Actinides Sorption on Glauconite Sand

Table 2. Percent Activity Repartition in the Analytical Train

	plutonium	americium
ultrafilter	12.6	89.2
cation resin	52.4	10.7
anion resin	34.7	0.1
neutral fraction	0.3	-
ultrafilter	15.1	90.7
anion resin	83.3	6.0
cation resin	0.3	3.3
neutral fraction	1.3	-

The distribution of transuranic nuclides leached from the glass results from various phenomena, the most important being colloidal filtration, ion exchange and chemical or physical adsorption. The situation is also complicated by the possibility of the coexistence of different chemical forms and valence states. However the results showed that for plutonium and americium over 90% of the retained activity is fixed in the initial top 20 mm; for neptunium the retention is lower and is spread over about twice this soil thickness.

With the aim of determining the main processes which affect the distribution of activity, the leaching solution was passed through an analytical train shown in Fig. 3 made up of a membrane filter, two ion exchange columns and a fraction collector for neutral species. In Table 2, are given the results of the distribution of plutonium and americium. These are based on the average of three experiments each carried out over a 34 day period. Results are given for two cases where the cation and anion resins are inverted in order to determine the eventual filtration effects of the uncharged species on the first column. The soluble charged fraction is considered to be that found on the second column; 34.7% and 0.3% are negatively and positively charged respectively for plutonium. The percentage of uncharged actinide is obtained by subtracting the soluble fraction from the total adsorbed on the corresponding initial column. The data show, in fact, that between 48-52% of plutonium particles seems to be uncharged and fixed by filtration. The largest "soluble" fraction, however, behaves as an anionic species and only a negligible cationic fraction was detected.

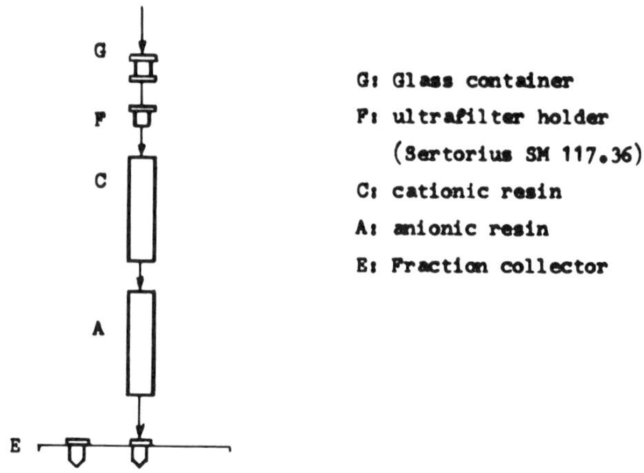

Fig. 3. Analytical Train for Studying Physico-chemical Forms of Glass Leachant

If these results are compared with those obtained with the soil, we would expect that, as sandy clays normally have a negligibly small anionic exchange capacity, continuous percolation of the contaminated solution would eventually saturate the soil with respect to its anionic exchange capacity. This in fact appears to be the case as a small anionic fraction (0.03%) was found to be released almost immediately after the start of the experiment. We, therefore, suggest that the major anionic fraction existing in the leachant (34.7%) is probably not an anionic "soluble" species but a negatively charged microcolloid being retained by the soil column.

For americium it can be seen that the largest amounts of activity are retained on the membrane filter. Of the remaining, some 3% of the total appears to be soluble, having a positive charge, and the remaining 6% is retained by filtration on the resin. Negatively soluble charged forms account for less than 0.1% of the total. Comparison with the soil shows that also in the case of americium this anionic soluble species is not retained by the column. The soluble anionic fractions observed for both plutonium and americium are considered to be probably carbonate complexes, while the species retained in the case of plutonium by the soil, are probably microcolloids which at high pH values are negatively charged (5,6,7).

CONCLUSION

The results presented in this paper are the first output of a new program carried out at the Joint Research Centre of Ispra on underground migration of actinides related to risk evaluation of the geological waste disposal option. Work is underway in order to better understand the phenomena observed in these laboratory experiments. Clearly more sophisticated analytical techniques are needed; these being developed and tested (8). Nevertheless these initial results already demonstrate a certain number of phenomena; firstly, that the soil matrix studied effectively retains the transuranic nuclides.

However the more effective processes seem not to be related to the high effective filtering ability of sandy clays. Further the data appears to show that both for plutonium and americium cationic species are only present at very low levels and negatively charged species appear, for plutonium at least, to be the main mobile form. For both plutonium and americium a soluble anionic form having a concentration of 10^{-13}M was detected which was not retained by the columns. Under the present experimental conditions, this concentration is about four orders of magnitude lower than the Maximum Permissible Concentration for drinking water.

REFERENCES

1. F. Girardi, A. Avogadro, G. Bertozzi, M. d'Alessandro, F. Lanza and N. C. Murray, IAEA-SM-243/161, Int. Atomic Energy Agency, Vienna (1979).
2. A. Saltelli, A. Avogadro and G. Bertozzi, Proc. of NEA/CEC Workshop on the Migration of Long-lived Radionuclides in the Geosphere, Brussels 29-31 January 1979, OECD Nuclear Energy Agency Publications (1979).
3. A. Avogadro, C. N. Murray and A. De Plano, Proc. of NEA/CEC Workshop on the Migration of Long-lived Radionuclides in the Geosphere, Brussels 29-31 Janauary 1979, OECD Nuclear Energy Agency Publications (1979).
4. R. Heremans, CEN-Mol (Belgium), personal communication.
5. Yu. P. Davydov, Radioknimiya 9, 99-109 (1967).
6. Yu. P. Davydov and Yu. I. Bondar, J. of Radioanal. Chem. 51, 325-335 (1979).
7. I. E. Starik, et al., Radioknimiya 9, 109-118 (1967).
8. C. N. Murray, A. Avogadro and G. Lazzari, Proc. of 2nd Int. Symp. of Radioecology, Cadarache, France, 19-22 June (1979) (in press).

POSSIBLE RETENTION OF IODINE IN THE GROUND

B. Allard, B. Torstenfelt, K. Andersson and J. Rydberg

Department of Nuclear Chemistry
Chalmers University of Technology
S-412 96 Göteborg, Sweden

INTRODUCTION

At the eventual release of radionuclides from a deep underground repository for high level waste, the only anionic fission product would be iodine, assuming reducing conditions (1-3). The major species would be I^- although leaching experiments on tuff from the Nevada Test Site have indicated that other species may be present (4). The retention of these anionic species would be low in most bedrocks (5-7). In this paper, the sorption of iodide on various minerals and compounds is discussed as well as the prospect of using common geologic media as iodide retarding backfill materials in an underground waste repository.

IODIDE CHEMISTRY IN THE GROUNDWATER/BEDROCK SYSTEM

Deep anoxic groundwaters in crystalline igneous rocks are reducing (8), and redox potentials below -0.12 V at pH 8 would be expected due to the presence of ferrous minerals in the rock (9,10). For the equilibrium between iodine species of different oxidation states at the same activity, the redox potential (E,V) will be given by the following expressions (calculated from standard potentials given in Ref. 11):

$$I^-/IO_3^- : \quad E = 1.09 - 0.0592 \text{ pH}$$

$$I^-/IO^- : \quad E = 1.32 - 0.0592 \text{ pH}$$

$$I^-/I_2(s): \quad E = 0.536 + 0.0592 \text{ pI}^-$$

Thus, the prevailing valence state in anoxic groundwater at pH 7-9 would be (-1), and only reactions of I^- are considered in this study. Significant retention of I^- could be expected either in groundwaters containing metal ions capable of forming strong cationic complexes or neutral compounds of low solubility or by chemisorption on minerals containing these same metals. In Table 1 some formation constants for metal iodides are given (12). None of the metal ions of Table 1 would normally be found to any significant extent in groundwaters, and none of the common cations in natural waters would form strong complexes with I^-. Another mechanism for sorption of I^- in the rock/groundwater system would be by anion exchange.

CHOICE OF GEOLOGIC MEDIA AND WATER

For the experiments minerals containing metal ions which form insoluble iodides, silicate minerals with different anion exchange capacities, and some metal hydroxides with large surface to volume ratios were chosen. (C f Table 2).

According to Table 1, minerals containing Hg or Bi would be expected to sorb iodine strongly. Minerals containing Cu(I), Ag(I), Cd(II), or Pb(II) would also show strong sorption. High sorption has previously been reported for chalcopyrite (0.21 m^3/kg) and galena (0.02 - 0.05 m^3/kg) (13). In the present study, chalcopyrite, galena, cinnabar and artificial cerargyrite (AgCl) were used.

The anion exchange capacity (with respect to Br^-) of some common rock-forming and secondary minerals (i.e. clays) of igneuos rocks are given in Table 3 (14). The anion exchange is low for all the minerals - rarely above 1 meq/kg. The highest values are obtained for serpentine and olivine. The comparatively high values for these minerals may be due to substitution of Fe(II) and Mg(II) by Fe(III) in the lattice, leading to an increased net surface charge. The clay minerals would be expected to exhibit some anion exchange properties due to the presence of hydroxy groups in the lattice. In this study, some silicates from different silicate groups were chosen. Quartz was used as an "inert" reference.

Among the hydroxides, the only naturally occuring one used was limonite. The hydroxides are expected to show some sorptive properties due to the high surface to volume ratio. The lead hydroxide is also capable of chemisorption.

An artificial groundwater with the initial composition 165 ppm Na^+, 3.9 ppm K^+, 18 ppm Ca^{2+}, 4.3 ppm Mg^{2+}, 123 ppm HCO_3^-, 9.6 ppm SO_2^{2-}, 70 ppm Cl^-, 12 ppm total SiO and pH 8.2 was used for the sorption measurements (15). Measurements were also performed in 4 M NaCl in order to study the effect of an increased anion concentration.

Table 1. Complex Formation Constants (log K_1) and Solubility Products (log K_s) for Metal Iodides at Zero Ionic Strength

Metal	log K_1[a]	log K_s[a]	Metal	log K_1[a]	log K_s[a]
Cu^+	(6)	-12.0	Hg_2^{2+}	(12)	(-29)
Ag^+	(6.8)	-16.08	Pb^{2+}	1.92	-8.10
Cd^{2+}	2.28	(-8)	Bi^{3+}	(2.4)	(-21)
Hg^{2+}	(12)	-28.33			

[a]Extrapolated or estimated values in paranthesis

EXPERIMENTAL

The sorption of I^-, in groundwater and 4 M NaCl, was measured as a function of pH for the minerals and compounds of Table 2 using a batch technique (15). The minerals were obtained from the Department of Geology, Chalmers University of Technology and from Ward's Natural Science Establishment. The purity of the minerals was checked by X-ray diffractometry only. The anion exchange capacity of the minerals was measured by a batch equilibration technique

Table 2. Solid Materials Used for Iodide Sorption Studies

Chemisorbing Minerals	Silicate Minerals
Cinnabar (HgS)	Olivine ($(Mg,Fe)_2SiO_4$)
Chalcopyrite ($CuFeS_2$)	Attapulgite ($Mg_{2,5}(H_2O)_2(OH)(Si_4O_{10})$)
Galena (PbS)	Serpentine ($Mg_6(Si_4O_{10})(OH)_8$)
Silver chloride (AgCl)[a]	Halloysite ($Al_4(Si_4O_{10})(OH)_8$)
	Montmorillonite ($(Al,Mg)_8(Si_4O_{10})_3(OH)_{10}aq$)
Hydroxides	Quartz (SiO_2)
Limonite (FeO(OH)aq)	
Ferric hydroxide ($Fe(OH)_3aq$)[a]	
Aluminum hydroxide ($Al(OH)_3aq$)[a]	
Lead hydroxide ($Pb(OH)_2$ aq)[a]	

[a]Prepared from or bought as reagent grade chemicals

Table 3. Anion Exchange Capacities (AEC) of Some Common Mineral

Class	Mineral	AEC[a], meq/kg
1. Sulfides	Chalcopyrite	0.5
	Pyrite	< 0.3
2. Oxides, hydroxides	Magnetite	0.4
	Limonite	0.6
	Gibbsite	0.5
	Hematite, corundum	< 0.3
3. Halides,	Fluorite	0.2
Carbonates,	Calcite, dolomite	< 0.3
Sulfathes,	Anhydrite	< 0.1
Phosphates	Apatite	< 0.1
4. Silicates		
a. Nesosilicates	Olivine	1.8
	Zircon	0.7
	Sphene	0.8
	Garnet (almandine)	0.4
b. Sorosilicates	Epidote	0.5
c. Cyclosilicates	Beryl	1.4
d Inosilicates	Augite, hornblende	< 0.2
	Attapulgite	0.7
e. Phyllosilicates	Kaolinite	0.5
	Serpentine	6.9
	Montomorillonite	0.9
	Halloysite	1.5
	Muscovite	0.6
	Biotite, chlorite	< 0.3
f. Tectosilicates	Anorthite	0.7
	Microcline, albite, bytownite, quartz	< 0.3

[a]For Br^- at pH 8

(16). The Fe-, Al- and Pb-hydroxides were prepared from reagent grade chemicals by precipitation at neutral pH from aqueous solutions. The precipitates were washed and dried at 150°C. Reagent grade AgCl was used. The solids (except the clays) were crushed in a mortar (to avoid iron contamination) and sieved. Only the size fraction 0.045-0.063 mm was used. Size fractioning was not performed for the clay minerals or the amorphous, artifically prepared compounds. All solids were prewashed with groundwater solution for five days. The solid phase was allowed to settle, the washing solution was discharged and new water was added. To this water 0.1 ml $Na^{131}I$-solution was added. The concentration of ^{131}I in the water phase was 10^{-12} M. In the case of 4 M NaCl the reagent grade NaCl had a maximum content of iodine of 0.00005%, which corresponds to 10^{-6} M I^- in 4 M solution, which may affect the sorption behavior. The solid/liquid ratio was 0.5 g/ 45 ml or 0.2 g/45 ml in all experiments. After a contact time of 24 h at room temperature (22 ± 2°C) under which the sample was agitated, the aqueous phase was separated from the solid by centrifugation (1h at 4000 g). Aliquots of the solutions were withdrawn and the remaining iodide concentration in solution was determined from activity counting using a well-type NaI scintillation counter. The pH of the solution was adjusted to a new level by the addition of 0.1 M NaOH or HCl and a new equilibration-sampling-counting cycle was performed with the same batch. Distribution coefficients (K_d) were calculated from:

$$K_d = (A_o - A)V/(Am)$$

where A_o = acitivity of water without mineral, A = activity of water with mineral, V = m^3 liquid and m = kg mineral.

RESULTS AND DISCUSSION

The measured sorption data (in terms of K_d) are given in Fig. 1-3. The chemisorbing agents all show a significant sorption. For naturally occurring minerals the highest K_d-values are obtained in groundwater. The order of K_d-values of the minerals follows the solubilities of the corresponding iodides. The increase in K_d with decreasing pH may be explained by an increased solubiltiy of the minerals (cf. ref 17). Significantly higher sorption was observed in groundwater than in 4 M NaCl, except for AgCl. All of the metals in the minerals would also form chloride complexes of low solubility (12). The Fe- and Al-hydroxides show a measurable sorption (K_d-values 0.01 - 0.02 m^3/kg). No significant difference between groundwater and NaCl solution is observed. A slight increase in K_d with decreasing pH exists. Sorption by Pb-hydroxide is high in groundwater, but is significantly lower in 4 M NaCl, just as for the chemisorbing agents.

A considerable sorption is obtained on some of the silicates. The expected high value of serpentine and olivine, due to the high AEC is not observed. Neither is the sorption on serpentine affected by the salt concentration. Olivine, however, acts more like an ion exchanger. At low pH the highest K_d-value is obtained for attapulgite which has a large surface to volume ratio (125-250 m^2/g) (18). Here the pH-dependence is considerable. In chloride solution the sorption on attapulgite is comparable to that of the other silicates, while the value of montmorillonite increases. A possible explanation is that montmorillonite flocculates in the presence of electrolytes while attapulgite is not affected. The addition of Ag^+ to the water significantly enhances the sorption even on an "inert" sorbent as quartz (K_d-values of 0.005-0.01 m^3/kg). This could be due either to formation and subsequent sorption of cationic (or neutral) complexes or to the formation and chemisorption on a surface coating of metal hydrolysis products on the solid. The formation of higher oxidation states of iodine, although improbable, can not completely be excluded in the presence of the various minerals. Any effects of iodide impurities in the NaCl solution have not been considered.

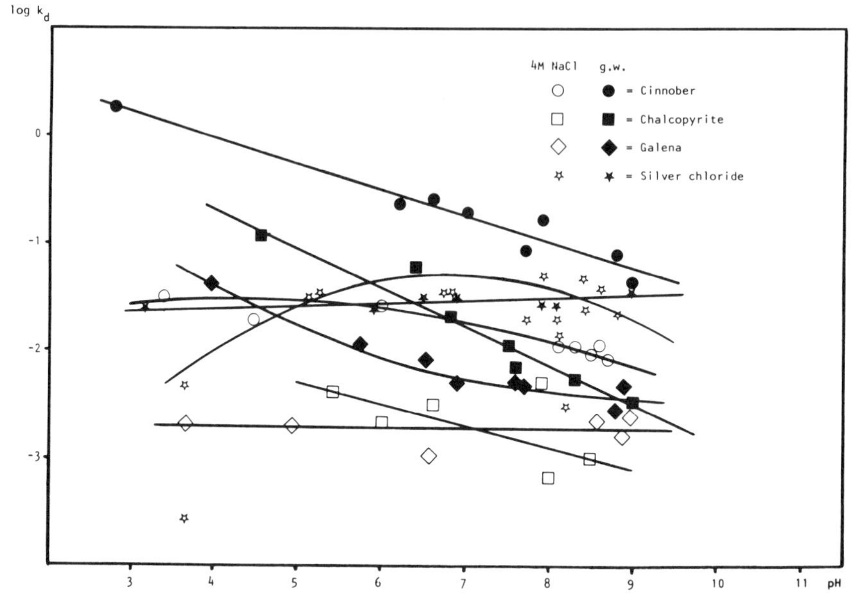

Fig. 1. Distribution Coefficients (log K_d, m^3/kg) vs. pH for I^- on Chemisorbing Agents

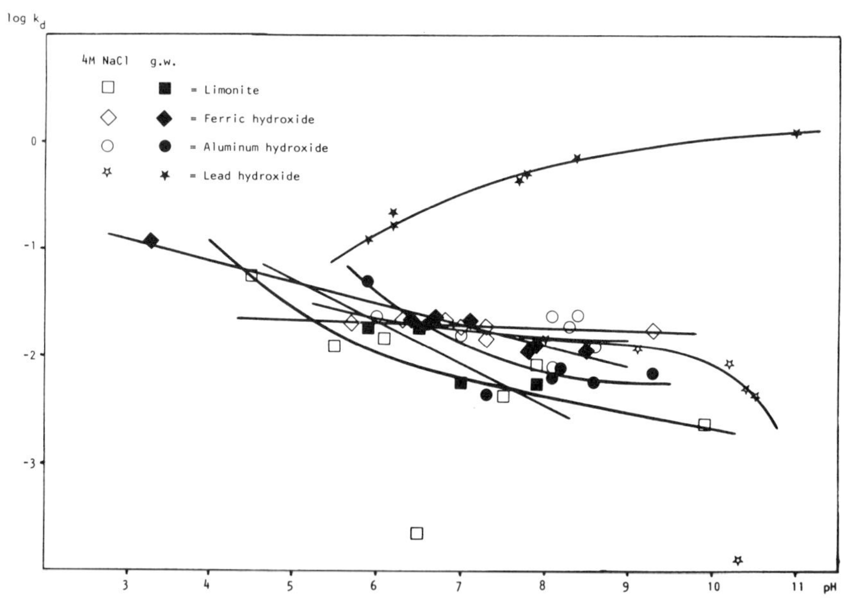

Fig. 2. Distribution Coefficients (log K_d, m^3/kg) vs. pH for I^- on Metal Hydroxides

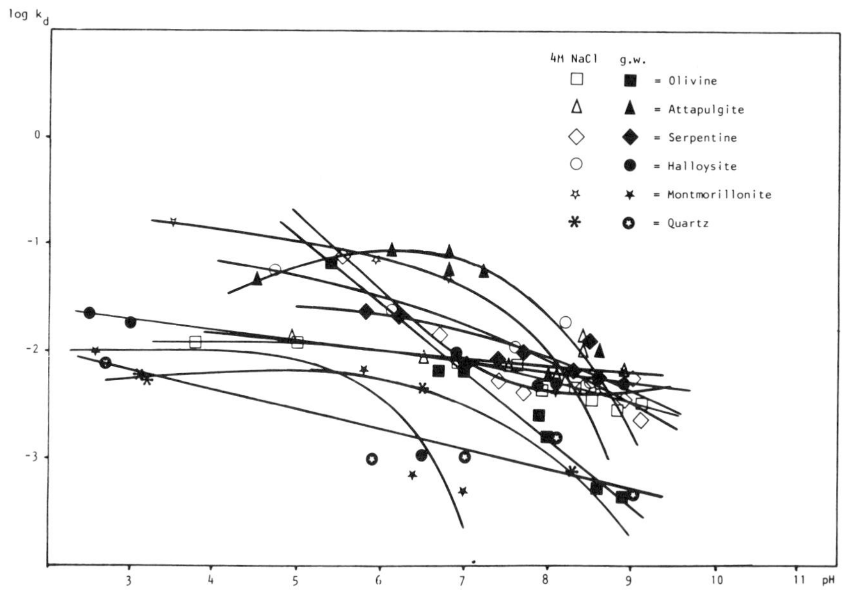

Fig. 3. Distribution Coefficients (log K_d, m^3/kg) vs. pH for I^- on Silicate Minerals

CONCLUSIONS

Significant iodide sorption occurs not only on the chemisorbing minerals containing Hg, Cu(I), Pb or Ag but also on some hydroxides and silicates with large surface to volume ratio. By the addition of such agents to e.g. a montmorillonitic buffer material, retention factors (the ratio between the velocity of a noninteracting specie and the velocity of the nuclide in question) in the 100-1000 range may be feasible.

The evident drawback from a practical point of view connected with the use of Cu-, Pb- or Ag-compounds as backfill additives is their potential values for other purposes. If, however, a metallic canister material containing Pb or Cu will be required in a future high level waste repositor, which has been suggested as possible alternatives (18), the release of Pb^{2+} and Cu^+ (monovalent under the conditions to be expected in deep groundwater) would significantly enhance the sorption of iodide even in the absence of specific iodide sorbing backfill material.

ACKNOWLEDGEMENTS

This work is supported by the National Council for Radioactive Waste Management (PRAV). Some of the minerals were provided by Dr. B. Ronge, Department of Geology, Chalmers University of Technology. Experimental help by Ms. Wanda May is gratefully acknowledged.

REFERENCES

1. Handling of Spent Nuclear Fuel and Final Storage of Virtified High Level Reprocessing Waste, Kärnbränslesäkerhet (KBS), Stockholm (1977).
2. B. Allard, H. Kipatsi and B. Torstenfelt, "Technetium: Reduction and Sorption in Granitic Bedrock," Radiochem. Radioanal. Lett. 37, 223 (1979).
3. E. A. Bondietti and C. W. Francis, "Geologic Migration Potentials of ^{99}Tc and ^{237}Np," Science 203, 4387 (1979).
4. K. Wolfsberg, Sorption Desorption Studies of Nevada Test Site Alluvium and Leaching Studies of Nevada Test Debris, LA-7216-MS Los Alamos Sci. Lab. (1978).
5. J. F. Relyea, L. L. Ames, R. J. Serne, R. W. Fulton and C. D. Washburne, "Batch K_d Determinations with Common Minerals and Representative Groundwaters," Batelle Pacific Northwest Lab., Progr. Rep. (1977).
6. B. Allard, H. Kipatsi and J. Rydberg, Sorption of Long-lived Radionuclides on Clay and Rock, Technical Rep. 55, Kärnbränslesäkerhet (KBS) Stockholm (1977).
7. S. Fried, A. M. Friedman, D. Cohen, J. J. Hines and R.G. Strickert, The Migration of Long-lived Radioactive Processing Wastes in Selected Rocks, Ann. Rep. Project AN0115A, FY 1977, ANL-78-46, Argonne National Lab. (1978).
8. Handling of Spent Nuclear Fuel and Final Storage of Unreprocessed Spent Fuel, Kärnbränslesäkerhet (KBS), Stockholm (1978).
9. I. Grenthe, "Determinations of Redox Potentials in Groundwater from Stripa and Finnsjön," App. 5 in Copper as Encapsulation Material for Unreprocessed Nuclear Fuel Waste, Technical Rep. 90, Kärnsbränslesäkerhet (KBS), Stockholm (1978).
10. B. Allard, J. Rydberg, H. Kipatsi and B. Torstenfelt, "Disposal of Radioactive Waste in Granitic Bedrock," in Radioactive Waste in Geologic Storage (S. Fried, ed. ACS Sumposium Series 100, American Chemical Soc., Washington 1979).
11. W. M. Latimer, The Oxidation States of the Elements and Their Potentials in Aqueous Solutions (Prentice-Hall, New York 1952).
12. R. M. Smith and A. E. Martell, Critical Stability Constants Vol. 4: Inorganic Complexes (Plenum Press, New York 1978).
13. D. Rancon and J. Rochon, "Retention des Radionuclides a Vie Longue par Divers Materieaux Naturels", Paper presented at "Workshop on the Migration of Long-lived Radionulcides in the Geosphere," OECD/NEA, Brussels, (1979).
14. B. Allard and G. W. Beall, work in progress.
15. B. Allard and G. W. Beall, "Sorption of Americium on Geologic Media," J. Environmental Sci. and Health, in press (1979).
16. C. Francis and D. F. Grigal, Soil Sci. 112, 17 (1971).
17. R. M. Garrels and C. L. Christ, Solutions, Minerals and Equilibria (Freeman, Cooper and Co., San Francisco, 1965).
18. W. L. Haden and I. A. Schwindt, "Attapulgite, Its Properties and Applications," Ind. Eng. Chem. 59, 9 (1967).

ASPHALT EMULSION SEALING OF URANIUM MILL TAILINGS

J. N. Hartley, P. L. Koehmstedt, D. J. Esterl

Battelle Pacific Northwest Laboratory
Richland, WA 99352

ABSTRACT

The use of asphalt emulsion to contain radon and radium in uranium tailings is being investigated at the Pacific Northwest Laboratory. Results of these studies indicate that a radon flux reduction of greater than 99% can be obtained using either a poured-on/sprayed-on seal (3.0 to 7.0 mm thick) or an admix seal (2.5 to 15.2 cm thick) containing about 18 wt % residual asphalt. A field test was carried out at the Grand Junction tailings pile in order to demonstrate the sealing process. A reduction in radon flux ranging from 4.5 to greater than 99% (76% average) was achieved using a 15.2 cm (6 in.) admix seal with a sprayed-on top coat. A hydrostatic stabilizer was used to apply the admix. This was followed by compaction to form the radon seal. Overburden was applied to provide a protective soil layer over the seal. Included in part of the overburden was a herbicide to prevent root penetration.

INTRODUCTION

The milling of uranium ore produces large quantities of waste (mill tailings) which remain potentially hazardous for a very long time due to the long half-lives of the radionuclides present. The two potentially hazardous radioactive decay products are radium-226 (half-life 1620 years) a solid and radon-222 (half-life 3.8 days), a radioactive gas which is considered to present the most significant exposure risk.

Based on projected U.S. nuclear generating capacity, 490 million metric tons (MT) of tailings will be produced by the year 2000 using conventional milling (1). These tailings would be in addition to the 107 million MT of tailings at currently active mill sites at the end of 1977 and 24 million MT of tailings at inactive sites. Proposed requirements for covering uranium tailings disposal include placing no less than 3 m (10 ft) of soil cover material over the tailings (1). This soil cover must not include mine waste or rock that contains elevated levels of radium. This technique might minimize human exposure from inhalation and ingestion, but it is not considered a totally satisfactory solution based on economics and the availability of cover material.

An alternative approach would be to place a cost-effective cover material that would reduce radon emanation to background levels and remain stabile for at least 1000 years. The Pacific Northwest Laboratory (PNL) is working on such an alternative (2, 3). The Department of Energy has contracted PNL to evaluate the use of asphalt emulsion sealants to retain radium in uranium tailings and to provide a barrier to prevent radon emanation to the atmosphere. In order for a stabilization or sealing material to last for millenia it must be inert, remain pliable, and not be affected by its surrounding environment. Since no materials have been tested for greater than 19 to 100 years, we cannot provide long-term stability data.

In the case of asphalt emulsion, we do know that <u>asphalt</u>, the primary constituent of asphalt emulsion, is present in very old (4800-5400 years) ornaments, figurines, statues, ceremonial baths and pavements which have survived the test of time. These objects have been excavated and recovered in excellent condition attesting to the potential long term stability of asphalt under anaerobic burial conditions.

This paper discusses the progress of this project including laboratory and field studies and summarizes the status of the sealing procedure for controling radon release from uranium tailings.

ASPHALT EMULSION

Asphalt emulsion consists of asphalt, water and an emulsifier (surface active agent) which are combined together in a colloid mill to form a homogeneous mixture of small asphalt droplets suspended in water (4). The quality of asphalt and water used to make the emulsion are very important. However, the most important component of any asphalt emulsion is the emulsifier (surfactant). To be an effective emulsifier for asphalt, the surfactant must be water soluble and must possess a proper balance between hydrophilic and hydrophobic properties. When used in combination with an acceptable asphalt, a good quality water and adequate mechanical mixing, the emulsifier is the major factor which influences initial emulsification, emulsion stability and ultimate field performance.

The only emulsions considered for sealing tailings are cationic asphalt emulsions. The positively charged surface of the asphalt droplets adhere to the negatively charged tailings. The surface charge (zeta potential) of cationic asphalt emulsions ranges from +12 to +130 mV. The choice of cationic emulsion depends on the surface area (particle size distribution) present in the material to be sealed. Different particle size distributions may require different choices of cationic emulsion in order to obtain the proper bonding, set time and penetration.

LABORATORY STUDIES

The overall objective of the laboratory studies is to investigate various asphalt emulsion sealants to contain radon and radium in the uranium tailings. Characterizing uranium tailings, formulating the seal and measuring radon diffusion are the primary activities. Uranium tailings are analyzed for physical and chemical characteristics that could have influence on formation of a radon-tight seal including particle size, chemical composition, pH and moisture content. Particle size (surface area) of the tailings varies widely and can have a significant effect on seal formation. For example, the high surface area material (slimes) contain considerable clay-like material which is very difficult to coat with the emulsion. Also, some tailings contain a very narrow range of particle size which makes compaction a potential problem. Moisture content and pH have an effect on the bonding mechanism. For example, very dry soil can dehydrate the emulsion prematurely.

Laboratory Radon Measurements

Radon diffusion measurements are performed to determine the effectiveness of various asphalt emulsion in producing a radon seal. The primary system used to test the seals is shown in Fig. 1. Radon gas is passed below the seal in addition to the radon generated by the tailings. This allows for a greater concentration of radon below the seal. Radon diffusion measurements are usually run for up to two weeks unless a major leak occurs. Any radon that diffuses through the seal is transported by the continuous nitrogen flow and is adsorbed on activated carbon in a dry ice-alcohol bath to improve its adsorption efficiency to greater than 99%. After the specified test time, the carbon is removed, sealed and counted. The radon content of the carbon is indirectly determined by counting the gamma activity of the Bi-214 daughter product. From this data the radon fluxes are determined.

Table 1. Radon Flux Reduction Using Poured-On Asphalt Emulsion Seals

Sample	Seal Thickness, mm	% Flux Reduction
Vitro-Site at Salt Lake City, UT	9	99.9%
Ambrosia Lake, NM	9	99.9%
Mexican Hat, UT	6	99.3%
Monument Valley, AR	9	99.9%
Shiprock, NM	9	99.9%
Tuba City, AR	9	99.9%
Falls City, TX	9	99.9%
Grand Junction, CO - 1	9	99.9% Pressurized Seal
Grand Junction, CO - 3	9	99.9% Pressurized Seal
Grand Junction, CO - 4	9	99.9% Pressurized Seal

Laboratory Results. Examples of radon flux reductions obtained by using poured-on seals are tabulated in Table 1. All flux reductions were greater than 99.3%. However, poured-on seals are not adequate for the field, since they do not have enough mechanical stability. Therefore, admixtures of Grand Junction tailings and asphalt emulsion were tested using the pressurized radon diffusion tests. Admixes containing from 10 to 18 wt % residual asphalt were compacted at about 5.6 kg/cm^2 (80 psi) and tested. Admixes containing 10 to 12 wt % residual asphalt stopped neither water vapor or radon, a 12 to 14 wt % residual asphalt admix only sealed out water vapor. Starting at 14 wt % residual asphalt, a marked reduction in radon flux is noted. At an 18 to 20 wt % residual asphalt content, greater than 99.9% reductions in radon fluxes were obtained. Based on these results, it was concluded that an 18 to 20 wt % admix seal would be used for the field test at Grand Junction.

FIELD STUDIES

The overall objective of the field studies is to demonstrate the effectiveness of a stabilization or sealing procedure using asphalt emulsion to contain radon. Techniques for applying the asphalt emulsion seal are being investigated at selected tailings sites. The objectives of the field tests are to obtain sufficient operating data to evaluate the technical and economic feasibility of the most promising application techniques. During June 1979, a field test was conducted at the Grand Junction tailings site. The primary emphasis of this test was to test a stabilization procedure consisting of site preparation (contouring, watering, compacting), asphalt emulsion application using a hydrostatic stabilizer and compactor to form the seal and overburden application.

Fig. 1. Operation of Radon Diffusion Test Apparatus

A 82.5 m x 82.5 m (275 ft X 275 ft) site was prepared for the sealing test by removing a 5 cm to 15.2 cm (2 to 6 in.) cover of overburden that contained too much clay to make an adequate seal, then contouring the site for drainage, watering and compacting. Following some preliminary seal application equipment tests, a 4328 m^2 (1.1 acre) area was sealed using a BOMAG MPH100 Hydrostatic Stabilizer as shown in Fig. 2. A few hours after the emulsion was applied, the tailings-emulsion mixture was compacted using a vibratory compactor to form an admix seal.

Several application rates were tried including a double pass application where about 8 wt % residual asphalt was applied on the first pass followed by an additional 8 wt % residual asphalt on the second pass. The BOMAG traveled between 9.1 and 12.2 m/min (30 and 40 ft/min) depending on the desired emulsion application rate. For example, an application rate of about 36.2 l/m^2 (8 gal/yd^2) was used in the small test plot with the BOMAG traveling at about 11.6 m/min (38 ft/min). After application and including time for the majority of the water to evaporate, the admix was compacted 4 to 6 times using a vibratory compactor to form the admix seal. The next step was to apply a 3.2 l/m^2 (0.7 gal/yd^2) fog seal (Fig. 3.) to fill any micro cracks that might have formed during compaction. Some of the problems encountered with the application procedure are:

1) Depth Control - Due to insufficiently compacted tailings, it was difficult to control admix depth; thus, the residual asphalt content varied from about 9 to 15 wt %. We were attempting to obtain about an 18 wt % residual asphalt in order to obtain a suitable seal. This problem can probably be overcome by some equipment modifications and improved compaction.

2) Lack of Water - Additional water application to the tailings would have made it easier to compact and easier for equipment to travel on. Very little water was applied over the admix seal prior to fog sealing because the water truck pump failed. This resulted in poor bonding between the admix and the fog seal. In future tests, additional water will be available.

Equipment modifications to both the hydrostatic stabilizer and distributor truck will be investigated during 1980 along with alternative application techniques such as a pug mill and paver.

Fig. 2. BOMAG MPH100 Hydrostatic Stabilizer and Distributor Truck Applying Asphalt Emulsion to Tailings

Field Radon Flux Measurements

The effectiveness of the asphalt emulsion seal was determined by measuring radon exhalation from the test area before and after the seal was applied. Nitrogen is passed over the tailings or seal picking up the radon and carrying it to an activated carbon canister kept in a dry ice/alcohol bath. After a specified time, the activated carbon is removed, allowed to reach equilibrium and counted. Radon fluxes are determined from this data.

Radon flux from the Grand Junction tailings pile test site prior to sealing ranged from 12 to 2400 pCi/m^2 s. The average flux was 270 pCi/m^2 s, while the mean was 73 pCi/m^2 s. The large range of fluxes is due to a mixture of slimes and sand (slimes generally have a higher concentration of radium). Radon fluxes after the seal was applied were reduced from 4.5 to 99.9% with average flux reduction of 76%. Considering that we did not obtain the residual asphalt required for a total seal (18 to 20 wt %) the test was a success. In most cases, areas covered with fog seal were sealed. Areas where the seal failed indicated an average 104% increase in flux. This is caused by the laterial diffusion of radon to the crack or breach in the seal; however, this increase in flux would not have a significant effect on overall radon concentration above the seal.

Fig. 3. Asphalt Emulsion Fog Seal Application over Admix Seal

CONCLUSIONS

The general conclusions that can be made as a result of both laboratory and field tests are as follows:

1) Cationic asphalt emulsion can be used effectively to stop radon exhalation from uranium tailings by either pouring/spraying-on or admixing with the tailings. Proper selection of the emulsion depends on the physical-chemical properties of the tailings or soil to be sealed.

2) Both laboratory and field tests indicate the potential for a flux reduction of greater than 99%. Field test radon reduction at the Grand Junction tailing test site averaged 76%.

3) An admix containing more than 16 wt % (about 18 wt %) residual asphalt is required to achieve a total seal.

4) Laboratory cyclic freeze/thaw tests (0 to 22°F) did not affect seal integrity in poured-on seals 3.0 to 7.0 mm thick.

5) Exposing laboratory seals to 10^5-R gamma radiation did not have any effect on seal integrity.

6) Long term stability of the seal is affected by the nature of the environment surrounding the seal. For example, if the seal were in an anaerobic environment, oxidation would be no problem and the seal should last for thousands of years.

7) Maintaining overburden over the seal provides erosion control. Revegetation or a rip-rap (rock) cover could be used to prevent soil erosion.

REFERENCES

1. U. S. Nuclear Regulatory Commission, Draft Generic Environmental Impact Statement on Uranium Milling, NUREG-0511, Vol. 1 (1979).
2. P. L. Koehmstedt, J. N. Hartley and D. K. Davis, Use of Asphalt Emulsion Sealants to Contain Radon and Radium in Uranium Tailings, BNWL-2190, Pacific Northwest Laboratory, Richland, WA (1977).
3. J. N. Hartley, P. L. Koehmstedt and D. J. Esterl, Asphalt Emulsion Sealing of Uranium Mill Tailings, (In progress) Pacific Northwest Laboratory, Richland, WA (1979).
4. J. N. Dybalski, The Chemistry of Asphalt Emulsion, presented at 55th Annual Meeting Transportation Research Board, (1976).

WASTE ISOLATION PERFORMANCE ASSESSMENT--A STATUS REPORT

H. C. Burkholder

Office of Nuclear Waste Isolation
Battelle Memorial Institute

INTRODUCTION

The assessment of system performance following emplacement of the waste is a key activity in the development of nuclear waste isolation systems. The assessment results provide a technical basis for decisions concerning site selection, repository design, package design, and system acceptability, and they bear directly on the central issue of nuclear waste isolation, "Do the proposed isolation systems sufficiently protect humans from the waste?" The objective of this paper is to describe the performance assessment problem, outline general approaches to its solution, and describe the status of the development, verification, and application of performance assessment technology. The objective of nuclear waste isolation performance assessment is to evaluate the effectiveness of nuclear waste isolation systems in preventing adverse radiological effects to present and future humans and their environments.

ISOLATION SYSTEM, DRIVING FORCES, AND PHENOMENA

A general nuclear waste isolation system is described pictorially in Fig. 1. As shown, the general system is divided into many components (waste form, stabilizer, etc.), not all of which are necessarily present in every isolation system. The components of the isolation system can be combined into groups in a way convenient for analysis:

1) waste package (waste form, stabilizer, container, overpack, and migration retardant)
2) repository (backfill, host rock, interstitial fluids, and shaft seal)

3) geosphere (surrounding geologic media, groundwater, and natural resources)
4) biosphere (land, surface water and atmosphere).

Although the system described in Fig. 1 is oriented toward a mined repository, other terrestrial disposal concepts, such as deephole or seabed, can be accommodated with only slight modification. The general isolation system shown in Fig. 1. contains many barriers and resistances to the release and transport of radionuclides from the waste to man. The barriers (i.e., stabilizer, canister, overpack, emplacement geology, and shaft seal) act to prevent the initiation of any radionuclide release from the waste form, and the resistances (i.e. waste form, migration retardant, backfill and surrounding geologies (act to control the rate of release or to retard the transport of the radionuclides after the release has begun. There are a number of possible concepts and materials for each of the barriers and resistances. Thus, isolation system developers have considerable flexibility in providing systems which sufficiently protect humans from the waste.

The general driving forces for change in the isolation system can be divided into four classes (cosmic, terrigenic, human, and waste and repository) depending on their origin. Each of these classes can be further divided into the several subclasses depicted in Fig. 1.

Individual phenomena which could result from the effect of these driving forces on the isolation system are listed in Table 1. Because the phenomenon resulting from the cosmic and terrigenic driving forces are strongly interactive, they are grouped together under the term natural phenomena. These groups of phenomena contain both events and processes. The events occur relatively instantaneously on the geologic time scale, and thus in a given time period they either occur or do not occur. The processes, on the other hand, occur continuously with time at some rate once they have been initiated.

IMPORTANT CHARACTERISTICS OF THE ASSESSMENT PROBLEM

Four characteristics of nuclear waste isolation performance assessment are particularly important. Two are apparent from examining Fig. 1. The first characteristic is complexity. The system has many interrelated components and driving forces and the assessment technology must analyze over positional scales from microns (e.g., canister corrosion phenomena) to hundreds of thousands of kilometers (e.g., tectonic phenomena) and time scales from microseconds (e.g., radiation damage phenomena) to hundreds of thousands of years (e.g., geosphere transport phenomena). The second characteristic is uncertainty. Because of limited understanding of some of

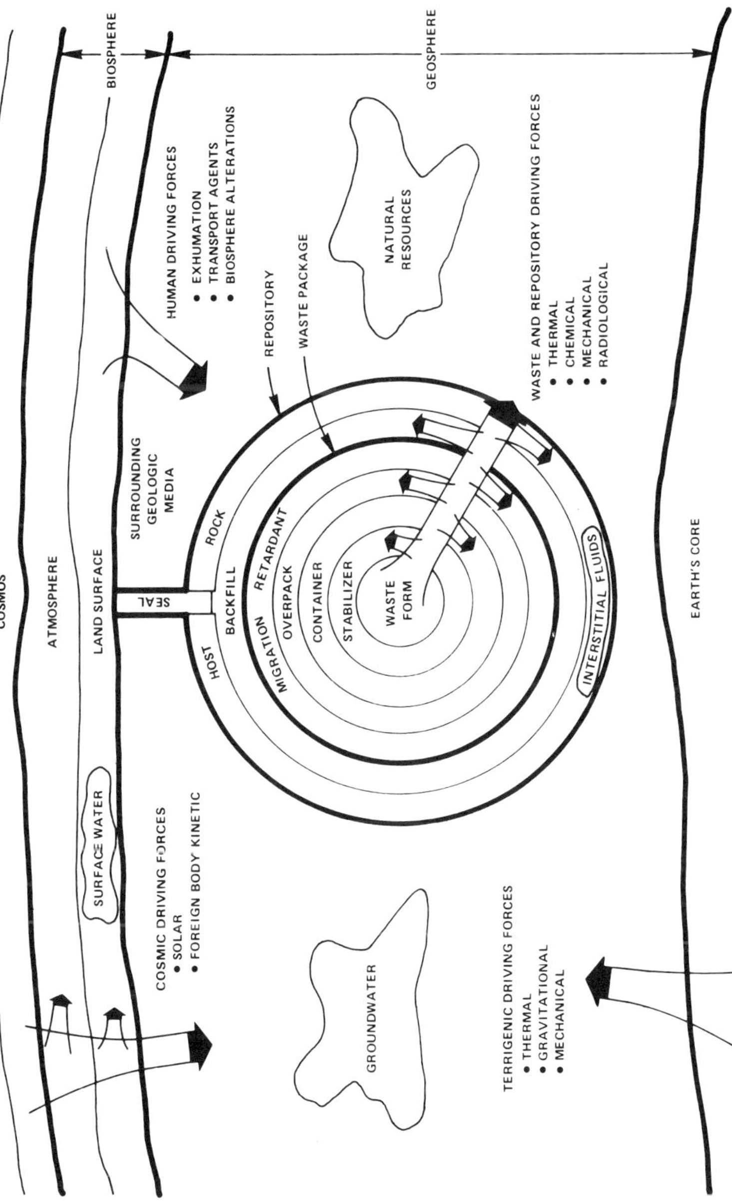

Fig. 1. Waste Isolation System and Driving Forces

Table 1. Potentially Important Phenomena

Natural	Human-Caused	Waste and Repository-Caused
Climatic Fluctuations	Improper Design/Operation	Thermal Effects
Glaciation	Shaft Seal Failure	Differential Elastic Response
	Improper Waste Emplacement	Non-Elastic Response
Denudation and Stream Erosion		Fluid Pressure Changes
	Undetected Past Intrusion	Fluid Migration
Magmatic Activity		
	Undiscovered Boreholes	Chemical Effects
Extrusive	Mine Shafts	
Intrusive		Geochemical Alterations
	Inadvertent Future Intrusion	Corrosion
Epeirogenic Displacement		Waste Package - Geology Interactions
	Archeological Exhumation	
Igneous Emplacement	Weapons Testing	Biochemical Gas Generation
Isostacy	Non-Nuclear Waste Disposal	
	Resource Mining (Mineral, Hydrocarbon, Geothermal, Salt)	Mechanical Effects
Orogenic Diastrophism		
	Storage of Hydrocarbons or Compressed Air	Local Fracturing
Near-Field Faulting		Canister Movement
Far-Field Faulting		Dewatering
Diapirism		
Diagenesis	Intentional Intrusion	Radiological Effects
Static Fracturing	War	Material Property Changes
	Sabotage	Radiolysis
Surficial Fissuring	Waste Recovery	Criticality
Impact Fracturing		Decay Product Gas Generation
Hydraulic Fracturing	Transport Agent Introduction	
Meteorites	Irrigation	
	Reservoirs	
Dissolutioning	Intentional Artificial Groundwater Recharge or Withdrawal	
Sedimentation	Chemical Liquid Waste Disposal	
Flooding		
	Biosphere Alteration	
Undetected Features		
	Establishment of New Population Center	
Faults, Shear Zones	Climate Modification	
Breecia Pipes		
Lava Tubes		
Gas or Brine Pockets		

the potentially relevant driving forces and resulting phenomena, limited accuracy of data, and limited ability to characterize the system without destroying its function, the assessment technology must estimate the uncertainty of the predicted results. The other two characteristics are apparent from the anticipated assessment applications. The assessment results will support decisions which will be scrutinized in technical, legal and political arenas. Therefore the auditability and comprehensiveness of the assessment logic are particularly important if the assessment results are to positively affect the decision process.

GENERAL PERFORMANCE ASSESSMENT APPROACHES

Waste isolation performance assessments must analyze either the driving forces themselves or the phenomena that result from their action to predict isolation system behavior. The work matrix for waste isolation performance assessment is shown in Fig. 2. The components from Fig. 1. are listed horizontally and the driving forces and phenomena are listed vertically. Those components where the effect of particular driving forces and phenomena may be important are denoted by an "X". As shown, the work matrix can be divided into two parts, the near-field and the far-field. Far-field performance assessment primarily involves modeling the effects of the external driving forces and phenomena on the geosphere and the biosphere. Near-field performance assessment primarily involves modeling the effects of the internal driving forces and phenomena on the waste package and the repository. Also as shown in Fig. 2., far-field performances assessment involves modeling the external driving forces and phenomena in portions of the repository, and near-field performance assessment involves modeling the internal driving forces and phenomena in portions of the geosphere.

Waste isolation performance assessment can be approached from either the traditional risk analysis or the classical systems analysis points of view. Traditional risk analysis describes system behavior as a series of events which lead to releases of radionuclides (release scenarios). The consequences of these scenarios are then analyzed and doses to humans are estimated. Probabilities can be determined for the release scenarios and these probabilities can be multiplied by the resulting dose consequences to provide risk values. Classical systems analysis describes system behavior as a combination of processes that lead to human doses, and the concepts of scenarios, consequences, probabilities and risk are not used. Because waste isolation performance assessment involves analysis of both events and processes, the traditional risk analysis approach must be modified to include processes, and the classical systems analysis approach must be modified to include events. Fig. 3. and 4. show conceptual organizations for waste isolation performance assessments based on these two general approaches.

		DRIVING FORCES/PHENOMENA \ PHASES	BIOSPHERE: ATMOSPHERE	LAND SURFACE	SURFACE WATER	GROUNDWATER	NAT. RESOURCES	GEOSPHERE: SURROUNDING GEOL.	SHAFT SEAL	REPOSITORY: HOST ROCK	INTER. FLUIDS	BACKFILL	MIGRATION RETARD.	WASTE PACKAGE: OVERPACK	CONTAINER	STABILIZER	WASTE FORM
EXTERNAL	COSMIC	SOLAR	X	X	X	X	X	X	X	X							
		FOREIGN BODY KINETIC	X	X	X	X	X	X	X	X							
	TERRIGENIC	THERMAL				X	X	X	X	X							
		GRAVITATIONAL				X	X	X	X	X							
		MECHANICAL				X	X	X	X	X							
	HUMAN	EXHUMATION				X	X	X	X	X							
		TRANSPORT AGENTS			X	X	X	X	X	X							
		BIOSPHERE ALTERATIONS	X	X	X	X	X	X	X	X							
INTERNAL	WASTE/REPOS.	THERMAL				X	X	X	X	X	X	X	X	X	X	X	X
		MECHANICAL						X	X	X	X	X	X	X	X	X	X
		CHEMICAL							X	X	X	X	X	X	X	X	X
		RADIOLOGICAL								X	X	X	X	X	X	X	X

FAR-FIELD NEAR-FIELD

Fig. 2. Waste Isolation Performance Assessment Work Matrix

In the modified risk analyses approach (Fig. 3.), the assessment activities for each group of components are divided into scenario analysis and consequence analysis. Scenario analysis involves investigation of how the containment barriers are breached by the combined action of the driving forces and resulting phenomena (events and processes). Consequence analysis involves investigation of how the release and tranport resistances are overcome by the driving forces and resulting phenomena once the barriers have failed. As suggested by the "state change" arrows, the scenario analysis activities are highly interactive. Because system components sometimes act as both barriers and resistances, the scenario analysis must continue to provide information about the barrier states to the consequence analysis after release from the waste form has occurred.

In the modified systems analysis approach (Fig. 4.), the assessment activities for each group of components are divided into process

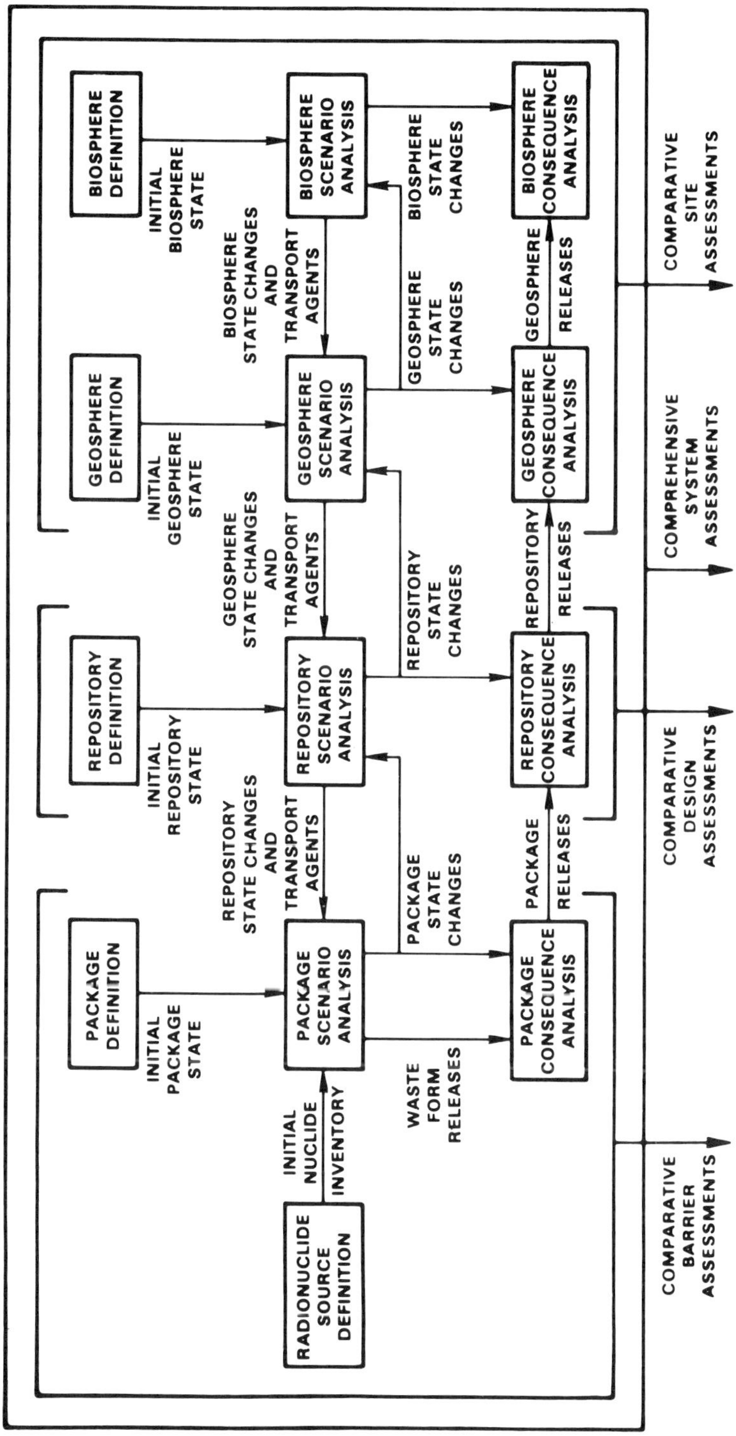

Fig. 3. Waste Isolation Performance Assessment - Modified Risk Analysis Approach

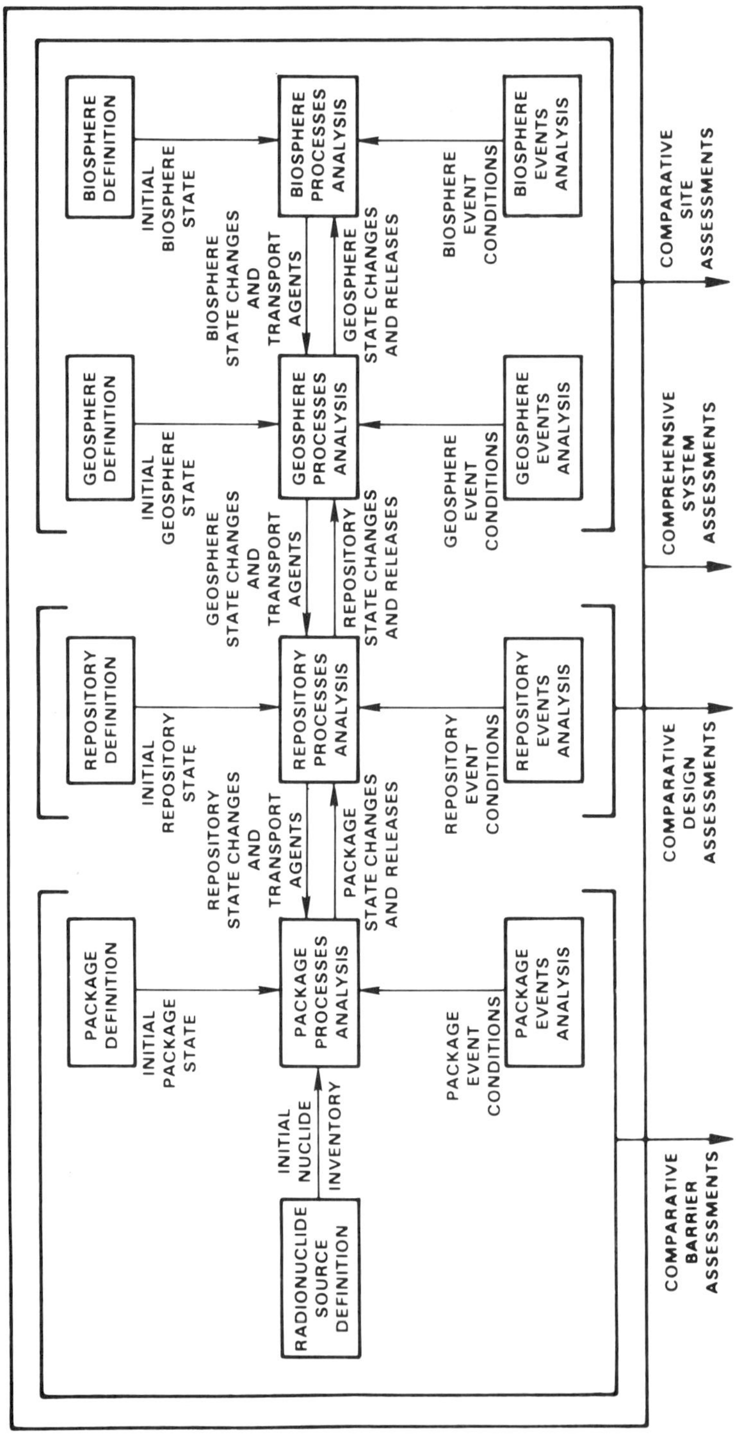

Fig. 4. Waste Isolation Performance Assessment - Modified Systems Analysis Approach

analysis and event analysis. Process analysis involves investigation of how the containment barriers are breached and the transport resistances are overcome by the combined action of the driving forces and resulting processes. Event analysis involves investigation of how the barriers are breached and the transport resistances are overcome by the combined action of the driving forces and resulting events. The results of the stochastic event analyses are superposed on the deterministic process analyses to predict overall system performance. As suggested by the "state change" arrows, the process analyses are highly interactive.

In principle both the modified risk analysis and modified systems analysis approaches can be successful in predicting waste isolation system performance because they both provide a logic which includes all of the elements necessary for a comprehensive assessment. However, the risk analysis approach seems more appropriate for far-field analysis because of the large number of phenomena that are currently modeled as events, and the systems analysis approach seems more appropriate for near-field analysis because of the large number of phenomena that are modeled as processes. Which of these approaches is more useful and efficient for the entire problem and particular applications remains to be determined, but an evolution toward the modified systems analysis approach can be expected as the performance assessment technology matures.

TECHNOLOGY DEVELOPMENT AND VERIFICATION

The development of performance assessment modeling technology occurs in three stages:

1) individual model development
2) coupled model development
3) individual and coupled model integration.

At all of these stages the models are validated/verified to the extent possible.

The procedure for developing and validating/verifying performance assessment models follows the scientific method. First, available information about the isolation system phenomena is collected and organized. This information comes from observation of natural behavior, laboratory experimentation, and field testing. Based on this understanding, hypothetical models are next developed. These models can be empirical (i.e., based on direct use of the observations) or theoretical (i.e., based on interpretation of the observations according to known physical laws). Finally the hypothetical model predictions are compared against additional observations of the isolation system phenomena under conditions different than those used for development of the models. If the comparison is unfavorable, the hypothetical models are modified and tested again. The iterative process of model modification and verification testing continues until a satisfactory agreement between prediction and observation is obtained. Peer review of this process leads to model acceptance.

The details of the development and validation/verification process for models is dependent on the type of phenomena being modeled. In general, the far-field phenomena occur over large position and time scales and are relatively uncertain. The far-field scenario analysis models can be developed only by extrapolation of past behavior into the future and are inherently impossible to completely verify by direct observation in the present. Hence general agreement by review of the models by knowledgeable scientists and engineers must suffice. The far-field consequence analysis and near-field models can be developed by natural observation and laboratory experiments and verified by short term experiments or field testing. The individual and coupled models must be integrated into a modeling system. The waste isolation performance assessment modeling system cannot be verified because isolation system behavior cannot be observed or tested over the time period of interest. However, data collected during the period of repository operation (approximately 40 years) can provide some verification of the near-field portion of the analysis system.

TECHNOLOGY APPLICATIONS

There are four general applications for isolation system performance assessment technology: 1) the evaluation and selection of disposal sites, 2) the evaluation and selection of repository designs, 3) the evaluation and selection of waste package designs, and 4) the qualification and licensing of the disposal system for waste emplacement and decommissioning. Each of these applications supports an important decision or decisions which must be made in the development of a nuclear waste isolation system. For each of these applications the assessment technology can be used (at least in principle) in either of two modes, specification or comparison. In the specification mode performance assessments work backward from known acceptability standards to determine sets of performance specifications on system parameters (e.g., thickness of emplacement geology, emplacement spacing, or canister corrosion rate). In the comparison mode performance assessments work forward from the parameters of existing systems to determine and compare their performance. Thus, in the specification mode performance assessments are used to determine limiting characteristics for the present state of the isolation system so that future performance will be acceptable, and in the comparison mode performance assessments are used to predict future performance from the characteristics of present systems so that these systems can be compared against one another or an acceptability standard.

Applications in the specification mode are more difficult than those in the comparison mode because of the inverted time sequence of the analysis and the lack of solution uniqueness (multiple sets of specifications will meet any acceptability standard). In addition, applications in the specification mode occur earlier in the isolation system development process than do applications in the comparison mode. As a result, the information required for a quantitative analysis of the system may not yet exist, and meaningful specification calculations may not be possible. Thus, the value of using waste isolation performance assessement technology in the specification mode is questionable. Application in the comparison mode are discussed in more detail below.

The choice between candidate sites requires a comparative evaluation of the expected performance of the emplacement geology, surrounding geologic media, and the biosphere, in preventing and retarding the release and transport of the radionuclides in the waste for a reference repository design and waste package design. As shown in Fig. 3. and 4., such an evaluation uses primarily the geosphere and biosphere portions of the overall assessment technology together with data from the site exploration activities and conditions from a reference analysis of package and repository performance. The choice between candidate repository designs requires a comparative evaluation of the expected performance of the repository for a reference waste package and site. This evaluation uses primarily the repository part of the assessment technology together with data from design testing activities and conditions from a reference analysis of site and package performance. The choice between candidate waste packages requires a comparative evaluation of the expected performance of the package for a reference site and repository design. This evaluation uses primarily the waste package part of the assessment technology together with data from the package testing activities and conditions from a reference analysis of repository and site performance. The demonstration that projected isolation system performance does not cause radiation doses greater than acceptable limits is necessary before the isolation system can be licensed. This requires use of the total assessment technology and the total data base.

TECHNOLOGY DEVELOPMENT AND VERIFICATION STATUS

At present there is significant variation in the development and verification status of the parts of waste isolation performance assessment technology (1-6). Consistent with the previous discussion, the far-field technology status is described in the context of the modified risk analysis approach, and the near-field technology status is described in the context of the modified systems analysis approach. In far-field (geosphere and biosphere) scenario analysis, preliminary models have been developed for a number of natural phenomena (e.g., climatic change). Some of these models are deterministic, some are probabilistic, and some are both. There have been preliminary integrations of some of the natural phenomena models together into far-field scenario modeling systems, and these systems have received limited peer review by the geoscience community. Only a relatively small amount of modeling work has been reported for the human-caused phenomena. Because humans have a large capacity to affect the isolation system, an increased emphasis on human-caused phenomena modeling is needed.

The far-field consequence analysis technology is probably the best developed of the performance assessment technologies. Hydrologic models have been completely developed and verified for porous media, but only partly developed and verified for fractured media. However, the porous media models may turn out to be sufficiently accurate for fractured media, if the application is done properly. The radionuclide contaminant transport models are generally less mature than the hydrology models, and improved mechanistic modeling of sorption phenomena may be needed. Short term field verification of the contaminant transport models have been performed for weakly sorbed fission products, but those done for the strongly sorbed actinides have been somewhat limited. Long term verifications for all nuclides are impossible, but some confidence may be gained from studies of transport away from mineral deposits and natural reactors. The biosphere transport-dose models are generally developed and verified sufficiently for waste isolation performance assessment application. An exception may involve the modeling of the long term accumulation and dispersion processes associated with the chronic release of radionuclides into the biosphere over long time periods. Some work has been completed in this area, but more may be needed. Far-field consequence analysis modeling seems to be receiving the proper degree of emphasis at the present time.

In contrast to the far-field models which are mostly "phenomenological", many of the near-field models are "fundamental" and predict system performance according to the known conservation and constitutive equations of physical and chemical science. Thus, there has been a historical difference in approach to modeling the far-field and the near-field, and this may make integration of the two technologies difficult. In order to have an integrated technology, either the near-field technology must be simplified to the "phenomenological" level, the far-field technology must be enhanced to the "fundamental" level, or the outputs of both must be converted into forms which can be input to each other. This latter alternative may be the only practical approach. Some models (e.g., thermomechanical) have been developed for the waste and repository driving forces; others have been developed for other assessment problems (e.g., in situ leaching of ores and geothermal and petroleum reservoir engineering); and some models that may be necessary for near-field analysis do not exist (e.g., waste-canister-geologic media reactions). The degree of verification of near-field models varies. Some models (e.g., thermal) have been verified for both porous and fractured media, while others (e.g., thermomechanical) have been verified for porous but not fractured media. Some work has been done to develop coupled near-field analysis models, but no effort has been made to develop an integrated near-field modeling system. An increased effort is needed in the near-field analysis area.

The foregoing suggests that: 1) the development and verification of individual models is highly advanced, 2) the development and verification (to the extent possible) of coupled models is moderately advanced, and 3) the integration of the modeling technology needs increased emphasis. The need for additional work is generally greater in the near-field than in the far-field. Although not within the scope of this paper, the need for performance assessment data is generally greater than that for performance assessment models themselves. However, the most important technical need concerning waste isolation performance assessement is a comprehensive and quantitative analysis to set research and development priorities in the model development, model verification, and data gathering areas. All three of these activities can go on forever and a means must be developed to focus available resources on critical assessment areas and to determine when research in non-critical areas can be discontinued. For example, one can question: 1) the extent to which fundamental models for the various sorption and leaching phenomena are really needed, 2) the degree of themomechanical model verification really necessary, and 3) whether the permeability of an almost impermeable rock needs to be known to more than conservative limits. At present these kinds of questions can only be answered in a qualititative way, and decisions relative to them are open to criticism for arbitrariness.

The need to rationally set research priorities can be satisfied by performing sensitivity and uncertainty analyses on the performance modeling system to determine: 1) the important assessment parameters, 2) the uncertainty levels present in our current knowledge about those parameters, and 3) the residual uncertainty levels expected to be present after all model development, model verification, and data gathering activities are completed. These sensitivity and uncertainty analyses need to be performed on the integrated performance modeling system as soon as possible. In this way the unimportant, the certain, and the impossible can be differentiated from the necessary when allocating waste isolation technology development and verification resources.

TECHNOLOGY APPLICATION STATUS

There have been a number of reported applications of performance assessment technology to evaluate both generic and specific isolation systems (and others have been completed but yet not reported). (1,5, 7-15). These analyses have emphasized the far-field portion of the assessment technology, and they collectively suggest that for "reasonable" isolation systems individual doses from possible but unlikely release scenarios are on the order of natural background or less. However, none of these applications have been for a system with a well developed data base. Hence, the ranges of uncertainty about the calculated doses are relatively large. Thus, although the results to date suggest that adequate isolation systems can be put into operation, the location and detailed development of specific systems remains to be completed.

REFERENCES

1. S. E. Logan and M. C. Berbano, EPA 520/6-78-005, U.S. Environmental Protection Agency, Washington, D.C. (1978).
2. J. E. Campbell, SAND 78-0029, Sandia Laboratories, Albuquerque, NM (1978).
3. R. A. Heckman and T. R. Donich, UCRL-52547, Lawrence Livermore Laboratory, Livermore, CA (1978).
4. H. C. Burkholder, PNL-2642, Pacific Northwest Laboratory, Richland, WA (1979).
5. A. D. Little, Inc., EPA-520/4-79-007, U.S. Environmental Protection Agency, Washington, D.C. (1979).
6. A. Brandstetter, PNL-2874, Pacific Northwest Laboratory, Richland, WA (1979).
7. H. C. Claiborne and F. Gera, ORNL-TM-4639, Oak Ridge National Laboratory, Oak Ridge, TN (1974).
8. H. C. Burkholder, "Incentives for Partitioning High Level Waste", <u>Nuclear Technology</u> <u>31</u>, 202 (1976).
9. F. Girardi, G. Bertozzi, and M. D'Alessandro, EUR 5902.e, Commission of the European Communities, Ispra, Italy (1977).
10. G. de Marsily, "Nuclear Waste Disposal: Can the Geologist Guarantee Isolation?" <u>Science</u> <u>197</u>, 519 (1977).
11. M. D. Hill and P. H. Grimwood, NRPB-R69, National Radiological Protection Board, Harwell, U.K. (1978).
12. Nuclear Fuel Safety Group, <u>Handling of Spent Nuclear Fuel and Final Storage of Vitrified High Level Reprocessing Waste</u>, Karn-Bransle-Sakerhet, Stockholm, Sweden (1978).
13. L. E. Berman, UCRL-13917, The Analytic Sciences Corporation, Reading, MA (1978).
14. <u>Waste Isolation Pilot Plant-Draft Environmental Impact Statement</u>, DOE/EIS-0026-D, U.S. Department of Energy, Washington, D.C. (1979).
15. <u>Management of Commercially Generated Radioactive Waste</u>, DOE/EIS-0046-D, U. S. Department of Energy, Washington, D. C. (1979).

ION MIGRATION IN OCEAN SEDIMENTS: SUBSEAFLOOR RADIOACTIVE WASTE DISPOSAL*

H. E. Nuttall, A. K. Ray and E. J. Davis

Department of Chemical and Nuclear Engineering
University of New Mexico
Albuquerque, NM 87131

ABSTRACT

In this study of seabed disposal, analytical ion transport models were developed and used to elucidate ion migration through ocean sediments and to study the escape of ions from the ocean floor into the water column. An unsteady state isothermal diffusion model was developed for the region far from the canister to examine the effects of ion diffusion, adsorption, radioactive decay, sediment thickness and canister position. Analytical solutions were derived to represent the transient concentration profiles within the sediment, ion flux and the ion discharge rate to the water column for two types of initial conditions: instantaneous dissolution of the canister and constant canister leakage. Generalized graphs showing ion migration and behavior are presented.

INTRODUCTION

The geologic disposal of nuclear waste in subseafloor sediments is one of the currently active waste management methods being studied by the USDOE (1). The current scheme, as illustrated in Fig. 1., requires the placement of high level waste canisters or spent fuel pins in the thick clay sediments located in selected regions of the oceans. Seabed nuclear waste disposal may be viewed as an isolation technique in which a set of sequential barriers are used to isolate radioactive nuclides until the nuclides have decayed to harmless

*This work supported by the United States Department of Energy under Contract DE-ACD4-76-DP00789.

elements or are released slowly in safe concentrations to the environvironment (2). Although several barriers are present, such as the solidified waste, the canister shell and the water column, the primary barrier is the geological formation or the material surrounding the canister in the case of canister over-packing. The near field and far field problem of simultaneous heat and ion transport was studied by Nuttall (3) using the large numerical code. In addition, simplified analytical models were developed to study the far field ion diffusion problem. A more complete analytical treatment of the far field problem was made by Nuttall and Talbert (1). Detailed analytical studies of near field heat transfer problems were performed by Schimmel and Hickox (4). The current study presented in this paper treats the far field problem in a general analytical form. Generalized dimensionless solutions are presented for the dynamic concentration profiles, ion flux and ion discharge rate to the water column.

THEORETICAL ANALYSIS

Consider the system shown in Fig. 1, which depicts a canister imbedded in a porous sedimentary layer bounded at $z = 0$ by an impermeable medium and bounded at $z = \ell$ by the ocean floor. The center of the canister is located at the arbitrary plane $z = \ell_1$ of the sedimentary layer, and we shall measure the radial position from the centerline of the canister. If the transport of the ionic species is governed by diffusion through the porous medium the concentration of the radioactive species is described by the unsteady state diffusion equation,

$$\frac{\partial c}{\partial t} = D_{eff}\left(\frac{1}{r}\frac{\partial}{\partial r}\left(r\frac{\partial c}{\partial r}\right) + \frac{\partial^2 c}{\partial z^2}\right) - \lambda c + S(r,z,t), \tag{1)}$$

where D_{eff} = (molecular diffusivity)/(retardation factor) = D/R^*, and λ is the radioactive decay constant of the decaying species. The source term, $S(r,z,t)$, depends on the rate of release of radioactive material, and we shall examine two possible rates of release below. The waste form, as represented by this source term, is assumed to be part of the geomedia. The initial and boundary conditions on Equation 1) are

$$c(r,z,0) = \frac{\partial c}{\partial r}(0,z,t) = \frac{\partial c}{\partial r}(\infty,z,t) = \frac{\partial c}{\partial z}(r,0,t) = 0, \tag{2)}$$

and $$c(r,\ell,t) = F(r,t), \tag{3)}$$

where $F(r,t)$ depends on the rate and mechanism of transport between the sediment and the ocean. To develop the general solution we shall treat this function as arbitrary.

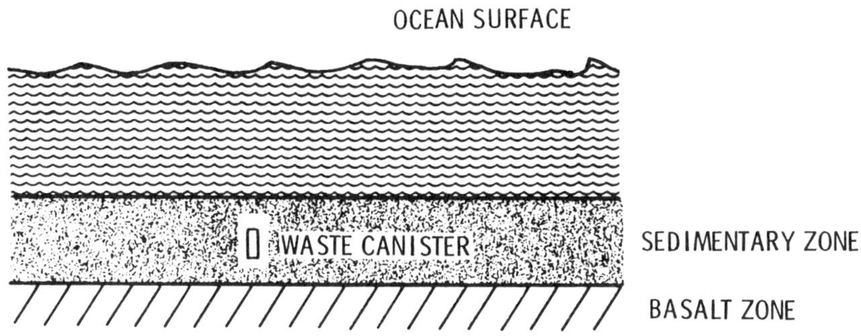

Fig. 1. Schematic Diagram of Seabed Disposal Concept

Now introduce the dimensionless quantities $\phi = c/c^*$, $R = r/\ell$, $\eta = z/\ell$, $\tau = tD_{eff}/\ell^2$ and $N_\lambda = \lambda\ell^2/D_{eff}$, where $c^* = 6M/\pi\ \ell^3$ is a reference concentration and let $\tilde{\phi} = \phi e^{N_\lambda \tau}$ to yield the following boundary value problem

$$\frac{\partial\tilde{\phi}}{\partial\tau} = \frac{\partial^2\tilde{\phi}}{\partial R^2} + \frac{1}{R}\frac{\partial\tilde{\phi}}{\partial R} + \frac{\partial^2\tilde{\phi}}{\partial\eta^2} + \frac{\tilde{S}(R,\eta,\tau,)\ell^2}{c^*\ D_{eff}} \tag{4}$$

with boundary conditions

$$\tilde{\phi}(R,\eta,0) = \frac{\partial\tilde{\phi}}{\partial R}(0,\eta,\tau) = \frac{\partial\phi}{\partial R}(\infty,\eta,\tau) = \frac{\partial\tilde{\phi}}{\partial\eta}(R,0,\tau) = 0 \tag{5}$$

$$\tilde{\phi}(R,1,\tau) = \tilde{F}(R,\tau), \tag{6}$$

where $\tilde{S}(R,\eta,\tau) = S(R,\eta,\tau)\ e^{N_\lambda\tau}$ and $\tilde{F}(R,\tau) = F(R,\tau)\ e^{N_\lambda\tau}$

Applying integral transform methods (the Fourier integral transform in η and the Hankel transform in R) we obtain the general solution

$$\tilde{\phi}(R,\eta,\tau) = \sqrt{2}\ e^{-N_\lambda\tau} \sum_{p=1}^{\infty} \cos\alpha_p\eta \int_{\beta=0}^{\infty} \beta J_0(\beta R)\ e^{-(\beta^2+\alpha_p^2)\tau}$$

$$\int_{\tau'=0}^{\tau} e^{(\beta^2+\alpha_p^2)\tau'}\ G(\beta,\alpha_p,\tau')d\tau' d\beta, \tag{7}$$

where $$G(\beta,\alpha_p,\tau) = \frac{\hat{S}(\beta,\alpha_p,\tau)\ell^2}{c^*D_{eff}} - \frac{dK(\alpha_p,1)}{d\eta}\hat{F}(\beta,\tau), \qquad 8)$$

and $\hat{S}(\beta,\alpha_p,\tau)$ and $\hat{F}(\beta,\tau)$ are the Hankel transforms of $\overline{S}(R,\alpha_p,\tau)$ and $\tilde{F}(R,\tau)$, and $\overline{S}(R,\alpha_p\tau)$ is the Fourier transform of $\tilde{S}(R,\eta,\tau)$ defined by

$$\overline{S}(R,\alpha_p,\tau) = \int_0^1 K(\alpha_p,\eta')\tilde{\phi}(R,\eta',\tau')d\eta', \qquad 9)$$

where $K(\alpha_p,\eta) = \sqrt{2}\cos\alpha_p\eta$, and $\alpha_p = (2p-1)\frac{\pi}{2}$, $p = 1,2,3...$ 10)

The Hankel transform is defined by

$$\hat{S}(\beta,\alpha_p,\tau) = \int_{R'=0}^{\infty} R'J_0(\beta R')\overline{S}(R',\alpha_p,\tau)dR', \qquad 11)$$

Equation 7) represents the general solution for any source term, $S(r,z,t)$, and any boundary condition $F(r,t)$ at the ocean floor. Since the canister can be expected to be small compared with the dimensions of the sedimentary layer we may write the source term as a point source,

$$S(r,z,t) = \frac{m(t)}{2\pi r}\,\delta(r-0)\,\delta(z-\ell_1), \qquad 12)$$

where $m(t)$ is the leakage rate, δ represents the Dirac delta function, and ℓ_1 is the vertical position of the canister. Furthermore, if the rate at which the radioactive species diffuses to the ocean floor is very small compared with the rate of transport in the sea water we may assume $F(r,t) = 0$. Now two special cases for the release rate are of particular interest.

Case I. Instantaneous Release

If the canister should break down suddenly, and all of the radioactive species are released, we may write

$$m(t) = M\delta(t-0), \qquad 13)$$

where M is the total mass of the radioactive species at the time of release, which is related to the initial mass, M_0, at the time of implantation of the canister by

$$M = M_0e^{-\lambda t_1} \qquad 14)$$

In this case the solution for $\phi(R,\eta,\tau)$ becomes

$$\phi(R,\eta,\tau) = \frac{M}{2\pi\ell^3 c^*\tau} e^{-(N_\lambda\tau+R^2/4\tau)} \sum_{p=1}^{\infty} \cos\alpha_p\eta \cos\alpha_p\eta_1 e^{-\alpha_p^2\tau} \quad 15)$$

Choosing $c^* = 6M/\pi\ell^3_{ref}$, where ℓ_{ref} is an arbitrary reference length, we have

$$\phi(R,\eta,\tau) = \left(\frac{\ell_{ref}}{\ell}\right)^3 \frac{e^{-(N_\lambda\tau+R^2/4\tau)}}{12\tau} \sum_{p=1}^{\infty} \cos\alpha_p\eta \cos\alpha_p\eta_1 \, e^{-\alpha_p^2\tau} \quad 16)$$

where $\eta_1 = \ell_1/\ell$. The flux of mass at the sediment/ocean interface is given by

$$J_s = -\frac{\partial\phi}{\partial\eta}\bigg|_{\eta=1} = \left(\frac{\ell_{ref}}{\ell}\right)^3 \frac{e^{-(N_\lambda\tau+R^2/4\tau)}}{12\tau} \sum_{p=1}^{\infty} \alpha_p \sin\alpha_p \cos\alpha_p\eta_1 \cdot e^{-\alpha_p^2\tau}, \quad 17)$$

and the total discharge rate of the radioactive species to the sea water is

$$Q = \frac{\pi}{3}\left(\frac{\ell_{ref}}{\ell}\right)^3 \sum_{p=1}^{\infty} \alpha_p \cos\alpha_p\eta_1 \sin\alpha_p \, e^{-N_\lambda\tau} \, e^{-\alpha_p^2\tau} \quad 18)$$

Case II. Constant Leakage Rate

For a constant rate of leakage we have

$$m(t) = R_0 e^{-\lambda\tau}\left[1-U_{t_1}(t)\right], \quad 19)$$

where $t_1 = M_0/R_0$ = the time required for total discharge,

and $U_{t_1}(t)$ = the unit step function

In this case the concentration distribution is given by

$$\phi(R,\eta,\tau) = \left(\frac{\ell_{ref}}{\ell}\right)^3 \frac{e^{-N_\lambda\tau}}{12\tau_1} \sum_{p=1}^{\infty} \cos\alpha_p\eta \cos\alpha_p\eta_1 \sum_{n=0}^{\infty} \frac{(-\alpha_p^2)^n}{n!} F_n(R,\tau) \quad 20)$$

$$\text{where } F_n(R,\tau) = \tau^n E_{n+1}\left(\frac{R^2}{4\tau}\right) - U_{\tau 1}(\tau) \cdot (\tau-\tau_1)^n E_{n+1}\left[\frac{R^2}{4(\tau-\tau_1)}\right] \qquad 21)$$

$$\text{and } E_{n+1}(z) = \frac{(-z)^n}{n!} \ln z + \psi(n+1) \sum_{\substack{m=0 \\ m\neq n}}^{\infty} \frac{(-z)^m}{(m-n)m!}, \qquad 22)$$

and $\psi(1) = -\gamma, \psi(n+1) = -\gamma + \sum_{m=1}^{n} m^{-1}$, for $n>1$, where $\gamma = 0.5772...$

The flux, J_s is given by

$$J_s = \left(\frac{\ell_{ref}}{\ell}\right)^3 \frac{e^{-N_\lambda \tau}}{12\tau_1} \sum_{p=1}^{\infty} \alpha_p \sin\alpha_p \cos\alpha_p \eta_1 \sum_{n=0}^{\infty} \frac{(-\alpha_p^2)^n}{n!} F_n(R,t) \qquad 23)$$

The total discharge rate to the ocean floor is

$$Q = \frac{\pi}{6\tau_1}\left(\frac{\ell_{ref}}{\ell}\right)^3 e^{-N_\lambda \tau} \sum_{p=1}^{\infty} \alpha_p \sin\alpha_p \cos\alpha_p \eta_1 \sum_{n=0}^{\infty} \frac{(-\alpha_p^2)^n}{n!} \times$$

$$\int_0^\infty F_n(R,t) R\,dR \qquad 24)$$

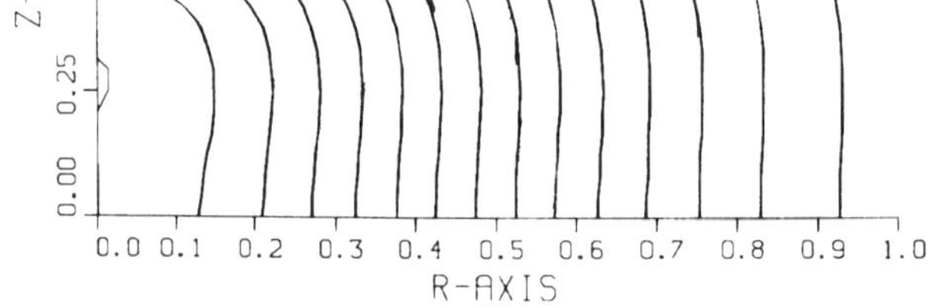

Fig. 2. Axisymmetric Concentration Contours (τ=0.1, η_1=0.5, N_λ=0)

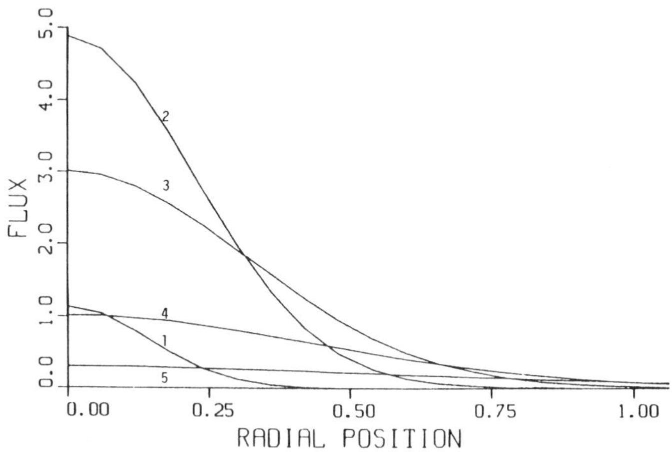

Fig. 3. Dimensionless Ion Flux at the Sediment/Water Interface (1 τ =0.01, 2 τ =0.025, 3 τ =0.05, 4 τ =0.1 and 5 τ =0.2). N_λ = 0.0

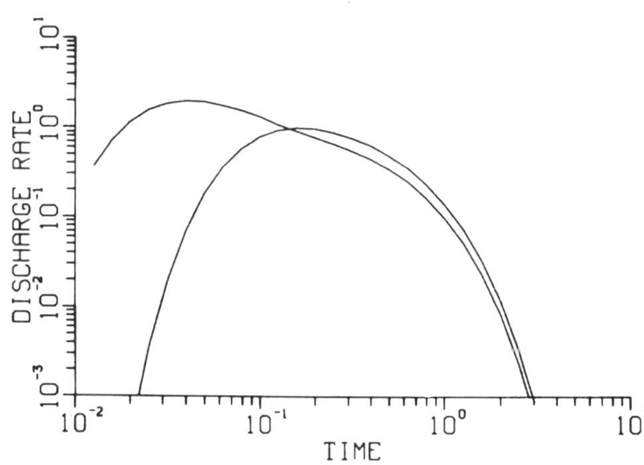

Fig. 4. Dimensionless Discharge Rate (1 η_1=0.5, 2 η_1=0). N_λ=0.0

MODEL APPLICATIONS

To show the nature of the ion transport predicted from the above analyses, the dimensionless solution for instantaneous dissolution of the canister was studied as a worst case. The analytical solution for the concentration of an ionic species within the sediment, Equation 16), was evaluated at two dimensionless times, τ =0.01 and τ =0.1 for η_1=0.5 and for N_λ=0. This represents a severe case in which a nondecaying species is migrating from the dissolved canister. A typical set of axisymmetric concentration contours in the dimensionless r-z plane is illustrated in Fig. 2. The figure corresponds to a dimensionless time following the peak breakthrough of the species at the sediment/ocean interface. The profiles in Fig. 2 show the effect of the zero flux boundary concentration at the bottom of the sediment layer. The maximum ion concentration remains below the original position of the canister. In Fig. 3 the axisymmetric dimensionless flux rate is plotted versus dimensionless radial position for several dimensionless times. As expected, the higher flux rates would be limited to the area directly above the canister. The two discharge curves illustrated in Fig. 4 show the effect of placing the canister deeper in the sediment formation (a factor of two deeper in this case). The Log/Log coordinates used tend to obscure the fact that the area under the two curves must be equal and that the curves decrease very slowly with time. Because the solutions were derived using dimensionless variables, the graphic results show general trends which are independent of initial concentrations, effective diffusivity, adsorption coefficient, and sediment thickness.

REFERENCES

1. W. P. Bishop, First-Year Report, SAND74-0410 (1975). D. M. Talbert, Seabed Disposal Program Annual Report, SAND76-0256, (1976); Seabed Disposal Program Annual Report, SAND77-1270, (1977); Seabed Disposal Program Annual Report, SAND78-1359, (1979); All four reports published by Sandia Laboratories, Albuquerque, NM (1979).
2. D. R. Anderson, Nuclear Waste Disposal Subseabed Geologic Formations: the Seabed Disposal Program, SAND78-2211, Sandia Laboratories, NM (1979).
3. H. E. Nuttall, Preliminary Seabed Disposal Modeling Studies, SAND78-1651, Sandia Laboratories, Albuquerque, NM (1979).
4. W. P. Schimmel, Jr., and C. E. Hickox, Application of Thermal Conduction Models to Deep Sea Disposal for Radioactive Wastes, SAND77-0752, Sandia Laboratories, Albuquerque, NM (1978).

RADIOACTIVE WASTE DISPOSAL INTO A PLASTIC CLAY FORMATION: PROBABILISTIC ASSESSMENT OF THE GEOLOGICAL CONTAINMENT

Marco D'Alessandro and *Arnold Bonne

Commission of European Communities, J.R.C., Ispra, Italy
*Centre d'Etude de l'Energie Nucléaire, Studiecentrum voor Kernenergie, Mol, Belgium

INTRODUCTION

This paper gives a first approach of the probabilistic safety analysis of a potential nuclear waste repository to be mined into the Boom Clay formation underlying the nuclear plants of Mol (Belgium). The study has been performed through a close collaboration between the J.R.C. of Ispra and the CEN/SCK of Mol.**

The J.R.C. of Ispra has developed a safety assessment methodology for nuclear waste disposal into geological formations. This methodology assesses the reliability of the geological containment in probabilistic terms through the application of the fault tree analysis and has been applied to two conceptual repositories mined into salt formations (1). In the present work, the same methodology is applied to a potential specific site to test the applicability of the methodology. As a result, important parameters and problem areas have been identified where further research work should be performed.

F T A Application

The evaluation of the geological containment can be performed by answering the two following questions: 1) How can the barrier fail? and 2) What is the probability of failure?

**CEN/SCK performs under contract with Commission of the European Communities an R & D program on disposal of conditioned radioactive waste in a clay formation.

Fault tree analysis (FTA) consists in the graphical representation of the possible ways by which a given system could cease to perform its function. This failure constitutes the "top event" of the tree. One then goes backward through a cascade of intermediate events to sets of elementary events, the "primary events." Each event or set of events whose occurrence is sufficient so that the top event may be reached constitutes a cut set.

Because of its binary logic, the fault tree approach is well suited for analyzing events which can either occur or not. In order to analyze slow continuous processes, this probabilistic approach seems less suited. However, it can be utilized also in this case by sharing the processes into classes characterized by different rates, each one being distinguished on the basis of a particular probability value. Therefore, we have classified as 1st order phenomena all those processes which are capable by themselves to cause the top event. The same phenomena, acting in a less severe way so that the concurrence of other phenomena is required to cause the top event, have been indicated as 2nd phenomena.

The FTA needs to be computer assisted, and a set of special codes able to calculate the probability of the top event (system unavailability), starting from the values of primary events, has been developed at J.R.C. Ispra. The code Bruna has been chosen from among these codes for the present work. As input, this code can accept a histogram with up to ten probability values, each with its own weighting. As output, the code gives corresponding histogram of the system unavailability and the relative weight of each cut set, i. e., the ratio between the cut set probability and the overall top event probability. This treatment permits, firstly, the use of weighted ranges of probability values instead of single-point values; further, it allows a check to be made of how much the uncertainty concerning the probability values of each primary event can affect the overall unavailability of the top event itself. Finally, the weight associated to each cut set permits the importance of various phenomena to be recognized.

It is worth noting that, at present, geology is a science with a limited predictive capability since it deals with past history of the earth rather than with its future evolution. However, the acquisition of probability figures can be reached on the basis of extrapolation of geological behavior into the future.

Site Description and Repository Dimensions

The clay formation of Boom belongs to the Rupelian age of the Oligocene and is covered essentially by Miocene, Pliocene and Pleistocene sands. In the Mol region, the clay of 109 m thickness is situated between -160 to -269 m below land surface. Neogenic sands above the clay formation contain an aquifer which constitutes an important water resource for the region; the water table is shallow (1-2 m). Below the clay, several aquifers exist which are not exploited in the Mol region. However, further south they are shallower and are utilized for industrial and domestic purposes.

In the conceptual design, the subsurface portion of the repository (apart from the access and ventilation shafts) consists of a horizontal network of circular galleries and storage holes excavated at a depth of about 220 m. Due to the huge pressure in a plastic clay body at such a depth, no hole can be kept open without a lining. Therefore, the waste will be separated from the geologic medium by a thick cast iron lining. For an assumed total electronuclear power production of 300 GW(e), the overall surface needed by the network of galleries can be evaluated to be about 2 km^2, while the actual area occupied by the holes and galleries will account for only 10% of the total.

Considered Events and Phenomena

The overall spectrum of possible events which could lead to geological segregation failure can be divided into different groups according to their suitability for a probabilistic treatment. In fact, other than in the case of meteoritic impact, the importance of which is intuitively negligible, any other failure case taken into account in our model cannot be considered as a random phenomenon suitable for probabilistic treatment without well defined limitations and starting assumptions. The following three main groups of events were identified:

A) Natural processes which are independent of repository existence. Such processes are linked to local evolving geological conditions, but they can be treated in probabilistic terms when the area and the time span in which they are expected to occur are limited enough so that steady geological conditions can be assumed.

The events of this kind considered in constructing the fault tree are the following*: 1) presence of ground water above; 2) presence of pressurized ground water below; 3) migration with retention (outside the clay); 4) migration without retention (outside the clay); 5) 3rd order fault; 6) 2nd order fault; 7) 1st order fault; 8) extrusive magmatic activity; 9) explosive magmatic activity; 10) diapirism; 11) glacial overburden (causing loss of plasticity); 12) Permafrost; 13) 1st order glacial action; 14) 2nd order glacial action; 15) subsidence; 16) 1st order stream erosion; 17) 2nd order

*In Table 1, the events required to be shared by different orders are described.

Table 1. Definition of Classes of Events in Relation to the Magnitude of the Consequences

Symbol	Definition	Consequences
E 5	3rd order fault (Displacement lower than 10 m)	Breaching of clay if preceded by loss of clay plasticity.
E 6	2nd order fault (Displacement greater than 10 m)	Breaching of clay and radioactivity release to ground water.
E 7	1st order fault (Displacement greater than 200 m)	Direct exhumation of waste and radioactivity release to land surface.
E13	1st order glacial action	Removal of the entire sedimentary cover (200 m) and release to land surface.
E14	2nd order glacial action	Removal of 100 m of sedimentary cover, able to affect the radio-elements partially uplifted.
E16	1st order stream erosion	Downcutting of the sedimentary cover up to the repository and release to land surface.
E17	2nd order stream erosion	Downcutting of the sedimentary cover up to the depth of 100 m.
E18	1st order denudation	Deepening of the sedimentary basin sufficient to cause a release to land surface.
E19	2nd order denudation	Deepening of the sedimentary cover affecting the radioelements partially uplifted.
E22	1st order human activities	Mining operations able to directly affect the repository and causing release to land surface.
E23	2nd order human activities	Mining operations able to exhume the partially uplifted radionuclides.
E24	1st order meteorite	Direct exhumation of waste and release to atmosphere.
E25	2nd order meteorite	Fracturing of geological formation and release to ground water.

stream erosion; 18) 1st order denudation; 19) 2nd order denudation; 24) 1st order meteorite (causing release to atmosphere); 25) 2nd order meteorite (causing release to ground water).

B) Changes of initial geological and/or lithological conditions which are caused by the presence of repository and hot waste into clay bed, e.g., thermal and radiation effect on the rock, mechanical stresses arising from repository construction, chemical interactions between waste and host rock. These changes are purely deterministic phenomena, and it is not possible to treat them in probabilistic terms; therefore, they have been omitted in the fault tree.

C) Human actions. The consequences of human intervention are difficult not only to place within the boundaries of random events but also to predict by any method; they seem basically unquantifiable. On the other hand, the possible interaction between future mankind and the lithosphere appears so important in respect of repository safety evaluation that it is not reasonable to neglect this aspect in our model. Therefore, the human actions will be included in the fault tree and treated in a probabilistic way. Nevertheless, we are aware of the weakness due to the presence of such unpredictable events in a general model having the aim of assessing, in probabilistic terms, the geological evolution of a given region. The following human actions have been considered in the model: 20) pumping of contaminated ground water; 21) drilling exposing water to aquifers; 22) 1st order human activities causing waste rise; 23) 2nd order human activities causing waste rise.

Receptor Definition

We analyzed the release from the repository to three receptors in the biosphere: ground water, land surface and air. A geochemical retention barrier separates the land surface from the ground water. The third receptor has been considered separately since it constitutes a particularly important pathway because of the relatively large hazard of plutonium inhalation. Consequently, for each receptor a single fault tree has been developed.

Time Frame Choice

The failure probabilities have been evaluated over four different time periods in order not only to get an idea of the probability trend as a function of time but, more important, to identify the event sets which, on any time interval, constitute the contributors of major importance. The four time spans considered are 2,000, 25,000, 100,000 and 250,000 years. The first one has been chosen as representative of a historical time period over which the present way of life of mankind can be reasonably extrapolated, while geological phenomena have small likelihoods of occurrence; further, this time period is largely sufficient to allow the fission products (^{99}Tc and ^{129}I excepted) to decay completely. The second one is long enough for a few important climatic effects to occur, such as glaciations and permafrost. The third one could cover completely the entire evolution of an ice age, where geological events of some importance begin to have the potential to affect the repository. The last one has been chosen because the radioactivity decrease is extremely slow after 250,000 years. Furthermore, reliable geologic predictions over such long time frames are beyond the present capability.

Release to Ground Water: Fault Tree Description and Results

In this model, the phenomena able to cause a ground water contamination are examined. It should be noted that the occurrence of such an event does not necessarily involve environmental pollution, since this latter requires the geochemical retention barrier to be bypassed. In the tree (Fig. 1.), the possibility of a release only to the upper aquifer (left branch, G2) has been separated from a release which could occur equally to the upper or lower aquifer (right branch, G4); a release only to the lower aquifer has been neglected since events able to cause a contamination only to underlying ground water do not seem conceivable in the geological context of the site. In both branches, the ground water presence is required as a necessary condition (G3, G6). Events that could cause release to the upper aquifer are 2nd order meteoritic impact or diapiric uplift. Release to either aquifer can be due to faulting phenomena, permafrost or drillings by man. Among these, faulting phenomena have been split into two different events (E5, E6) according to the size of their displacements (see Table 1).

The results of the computer treatment are shown in Fig. 2. The phenomenon governing the top event probability curve is the 2nd order fault, i.e., a sudden displacement greater than 5 ~ 10 m. This event for a time span of 2,000 years accounts for 50% of the total top event probability, while the remaining 50% is due to human actions. The faulting weight increases rapidly at 25,000 years because of a diminishing importance of human actions. Beyond this time, the faulting weight remains constant until 100,000 years and increases slightly over the longest time considered. Other phenomena of some importance are the loss of plasticity due to a glacial overburden followed by 3rd order faulting and the effects of permafrost. Both phenomena present a constant trend of importance although never accounting for more than 10% of the total. In conclusion, it should be noted that the activities of future generations are likely to be dominant causes of breaching the geological containment in the short term. Geological phenomena require longer time periods.

It is worth emphasizing that for assigning probability values to primary events, different processes have been used (2). Concerning the 2nd order fault, by far the most important event, we have extrapolated in the future the quaternary fault frequency detected by a graphical interpretation of Landsat images (photosatellites) of the region. Therefore, among all the parameters governing the probability figure of such event, the age of fault network teledetected on the Campine as well as the distinction between 1st and 2nd order fault (i.e., the highest displacement which can be plastically absorbed by clay) appear to be of prime importance. In order to improve not only a specific probabilistic evaluation but also the overall risk assessment of geological disposal, it would be very useful to obtain a better understanding of these aspects.

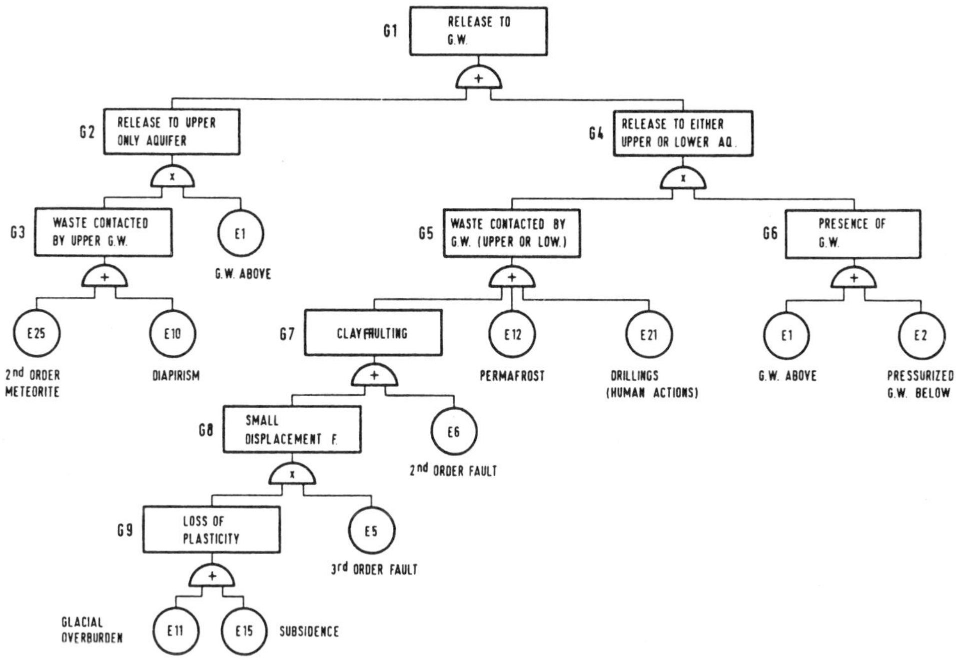

Fig. 1. Release to Groundwater

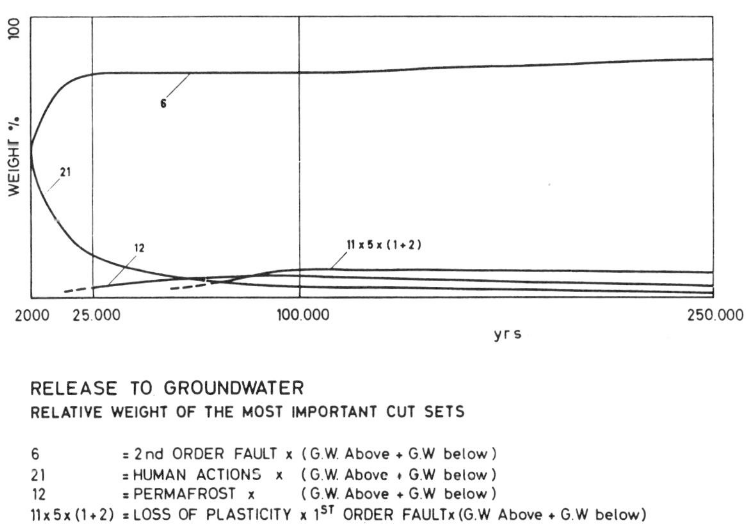

Fig. 2. Relative Weight of the Most Important Cut Sets

Release to Land Surface

The tree (Fig. 3.) consists of three main branches, each of them illustrating a particular way of geological containment failure. In the first branch of the left-hand side, the uplift of waste can be direct (human actions, huge displacement fault, magmatic activity) or due to ground water transport (G3). This latter occurrence is subject to two conditions: the top event of the 1st fault tree must be reached (E50) and, subsequently, the retention barrier must be bypassed through exploitation of contaminated ground water or through upward migration for a sufficiently long time interval. In the second branch (G5), exogenetic phenomena capable of removing 220 meters of sediment overlying the repository are considered: stream erosion, denudation and glacial erosion are here called "1st order phenomena" (see Table 1). The last branch on the right hand side takes into account the following sequence of events: a failure of the geological barrier is followed by ground water contamination; the ground water flows upward (conservative assumption) through glauconitic sands which retain the radionuclides. Subsequently, endogenetic or exogenetic phenomena as well as human actions acting on contaminated levels finally cause a release to land surface. In this case, these events are called "2nd order phenomena" as they act in a less severe way compared to the previous example, since contaminated levels are assumed shallower than the repository.

As far as the results are concerned (Fig. 4.), in the initial 2,000 years the only event having a probability different from zero is the "Human Activities," the weight of which accounts for 100% of the top event probability. Within 25,000 years, the weight of this event decreases, but it still maintains the highest importance with a figure of 57%. Over longer time spans the "migration of contaminated ground water" rapidly gains importance; this cut set assumes that the top event of the 1st fault tree has already occurred. Further, phenomena related to new glacial age occurrence appear to be of some importance. From this study, we can see that over short time intervals, once again the human actions appear as the most important way of release. Within more extended time periods, the results emphasize the importance of the soil retention capability.

Direct Release to Atmosphere

This very simple model takes into account only two events: the fall of a giant meteorite on the repository area and an explosive magmatic extrusion directly affecting the waste. Because of the linear increasing of event probablilities as a function of time and the rather simple structure of the fault tree, any further tree description and comments about the results have been omitted here.

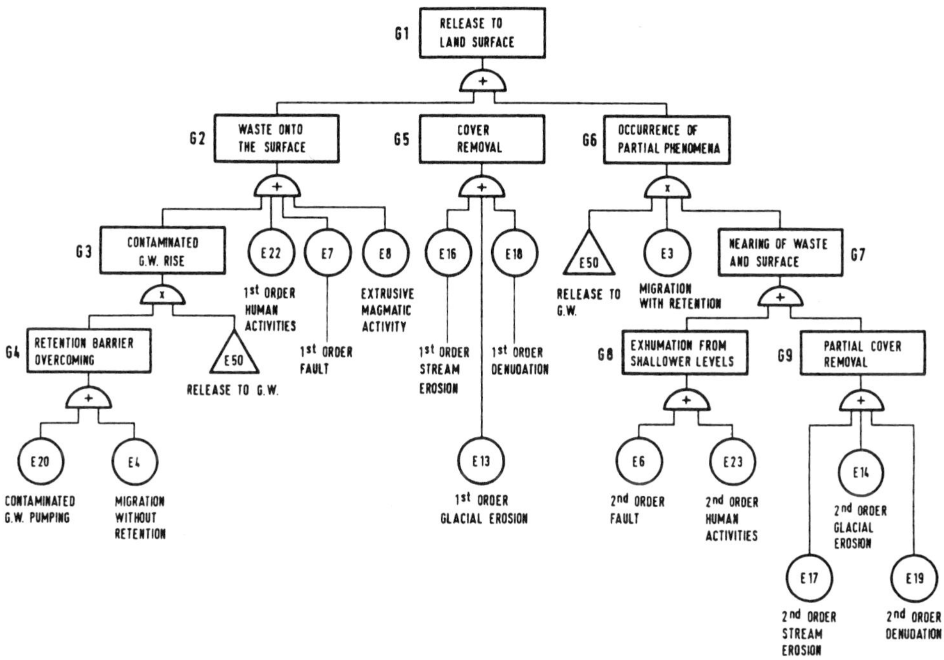

Fig. 3. Release to Land Surface

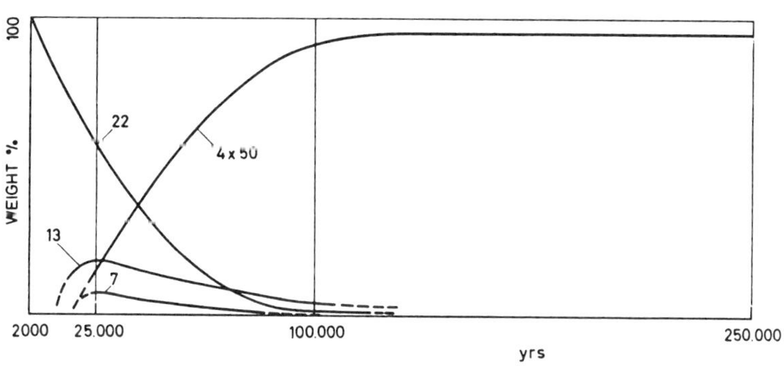

RELEASE TO LAND SURFACE
RELATIVE WEIGHT OF THE MOST IMPORTANT CUT SETS

22 = 1ST ORDER HUMAN ACTIONS
4 x 50 = RELEASE TO G.W. x MIGRATION WITHOUT RETENTION
13 = 1ST ORDER GLACIAL ACTION
7 = 1ST ORDER FAULT

Fig. 4. Relative Weight of the Most Important Cut Sets

CONCLUSIONS

The Fault Tree binary logic shows some drawbacks in analyzing processes which occur continuously at some finite rate. However, this application to a real site has shown that this methodology can be utilized provided some basic hypothesis be made to assess probabilities of all geological events which could lead to breaching of geological containment (3). Aspects and properties of the geological formation as well as problem areas that govern the release probabilities and the overall concept acceptability have been identified:

1. Intrinsic properties of the host rock in relation to various geological phenomena. The plasticity and impermeability features should be investigated as a function of the severity degree of the different phenomenon which can affect the formation. For instance, the greatest faulting phenomena capable of being plastically absorbed by a given clay layer should be clearly identified in order to group faulting phenomena in two classes according to their capability to breach or not to breach the plastic layer.

2. Intrinsic properties of the host rock in relation to waste repository presence. Intrinsic properties of the formation could change due to events and processes induced by the presence of the repository itself (e.g., thermal and radiation effects, mechanical disturbances, sealing and backfilling operations). Their occurrence is certain in some cases, but the reversibility of the effects and the possible consequences are rather unknown. Therefore, they should be carefully investigated.

3. Human actions. It is rather intuitive that they might influence geologic containment, but it is very difficult to insert them in a probabilistic model. Due to the rather high weight of human activities within shortest time spans, it would be worthwhile to reduce the uncertainty through some kind of predictive sociological studies.

REFERENCES

1. G. Bertozzi, M. D'Alessandro, F. Girardi and M. Vanossi, "Safety Assessment of Radioactive Waste Disposal into Geological Formations: A Preliminary Application of Fault Tree Analysis to Salt Deposits," Nucl. Sci. Tech., ERU 5901 EN (1978).
2. M. D'Alessandro and A. Bonne, Radioactive Waste Disposal into a Plastic Clay Formation: Probabilistic Assessment of the Geological Containment, Report EUR (in preparation).
3. M. D'Alessandro, C. N. Murray, G. Bertozzi and F. Girardi, "Probability Analysis of Geological Processes: A Useful Tool for the Safety Assessment of Radioactive Waste Disposal," Radioactive Waste Management (in press).

FINITE MEDIUM GREEN'S FUNCTION SOLUTIONS TO NUCLIDE TRANSPORT IN POROUS MATERIAL*

Steven G. Oston

The Analytic Sciences Corporation
Six Jacob Way
Reading, MA 01867

INTRODUCTION

It is important to develop mathematical models to estimate the release of buried radioactive waste to the biosphere and its potential impact on man. A necessary link in simulating this process is a model that accurately represents the subsurface transport. Among the analytical techniques used to predict the transport of nuclides in porous materials is the Green's function approach (1), i.e., the response characteristics of a geologic pathway to an impulse function input. To date, the analyses all have set the boundary conditions needed to solve the 1-D transport equation as though each pathway were <u>infinite</u> in length. This paper critically examines the effects that the infinite pathway assumption has on Green's function models of nuclide transport in porous media. It treats the more difficult problem of obtaining suitable Green's functions for <u>finite</u> pathways whose dimensions may not be much greater than the diffusion length. Such pathways may occur in multipath representations of a waste repository, for example.

TRANSPORT EQUATION, INITIAL AND BOUNDARY CONDITIONS

The 1-D transport equation in terms of mass flux, q, of a nuclide is:

*This work supported by Lawrence Livermore Laboratory Subcontract 2280709.

$$\frac{\partial q}{\partial t} + \frac{V}{B}\frac{\partial q}{\partial x} = \alpha \frac{V}{B} \frac{\partial^2 q}{\partial x^2} \qquad 1)$$

with V = interstitial water velocity

B = retardation factor based on an ion exchange distribution coefficient, K_d.

B = 1 + $K_d \rho/\varepsilon$, where ρ and ε are, respectively, the bulk density and porosity of the geologic medium.

α = longitudinal dispersion length such that αV is the coefficient of dispersion.

Solution of Eq. 1) requires two boundary conditions and an initial condition. Consider the case where no radionuclides are present in a pipe until time t = 0 when a unit amount of material is instantaneously injected at the upstream end of the pipe ($x = -\ell$), as depicted in Fig. 1. This situation implies an initial condition and boundary condition of

$$q(x,0) = 0 \quad \text{and} \qquad 2)$$

$$q(-\ell,t) = \delta(t) \qquad 3)$$

where $\delta(t)$ is a delta function in time. At the downstream end of the pipe (x = 0) it is necessary to match both mass flux, q, and concentration, c, with those of succeeding pipe.

The exact matching of boundary conditions at the interface between continuous pipes can involve tedious mathematics. A much more tractable problem is one in which the various pipes are mathematically isolated from one another. That is, by suitably choosing limiting forms of the boundary conditions, the solutions to Eq. 1) for each region are effectively <u>decoupled</u>. This method is satisfactory if the limiting forms of the boundary conditions are justified on physical grounds.

For the present, two different situations at x = 0 will be considered. In the first case the pipe in question empties into a much larger body of water, such as an aquifer, where there is considerable dilution of the waste. The following relationship between mass flux and concentration at any point is invoked:

$$U \frac{\partial q}{\partial x} + Q \frac{\partial c}{\partial t} = 0 \qquad 4)$$

where U = V/B and Q = water volumetric flow rate. 5)

When the radionuclide concentration is very small in the medium that follows, e.g., an aquifer, then both c and $\frac{\partial c}{\partial t}$ are small at the boundary. Thus, for the limiting case, a suitable boundary condition at x = 0 is:

$$\frac{\partial q}{\partial x}(0,t) = 0 \tag{6}$$

Qualitatively, the zero gradient boundary condition is tantamount to ignoring advection in the vicinity of the x = 0 boundary and allowing nuclide flow across the boundary via hydrodynamic dispersion only. This argument follows from neglecting the second term in Eq. 1). The second boundary condition deals with an opposite case, that of the pipe in question emptying into a very much smaller pipe. Obviously, the velocity increases rapidly at the boundary owing to the constriction. The appropriate boundary condition in this limiting case of infinite velocity at the boundary is obtained by assuming "slug flow". Slug flow depicts the flow as moving without axial mixing; that is, hydrodynamic dispersion is neglected. The mathematical description of slug flow at x = 0 is:

$$\frac{\partial q}{\partial t}(0,t) + U\frac{\partial q}{\partial x}(0,t) = 0 \tag{7}$$

This equation can be obtained from Eq. 1) by assuming its right-hand side is negligible.

TRANSPORT EQUATION SOLUTIONS

A method of solving the 1-D transport equation is by the Laplace transform technique. The transformed equation, making use of initial condition Eq. 2) is:

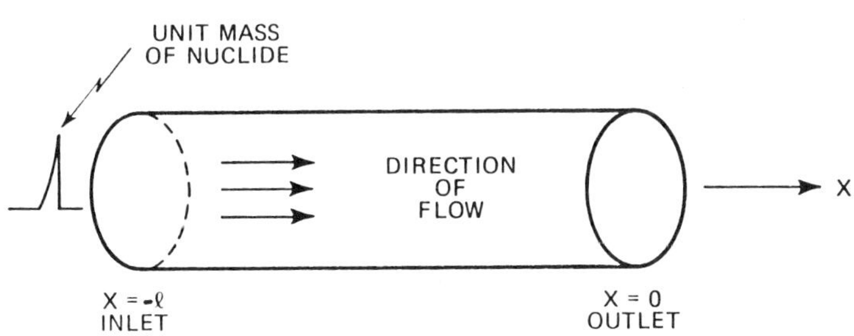

Fig. 1. A One-Dimensional Pipe

$$s\bar{q} + U\bar{q}_x = \alpha U\bar{q}_{xx} \tag{8}$$

In this notation, $\bar{q}$ is the transform of q. Each subscript x denotes a differentiation with respect to x; s is the Laplace tranform variable such that

$$\bar{q}(s) = \int_0^\infty e^{-st}\, q(t)\, dt \tag{9}$$

We are interested in finding the mass flux at x = 0, the end of the pipe, given a delta function input at the pipe inlet. The transform of that response function in the pure diffusion boundary case is

$$\bar{q}(0) = \frac{Ke^{\ell/2\alpha}}{K \cosh K\ell + \frac{1}{2\alpha} \sinh K\ell} \tag{10}$$

where

$$K = K(s) = \frac{1}{2\alpha} \sqrt{1 + \frac{4\alpha s}{U}} \tag{11}$$

For the slug flow boundary conditions the response function is

$$\bar{q}(0) = \frac{UKe^{\ell/2\alpha}}{UK \cosh K\ell + (s + \frac{U}{2\alpha}) \sinh K\ell} \tag{12}$$

Inversion of Equations 10) and 12) is accomplished by the method of residues, giving solutions in the time domain in the form of an infinite series. The mass flux at x = 0 is found to be

$$q(0,t) = \frac{U}{2\alpha}\, e^{\ell/2\alpha}\, e^{-Ut/4\alpha} \tag{13}$$

$$\times \sum_n \frac{\xi_n^2 \cos\left(\frac{\xi_n \ell}{2\alpha}\right) e^{-U\xi_n^2 t/4\alpha}}{\left[\frac{\lambda}{2}(1 + \xi_n^2) - 1\right]\left[\frac{\ell}{2\alpha} + \lambda \sin^2\left(\frac{\xi_n \ell}{2\alpha}\right)\right] - \cos^2\left(\frac{\xi_n \ell}{2\alpha}\right)}$$

The quanitities ξ_n are obtained from the transcendental equation

$$\tan\left(\frac{\xi_n \ell}{2\alpha}\right) + \frac{\xi_n}{1 - \frac{\lambda}{2}(1 + \xi_n^2)} = 0 \tag{14)}$$

The parameter λ takes on the values $\lambda = 0$ for pure diffusion at $x = 0$ and $\lambda = 1$ for slug flow at $x = 0$.

Equation 13) with $\lambda = 0$ or 1 is the Green's function for a finite, one-dimensional medium. Currently, semi-infinite (2) or infinite medium Green's functions are employed to estimate nuclide transport. The semi-infinite medium Green's function is obtained from the transport equation, which is solved for an infinitely long pipe with an inlet at $x = 0$. A unit impulse of material is injected at $t = 0$; thus,

$$q(0,t) = \delta(t) \tag{15)}$$

The second boundary condition invoked is that the mass flux must be zero at $x = \infty$.

$$q(\infty,t) = 0 \tag{16)}$$

As an initial condition Eq. 2) is applicable. The mass flux at a distance ℓ from the source is given by

$$q(\ell,t) = \frac{\ell}{[4\alpha U\pi t^3]^{1/2}} \exp\,[-(\ell - Ut)^2/4\alpha Ut] \tag{17)}$$

The infinite medium Green's function (1) utilizes the following boundary conditions:

$$q(+\infty,t) = 0 \quad \text{and} \tag{18)}$$

$$q(-\infty,t) = 0 \tag{19)}$$

The initial condition corresponds to an instantaneous injection of a unit mass at $x = 0$; that is

$$q(x,0) = U\delta(x) \tag{20)}$$

where $\delta(x)$ is the delta function. The resultant Green's function evaluated at $x = \ell$ is

$$q(\ell,t) = \frac{U}{\sqrt{4\pi\alpha t}} \exp\,[-(\ell - Ut)^2/4\alpha Ut] \tag{21)}$$

NUMERICAL RESULTS

The various Green's functions have been numerically evaluated for a single pipe to compare their important features as a function of problem parameters. Fig. 2. is a typical plot of mass flux vs. time at a fixed position a distance ℓ from the pipe inlet at which a unit impulse of material is injected at t = 0. The coordinates are normalized so that all pulses have a unit area.

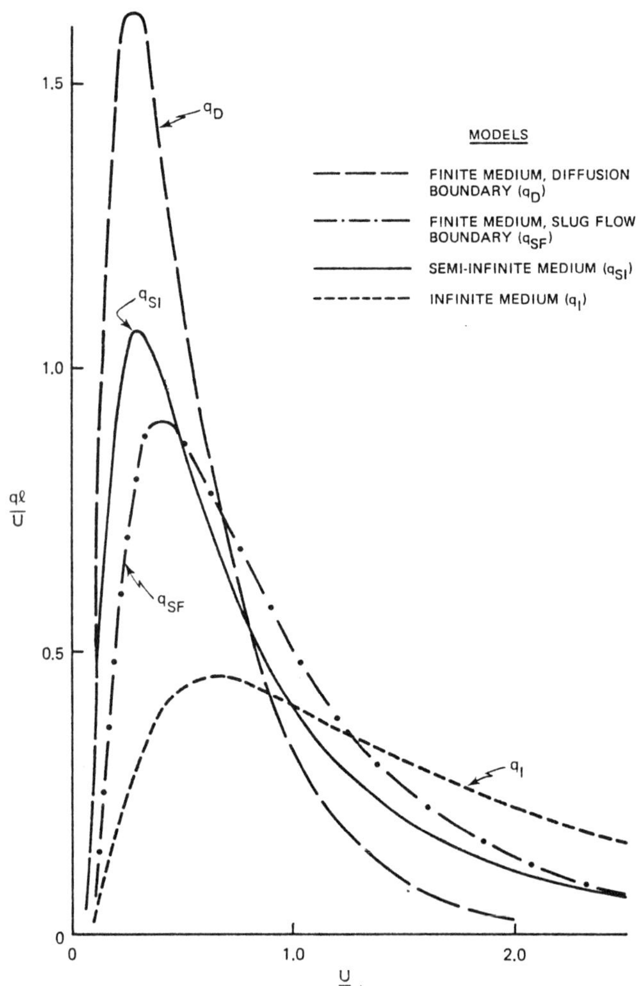

Fig. 2. Nuclide Flux Versus Time Histories for ℓ/α = 2

Fig. 2. compares the four Green's functions for a pipe with $\ell/\alpha = 2$. The diffusion curve q_D has a considerably higher peak than the slug flow peak, a faster rise time, and an earlier arrival time. The slug flow curve is characterized by a longer tail. The diffusion model represents a pipe joining a larger body of water with a much lower nuclide concentration. The large body of water will tend to draw the waste by diffusion rapidly out of the pipe. In contrast, the slug flow model represents a pipe joining a small area pipe. At the interface the flow resistance is high. Hence, nuclide transport across the first pipe is slower. The semi-infinite model (q_{SI}) occupies an intermediary position. The transit time of the peak approximates that the diffusion model, although the overall shape is somewhat closer to the slug flow model. The q_I result for the infinite pipe has by far the lowest peak and is the most delayed in time of the four Green's functions. This model allows backward diffusion from the injection point, i.e., in a direction opposite from the bulk flow, and therefore the pulse is wider than the other three models.

Flux-time histories have been computed (3) for ℓ/α ratios up to 24 where the four models become virtually identical and approach Gaussian distributions centered about $t = \ell/U$. In the limit of long pipes ($\ell/\alpha = 10$) there is little sensitivity of pulse shape to selection of model.

INTERPRETATION OF MODELS

Given a one-dimensional pipe with nuclide transport, four response functions have been presented that allow the flux at the pipe outlet to be computed from a delta-function input at the pipe inlet. The choice of the appropriate Green's function is now addressed in terms of inequalities involving ratios of physical parameters of two consecutive pipes that determine the appropriate Green's function selection for the first pipe. For the q_D model to be valid, it is necessary that either

$$Q_2 \gg Q_1 \text{ and } U_1 \cong U_2 \qquad 22)$$

$$\text{or} \quad U_1 \gg U_2 \text{ and } Q_1 \cong Q_2$$

where the subscript 1 indicates the pipe of interest, and the subscript 2 indicates the following pipe. For the q_{SF} model to be valid, it is necessary that

$$U_2 \gg U_1 \text{ and } Q_1 = Q_2. \qquad 24)$$

The semi-infinite medium Green's function, of course, is never strictly applicable to finite length pipes, but may be the best choice when

$$Q_1 \cong Q_2 \text{ and } U_1 \cong U_2 \qquad 25)$$

or for pipes long compared to the diffusion length α when injection is at an endpoint. The infinite medium Green's function q_I is seemingly the most advisable choice to describe the aquifer or any long pipe when material is injected somewhere within the pipe boundaries, but not at an endpoint.

CONCLUSIONS

It has been shown elsewhere (3) that the above inequalities for the q_D and q_{SF} models can indeed be obeyed for a multi-pipe representation of an unflawed salt repository. For instance, the q_D model appears useful to describe flow from the repository through various rock formation layers of increasing porosity to an aquifer. The q_{SF} model, on the other hand, may be valid to describe flows through tunnel and shaft fracture zones of decreasing area. These conclusions are based on the limited extent to which important geologic parameters are known.

It has also been demonstrated (3) that the transforms of the finite medium Green's functions are analytic. Thus the use of finite medium Green's functions has computational advantages over the non-analytic infinite or semi-infinite Green's functions, in particular for application to the solution of problems involving stacked pipes, i.e., a series of connected pathways.

Because of the ease of treatment of stacked pipes and their more accurate representation of short pathways, finite medium Green's functions represent an improvement over current waste transport models and extend the range of applicability of the Green's function approach to solving radionuclide transport problems.

REFERENCES

1. L. E. Berman, Analysis of Some Nuclear Waste Mangement Options, The Analytic Sciences Corporation, Report TR-1103-1-1, USDOE Report UCRL-13917 (1978).
2. A. M. Kaufman, Waste Pipe Calculus, Lawrence Livermore Laboratory, Report UCID-17953 (1978).
3. S. G. Oston, Finite Medium Green's Function Solutions to Nuclide Transport in Porous Media, The Analytical Sciences Corporation, Report TR-1485-2, USDOE Report UCRL-15040 (1979).

PRELIMINARY RATE EXPRESSIONS FOR ANALYSIS OF RADIONUCLIDE MIGRATION RESULTING FROM FLUID FLOW THROUGH JOINTED MEDIA*

Kenneth L. Erickson

Sandia Laboratories**
Albuquerque, NM 87185

ABSTRACT

A theoretical and experimental basis is being developed for analysis of radionuclide transport in jointed geologic media. Batch equilibration and rate experiments involving samples of Eleana argillite and Tertiary silicic tuffs in contact with solutions containing Cs, Sr, or Pm indicated that most radionuclide sorption was associated with the surfaces of very small intergranular regions and that the rate of sorption was controlled by diffusion of the nuclides into such regions. Based on these experimental results, the continuity equations for radionuclides in the mobile and immobile phases were reduced to a model analogous to Rosen's equations for packed beds and were solved similarly. Using the model and experimental data, limited radionuclide transport analyses were made which indicated that important parameters controlling transport include the intergranular porosity and nuclide penetration depth, fracture plate spacing and length, fluid velocity and sorption distribution coefficient. Many of these parameters represent physical quantities or processes which can be quantified in the laboratory. However, fluid velocities and fracture plate spacings and lengths must be obtained from the field and methods must be developed to establish reliable bounds for such field-determined parameters.

*This work supported by the U. S. Department of Energy (DOE), under Contract DE-AC04-76-DP00789.
**A DOE facility.

INTRODUCTION

Current literature regarding analysis of radionuclide transport in jointed geologic media is very limited. This situation has probably partly resulted because finite rates of mass transfer between phases must be considered. This paper summarizes initial work conducted to develop theoretical and experimental approaches which adequately account for finite mass transfer rates between phases and which will provide a basis for analyzing radionuclide transport in jointed geologic media. In the following discussion, the theoretical and experimental approaches are illustrated and then used to develop a transport model for relatively simple systems. The model and experimental data are used for limited transport analyses which identify important parameters and some possible deficiencies in current methods of parameter evaluation.

THEORETICAL APPROACH

For purposes of discussion, the solution phase which is subject to bulk motion has been referred to as the mobile phase, and the solid phase, which includes voids containing stagnant solution, has been referred to as the immobile phase and was idealized as consisting of distinct permeable and impermeable regions. The permeable regions were defined as the fractures, pores, and intergranular regions containing stagnant solution, and the impermeable regions were defined as the crystal grains.

Assuming that the mobile phase is an aqueous solution in laminar flow, the material balance for a dissolved component is

$$\frac{\partial C}{\partial t} = -\vec{\nabla}\cdot\vec{v}C - \vec{\nabla}\cdot\vec{J} + \sum R_j \qquad 1)$$

where C is the component's concentration; $\vec{J}$ is the component's flux relative to the solution's mass average velocity $\vec{v}$; R_j are terms representing rates of production and/or depletion due to homogeneous chemical reactions and/or nuclear decay and t is time. The initial condition assumed for the mobile phase is that $C(X_1,X_2,X_3,0)$ is a known function of the general coordinates X_i. Two commonly used boundary conditions are sorption equilibrium at the interface between phases and continuity of the radionuclide flux across the interface.

For the permeable regions of the immobile phase, the material balance is

$$\frac{\partial C}{\partial t} = -\vec{\nabla}\cdot\vec{\hat{J}} + \sum \hat{R}_j \qquad 2)$$

where $\hat{C}$ is the local concentration; $\vec{\hat{J}}$ is the diffusional flux, and $\hat{R}_j$ are terms representing rates of production and/or depletion due to homogeneous chemical reaction and/or nuclear decay. Similar initial and boundary conditions as given for Eq. 1) also apply to Eq. 2). Futhermore, at the interface between permeable and impermeable regions, the conditions of sorption equilibrium and continuity of the radionuclide flux also apply.

EXPERIMENTAL APPROACH

The objective of the experimental program is to identify the important sorption mechanisms and the important homogeneous and heterogeneous chemical reactions and to obtain sufficient data so that these phenomena can be described quantitatively using appropriate mathematical expressions. A basic approach for achieving this objective is illustrated in the following discussion of experimental results.

Initial batch equilibration and rate experiments have been conducted using samples of Tertiary silicic tuffs which were contacted with simulant groundwater solutions containing cesium, strontium and promethium and also using samples of Eleana argillite contacted with solutions containing cesium (1). The argillite experiments have been most amenable to analysis, and the results from those experiments are summarized below, so that the development of rate expressions can be illustrated in the next section.

The Eleana argillite primarily consisted of quartz and layered silicate minerals, kaolinite, pyrophyllite and mixed layer "illite." Samples of the argillite generally contained some macroscopic fractures, and scanning electron micrographs indicated that the surface of the samples contained intergranular regions having thin, approximately rectangular cross-sections and wall spacings on the order of 0.1 microns and less. The porosity associated with these regions appeared to be between one and ten percent. For 30 to 60 mesh particles, surface areas determined by the BET method were about 9 m^2/gm.

Based on the solution chemistry of cesium and of the constituents of the groundwater solutions, it was felt that if the surface of the immobile phase remained unaltered, no homogeneous and no heterogeneous chemical reactions other than sorption should be involved. The sorption equilibrium isotherm at room temperature determined from batch equilibration experiments using 30 to 60 mesh particles, was found to be represented by

$$\hat{\hat{C}} = (0.05\ C\ \text{cm})/(1 + (5 \times 10^7\ \text{cm}^3/\text{mole})\ \hat{C}) \qquad 3)$$

for values of $\hat{C}$ between 10^{-2} and 10^{-10}M. The term $\hat{\hat{C}}$ represents the nuclide concentration associated with the surfaces of the impermeable regions, and it has been assumed that at the interface between phases $C = \hat{C}$. The distribution coefficient K is defined as $\hat{\hat{C}}/C$.

Sorption rate data were obtained from independent batch experiments using argillite tablets which were approximately 1/4 inch thick by 2-3/8 inch in diameter and were contacted with well-mixed cesium solutions. At the conclusion of these experiments, the tablets were examined using autoradiography. Based on autoradiographs from intact and cleaved tablets, it appeared that most sorption occurred in intergranular regions having effective lengths of 0.01 cm or less, which was consistent with the surface area measurements and estimated intergranular porosities.

RADIONUCLIDE TRANSPORT MODEL

Analysis of Rate Data

Based on the literature concerning ion-exchange kinetics (2), the rate of cesium sorption should have been limited by diffusion rather than by the intrinsic rate of a "sorption reaction." For the batch rate experiments, the boundary layer thickness in the mobile phase should have been on the order of 10^{-2} to 10^{-3} cm, and for the tablets, it appeared reasonable to expect that diffusion into the relatively long, narrow intergranular regions would primarily dominate the rate of mass transfer across the interface between mobile and immobile phases.

Referring to Fig. 1., equations describing mass transfer in the intergranular regions were developed (3) by assuming local sorption equilibrium between the bulk stagnant solution and impermeable surface, constant-valued parameters, continuous physical and chemical properties, by assuming that solution-phase nuclide concentrations were sufficiently dilute so that Fick's law was valid, that diffusion was essentially one-dimensional, and that there was no surface diffusion. The resulting flux expression describing mass transfer across the interface between phases was

$$\vec{n}\cdot\vec{J} = \vec{n}\cdot\vec{\hat{J}} = -(D/a_t)(\partial\hat{C}/\partial X_1) = -(\overline{D}/a_t)(\partial\overline{C}/\partial X_1) \qquad 4)$$

where a_t is a tortuosity factor, D is the diffusion coefficient for the nuclide in the solution; $\overline{C} = \overline{K}C$, $\overline{D} = D/\overline{K}$; $\overline{K} = 1 + a_c hK$, h is the ratio of the perimeter length to the area of a simple characteristic cross-section for the intergranular regions, and a_c is a roughness factor. The material balance for the bulk of the intergranular regions is

$$\partial\overline{C}/\partial t = (\overline{D}/a_t^2)\ \partial^2\overline{C}/\partial X_1^2 \qquad 5)$$

with $C(X_1,0) = 0$, and $\partial\hat{C}(\overline{H},t)/\partial X_1 = 0$. The $\hat{R}_j$ terms in Eq. 2) have been assumed negligible, and $\overline{H}$ is a characteristic depth to which the intergranular regions penetrate the immobile phase.

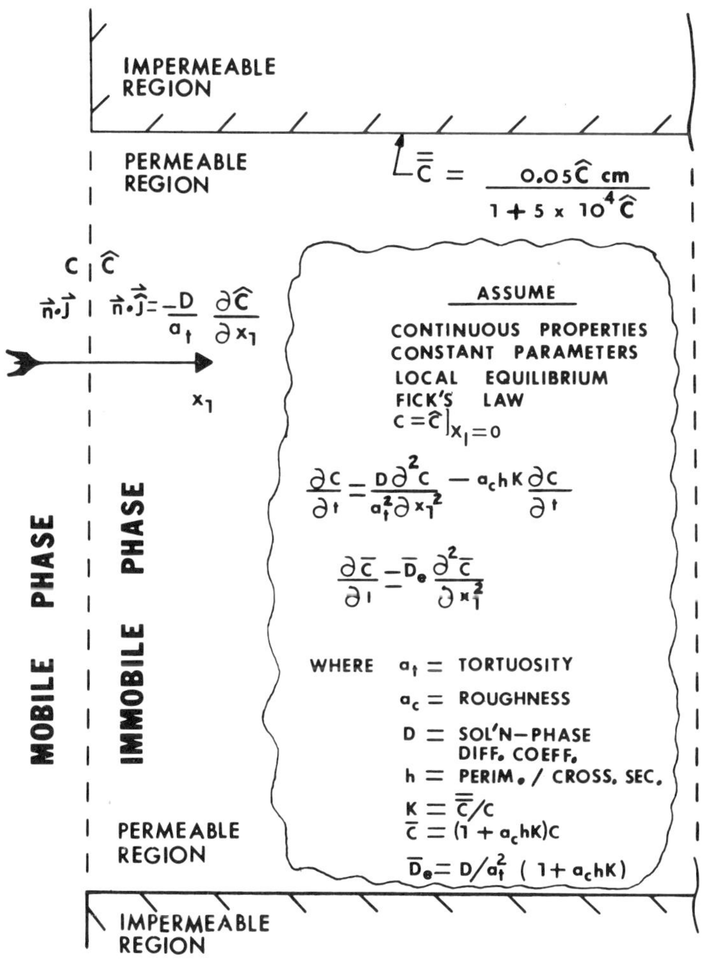

Fig. 1. Schematic Representation of Mass Transfer in Intergranular Regions

To describe a batch rate experiment, Eq. 5) is solved simultaneously with Eq. 6) below

$$(V/apAa_t\overline{K})\ (\partial C/\partial t) = (\overline{D}/a_t^2)(\partial\overline{C}/\partial X_1)\,|_{\text{interface}} \qquad 6)$$

where $C(0) = C_o$ = a constant; A is the gross geometrical surface area of the tablet; a is a roughness factor for the tablet's surface; p is the interfacial intergranular porosity, and V is the volume of the mobile phase. The solution to Eqs. 5) and 6) can be obtained from Crank (4). Some typical data from batch rate experiments with argillite tablets and cesium solutions are shown in Fig. 2. Analogous data calculated using the given parameter values are shown by the solid line. The agreement between calculated and experimental data is not entirely quantitative. However, when inhomogeneities in and variations between rock samples are considered, as well as the purely representative nature of some of the parameters, and when initial effects due to diffusion through the boundary layer in the mobile phase are also considered, the agreement between calculated and experimental results appears sufficiently good to support the validity of Eqs. 5) and 6) and the parameter values used.

Model Development and Solution of Equations

Based on the preceding analyses, a fundamental transport model for cesium solutions in laminar, one-dimensional flow through linear rectangular fractures in Eleana argillite was obtained by: assuming constant, uniform dimensions and constant, continuous properties for the fractures, replacing the parabolic velocity profile by an appropriate average, assuming that the fracture wall spacing H is sufficiently small so that diffusion in the X_1-direction in the mobile phase can be approximated as a quasi-steady-state process, neglecting terms due to nuclear decay, and assuming constant-valued parameters. The result is

$$\partial C/\partial t + v(\partial C/\partial Z) = -\,(1/M)(\partial Q/\partial t) \qquad 7)$$

where $Q = \int_o^{\overline{H}} \overline{C}(X_1,Z,t)dX_1$; $C(0,t)/C_o = 1$; $C(Z,0) = 0$, and $Q(Z,0) = 0$.

$$\partial\overline{C}/\partial t = \overline{D}_e\ \partial^2\overline{C}/\partial X_1^2 \qquad 8)$$

with $\overline{C}(X_1,Z,0) = 0$, and $\partial\overline{C}(\overline{H},Z,0)/\partial X_1 = 0$

$$\partial Q/\partial t = (1/R_M)(C - \overline{C}_s/\overline{K}) \qquad 9)$$

The term C is now the average concentration for the rectangular cross-section; C_o is the constant mobile phase concentration entering the fracture; $\overline{C}_s$ is the value of $\overline{C}$ at the interface between mobile and immobile phases; $\overline{D}_e = \overline{D}/a_t^2$; $M = H/2ap\overline{H}$; $R_M = H\overline{H}/4D$, and Z is the spatial coordinate parallel to the direction of bulk fluid motion.

Eqs. 7), 8), and 9) can be solved analogously to Rosen's (5) development for packed beds. The result is identical in form to Rosen's solution. The numerical differences which exist are small and should be masked by the inherent uncertainties associated with the parameter values, and for purposes of making preliminary calculations, the results given by Rosen (6) should be applicable. For values of $2ap\bar{D}_e\bar{K}Z/H\bar{H}v$ on the order of 50 and greater, the solution approaches the following asymptotic expression

$$U(\bar{Z},\bar{t}) = 1/2 + (1/2)\mathrm{erf}((2Y/\bar{X}) - 1)/2((1 + 3g)/3\bar{X})^{1/2} \qquad 10)$$

where $U = C/C_0$; $\bar{t} = t-Z/v$; $\bar{Z} = Z/Mv$; $\bar{X} = 2ap\bar{D}_e\bar{K}Z/H\bar{H}v$; $Y = (D/2\bar{H}^2a_t^2K)\bar{t}$, and $g = H/4\bar{H}$.

Discussion

If $t_{0.01}$ is defined as the elapsed time required for U to reach a value of 0.01 for a given value of Z and if $t_{0.5}$, $v_{0.01}$, and $v_{0.5}$ are defined analogously, then from Eq. 10) and appropriate values

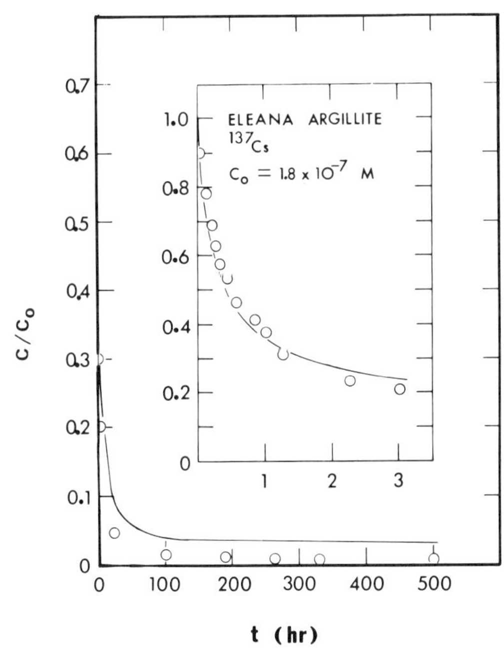

PARAMETER	VALUE USED	ESTIMATED RANGE
a	1	1 - 2
a_c	$\sqrt{2}$	1 - 4
a_t^2	2	1 - 4
D	2×10^{-5} cm^2/sec	---
$\bar{H}$	5×10^{-3} cm	$10^{-3} - 10^{-2}$ cm
h	6×10^{6} cm^{-1}	*
K	5×10^{-2} cm	---
$\bar{K}$	4×10^{5}	---
p	3×10^{-2}	$10^{-2} - 10^{-1}$

* CALCULATED SO AS TO BE CONSISTENT WITH SURFACE AREA MEASUREMENTS AND VALUES USED FOR a_c, a_t, H, AND p

Fig. 2. Comparison of Experimental (Circles) and Calculated (Solid Line) Rate Data Using the Parameter Values Given of the error function it can be shown that as Z/v becomes sufficiently large (7),

$$v_{0.5} = vM/(M + \bar{K}) \approx v_{0.01} \quad \text{11a)}$$

$$t_{0.5} = (1 + \bar{K}/M)Z/v \approx t_{0.01} \quad \text{11b)}$$

Conversely, as Z/v becomes small, $V_{0.5}$ and $v_{0.01}$ approach v. Referring to Rosen's numerical results, if $g/\bar{X}$ is much less than 0.2 and $\bar{X}$ is greater than 50, Eq. 11a) applies. For Eq. 11b) to apply, $\bar{X}$ must be much greater than 50, say on the order of 500. This leads to the following criteria for application of Eq 11b):

$$Z/v > 250H\bar{H}/ap\bar{D}_e\bar{K} \quad \text{11c)}$$

As also shown by Rosen, for t = Z/v, the value of $\bar{t}$ is zero, and

$$U(\bar{Z},0) = \exp\ (-Z/vMR_M) \quad \text{13)}$$

which leads to the following criteria for when the value of $t_{0.01}$ can be greater than Z/v

$$Z/v > 2.3H^2/4apD \quad \text{14)}$$

RADIONUCLIDE TRANSPORT ANALYSIS

Model Evaluation

Some initial experiments have been conducted in which stable cesium solutions ($C_0 \approx 10^{-3}$ M) have been pumped through artificially prepared fractures in argillite samples. These experiments were primarily designed to develop sample preparation techniques and quantitative analysis of the data has been complicated by the non-linear nature of the sorption isotherm. However, qualitative analysis of the data has been encouraging.

General Analyses

The equations and discussion of the preceding section are summarized in Fig. 3., which represents the superposition of independent plots of Eq. 11c) and Eq. 14). For values of $4apD/H^2$ and Z/v in regions 3 and 4, the breakthrough time t_B at which the value of U equal to 0.01 or greater appears at distance Z is equal to Z/v. For values of $4apD/H^2$ and Z/v in regions 1 and 6 and values of $ap\bar{D}_e\bar{K}/H\bar{H}$ and Z/v in regions 1, 2 and 3, t_B is given by

$$t_B \approx (1 + 2ap\bar{H}\bar{K}/H)Z/v \quad \text{15)}$$

For other combinations of parameter values the breakthrough time t_B is between Z/v and that given by Eq. 15).

The above conditions for the applicability of Eq. 15) essentially define the parameter values required to apply the local equilibrium assumption to transport models for jointed media. Furthermore, many of the required parameters represent physical quantities or processes which can be quantified in the laboratory. However, parameters such as v and H must be determined from field measurements. Current measurement techniques are crude and introduce large uncertainties into the parameter values. The importance of these uncertainties can be illustrated by choosing a typical hydraulic gradient $\Delta P/Z$ and then calculating the maximum value of H which will allow some minimum value of t_B at selected distances L from a radionuclide source. For purposes of illustration, parameter values are taken from Fig. 2.; t_B is taken as 10^6 years, and $\Delta P/Z$ as about 10 feet of water per mile. The largest maximum values of H will occur when Eq. 15) is valid, and if the value of $2ap\bar{H}\bar{K}/H$ is much greater than unity,

$$H_{max} = (2ap\bar{H}\bar{K}/t_{0.01} \times 10^4 \text{ m cm}^{-2} \text{ day}^{-1})^{1/3} L^{1/3}$$

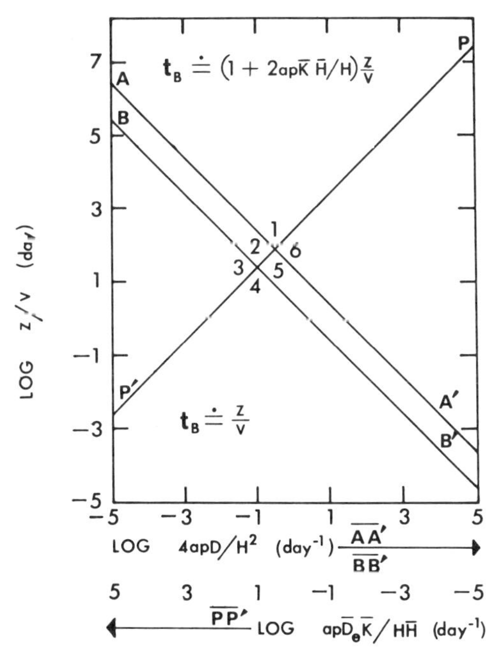

Fig. 3. Effect of Parameter Values on t_B

For axial distances of 30 and 10^4 meters, the resulting values of H_{max} are about 10^{-3} and 10^{-2} cm, respectively, and are on the same order as some values estimated for actual joint widths. Therefore, the uncertainties in such estimates must be reliably determined (3).

CONCLUSIONS

The initial work conducted to develop a theoretical and experimental approach for analysis of radionuclide transport in jointed geologic media has been encouraging. The transport model developed for relatively simple systems indicated that important parameters controlling radionuclide transport include the intergranular porosity (p), penetration depth ($\bar{H}$), the fracture plate spacing (H), length (Z), the fluid velocity (v) and the sorption distribution coefficient (K). Values determined for these parameters can be used to determine when the assumption of local equilibrium between bulk phases can be applied to radionuclide transport in jointed geologic media. Furthermore, many of the parameters represent physical quantities or processes which can be characterized in the laboratory. However, parameters such as joint widths (H) and fluid velocities (v) must be obtained from the field, and methods must be developed to establish reliable bounds on the uncertainties in the values of such field-determined parameters.

REFERENCES

1. K. L. Erickson, to be published.
2. F. Helfferich, Ion Exchange (McGraw-Hill, New York 1962).
3. K. L. Erickson," A Fundamental Approach to the Analysis of Radionuclide Transport Resulting from Fluid Flow Through Jointed Media," SAND80-0457, Sandia Laboratories, Albuquerque, NM 87185 (1980).
4. J. Crank, The Mathematics of Diffusion (Clarendon Press, London 1975).
5. J. Rosen, "Kinetics of a Fixed-Bed System for Solid Diffusion into Spherical Particles," J. Chem. Phys. 37, 387 (1952).
6. J. Rosen, "General Numerical Solution for Solid Diffusion in Fixed Beds," Ind. Eng. Chem. 46, 1590 (1954).
7. T. Sherwood, R. Pigford, C. Wilke, Mass Transfer (McGraw-Hill, New York 1975).

THE COMPLEX RELATIONSHIP BETWEEN GROUNDWATER VELOCITY AND CONCENTRATION OF RADIOACTIVE CONTAMINANTS

F. E. Kaszeta and F. W. Bond

Water and Land Resources Department
Battelle, Pacific Northwest Laboratories
Richland, WA 99352

INTRODUCTION

The program for Assessment of the Effectiveness of Geological Isolation Systems (AEGIS) is concerned with investigating methods and techniques of using geologic formations to store high level radioactive wastes and/or spent fuel from commercial nuclear power reactors. One task under AEGIS is to develop and implement hydrologic and transport models to predict the groundwater transport of radioactive contaminants. A subtask under model development and implementation is to perform a sensitivity analysis of the model results. The analysis was performed for a hypothetical geologic repository located in bedded salt. Groundwater velocities in the quifer were determined with the two-dimensional Variable Thickness Transient (VTT) model (1) based on the finite difference method. The velocity profile from the VTT model, along with data on dispersion, half-life, and leach rate was input to the one-dimensional Multi-component Mass Transport (MMT) model (2) to predict contaminant concentrations entering the bioshpere.

This paper uses the results from the MMT model to examine the complex interrelationship between groundwater velocity and contaminant dispersion, decay, and retardation with regard to their influence on the contaminant concentration distribution as it travels through the geosphere to the biosphere. Others (3,4,5) have performed sensitivity analyses; but for the most part, they have studied the sensitivity of each parameter individually. This paper, on the other hand, also addresses the combined or total effect of the parameters as they influence and interact with one another. Results are presented with regard to how the contaminant concentration distribution, whose transport through the groundwater flow system is governed by the groundwater velocity, contaminant retardation, and leach rate, is distorted during transport by the combined effects of dilution, dispersion and decay.

In order to better illustrate this combined effect, two cases are presented: 1) a hypothetical case assuming no retardation, no dispersion, and no decay; and 2) an example case with set parameter values. For purposes of this study, the source term is assumed to be independent of the regional hydrologic conditions. As a result, the conclusions drawn are not necessarily valid for breach scenarios where the source term is solubility-limited and flow past the waste is dependent on regional hydrologic conditions.

LEACHING AND TRANSPORT OF THE CONCENTRATION DISTRIBUTION

Transport begins when the contaminant is leached from the repository. Although the major characteristics of the contaminant concentration distribution are inherited during leaching, the distribution is modified throughout transport. The source term distribution is defined as the concentration distribution at the time of complete leaching. Dilution, dispersion and decay distort the concentration distribution during leaching, characterizing the shape of the source term distribution. Because the hydrologic and transport models assume that dilution occurs only during leaching, modification of the concentration distribution after completion of leaching is due only to dispersion and decay.

HYPOTHETICAL CASE

The hypothetical case uses a specific leach rate and groundwater velocity but assumes a hypothetical non-decaying, non-dispersing contaminant with no retardation properties. The shape of the concentration distribution is determined by the leach rate and the groundwater velocity. The following values were used:

leach rate = 1.83×10^{-3} g-leached/cm^2 of surface area/yr for 18,500 years, resulting in a hyperbolic source term versus time as surface area decreases

groundwater velocity = 4.2 ft/day (corresponding to a hydraulic conductivity of 100 ft/yr)

An initial contaminant inventory of 1,000 curies and a path length of 18,600 feet are used for all analyses.

The source term distribution is characterized by the rate of contaminant release whether instantaneous, continuous, or hyperbolic. Rapid leach rates result in shorter release times causing narrower, more spiked source term distributions. Similarly, slow leach rates result in longer release times, causing broader source term distributions with lower peak concentrations.

As groundwater velocity increases, the contaminant is released into a larger volume of water resulting in greater dilution and, therefore, a lower peak concentration. Dilution in the aquifer is a function of groundwater velocity alone. Without retardation, dispersion, or decay, for any distance traveled or for any point in time after complete leaching, regardless of the groundwater velocity, the source term distribution maintains its original shape as it moves along the flow path with the groundwater. Peak concentration at exit to the biosphere decreases with increased velocity as a result of dilution in the aquifer (see Fig. 4). For this study it was assumed that the source term is not affected by the groundwater velocity.

EXAMPLE CASE

Under actual field conditions, dilution, dispersion, and decay influence the concentration distribution. Groundwater velocity and retardation, which alter the rate of contaminant transport, complicate their effect. The contaminant I-129 with a groundwater velocity and leach rate identical to the hypothetical case is used to illustrate the example case. Additional transport parameters are defined for I-129 as follows:

retardation factor = 18.5 (distribution coefficient = 0.5 ml/g soil-to-solution ratio = 35 g/ml)

dispersion length = 10 ft

half-life (decay) = 1.59×10^7 yrs

The individual relationships between groundwater velocity and dispersion, decay and retardation are discussed first, followed by a discussion of the combined or total effect of all the parameters with regard to groundwater velocity.

Dispersion length refers to the total amount of dispersion occurring for a specific distance transported, both during and after release. The amount of dispersion is determined by the total distance traveled, not the rate of transport; but its influence on peak concentration is dependent upon the shape of the concentration distribution. For a hyperbolic or spiked source term, as would result under conditions of low groundwater velocities, dispersion can significantly reduce the peak and spread the base of the source term distribution. In comparison, when the source term is constant (i.e., square-shaped release), as would result at high groundwater velocities, dispersion rounds the leading and trailing edges of the source

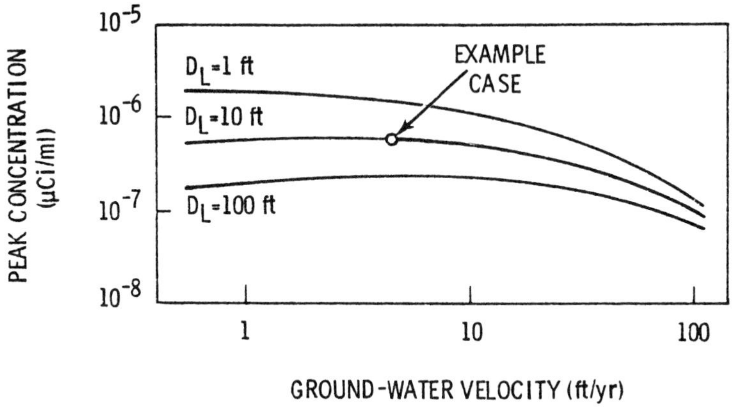

Fig. 1. The Effect of Dispersion Length, D_L, on Peak Concentration for Varying Groundwater Velocities.

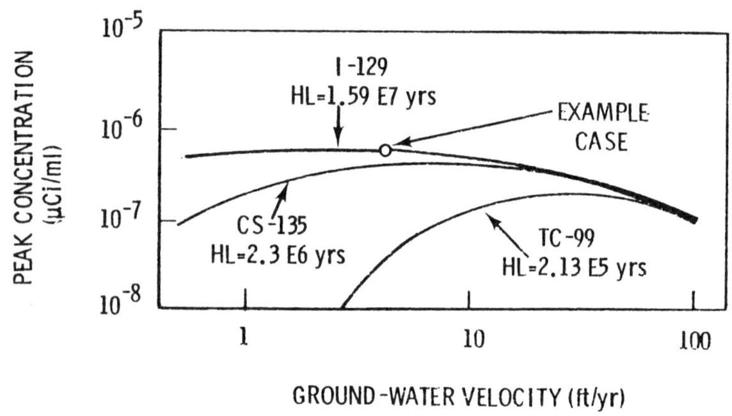

Fig. 2. The Effect of Contaminant Decay on Peak Concentration for Varying Groundwater Velocities

term distribution but has little or no influence on reducing the peak. This relationship is illustrated in Fig. 1. At low velocities, dispersion is the primary factor reducing the more peaked concentration distribution. At high velocities, dilution is the primary factor reducing the peak of the more square-shaped distribution. Fig. 1 also illustrates that, for the example case, an increase in dispersion length reduces peak concentration noticeably regardless of groundwater velocity.

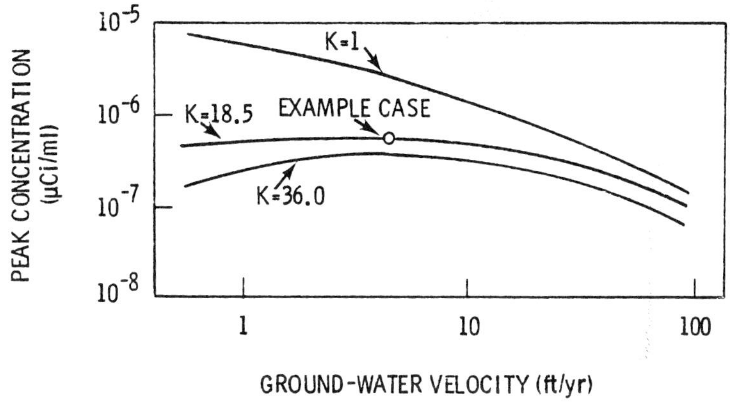

Fig. 3. The Effect of Retardation Factor, K, on Peak Concentration for Varying Groundwater Velocities

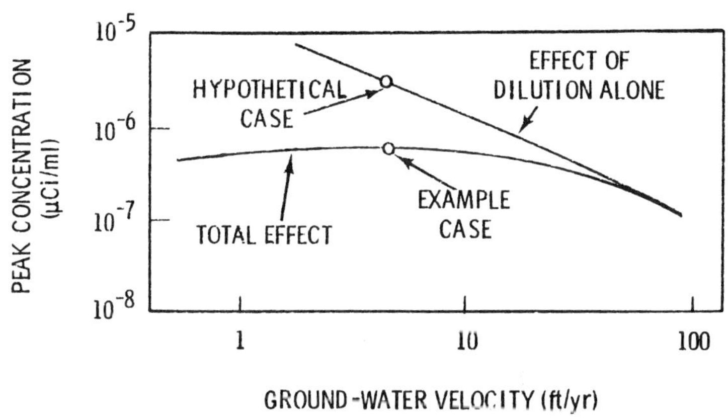

Fig. 4. Comparison of the Hypothetical Case to the Example Case for Varying Groundwater Velocities

Contaminant decay refers to the reduction in contaminant concentration that results from the half-life decay process and is a function of transport time alone. The influence of radioactive decay on the concentration distribution both during and after release is to act as a scaling factor on its overall size. It does not affect the time of peak concentration. Longer leach times and transport times allow greater time for contaminant decay; therefore, lower groundwater velocities, which lengthen transport time, will reduce peak concentration more significantly for contaminants with shorter half-lives (Fig. 2).

Retardation factor is measured of the rate of movement of the contaminants in relation to the groundwater. Retardation alters the contaminant transport time, affecting both decay and dispersion. A high retardation factor retards the movement of the contaminants in relation to the groundwater resulting in a narrower source term distribution. Because total transport time is altered, the amount of contaminant decay is also affected. For the example case, greater retardation reduces peak concentration regardless of groundwater velocity but is more significant for lower velocities because of the longer transport times encountered (Fig. 3).

The total effects of dispersion, dilution, decay and retardation at different groundwater velocities are shown in Fig. 4. This figure illustrates that a maximum peak concentration, the threshold concentration, occurs at a specific groundwater velocity where the effects of dilution balance those of dispersion and decay.

At lower velocities the influences of dispersion and decay exceed those of dilution in reducing peak concentration. At higher velocities, however, the effects of dilution dominate. As the influences of dispersion and decay increase, their effects outweigh that of dilution, resulting in the threshold concentration occurring at a higher groundwater velocity. These results may be valid for those cases when the source term is independent of regional hydrologic conditions since solubility-limited leaching was not considered.

SUMMARY AND CONCLUSION

The rate of transport of contaminants through the geosphere is governed by groundwater velocity, leach rate, and contaminant retardation. The dominant characteristics of the contaminant concentration distribution are inherited during leaching and modified during transport by dilution, dispersion and decay. For a hypothetical non-decaying, non-dispersing contaminant with no retardation properties, the shape of the source term distribution is governed by the groundwater velocity (dilution) and leach rate. This distribution remains unchanged throughout transport. Under actual conditions, however, dispersion, decay and retardation modify the concentration distribution during both leaching and transport. The amount of dispersion is determined by the distance traveled, but it does have a greater peak-reducing influence on spiked distributions than square-shaped distributions. Decay acts as an overall scaling factor on the concentration distribution. Retardation alters the contaminant travel time and therefore indirectly influences the amount of dilution, dispersion and decay.

Simple relationships between individual parameters and groundwater velocity as they influence peak concentration do no exist. For those cases where the source term is not solubility-limited and flow past the waste is independent of regional hydrologic conditions, a threshold concentration occurs at a specific groundwater velocity where the effects of dilution balance those of dispersion and decay.

ACKNOWLEGEMENTS

This research was supported by the Waste Isolation Safety Assessment Program (WISAP) conducted by Pacific Northwest Laboratory. This program was sponsored by the Office of Nuclear Waste Isolation, which is managed by Battelle Memorial Institute under contract EY-76-C-06-1830 with the U. S. Department of Energy. On October 1, 1979, WISAP became the Assessment of Effectiveness of Geologic Isolation Systems (AEGIS) Program and the Waste/Rock Interaction Technology (WRIT) Program. This report is issued by AEGIS.

REFERENCES

1. K. L. Kipp, A. E. Reisenauer, C. R. Cole and C. A. Bryan, Variable Thickness Transient Groundwater Flow Model Theory and Numerical Implementation, BNWL-1703, Battelle, Pacific Northwest Laboratory, Richland, WA (1976).
2. S. W. Ahlstrom, H. P. Foote, R. C. Arnett, C. R. Cole and R. J. Serne, Multicompnent Mass Transport Model: Theory and Numerical Implementation (Discrete-Parcel-Random-Walk Version), BNWL-2127, Battelle, Pacific Northwest Laboratory, Richland, WA (1977).
3. M. O. Hill, Analysis of the Effect of Variations in Parameter Values on the Predicted Radiological Consequences of Geologic Disposal of High Level Waste, NPPB report (to be published).
4. H. C. Burkholder and C. DeFigh-Price, Diffusion of Radionuclide Chains Through an Adsorbing Medium, BNWL-SA-5787, Battelle, Pacific Northwest Laboratory, Richland, WA (1977).
5. M. A. Marino, "Distribution of Contaminants in Porous Media Flow," Water Resources Research, 10, 1013 (1974).

KINETIC EFFECTS ON RADIONUCLIDE TRANSPORT OF GROUNDWATER*

Herman S. Levine

Sandia Laboratories**
Albuquerque, NM 87185

INTRODUCTION

A realistic description of the fluid transport of nuclides through complex stratified geomedia is needed to evaluate the risk associated with the underground storage and disposal of radioactive wastes. In the current approach by others to the problem (1-4), groundwater flow is treated separately from contaminant transport since it is assumed that radionuclides would be present only in trace quantities and, therefore, not alter bulk fluid properties. The transient flow of groundwater through porous media is generally understood (1, 5) and descriptions of fluid velocity can be found. Retardation of the movement of a dissolved radionuclide by a porous medium relative to groundwater flow is then approximated by using well established theories of ion exchange separation processes. Decay chains are also included to account for production and loss of nuclides through radioactive decay as a function of time or distance from the source.

This current transport model does provide a description of the migration of dissolved radioactive waste from a repository utilizing the limited data on and understanding of geochemical interactions presently available from simplified experimental procedures. Studies have shown, however, that the interactions of radionuclides with geomedia are highly variable and complex and that the simple linear equilibrium sorption process currently used is too incomplete to

*This work supported by the U. S. Department of Energy (DOE) under contract No. DE-AC-04-76DP00789.
**A U.S. DOE facility.

adequately model all mechanisms affecting the transport of radioactive waste. Other relevant phenomena and rate effects have been considered as potential contributors to the transport process, but there is currently insufficient data to evaluate them and none have yet been included in any model (4). Some understanding of the potential role played by these other phenomena and rate effects is needed. A more general model of the transport problem is presented here with special emphasis on the effects of sorption kinetics on the form of the transport equation.

THEORY

The migration of soluble nuclides through a porous adsorbing solid medium is basically described by a continuity equation for each component in each phase. Consider a unit representative elementary volume of material of solid density ρ and average porosity ε. Groundwater of unit density (assumed) flows through the connected pores of the element with a known constant average pore velocity μ_o.* Generalization to a variable pore velocity can, in principle, be made. The final conclusions should not be altered. The average concentration of the trace components is $\overline{C}_i$ per unit volume of liquid and $\overline{Q}_i$ per unit mass of solid. While the latter is the conventional concentration definition for the sorbed material, only a surface concentration is implied here. Let

$$C_i = \varepsilon \overline{C}_i \quad \text{and} \quad Q_i = (1-\varepsilon)\rho\overline{Q}_i \tag{1}$$

which transforms the two different concentrations to a common system volume basis, and

$$T = \mu_o t \tag{2}$$

expresses time in terms of the distance traveled by the liquid moving at constant velocity μ_o. The transformed diffusion flux of component i in a moving liquid is

$$f_i = -\frac{D_i}{\mu_o}\frac{\partial C_i}{\partial x} + C_i \tag{3}$$

for unidirectional flow where D_i is a dispersion coefficient (5). Then mass conservation in our tranformed space requires that

$$\frac{\partial C_i}{\partial T} + \frac{\partial f_i}{\partial x} = r_i + \frac{\lambda_{i-1}}{\mu_o} C_{i-1} - \frac{\lambda_i}{\mu_o} C_i + \sum_j A_{ij} \tag{4}$$

* Generalization to a variable pore velocity can, in principle, be made. The final conclusions should not be altered.

and $$\frac{\partial Q_i}{\partial T} = R_i + \frac{\lambda_{i-1}}{\mu_o} Q_{i-1} - \frac{\lambda_i}{\mu_o} Q_i + \sum_j B_{ij} \qquad 5)$$

for component i in each of the phases subject only to the condition

$$r_i + R_i = 0 \qquad 6)$$

where the λ's are first order radioactive decav constants and (i-1) is a precursor.

Equations 4) to 6) provide a general statement of the nuclide transport problem in terms of the kinetics of chemical and physical interaction processes. In Equations 4) and 5) the net accumulation of i in either phase is equated to a series of sources and sinks on the right; i.e., production terms for all chemical and physical processes that affect C_i. The first three account for the kinetics of sorption processes, nuclide generation from a precursor, and nuclide decay, respectively. The A_{ij} and B_{ij} terms (for j processes) deal with the kinetics of other contributing phenomena such as complex ion and atypical species formation, precipitation, oxidation/reduction and polymerization reactions, diffusion from the surface to the interior of the solid, etc. Equation 6) is required to insure that any change of i in the liquid by a sorption reaction causes an equal but opposite change in the solid. Similar restraints may be required for some of the other production processes. Significantly, the liquid flow velocity, μ_o, is found to be a natural yardstick for measuring the relative magnitude and importance of all the production terms. This will be explained later.

Kinetic expressions for the sorption process are of the form

$$R_i = R_i(C_i, Q_i) \qquad 7)$$

depending on the mechanisms involved. For an exchange between univalent ions Amundson (6) and Thomas (7) found that, in terms of the present nomenclature,

$$R_i = \frac{K_1}{\mu_o} Q_i + \frac{K_2}{\mu_o} C_i(Q_o - Q_i) \qquad 8)$$

where K_1, K_2 and Q_o are constants. It has been referred to as adsorption with Langmuir-type kinetics. A simpler expression

$$R_i = \frac{k_i}{\mu_o}\left[\left(\frac{1-\varepsilon}{\varepsilon}\right)\rho K_{d_i} C_i - Q_i\right] \qquad 9)$$

was referred to as adsorption with linear kinetics (6,8) and also deals with an exchange between univalent ions where k_i is a desorption rate constant. Since in the limit $R_i = 0$, there will be an equilibrium value of $Q_i \equiv Q_i^*$ for each C_i,

$$Q_i^* = \frac{(1 - \varepsilon)}{\varepsilon} K_{d_i} C_i \qquad 10)$$

so that from Equation 9),

$$Q_i = Q_i^* - \frac{\mu_o}{k_i} R_i . \qquad 11)$$

More generally, the exchange may involve polyvalent ions such that

$$Q_i^* = \frac{(1 - \varepsilon)}{\varepsilon} K_{d_i} C_i^n \qquad (n<1) \qquad 12)$$

which is generally referred to as a Freundlich isotherm. Several sorption mechanisms may occur simultaneously in geomedia because of the complex nature of the material. The nonlinear Freundlich isotherm in Equation 12) should be generally valid in limited concentration ranges. The linear isotherm in Equation 10), however, is generally used in current codes. Using Equations 4), 5), 6), and 11), it is found that

$$\frac{\partial Q_i^*}{\partial T} + \frac{\partial C_i}{\partial T} + \frac{\partial f_i}{\partial x} + \frac{\mu_o}{k_i}\frac{\partial G_i}{\partial T} =$$

$$\frac{\lambda_{i-1}}{\mu_o}(C_{i-1} + Q_{i-1}) - \frac{\lambda_i}{\mu_o}(C_i + Q_i) + \sum_j (A_{ij} + B_{ij}) \qquad 13)$$

where

$$G_i = \frac{\partial C_i}{\partial T} + \frac{\partial f_i}{\partial x} - \frac{\lambda_{i-1}}{\mu_o} C_{i-1} + \frac{\lambda_i}{\mu_o} C_i - \sum_j A_{ij} \qquad 14)$$

Two special cases of Equation 13) are of interest which depends on extremes in the value of μ_o/k_i. The quality μ_o/k_i is a measure of the velocity of fluid flow relative to that of the sorption kinetics. If $k_i \gg \mu_o$ such that $\mu_o/k_i \to o$, then

$$Q_i = Q_i^* = \frac{(1-\varepsilon)\rho}{\varepsilon} K_{d_i} C_i^{\,n} \qquad 15)$$

Define retardation coefficients as

$$K_i = 1 + \frac{(1-\varepsilon)\rho}{\varepsilon} K_{d_i} n C_i^{\,n-1}$$

$$K_i' = 1 + \frac{(1-\varepsilon)\rho}{\varepsilon} K_{d_i} C_i^{\,n-1} \qquad 16)$$

Equation 13) then becomes

$$K_i \frac{\partial C_i}{\partial T} + \frac{\partial f_i}{\partial x} = \frac{\lambda_{i-1}}{\mu_o} K_{i-1}' C_{i-1} - \frac{\lambda_i}{\mu_o} K_i' C_i + \sum_j (A_{ij} + B_{ij}) \qquad 17)$$

If, futhermore, $n = 1$ and $\sum_j(A_{ij} + B_{ij}) = 0$ then Equation 17) simplifies further to a form equivalent to that used in the current transport model discussed earlier. The current model then is just a very special and limited case of the more general treatment. It is found, if (μ_o/k_i) is relatively small, that $Q_i \approx Q_i^*$. Since $(\partial G_i/\partial T)$ involves higher order derivatives, the product $(\mu_o/k_i)(\partial G_i/\partial T)$ should be significant only in the region of the leading or following edge of the concentration wave passing through the geomedia. Only minor deviations from Equation 17) would then be observed.

If (μ_o/k_i) is relatively large, then $Q_i \neq Q_i^*$ and Equation 13) cannot be simplified much further. While it is not clear how large (μ_o/k_i) must be to be significant, the nature and extent of the departure from the current transport model depends on the magnitude of (μ_o/k_i). If $\mu_o \sim 1.5$ cm/day, then $(\mu_o/k_i) \geq 1$ if the sorption process is less than 90 percent complete in one day. Many laboratory K_d measurements show that much longer periods of time are often required for equilibration (9). For this given set of conditions, some noticeable changes in nuclide migration behavior should be observed. The parameter (μ_o/k_i) therefore appears to have an impact on the form of transport equations. It should, ideally, be nondimensionalyzed. This might be accomplished by introducing a characteristic length such as the pore volume to surface area ratio.

SUMMARY AND CONCLUSIONS

A general description of the transport of radionuclides in porous geomedia was formulated. A mass conservation equation for each phase and kinetic expressions for all the sources and sinks in the system were used. Although the system of equations formulated is relatively complex, the kinetic approach taken provides a more complete description of the transport process. This formulation may be simplified to the current equilibrium sorption model by assuming ion exchange kinetics are more rapid than bulk fluid flow, adsorption follows a linear isotherm, and all other geochemical interactions - remineralization, preciptiation, etc. - may be ignored. The more complete formulation presented here, however, may be more realistic in modeling radionuclide transport processes where adsorption isotherms are nonlinear and ion exchange kinetics are extremely slow. Further, experimentally determined rate constants and hydrologic data may be used in this formulation to identify rate determining steps and ranges for system parameters that may control the radionuclide transport process.

REFERENCES

1. R. T. Dillon, R. B. Lantz and S. B. Pahwa, Risk Methodology for Geologic Disposal of Radioactive Waste: The Sandia Waste Isolation Flow and Transport (SWIFT) Model, SAND78-1267, Sandia Laboratories, Albuquerque, NM 87185, (1978).
2. D. H. Lester, G. Jansen and H. C. Burkholder, "Migration of Radionuclide Chains Through an Adsorbing Medium", Adsorption and Ion Exchange AICHE Symposium Series No. 152, 71, 202 (1975).
3. R. D. Arnett, Conceptual and Mathematical Modeling of the Hanford Groundwater Flow Regime, ARH-ST-140, Atlantic Richfield Comp., Hanford, WA (1976).
4. H. C. Burkholder, Waste Isolation Safety Assessment Program Technical Progress Report for FY-77, PNL-2642, Battelle Northwest Laboratory, Hanford, WA (1979).
5. J. Bear, Dynamics of Fluids in Porous Media (American Elserver, 1975).
6. N. R. Amundson, "Mathematics of Adsorption in Beds. II," J. of Phys. Chem. 54, 812 (1950).
7. H. C. Thomas, "Heterogeneous Ion Exchange in a Flowing System," J. Am. Chem. Soc. 66, 1664 (1944).
8. H. C. Thomas, Ann. N.Y. Acad. Sci. 49, 161 (1948).
9. R. G. Dosch and A. W. Lynch, Interaction of Radionuclides with Argillite from the Eleana Formation on the Nevada Test Site, SAND78-0893, Sandia Laboratories, Albuquerque, NM (1979).

THE EFFECT OF VARIATIONS IN PARAMETER VALUES ON THE PREDICTED RADIOLOGICAL CONSEQUENCES OF GEOLOGIC DISPOSAL OF HIGH-LEVEL WASTE

Marion D. Hill

National Radiological Protection Board
Harwell, Oxfordshire, United Kingdom

INTRODUCTION

A preliminary assessment of the radiological consequences of geologic disposal of high level waste carried out at the National Radiological Protection Board identified several areas where further research is required before this disposal option can be fully evaluated (1). This paper summarizes an analysis of the sensitivity of the results of the preliminary assessment to the assumptions made and the values of the parameters used (2). The purpose of the analysis was to clarify research objectives rather than to derive the possible range of potential doses. The numerical results must be viewed in this context and should not be regarded as representative of those which would be obtained for a specific repository.

Both the preliminary assessment and the sensitivity analysis deal only with disposal of vitrified reprocessing waste. The studies are essentially generic and not site or rock specific. However, where detailed assumptions are required, a repository in hard crystalline rock is considered. Only the consequences of groundwater contacting the waste and transporting radionuclides to the biosphere are dealt with in the sensitivity analysis since the preliminary assessment showed that this is the most likely release mechanism. The assumptions and parameter values used in the preliminary assessment and the reference case are given in Table 1. The approach adopted in the sensitivity analysis was to vary each of the parameters in turn, holding the others constant. This facilitates interpretation of the results. The effects of the assumptions made were examined by reference to the results of the parametric variations. Throughout the analysis, the release, geosphere transport and biosphere models used are those described in the preliminary assessement

(1). Only doses to individuals via drinking water were calculated. The sensitivity of the results to variations in the assumptions and parameters used in modelling the transport of radionuclides through the biosphere was not determined. Previous work showed that uncertainties about the behavior of waste in the repository and transport of radioactivity through the geosphere are likely to have a greater influence on predicted doses than the assumptions made about dilution and dispersion in the biosphere. It is, therefore, more important from the point of view of making recommendations for research to assess the sensitivity of the results to the assumptions and parameters used in the release rate and radionuclide migration calculations.

REFERENCE CASE

The quantity of waste considered is that arising from the generation of 330 GW(e)y electrical energy. Details of the inventory of

Table 1. Assumptions and Values of Parameters Used in the Reference Case

Assumptions

1. Circulating groundwater enters the repository approximately 1000 years after disposal.

2. The waste canisters are not intact at the time of groundwater ingress and do not inhibit leaching.

3. Groundwater contacts all the waste at the same time.

4. The vitrified waste has fractured into pieces of diameter 10 cm.

5. The dominant long term leaching mechanism is dissolution of the waste.

6. The sorption constants of radionuclides and the groundwater flow parameters are constant along the whole of the flow path from the repository to the biosphere.

Parameters

1. Leach rate of vitrified waste = 10^{-5}g cm^{-2} d^{-1}
2. Length of groundwater flow path = 10 km
3. Groundwater velocity = 0.3m d^{-1}
4. Dispersion coefficient = $10m^2d^{-1}$
5. Effective flow rate of freshwater body = 0.3 m^3 s^{-1}
6. Individual daily intake of drinking water = $2x10^{-3}$ m^3

radionuclides and the values used for dose* per unit intake of ingested activity are given in Reference (2).

The calculations show that four radionuclides give rise to the highest peak annual individual doses via drinking water. These are: ^{237}Np (6×10^{-2} Sv y^{-1}), ^{129}I (8×10^{-5} Sv y^{-1}), ^{99}Tc (7×10^{-5} Sv y^{-1}) and ^{226}Ra (3×10^{-5} Sv y^{-1})$^{+}$. The remaining radionuclides give rise to peak annual individual doses below 10 μ Sv. These results are similar to those obtained in the preliminary assessment but are not identical to them due to changes in dosimetric data and models. These changes are discussed in more detail in reference (2). The major difference is that in calculated dose due to ^{237}Np. This arises from the use of a value of 10^{-2} for Fi (the fraction of ingested activity transferred from the gastro-intestinal tract to blood) (3). This value is based on the limited evidence currently available and is subject to considerable uncertainty.

VARIATION OF THE LEACH RATE OF THE WASTE

Most reported values for the leach rates of glasses to be used to incorporate radioactive waste range from about 10^{-3}g cm^{-2} d^{-1} to 5×10^{-11}g $cm^{-2}d^{-1}$. The higher values were obtained from measurements on borosilicate glasses at temperatures close to the limit suggested for a geologic repository in hard rock (about 100°C). Although leach rates would be higher under extreme conditions of temperature and pH, the value of 10^{-3}g cm^{-2} d^{-1} probably represents the maximum realistic leach rate under expected repository conditions. The leach rate of 5×10^{-11} g cm^{-2} d^{-1} was obtained from an experiment in which blocks of nepheline syenite glass were buried at shallow depth beneath the water table. The values measured in this long term experiment are much lower than those normally reported for laboratory tests but it is not clear whether they would be applicable under repository conditions. At the ambient temperatures expected at repository depths, the minimum reported leach rates for borosilicate glasses are typically of the order of 10^{-7}g cm^{-2} d^{-1}.

For the purpose of investigating the sensitivity of the results to variations in leach rate a range of 10^{-3} to 10^{-7} g cm^{-2} d^{-1} was used. As indicated above this does not represent the extreme range of measured values but it is sufficiently wide to examine the effects of variations in this parameter. With the assumption that the waste is fractured into pieces of 10 cm diameter and has a density of 2.6×10^{3} kgm^{-3}, this corresponds to dissolution times of $35-3.5 \times 10^{5}$ years.

*Throughout this paper the term "dose" is used to refer to the effective dose equivalent.

$^{+}$ 1 Sv = 100 rem

The peak annual doses due to representative radionuclides are shown in Table 2. The results show that in general, the peak doses due to those radionuclides which are poorly sorbed on geologic media are strongly dependent on leach rate. Over most of the range of leach rates considered the peak doses due to ^{99}Tc and ^{129}I are approximately proportional to leach rate. This shows that when the time taken for radionuclides to be released from the waste substantially exceeds their transit time from the repository to the freshwater body, the leach rate becomes the principal factor controlling peak doses. In the case of the actinides and their daughter products this effect is complicated by chain decay. For example, the variations in peak annual doses due to ^{226}Ra follow closely those in doses due to its precursors ^{230}Th and ^{234}U. Thus, although on the basis of its sorption constant the peak dose due to ^{226}Ra would be expected to depend on leach rate, in fact it is almost independent of it over the range considered.

VARIATIONS IN GEOSPHERE TRANSPORT PARAMETERS

Groundwater Velocity and Pathlength

The groundwater velocity and the length of the flowpath from the repository to the biosphere are both site specific parameters which will depend on the hydrologic conditions existing at a site when groundwater ingress occurs. They will be affected by the events and processes leading to failure of geologic containment. Although the repository will be sited in an area where there is little or no groundwater flow, it is assumed that hydrologic conditions around the repository change and circulating groundwater enters it. The groundwater velocity and pathlength will depend on factors such as the extent to which the rock around the repository has become fractured, the groundwater flow pattern in the area and thermal effects on water movement. The ranges of values considered for these parameters are therefore large, spanning those which might be encountered in various types of geologic formations in the UK.

The results of the calculations for a selection of radionuclides are given in Table 2. They show that increasing the pathlength and decreasing the groundwater velocity both lead to a decrease in peak annual doses. The decreases become very large when the transit time of the radionuclide through the ground exceeds its half-life except in cases where the peak dose is determined by in-growth from its precursors. When radioactive decay does not influence peak dose (i.e. when the half-life is very long) its variation with groundwater velocity and pathlength depends on the sorption constant of the radionuclide. The principal effect here is dispersion.

Table 2. Variation of Peak Annual Individual Doses with Leach Rate and Geosphere Transport Parameters

Radionuclide/ Parameter	Peak annual dose (Sv)				Time to peak (years after leaching begins)			
Leach rate ($gcm^{-2}d^{-1}$)	10^{-3}	10^{-5}*	10^{-6}	10^{-7}	10^{-3}	10^{-5}*	10^{-6}	10^{-7}
^{99}Tc	2×10^{-3}	7×10^{-5}	8×10^{-6}	8×10^{-7}	1×10^{2}	2×10^{2}	2×10^{2}	2×10^{2}
^{129}I	2×10^{-3}	8×10^{-5}	8×10^{-6}	8×10^{-7}	1×10^{2}	2×10^{2}	2×10^{2}	2×10^{2}
^{135}Cs	3×10^{-7}	3×10^{-7}	3×10^{-7}	1×10^{-7}	9×10^{4}	9×10^{4}	1×10^{5}	1×10^{5}
^{226}Ra	3×10^{-5}	3×10^{-5}	3×10^{-5}	3×10^{-5}	1×10^{6}	1×10^{6}	1×10^{6}	1×10^{6}
^{234}U	1×10^{-8}	1×10^{-8}	1×10^{-8}	9×10^{-9}	1×10^{6}	1×10^{6}	1×10^{6}	1×10^{6}
^{237}Np	1×10^{-1}	6×10^{-2}	1×10^{-2}	1×10^{-3}	1×10^{4}	1×10^{4}	1×10^{4}	1×10^{4}
Groundwater pathlength (km)	1	10*	100	10^{3}	1	10*	100	10^{3}
^{99}Tc	8×10^{-5}	7×10^{-5}	6×10^{-5}	2×10^{-5}	2x10	2×10^{2}	1×10^{3}	1×10^{4}
^{129}I	8×10^{-5}	8×10^{-5}	6×10^{-5}	2×10^{-5}	2x10	2×10^{2}	1×10^{3}	1×10^{4}
^{135}Cs	2×10^{-6}	3×10^{-7}	3×10^{-8}	3×10^{-10}	1×10^{4}	9×10^{4}	9×10^{5}	8×10^{6}
^{226}Ra	8×10^{-4}	3×10^{-5}	2×10^{-6}	2×10^{-7}	1×10^{5}	1×10^{6}	1×10^{7}	1×10^{8}
^{234}U	6×10^{-7}	1×10^{-8}	6×10^{-10}	6×10^{-11}	1×10^{5}	1×10^{6}	1×10^{7}	1×10^{8}
^{237}Np	2×10^{-1}	6×10^{-2}	7×10^{-3}	2×10^{-4}	2×10^{3}	1×10^{4}	9×10^{4}	9×10^{5}
Radionuclide/ Parameter	Peak annual dose (Sv)				Time to peak (years after leaching begins)			
Groundwater velocity (md^{-1})	3	0.3*	0.03	0.003	3	0.3*	0.03	0.003
^{99}Tc	8×10^{-5}	7×10^{-5}	5×10^{-5}	8×10^{-6}	2x10	2×10^{2}	1×10^{3}	6×10^{3}
^{129}I	8×10^{-5}	8×10^{-5}	5×10^{-5}	8×10^{-6}	2x10	2×10^{2}	1×10^{3}	6×10^{3}
^{135}Cs	3×10^{-6}	3×10^{-7}	2×10^{-8}	4×10^{-10}	1×10^{4}	9×10^{4}	8×10^{5}	3×10^{6}
^{226}Ra	9×10^{-4}	3×10^{-5}	1×10^{-6}	7×10^{-8}	1×10^{5}	1×10^{6}	1×10^{7}	6×10^{7}
^{234}U	9×10^{-7}	1×10^{-8}	4×10^{-10}	2×10^{-11}	1×10^{5}	1×10^{6}	1×10^{7}	6×10^{7}
^{237}Np	2×10^{-1}	6×10^{-2}	5×10^{-3}	7×10^{-5}	1×10^{3}	1×10^{4}	9×10^{4}	5×10^{5}
Sorption constant +	Reference	Maximum	Minimum		Reference	Maximum	Minimum	
^{99}Tc	7×10^{-5}	4×10^{-8}	7×10^{-5}		2×10^{2}	3×10^{5}	2×10^{2}	
^{129}I	8×10^{-5}	2×10^{-6}	8×10^{-5}		2×10^{2}	1×10^{5}	2×10^{2}	
^{135}Cs	3×10^{-7}	3×10^{-6}	5×10^{-9}		9×10^{4}	1×10^{4}	3×10^{6}	
^{226}Ra	3×10^{-5}	3×10^{-8}	8×10^{-5}		1×10^{6}	9×10^{7}	8×10^{3}	
^{234}U	1×10^{-8}	8×10^{-11}	2×10^{-5}		1×10^{6}	9×10^{7}	1×10^{4}	
^{237}Np	6×10^{-2}	1×10^{-6}	8×10^{-2}		1×10^{4}	8×10^{6}	1×10^{3}	

* Reference value + See reference (2) for values

Sorption Constants (Retardation Factors)

Throughout the calculations it is assumed that conditions of linear equilibrium sorption prevail. In this case, the sorption constant (K) is the ratio of groundwater velocity to radionuclide migration velocity. To investigate the effects of variations in sorption constant calculations were carried out for a number of fission products and actinides, including those radionuclides which, on the basis of previous calculations, give rise to the highest peak doses. For each radionuclide, the range of sorption constants considered is representative of values reported in the literature. Results for a selection of radionuclides are given in Table 2.

The results show that, in general, as the sorption constant increases the peak dose due to the radionuclide decreases and the peak occurs at a later time. However, while the increase in the time at which the peak occurs is approximately proportional to the change in K value this is not the case with the decrease in peak dose. The difference between peak doses calculated using K values of 1 and 100 is small. Large decreases in peak dose occur when the transit time of the radionuclide becomes comparable to or greater than its radioactive half-life. These effects are complicated by in-growth in the case of the actinides and their daughter products. Changes in the sorption constant of parent radionuclides affect not only the peak dose due to the radionuclide itself, but also those due to its daughters.

Dispersion Coefficient (Dispersivity)

Dispersion is a combination of purely mechanical effects (variations in groundwater velocity about the mean value, due to inhomogeneities in the geologic medium) and physicochemical effects (molecular diffusion resulting from concentration gradients). The ratio of dispersion coefficient to groundwater velocity is the characteristic dispersive length of the geologic medium and is a measure of its heterogeneity. Groundwater may flow through different geologic media on its path to a freshwater body. However, the geosphere transport model assumes a constant dispersion coefficient so it was not possible to investigate the effect of variations in heterogeneity along the flow path. Instead a wide range of characteristic dispersive lengths was considered, the minimum being 3m and the maximum 3km. The results for variations in groundwater velocity indicated that the effects of dispersion may vary in magnitude at different velocities. Therefore, at each of the groundwater velocities, a range of dispersion coefficients corresponding to the characteristic dispersive lengths was used.

The results of the calculations are given in detail in Reference 2. They show that the effect of dispersion on peak dose varies with the groundwater velocity and the sorption characteristics of the radionuclide. At the higher groundwater velocities peak doses are determined principally by convective flow; dispersion plays only a minor role. At the lower groundwater velocities varying the dispersion coefficient has more significant effects on peak doses due to strongly sorbed radionuclides. If the transit time of the radionuclide is substantially greater than its half-life an increase in dispersion leads to an increase in peak dose. This is due to the decrease in transit time caused by increasing dispersion allowing less time for radioactive decay. Where the transit time of the radionuclide is less than its half-life, increasing the dispersion coefficient results in a relatively small decrease in peak dose.

TIME OF GROUNDWATER INGRESS

The sensitivity of calculated peak doses to the time at which groundwater ingress is assumed to occur can be inferred from the results described in previous sections and consideration of the time variation of the inventory of principal radionuclides in the waste. The radionuclides which give rise to the highest doses are either very long-lived or grow in from very long-lived parents so that their inventories effectively remain constant for long times after disposal e.g. ^{99}Tc, ^{129}I, ^{237}Np, ^{226}Ra. The variation of the inventory of radionuclides with time will only have a noticeable effect on peak doses if the time taken for the waste to dissolve and to migrate from the repository to the biosphere (for convenience this will be called the transport time) is shorter than the time at which groundwater ingress occurs. In all other cases, the effects of radioactive decay during leaching and transport through the geosphere will outweigh those occurring prior to entry of groundwater into the repository.

It can therefore be concluded that, considering only the time variation of the inventory of radionuclides, peak doses will depend on the time at which groundwater ingress occurs if either:

a) groundwater ingress occurs within about 300 y after disposal (when the shorter-lived fission products such as ^{90}Sr and ^{137}Sc have decayed to very low levels) and the transport times of the shorter-lived radionuclides are less than approximately 300 y; or

b) groundwater ingress occurs at a time when the inventory of radionuclides which are of most concern is beginning to decay significantly (for ^{237}Np, ^{99}Tc and ^{129}I more than about 10^6 y), and the transport times are shorter than the time at which leaching begins.

Conditions within the repository and the integrity of the vitrified waste and the canisters will all vary with time after disposal. The rate of release of radionuclides into groundwater is therefore time dependent. At the present state of knowledge, the exact nature of this time dependence is difficult to predict. However, it has been shown that the leach rate of waste increases with increasing temperature. Hence, if groundwater ingress occurs within a few hundred years of disposal, the leach rate would be greater than if it occurred at times when the shorter-lived fission products have decayed and the temperature within the repository has returned to the ambient level. This conclusion does not take into account possible reductions in the integrity of the waste, but available data indicate that these are likely to be small over this time period, and probably are very much longer ones.

The above discussion shows that if groundwater enters the repository during the period when the inventory of radionuclides in the waste and the leach rate remains constant, peak doses will be independent of the time of ingress. They will, however, still depend on the geosphere transit time of radionuclides which will be affected by the events and processes leading to water ingress. If groundwater enters the repository during the first few hundred years after disposal, peak doses due to poorly sorbed radionuclides (such as ^{99}Tc and ^{129}I) will be quite strongly dependent on the ingress time because they depend on the rate of release from the waste. The peak doses due to more strongly sorbed radionuclides will only be affected by the by the water ingress time if the events and processes leading to failure of geologic containment have created short flowpaths to the biosphere or high groundwater velocities. At very long times after disposal, the very long-lived radionuclides begin to decay to low levels and the transport times of all radionuclides are likely to be shorter than the time at which water ingress occurs. Peak doses will therefore depend on the time of ingress because they will be determined by the inventory of radionuclides in the waste.

DISCUSSION AND CONCLUSIONS

The results summarized in Sections 3 and 4 show that over the range of parameters considered, the geosphere transit times of almost all the radionuclides (^{99}Tc and ^{129}I are the only notable exceptions) are greater than the leach time. Peak doses are strongly dependent on groundwater velocity, pathlength and sorption constants, but are only marginally affected by the rate of release of radionuclides into groundwater. It can therefore be concluded that the main barrier to the return of activity to man's environment is that provided by slow migration through the geosphere while the leach resistance of the waste form is relatively unimportant.

This conclusion has implications for research on the leach rate of vitrified waste, the development of extremely leach resistant waste forms and investigation of buffering materials which could affect leach rates. It is obvious that measurements of short term leach rates are of no value for the assessement of potential doses. Research should be directed towards the prediction of leach rates over long time periods under conditions representative of those expected in a repository. Development of a more highly leach resistant waste form would not appear to be warranted. Similarly, the alternative approach of surrounding waste canisters with a buffer material which causes local alterations in groundwater chemistry, thus reducing leach rate must be viewed with caution. If further research confirms that only a few radionuclides migrate rapidly with groundwater, it may be preferable to surround waste canisters with a material which preferentially sorbs these radionuclides rather than attempting to reduce the rate of release of all radionuclides. Additional calculations (2) showed that materials of very high sorptive capacity could reduce peak doses. As discussed in the previous section, calculated peak doses are largely insensitive to the time at which circulating groundwater enters the repository if this time is greater than a few hundred years after disposal and less than the time required for the very long-lived radionuclides to decay. This result can be used to define preliminary criteria for engineered barriers to radionuclide migration. For example, encapsulating waste in canisters which are designed to remain intact for periods as long as 10^5 years is unlikely to lead to a greater reduction in peak doses than the use of canisters with a mean life of a few hundred years.

The conclusions concerning the importance of the geologic barrier to radionuclide migration and the insensitivity of potential doses to the time at which water ingress occurs are to a large extent contingent on geosphere transit times of radionuclides being long. The shortest pathlength from the repository to the biosphere is upwards through the back-filled tunnels and access shafts. Groundwater may move in this direction under the influence of the thermal gradient in the rock if water ingress occurs at a sufficiently early time or if hydrologic conditions change to provide a potential for upward water movement. It is therefore important that the material used for backfilling should retain its structural integrity, offer high resistance to water flow and, ideally, have a high sorptive capacity for the radionuclides of major concern in terms of potential doses.

The importance of sorption of radionuclides on geologic media was highlighted in the preliminary assessment (1) and is confirmed by the sensitivity analysis. It is the main factor determining which radionuclides give rise to the highest peak doses, in many cases its effects override differences in radiotoxicity and inventory of radionuclides in the waste. Throughout the analysis, it is the radionuclides which are assumed to be poorly sorbed (^{237}Np, ^{226}Ra, ^{129}I and ^{99}Tc) which dominate peak doses. Although it has been shown that the geosphere transit time is the fundamental parameter controlling doses, it should be noted that the range of reported sorption constants is larger than those of groundwater velocity and pathlength. Uncertainties in sorption constants are therefore the major source of uncertainty in the prediction of doses.

Finally, it can be seen from the results obtained by varying the dispersion coefficient that dispersion can have a significant effect on peak doses. When the geosphere transit times of radionuclides are greater than their radioactive half-lives, dispersion increases peak doses over those predicted considering convection alone. Hence, models of radionuclide migration which neglect dispersion may underestimate peak doses and the assumption of a low dispersion coefficient is not necessarily conservative.

Note

The work described in this paper was carried out under contract to the United Kingdom Atomic Energy Authority, Harwell, on behalf of the Commission of the European Communities.

REFERENCES

1. M. D. Hill and P. G. Grimwood, Preliminary Assessment of the Radiological Protection Aspects of Disposal of High-Level Waste in Geologic Formations, NRPB-R69, (1978).
2. M. D. Hill, Analysis of the Effect of Variations in Parameter Values on the Predicted Radiological Consequences of Geologic Disposal of High-Level Waste, NRPB-R86, (1979).
3. N. Adams, B. W. Hunt and J. A. Reissland, Annual Limits of Intake of Radionuclides for Workers, NRPB-R82, (1978).

SENSITIVITY STUDIES OF THE SWIFT RADIONUCLIDE TRANSPORT MODEL

James E. Campbell, Brenda S. Langkopf, Ronald L. Iman
and Mark Reeves*

Sandia Laboratories, Albuquerque, NM,
*INTERA Environmental Consultants, Houston, TX

INTRODUCTION

Risk analysis is a vital part of any safety assessment of a proposed nuclear waste depository. Risk analysis results are expressed in terms of radiation-induced health effects and are calculated using mathematical models. These models describe the following processes: 1) radionuclide migration in groundwater from the depository to a discharge point in the surface environment, 2) radionuclide migration and accumulation in the surface environment, 3) human uptake and 4) radiation dosimetry and health effects. There are large uncertainties associated with many of the input variables to these models so that ranges, rather than point values, are often specified. Consequently, the risk estimates themselves have large uncertainties. Therefore, point estimates of risk, without an indication of the possible distribution of results, are probably meaningless and may be misleading.

Given the importance of quantifying uncertainties in risk estimates, it is clearly important to identify those input variables which are major contributors to uncertainty in model output. It is also important to establish techniques to generate model output distributions with a limited number of computer runs for assumed ranges and distributions on model input variables. Such is the purpose of a sensitivity analysis. The work reported here is part of a larger program funded by the Nuclear Regulatory Commission to develop a risk assessment methodology for geologic disposal of radioactive waste. Two important parts of this methodology were used in this study, namely a numerical geosphere transport simulator and statistical techniques for analyzing transport calculations. The transport simulator is the Sandia Waste Isolation Flow and Transport (SWIFT) code

(1). It is a three-dimensional, finite-difference model which solves conservation equations for fluid flow, heat transport, (possibly non-trace) solute mass and radionuclides in trace quantities. The statistical techniques (2) were designed to reduce the number of transport calculations while providing a reliable identification of important input variables; i.e. those variables which explain a significant amount of variation in the model output. These models are discussed further in Section III. To demonstrate the application of the transport and statistical methods, we have chosen a reference site and a particular depository breachment scenario. The physical problem is described in the next section, and results from the sensitivity analysis are presented in Section IV.

PHYSICAL PROBLEM

The sensitivity analysis described in this paper was applied to a strictly hypothetical system. This system consists of both a reference site containing a nuclear waste depository (3) and a disruptive-event scenario for the possible release of radioactive materials from the depository to the biosphere. The reference site is located in a symmetrical upland valley and is shown in cross-section in Fig. 1. The crest of the ridge surrounding the valley is at an elevation of 6000 feet, and the crest is a surface- and groundwater divide so that the only water moving in the valley falls in the valley itself. The valley is drained by a major river, River L, which is at elevation 2500 feet. Tributaries to River L exist, such as River U, but they are normally dry. The valley receives 40 inches of rainfall per year, of which 16 inches are lost by evapotranspiration and the remaining 24 inches recharge the groundwater system. The geology of the site is also shown in Fig. 1. The valley is underlain by crystalline bedrock which crops out only over a narrow width at the ridge crest surrrounding the valley. This bedrock is assumed to be impermeable to groundwater flow. Over the bedrock is a sequence of sedimentary rock as shown in Fig. 1. Nominal values of all hydraulic and thermal properties are given in Reference 3.

One disruptive event which conceivably could release radioactive waste from a depository such as that of the reference site is the formation of a U-tube. Here we assume seal failures of a shaft (30 feet in diameter) on the up-dip side and of a borehole (3 feet in diameter on the down-dip side. The shaft, for example, could be a man-materials connection to the surface, and the borehole could be a ventilation pipe. The U-tube, as shown in Fig. 1, provides a flow path between the depository and the upper aquifer. The shaft and borehole are not extended through the upper shale and upper sand and gravel as there is no significant vertical hydraulic gradient. Numerical simulations have been carried out for the full two-dimensional system of Fig. 1 and for some selected three-dimensional systems in the near field of the depository using the SWIFT computer code. For

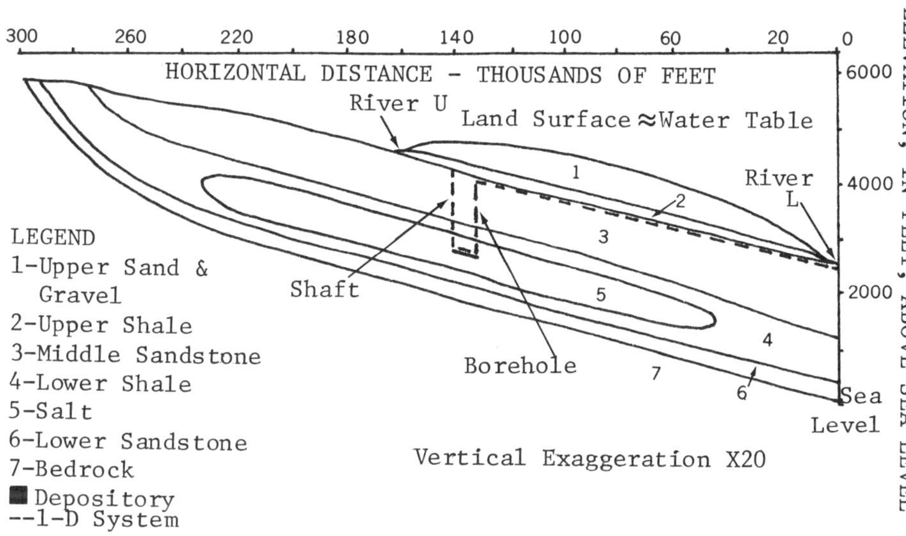

Fig. 1. Geologic Cross-Section Through Reference Site

the sensitivity analysis, however, a simplified one-dimensional representation of the full system and the U-tube was used. This representation is shown as a dashed line in Fig. 1. A simulation of the full system indicated that the radionuclide migration path could be represented by two one-dimensional segments, one for the borehole and one for the aquifer. These segments had constant but different flow velocities. Radionuclides were introduced into the system from an assumed inventory and leach rate.

As indicated in the Introduction, fluid flow and radionuclide transport are controlled by hydrological and geochemical variables whose values are not generally well known for any given site. For the reference site, appropriate estimates of these uncertainties were provided by earth scientists from Sandia Laboratories, the U. S. Geological Survey and INTERA Environmental Consultants. The results of this effort are shown in Table 1 for the simplified ^{238}U chain, $^{238}U \rightarrow {}^{234}U \rightarrow {}^{230}Th \rightarrow {}^{226}RA$. The ranges chosen are somewhat global in nature. Measurements should provide more narrow ranges for a specific candidate disposal site.

DESCRIPTION OF STATISTICAL METHODS

For the physical problem described above, we present results for the total (integrated) radionuclide discharge in the 10^6-year time period following depository closure. The pertinent questions are:

1. Which variables, of those listed in Table 1, have a significant effect upon the total discharge and what is their relative importance?

2. What relationships exist between the distributions of model input and model output?

Conceptually, the manner in which the first question is answered is based on a simple geometrical picture such as the one shown in Fig. 2. The dependent variable Y is given as a function of the two independent variables X_1 and X_2 over their respective ranges. Surface S is called the response surface. In general, the value of the dependent variable Y is known only at certain discrete points (i.e. for particular values of X_1 and X_2) as indicated by the crosses in Fig. 2. In the physical problem addressed in this paper, the discrete points are known as a result of transport calculations with the SWIFT code. The fitting of the response surface by stepwise regression is of central importance to the sensitivity analysis as this process selects the important independent variables. Standardized regression coefficients derived from the regression analysis characterize the degree of dependence of model output on model input variables. These coefficients provide a basis for determining the relative ordering of the independent variables. Thus, the first question posed at the beginning of this section may be answered through response-surface construction. Although simple in concept, response surface construction involves several statistical procedures. There is, for example, preliminary variable assessment using partial correlation coefficients, stepwise regression for construction of the response surface, protection against overfit, use of

Table 1. Parameter Ranges and Distributions

Variable*	Range	Distribution
ϕ_s	0.005 - 0.20	Log Normal
ϕ_a	0.05 - 0.30	Normal
K_s	0.01 - 50 ft/d	Log Normal
K_a	1.0 - 50 ft/d	Log Normal
α	45 - 500 ft	Uniform
K_d(U)	10^{-2} - 10^5 cm^3/g	Log Uniform
K_d(Th)	10^{-2} - 10^5 cm^3/g	Log Uniform
K_d(Ra)	10^{-2} - 10^3 cm^3/g	Log Uniform
τ	10^3 - 10^7y	Log Uniform

*Variable definitions:
ϕ = porosity, K = hydraulic conductivity, α = dispersivity,
K_d = distribution coefficient, τ = leaching time of a constant-rate leach model, s = shaft and borehole, a = aquifer.

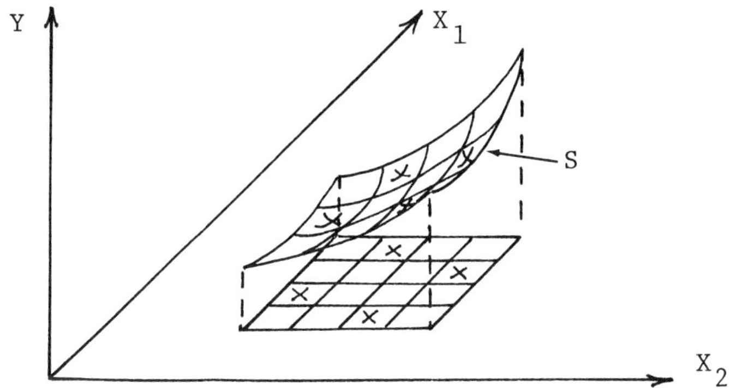

Fig. 2. Conceptualization of a Response Surface

t-tests to establish significance levels for the independent variables and various tests to examine the effect of input variable distribution assumptions on the output variable distribution. Details of these procedures can be found in Reference 3 and in references cited therein. Here, however, we wish to highlight two relatively new statistical techniques which were used.

The first technique is Latin Hypercube Sampling (LHS) (4) which minimizes, relative to other commonly used schemes, the number of model calculations required for response surface construction. (For the numerical radionuclide transport simulator used here, the calculations can be quite expensive.) LHS assures selection of points from the full range of input variables and the fitted response surface yields an unbiased estimate of cumulative distribution functions and mean values for the model output. It is for these reasons that LHS is being used for sensitivity and risk analysis. A brief description of LHS is provided in the following discussion. First the independent variable ranges X_i are divided into intervals I_{ji} based on equal probablility. Values T_{ji} are then randomly selected from each of these intervals (indicated by stars below). An input vector is formed by randomly selecting the T_{ji} without replacement for each variable. For the simple four-point example of Fig. 2. each variable range is divided as shown below,

and the four input vectors of independent variables are $T_1 = (T_{31}, T_{12})$, $T_2 = (T_{11}, T_{22})$ and $T_3 = (T_{41}, T_{32})$, $T_4 = (T_{21}, T_{42})$.

The second technique from the response surface methodology which we wish to highlight is rank regression (5). The motivation here stems from the nonlinearity which is frequently present in the dependence $Y = f(X_1, X_2, \ldots)$ of the basic response variable upon the independent variables. If, however, the dependence is monotone, as it frequently is for transport calculations, then it may be linearized with the rank transformation. The concept of a rank transform is simple in that each variable (independent and dependent) is replaced by its corresponding rank. For example, the smallest value of X_i is replaced by the rank 1, the next smallest by the rank 2, and so on. Thus the original discrete valued function is replaced by the rank-transformed function $R(Y) = f[R(X_1), R(X_2), \ldots]$. It is this transformed function which is fitted with a response-surface using stepwise regression and which is used to establish the significance and relative importance of the independent variables.

To answer the second question posed at the beginning of this section, i.e., the relationship of distributions of output variables to the assumed distributions of input variables, we can either propagate new distributions through the fitted response surface or reweight the base set of Latin hypercube input vectors as outlined in Reference 6 to reflect different distribution assumptions. This aspect of the analysis thus provides a means of assessing the sensitivity of model output to the distribution assumptions of the input variables. The utility of this procedure is demonstrated in the next section for the physical problem discussed in Section II.

RESULTS AND CONCLUSIONS

To identify the important input variables for the system described in Section II and to establish relative importances of these quantitites, stepwise regression analysis was performed on the total discharge to River L over a fixed time (10^6 years). The important input variables for total discharge are shown in Table 2. (In-

Table 2. Important Input Variables for Total Discharge

	^{238}U	^{234}U	^{230}Th	^{226}Ra
	K_d (U)	K_d (U) τ K_s ϕ_s	K_d(Th) τ K_s ϕ_s	K_d(Ra) K_s
N	30	28	39	47
R^2	0.657	0.850	0.909	0.685

tegrated discharge for shorter time periods and peak discharge rate were also evaluated as dependent variables. Important input variables were, with few exceptions, identical to those given in Table 2.)

The number N of nonzero values for each isotope which were used in the stepwise regression analysis are shown in the table. Regression analysis was performed both with and without zero values. Although the important input variables were similar in the two cases, inclusion of zero values tended to obscure relationships between independent and dependent variables. The R^2 values, if multiplied by 100, represent the percent of the output variation (expressed in ranks) which is assigned to the input variables selected by stepwise regression. For each isotope, the distribution coefficient K_d ranked as the most important variable. The time of leach for a constant leach rate model, hydraulic conductivity in the shaft K_s, and shaft porosity ϕ_s enter as less important variables.

Fig. 3. shows cumulative distributions of total discharge for ^{238}U. The solid line represents results from the original SWIFT calculations. Although only 30 nonzero values were used to fit the response surface, the agreement is quite good. The agreement between SWIFT and the response surface curves for ^{234}U, ^{239}Th and ^{226}RA (not shown) is of similar high quality. Fig. 4. shows cumulative distributions of total discharge for ^{238}U where zero values have been in-

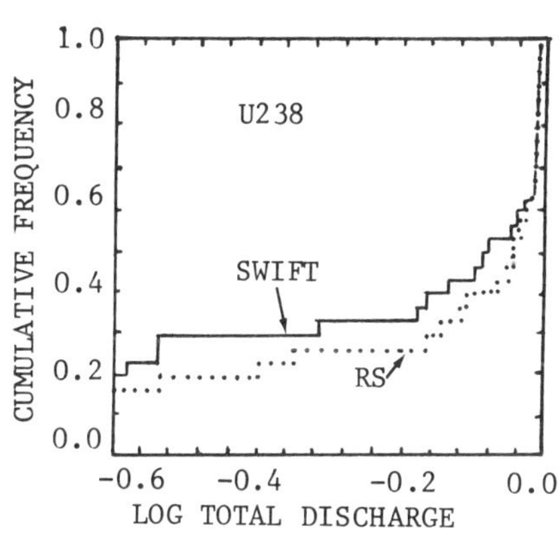

Fig. 3. Cumulative Distributions of Total Discharge Relative to Source. The Solid Line Represents Results From SWIFT. The Dotted Line Represents Results From Fitted Response Surface. Calculations Which Produced Zero Discharge are Excluded.

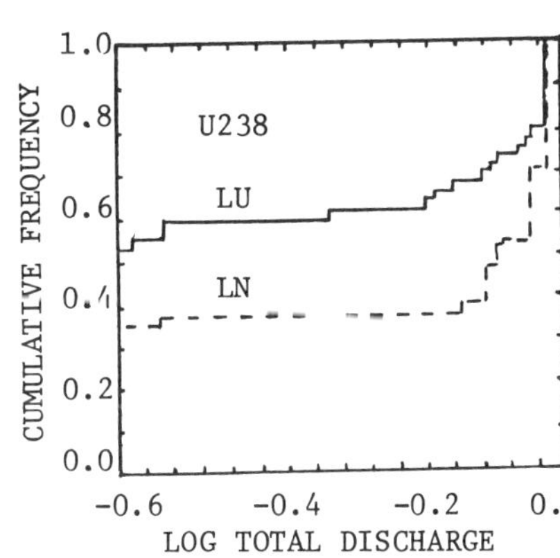

Fig. 4. Cumulative Distributions of Total Discharge Relative to Source. The Solid Line Represents Results from SWIFT with Log Uniform Distributions for K_d Values. The Dashed Line was Obtained by Reweighting SWIFT Results to Represent Log Normal Distributions of K_d Values.

cluded. These results are included to show the sensitivity of SWIFT output to the distribution assumption for the distribution coefficient (K_d). The curve labelled LU represents the original SWIFT results where a log-uniform distribution was used for the K_d values. To obtain the curve labelled LN, the original input vectors were reweighted to represent a log-normal distribution for the K_d values.

In conclusion, we note four things: 1) Although we have used only nine independent variables in this case, the methods are quite efficient for much larger sets of input variables (7). Therefore, our simulation and statistical techniques appear to be applicable to a field-scale waste-isolation problem. 2) Response surfaces have been constructed with less than 50 calculations by the geosphere-transport simulator. 3) The dispersivity is unimportant because total releases are used, and therefore, details of the shape of the breakthrough are unimportant. 4) The dominance of the distribution coefficient in each case is reasonable in view of the rather large ranges used for this quantity.

REFERENCES

1. R. T. Dillon, R. B. Lantz, and S. B. Pahwa, Risk Methodology for Geologic Disposal of Radioactive Waste: The Sandia Waste Isolation Flow and Transport (SWIFT) Model, SAND78-1267, Sandia Laboratories, Albuquerque, NM (1978).
2. R. L. Inman, J. C. Helton and J. E. Campbell, Risk Methodology for Geologic Disposal of Radioactive Waste: Sensitivity Analysis Techniques, SAND78-0912, Sandia Laboratories, Albuquerque, NM (1978).
3. J. E. Campbell, Risk Methodology for Geologic Disposal of Radioactive Waste: Interim Report, SAND78-0029, Sandia Laboratories, NM (1978).
4. M. D. McKay, W. J. Conover and R. J. Beckman, "A Comparison of Three Methods for Selecting Values of Input Variables in the Analysis of Output from a Computer Code," TECHNOMETRICS (to be published)
5. R. L. Iman and W. J. Conver, "The Use of the Rank Transformation in Regression," TECHNOMETRICS 21 (to be published).
6. R. L. Iman and W. J. Conover, "Small Sample Sensitivity Analysis Techniques for Computer Models with Application to Risk Assessment," submitted to Communications in Statistics.
7. J. C. Helton and R. L. Iman, Risk Methodology for Geologic Disposal of Radioactive Waste. Sensitivity Analysis of Pathways Model, SAND79-1393, Sandia Laboratories, Albuquerque, NM (1980).

DEVELOPMENT OF SCENARIOS FOR THE LONG-TERM RELEASE OF RADIONUCLIDES FROM THE PROPOSED WASTE ISOLATION PILOT PLANT IN SOUTHEASTERN NEW MEXICO*

Felton W. Bingham and George E. Barr

Sandia Laboratories
Albuquerque, NM 87185

INTRODUCTION

The Waste Isolation Pilot Plant (WIPP) is a project of the U.S. Department of Energy. At the plant, radioactive waste produced in defense programs will be buried in bedded-salt formations lying deep below the Delaware basin in southeastern New Mexico. The principal contaminants in this waste will be transuranic elements; little of the waste will contain fission products. Because the time required for isolating these long-lived elements is still under discussion, the safety analyses of the WIPP repository must deal with conditions in the far distant future, at times as long as hundreds of thousands of years. The analyses must, in fact, estimate the consequences of unexpected releases of waste from the repository at such distant times. Computer codes for predicting these consequences are available, but they must be supplied with descriptions of the mechanisms that can be assumed to release the waste. Users of these codes, in other words, must develop scenarios--hypothetical sequences of events that could conceivably release waste from a repository.

Approximately three years ago, the WIPP staff began to prepare detailed safety analyses. Since it was clear that they would need descriptions of release mechanisms, the study group began their work by examining hypothetical releases. This examination, the first step in the safety analyses, had four purposes:

1. To decide which of the many conceivable processes for the release of waste are possible at the WIPP site.

*This work supported by the U.S. Department of Energy.

2. To point out which of these processes need further experimental study and which of them can be eliminated from modeling.
3. To recommend the most probable processes for modeling early in the safety-analysis program.
4. To group the processes into classes, each represented by a single basic computer model.

To meet these objectives, the group used the methods described in this paper.

PRELIMINARY ANALYSES

Like some other investigators of the geologic disposal of waste (1, 2), the WIPP study group began their work by constructing a primitive fault tree. One part of the tree described events that can release waste from the repository. This part had three major branches because release can occur only if three events take place: waste must become available for release, a force must move the waste, and a path through which the waste can move must be created. Following the usual methods of fault-tree construction, the group expanded these branches by listing specific faults and then expanding these faults until the branches reached well-defined basic events. A computer code determined the cut sets of the tree and calculated the probabilities of their occurrence from probabilities attached to each fault.

The group gained five important insights from constructing the fault tree. First, assigning accurate probabilities to the tree branches proved to be extremely difficult. It was necessary, for example, to estimate the future probability of geologic events that had previously occurred at the WIPP site only rarely or never in the history of the earth; to make such guesses from the limited data available would stretch not only the predictive ability of geology but also the applicability of probability theory to predictions of the future. To estimate the condition of buried waste after thousands of years would require more extensive data than those now available. And to predict the structure of human society more than a few hundred years into the future would require data that no experiments will provide. For reasons like these, the WIPP group felt it prudent to avoid safety-analysis methods that lean heavily on probabilities.

Second, the number of cut sets available from the simple fault tree was extremely large. The tree would have become even less manageable if it had been expanded until the group felt confident that it was complete. Furthermore, the relative importance of the cut sets was severely shifted by the necessary inclusion of events for which no probability data exist. Because such events had to be given unit probability by default, the cut sets containing them had artificially high probabilities and, therefore, artificial importance.

Third, the tree afforded the group only a limited capability for discovering and handling competing risks. The cut sets derived from the tree cannot all occur simultaneously; some events are mutually exclusive. The standard computer techniques for analyzing fault trees do not handle these competing cut sets; careful interpretation of the computer output is required, and when the number of cut sets is large, that interpretation must be lengthy and complex.

Fourth, the tree had to include events that depend on time in widely different ways. For this reason, the complete tree would have had to contain conditional statements and probabilities. Even a careful insertion of these conditions would have produced, in practice, a separate, complex fault tree describing events at each future time for which predictions are needed.

Fifth, assigning probabilities was not an appropriate way to handle the events whose future occurrence can be tested experimentally. Probabilities have little meaning when attached to deterministic events like those whose occurrence depends, for example, solely on waste form or on interactions between the waste and the medium in which it is buried.

These five realizations led the WIPP study group to feel that a straightforward continuation of their fault-tree analysis would not provide the scenarios they needed in the time available for guiding experimental programs and plant design. Although the fault-tree analysis would eventually produce a risk assessment derived from first principles, the need for detailed probabilities, for other fundamental data, and for complex interpretations would require studies lasting for years.

CONSTRUCTION OF EVENT TREES

The group therefore switched their attention to the construction of event trees. The cut sets of an event tree are scenarios. Each event tree begins with a breaching event--an event assumed to disrupt the repository so severely that subsequent events can move waste through the breach. The tree also displays these subsequent events, all of which lead to a final event that brings radioactive waste into contact with people. Each of the subsequent events specifies, at least implicitly, a mechanism for moving the waste through the breach and the response of the salt to the breach. In addition to the fault tree, the group used several sources of information for constructing the event trees: local experience in drilling and mining, the results of experiments investigating the interactions between waste and salt, and discussions with consultants and staff in hydrology, geophysics, and mining engineering.

Contents of Event Trees

The study group began their construction of event trees with the breaching and nuclide-transport mechanisms in the original fault tree. It was not necessary, however, to carry the derivation of basic events to the level of detail required for full fault-tree analysis. Instead, the group simply specified breaching events that they knew were physically possible in the Delaware basin; they did not specify in detail how the breaching events might come about. This practice took advantage of the fact that many cut sets begin with similar breaching events; unless there was some insight to be gained from further expansion, the basic events contributing to the breach could remain lumped together.

The group continued to construct event trees until they felt confident that the trees included all the breaching mechanisms that could be important in the Delaware basin. Since the group were constructing scenarios only for analyses of the WIPP repository, it was not necessary for them to produce event trees describing all the releases that might be possible at other repository sites. Some of the event trees are unique to the WIPP site. A compilation of the event trees, along with a discussion of their derivation, appears in a summary document (3).

The final set of event trees encompassed 19 breaching events, most of which gave rise to several different scenarios. Each scenario, when fully defined for computer modeling, specifies an event assumed to breach the repository, a time at which the breach occurs, a mechanism for moving radionuclides through the breach, the response of the burial medium to the breach, the radionuclides present in the waste, and the physical and chemical condition of the waste.

Relative Ranking of Event Trees

By constructing the set of event trees, the study group fulfilled the first of the four purposes--to pick physically possible release scenarios from the many conceivable ones. They had, in fact, also fulfilled the second purpose because the construction of the trees had identified many processes that were not completely understood. To satisfy the third purpose--the identification of the scenarios that should be modeled earliest--the group turned to the problem of assigning probabilities to the events in the event trees.

Although they felt that estimates of <u>absolute</u> probabilities for these events must be highly uncertain, the group felt that it should be possible to assign meaningful <u>relative</u> probabilities simply by making the same sorts of assumptions when estimating each probability. The group made these assignments reluctantly because of the possibility that other people might erroneously interpret them as absolute probabilities. The relative probabilities are not predictions of the distant future, and they are probably useful only for comparisons among the WIPP scenarios.

Accurate absolute probabilities are not necessary for the WIPP long-term safety analyses, because the computer modeling implicitly assumes unit probability for each modeled event. If the computer analysis should predict unacceptable consequences from a hypothetical release, an accurate derivation of absolute probabilities would become necessary, for it would then be essential to know whether the risk of the release might also be unacceptable. But so long as the consequences appear to be acceptable under the assumption of unit probability, there is little need for a detailed study of absolute probabilities.

In assigning relative probabilities, the group was guided by previous attempts to assign probabilities, by discussions with persons experienced in waste-management analyses, and by data collected especially for their study. When no data or previous assignments were available, the group debated among themselves, attempting always to base their estimates on uniform assumptions and on similar arguments. After ranking the scenarios according to the relative probabilities, the group was able to recommend the scenarios that appeared most likely.

WIPP SCENARIOS

The group was now ready to addresss the fourth purpose of their study--grouping the scenarios into classes for computer modeling. The group found that they could collect the scenarios into three well-defined classes, each characterized by a mechanism for radionuclide transport. If a single computer model can serve to represent a class of several scenarios, the modeling task can, of course, be greatly simplified. Furthermore, the detailed computer analysis of each model can use bounding parameters; the predicted consequences are then bounds to the consequences that can be expected from each scenario in the class. For example, the computer model for scenarios involving water flow contains parameters describing the rate of flow through the breached repository; if this rate is taken in successive computer runs to be the maximum and the minimum rate that is physically possible at the WIPP site, the predicted consequences of the scenario will bound the effects of the flowing water. If the bounds to the consequences of an entire class of scenarios are determined in this way, it is not necessary to model each of the scenarios in detail. Moreover, this method of finding an envelope that bounds the scenario parameters avoids any need to assume the worst conceivable conditions for each event in the event tree--conditions that often cannot physically exist together in the same chain of events.

This practice of grouping scenarios into classes and computing their bounding consequences was the study group's method of dealing with competing risks and the large number of cut sets that can be derived from event trees. The event trees do not need to be expanded until all possible fundamental events have been included. A tree requires expansion only until it permits a bounding computer model to be constructed or until it can be shown to arise from a single event that experiments can prove impossible.

The WIPP scenarios fall into three groups: those that expose waste directly to the biosphere, those that begin with water flow between two aquifers at the site, and those that begin with water flow from only one aquifer.

The first group includes four separate scenarios. Only one was recommended for extensive modeling: drilling directly into the repository and bringing radioactive material to the surface.

The scenarios in the second group depend on the documented existence of water-bearing formations above and below the repository horizons. The breaching events connect these two aquifers, allowing water from the lower aquifer to flow through the repository and into the upper aquifer. As the flowing water dissolves the salt formation, it attacks the waste and carries radionuclides through the upper aquifer to the biosphere at a natural discharge point approximately 22 kilometers from the site. Several events could conceivably produce such a connection--extensive faulting or the failure of seals around boreholes, for example.

Scenarios in the third group begin with breaching events that allow water to flow from the upper aquifer to the repository. After such a hydraulic connection has been established, three mechanisms--convection, diffusion, and pressurization--might move waste from the repository to the aquifer. A convective cell set up by heat from the waste is not possible in the WIPP repository because little heat-producing waste will be emplaced there. Waste could, however, diffuse upward to the aquifer through a static pool of water filling the connections. And if the waste generates gases after burial, the gas pressure might force waste through the connections and to the aquifer. Several events listed in the other two scenario groups--faulting or drilling, for example--could produce these scenarios. In addition, the study group formed scenarios based on the arrival of an underground dissolution front, on the flooding of potash mines in the area near the repository, and on extensive cracking induced in the rock above the repository by any of several mechanisms.

At 1000 years after emplacement, the scenarios with the highest relative probabilities are those that begin with drilling. At 10,000 years, they begin with the cracking and seal-failure events as well as some of the drilling events. At 100,000 and 1 million years, cracking and the approach of the underground dissolution front dominate the breaching events. Each of these scenarios can be represented, at least for predicting upper limits to its effects, by one of four general models: direct drilling into the waste, a hydraulic connection between the upper and the lower aquifers, a hydraulic connection that supports no flow but allows the diffusion of radionuclides to the upper aquifer, and a hydraulic connection that allows water to flow from the upper aquifer, through the repository, and back to the aquifer. The last of these four models describes a bounding condition representing the most extensive radionuclide transport that a single aquifer could produce.

The two models that connect the repository to a single aquifer both describe the third group of scenarios. Because diffusion and flow require different mathematical treatments, it proved useful to construct these two separate models to bound the consequences of the scenarios in the third group.

Although none of the scenarios appeared likely within 1 million years at the WIPP site, the group recommended that these four basic models be set up for computer analysis. The current analyses of the long-term effects of the WIPP repository are derived from these models.

COMMENTS

Because the results of the modeling were reported at this meeting last year, they will not be repeated in this presentation. It is worth repeating, however, that the predicted radiation doses generally fall far below the doses delivered by natural background radiation.

It is important to point out that the methods used by the WIPP study group may not be appropriate for the construction of scenarios that might occur at other proposed repository sites. This observation is particularly true for generic studies that do not confine their attention to a well-defined and well-characterized site. Many of the most important assumptions of the WIPP study were possible only because they were supported by the extensive data describing the WIPP site and the surrounding region.

REFERENCES

1. S. E. Logan, A Technology Assessment Methodology Applied to High-Level Radioactive Waste Management, Ph. D. thesis, University of New Mexico, Albuquerque, NM (1974).

2. R. Proske, Systemstudie "Radioactive Abfalle in der Bundesrepublik Deutschland," Teil 6: Beitrage zur Risikoanalyse eines Hypothetischen Endlagers in Referenzsystem "Lagerung und Verfestigung von HAW aus einer Weideraufarbeitungsanlage," KWA 1214 a. Gellschaft fur Strahlen- und Umweltforschung/Institut fur Tieflagerung, Clausthal-Zellerfeld, Federal Republic of Germany (1976).
3. F. W. Bingham and G. E. Barr, Scenarios for Long-Term Release of Radionuclides from a Nuclear-Waste Repository in the Los Medanos Region of New Mexico, SAND78-1730, Sandia Laboratories, Albuquerque, NM (1978).

A STOCHASTIC MODEL OF RADIONUCLIDE MIGRATION FROM NATURAL AND ENGINEERED REPOSITORIES

Anne S. Kiremidjian and Paul Kruger

Civil Engineering Dept.
Stanford University
Stanford, CA 94305

INTRODUCTION

A stochastic model is being developed to analyze available migration data from existing radionuclide repositories, as a tool for assessing the adequacy of geologic formations as long term repositories of high level nuclear wastes. Since factual data on long term disposal of commercial nuclear wastes do not exist, the data needed for licensing and regulatory decisions on nuclear waste management must come from controlled laboratory and field experiments and from indirect sources. Much information is already available from the former. For example, considerable efforts are underway in the DOE-sponsored program to develop methods for forecasting radionuclide migration. Two major programs are the general models being developed at Sandia Laboratories (1) and at Battelle Pacific Northwest Laboratories (2). The Sandia program is a mathematical model to estimate the long term stability of engineered repositories, while the PNL model is focused on repository performance assessment (acronymed WIPAP). The Sandia program uses analytical approaches to estimate the geologic stability based on probabilistic models and numerical simulation, and are being applied to idealized waste repositories. The PNL program is divided into two parts, 1) release scenarios and 2) consequence analysis, the release scenarios are analyzed by a number of techniques, such as lumped-parameters simulation simulation, Monte Carlo simulation, and fault/event tree analyses. Stottlemeyre et al.(3) described a simplified one-dimensional model to simulate the system response to potential disruptive phenomena over time. The model is being developed to expand its capability. Although it now operates in a deterministic manner, based on user selected scenarios, it is planned to develop the model for stochastic operation. The consequence analysis involves deterministic modeling of the processes transporting the release radionuclides through dose commitment.

A larger source of indirect data is also available from many forms of natural, inadvertent, and engineered geologic repositories of radioactive materials. These repositories occur with widely diverse radiochemical composition, engineered or natural form of emplacement, and in various geologic and hydrologic environment. They include natural repositories of heavy elements, repositories created by underground nuclear explosions, and engineered repositories from planned burials or inadvertent leaks from surface storage facilities. Since existing data from such repositories are likely to be sparse and in some cases the physical processes responsible for migration are not fully understood, a stochastic approach to the modeling of the migration is warranted. The overall model includes two types of evaluations. The first involves the identification of the sequences of events which lead to the repository conditions necessary for the initiation of radionuclide migration and evaluation of their probabilities of occurrence. The second considers the radionuclide transport into the surrounding media as a function of time and space. A second major effort is identifying the parameters to be used in characterizing repository and migration parameters. A method for consolidation, sorting and statistical evaluation of migration data is developed. The resulting attribute matrices will serve as input parameters to the stochastic model. Since data for many of the parameters are sparse, Bayesian statistical techniques are employed for their evaluation. Preliminary attribute matrices have been developed for four test cases: the OKLO nuclear fission reactor in Gabon, Africa; the many underground nuclear explosions at the Nevada Test Site; the leaked radionuclides at the Hanford Works; and the buried glass block experiments at Chalk River, Canada.

DEVELOPMENT OF MODEL PARAMETERS

Forecasting the migration or retention of radionuclides from engineered nuclear waste repositories over long future time periods depends on our ability to properly identify the physical characteristics of such repositories and model the dynamic influences which may change these characteristics. The initial state of a repository is described by the following three model evaluation groups: 1) waste and emplacement form characteristics, 2) repository characteristics, 3) migration characteristics. All attributes that contribute to the retention or migration of radionuclides fall into one of these three groups, which are further divided into specific states and attributes. Table 1 lists the states and attributes for each of the parameters to be used in the evaluation model. For example, under the model parameter named Waste and Emplacement Form, the state is radionuclide properties further described by seven attributes. The state

Table 1. Matrix of Repository Characteristics

1. Waste and Emplacement Form Characteristics	2. Repository Characteristics
Radionuclide Properties	Physical State
Isotope	Depth of Burial
Half-live	Volume
Concentration	Stratigraphy
Chemical State	Geothermal Gradient
Physical State	Physical Properties
Solubility	Composition
Volatility	Volume
Emplacement Form	Age
Composition	Microfracture Density
Temperature	Macrofracture Density
Volume	Young's Modulus
Leach Rate	Compressive Strength
Time of Emplacement	Poisson's Ratio
	Density
3. Migration Characteristics	Thermal Conductivity
	Specific Heat
Hydrogeology	Melting Temperature
Total Porosity	Velocity Ratio
Effective Porosity	Seismicity
Hydraulic Conductivity	Earthquake History
Intrinsic Permeability	Earthquake Magnitude
Hydraulic Gradient	Fault Locations
Transmissivity	Type of Faults
Storage Coefficient	Fault Activity
Pore Velocity	Ground Motion
Total Discharge	Tectonics
Hydrology	Plate Movement Rate
Rainfall	Plate Movement Type
Infiltration Capacity	Volcanics
Location of Water Table	Caldera Dimensions
Fluvial Activity	Volcanic Cycle
Recharge and Discharge	Magnetic Movement
Geochemistry	Glaciation
Distribution Coeff.	Cycle
Acidity	Dimensions
Redox Potential	Overburden Density
Hardness	
Fluid Temperature	
Fluid Density	
Chem. Equilibrium Const.	
Retardation Factor	

and attributes listed under Repository Characteristics pertain to the probability of water intrusion. Under Migration Characteristics the states, hydrology, hydrogeology and geochemistry, affect radionuclide retention and migration once groundwater has entered the repository. The matrix formulation is a compact and concise way to organize the attributes for easy reference and modeling purposes. These matrices also allow for the coupling between data consolidation and evaluation and migration modeling. As an example, Migration Characteristics has been put into matrix form in Fig. 1.

The state vector θ_{11} refers to hydrogeology, and is described by η = total porosity, θ = effective porosity, K = hydraulic conductivity, k = intrinsic permeability, I_h and I_v = hydraulic gradients, T = transmissivity, S = storage coefficient, v = pore velicity, and Q_{gw} = total discharge. The elements of the state vector $\underset{\sim}{\theta}_{12}$, hydrol-

$$\text{Migration Characteristics} = \begin{bmatrix} \underset{\sim}{\theta}_{i1} \\ \underset{\sim}{\theta}_{i2} \\ \underset{\sim}{\theta}_{i3} \end{bmatrix} = \begin{bmatrix} \eta & \theta & K & k & I_h & I_v & S & v & Q_{gw} \\ R_a & f_p & D_a & V_a & Q_r & RCh & DCh & & \\ K_D & pH & Eh & Hd & T_f & \rho_f & K & R & \end{bmatrix}$$

Fig. 1. Migration Characteristics in Matrix Form, in Which $\underset{\sim}{\theta}_{i1}$ is the State Vector of Hydrogeology, $\underset{\sim}{\theta}_{i2}$ is the State Vector of Hydrology, and $\underset{\sim}{\theta}_{i3}$ is the State Vector of Geochemistry.

ogy, are R_a = rainfall, f_p = infiltration capacity, D_a = depth to aquifer, V_a = dimensions of aquifer, Q_r = flexural activity, RCh = recharge point, and DCh = discharge point. The geochemistry vector, $\underset{\sim}{\theta}_{i3}$, is described by K_D = distribution coefficient, pH = acidity, Eh = potential; Hd = hardness, T_f = temperature of fluid, ρ_f = density of fluid, K = equilibrium constant, and R = retardation factor. The subscript i for each state vector $\underset{\sim}{\theta}$ refers to the time at which these properties are observed. The matrix formulation is used in the identification and sorting of pertinent data from existing repositories.

STOCHASTIC MODEL DESCRIPTION

The two major aspects of stochastic modeling for radionuclide migration identified in the introduction are the event path analysis and the basis for transport in geomedia. Developments have already been made (4,5) in the identification of the primary causes for disruption of the repository environment and the sequences of following events. Although all possible sequences of events should be initially identified, it is clear that some of the event paths which can be included in a generic model will not apply to specific repository types. In that sense, the general event path model can be said to be independent of repository site characteristics and waste form.

The major steps in the current model are shown in Fig. 2. The geologic and hydrologic regime of the region immediately surrounding a repository (Fig. 2a.) is characterized through the matrix of attributes (Fig. 2b.). Many of the attributes cannot be described by precise physical constants and thus are represented in probabilistic form (Fig. 2c.), although not all attributes are of stochastic nature. Given the repository regime, the various sequences of events can be identified (Fig. 2d.) and used in a simulation procedure (Fig. 2g.).

The dynamic influences which can lead to the initiation of migration are faulting, flooding, volcanogenic effects, meteororic impaction, glaciation, uplift, and quasi-static water intrusion. These are identified in Fig. 3. as initial extreme events. They are

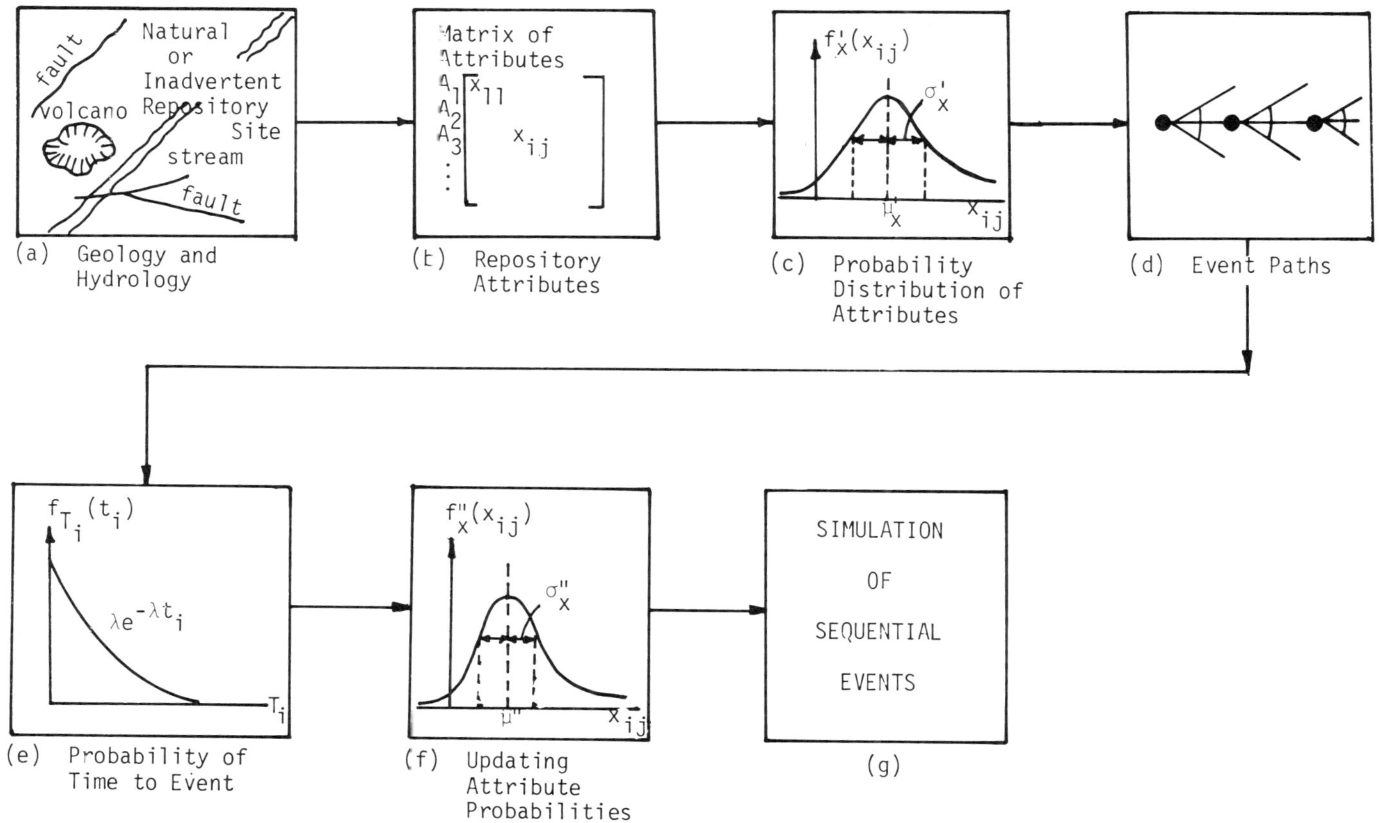

Fig. 2. Event Path Model

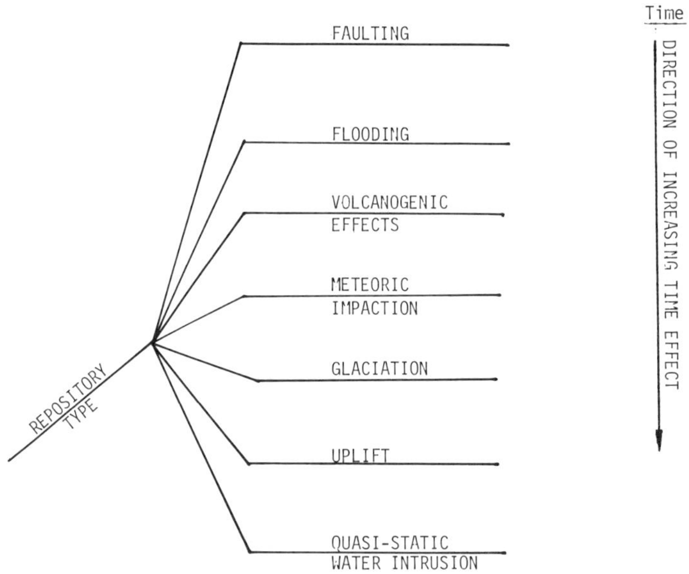

Fig. 3. Initial Extreme Events

shown in the order of increasing time of forecast. Thus, if a long-time period forecast is considered ($\sim 10^6$ yrs.), then glaciation and uplift will have important contribution to the model, while if only 1,000 yr. forecast is used, their probability of occurrence will be too low to be of significance. Another key aspect in the model is the sequence in which extreme events may occur. For example, major flooding following a major earthquake has more severe consequences than an earthquake occurring after a major flood has taken place. In order to represent this aspect properly, it is necessary to consider the time between events as shown in Fig. 2e. Simulation of the time between events will insure the proper sequences of events. This type of analysis is in contrast to the fault tree analysis where only the occurrence and non-occurrence of events are simulated without considerations for the time between their occurrence.

After the occurrence of each extreme event, some of the repository characteristics will change. Thus, an important provision in the model is the continuous updating of probability distributions of attributes (Fig. 2f). The Bayesian statistical techniques provide the necessary tools for this modification of parameter probabilities. Currently computerized procedures are being tested for the event path model. Their sensitivity are being evaluated using partial data from existing natural and inadvertent repositories. The event path model forms the basis for the radionuclide migration model. At any given time period of forecast, migration or retention of radionuclides will depend on the repository characteristics altered by the occurrence of extreme events which have taken place during that time. Thus, a migration model has to be considered in conjunction with the event path model. This aspect of the overall stochastic model is currently under investigation.

ACKNOWLEDGEMENT

This research is partially supported by the Office of Nuclear Research, United States Nuclear Regulatory Commission.

REFERENCES

1. J. T. Campbell, Risk Methodology for Radioactive Waste Disposal in Geologic Media, SAND 78-0029, Sandia Laboratories, Albuquerque, NM (1979).
2. H. C. Burkholder, "Waste-Isolation Performance Assessment Program," Proc. National Waste Terminal Storage Program Information Meeting, OWNI-62 (1979).
3. J. A. Stottlemeyre, G. M. Petrie and M. F. Mullen, "Computer Enhanced 'Release Scenario' Analysis for Nuclear Waste Repository," Scientific Basis for Nuclear Waste Management 1 (G. J. McCarthy, ed; Plenum Press, New York, NY 1979).
4. P. Kruger and A. S. Kiremidjian, "Evaluation of Retention and Migration Data from Natural and Inadvertent Repositories," Progress Report to NRC Project 04-78-266, Stanford University CE42-78-4, (1979).
5. P. Kruger and A. S. Kiremidjian, "Evaluation of Retention and Migration Data from Natural and Inadvertent Repositories," Progress Report to NRC Project 04-78-266, Stanford University CE42-78-5 (1979).

AN APPLICATION OF GEOMETRIC PROBABILITY TO THE EXISTENCE OF FAULTS IN ANISOTROPIC MEDIA

Robert M. Cranwell* and Fred A. Donath+

*Sandia Laboratories, Albuquerque, NM and
+University of Illinois, Urban, IL

INTRODUCTION

When considering the possibility of release of radioactive waste from the confines of a repository in layered geologic media, it is generally accepted that faulting presents a potential threat to the long term isolation of this waste from circulating groundwater. A fault can penetrate either the isolating barrier media or the waste repository itself to form a potential conduit for groundwater to make contact with the radioactive waste.

Three primary aspects of faults which relate to their potential for degradation of a repository site are: 1) the possibility of an existing but undetected fault intersecting the repository site, 2) the potential for a new fault occurring and propagating through the repository site and 3) the ability of any such fault to transmit groundwater. Given that a fault might be present in the region surrounding the site, the probability that it intersects the site depends primarily on its orientation and on the density of faulting in the area. Once these parameters are known, a model can be developed to determine the probability that an existing but undetected fault will intersect the repository site.

Similar techniques can be used to estimate the potential for new faults occurring and intersecting the repository site, or intersection from progagation along existing faults. However, additional data including in situ stress measurements and records of seismic activity would be needed. One can estimate the stress level at which the strength in the surrounding media will be exceeded, and thus determine a time-dependent probability of movement along a pre-existing fault or of a new fault occurring, from a predicted rate of change in local stresses. In situ stress measurements taken at intervals of time could aid in determining the rate of stress change in the surrounding media, although measurable changes might not occur over the available period of observation. In situ stress measurements might also aid in assessing the ability of existing faults to transmit fluids.

PROBABILISTIC TREATMENT

Assume that initially - i.e., at time of repository closure - the mean density of faults existing in a region R surrounding the repository site is λ_0 per unit area. Assume further that new faults appear in this region according to a nonstationary Poisson process with mean rate $\lambda_1(t)$ per unit area per year, where t indicates the time-dependent rate of formation of new faults. Then, the mean density of faults existing in the region R at some time t following closure of the repository site can be represented by

$$\lambda(t) = \lambda_0 + \int_0^t \lambda_1(\tau)\, d\tau \,. \qquad 1)$$

From equation 1) the probability of exactly N faults existing in the region R by time t is given by

$$P(N, t) = \frac{[\lambda(t)A]^N}{N!} \exp\,[-\lambda(t)A], \qquad 2)$$

where A is the area of region R. Let p denote the conditional probability that, if a fault exists in the region R, it will intersect the repository site. Then the probability that at least one fault intersects the repository site in the time interval (0, t) is given by

$$P = 1 - e^{-\lambda(t)pA}. \qquad 3)$$

From equation 3) we see that to determine a value of P we need to know values of $\lambda(t)$ and p. $\lambda(t)$ can be evaluated once a specific site for the waste repository has been selected. Geologic and historic records of tectonic and seismic activity, determination of ages of existing faults, and in situ measurements of local stresses can all be used to arrive at a represention for $\lambda(t)$.

To arrive at a representation for p which takes into account the presence of anisotropy in the surrounding media, we consider faults in the region R as anisotropic random lines in a plane (1) and examine the measure of those lines intersecting the repository site. The position of a line is described in terms of the coordinates (r, α), where r is the signed distance from the origin 0 to the line and α is the angle between a reference axis and the perpendicular to the line (see Fig. 1. below). The position of a random line is thus determined by the values of the random variables r and α. We can denote by $F(\alpha)$ the cumulative distribution function of α and assume that the distribution of the orientation is translation-invariant; that is, the orientation of a random line is invariant under translations of the coordinate axes. Then for any given α the density of r is dr, and the combined density of r and α is $drdF(\alpha)$. The measure m(W) of all positions of a random line satisfying some condition W can be written as

$$m(W) = \iint drdF(\alpha), \tag{4}$$

where the integration covers all positions of the random line such that condition W is satisfied. Note that in the isotropic case this measure becomes $\frac{1}{\pi}\, drd\,\alpha$. We can define the thickness $T(\alpha)$ of a set S in the direction α to be the length of the projection of S onto a line with direction α. For example, the thickness of a circle of radius r is 2r, and the thickness of a rectangular region of dimensions s_1 and s_2 is $T(\alpha) = s_1|\cos\alpha| + s_2|\sin\alpha|$. From the definition of $T(\alpha)$ we have that the integral $\int dr$ of all positions of a line with perpendicular direction α such that it intersects the set S is $T(\alpha)$. From this and equation 4) the measure m(S) of all positions of a random line such that it intersects a (measurable) set S in the plan is given by

$$m(S) = \int T(\alpha)\, dF(\alpha). \tag{5}$$

Equation 5) can be used to obtain an expression for the probability that a random line intersects a set S_1 given that it intersects another set S_0. It is necessary to consider conditional probabili-

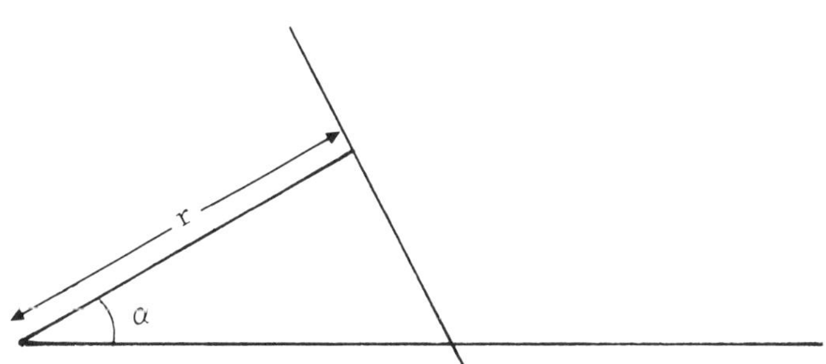

Fig. 1.

ties because the unconditional probability that a line, random over the whole plane, will intersect any bounded set S is zero.

If we assume that a random line intersects a measurable set S_0, then from equation 5) we have that the probability that it also intersects a subset S_1 of S_0 is given by

$$P[\text{line intersects } S_1 | \text{intersects } S_o] = \frac{\int T_1(\alpha)dF(\alpha)}{\int T_0(\alpha)dF(\alpha)}, \qquad 6)$$

where $T_0(\alpha)$ and $T_1(\alpha)$ are the thicknesses of S_0 and S_1, respectively, in the direction α.

Viewing the repository site area as a region R_1 which is a subset of a larger region R_0 containing a collection of faults with a definite orientation, we can use the above arguments to arrive at an estimate of the parameter p.

APPLICATION

Within the upper few hundred meters of the earth's crust, at depths at which high-level nuclear waste might be emplaced, one of the three principal stresses is likely to be vertical or very steeply plunging. Therefore, there are three ideal orientations of potential planes of faulting, defined by which of the three principal stresses is vertical (Fig. 2.)(2). A recent review of the state of stress in the earth's crust (3) indicates that at depths less than 1 km the maximum principal stress commonly is hoizontal and the minimum principal stress is vertical. For purposes of illustration, we will assume this orientation to exist at a hypothetical repository site, although we note that this might not be representative of conditions in bedded or domed salt. This stress orientation favors the development of low-angle reverse faults (see Fig. 2b.) which, if unaffected by the presence of anisotropy, would dip about 30 degrees, since in isotropic media this is the inclination of a fault to the maximum principal stress responsible for it. However, the presence of planar anisotropy can cause this angular relationship to vary from a few degrees to more than 45 degrees (4).

We shall assume that, because of the presence of anisotropy, the dip angle of existing faults or of newly formed faults in a region R_0 containing the repository region R_1 tend to be uniformly distributed over the range from 10 degrees to 60 degrees. We shall assume further that the regions R_0 and R_1 are rectangular regions with dimensions s_{01}, s_{02} and s_{11}, s_{12}, respectively. Because the dip angle of the faults are uniformly distributed over the range from 10 degrees to 60 degrees, the perpendicular direction α of

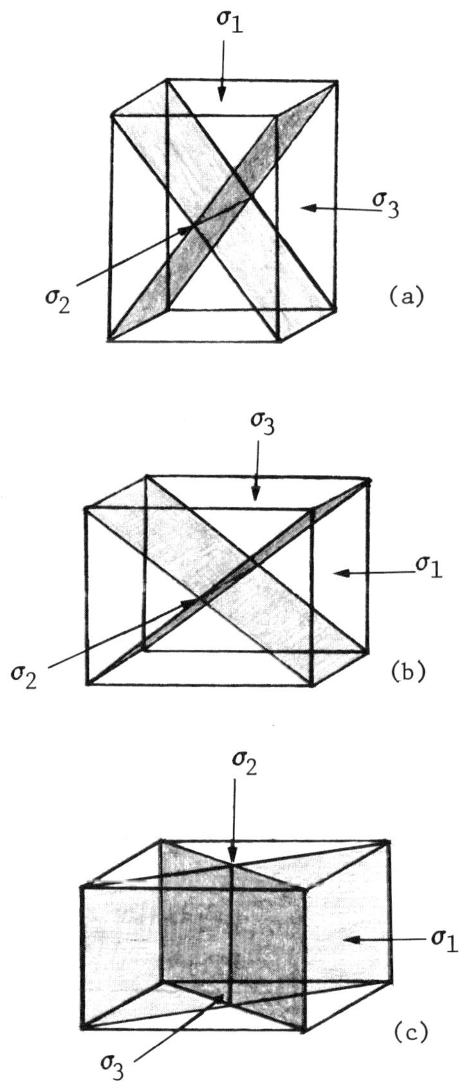

Fig. 2. Fault planes (shaded) and principal stress directions for one principal stress vertical. a) High angle-normal faults with σ_1 vertical, b) Low angle-reverse faults with σ_3 vertical, and c) Strike-slip faults with σ_2 vertical.

these fault lines is uniformly distributed over the range from 30 degrees to 80 degrees. Thus,

$$m(R_0) = \int T_0(\alpha)dF(\alpha) = \int_{30}^{80} \frac{1}{50}(s_{01} \cos \alpha + s_{02} \sin\alpha)\, d\alpha$$

$$= \frac{1}{50}[s_{01}(.48) + s_{02}(.7)].$$

Similarly,

$$m(R_1) = \int T_1(\alpha)dF(\alpha) = \frac{1}{50}[s_{11}(.48) + s_{12}(.7)].$$

Thus, the probability that a fault intersects the repository site R_1, given that it intersects the region R_0, is

$$P = \frac{m(R_1)}{m(R_0)} = \frac{s_{11}(.48) + s_{12}(.7)}{s_{01}(.48) + s_{02}(.7)}$$

Although we have assumed a uniform distribution of dip angles, it should be recognized that this technique could be used with other assumed distributions.

FORMATION OF NEW FAULTS

Whether a new fault will form or renewed movement will occur along a pre-existing fault in anisotropic media can be determined by a linear failure criterion of the form

$$\tau = \tau_o + \sigma\mu \qquad 7)$$

where τ and σ are the shear and normal stresses on the potential fault plane or, on a pre-existing fault plane, and μ is the tangent of the angle of internal friction for intact anisotropic rock or the coefficient of sliding friction for displacement on an already present fault surface. The term τ_o is the cohesive strength for intact rocks or the shearing resistance on an existing fault.

For intact anisotropic rock, τ_o and μ vary systematically with the inclination β of the anisotropy to the direction of maximum principal stress according to the relationship

$$\tau_o = \underline{a} - \underline{b} \cos 2(\gamma - \beta) \tag{8}$$

and

$$\mu = \underline{c} - \underline{d} \cos 2(\gamma - \beta) \tag{9}$$

where γ is the orientation of β for with τ_o is a minimum (5). For faults that develop parallel to the anisotropy or for pre-existing faults the values of τ_o and μ are constant, although it should be noted that the values of these parameters might vary with environmental conditions such as temperature.

The formation of a new fault or the displacement along an existing fault can thus be predicted to occur when the existing state of stress in the rocks satisfies one or the other of the stated criteria for fault formation or for sliding, i.e., when

$$\tau - \tau_o \geq \mu\sigma$$

where

$$\tau = (\sigma_1 - \sigma_3) \sin\theta \ \cos\theta \tag{10}$$

and

$$\sigma = \sigma_1 \sin^2\theta + \sigma_3 \cos^2\theta \tag{11}$$

and θ is the angle between the fault and the direction of σ_1, the maximum principal stress. Thus, once the statistical distribution of τ and σ are determined, one can arrive at an estimate of the probability of a new fault forming or of movement occurring along a pre-existing fault. The methods described in the earlier part of this paper can be used to arrive at an estimate of the probability that the fault will intersect the waste repository site.

Because there is a functional relationship between τ and σ, one need only determine the distribution function for one of the variables, say σ. Once this is known, the distribution function of τ can easily be determined in the following manner. Assume that σ is a continuous random variable with density $f(\sigma)$ and that $f(\sigma) > 0$ for $a < \sigma < b$. Then the density $g(\tau)$ of τ is given by

$$g(\tau) = f\left(\frac{\tau - \tau_o}{\mu}\right) \frac{1}{\mu}$$

Similar techniques apply if σ is a discrete random variable.

For a given state of stress in the earth's crust, the criterion for development of a new fault might not be satisfied by the existing state of stress but the criterion for sliding could be met for certain orientations of the sliding surface. Thus, for this situation the problem is to identify the range of orientations that satisfy the sliding criterion and to determine the probability that an existing fault surface falls within this range. This probability could be determined from knowledge of the existing stress field and orientation of the anisotropy in the region being considered, or from collected data concerning the surface deformation of the fault being considered (6).

REFERENCES

1. L. A. Santalo, Integral Geometry and Geometric Probability, (Addison-Wesley, 1976).
2. E. M. Anderson, The Dynamics of Faulting (2nd Ed.), (Oliver and Boyd, Edinburgh, 1951).
3. A. McGarr and N. C. Gay, "State of Stress in the Earth's Crust", Ann. Rev. Earth and Planetary Sci. 6, 405-436 (1978).
4. F. A. Donath, "Fundamental Problems in Dynamic Structural Geology", The Earth Sciences: Problems and Progress in Current Research (T. W. Donnelly, ed; Univ. of Chicago Press, 83-103, 1963).
5. F. A. Donath, "Effects of Cohesion and Granularity on Deformational Behavior of Anisotropic Rock", Studies in Mineralogy and Precambrian Geology (B. R. Doe and D. K. Smith, ed; Geological Society of America Memoir 135, 95-128 (1972).
6. J. B. Walsh, "Dip Angle of Faults as Calculated from Surface Deformation", Journal of Geophysical Research 74 (8), 2070-2080 (1969).

GEOLOGIC WASTE DISPOSAL AND A MODEL FOR THE SURFACE MOVEMENT OF RADIONUCLIDES

Jon Helton, Ron Iman, Jack Brown*, Steve Schreurs**

Sandia Laboratories, Albuquerque, NM 87185
*Auburn University
**U.S. Nuclear Regulatory Commission

INTRODUCTION

In the following, a model for the surface movement of radionuclides is presented. This model, which is referred to as the Pathways Model, was constructed at Sandia Laboratories in a project funded by the Nuclear Regulatory Commission to develop a methodology to assess the risk associated with the geologic disposal of high level radioactive waste. The methodology development involves work in two major areas: a) models for physical processes and b) statistical techniques for the use and assessment of these models. The following presentation of the Pathways Model will involve topics from both areas. Additional information on the entire project can be obtained from its interim report (1). The work introduced in this paper is described in greater detail in various referenced technical reports; these reports should be consulted for more detailed information than can be provided here.

A model for the surface movement of radionuclides in the context of geologic waste disposal should permit consideration and assessment of the many factors that may affect both radionuclide distribution in the environment and our ability to predict such distribution. Desirable features include the following. First, due to: a) the possible diversity of specific waste disposal sites, b) the changes which may occur at a particular site during the long term periods involved in radioactive waste disposal and c) the potential assessment and comparison of different sites, the model should be able to represent radionuclide movement in different physical settings. Second, the model should be able to represent radionuclide accumulation and possible reconcentration as well as dispersion and dilution. Third, as distribution coefficients and other factors influencing radionuclide movement are inexactly known, the model should be amenable to sensitivity analysis to determine a) the relative importance of individual variables and b) the effects of uncertainty in knowledge of these variables. Fourth, as potential release periods in radioactive waste management are lengthy, it should be possible to determine the asymptotic behavior of the model. Fifth, as river discharge and other factors influencing radionuclide movement are functions of time (actually, stochastic processes), it should be possible to determine what effects the use of average values and other simplifying assumptions has on model predictions.

There are many models that represent the environmental movement of radionuclides. Recent surveys are provided by References 2 and 3. However, after considering existing models, it was decided that it would be more productive to construct a model tailored to the specific needs of the waste disposal project than to adapt an already existing model to those needs. Of the existing models, the Pathways Model has the greatest conceptual similarity with the HERMES model (4).

THE PATHWAYS MODEL

For the environmental movement of radionuclides in the context of geologic waste disposal, there are two processes involved with distinctly different time scales. The first process is the long term distribution and accumulation of radionuclides in the surface environment. Radionuclide release to, movement through and removal from the environment will generally be very slow. The second process is the movement of radionuclides from the surface environment to man. Here the time scale is much shorter. Therefore, the Pathways Model is divided into two submodels. One of these, the Environmental Transport Model, represents the long term distribution and accumulation of radionuclides in the environment. The other, the Transport to Man Model, represents the movement of radionuclides from the environment to man.

In the Environmental Transport Model, radionuclide movement is represented with a compartment model. With this approach, radionuclides in an area under consideration are divided into a number of "compartments", and then radionuclide movement between these compartments is represented by a system of linear differential equations. The basic idea is to place radionuclides which are in physical regions with different characteristics in different compartments and then to determine the distribution which results from radionuclide decay and from radionuclide movement between regions.

The Environmental Transport Model uses an algorithm to construct the radionuclide transport equations described in the preceding paragraph. The basis of this algorithm is a building block called a zone. A zone corresponds to a physical region in an area to be modeled. Further, each zone has four subzones (ground water, soil, surface water and sediment), and various movements of water and solid material associated with these subzones are possible. It is the movements of water and solid material which are assumed to be responsible for radionuclide transport. In such movements, it is assumed that radionuclides are partitioned between the liquid and solid phases on the basis of distribution coefficients. A system of transport equations incorporating decay chain characteristics and regional diversity is constructed by linking a suitable number of these zones together. In this construction, each subzone has one uniformly mixed compartment associated with it for each radionuclide in the decay chain under consideration. A single zone is represented in Fig. 1., and the linkage of a number of zones to represent the movement of a decay chain is represented in Fig. 2. The computer program which implements the Environmental Transport Model actually permits a greater variety of flows between zones than indicated in Fig. 2.

In the Transport to Man Model, radionuclide movement is represented by a model based on simple food chains and concentration ratios. With this approach, food chains which originate in an affected area and lead to man are determined. Then the radionuclide concentration in an affected food type is determined by the use of concentration ratios, and the amount of radionuclide consumed follows from the amount of food consumed and the radionuclide concentration in that food. This approach is the same as currently employed by the Nuclear Regulatory Commission (5). The Environmental Transport Model is used to provide radionuclide concentrations in soil and water which serve as input to the Transport to Man Model. A more detailed description of the Pathways Model is provided in Chapter 4 of the project's interim report (1). Further, a user's manual which de-

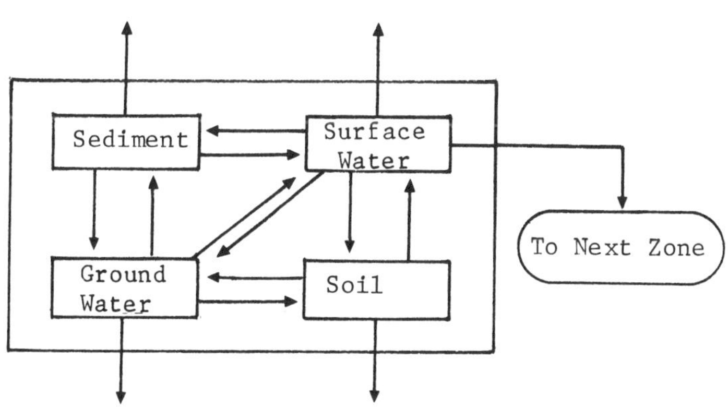

Fig. 1. Division of a Zone into Subzones. Arrows represent possible movements of water and solid material; those arrows above and below Fig. 1. represent movements to sinks.

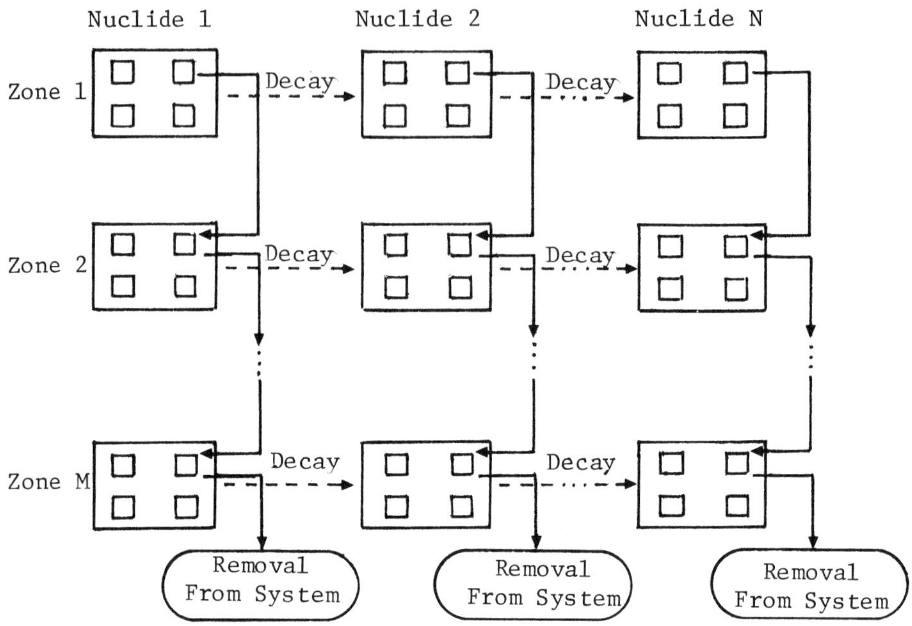

Fig. 2. Flows Associated with the Movement of a Chain of N Radionuclides Through a System of M Zones. Fig. 1. represents the flows associated with a single zone in greater detail.

scribes the model and the computer program which implements it is also available (6).

It is felt that the division of the Pathways Model into the submodels Environmental Transport and Transport to Man and the construction of the Environmental Transport Model on the basis of compartments provide the capability to represent different types of radionuclide movement and also to assess both radionuclide dilution and possible reconcentration. By suitably defining the zones and subzones associated with the Environmental Transport Model and organizing the operation of this model in conjunction with the Transport to Man Model, it is possible to represent a variety of different and possibly diverse sites and also to represent different sets of conditions which could exist through time at a specific site. The computer program which implements the Pathways Model is written to permit, and indeed require, significant user involvement in its operation.

SENSITIVITY ANALYSIS

For the Pathways Model, an approach to sensitivity analysis based on regression analysis has been found to be productive. The

Table 1. Stepwise Regression Results of Variables Affecting U234 Concentration in a Flood Plain

Step	Variable	Rank R^2	Raw R^2	St. Reg. C.
1	B43	0.614	0.368	0.745
2	B42	0.796	0.665	1.283
3	B42*B42	0.849	0.669	-0.866
4	B10*B43	0.878	0.702	0.254
5	B12*B43	0.901	0.812	-0.189
6	B4*B7	0.914	0.835	0.118

overall approach is described in Reference 7, and applications of the approach to the Environmental Transport Model are described by References 8 and 9. Basically, the idea is: a) to start with a set of input variables with selected ranges and distributions, b) to select model inputs from these variables according to their ranges and distributions, c) to generate model output with selected inputs and d) to assess the relationship between model input and output by stepwise regression. Special techniques found to be useful include: a) Latin hypercube sampling to select values of input variables, b) the rank transform to reduce the effects of nonlinearity in the relationships between model input and output and c) the PRESS (predicted error sum of squares) criterion to indicate overfit during regression analysis.

An example of a sensitivity analysis calculation is contained in Table 1. This example is taken from Table 6.10 of Reference 8 and indicates variables affecting the buildup of ^{234}U in a flood plain. It is assumed that ^{234}U is entering a river and that various processes are acting to move it into the flood plain. The site and the assumptions underlying its derivation are described in Reference 8. Approximately 20 independent variables were involved in the regression analysis which generated Table 1. The following variables actually appear in the table: B4--average soil subzone thickness; B7--a conversion factor such that the product of B7 and the pore volume of the soil subzone is equal to the annual water exchange between the soil subzone and the surface water subzone; B10--a conversion factor such that the product of B10 and the mass of solids contained in the soil subzone is equal to the annual solid exchange between the soil subzone and the surface water subzone; B12--the regional erosion rate; B42--the distribution coefficient for ^{234}U in the soil subzone; and B43--the distribution coefficient for ^{234}U in the surface water subzone.

The stepwise regression analysis which generated Table 1 was based on rank transformed data. The first two columns indicate the variables and cross products in order as selected by the regression analysis. The third and fourth columns contain R^2 values based on rank transformed data and interpolations from rank transformed data to raw (i.e., untransformed) data, respectively. The Rank R^2 indicates how successful the regression model is in predicting the relative importance of variables in influencing the Environmental Transport Model while the Raw R^2 indicates how successful the regression model is in duplicating the Environmental Transport Model. The fifth column contains standardized regression coefficients for the individual variables in the regression model at the last step of the regression analysis. The absolute value of these coefficients provides a ranking of the relative importance of the individual variables in influencing the model. Many regression analyses of the form indicated here are presented in References 8 and 9.

ASYMPTOTIC BEHAVIOR

It is possible to show that, for a fixed rate of radionuclide input, the radionuclide concentrations as predicted by the Environmental Transport Model will increase monotonically to asymptotic limits. For a solution y(t) to the radionuclide transport equations generated in the Environmental Transport Model, the expressions asymptotic behavior and asymptotic limit are used in reference to the limit of y(t) as $t \to \infty$. As the time required in many situations to reach steady state in the surface environment may be short relative to other processes involved in the release of radionuclides to the surface environment (i.e., ground water transport), it may be desirable to use asymptotic values for radionuclide concentrations in the surface environment rather than calculating and using their values as functions of time. For the same zone and variables used in the generation of Table 1, the variation in time to 90 percent steady state and concentration at steady state for an assumed input rate of 1 atom/yr of ^{245}Cm to the surface water subzone is represented in Table 2. This table is adapted from Tables 4.4 through 4.7 of Reference 9; this report contains numerous such compilations as well as sensitivity analyses similar to the one presented in Table 1 for the variables influencing time to 90 percent steady state and concentration at steady state. As the Environmental Transport Model is linear with respect to rate of radionuclide input, such calculations can be scaled to represent other input rates.

TIME VARYING PARAMETERS

In the Environmental Transport Model, as will probably be the case for any model used to represent the surface movement of radionuclides, many parameters are based on "averaged" water and solid flow rates. Often, mean annual flow rates are used whereas the flow rates are (at best) periodic functions or (more realistically) periodic stochastic processes. It is desirable to assess the effect that the replacement of parameters with their average values has on model predictions. Such a study has been initiated for the Environmental Transport Model (10).

Table 2. Time to 90 Percent Steady State and Concentration at Steady State for an Assumed Input to the Surface Water Subzone of 1 Atom/Yr of ^{245}Cm. Time is in years; surface water concentration is in atoms/L; all other concentrations are in atoms/kg.

		Ground Water	Soil	Surface Water	Sediment
Time to	Minimum	.10E+03	.10E+02	.10E+01	.10E+02
90%	Maximum	.28E+05	.51E+04	.27E+03	.44E+03
Steady	Mean	.11E+05	.34E+03	.20E+02	.72E+02
State	Std. Dev.	.10E+05	.43E+03	.33E+02	.65E+02
Conc. at	Minimum	.12E-16	.14E-13	.35E-13	.42E-12
Steady	Maximum	.13E-09	.38E-09	.94E-13	.45E-09
State	Mean	.30E-11	.37E-10	.54E-13	.75E-10
	Std. Dev.	.11E-10	.58E-10	.14E-13	.87E-10

That study investigates the effects of multiplying the coefficient matrix for the radionuclide transport equations by a periodic function and by a periodic stochastic process. The function and the stochastic process were picked to mimic observed hydrologic patterns. For the periodic hydrologic process, it was possible to calculate bounds on the difference between predictions made with the untransformed and transformed coefficient matrix; for the stochastic hydrologic process, there was less success at obtaining analytic assessments of the untransformed and transformed cases, and computer simulations were used to obtain an indication of their differences. Overall, the variation was not large between concentrations calculated with averaged yearly values and with either periodic or periodic stochastic perturbations of these values. This is especially true as the uncertainty in some of the other parameters which are used as model input (particularly distribution coefficients) can produce much larger variation in model predictions.

REFERENCES

1. J. E. Campbell, R. T. Dillon, M. S. Tierney, H. T. Davis, P. E. McGrath, F. J. Pearson, H. R. Shaw, J. C. Helton and F. A. Donath, Report No. SAND78-0029, Sandia Laboratories, Albuquerque, NM (1978).
2. F. O. Hoffman, C. W. Miller, D. L. Shaeffer and C. T. Garten, "Computer Codes for the Assessment of Radionuclides Released to the Environment," Nuclear Safety 18, 343-354 (1977).
3. D. L. Strenge, E. A. Watson and J. G. Droppo, Report No. BNWL-B-454, Battelle Pacific Northwest Laboratories, Richland, WA (1976).

4. J. F. Fletcher and W. L. Dotson (Compilers), Report No. HEDL-TME-71-168, Hanford Engineering Development Laboratory, Richland, WA (1971).
5. U.S. Nuclear Regulatory Commission, "Calculation of Annual Doses to Man from Routine Releases of Reactor Effluents for the Purpose of Evaluating Compliance with 10 CFR Part 50, Appendix I," Regulatory Guide 1.109, Revision 1 (1977).
6. J. C. Helton and P. C. Kaestner, Report No. SAND78-1711, Sandia Laboratories, Albuquerque, NM (1980).
7. R. L. Iman, J. C. Helton and J. E. Campbell, Report No. SAND78-0912, Sandia Laboratories, Albuquerque, NM (1978).
8. J. C. Helton and R. L. Iman, Report No. SAND79-1393, Sandia Laboratories, Albuquerque, NM (1980).
9. J. C. Helton, J. B. Brown and R. L. Iman, Report No. SAND79-1908, Sandia Laboratories, Albuquerque, NM (1980).
10. J. B. Brown and J. C. Helton, Report No. SAND79-1909, Sandia Laboratories, Albuquerque, NM (1980).

REGIONAL THERMOHYDROLOGICAL EFFECTS OF AN UNDERGROUND REPOSITORY FOR NUCLEAR WASTES IN HARD ROCK*

J. S. Y. Wang and C. F. Tsang

Lawrence Berkeley Laboratory
University of California
Berkeley, CA 94720

INTRODUCTION

The principal mechanism for underground migration of radionuclides is transport with groundwater. In the selection and assessment of suitable sites as potential repositories of nuclear wastes, it is important to predict the effects of heat generated by the wastes upon the water movement between the repository and the biosphere. With the absence of relevant engineering experience and the limitation on duration of in situ testing, simulation of global and long term thermal responses of the rock mass in hypothetical models is the main method for the understanding of the mechanism and the identification of the sensitive parameters controlling groundwater flow. This paper presents results of two sets of calculations: 1) Regional temperature effects are studied for different waste forms, repository dimensions and rock formations. The type of reprocessing treatment and the length of cooling period of the nuclear wastes before emplacement into the repository are found to be two of the more important factors. 2) Thermally induced fluid flow is calculated assuming a simple two-fracture system to make a "worst case" estimate of the transit time of water from repository to ground surface. Recharge capacity from the surrounding formation is found to be a controlling factor.

*Work done under University of California Contract W-7405-ENG-48 for the U. S. Department of Energy.

TEMPERATURE FIELD

To study the long term regional changes in temperature in the rock mass the repository is idealized to be a flat circular disk loaded uniformly with nuclear wastes at time t=0. The repository is assumed to be at a depth D = 500 m below surface in granite and to have a radius R = 1500 m. The principal mode of heat transfer in the rock mass is assumed to be by heat conduction. This has been shown to be a good assumption based on recent Stripa data analysis (1). The temperature rise is calculated by integrating the instantaneous disk source function (2) over the heat power function of the nuclear wastes (3). The presence of a boundary of constant temperature at the ground surface is corrected by the method of image. The thermal diffusivity for granite is assumed to be 1.15×10^{-6} m^2/sec and the thermal conductivity is 2.5 W/m/°C.

The temperature changes are proportional to the waste density, or equivalently the initial power density with which the repository is loaded. Besides this simple proportionality, the temperature field, especially the long term behavior, is sensitive to the decay characteristics of the heat power functions of the wastes. The time dependence in power is determined by the waste composition. The nuclear wastes contain short lived fission products and the long lived actinides. The fission products generate most of the heat at early life of the wastes and decay rapidly after discharge from the reactor. Accordingly, the effects of a period of near surface cooling of the wastes before burial in a repository may be expected to be significant. Fig. 1. illustrates the effects of pre-emplacement surface-cooling times of spent fuel discharged from a pressurized water reactor on the repository temperatures. Although the temperature rises are calculated on a uniformly loaded model, it can be concluded from these results that the period for which the wastes are cooled before emplacement is an important factor in determining the thermal effects.

Fig. 1. also indicates that the temperature at the repository reaches a maximum after a period of less than 100 years and thereafter decays very slowly over a period of thousands of years. After 100 years, most of the fission products have decayed and the actinides and their daughters are the main heat source. Therefore, the reprocessing treatment of spent fuel to recover the uranium and plutonium may affect the long term thermal field significantly. To study the far field effects and, in particular, those at the surface, the maximum temperature gradient at the epicenter above the repository has been calculated for a variety of conditions (4). Fig. 2 shows the large differences in surface gradient between the spent fuel and the reprocessed wastes for a number of different rock formations. Either form of the wastes is assumed to be buried 10 years after discharge from the reactor at an initial loading density in the repository of 10 W/m^2. With the same initial power density, the

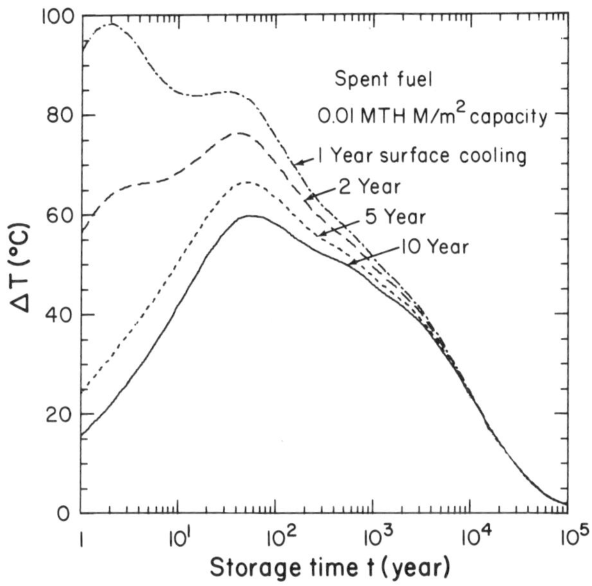

Fig. 1. Effects of Surface Cooling on the Repository Temperatures with Waste Density of 0.01 Metric Ton of Uranium per m^2

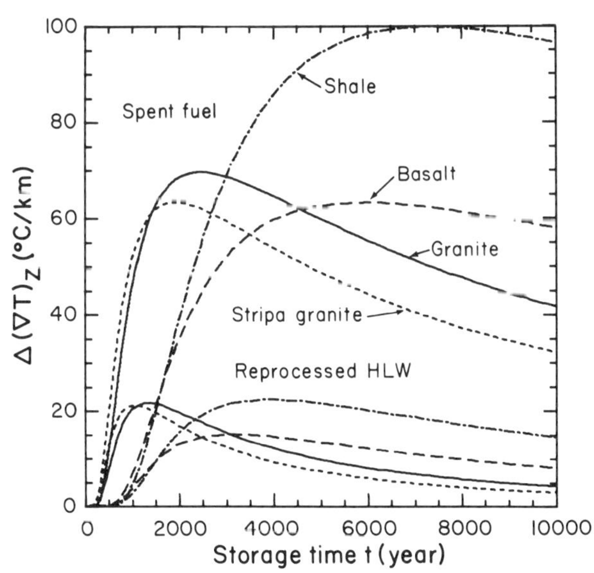

Fig. 2. Effects of Waste Types and Rock Formations on the Ground Surface Thermal Gradient with Initial Power Density of 10 W/m^2

accumulated heat generated by the spent fuel is much larger than the reprocessed wastes. From Fig. 2., it can also be seen that very different thermal fields are expected for different rock formations.

BUOYANT FLOW IN TWO-FRACTURE MODEL

Variations in the temperature of the rock mass cause changes in the density and viscosity of water in the rock fracture system and induce buoyant flows. The permeability of many crystalline and argillaceous rocks arises mainly from the hydraulic conductivity of fractures. The concern on fracture flow is related to the important question of the possibility of waste components being carried by the fracture water from the repository to the surface. A porous medium model to represent fractured rock mass with multiple fracture sets has been used in generic studies of granite formation (5). In this paper, a simple model is used that comprises a horizontal fracture at the depth of repository connecting a recharge zone to the repository, which is also connected to a discharge zone in the opposite direction, and intersecting a vertical fracture from the center of the repository to the surface as shown in Fig. 3. Before the repository is loaded and the rock mass subjected to changes in temperature, it is assumed that the original groundwater flow is horizontal from recharge zone to discharge zone. As the rock temperature increases, the water initially at the depth of the repository will move upward in the vertical fracture by the buoyancy forces.

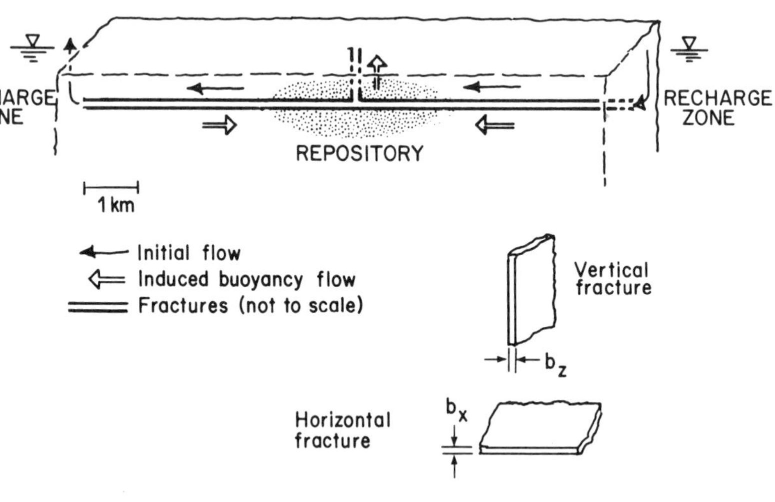

Fig. 3. Two-Fracture Model for Simulating Buoyant Flow

From Darcy's law and Boussinesq approximation, the one-dimensional flows within the fractures can be derived (4). The recharge and discharge zones are assumed to be far away from the repository and they are maintained at normal hydrostatic pressure and ambient temperature. Before the emplacement of wastes, an ambient temperature field of 20°C at ground surface and a normal geothermal gradient of 30°C/km is assumed. The buoyancy force is proportional to the density contrast between the heated water in the vertical fracture and the cooled water in the recharge and discharge zones. The large difference in the long term temperature fields between spent fuel and reprocessed waste results in large difference in the vertical flow as illustrated in Fig. 4. In Fig. 4., the upward movement of water initially at the depth of the repository is plotted as a function of time. The results in two cases, one with the repository at a depth of 500 m and the other at a depth of 1000 m are compared. There is little difference due to the change in depth. Essentially, the flow of water as a result of buoyancy depends upon the average temperature of the water throughout the length of the vertical fracture. The more important factor affecting the buoyant flow of groundwater is the ratio between the distance L from repository to the recharge zone and the depth D of the repository. For the particular case illustrated in Fig. 4., the repository is assumed to be located a distance, L = 5000 m, midway between the recharge and distance zone. The vertical and horizontal fractures are assumed to have the same aperture $b_z = b_x = 1\mu m$.

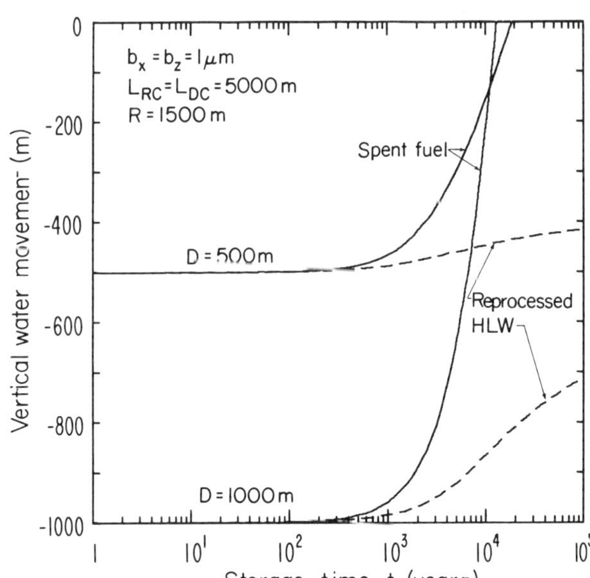

Fig. 4. Effects of Waste Types and Repository Depths on the Water Movement along the Vertical Fracture

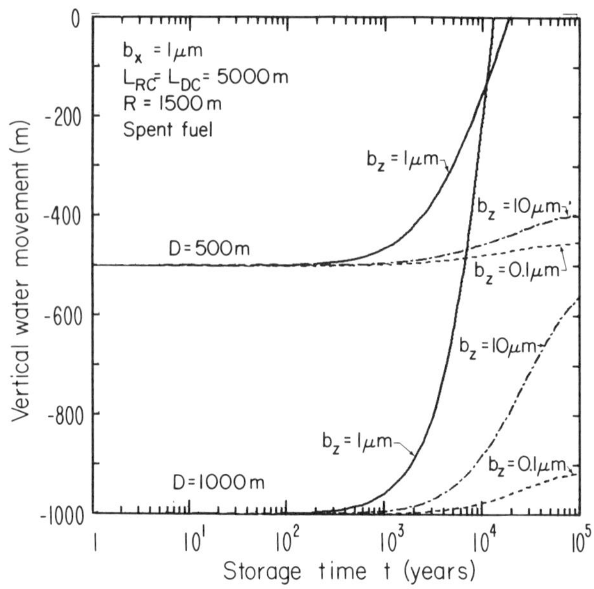

Fig. 5. Effects of Vertical Fracture Apertures on the Water Movement Along the Vertical Fracture

In addition to the temperature change and hydrologic connecting distances L and D, the buoyant flow depends sensitively on the apertures and permeabilities of the fractures. For a fracture with aperture b, the permeability $k = b^2/12$ for laminar fracture flow is assumed (6,7) in the calculation. If the horizontal fracture representing the recharge path from the surrounding formation has a given finite aperture, the dependence of the buoyant flow on the vertical aperture is of great interest. In Fig. 5., the results of a constant horizontal aperture b_x = 1 μm and with a range of vertical apertures b_z = 10 μm, 1 μm and 0.1 μm are shown. The movement of the groundwater in the vertical fracture is significantly slower both for the case of $b_z = 10\, b_x$ and of $b_z = 0.1 b_x$ than with $b_z = b_x$. For nonzero distance L and a finite horizontal aperture b_x, the buoyant flow does not become infinite as the vertical fracture aperture b_z increases. On the contrary, the buoyancy flow decreases as b_z increases. This can be easily explained by the fact that a large vertical fracture with storage capacity reduces fluid flow velocity. Thus, the buoyant flow in the vertical fracture is controlled by the finite recharge capacity through the horizontal fracture.

DISCUSSION AND CONCLUSION

Although the thermo-hydrologic model used for the analysis presented in this paper is very simple, it should possess the same physical behavior as that of the more complex repository systems. Accordingly, it should provide a good insight into the dynamics of the thermally-induced groundwater flow, and illustrate the sensitivity of this flow to various parameters. The actual numerical results should be considered as no more than an order of magnitude estimations. It must be pointed out also that the transport of nuclides does not take place at the same rate as that of the groundwater. Nuclide transport is retarded in a certain degree by various physical and chemical processes, such as sorption.

The calculations reported in this paper suggest that the magnitude of the thermally-induced buoyant groundwater flow carrying toxic wastes from a repository to the biosphere depends upon many factors. Of these, the aggregate increase in the temperature of the rock mass containing the repository is one of the most important. This temperature is affected by the design of the repository, the kind of nuclear waste buried in it and the period for which this waste has been cooled near surface before burial. Significant differences exist between reprocessed waste and spent fuel in respect of the degree to which the rock mass is heated and hence the time taken for groundwater to reach the surface by buoyant flow. The heat capacity of different kinds of rocks has a significant effect on buoyant groundwater flow also. Finally, the bouyant groundwater flow also depends strongly upon the ratio of the hydraulic transmissivities of the vertical and horizontal fractures.

REFERENCES

1. M. Hood, "Some Results from Field Investigation of Thermomechanical Loading of Rock Mass when Heater Canisters are Emplaced in Rock," 20th U. S. Symposium on Rock Mechanics, Austin, TX, June (1979).
2. H. S. Carslaw and J. C. Jaeger, Conduction of Heat in Solids (Oxford at the Clarendon Press, 2nd Ed. 1959).
3. R. A. Kisher, J. R. Marshall, D. W. Turner and J. E. Vath, Nuclear Waste Projections and Source-term Data for FY 1977, Y/OWI/TM-34, Office of Waste Isolation, Oak Ridge National Laboratory, Oak Ridge, TN, April (1978).
4. J. S. Y. Wang, C. F. Tsang, N. G. W. Cook and P. A. Witherspoon, A Study of Regional Temperature and Thermohydrological Effects of an Underground Repository for Nuclear Waste in Hard Rock, LBL-8271, Lawrence Berkeley Lab., U. of California, Berkeley (1979).
5. Groundwater Movements Around a Repository, KBS Teknisk Rapport 54:01-06, Karnbransksakerhet, Stockholm, Sweden, Sept. (1977).

6. H. Lamb, Hydrodynamics (Cambridge University Press, New York, 6th Ed. 1932).
7. P. A. Witherspoon, J. S. Y. Wang, K. Iwai and J. E. Gale, Validity of Cubic Law for Fluid Flow in a Deformable Rock Fracture, LBL-9557, Lawrence Berkeley Lab., U. of California, Berkeley (1979).

ENVIRONMENTAL EFFECTS FROM DISPOSAL OF DEFENSE HIGH-LEVEL RADIOACTIVE WASTES IN A GENERIC GEOLOGIC REPOSITORY

N. Chipman, G. Simpson, W. Rodger*, R. Frendberg* and H. Lawroski**

Exxon Nuclear Idaho Company, Idaho Falls, ID
*Nuclear Safety Associates, Bethesda, MD
**Consultant, Idaho Falls, ID

INTRODUCTION

High level radioactive wastes generated during the reprocessing of defense-related nuclear fuels are presently stored as calcined solids at the Idaho Chemical Processing Plant (ICPP). This method of handling nuclear waste has proved to be both safe and cost effective. However, for disposal, the waste form and/or disposal location may need to be altered.

Many plans for nuclear waste management can be advanced; however, the preferred alternatives for the ICPP waste are in these general categories:

1. Continue current operations.

2. Modify the waste form and dispose at the Idaho National Engineering Laboratory (INEL).

3. Modify the waste form and dispose at an off-site geologic repository.

4. Remove actinides from the waste, dispose of the actinide-depleted waste at the INEL, and dispose of the actinide waste at an off-site geologic repository.

The environmental impacts of disposing of the waste at the INEL and transporting it to an off-site repository have been previously reported (1). The radiological consequences of disposing of the wastes in a generic geologic repository are reported here. The calculations are based on the scenarios used in the preparation of the Draft Environmental Impact Statement, Management of Commercially Generated Radioactive Waste (2). Calculations were completed for both the calcine and glass waste forms. Calcine will be "stabilized" by a heating process to remove residual nitrates and/or water; all other properties remain the same. The glass waste form has two sub-options, the actinide fraction of the waste made into a glass and the calcine made into a glass.

METHODOLOGY

The scenarios considered for release from the repository and their estimated probabilities are:

1) Canister drop down the repository shaft ($7x10^{-7}$ events/yr)
2) Meteorite impact ($2x10^{-13}$ events/yr)
3) Fault fracture and flooding ($2x10^{-13}$ events/yr)
4) Drilling into the repository ($5x10^{-7}$ events/yr)
5) Solution mining in a salt formation ($1x10^{-6}$ events/yr).

Possible maximum individual doses (maximum event consequences) and population doses were calculated for each waste form and for sixteen time periods extending to 100 million years after disposal. Only three time periods, 1,000, 10,000 and 1 million years, are reported here. Population doses were integrated over time and were calculated to include probability, radioactive decay, population and time period of concern. Population doses are cumulative; therefore, doses calculated for shorter periods of time are included in those for longer periods of time. Most scenarios could result in small doses to an assumed two million people living within 80 km of the repository. The drilling scenario would result in exposures to a limited number of people (25 were assumed). Forty million people could be exposed by eating contaminated salt from the solution mining scenario. Scenarios for fault and flooding, drilling and solution mining are assumed not to start until 1,000 years after the repository is closed. Details of the assumptions and calculations can be found in Reference 3.

SOURCE TERMS

Worst case source terms for the calculations were determined by considering the inventories of all consequential isotopes in the waste and allowing decay and in-growth of daughter products. These include all wastes projected to be calcined through the year 2020. At the base year of 2020, the total quantity of waste is $2x10^7$ kg ($4x10^7$ lb) calcine (or equivalent if in another waste form). The concentrations of radionuclides used in all calculations are shown in Table 1.

Table 1. Initial Radionuclide Concentrations of Input Solid Wastes (4), All Options

Nuclide	Activity (Ci/kg)	Nuclide	Activity (Ci/kg)	Nuclide	Activity (Ci/kg)
^{79}Se	6.4E-05*	^{87}Rb	3.6E-09	^{90}Sr	1.3E+01
^{90}Y	1.3E+01	^{93}Zr	3.1E-04	^{93m}Nb	7.5E-05
^{99}Tc	2.1E-03	^{106}Ru	9.7E-01	^{106}Rh	9.7E-01
^{107}Pd	2.0E-06	^{126}Sn	3.2E-05	^{126m}Sb	3.2E-05
^{126}Sb	3.2E-05	^{134}Cs	3.3	^{135}Cs	7.5E-05
^{137}Cs	1.3E+01	^{137m}Ba	1.2E+01	^{144}Ce	8.2
^{144}Pr	8.2	^{144}Nd	0.0**	^{147}Pm	1.2E+01
^{147}Sm	0.0**	^{151}Sm	1.7E-01	^{154}Eu	1.8E-01
^{226}Ra	0.0**	^{230}Th	0.0**	^{233}Pa	0.0**
^{233}U	1.2E-12	^{234}U	4.3E-10	^{235}U	1.8E-09
^{236}U	1.0E-08	^{237}U	4.8E-12	^{238}U	1.0E-14
^{237}Np	4.8E-08	^{239}Np	0.0**	^{238}Pu	7.0E-02
^{239}Pu	7.0E-04	^{240}Pu	6.5E-04	^{241}Pu	1.6E-01
^{242}Pu	1.8E-06	^{241}Am	9.8E-04	^{243}Am	8.3E-06
^{242}Cm	6.5E-04	^{244}Cm	5.2E-04		
Total	8.5E+01				

*6.4E-05 = 6.4×10^{-5} = 0.000064.
**Initial input is 0.0, but they appear later as daughter products.

RESULTS

Possible environmental impacts from disposal of ICPP wastes in a geologic repository are evaluated and compared for actinide glass, stabilized calcine and glass waste forms in Table 2. For disposing the actinide fraction of the waste as a glass, the highest maximum event consequence is 10,000 mrem as a result of intrusion into the waste by drilling. A meteorite impact on the repository would result in a 2000 mrem exposure. The other scenarios would result in insignificant maximum event consequences. The largest population dose comes from the solution mining scenario over long time periods. The 69,000 man-rem dose is due mainly to the large number of people exposed (40 million) and the long time span. No individual would receive a dose higher than three mrem. The other scenarios would not cause any significant population doses over any time period. Over any time period, the largest total dose is only 0.000027 percent of the total dose to the same population due to background radiation.

For disposing the entire waste inventory as a calcine, the highest maximum consequence is 5600 mrem as a result of a meteorite impact. The drilling and fault-flooding scenarios would result in individual doses of 640 and 160 mrem, respectively. The solution

Table 2. Radiological Consequences of Repository Disposal

Release Scenario	Maximum Event Consequence (mrem/event)	Population Dose (man-rem) For Periods Ending At 1,000 yr	10,000 yr	1 million yr
Actinide Glass				
Canister Drop	0.003	$1.6x10^{-6}$	$1.6x10^{-6}$	$1.6x10^{-6}$
Meteorite Impact	2,000	$1.6x10^{-6}$	$2.7x10^{-6}$	$7.5x10^{-6}$
Fault & Flooding	11	0	$1.3x10^{-12}$	$2.2x10^{-6}$
Drilling	10,000	0	0.21	74
Solution Mining	3.0	0	72	69,000
TOTAL		$3.2x10^{-6}$	72	69,000
Stabilized Calcine				
Canister Drop	0.002	$1.0x10^{-6}$	$1.0x10^{-6}$	$1.0x10^{-6}$
Meteorite Impact	5,600	$2.6x10^{-6}$	$3.8x10^{-6}$	$8.5x10^{-6}$
Fault & Flooding	160	0	$150x10^{-9}$	$32x10^{-6}$
Drilling	640	0	0.06	2.0
Solution Mining	3.0	0	72	69,000
TOTAL		$3.6x10^{-6}$	72	69,000
Glass				
Canister Drop	0.00007	$34x10^{-9}$	$34x10^{-9}$	$34x10^{-9}$
Meteorite Impact	5,600	$2.6x10^{-6}$	$3.8x10^{-6}$	$8.5x10^{-6}$
Fault & Flooding	11	0	$1.3x10^{-12}$	$2.2x10^{-6}$
Drilling	430	0	0.04	1.3
Solution Mining	3.0	0	72	69,000
TOTAL		$2.6x10^{-6}$	72	69,000

mining and canister drop scenarios would cause much lower event consequences. The largest population dose (69,000 man-rem) again comes from the solution mining scenario because all the waste is released in this scenario. This dose is the result of the large population and long time span involved, not any significant individual doses. No other significant population doses would occur. The largest total dose over any time period is again only 0.000027 percent of the total dose to the same population due to background radiation.

For disposing the entire waste inventory as a glass, the highest maximum event consequence is 5600 mrem as a result of a meteorite impact. The drilling scenario would result in a 430 mrem exposure. No other significant event consequences would occur from the other scenarios. The largest population dose (69,000 man-rem) again comes from the solution mining scenario. The dose is again mainly due to the large population and long time span involved. No individual would receive a significant dose. All other population doses would be extremely low. The largest total dose over any time period is only 0.000027 percent of the total dose to the same population due to natural background radiation.

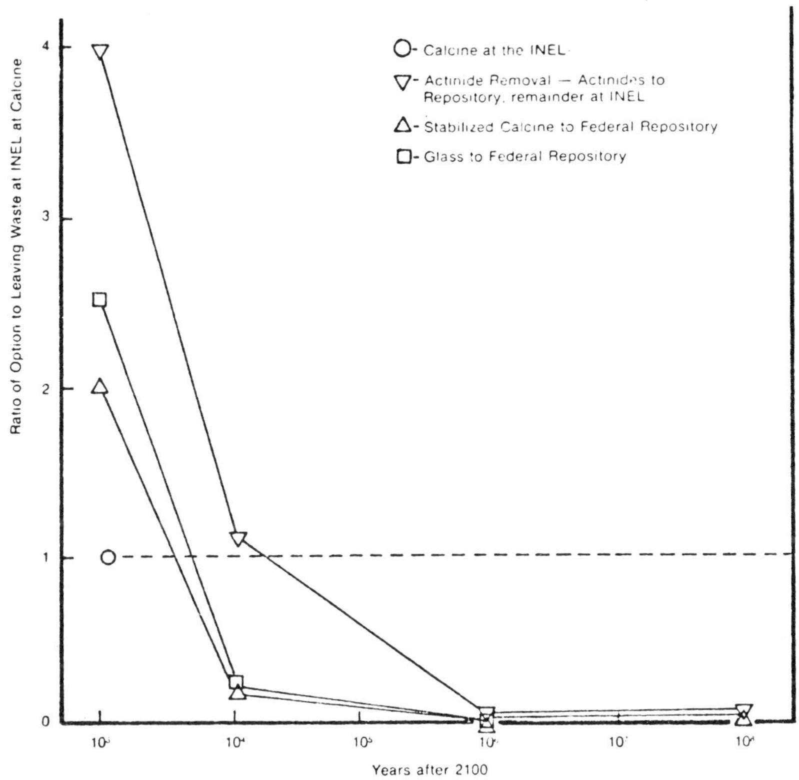

Fig. 1. Comparison of Population Dose Risks from Alternatives

SUMMARY AND CONCLUSIONS

The highest maximum event consequence for any scenario and waste form is 10,000 mrem for the drilling scenario with actinide glass. This dose is lower by factors of 16 and 23 for stabilized calcine and glass, respectively. The second highest event consequence is 5600 mrem for stabilized calcine and glass and 2000 mrem for actinide glass from a meteorite impact. Both events have very low probabilities of occurrence, and the resulting doses would cause very few, if any, health effects. The only significant population dose is 69,000 man-rem from the solution mining scenario for all three waste forms over the one million year period. The dose, however, is due to the large exposed population; no one individual would receive more than three mrem. The solution mining scenario would not exist at all if the repository were in any geologic formation other than salt.

Doses to individuals and to the population for each waste form and time period can be compared to the impacts of leaving the waste at the ICPP (1, 4). The two largest maximum event consequences for leaving the waste at the ICPP are 490,000 mrem resulting from a meteorite impact scenario and 100,000 mrem resulting from an airplane striking the facility ($1x10^{-8}$ events/yr). A maximum event consequence of 180,000 mrem results from an accident scenario while transporting the waste to a geologic repository. The total population dose for all activities including operational and occupational exposures, transportation exposures and exposures from intrusion and migration are shown as a ratio to the population dose from leaving the waste at the ICPP in Fig. 1. Because of the projected exposures required to further treat the waste and transport it to a geologic repository, thousands of years are necessary for the population dose from the geologic repository disposal alternative to equal or be less than the population dose from leaving the waste at the ICPP.

Once ICPP wastes are disposed in a geologic repository, no catastrophic or even dangerous doses could be received from these exposure scenarios. No clear choice between alternative waste forms, based on potential radiation exposure after geologic disposal, can be made from this analysis. Disposal in a geologic repository such as the generic one described here appears to be a safe method for management of ICPP high level wastes.

REFERENCES

1. N. A. Chipman and G. G. Simpson, Environmental Impact Analysis: Long Term Management of Defense High Level Radioactive Wastes, Idaho Chemical Processing Plant, Idaho National Engineering Laboratory, ACI 375 (1979).
2. U.S. Department of Energy, Draft Environmental Impact Statement Management of Commercially Generated Radioactive Waste, DOE/EIS-0046-D (1979).
3. N. A. Chipman and G. G. Simpson, Geologic Repository Release Studies: Long Term Management of Defense High Level Radioactive Wastes, Idaho Chemical Processing Plant, Idaho National Engineering Laboratory, ENI 105 (1979).
4. G. G. Simpson and N. A. Chipman, Source Terms for Environmental Impact Analysis: Long Term Management of Defense High Level Radioactive Wastes, Idaho Chemical Processing Plant, Idaho National Engineering Laboratory, ACI 359 (1978).

FAR FIELD THERMAL CALCULATIONS FOR THE WIPP SITE IN SOUTHEASTERN NEW MEXICO (SENM)*

S. B. Pahwa and J. R. Wayland

INTERA, Houston, Texas;
+Sandia Laboratories, Albuquerque, NM 87185

INTRODUCTION

The evaluation of the safe burial of heat producing radioactive waste in bedded salt requires estimates of the increased temperatures introduced into the surrounding geological strata. The specific problem of the far field (regional) temperature field calculation and the interaction with existing aquifers is considered for a specific site (the WIPP site in SENM).** The hydrologic transport of radionuclides in the geologic media is modeled by a finite-difference solution of the coupled pressure, energy, and mass transport equations. Thus, the same model can be used to investigate temperature increases in the aquifers and at the WIPP ground surface. Also, the dependence of the temperature changes with aquifer and thermophysical parameter variations will be studied. The purpose of this work is to perform a sensitivity analysis of emplacement scenarios upon the temperature gradients and heat flow of the near-surface geologic media.

MODELING

The proposed study area (Los Medanos) is located in Eddy and Lea Counties, southeastern New Mexico, about 27 miles east of Carlsbad.

WIPP Geology and Hydrology

The waste horizon in the Salado Formation is made up of halite with interbedded anhydrite, polyhalite, and clays. The Salado is

*This work supported by the U. S. Department of Energy
**The thermomechanical calculations are given in other reports

overlain by the Permian age Rustler Formation (300 to 400 ft thick), which consists mainly of interbedded anhydrite, dolomite, siltstones, clays, and some halite. There are two dolomite beds within the Rustler that range between 20 to 30 ft thick. The Magenta, the upper dolomite member, is dense, finely crystalline dolomite. The lower dolomite member, the Culebra, is vuggy with some anhydrite. These are the dominant water-bearing formations of the Rustler (2). The values of transmissivity in the formations immediately above the repository are from 1 to 20 ft^2 d. For model calculations, the two aquifers were combined and represented by one 40 ft vertical grid block. Above the Rustler Formation, the Dewey Lake Redbeds are composed of interbedded siltstones and sandstones that vary in thickness from 200 to 540 ft. The Dewey Lake is overlain in about half of the study area by the Triassic Santa Rosa sandstone. The surface is thinly covered by dune sands. Below the lower salt bearing formation (Castile), are several low permeability geological formations. This group of aquifers is called the Delaware Mountain Group and is modeled as one 1000 ft thick aquifer. Two 500 ft thick numerical grid blocks were used.

Sources

A range of possible thermal loading densities for the WIPP site has been investigated. Fission products producing the heat (^{137}Cs and ^{90}Sr) are adequately characterized by a 30-year half life. At relatively long times (tens of thousands of years) when heat released by the decay of fission products is no longer the major portion of the total source (when actinides become important), the assumption of 30-year half life for total heat decay is weak. To help establish bounds upon expected temperature fields, a series of test cases will be considered. These are not to be construed as being the WIPP, but only as representative test cases. The base case (A in Table 1) has a loading density of 100 kW/acre at 2700 ft and 30 kW/acre at 2100 feet. Both levels cover 1470 acres. As indicated in Table 1, three more cases are considered including a single-level repository of 20 acres, and thermal loading of 30 kW/acre at a depth of 2100 ft.

Boundary Condition at the Surface

Surface temperature measurements show approximately sinusoidal behavior as a function of time. Over the time periods of interest (hundreds to thousands of years), daily and seasonal variations in the surface temperature can be neglected. It is convenient to use a constant temperature boundary condition at the surface, but the adequacy of this boundary condition needs to be evaluated or justified. Physical boundary condition at the earth's surface is a combination of 1) radiation, 2) convection by air, and 3) heat removal by evaporation of water. It is a common practice in engin-

Table 1. Thermal Loadings of the Cases Modeled

Case	No. of Levels	Thermal Loading	Area
A	2	100 kW/acre @ 2700 ft 30 kW/acre @ 2100 ft	1470 acres
B	2	75 kW/acre @ 2100 ft 22.5 kW/acre @ 2100 ft	1470 acres
C	1	30 kW/acre @ 2100 ft	1470 acres
D	1	30 kW/acre @ 2100 ft	20 acres

eering to combine the effects of convection and radiation heat losses by a linear convection equation (3). Neglecting the heat carried by evaporation, the heat flux from the surface can be described by an effective heat transfer coefficient. Thermal resistance of air indicates heat transfer coefficient of 1.3 BTU/hr-ft^2-°F or higher. Thermal conductivity of the near-surface formation (Dewey Lake) at WIPP is roughly 1 BTU/hr-ft-°F. Therefore, the resistance of air represents less than 1 ft of Dewey Lake formation resistance. Air resistance is essentially negligible. The results obtained are valid except at the surface and very near the surface. For temperature increases at the surface, either the thermal resistance should be adjusted to the constant temperature boundary to include the air resistance, or interpret the calculated temperatures at rougly 1 ft depth as the surface temperature. A comparison of these two cases found them to be equivalent.

Numerical Discretization

There is one aspect of numerical modeling that must be clearly understood. The finite-difference equations solved are flux balances of flow and energy around each numerical grid block. Therefore, the calculated pressures and temperatures are average quantities over the grid block. Temperature profiles within the grid blocks are not calculated. If the thermal front moves to less than a grid block spacing information obtained by finite difference modeling may be of little value. The basic objective of this work has to calculate temperature changes in the Rustler and near the ground surface. The grid sizes used were 1000 ft in the horizontal direction, and varied from 40 ft near the surface to 110 ft at the repository levels to 500 ft below Castile in the vertical direction. The vertical grid spacing and associated physical properties are given in Tables 2 and 3. Assuming the grid block average values to be the same as block center values is an excellent approximation except in the source blocks. The calculated temperatures in the repository blocks are average repository temperatures.

Table 2. Vertical grid spacing

Geologic Formation	Depth From Surface (ft)	Thermal Conductor (BTU/ft/day F)*	Porosity	Horizontal Hydraulic Conductivity (ft/day)
Dewey Lake	0-550	Type 3	0.005	10^{-5}
Rustler	550-635	Type 2	0.005	10^{-5}
	635-675	Type 2	0.100	1.0
	675-760	Type 2	0.005	10^{-5}
Salado	760-2830	Type 1	0.005	10^{-5}
Castile	2830-3330	Type 1	0.005	10^{-5}
	3330-3630	Type 2	0.005	10^{-5}
	3630-4030	Type 1	0.005	10^{-5}
Delaware	4030-5030	Type 4	0.15	0.15
Mountain	5030-7430	Type 4	0.005	0.0

*See Table 3

REGIONAL TEMPERATURES

Temperatures at the Emplacement Horizons

The temperatures at the emplacement horizons all show strong dependence upon the thermal loading density (kW/acre). Emplacement horizons were represented by 110 ft thick layers. Calculated temperatures in these layers for the different loading scenarios are shown in Fig. 1. Note that the peak temperature is reached about 50 years after emplacement. It was assumed that the emplacement was instantaneous into a layer 110 ft thick. The detailed near field calculations are given in SAND78-0924 (4). If there is no water expansion or if flow in the aquifers is turned off, the impact upon the calculated temperatures is not significant.

Temperature of Aquifers

The temperature fields in the Rustler Formation are more accurately determined by this calculation. For the base case, the temperature increases can be as much as 20°F as shown in Fig. 2. The peak temperature rise occurs at about 2000 years after emplacement for all cases. Again, there is little or no impact from water expansion or if the slow flow of the aquifer is turned off.

Table 3. Thermophysical Properties

	Temp(F)	K (BTU/Ft-Day-F)	Specific Heat (BTU/Ft^3°F)
Rock Type 1	126.00	79.10	29.4
	180.00	66.60	
	378.00	54.10	
Rock Type 2	117.00	43.00	33.6
	207.00	3.30	
	369.00	27.80	
Rock Type 3	126.00	26.40	49.5
	180.00	22.20	
	378.00	18.00	
Rock Type 4	70.00	22.50	33.5
	150.00	22.50	
	400.00	22.50	

Surface Temperatures

Of considerable interest for environmental impact consideration is the temperature variation in the soil column just at the surface. The grid block chosen for this calculation has a 40 ft width at the surface. The temperatures are calculated at the midpoint of each node. The results are shown in Fig. 3. Note that the maximum temperature increase at 20 ft below the surface is, for the base case, about 1°F. The temperature increase for emplacement at 30 kW/acre is less than 0.5°F with a somewhat delayed peak value. Perhaps of

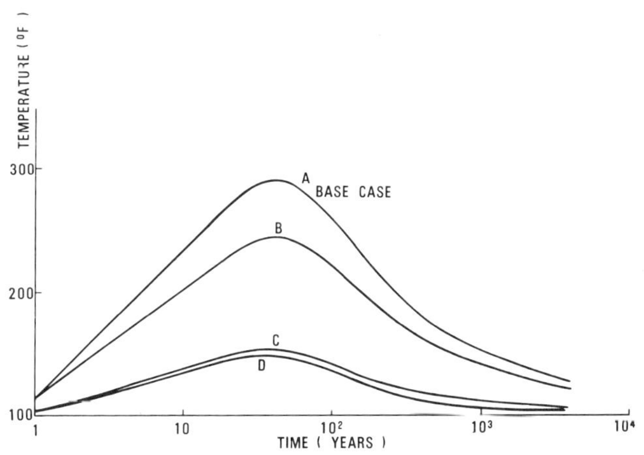

Fig. 1. High Level Waste Temperature Showing Effect of Different Heat Loads

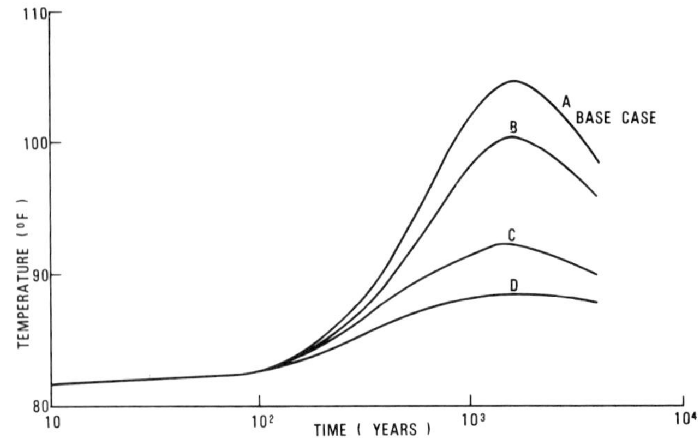

Fig. 2. Temperature in the Rustler Showing the Effect of Heat Loads

more interest are the variations that are introduced by uncertainty in the thermal conductivity. The results for the base case are shown in Fig. 4. as curve A (K increased to 2K) and curve C (K decreased to 1/2K). Only the conductivity in the Salado was varied. The change is as much as going from a 130 kW/acre, two-level, 1470 acre repository to a one-level, 20 acre, 30 kW/acre emplacement.

From the biologist's point of view, the changes in the top six inches of the soil column are most important. The near study state close to the surface allows linear interpolation to be used from the surface (Z = 0) to the first grid block center (Z = 20 ft). Note that the sinusoidal temporal variation has been ignored. The results at 6 inches are shown in Fig. 5. The very small maximum change (base case A) of about 0.05°F clearly indicated that there should be only a

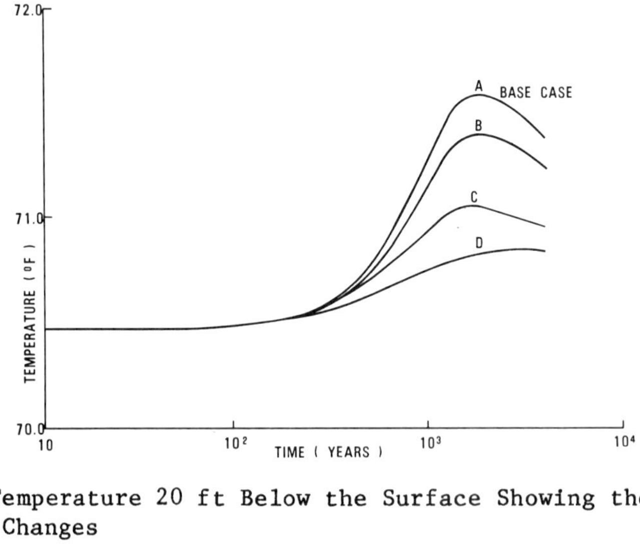

Fig. 3. Temperature 20 ft Below the Surface Showing the Effect of Heat Load Changes

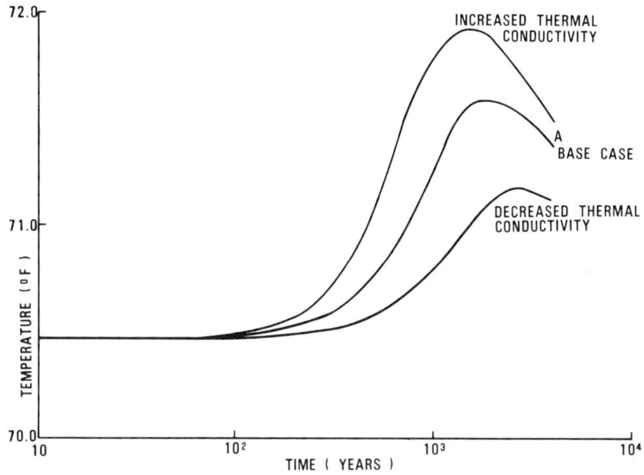

Fig. 4. Temperature 20 ft Below the Surface Showing the Effect of Thermal Conductivity

minimum increase. The maximum geothermal flux increase is 3.5 times the amount calculated for base case A. Another aquifer system that might be affected is in the Delaware Mountain Formation. The temperature increase in these aquifers is relatively small; from 0.3°F for the 30 kW/acre 20 acre case to 5.4°F for the base case of 130 kW/acre.

CONCLUSIONS

This idealized study indicates that there is little impact upon the far thermal field from water expansion or for stoppage of aquifer

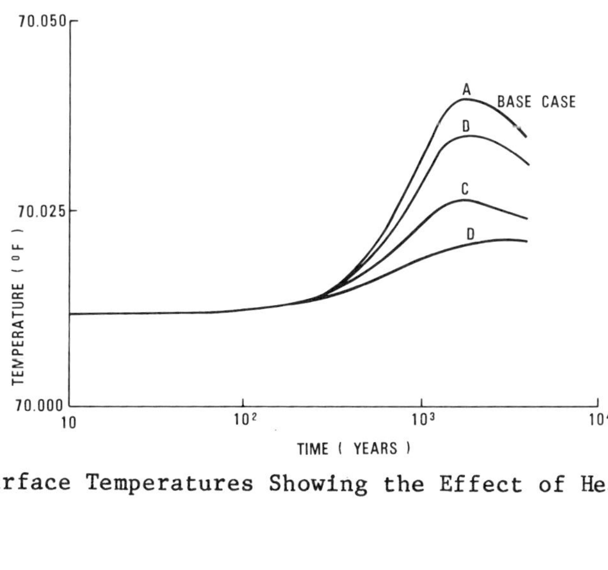

Fig. 5. Surface Temperatures Showing the Effect of Head Load Changes

flow. The variability of salt thermal conductivity in the Salado Formation can be as important to the magnitude of the temperature fields as the extremes proposed for thermal loading densities.

REFERENCES

1. S. J. Lambert and J. W. Mercer, SAND77-1401, Sandia Laboratories, Albuquerque, NM 87185 (1977).
2. J. W. Mercer and B. R. Orr, USGS Open File Report 77-123, Albuquerque, NM, February (1977).
3. D. Kern, Process Heat Transfer (McGraw-Hill Book Company, Inc., New York, 1950).
4. C. E. Sisson, SAND78-0924, Sandia Laboratories, Albuquerque, NM 87185 (1978).

TEMPERATURES GENERATED BY UNDERGROUND STORAGE OF DEFENSE WASTE CANISTERS*

M. H. Tennant

E. I. du Pont de Nemours and Company
Savannah River Laboratory
Aiken, SC 39801

INTRODUCTION

The plan for long term management of Savannah River high level liquid waste is that it will be calcined, combined with glass frit at high temperatures, and allowed to solidify in cylindrical steel canisters (2 ft in diameter, about 9 ft tall). The sealed canisters would then be inserted into cylindrical steel storage containers. After interim storage, the storage containers would be placed in an underground federal repository for long term storage. Although specifications for repository storage have not been established, it is assumed that a series of 15 to 50 foot-wide disposal rooms would be mined at depths ranging from 1000 to 3000 feet. Storage containers would be placed in holes drilled in the disposal room floors.

After container emplacement, the radioactive decay heat would be dissipated by conduction through the surrounding geologic formations to the earth's surface. From earth's surface, the energy would be transferred by convection to the atmosphere and ultimately to deep space by radiation. To ensure safe storage, the design selected for an underground repository must take into account temperature restrictions placed on the defense waste glass, containers and surrounding geology. Studies are planned at Savannah River Laboratory, in conjunction with Office of Nuclear Waste Isolation, to set temperature restrictions based on leach rates and canister compatibility with salt, water and salt brine. The primary motivation for this study was the need to determine realistic temperature ranges for

*Work done under USDOE Contract No. DE-AC09-76 SR0001.

leaching and compatibility tests associated with salt repositories. Determination of the limiting thermal restrictions will require consideration of other phenomena such as devitrification, decrepitation, salt brine migration and mine stability.

DISCUSSION

Methodology

Three models were used in this study. The models are classified as far-field, near-field and very near-field. Far-field refers to the geologic region extending hundreds of feet from repository, where the thermal response is sensitive to areal power densities, geologic properties and stratigraphy. Near-field refers to the geologic region within tens of feet of the cavity in which the canister is placed, where the thermal response is sensitive to the overlying mine shaft, shielding plugs, liners and local material properties. Very near-field refers to that region in the immediate vicinity of the canister, where the thermal response is sensitive to the size and content of any annuli between the canister cavity and waste glass.

All three models aid in repository design. Far-field models can be used to determine geologic temperatures affecting overlying aquifers and uplift. Near-field models can be used to determine temperatures affecting reconsolidation and retrievability in the vicinity of the canister cavity. Very near-field models can be used to determine waste glass and container temperatures affecting waste glass devitrification and leaching as well as canister compatibility.

The relationship between the three models is shown in Fig. 1. The far-field model assumes a thin cylindrical 833-acre repository* buried 2000 feet below the earth's surface. The repository itself is represented as a uniformly distributed heat source. The near-field model considers a narrow cylindrical slice from the center of the repository. It contains a single canister. Finally, the very near-field model considers the radial temperature rises across insulating annuli, and is independent of canister spacing.

The models used are conservative and simple. However, the added complexity and calculation times required for more detailed models is not presently justified because of the large uncertainties that currently exist in repository design, properties and waste canister power.

*Proposed size of a national repository. All Savannah River waste will probably require less than 150 acres.

General Assumptions

1. Repository Conditions:

The entire repository was assumed to be simultaneously loaded with defense waste canisters having the thermal power and decay rates consistent with Fig. 2. Five-year-old waste represents the freshest waste that might be processed. Most of the waste will be as much as 35 years old. No attempt was made to model mine shafts, shield plugs, liners, overlying aquifers, salt brine migration, groundwater movements, reconsolidation, or any conditions pertaining to a retrievable period. Moisture was considered only in the annulus surrounding a canister.

2. Initial Conditions:

At the time of burial, the temperature of the geology was assumed to be 16°C at the earth's surface and to increase 11°C for every 1000 feet below the earth's surface. Geothermal fluxes in the continental United States range from 53 to 350 Btu/yr-ft^2 resulting in geothermal gradients ranging from 3°C to 33°C per 1000 feet depth (1,2).

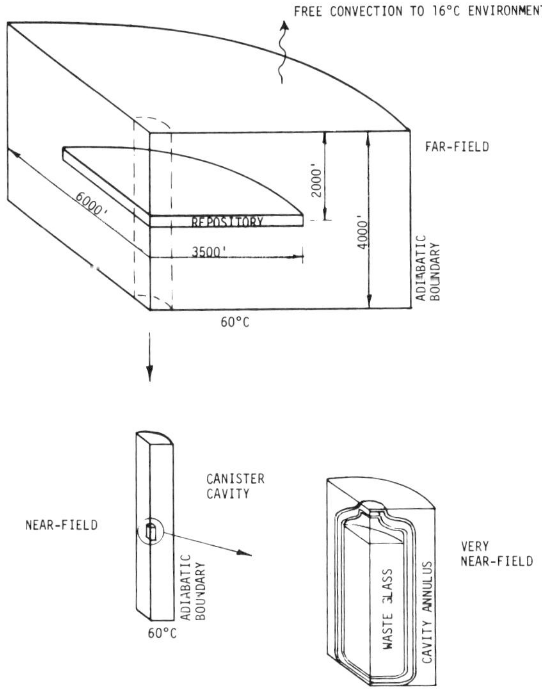

Fig. 1. Relationship Between Models

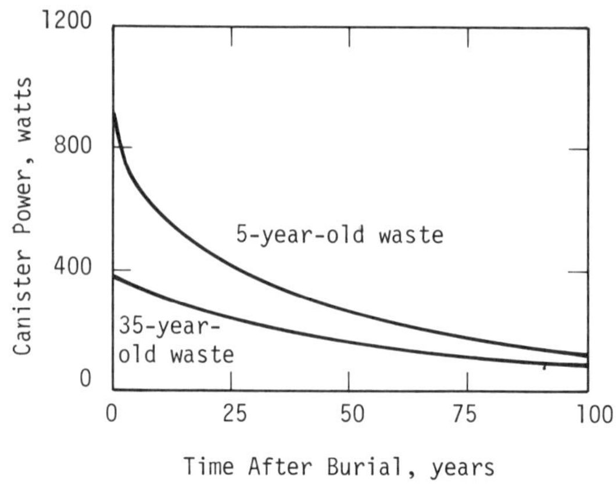

Fig. 2. Defense Waste Power Decay

3. Boundary Conditions:

The axial boundary conditions (Fig. 1.) were fixed by a temperature of 60°C at a depth of 4000 feet and convection (convective heat transfer coefficient of 1.0 Btu/hr-ft^2-°F) to a 16°C environment at the earth's surface. Radially the far- and near-field models required adiabatic boundary conditions at 6000 feet and half the canister spacing, respectively. The canister cavity temperature, required as a radial boundary condition for the very near-field model, was determined from the near-field calculation.

4. Stratigraphy:

The stratigraphy assumed for the salt repository is shown in Table 1. Stratigraphy at a given site may vary markedly from that assumed.

5. Thermal Properties:

The temperature dependent thermal properties used are given in Table 1. These properties are typical, but may be significantly different from the properties at a given site.

Solution Techniques

A cylindrical implicit transient version of the HEATING5 conduction code was used to calculate the temperatures for the far- and

Table 1. Material Properties

Salt Temperature, °F	Thermal Conductivity (K_s), BTU/hr-ft-°F
32	3.47
122	2.85
212	2.39
302	2.05
392	1.77

Repository

Layer	Depth, ft	Density, lbm/ft^3	Specific Heat, BTU/lbm-°F	Thermal Conductivity, BTU/hr-ft-°F
1	0-1000	512	0.22	0.4 K_s
2	1000-1825	471	0.22	0.57 K_s
3	1825-2165	430	0.22	K_s
4	2165-3000	471	0.22	0.57 K_s
5	3000-4000	512	0.22	0.4 K_s

Backfill, 30% free volume	Thermal Conductivity BTU/hr-ft-°F
Dry Crushed Salt	0.15
Wet Crushed Salt	1.20

near-field models. Simple one-dimensional conduction calculations were used for the very near-field calculations.

Results

Far-Field: After burial, the temperature of the surrounding geology rises as the repository dissipates heat. As the repository temperature peaks, all the geology within 500 feet at the repository is affected. As time continues, the repository temperature will fall and the volume of geology affected will slowly grow.

Near-Field: Figs. 3. and 4. show the near-field results. Fig. 3. shows that canister cavity temperatures tend to peak earlier with increased spacing. Fig. 4. shows a sharp decline in peak canister cavity temperatures as the spacing is increased from 10 to approximately 30 feet. After that, increased spacing does not appreciably lower peak canister cavity temperatures which are below 100°C.

Very Near-Field: Fig. 5. shows the temperature rise across crushed salt backfilled annuli for five-year-old waste. Dry annuli result in 8 times higher temperature rises then do wet annuli. For 35-year-old waste, temperature rises across the annuli would only be ~40% of the values given in Fig. 5.

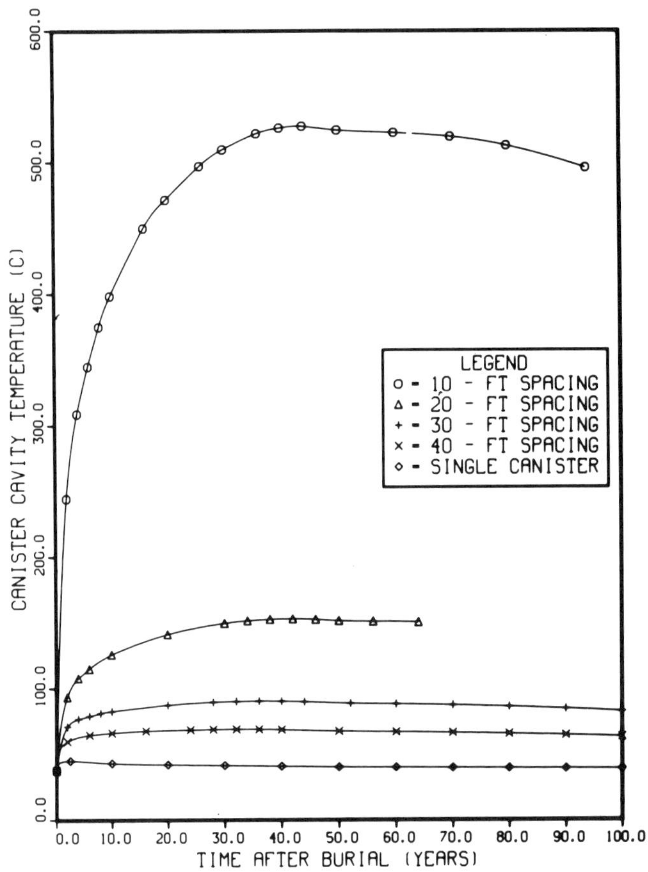

Fig. 3. Canister Cavity Temperature in a Salt Repository - 5-Year-Old Waste

CONCLUSIONS

1. For conditions conductive to leaching, i.e., moisture in contact with glass, the glass surface temperature would essentially be the canister cavity temperature because temperature rises across wet annuli are small. In turn, canister cavity temperatures are dominated by canister spacing for a given canister power decay curve. Actual canister powers may be as much as 40% less than those used for these calculations. Owing to the low heat generating characteristics of Savannah River waste, glass surface temperatures can easily be maintained below 100°C under leaching conditons.

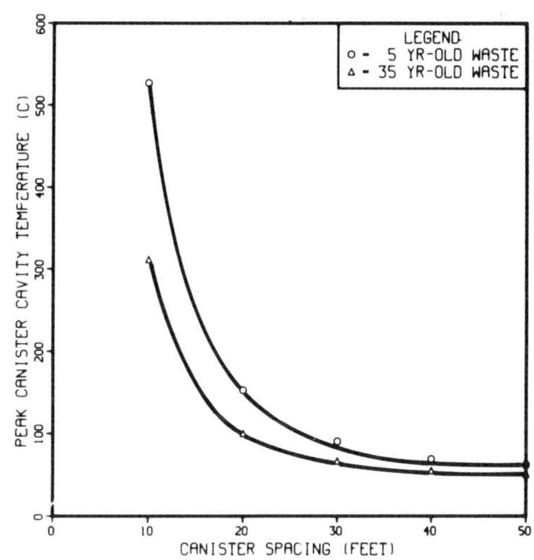

Fig. 4. Peak Canister Cavity Temperature in a Salt Repository

2. For dry salt in contact with the storage container, the temperature rises across the annulus may be large. The container temperature would tend to peak earlier for dry annuli than for wet annuli.

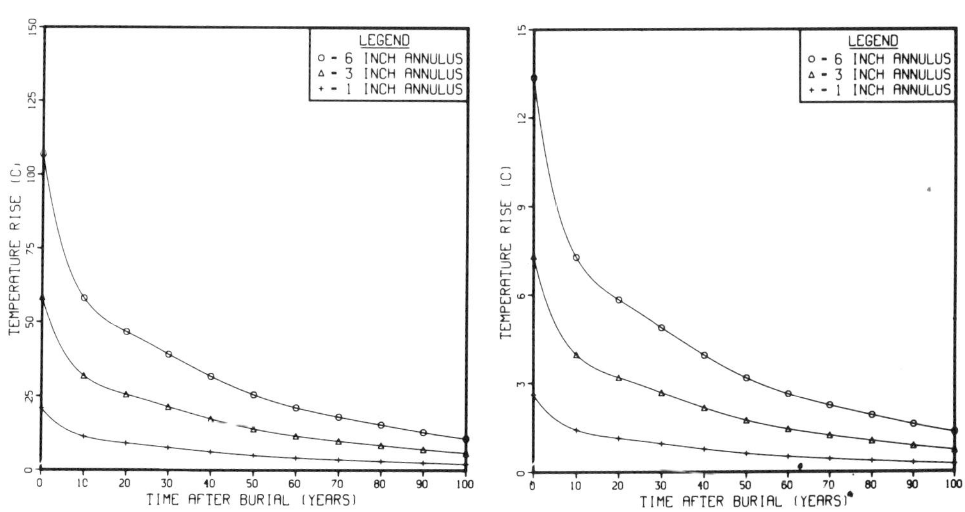

Fig. 5. Temperature Rises Across Dry and Wet Crushed Salt Backfill for 5-Year-Old Waste

3. When canister cavity temperatures peak, salt temperatures would be increased over normal approximately 500 feet in any direction extending from the repository.

REFERENCES

1. E. C. Robertson, The Nature of the Solid Earth, Chapter 19, (McGraw-Hill, Inc., NY, 1972).
2. B. Gutenberg, Physics of the Earth Interior, (Academic Press, NY, 124, 1959).

CALCULATED HYDROGEOLOGIC PRESSURES AND TEMPERATURES RESULTING FROM RADIOACTIVE WASTE IN THE ELEANA ARGILLITE*

R. R. Eaton[+], W. D. Sundberg[+], D. E. Larson[+] and M. P. Sherman[++]

[+]Sandia Laboratories**, P. O. Box 5800, Albuquerque, NM 87185; [++]Rochester Institute of Technology, Rochester, NY

ABSTRACT

The SHAFT 78 code (multi-dimensional, two fluid phases, porous medium) has been used to begin assessment of the consequences of nuclear waste burial in a 1000 acre repository emplaced in argillite. The repository is assumed to contain spent fuel [SF (UO_2)] at a loading of 150 kW/acre and to be located at a depth of 600 m. It was found that with perfect backfill, permeability = 1.0×10^{-7} darcys, a maximum fluid pressure in the repository of 770 bars existed at a time of 55 years after burial. Holding all other input variables constant, the maximum fluid pressure in the underground workings never exceeded the local lithostatic pressure when the permeability of the backfill material was increased to 1.0×10^{-1} darcys. The calculated temperature profiles are essentially independent of backfill permeability and porosity indicating that the heat flow is conduction dominated.

INTRODUCTION

Numerous one- and two-dimensional thermal studies have been previously made to assess the consequences of nuclear waste burial in various geologic media using heat conduction models to calculate temperature fields (1). For low permeability rocks, these models provide complete temperature fields near repositories for a wide range of materials and heat loadings. However, conduction models

*This work was supported by the U. S. Department of Energy under contract DE-AC04-76DP00789.

**A U.S. Department of Energy Facility.

cannot predict the fluid pressure (pore pressure) or fluid transport resulting from waste emplacement. Large fluid pressures occurring within a rock matrix as a result of such heating could have significant effects on the integrity of a repository, such as fracturing the rock. The magnitude of the fluid pressure fields resulting from waste emplacement in argillite is considered in this paper. The algorithm of the SHAFT 78 code used for repository analysis in this paper is based on mass and energy balance equations for two-phase flow in a saturated porous medium (2). Density and specific internal energy are used as independent variables. The water mass flux is modeled using the darcy formulation. In the solution method the rock rock volume is subdivided into a finite number of arbitrarily shaped volume elements each having a centered node point. Solutions are obtained at discrete time steps using Evans' semi-implicit iterative technique.

REPOSITORY ANALYSIS

The code has been applied to several repository problems. The material properties used for these studies are: argillite permeability (k) = 1.0×10^{-7} darcy, porosity (ϕ) = 9.1%, specific heat (C_p) = 1046.0 J/kg°C, thermal conductivity (K) = 2.7 J/m·s°C, bulk density (ρ) = 2530 kg/m^3. The initial (t = 0) temperatures and pore pressures were calculated assuming a water table at the earth's surface and 1.5 μcal/cm^{2}s geothermal heat flux. The two-dimensional nodalization for the repository is given in Fig. 1a with the repository located between $0 \le y \le 10$ and $0 \le R \le 1134$ m. The most critical input variable is the assumed argillite permeability. The assigned value of 1×10^{-7} darcy falls in the middle of the reported range for shales in general (1×10^{-5} to 1×10^{-8} darcys) (3) on the high permeability side of values reported from in situ testing at depth on the Savannah River reservation in South Carolina (3×10^{-7} to 4×10^{-9}) (4) and on the high permeability side of regional permeabilities backed out of groundwater modeling of the San Juan basin (3×10^{-7} to 3×10^{-8} darcys) (5). The assigned value is likely to be reasonable in considering fluid pressurization in many argillaceous rocks. This is especially true if no clay dehydration and contraction occurs near the boiling point of water in the rock being considered. Rocks totally lacking expandable clays should be directly amiable to the treatment used here. Other material properties are taken or approximated from values collected in support of the Eleana near-surface heater tests, operated by Sandia Laboratories, on the Department of Energy's Nevada Test Site (6).

In the first case modeled, no provision was made for considering different rock properties for the mine drift and shaft, i. e., the repository was assumed to contain perfect backfill. The isobars for this case are given in Fig. 2 at 50 years after burial. The important item to notice is the extremely high pressures near the reposi-

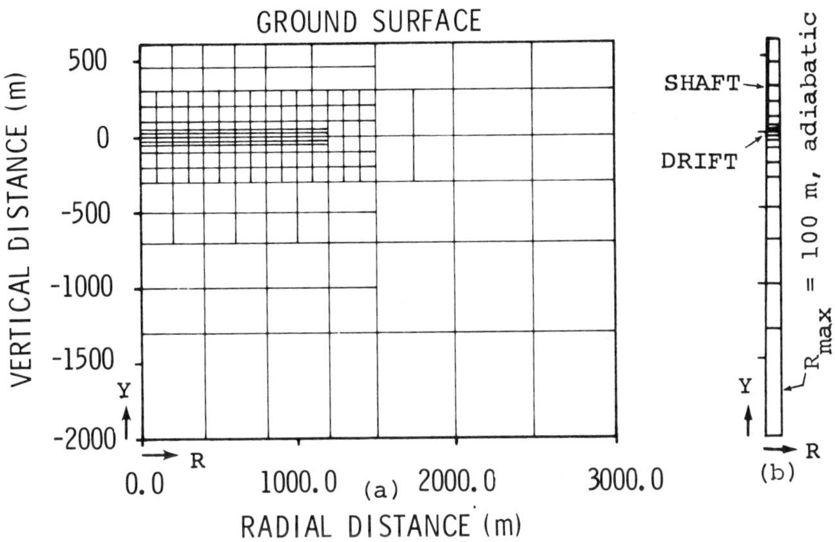

Fig. 1. Axisymmetric Grid Systems. Repository Located Between $0 \leq Y \leq 10$. (a) Perfect Backfill Geometry. (b) Quasi One-D Model Used to Analyze Imperfectly Backfilled Drift and Shaft Cases.

tory caused by the small temperature-independent permeability assumed for the argillite which restricts the transport of water. The model predicts no vaporization for this case. The maximum pressure (770 bars) for this case occurs at 55 years, at the centerplane of the repository disc and centerline (R = 0) of the model. The centerline pressure and temperature distributions at 55 years are given in Fig. 3. This curve shows that the local lithostatic pressure is exceeded by the fluid pressure for a distance of 205 m above the centerline by the repository. Such volatile overpressuring would likely not be stable since argillite appears to have a near-zero tensile strength in-situ. Thus, should such overpressuring occur, it is likely that most of the water in the overpressured zone would be released by vertical uplift of the overlying strata. Two factors affecting the result presented here for the perfect backfill limiting case must be kept in mind. First, perfect backfill has been assumed from time zero. Thus, all possibility of fluid venting into underground workings or access shaft has been eliminated, and fluid pressures calculated represent a maximum. Secondly, the argillite matrix is treated by SHAFT 78 as totally non-deformable. Thus, only thermal expansion of fluid in constant-volume pore space is treated, which should lead to an overestimate of the fluid pressures.

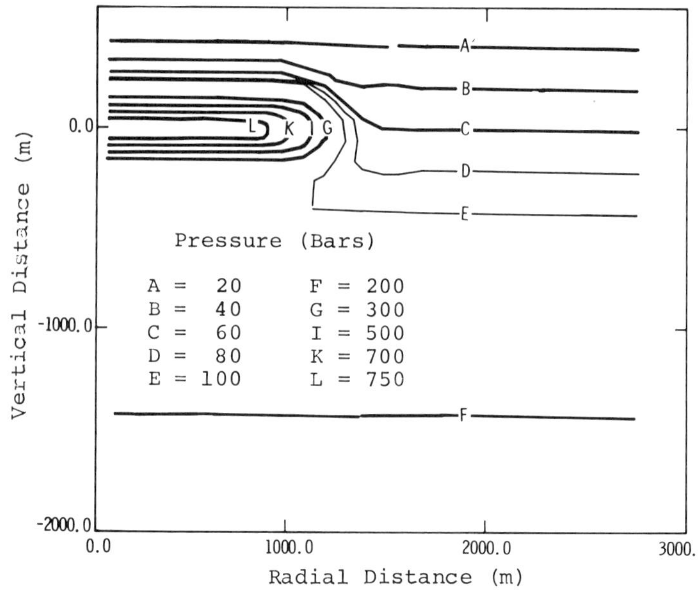

Fig. 2. Isobars for 1000 Acre Repository: 150 kW/acre, Spent Fuel, k = $1x10^{-7}$ darcy, ϕ = 9.1%, 50 years.

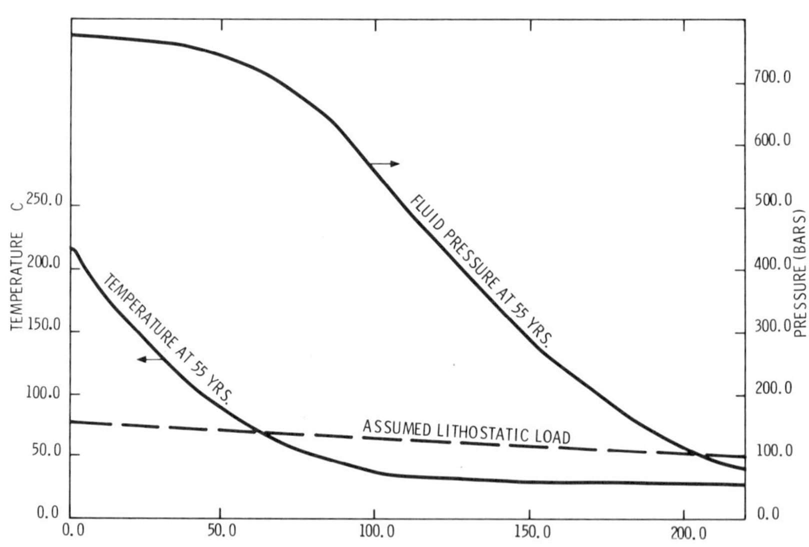

Fig. 3. Temperature and Pressure at R = 50.0 for 150 kW/acre, Spent Fuel, k = $1x10^{-7}$ darcy, Time = 55 years, ϕ = 9.1%.

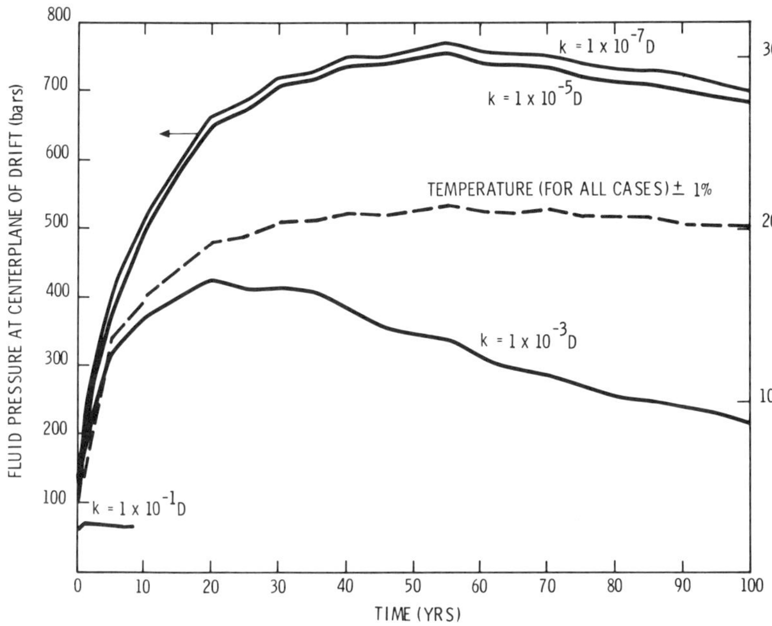

Fig. 4. Repository Pressure and Temperature (R = 50.0) for Several Mine Drift and Shaft Permeabilities; 150 kW/acre, Spent Fuel, Burial Depth = 600 m, ϕ = 9.1%.

In order to parametrically study the effects of varying the permeability of the material in the location of the mine shaft, $0 \text{ m} \leq R \leq 5 \text{ m}$, $10 \text{ m} \leq Y \leq 600 \text{ m}$, and drift $0 \leq R \leq 100 \text{ m}$, $0 \text{ m} \leq Y \leq 10 \text{ m}$, while leaving the argillite permeability outside these regions constant, a quasi one-dimensional model was used (see Fig. 1b) (7). The model was truncated at R = 100 m in order to keep the computer times to a reasonable limit. The repository loading of 150 kW/acre was used. Additional studies discussed in the full report show this assumption to be reasonable. This model will give a reasonable indication of the pore pressures that exist during the post-operational phase of a repository with fully saturated, imperfectly backfilled shaft and drift. During the operational phase of the repository, the mine shaft and drift could be partially saturated or filled with air. The fully saturated drift/shaft assumption made in the code will result in an overprediction of the argillite pore pressures. However, this overprediction will not be significant as long as the assigned drift/shaft permeabilities are high enough that fluid release from the argillite into the shaft and drift is not significantly hindered. Fig. 4 gives the results of this study. The drift/shaft permeability of $k = 1\text{x}10^{-5}$ darcy reduced the peak pressure by 2%, $k = 1\text{x}10^{-3}$ reduced it by 45% and

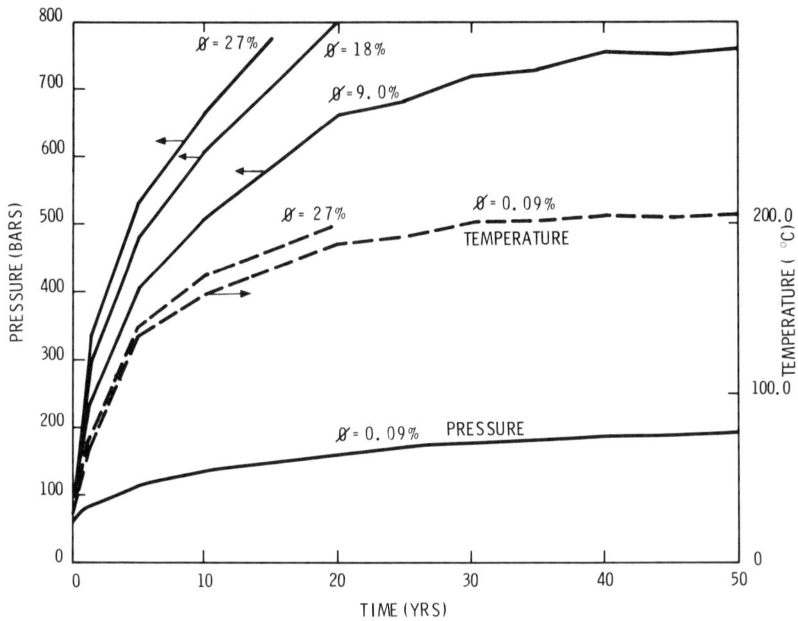

Fig. 5. Repository Pressure and Temperature (R = 50.0) for Several Mine Drift and Shaft Porosities; 150 kW/acre, Spent Fuel, Burial Depth = 600 m, k = 1×10^{-7} darcy.

k = 1×10^{-1} reduced it by 91%. For the latter case, the calculated fluid pressure is everywhere below the local lithostatic pressure, and the maximum argillite pressure 15 m below the drift was 130 bars (the local lithostatic pressure = 160 bars). The temperature history is nearly independent of permeability. Next, the porosity in the drift and shaft was varied from 0.09% to 27% while holding the permeability constant (k = 1×10^{-7} darcys). The increases in fluid pressures with increasing porosity (Fig. 5) are to be expected when one considers the greater mass of expanding water, for large porosity (ϕ) with no decrease in flow resistance (k).

The last case computed used the extreme values for drift/shaft permeability and porosity (k = 0.1 darcy and ϕ = 27%). Fig. 6 displays the vertical distribution of calculated fluid pressures in the argillite and drift at R = 50 m for 5, 10 and 15 years after burial. For this combination of permeability and porosity, the local lithostatic pressure is exceeded 35 m above the drift centerline at 15 years after burial.

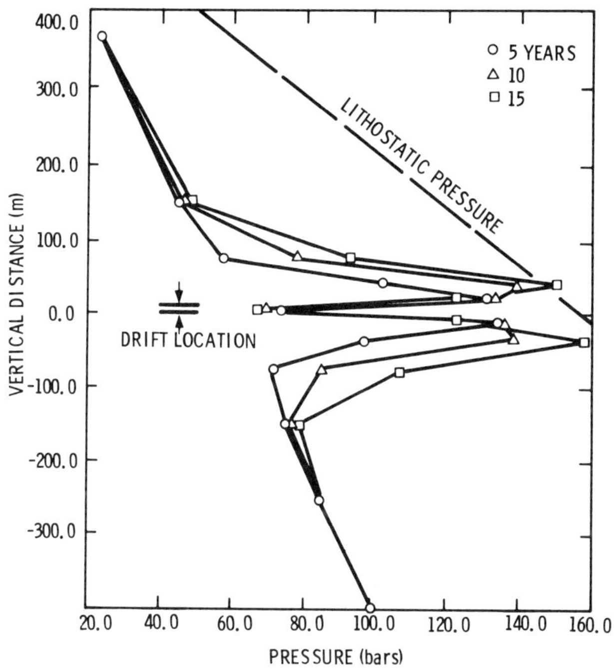

Fig. 6. Vertical Distribution of Pore Pressures for k = 0.1 darcy and ϕ = 27%, 150 kW/acre, Spent Fuel, Burial Depth = 600 m.

SUMMARY OF RESULTS

1. For the case of a repository with perfect backfill, the maximum repository pore pressure exceeds the local lithostatic pressure by a factor of four. This result, though specific to the assumed power density and fluid permeability, indicates that fluid release from the argillite into the repository may be a critical variable in disposal of waste in argillaceous rocks. This suggests that temperature-dependent permeability data must be used in such modeling as soon as possible.

2. Response of repository pressure levels to drift/shaft backfill permeability and porosity:

   a. Fluid pressure levels are a strong function of drift/shaft backfill permeabilities. An increase of six orders of magnitude in the backfill permeability (1×10^{-7} darcy to 0.1 darcy) assures that the maximum pressure encountered is below local lithostatic pressures at constant porosity (9%).

b. Increasing the drift/shaft backfill porosity at constant permeability could cause a major increase in fluid pressure.

3. Repository temperature is a weak function of backfill permeability and porosity in the range considered. Consequently, conduction models can be used to obtain reliable temperature distributions.

4. Fluid pressures near the repository are largely relieved by the presence of high-permeability backfill, thus reducing the concern of volatile overpressurization and the resultant possible degradation of the argillite. However, when both the permeability and porosity of the drift are increased, combinations of these parameters exist which will cause the calculated fluid pressure in the argillite to exceed the lithostatic pressure, Fig. 6. Certainly, more detailed canister and room-scale modeling must be done, including the use of temperature and pressure dependent permeability data, to assess in detail the areas where fluid overpressuring remains a possibility. Additionally, the large fluid pressures calculated with this code make it pertinante that this work be coupled with a rock mechanics code in order to obtain a better understanding of the conditions under which rock matrix failure might occur.

REFERENCES

1. D. F. McVey, "Near Field Interactions," presented at the 19th Annual ASME Symposium on Geological Disposal of Nuclear Waste, Albuquerque, NM (1979).
2. K. Pruess, et. al, SHAFT 78--A Two-Phase Multidimensional Computer Program for Geothermal Reservoir Simulation, LBL-8264, Lawrence Berkeley Laboratory, Earth Sciences Division, Berkeley, CA 94720.
3. K. Magara, "Permeability Considerations in Generation of Abnormal Pressures," Soc. Petr. Engr. 11, 236-242 (1971).
4. C. D. Marine, "Geohydrology of Buried Triassic Basin at Savannah River Plant, S. Carolina," Am. Ass. Petr. Geol. Bull. 58, 1825-1837 (1974).
5. F. Lyfort, USGS, Albuquerque, NM, personal communication, (1978).
6. D. F. McVey, A. R. Lappin and R. K. Thomas, Small-Scale Heater Tests in Argillite of the Eleana Formation at the Nevada Test Site, SAND79-0344, Sandia Laboratories, Albuquerque, NM (1979).
7. R. R. Eaton, W. D. Sundberg, D. E. Larson and M. P. Sherman, Status of SHAFT 78 With Respect to Modeling Radioactive Waste Burial in Eleana Argillite Including Calculations to Date, SAND79-1484, Sandia Laboratories, Albuquerque, NM (1979).

A SYSTEMATIC APPROACH FOR ASSESSMENT OF ALTERNATIVE NUCLEAR WASTE MANAGEMENT STRATEGIES

Gour-Tsyh Yeh

Environmental Sciences Division
Oak Ridge National Laboratory
Oak Ridge, TN 37830

ABSTRACT

A systematic approach for alternative nuclear waste management assessment is prescribed for evaluating the fate and transport of radwaste in the environment and to man. The approach consists of a system of flow dynamic and waste transport models for the physical environment, transfer models to biota, and dose models to man. Flow dynamic and waste transport models are included for computing spatial-temporal variations of the flow field and radwaste concentration in subsurface aquifers, surface water regimes, estuaries/oceans, and the atmosphere. Transfer models are linked to flow dynamic and waste transport models to study the uptake from and to the physical environment of the waste by aquatic and terrestrial biota. Dose models are used to compute both the radwaste input to man and its effects on man. With these series of computations, the radiation dose from alternative nuclear waste management strategies can be computed. Coupled with the exposure and effect analysis, the approach should facilitate the selection of the best management methodologies among many alternatives.

INTRODUCTION

One of the critical issues of nuclear power production and utilization is the safe management of potentially hazardous radioactive waste materials. Comprehensive assessments are needed to quantify effectiveness of each of the management alternatives, and thus increase public acceptance of nuclear energy. Safety assessments would inevitably include study of the transport, transformation, fate and pathway of the radwaste associated with nuclear energy.

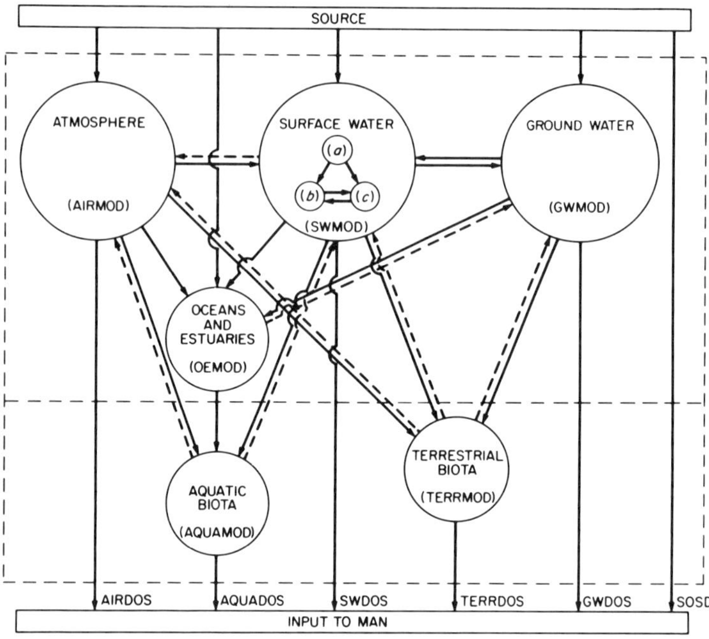

Fig. 1. Pathways of Radionuclides Released into the Environment to Man and the Associated Assessment Models

They would also be concerned with possible exposure and the consequent adverse health, ecological and abiota effects.

Environmental release of radwastes may occur in subsurface aquifers, surface water regimes, oceans/estuaries, the atmosphere, or even directly to biota. However, following the release, there will be redistribution of the wastes into all media. The important pathways of the radionuclides from source release to man are shown in Fig. 1. Accordingly, for a complete systematic approach, flow dynamic, waste transport, biological transfer, and dose models are needed to cover the entire biosphere. Flow dynamic models are required to simulate the flows of the carrier fluid. Waste transport models are necessary for predicting the concentration distribution. Biological transfer models are needed to simulate the uptake and release between the biota and physical environment. Dose models must be provided to compute radiation dose and its effect on man through external exposure and the inhalation, ingestion, immersion, submersion and other internal exposures. The availability and minimum requirements of these models will be discussed in the next section.

ASSESSMENT MODELS

For thorough assessment of alternative waste management strategies, a complete set of flow dynamic, waste transport, biological transfer, and dose models are required. Computer codes of biological transfer models (AQUAMOD and TERMOD) and dose models (AIRDOS. SWDOS, GWDOS, AQUADOS, TERDOS and SOSDOS) have been routinely used for several years to assess the radiological impact of environmental release of radionuclides (1-4). Many of the environmental transfer coefficients and other pertinent data required for a given simulation are already available. Hence, in this paper, I will be primarily concerned with flow dynamic and waste transport models in the physical environment of subsurface aquifers, surface water regimes, estuaries/oceans, and the atmosphere.

In order to predict the concentration distribution of radwastes in the physical environment, it is envisioned that at least 36 types of models listed in Table 1 are required. These include from steady one-dimensional to transient three-dimensional models of flow dynamics and waste transport in subsurface aquifers, surface water regimes, estuaries/oceans, and the atmosphere. Most of these models are available in the literature (5); but for a systematic approach, they must be adapted, modified and documented in a unique format so that the flexibility and utility of combining the models can be achieved. They can then be used separately to deal with site-specific problems or combinedly to handle large scale generic study. If the combination of models is employed, iterations must be used to obtain interfacial values of the media when the feedback beetween media is strong. When the feedback between media is weak, the output from one media may be used as input to the subsequent media and the excessive computation of iterations may be ignored.

Because of the severe consequence that may result from unsound management practice based on poor assessment, each of the models listed in Table 1 must be capable of yielding valid results when applied to any management proposal. To achieve this goal, each model must satisfy the following minimum requirements: (i) the problem must be logically formulated and expressed in mathematical equations to include as many processes as possible, (ii) the up-to-date and efficient computational algorithm must be divised and analyzed, (iii) the model must be validated by comparing the model results with well-known analytical solution, other type of numerical simulations, laboratory experiments, and/or field measurements, and (iv) demonstrative applications must be made. The processes governing the radwaste transport within a medium may include mainly the advection of carrier fluid, hydrodynamic dispersion, radioactive decay, and the medium-dependent processes. For example, adsorption by the soil matrix will occur in the subsurface aquifer. The adsorption and desorption by sediments, which themselves are transported along with the radionuclide by the fluid, can occur in the surface water regimes

Table 1. Flow Dynamics and Waste Transport Models Required for a Systematic Modeling Approach

		Model No.	Flow						Radionuclides						Sediments					
			S	T	1-D	2-D X-Y	2-D X-Z	3-D	S	T	1-D	2-D X-Y	2-D X-Z	3-D	S	T	1-D	2-D X-Y	2-D X-Z	3-D
Subsurface Aquifer (GWMOD)	Unsat-Sat.	1		X			X													
		2								X			X							
		3		X				X												
		4								X				X						
	Sat.	5		X		X														
		6								X		X								
	Unsat.	7		X	X															
		8								X	X									
Surface Water Regimes (SWMOD)	Runoffs	9		X		X														
		10								X		X				X		X		
	Streams/Rivers	11	X		X															
		12		X	X															
		13								X	X					X	X			
		14	X				X													
		15		X			X													
		16								X			X			X			X	
	Surface Impoundment	17	X			X														
		18		X		X														
		19								X		X				X		X		
		20	X					X												
		21		X				X												
		22								X				X		X				X
Oceans & Estuaries (OEMOD)		24		X	X															
		25								X	X					X	X			
		26		X		X														
		27								X		X				X		X		
		28		X			X													
		29								X			X			X			X	
		30		X				X												
		31								X				X		X				X
Atmosphere (AIRMOD)		32	X		X															
		33	X			X														
		34		X				X												
		35							X					X						
		36								X				X						

S = Steady state simulation
T = Transient state simulation
1-D = One-dimensional
2-D X-Y = Two-dimensional in the horizontal plane X-Y
2-D X-Z = Two-dimensional in the vertical plane X-Z
3-D = Three-dimensional

and the estuaries/oceans. Cloud scavenging and precipitational washout may take place in the atmosphere. On the other hand, transfer of wastes across media interfaces may be induced by various kinds of processes such as deposition, resuspension, aeration, seepage, infiltration, volatilization, evaporation and transpiration. All these processes must be included in the transformation of processes into equations. The computational algorithm may range from simple analytical and empirical methods to realistic sophisticated numerical techniques including finite difference, finite element, and Monte Carlo simulations. Analytical and empirical models are best utilized for analyzing the response of the media to the variation of model parameters. Insensitive parameters are then excluded from the sophisticated numerical based models.

EXAMPLE MODEL FOR SUBSURFACE AQUIFER

Six models (6-11), each representing an example of model type nos. 1, 2, 3, 4, 17 and 18 in Table 1, have been completed according to the minimum requirements described above. Continuing efforts are being made to have at least one model available for each of the remaining 30 model types in the same manner. FEMWATER (8) which is an example of model type no. 1, will be used to illustrate the model buildup and documentation. It is common practice to bury low level radwastes in shallow trenches above groundwater table. Under such circumstances, unsaturated groundwater flow is important in moving radionuclides to the water table and transporting them to

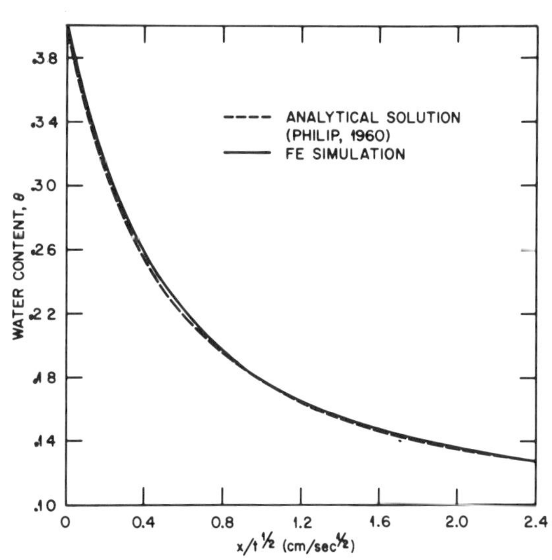

Fig. 2. Comparisons of Philip's Analytical Solution and the FEMWATER Simulation

other media through the saturated zone. Therefore, the model used to simulate the water movement must describe unsaturated as well as saturated flow. If the burial ground is located in humid climates, the model must allow for relatively large fluctuation of the water table elevation. The major input of water to the subsurface aquifer system is the infiltration by direct precipitation and indirect surrounding surface waters. The accuracy required in this simulation can only be obtained by numerical solution of the nonlinear differential equations of the modified Richards types and of Darcy's law. During the rainfall storm, the aquifer may or may not have the capacity to receive all the excess rainfall. During and after the storm, the water in the aquifer may move to the sloping air-land interface or to streams, rivers, or impoundments. These necessitate the consideration of Dirichlet, Neumann, and variable boundary conditions. Mathematical equations that govern the saturated-unsaturated flow and three types of boundary conditions are described in FEMWATER (8) which is a member of GWMOD in Table 1.

Because of its flexibility to handle complex boundaries and the ease in treating Neumann boundary conditions, the finite element method (FEM) is chosen to approximate the governing partial differential equation of continuity. To obtain a continuous flow field, FEM is also applied to the Darcy's law. After incorporating the boundary conditions, two matrix equations are obtained; one for the pressure field and the other for the velocity field. Six up-to-date alternative numerical schemes are used to solve the matrix equations for the pressure head and velocity components (8). Numerical experiments on a small laboratory-sized watershed given by Freeze (12) show that all six schemes yield comparable results and the expected behavior of the model are obtained.

FEMWATER is validated by comparing the model simulations with Philip's exact solution (13) for an ideal horizontal infiltration, with Freeze's transient finite difference model (12) for a hypotheti-

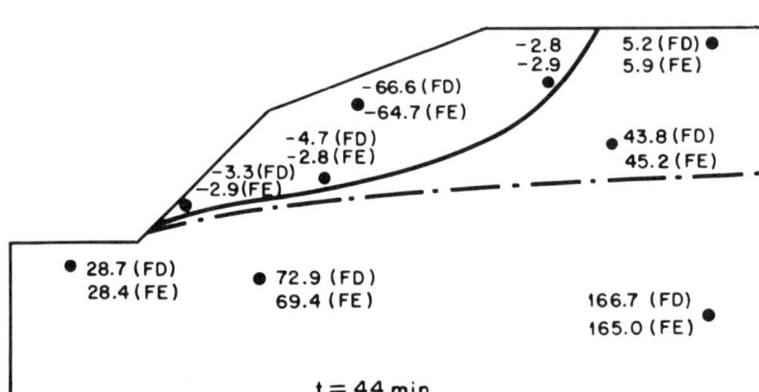

Fig. 3. Comparison of Freeze's·Finite-difference and FEMWATER Simulation

cal watershed, and with Hewlett and Hibbert's laboratory experiment (14). To compare the numerical simulation with Philip's exact solution (13), a diffusivity function of the moisture, θ, $D(\theta) = (1/2\theta - 1/4\sqrt{\theta})$ is used. Initial and boundary conditions are specified as: $\theta(x,0)=0.1$ and $\theta(0,t) = 0.4$, $\theta(\infty,t) = 0.1$. Fig. 2. shows the comparisons of moisture content profiles as simulated by FEMWATER and Philip's exact solution. Excellent agreement is obtained, thus the reliability of the computer code is assured. Freeze (12) employed a hypothetical watershed measured 6 X 3 -m with sloping surfaces (Fig. 3) to test his finite difference model. The watershed is assumed to be filled with highly permeable sands. The bottom, right, and left sides and the right sloping segment on the surface are considered impervious. The left horizontal segment on the surface is treated as a Dirichlet boundary. The left sloping and right horizontal segments on the surface are assumed to be variable boundaries to simulate the dynamic alteration of seepage and infiltration. Comparisons of simulated results at several points by FEMWATER and Freeze's finite difference model are also shown in Fig. 3. General agreement is obtained. Thus, the accuracy of the numerical algorithm can be asserted. Hewlett and Hibbert's experimental watershed is a 92 X 92 X 1400 -cm inclining concrete trough on a 40% slope. The structure was packed with a natural sandyloam soil to a bulk density of 1.3 for the experiment. Continuous measurement of outflow was taken for 150 days after the watershed was soaked by a sprinkling system for several days. The simulated outflow hydrograph by FEMWATER is compared with the measured values in Fig. 4. The agreement is considered acceptable.

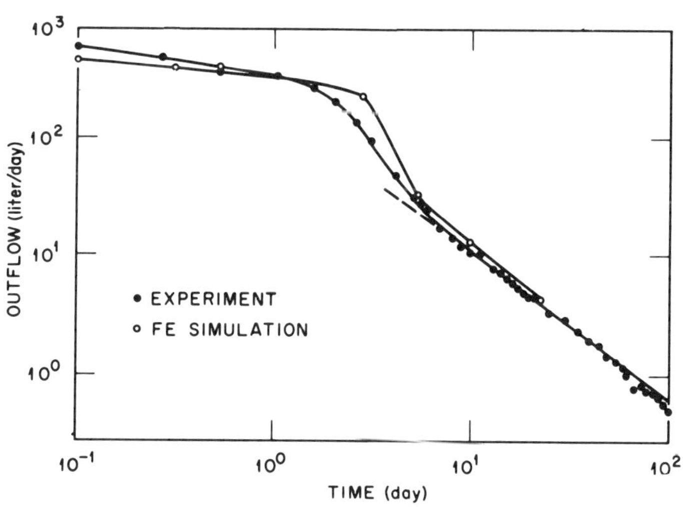

Fig. 4. Comparison of FEMWATER and Hewlett and Hibbert's Measured Hydrographs

CONCLUSION

A system of flow dynamic, waste transport, biological transfer, and dose models are prescribed for a systematic approach of assessing the effects of alternative nuclear waste management strategies on man. The minimum requirements for every flow dynamic and waste transport models are described. These minimum requirements are exemplified by a groundwater flow model, FEMWATER. FEMWATER is one of six models that have been completed satisfying the minimum requirements. Continuing efforts are being made to have at least one model available for each of the 36 model types prescribed in Table 1.

REFERENCES

1. R. S. Booth, S. V. Kaye, and P. S. Rohwer, Proc. Third National Symp. on Radioecology, Oak Ridge, TN, May 10-12, 1971, 877 (1971).
2. R. E. Moore, ORNL/TM-4687, Oak Ridge National Laboratory, Oak Ridge, TN (1975).
3. R. S. Booth and P. S. Rohwer, Fourth National Symp. on Radioecology, Oregon State University, Corvallis, OR, May 12-14 (1975).
4. E. Halfon and R. E. Bargmann, Proc. Fourth National Symp. on Radioecology, Corvallis, OR, May 12-14, 184 (1975).
5. P. T. Owen, N. S. Dailey, C. A. Johnson and F. M. Martin, ORNL/EIS-147, Oak Ridge National Lab., Oak Ridge, TN (1979).
6. G. T. Yeh, ORNL-5521, Oak Ridge Nat. Lab., Oak Ridge, TN (1979).
7. G. T. Yeh, ORNL-5522, Oak Ridge Nat. Lab., Oak Ridge, TN (1979).
8. G. T. Yeh and D. S. Ward, ORNL-5567, Oak Ridge Nat. Lab., Oak Ridge, TN (1979).
9. G. T. Yeh and D. S. Ward, ORNL-5601, Oak Ridge Nat. Lab., Oak Ridge, TN (1979).
10. G. T. Yeh, A Transient Three-dimensional Finite Element Model of Water Flow in a Complex Aquifer System, ORNL- , Oak Ridge Nat. Lab., Oak Ridge, TN (In Press).
11. G. T. Yeh, A Transient Three-dimensional Finite Element Model of Waste Transport in a Complex Aquifer System, ORNL- , Oak Ridge Nat. Lab., Oak Ridge, TN (In Press).
12. R. A. Freeze, Water Resources Research 8(3), 609 (1972).
13. J. R. Philip, Nature 185(4798), 233 (1960).
14. J. D. Hewlett and R. A. Hibbert, J. Geophy. Research 68(4), 1081 (1963).

A REVIEW OF THE ENVIRONMENTAL IMPACT OF MINING AND MILLING OF URANIUM ORES

J. M. Costello, D. R. Davy, F. C. R. Cattell,
J. E. Cook

Australian Atomic Energy Commission
Research Establishment, Private Mailbag
Sutherland, NSW, 2232, AUSTRALIA

INTRODUCTION

This review examines some of the environmental impacts of uranium mining and milling including use of natural resources, resulting effluents and the health of uranium workers and the general public. Impacts are expressed in terms of a gigawatt year of electricity (GWe y) generated in a light water reactor (LWR) using the uranium in a "once-through" fuel cycle.

Site specific impacts from uranium mining and milling depend upon many factors including local regulatory practices and standards, type of technology, meteorology and hydrology, and the regional population density. International recommendations on radiation protection in uranium mining have been made (1) and national codes of practice prepared (2). International guidelines for management of wastes from uranium mining and milling (3) have recently emphasized practices for the stabilization and disposal of uranium mill tailings(4).

Several generic environmental surveys, including data on uranium mining and milling for US conditions, give a range of data on the use of land, water and fossil fuels (5,6) assessments of chemical radiological and thermal effluents released to the biosphere (5,6,7), and radiological dose commitments and upper limit estimates of resultant health effects (6,7,8). Radiological dose commitments based on these data (5) have also been reviewed by UNSCEAR (9). The Final Generic Environmental Statement on the use of Recycle Plutonium in Mixed Oxide Fuel in Light Water Reactors (GESMO)(6) appeared more comprehensive than earlier work and contained revised estimates of radon emission during mining and milling. In this paper, GESMO (6) has been used as a major data source for comparison with estimates by UNSCEAR (9).

DESCRIPTION OF URANIUM MINING AND MILLING OPERATIONS

Uranium Mining

Uranium occurs as a low grade ore containing generally more than 0.1 wt % uranium. It may be mined by opencut, underground or solution techniques, depending on the location of the ore body. It may also be a by-product of residues in other mining operations, e.g., gold mine tailings. Opencut mining is used where the overburden and waste rock associated with the ore can be removed economically. Ore bodies deeper than about 100 m are generally mined by underground techniques which are more selective and produce less waste rock than opencut operations, but may lead to a lower uranium recovery. In solution mining, leachant is injected for in situ extraction of uranium and the solution is removed for purification. It may only be employed with horizontal orebodies lying on impervious rock strata below the water table, and requires extreme care to avoid contamination of groundwater (10), recoveries are about 65-70%(11). In 1975, almost 60% of uranium in the USA was mined by opencut and 40% by underground techniques; solution mining accounted for only a small percentage.

Uranium Milling

Uranium ores are crushed, ground and processed to recover most of the uranium as a crude oxide concentrate, U_3O_8, termed yellowcake, Uranium is leached either by sulphuric acid or sodium carbonate, the choice depending on the acidic or basic nature of the host rock.

Acid leaching is suitable for most uranium ores. Over 90% of the uranium content may be extracted by heating the crushed ore with sulphuric acid and an oxidant, washing the insoluble tailings by counter-current decantation, and transferring them to a retention pond, sometimes after neutralization. Uranium in the leach product is purified by solvent extraction or ion-exchange and precipitated as an alkali diuranate, frequently with ammonia. The product is filtered, dried, calcined and packaged.

Carbonate leaching is suitable for some ores in Western Australia, Canada and the US. Crushed ore is leached with sodium carbonate solution with air or oxygen as oxidant. The tailings are washed and stored and the extract is decanted, filtered, and the uranium selectively precipitated from dissolved contaminants with sodium hydroxide by carefully controlling pH. The slurry is thickened and centrifuged; sodium carbonate is regenerated by carbonation of the supernate; and the sodium diuranate dried and packaged or converted to ammonium diuranate for decomposition to U_3O_8.

BASIS FOR IMPACT ASSESSMENT

GESMO (6) assumed a proportion of 3 underground to 2 opencut mines with an ore grade of 0.085% uranium. This is much lower than earlier studies and for large deposits in Australia and Canada. A recovery of 90.5% uranium was assumed in a composite mill with 85% processed by acid leaching and the remainder by carbonate leaching. Mining and milling in semi-arid and remote areas of the U. S. was assumed. Generation of 1 GWe y required production of 260 t U in yellowcake from about 288 t U in ore. Nuclear power and fossil fuels were assumed to supply energy needs for processing in the ratio 1:2.

ENVIRONMENTAL IMPACTS

Use of Natural Resources in Uranium Mining and Milling

GESMO (6) predictions of land, water and fossil fuel requirements are given in Table 1. These are greater than predicted for higher grade deposits in Australia.

Land Use

Most mined land can be reclaimed by grading overburden heaps and converting excavations into lakes. Water pumped from mines may temporarily lower the water table. Most of the land area sequestered for

Table 1. Use of Natural Resources per GWe y

Resource	Mining	Milling	Resource	Mining	Milling
Land use	(ha)		Energy use		
Disturbed	34	3.8	Coal t	-	-
Committed	1.1	3.2	J equ.	-	-
			Gas J equ.	-	$3.2x10^{14}$
Water discharged	(m^3)		Fuel oil m^3	$1.9x10^3$	-
			J equ.	$7.0x10^{13}$	-
to air	-	$1.1x10^6$	Electricity	$7.3x10^{-4}$	$1.7x10^{-3}$
to water	-	-	GWe y		
to ground	$2.8x10^6$	-	J equ.	$2.3x10^{13}$	$5.4x10^{13}$
Total	$2.8x10^6$	$1.1x10^6$			

Table 2. Effluents Produced per GWe y

Effluent	Mining	Milling	Effluent	Mining	Milling
	(J)			(T Bq)	
Heat dissipated	$1.0x10^{14}$	$3.7x10^{14}$	Plant effluent to atmosphere		
Effluents from coal eqv. 2/3 electricity	(t)		Radon-222	220	41
Sludge	210	500	Radium-226	-	$1.2x10^{-4}$
SO_x to atmosphere	19	45	Uranium	-	$4.9x10^{-3}$
NO_x to atmosphere	15	35	Plant effluent to water	assumed nil	assumed nil
CO to atmosphere	0.30	0.70	Plant waste	(m^3)	
Particulates to atmosphere	0.93	2.1	Chemical compounds	-	-
Hydrocarbons	0.15	0.35	Mill tailings	-	$2.0x10^5$
Plant effluent to atmosphere	(t)		Radioactive solids	-	-
SO_x	21	0.14			
NO_x	17	28			
CO, NH_3, Fluorides	-	-			
Particulates	2.5	1.4			
Hydrocarbons	1.2	5.5			

milling is used for permanent storage of ore tailings. After rehabilitation, the land might be released for restricted uses. Process water may be obtained from mine seepage and recirculated via the tailings dam.

Wastes and Effluents

GESMO (6) estimates of effluents from uranium mining and milling are shown in Table 2. Opencut mining produces solid wastes of overburden and host rock. Average ratios of waste to ore are about 30 for US mines (5) but are 4 and 4.7 at two Australian deposits (12,13). Liquid effluents from mining consist of surface water runoff, seepage from ore stockpiles and waste rock heaps, and mine water ingress. These waters, which may contain dissolved minerals and members of the uranium decay chain, may be impounded and evaporated or used as process water for the mill. Uranium bearing areas frequently contain non-radioactive toxic mineralization which has been a source of environmental pollution (14).

The major chemical and radiological solid and liquid wastes are in the mill tailings slurry, which contains leached ore and precipitated minerals extracted from the ore, and waste solutions from treatment process chemicals. The slurry is pumped into a tailings pond, designed to retain solids and reduce seepage. Airborne wastes from mining and milling include dusts and radon gas from the mine, tailings piles, and mill operations, airborne chemical contaminants from the treatment process, and combustion products of petroleum.

HEALTH EFFECTS AND RADIOLOGICAL DOSE COMMITMENTS

Uranium mining and milling may have adverse effects on the health of workers and the general public. Health effects are not all equivalent; they may include accidents during construction and operation and in transport of materials, and effects from radiological, chemical and mineral pollutants. These health effects may be transient or permanent, trivial or fatal, localized or distributed, and immediate or delayed for many years after operations have ceased. The GESMO study (6) estimated radiological exposure to both uranium workers and the general US public from mining and milling uranium supporting a growing nuclear power program over a 26 year period. The resultant dose estimates expressed in person-sieverts per GWe y are listed in Table 3. Numerical estimates of radiological health effects, based on incidence of fatal cancers inferred from proportionality factors listed by the International Commission on Radiological Protection (15), are also shown.

Table 3. Dose Commitments and Inferred Cancer Mortality per GWe y

	Occupational		Public	
	Mining	Milling	Mining	Milling
	Person Sievert per GWe y*			
Total Body	2.5	1.1	6.3	1.2
G.I. Tract	2.5	0.5	0.3	0.06
Bone	3.7	5	20	5
Liver	2.6	0.5	6	1
Kidney	3.7	0.6	30	5
Thyroid	2.6	0.5	0.02	0.004
Lung	13.7	10	2	0.5
Skin	3	0.4	0.02	0.004
	Inferred cancer mortality deaths per GWe y**			
	0.08	0.04	0.1	0.02

* From Ref. 6 ** Estimated from ICRP risk coefficients (Ref. 15)

Uranium Workers

GESMO (6) estimates of the radiation dose to occupational workers from mining and milling inferred 0.12 cancer deaths per GWe y of power generated. An alternative estimate for uranium miners may be derived from data by UNSCEAR (9) and Pochin (8). From data on underground miners, UNSCEAR (9) indicated a risk of fatal lung cancer of 2×10^{-4} per Working Level Month (WLM) and an average of 1.3-1.9 WLM per year for miners. An additional risk of about $(0.5-1) \times 10^{-4}$ per person per year is incurred from external radiation at the mines. UNSCEAR assumed 50 miner-years and Pochin (8) 200 miner-years per GWe y. These data indicate a total of 0.02-0.08 cancer deaths for miners per GWe y, compared with 0.08 deaths per GWe y from GESMO data. Both Pochin and UNSCEAR assumed that the occupational radiation dose from milling operations was small compared with that from mining.

Public Health

GESMO (6) estimates of the radiation dose to the general public from mining and milling indicate an inferred 0.12 cancer deaths per

Table 4. Collective Dose Commitment to Public and Inferred Cancer Mortality Resulting from Mining and Milling (per GWe y)

	Person Sievert per GWe y						
	TBq Release	Segmental Bronchioles	Lung (Whole)	Bone Marrow	Bone Lining Cells	Gonads Whole Body	Cancer Mortality
During Mining							
Radon-222	220	30	7	0.05	0.05	0.05	0.01
Lead-210	0.11		1	0.6	2	0.5	0.01
After Mining*							
Radon-222	$4\ y^{-1}$	50	10	0.09	0.09	0.09	0.02
Lead-210	$0.002\ y^{-1}$		2	1	4	0.8	0.02
During Milling							
Radon-222 (+ Lead-210)	41	6	1	0.1	0.5	0.1	0.004
Dust*							3×10^{-4}
Uranium-238	4.8×10^{-3}		0.03	0.005	0.018		
Thorium-230	1.2×10^{-4}		0.01	0.007	0.13		
Radium-226	1.2×10^{-4}		0.007	0.007	0.007	0.007	
Lead-210	1.2×10^{-4}		0.001	7×10^{-4}	0.003		
After Milling* (no radon abatement)	$4\ y^{-1}$	50	10	1	4	0.9	0.04

* Incomplete (100 year dose commitment)

GWe y. Alternative estimates (UNSCEAR)(9) and modified for the greater radon-222 release assumed by GESMO (6) (Table 4) indicate 0.024 fatal cancers per GWe y in the general public, plus $8x10^{-4}$ deaths per year per GWe y resulting from continuing emission of radon gas from the tailings dam and mine after operations have ceased, if no measures are taken for radon abatement.

NON-RADIOLOGICAL HEALTH EFFECTS

Pochin (8) assumed the risks to miners from accidents and ill health resulting from exposure to dust and silica in underground mines as $1.5x10^{-3}$ and $5x10^{-4}$ deaths per person per year, indicating 0.075-0.3 deaths from accidents per GWe y, and 0.025-0.1 deaths per GWe y from inhalation of non-radioactive siliceous dusts. Pochin suggested 0.1 fatal accidents per GWe y in milling operations; he also deduced mortality rates of 0.25 and 0.003 per GWe y from construction work and transportation operations, respectively for the entire fuel cycle. This paper has arbitrarily assigned 30% of these to mining and milling, i.e., about 0.08 deaths per GWe y.

ADDITIONAL WASTE TREATMENT

A study (16) on reduction of environmental effects from uranium milling concluded that the most cost-effective reduction in radiological dose commitment could be obtained by a) minimizing airborne dispersal of tailings containing radioactive dusts by covering exposed wastes during mill operations and subsequently rehabilitating with a permanent stabilized earth cover; b) increasing the efficiency of dust collectors on some ventilation streams from ore crushing and yellowcake calcination and packaging; and c) reducing the long term release of radon gas from stored tailings by covering them with a thick layer of earth.

The importance of stabilization of mill tailings is receiving international attention (4). The long term erosion characteristics of above ground retention systems are uncertain; it was recommended that tailings from the Ranger Uranium Mine in North Australia be temporarily stored above ground in an engineered structure and returned to the opencut for coverage and revegetation after mining (17).

SUMMARY AND CONCLUSIONS

1. The actual environmental effects from uranium mining and milling will be site specific.

2. A generic estimate can be made of health effects, typical of the U S industry (Table 5). The number of premature deaths from occupa-

Table 5. Inferred Premature Fatalities per GWe y

	Cause of Fatalities			Total
	Radiation	Accidents	Other[(a)]	
Occupational[(b)]				
Mining	0.02-0.08	0.075-0.3	0.025-0.1	0.12-0.48
Milling	<0.01-0.04	0.1		0.11-0.14
Construction		0.08		0.08
Transport	0.001	0.001		0.002
Total(occupational)	0.03-0.12	0.26-0.48	0.025-0.1	0.31-0.7
Public[(b)]				
Mining[(c)]	0.02-0.1			0.02-0.1
Milling[(c)]	0.004-0.02			0.004-0.02
Total(occupational and public)	0.05-0.24	0.26-0.48	0.025-0.1	0.33-0.82

(a) Diseases not related to radioactivity.

(b) Lower values UNSCEAR, higher values GESMO.

(c) Continuing release after operation finished is assessed in text.

tional causes is estimated to be between 0.3 and 0.7 per GWe y of energy produced in an LWR with a once through fuel cycle. Of this number between 0.03 and 0.12 per GWe y may arise from cancer associated with radiation exposure and between 0.25 and 0.48 from accidents. The remainder may result from exposure to siliceous dusts.

3. The number of premature deaths among the general public per GWe y is estimated to be between 0.02 and 0.12 from cancer associated with radiation exposure. If no attention is given to reducing radon gas emanation from mines and mills which have finished production, the continuing number of inferred cancer mortalities may be about 0.008 deaths per year per GWe y.

4. By way of comparison, exposure to a natural background radiation level of 1 mSv y^{-1} (100 mrem y^{-1}) for one year may ultimately lead to an inferred 7-30 premature deaths from cancer per million of population, using the same limits of proportionality. In the US, at present rates of consumption, 1 million people use about 1.2 GWe y annually.

ACKNOWLEDGEMENT

This paper is a condensation of a report to the United National Environment Program (UNEP); the authors gratefully acknowledge permission to publish some of the findings.

REFERENCES

1. International Commission on Radiological Protection, Radiation Protection in Uranium and Other Mines, ICRP Publication 24, (Pergamon Press, Oxford, 1977).
2. Commonwealth Department of Health, Code of Practice on Radiation Protection in the Mining and Milling of Radioactive Ores, Canberra, ACT (1975).
3. International Atomic Energy Agency, Management of Wastes from the Mining and Milling of Uranium and Thorium Ores, Safety Series No. 44 (1976).
4. R. E. Cunningham, "Issues on Management Stabilization and Environmental Impacts of Uranium Mill Tailings," Proceedings of the Seminar on Management, Stabilization and Environmental Impact of Uranium Mill Tailings, OECD/NEA, Albuquerque (1978).
5. USAEC Fuels & Materials Directorate of Licensing, Environmental Survey of the Uranium Fuel Cycle, WASH-1248 (1974).
6. US Nuclear Regulatory Commission, Final Generic Environmental Statement on the Use of Recycle Plutonium in Mixed Oxide Fuel in Light Water Cooled Reactors, NUREG-0002, Vol. 3, Health, Safety & Environment, August (1975).
7. U.S. Environmental Protection Agency, Environmental Analysis of the Uranium Fuel Cycle, Part 1, Fuel Supply, EPA-520/9-73-003-B, October (1973).
8. E. E. Pochin, Estimated Population Exposure from Nuclear Power Production and Other Radiation Sources, NEA/OECD (1976).
9. United Nations Scientific Committee on the Effects of Atomic Radiation, Sources and Effects of Ionizing Radiation (1977).
10. US Environmental Protection Agency, State of the Art, Uranium Mining, Milling and Refining Industry, National Environmental Research Center, Office of Research and Development, June (1974)
11. J. N. Frank, Cost Model for Solution Mining of Uranium, US Energy Research and Development Administration, October (1976).
12. Ranger Uranium Mines Pty Ltd, Environmental Impact Statement, February (1974).
13. Queensland Mines Limited, Final Environmental Impact Statement, Nabarlek Uranium Project, January (1979).
14. D. R. Davy, (Ed.), Rum Jungle Environmental Studies, AAEC/E365, September (1975).
15. Recommendations of the International Commission on Radiological Protection, ICRP Publication 26 (Pergamon Press, Oxford, 1977).
16. M. B. Sears, R. E. Blanco, R. C. Dahlman, G. S. Hill, A. D. Ryan and J. B. Witherspoon, Correlation of Radioactive Waste Treatment Costs and the Environmental Impact of Waste Effluents in the Nuclear Fuel Cycle for Use in Establishing "As Low As Practicable" Guides - Milling of Uranium Ores, ORNL-TM-4903, Vol. 1, May (1975).
17. Ranger Uranium Environmental Inquiry, Second Report, Australian Government Publishing Service, Canberra, ACT (1976).

PROCESSING OF HIGH-TEMPERATURE SIMULATED WASTE GLASS IN A CONTINUOUS CERAMIC MELTER

S. M. Barnes, R. A. Brouns and M. S. Hanson

Chemical Technology
Pacific Northwest Laboratory
Richland, WA 99352

INTRODUCTION

The processing of spent nuclear fuel for defense and commercial purposes has generated a volume of high level liquid waste, which is being retained in storage tanks. The Pacific Northwest Laboratory (PNL), operated by the Battelle Memorial Institute under its contract with the Department of Energy, is developing suitable solidification processes for these and other radioactive wastes.

The development of continuous ceramic-lined melters for this purpose began in 1973. Until recently the emphasis has been placed upon low-melting (1000°-1050°C) borosilicate glasses (1,2). In November of 1978 a bench-scale, ceramic-lined melter was completed; this Joule-heated melter was designed to vitrify simulated high level waste compositions that may require melting cavity temperatures approaching 1500°C. The melter processes simulated high level waste and glass formers to produce a glass stream that can be directed into a storage canister or a waste-product forming device.

High-temperature waste forms exhibiting improved product durability have been developed and are undergoing evaluation (3). In addition, improved waste loadings are possible in certain high-temperature compositions.

Two high-temperature glass compositions have been processed in the bench-scale melter. These experiments were designed to test the behavior of the compositions in vitrification process equipment and to obtain samples for characterization and durability tests. The first test produced a borosilicate glass designed to vitrify the residual-liquid component of the defense waste stored at Hanford; the second produced a celsian glass-ceramic composition containing simulated high level commercial waste.

BENCH-SCALE MELTER

In the continuous-melting technology developed at PNL, molten glass is contained by refractory ceramic materials encased in a metallic shell. The ceramics are installed in increasingly corrosion-resistant layers. The composite wall design decreases the potential for glass leakage and increases the thermal insulation. Heat is generated in the melting cavity by passing an alternating electric current through the glass from one electrode to the other. The bench-scale melter builds upon this technology and adds several unique design features; among these are tin oxide electrodes, a multiple-entry feed system, and an airlift drain system.

Tin oxide was selected as the electrode material because of its inherent resistance to oxidizing glass corrosion at temperatures above 1100°C. The bench-scale melter utilizes Corhart* T-1185 sintered tin-oxide electrodes to form the end walls of the glass cavity, as shown in Fig. 1. An air-cooled baffle is bolted to the rear of the electrode wall to reduce the electrode operating temperature and to prevent glass leakage onto the electrical connections.

A twin-screw volumetric feeder delivers the feed materials to a diverter system, which distributes it across the glass surface via two feed-addition ports. This dual feed system is required to prevent unreacted materials from bridging across the long, narrow glass cavity of the bench-scale melter. The melting cavity is 20 cm wide, 22 cm deep, and 61 cm long. The overall melter dimensions are 1.3 m, 1.4 m, and 1.4 m, respectively.

To eliminate nearly all of the flexible service connections required by earlier tilt-to-interrupt-pour melter designs, an airlift draining system was incorporated into the bench-scale melter design, as shown in Fig. 1. This airlift drain design consists of a small-diameter ZGS** platinum tube inserted into a semi-circular riser cut into the glass-contact refractory. Air injection into the riser causes a localized glass-density decrease. The density decrease lifts the glass from the riser into the overflow drain.

---

*Corhart T-1185 is a product of Corhart Refractories Company, a Division of Corning Glass Works, Louisville, KT.

**ZGS platinum is a product of Matthey Bishop, Inc., Malvern, PA.

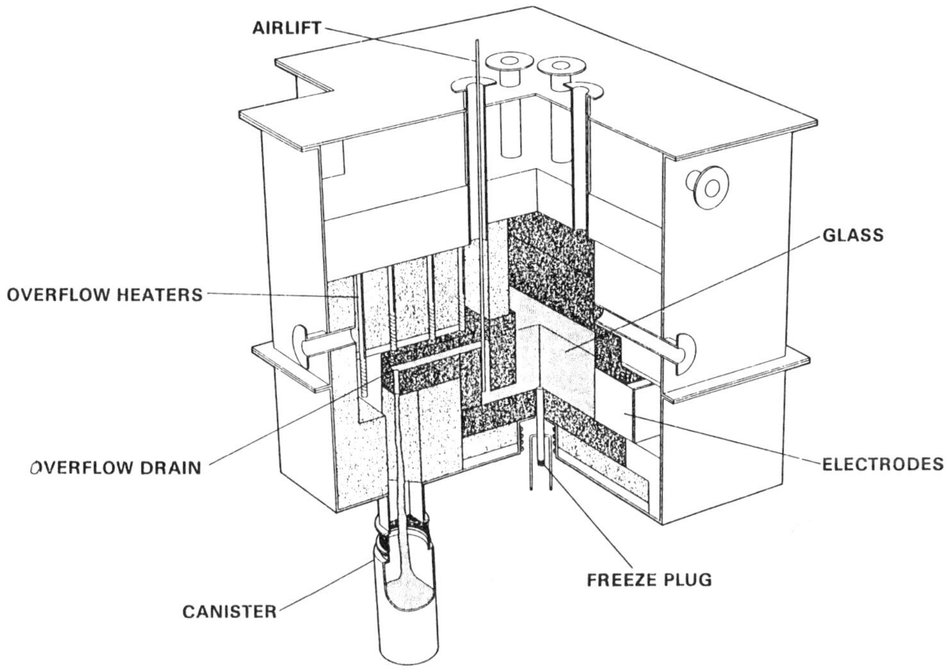

Fig. 1. Cut-away of Bench-Scaled Ceramic Melter

RESIDUAL-LIQUID GLASS TEST

Two test runs with the high-temperature residual-liquid composition were completed early in 1979. The composition of the glass, listed in Table 1, has a viscosity of 100 poise at 1200°C. Glass melting experience indicates that a glass viscosity near 100 poise results in good mixing and pouring characteristics.

Initial Glass Testing

The first of the two tests was a scoping run designed to examine the melting behavior of the composition in preparation for a 24-h

Table 1. Waste Glass Compositions

Glass Formers	Residual-Liquid, wt%	Glass-Ceramics, wt %
SiO_2	63.0	28.16
BaO	--	14.88
B_2O_3	10.0	6.40
Al_2O_3	5.0	12.88
CaO	--	4.00
TiO_2	--	4.00
ZnO	--	3.60
Li_2O	2.0	2.40
Na_2O	--	1.60
MgO	--	1.20
ZrO_2	--	0.80
Waste Oxides		
MoO_3	--	2.79
Fe_2O_3	--	2.52
Na_2O	17.3	2.47
Al_2O_3	0.82	--
TiO_2	0.90	--
ZrO_2	--	2.36
BaO	--	0.81
NiO	--	0.57
K_2O	--	0.56
SrO	0.52	0.51
Cs_2O	0.44	--
TeO_2	--	0.36
P_2O_5	--	0.28
Cr_2O_3	--	0.15
CoO	--	0.07
CaO	--	0.06
Ag_2O	--	0.03
Rare Earth Oxides [a]	--	6.45

[a] Commercial rare earth mix nominal oxide composition: 0.2% Y_2O_3, 24% La_2O_3, 48% CeO_2, 5% Pr_6O_{11}, 17% Nd_2O_3, 3% Sm_2O_3, 0.8% Eu_2O_3, and 2% Gd_2O_3.

run. The first run demonstrated an average feedrate of 9 kg/h with the glass temperature between 1120°C and 1175°C; foaming conditions in the melter ended the run after 6 hours. Foaming was caused both by overfeeding of the melter after a partial failure of the feed system and by the interaction of the initial glass composition in the tank with the new composition replacing it. Foaming caused by overfeeding has been reported in the literature (4).

When the foam receded, small partially fused bodies were observed floating on the glass surface. Chemical analysis showed these bodies to be rich in silica (approximately 80%). At that point the formation of these bodies was attributed to either a separation of the silica and fluxing agents during the foaming period or an insufficient glass processing temperature.

The glass temperature was increased for the second test. This temperature, as monitored by thermocouple inserted into the glass, was maintained above 1200°C for all but the first hour of the test. The maximum temperature noted was 1325°C.

During the second test run 190 kg of the feed composition was added to the melter at an average feedrate of 8 kg/h. The peak feedrate obtained for the entire experiment was 14 kg/h. The total residual-liquid glass production for the two tests was 200 kg.

As in the earlier, short term test there was some slight frothing caused by interactions of the old and new glass near the unreacted feed for the first few hours of the run. This effect slowly diminished as the new composition replaced the old. Within the first seven hours of the test a fully developed, stable cold cap with fluidizing vents similar to those experienced in other melters had formed (5). The fluidizing vents are a result of decomposition gases passing through and fluidizing regions of unreacted feed materials. No foaming other than the initial frothing was observed. Formation of the partially fused bodies occurred again in this 24-h run. These bodies slowly dissolved into the glass within two weeks. The formation of these bodies of fused silica was attributed to separation of the silica and fluxing agents caused by the nonhomogeneous feed mixture entering the melter. It is not believed that the formation of these bodies would interfere with a test of longer duration because the periodic addition of the fluxing agents would dissolve the covered bodies in the nonhomogeneous feed condition. Their formation would be prevented if the feed was well mixed.

Leaching

Several leaching tests were performed on the residual-liquid glass and a standard commercial waste glass composition (76-68). The leaching tests performed were a 72 h Soxhlet test, a 19 h pH-4 test and a 19 h pH-9 test. The leaching results are presented in Table 2. There were no crystalline phases present in the residual-liquor glass (6).

Concurrent with the work discussed in this paper, second-generation high temperature waste glass formulations exhibiting increased chemical durability are being developed at PNL. Increased levels of the network formers SiO_2 and Al_2O_3 and decreased levels of the network modifiers such as Na_2O, Li_2O and CaO are necessary to significantly improve leach resistance.

GLASS-CERAMIC TEST

During August of 1979, 300 kg of a celsian glass-ceramic developed for commercial waste was processed in the bench-scale melter. This composition, shown in Table 1, was designed to have a viscosity of 100 poise at 1200°C. The average feedrate for this run was 9 kg/h and the peak feedrate observed was 16 kg/h. No processing problems were encountered during the entire 35-h test. The glass produced in this run was cast into 30.5 cm-dia. by 30.5 cm-tall monoliths. Proper heat treatment of this glass produces the desired crystalline phases.

The leach rate and crystallinity data for this composition is not available pending the verification of the heat treatment schedule. The maximum crystal yield for celsian glass-ceramics has been estimated to be 35% by W. Lutze et al (7).

During this test an unplanned power outage provided an interesting insight into the processability of glass-ceramic compositions. Prior to this test, the high potential for glass-ceramics to devitrify and form insoluble crystals was a major concern. The buildup of precipitate during normal operation or in the event of an electrical power failure could potentially preclude operation and require

Table 2. Chemical Durability of a Standard Commercial Waste Glass and the Residual-Liquid Glass

	Percentage Weight Loss			
Waste Form	Soxhlet 72-h, 99°C	pH-4 19-h, 23°C	pH-9 19-h, 23°C	Soxhlet g/cm^2- d
Residual-Liquid	3.7	0.2	0.4	$4.75x10^{-5}$
Commercial (76-68)	5.8	0.17	0.2	$7.42x10^{-5}$

melter replacement. With the glass-ceramic composition in the melter, an 11 h power outage occurred resulting in a bulk-glass temperature decline from 1310°C to 620°C. At the lower temperature, melter recovery by joule heating was not possible. Visual inspection of the melting cavity confirmed that devitrification had occurred. An auxiliary heat source was used to heat the vapor space above the melting cavity; this heating raised the temperature of the devitrified glass-ceramic to a level at which joule heating was again practicable. This restart operation was completed within 17 hours. No detectable crystalline phases were present in the glass product samples obtained after the outage.

CONCLUSIONS

Recent operations have demonstrated that high-melting-point glasses and glass-ceramics can be successfully processed in joule-heated, ceramic-lined melters with minor modifications to the existing technology. Over 500 kg of simulated waste glasses have been processed at temperatures up to 1410°C. The processability of the two high-temperature waste forms tested is similar to existing borosilicate waste glasses.

High-temperature waste glass formulations produced in the bench-scale melter exhibit quality comparing favorably to standard waste glass formulations.

REFERENCES

1. J. L. Buelt, "A Review of Continuous Ceramic-Lined Melters and Associated Experience at PNL," Ceramics in Nuclear Waste Management CNF-790420, Technical Information Center, U. S. Dept. of Energy, Spring Field, VA, 107-113 (1979).
2. C. C. Chapman, J. L. Buelt, "The Use of a Continuous Melter in Immobilization of Radioactive Defense Waste," Presented at the American Nuclear Society Meeting in San Diego, CA, June 1978, PNL-SA-6867, Pacific Northwest Laboratory, Richland, WA (1978).
3. J. M. Rusin, "Alternative Waste Forms--A Comparative Study," Presented at the Materials Research Society Annual Meeting in Cambridge, MA, November 1979, PNL-SA-7769, Pacific Northwest Laboratory, Richland, WA (1979).
4. J. L. Buelt, "Liquid-Fed Ceramic Melter," Quarterly Progress Report--Research and Development Activities - High Level Waste Immobilization Program: April through June 1979 (J. L. McElroy, ed. PNL-3050-2 Pacific Northwest Laboratory, Richland, WA, in press).

5. C. C. Chapman, Vitrification of Hanford Wastes in a Joule-Heated Ceramic Melter and Evaluation of Resultant Canisterized Product, PNL-2904, Pacific Northwest Laboratory, Richland, WA (1979).
6. M. J. Kupfer, Vitrification of Hanford Radioactive Defense Wastes, RHO-SA-89, Rockwell Hanford Operations, Richland, WA (1979).
7. W. Lutze, "Characterization of Glass and Glass Ceramic Nuclear Waste Forms," Proc. of Symp. on Science Underlying Radioactive Waste Management Materials Research Society Annual Meeting Vol. 1, (G. J. McCarthy, ed., Plenum Press, New York, NY, 1979).

FIXATION OF MEDIUM-LEVEL WASTES IN TITANATES AND ZEOLITES: PROGRESS TOWARDS A SYSTEM FOR TRANSFER OF NUCLEAR REACTOR ACTIVITIES FROM SPENT ORGANIC TO INORGANIC ION EXCHANGERS

S. Forberg*, T. Westermark*, R. Arnek**, I. Grenthe**, L. Fälth+, and S. Andersson+

*Department of Nuclear Chemistry, **Department of Inorganic Chemistry, The Royal Institute of Technology, S-10044 Stockholm; +Inorganic Chemistry 2, Institute of Technology, S-22007 Lund, Sweden

INTRODUCTION

In order to limit the activity level in different water compartments of nuclear power stations, streams of water are passed through organic ion exchangers, mostly in mixed beds. After use, the organic resins from Swedish reactors are either incorporated in concrete blocks or dried, mixed with melted bitumen and poured into drums. In both cases, the end volume is larger than the bed volume of the exchanger. Considerable volumes have to be stored or disposed of in a safe way. The time to be considered is a few centuries, mostly due to the fission products ^{90}Sr and ^{137}Cs. There are reasons for concern about the long term stability and leaching resistance of the concrete blocks and bitumen as well. Inorganic ion exchangers could be sintered to products of small volume and high stability.

The inorganic exchangers considered for this purpose were zeolites and titanates. The zeolites have high selectivity and good stability in neutral aqueous solutions. They can be melted to glass or sintered to ceramics. Titanates can be produced with good quality as ion exchangers (1,2,3). They can be sintered to extremely stable products (4). Leakage of counterions from these exchangers prohibit direct application in reactors. Results from work on the transfer of activities from organic to inorganic exchangers is the subject of this paper (5,6,7).

PURPOSE AND MEANS OF THE PRESENT STUDY

The purpose of this work was to reduce the radiotoxicity of spent organic resins to a matter of short term concern, say a few decades, and to solidify extracted activities in stable inorganic substances. ^{90}Sr and ^{137}Cs constitute, for a few centuries, the major radiotoxins in this waste. The transfer of 99.9% of the Sr and Cs was consequently established as the prime goal. Any reduction of the activity of the ^{60}Co remaining in the resin would, of course, further shorten the required isolation period. A substantial reduction of other short lived gamma emitters would render the handling of the resin more convenient. A secondary goal was that the final volume of transferred activities should be as small as possible, due to the cost of disposal space. A prerequisite was, however, that the final form must be stable enough to permit disposal "forever".

The obvious medium for the transfer of ions from one exchanger to another is an aqueous solution. Mineral acids were tried as eluants but they were found to be inferior to complexing agents, especially when the overall function of the process was considered. The use of a complexing agent in a circulating solution permitted a closed system with essentially two steps: elution of the organic resin and sorption in inorganic ion exchangers. Some zeolites sorbed cesium excellently and sorbed complexed divalent cations very poorly. Titanate sorbed the latter ions much more efficiently than the zeolites did. Titanate sorbed cesium less efficiently than it did divalent cations. Hence, a dual sorbent system was the obvious choice.

DESCRIPTION AND PERFORMANCE OF A BENCH-SCALE SYSTEM

The main features of the system used are shown in Fig. 1. The system was sized for 1 litre of organic resin at a time. For practical reasons, the resin was divided into two beds. The whole system was placed behind a radiation shielding wall of 10 cm of lead in a fume hood, designed to house a total activity of about 3×10^{10} Bq. The activities encountered were considerably lower and could conveniently be handled with full safety.

The coarse-grained organic ion exchanger came from a boiling water reactor (OKG). A large fraction had another, undesirable origin than the parallel cleaning circuit. Although great care was taken when packing the columns, the pale, light anion beads always rose and became concentrated at the top of the beds.

The complexing agent used was disodium tartrate. The preferred concentration was 0.3 M, the preferred pH 5.5. Sodium salts of other tri-(citric), di- and monobasic organic acids were tried and found to

Spent organic ion exchanger (mixed bed) Grain sizes: 300-1200 μm Eluted:		Zeolite (mordenite) < 63 μm Sorbed:	Titanate 250-500 μm		Weak cation exchanger (carboxylic acid) in H^+/Na^+ form for control of pH
99.9%	Sr		"All"*		
99.9%	Cs	"All"*			
90 %	Co	Traces	80%	10%	
95 %	Zn	Traces	90%	5%	

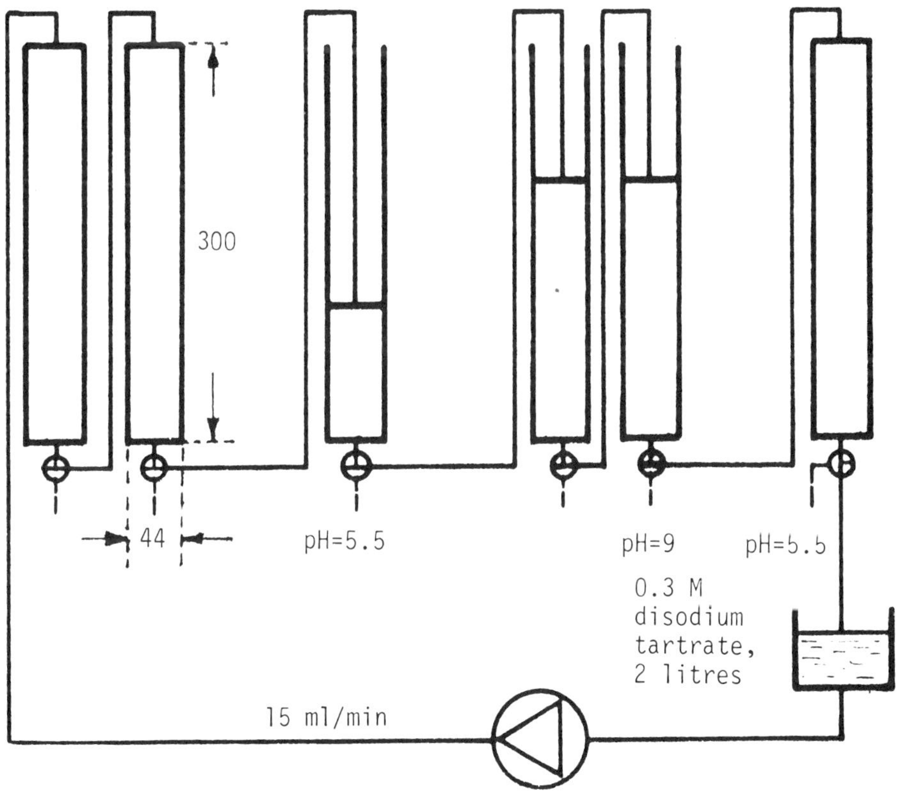

Fig. 1. The Columns and Beds, to Scale, With Connections and Circulating System. The Flow Rate in the Beds is ~ 1 cm/min.

---

* See text

be inferior (7,8). In order to avoid the formation of gas bubbles (CO_2) in the columns, the resins were preconditioned by the addition of tartrate to a stirred slurry in an open vessel. The slurry turned acidic due to the exchange between sodium and hydrogen ions and any carbonate in the anion exchanger was evolved as carbon dioxide. In a large-scale (say 1 m^3) system, this pretreatment seems somewhat inconvenient. Therefore, means of eluting the carbonate as bicarbonate ions in solution were tried. The obvious method was to control the pH by adding ammonia to the first volume of the influent. The moderate and temporary increase of the pH value at the front of the eluant rendered ^{60}Co and other corrosion products less accessible for elution. Formation of scarcely soluble hydroxides or ferrites was thought to be the mechanism. A less efficient transfer would be the final result. The flow rate, 1 cm/min in the columns, corresponded to one bed volume (in two columns) of the resin per hour. The processing time was about 20 h.

ANALYSES AND RESULTS

Although the organic resin was eluted as a mixed bed, due to their segregation during packing and the non-representative composition of the mixture, the anion and cation exchangers were analysed separately. After the water had been displaced with ethanol, trichloroethene was added gradually, and floating fractions were skimmed off and analysed by means of a Ge(Li) spectrometer. The different fractions were distinguished and characterized subjectively by the laboratory worker on the basis of their behavior during separation. The activities of such fractions from the ion exchanger before and after 18 h of continuous elution with 0.3 M tartrate are listed in Table 1. Besides the major, underlined fractions of anion and cation exchanger, there were minor fractions, $< 10\%$ by weight, and very small ones, $<1\%$ by weight. The activity of each fraction was weighted with the factor 1, 0.1 or 0.01 before listing it in Table 1 in order to tabulate figures of each fraction close to the activity per kg of the group of "anionic" or "cationic" exchangers within each sample.

The activity of the resin was measured 4 months after delivery from the reactor station. Because ^{110m}Ag interfered with the gamma spectrum of ^{137}Cs, the pure peak at 605 keV from ^{134}Cs was used for the evaluation. The activity ratio between ^{137}Cs and ^{134}Cs was determined from their spectra in a sample of mordenite and found to be 1.7. The upper limits given in Table 1 represent a confidence level of 99.5% (net, non-significant value + 3 SD). Where "not detected" is indicated, the detection limit varies with the intensity and composition of the spectrum and with the time available. ^{90}Sr was separated chemically, with ^{85}Sr as a tracer for yield determination, and analysed by means of cerenkov radiation from ^{90}Y growing-in in an aqueous solution.

Table 1. Activities of dry (105°C) fractions of ion exchangers in MBq/kg (before weighting). Each sample of organic resin was separated by skimming off fractions - in the sequence given - while adding trichloroethene to a slurry in ethanol. The fractions were classified as "anionic", A, or "cationic", C, according to their visual appearance. The major part of the mixed-bed was cationic. The major fractions of anion and cation exchangers are underlined. The activities of minor fractions are weighted by the factors (w) listed. These factors correspond approximately to the weight fractions within the anion or cation exchangers. The toxicity indices indicate the relative radiotoxicity in water according to ICRP Publ. 2 and 6(9) - the figures correspond to $(MPC)_w$ (1 mCi = 37 MBq) for 168 h/week (occupational). nd = not detected.

Nuclide	^{137}Cs	^{90}Sr	^{60}Co	^{125}Sb	^{54}Mn	^{110m}Ag	^{65}Zn	^{58}Co
Half-life, Y	30	28	5.2	2.7	0.9	0.7	0.7	0.2
Toxicity, m^3/mCi	5	250	2	1	1	3	1	1
1 Non-Eluted Resin								
1 A w = 0.1	1.5		30		1.5	1.5	14	
2 " " 1	12.6		172		8.5	8.3	83	
3 " " 0.1	170		200		2	20	300	5
4 " " 0.01	0.7		70		30	3.5	25	
5 C " 0.1	6.0		9		2.5	1.5	12	
6 " " 1	1660		3200		97	210	6800	105
7 " " 0.01	nd		200	25	3.5	nd	50	
A-C (Mixed bed)*		41						
2 Eluted Resin, Top of Column 1								
1 A w = 0.1	nd		200	3	9	30	70	
2 " " 1	"		69	1	3.4	25	27	
3 " " 0.1	"		12	0.2	0.4	0.5	4	
4 C " 1	"		85	0.3	4.3	61	29	
3 Eluted Resin, Average of Column 1								
1 A w = 0.1	1.0		300		10	130	110	
2 " " 1	2.1	0.008	57	0.2	2.2	55	43	
3 A/C" 0.1	0.4		20		0.9	10	8	
4 C " 1	2.6	0.003	49		nd	176	53	
5 " " 0.1	0.4		15		0.3	20	20	0.3
6 " " 0.01	nd		16	25	3	nd	60	
4 Titanate Bed (w = 1)								
1 At 0 cm depth	nd		3200		700	400	10000	nd
2 0-10 " "	"	**"all"	6400		nd	nd	15800	160
3 10 " "	"		2300		"	"	1380	230
4 10-20 " "	"		1160		"	"	670	39
5 20 " "	"		21		2.6	"	6	1

* A:C ~ 1:4 **See text

Table 2. Content of given elements in g/kg of dried resin before and after elution. Before drying resins, the exchangers were separated by fractional sedimentation.

	Co	Cr	Cu	Fe	Mn	Ni	Zn
Anion exchanger							
before elution	<0.16	1.3	0.8	1.6	0.9	<0.16	0.9
after elution	<0.1	1.4	2.3	0.9	0.5	<0.1	<0.1
Cation exchanger							
before elution	<0.1	0.1	7.8	0.5	4	<0.1	6
after elution	<0.05	0.05	1.6	0.14	0.3	<0.05	<0.05

The amounts of corrosion products are also of interest. Eluted and non-eluted resins were separated in anion and cation exchangers by upward pulsation of water in a column, dried, dissolved and analyzed by atomic absorption. The results are presented in Table 2. The eluted resin for this table was not the same as that for Table 1.

Sorption in mordenite is presented in Table 3. In this case, the effluent from two bench-scale experiments had passed through the bed measured. Sorption in the foremost titanate column is presented in Table 1, where the resin fractions and the titanate are from the same experiment. The indication that "all" ^{90}Sr of the effluent from the zeolite column in Fig. 1. is sorbed in the upper half of the titanate bed is a conclusion from parallel sorption experiments in which ^{85}Sr, added as a tracer, was quantitatively sorbed in a fraction of the column length necessary for sorption of ^{65}Zn (8).

DISCUSSION AND CONCLUSION

It is evident from Table 1 that the goal of removal of 99.9% of the ^{90}Sr and the ^{137}Cs from the organic resin has been achieved. The quantitative sorption of the ^{90}Sr in the titanate has been discussed above. The quantitative sorption of ^{137}Cs from the influent in the mordenite bed is indicated by Table 3 and by the fact that ^{137}Cs was not detected in the titanate (Table 1). The elution of 90% of the ^{60}Co indicated in Fig. 1. has been surpassed in other experiments, where the total activity of the bed was measured by direct spectrometry before and after the elution. The same was found for ^{65}Zn. Hence, the overall function indicated in Fig. 1. has been obtained. The figures for the migration of ^{60}Co and ^{65}Zn through the inorganic exchangers are rather conservative.

Further elution with tartrate is not an effective way of reducing the remainder of the gamma dose rate from the resin. After the elution, the dose rate is dominated by ^{110m}Ag and by other corrosion products outside the cation exchanger. None of these categories are easily eluted with tartrate. Nonionic activities normally forming divalent cations are thought to be in particle form or CRUD,

Table 3. Sorption in mordenite - relative counting rates in <u>pulses/min</u>, from some gamma lines at different depths of a zeolite bed after passage of the effluent from organic resin beds of a total height of 120 cm.

	Nuclide	^{134}Cs	^{65}Zn	^{60}Co
	Gamma energy, keV	605	1115	1332
Distance	0 - 2.5 cm	15000	170	220
from the top	3.5 - 6.0 cm	7 300	45	96
of the bed	8.5 - 11.0 cm	104	32	15

either retained by the mixed-bed resin or formed by precipitation during the elution. Some elements were transferred from the cation to the anion exchanger during the elution, e.g. copper in Table 2 and ^{110m}Ag in Table 1. Quantitative separation of the anion from the cation exchanger previous to the elution might facilitate a more quantitative elution. The activity of the ^{137}Cs and a part of the ^{60}Co in the major fraction of the non-eluted anion exchanger listed in Table 1 might be explained by improper separation -- 0.8% of the cation exchanger following the anion exchanger would be sufficient. In such a case, the elution of a completely separated anion exchanger might not be necessary with respect to the radioactive species discussed here.

FIXATION OF MLW IN ZEOLITES AND TITANATES

The appearance in Table 1 of more activities outside the major fractions in the sample from the average of the bed than in that from the top of the bed might indicate a migration of CRUD during the elution and also the possibility of separating it mechanically, e.g., by means of ultra-sound, from the ion exchanger. In that case, the residual activity of the resin could be further reduced. The CRUD could be mixed with the inorganic ion exchangers and sintered together with them.

The bed volumes of the spent zeolite and the spent titanate may be about 5% and 30% of the bed volume of the organic resin treated. This means that the volume after proper sintering will be considerably less than 10% of that of the organic resin. A research program on the sintering procedure and the stability and crystallographic data of the resulting products has been started, involving both hot isostatic pressing (10), simpler hot pressing and sintering of zeolites and titanates together and separately (11). In order to obtain the best stability, the titanates will be treated both as they are obtained from the process and after elution of remaining sodium. Incineration in a fluidized bed is one approach for conditioning the eluted resin. The ashes could be compacted to a stable product of small volume.

ACKNOWLEDGEMENT

We are grateful for the technical assistance we received from concerned members of each department and of Studsvik Energiteknik AB; to the Nuclear Power Station at Oskarshamn (OKG) for delivery of the organic resin and for performance of AA analysis; and to the staff of the sponsor for encouragement and discussion. The work was sponsored by the National Council for Radioactive Waste.

REFERENCES

1. S. Forberg and P.-I. Olsson, Solidification of High Level Radioactive Wastes with Methods Satisfying Requirements: Synthesis and Characterization of Titanate Ion Exchangers for Sorption of Fission Products, Well Suited for Final Sintering, TRITA-KKE-7605 (in Swedish, summary in English), the Royal Institute of Technology, Stockholm (1976).
2. S. Forberg and P. -I. Olsson, "Method of Preparing Titanates Suitable as Ion-Exchange Material," U. S. Patent 4,161,513 (1979).
3. S. Forberg and P. -I. Olsson, "Method of Preparing Titanates Suitable as Ion-Exchange Material," (in Swedish), Swedish Patent 116-7614116-7 (1979).
4. S. Forberg, T. Westermark, H. Larker and B. Widell, "Synthetic Rutile Microencapsulation: A Radioactive Waste Solidification System Resulting in an Extremely Stable Product," Scientific Basis for Nuclear Waste Management, Vol. 1, (G. J. McCarthy, ed; Plenum Press, NY, 201-5, 1979).
5. A. Hultgren (ed), Research Report, The National Council for Radioactive Waste, Fack, S-102 40 Stockholm, June (1977).
6. A. Hultgren (ed), Research Report, The National Council for Radioactive Waste, Fack, S-102 40 Stockholm, July (1978).
7. A. Hultgren (ed), Research Report, The National Council for Radioactive Waste, Fack, S-102 40 Stockholm, July (1979).
8. R. Arnek and S. Forberg, A System for the Transfer of Long Lived Radioactive Nuclides from Spent Resins to Zeolites-Titanates, Report PRAV 3.19, National Council for Radioactive Waste, Fack, S-102 40 Stockholm, Sweden (1979).
9. The International Commission on Radiological Protection, Publ. 2 (1959) and 6 (1962). See also publication 26 and 30.
10. H. T. Larker, "Hot Isostatic Pressing for the Consolidation and Containment of Radioactive Waste," Scientific Basis for Nuclear Waste Management, Vol. 1, (G. J. McCarthy, ed; Plenum Press, NY, 207-210, 1979).
11. J. K. Johnstone, T. J. Headly, P. F. Hlava and F. V. Stohl, "Characterization of a Titanate Based Ceramic for High Level Nuclear Waste Solidification," Scientific Basis for Nuclear Waste Management, Vol. 1, (G. J. McCarthy, ed; Plenum Press, NY, 211-217 (1979).

DEVELOPMENT OF AN ION-EXCHANGE PROCESS FOR REMOVING CESIUM FROM HIGH-LEVEL RADIOACTIVE LIQUID WASTES*

P. K. Baumgarten, R. M. Wallace, D. A. Whitehurst and J. M. Steed

E. I. du Pont de Nemours & Co., Inc.
Savannah River Laboratory
Aiken, SC 29801

INTRODUCTION

A process is now being developed to solidify and isolate the biologically hazardous radionuclides from approximately 20 million gallons of alkaline high level radioactive waste accumulated at the Savannah River Plant (SRP) and now being stored in large underground tanks. The waste consists mainly of liquid waste supernate, a damp sodium salt cake and a gelatinous, insoluble sludge. The process involves dissolving the salt cake in water, separating it from the sludge, washing the sludge and adding the washings to the dissolved salt cake. The aqueous portion is then treated by ion exchange to remove cesium-137, plutonium and other actinides and soluble strontium-90. After elution, the cesium and actinides are concentrated by evaporation and immobilized on zeolite. The strontium is similarly concentrated. The cesium-zeolite and strontium concentrate are combined with washed sludge for calcination. The calcined mixture is melted with glass frit to form a leach resistant glass. The final glass is encased in 0.8 m^3 metal canisters. The decontaminated aqueous waste is concentrated by evaporation and returned to cleaned underground tanks for storage as a salt cake.

Bench scale demonstration and initial process studies of the cesium ion-exchange step were reported by Wiley (1) at the Savannah River Laboratory (SRL). Decontamination factors (concentration in

*The information contained in this article was developed during the course of work under Contract No. AT(07-2)-1 with the U.S. Department of Energy.

feed/concentration in product) of over 100,000 were reported. The tests were made in shielded remote cells ("high-level caves") on actual wastes from tanks, adjusted to standard reference compositions and slurried with radioactive sludge to assure equilibrium composition. The resin used was Duolite* ARC-359, a gel-type sulfonated phenol-formaldehyde cation exchange resin which had proved in the past (2) to be particularly selective to cesium at high pH, even in solutions with large sodium concentrations. Duolite ARC-359 resin is a refined version of Duolite C-3 resin used in some of the early tests.

Additional work is now going forward to develop the basic data needed for the design of plant columns and to demonstrate reliable, day-to-day performance. Process features are: 1) two resin columns in series with continuous monitoring of cesium concentrations between the columns (cesium is too dilute in the second column effluent in the face of relatively high concentrations of short-lived ruthenium-106); 2) downflow feed (exhaustion) and upflow elution and regeneration, followed by water rinses; 3) periodic backwashing to remove fines, correct channeling and to regrade the resin; and 4) valveless operation to provide a low-maintenance installation for remote canyon (protective barrier) operation. The countercurrent mode produces better decontamination and uses less eluant/regenerant than concurrent operation (3). Separate elution and regeneration are required according to the simplified equations:

Feed: $R\ Na + Cs^+ \rightarrow R\ Cs + Na^+$

Elute: $R\ Cs + NH_4^+ \rightarrow R\ NH_4 + Cs^+$

Regenerate: $R\ NH_4 + Na^+ \rightarrow R\ Na + NH_4^+$

where R is the resin. The eluant is a solution of 2M ammonium hydroxide-2M ammonium carbonate with a pH of 9.5. The regenerant is 2M NaOH.

Included in the program were: 1) developing equilibrium and diffusion data on Duolite ARC-359 resin and correlating the effects of hydroxyl and cesium ion concentrations; 2) operating two 7.5-cm diameter x 1-m high columns to determine effects of process variables such as feed rate and composition, resin batch and backwashing.

RESIN SODIUM CAPACITY

Duolite ARC-359 resin has two types of replaceable hydrogens: strong acid hydrogens associated with the sulfonic acid group and

*Registered trademark of Diamond Shamrock Corp., Cleveland, OH.

very weakly acidic phenolic hydrogens. The capacity of the sulfonic acid hydrogens was determined to be 2.5 gram equivalents per kg of resin. The excess capacity due to the phenolic groups was measured by equilibrating samples of resin originally in the sodium sulfonate form with sodium hydroxide solutions. The amount of acid consumed when the resins were returned to that form was determined with a mixture of sodium nitrate and dilute nitric acid. Excess capacities were found to be quite high (Table 1) and are more than double those of the sulfonate groups at the high hydroxide concentrations typical of the expected feed composition.

EQUILIBRIUM CESIUM DISTRIBUTION

Previous work by Wiley and Wallace (4) showed that the cesium distribution coefficient, K_D, between Duolite ARC-359 resin and aqueous solutions was a function of sodium and hydroxide ion concentrations. The present work revealed that it also depended on cesium concentration.

Distribution coefficients were measured at room temperature for solutions of nearly constant sodium ion concentration (about 5.75M) but varying hydroxide (0 to 5.75M) and cesium (10^{-9} to 0.1M) ion concentrations. Typical results are given in Fig. 1. The data were fitted to equations based on models which assumed that two or three different types of sites participated in ion exchange. Each type of site was assumed to have its own capacity and selectivity. A three-site equation fit satisfactorily over the whole concentration range.

Table 1. Excess Capacity of Duolite ARC-359 Resin in Alkaline Solutions

NaOH Concentration, Moles/Liter	Excess Na^+ on Resin, Equivalents/kg of Dry Resin[a]
0.0047	0.776
0.0114	1.58
0.0492	2.02
0.0987	2.59
0.1926	3.13
0.4972	3.77
0.9643	4.46
1.975	5.31
4.979	7.97

[a]One kg of dry resin has an expanded bulk wet volume of 340 mL in 1 molar NaOH.

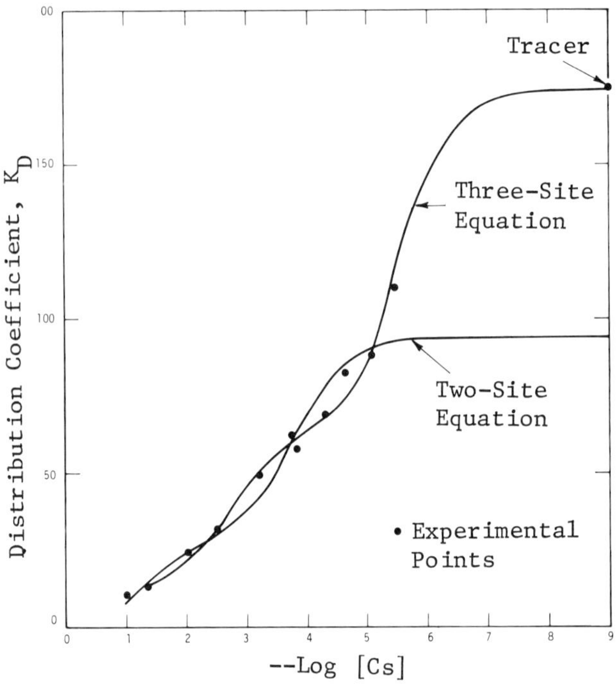

Fig. 1. Distribution Coefficient vs. Cesium Concentration for Duolite ARC-359 Resin in 4.7M $NaNO_3$ - 0.95M NaOH

The equation derived for the three-site model is:

$$K_D = \frac{C_1K_1}{1+K_1[Cs]} + \frac{C_2K_2}{1+K_2[Cs]} + \frac{C_3K_3}{1+K_3[Cs]}$$

where [Cs] is the cesium ion concentration in solution; K_1, K_2 and K_3 are the equilibrium quotients (selectivity coefficients) of the three sites; and C_1, C_2 and C_3 are their capacities.

The C and K values could be fit to sodium hydroxide concentrations between 0.56M and 5.75M (a total of 84 data points) by using a single set of capacity (C) values and assuming that each of the selectivity coefficient (K) values was proportional to the square of hydroxide concentration [OH]. The final values were:

$$K_1 = 6.04 + 1.55\ [OH] + 1.31\ [OH]^2$$

$$K_2 = 573 + 122\ [OH] + 197\ [OH]^2$$

$$K_3 = 2.88 \times 10^5 + 2.52 \times 10^4\ [OH] + 9.32 \times 10^4\ [OH]^2$$

$C_1 = 1.83$

$C_2 = 5.98 \times 10^{-2}$

$C_3 = 3.44 \times 10^{-4}$

KINETICS

An important factor in the ion exchange process is the rate at which ions diffuse from solution into the resin. Diffusion parameters were measured at various flow rates by passing simulated supernate solution containing cesium-137 tracer through a shallow layer of resin (~0.05 g). Flow was stopped after a prescribed time and the residual solution was removed rapidly. The resin was then washed with water and counted for gamma activity. The diffusion parameter (D/a^2) was calculated from the equation:

$$D/a^2 = \frac{1}{\pi t}\left(1 - \sqrt{1 - \frac{\pi}{3} U}\right)^2$$

where $\underline{D}$ is the diffusion coefficient, $\underline{a}$ is the particle radius (assuming spherical shape) and $\underline{U}$ is the fractional attainment of equilibrium (5).

A reasonably constant value of the parameter, namely $13.2 \pm 0.7 \times 10^{-6}$ sec^{-1} was obtained for times ranging from 0.5 to 2.0 hours for a 1978 batch of Duolite ARC-359 resin at flow rates of 0.57 and 0.88 cm/min (27 and 42 mL/hr). The parameter at a lower flow of 8.8 mL/hr was 10.2 sec^{-1}, which suggests that film diffusion was a controlling factor in that case.

PROCESS STUDIES

Apparatus, Solutions and Analysis

The equipment for doing the process studies consists of two glass columns in series. Above each 7.5-cm diameter x 1-m tall resin column is a 0.9-m backwash column of the same diameter connected with it through a 2.5-cm i.d. ball valve and pipe. The valve is closed except during backwash. The feed solution is a filtered synthetic waste liquor dosed with nonradioactive cesium-133.

Product streams are sampled in an isolation box to avoid contamination. Effluents containing less than 1 part per million cesium are analyzed by isotope dilution-mass spectrometry using a special technique within the instrument in which the cesium is separated from the large quantity of sodium salts. Solutions containing cesium

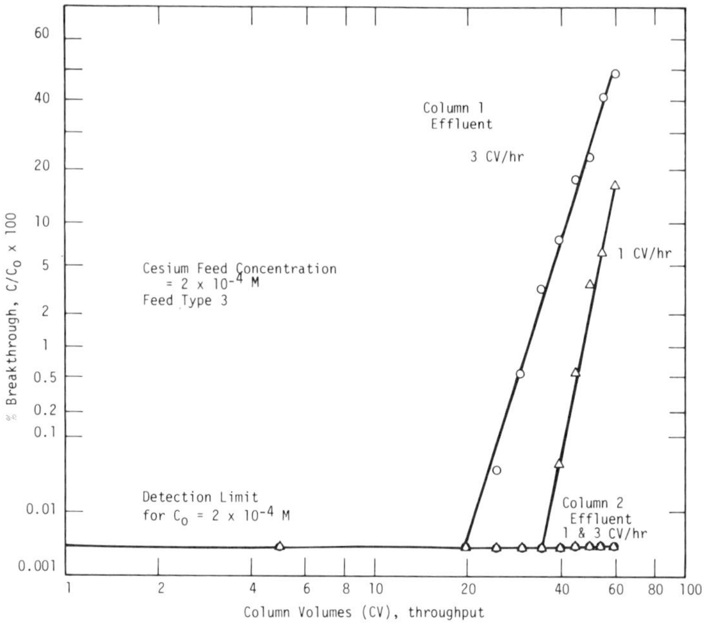

Fig. 2. Process Tests -- Typical Breakthrough Curves for 1978 Resin

above 1 ppm, i.e., mainly for first column effluent, elution and regeneration streams, are analyzed by flame emission in a Jarrell-Ash spectrophotometer. Because the mass spectrometer detection limit is about 1 part per billion, or about 7.5 x 10^{-9} molar, and feed concentrations are 2 to 8 x 10^{-4} molar, a decontamination factor (DF) of between 10^4 or 10^5 can be demonstrated.

Cesium Breakthrough Curves with 1978 Resin

Three ARC-359 resin batches manufactured in 1973, 1977 and 1978 were used. The best process results were obtained with 1978 resin and equaled or exceeded the results Wiley (1) had demonstrated with 1973 resin. As shown in Fig. 2, the goal of 0.01% breakthrough (DF = 10,000) was reached in Column 2 effluent after >60 column volumes* (CV) throughput at feed rates of both 1 and 3 CV/hr. 53% breakthrough was obtained in Column 1 effluent after 60 CV's at 3 CV/hr and after >60 CV's at 1 CV/hr. The 53% breakthrough point corresponded to the effective stoichiometric feed volume, i.e., the feed volume at which the resin would be in equilibrium with the concentration of cesium in the feed solution (6).

*Column volumes are based on a single column.

Percent breakthrough (100x effluent concentration, C/feed concentration, C_o) plots (Fig. 2.) as a straight line vs. throughput (column volumes of solution fed) on log-probability paper, once the effluent concentration exceeds the mass spectrometer detection limit. The straight line plot would be expected from J-function prediction as discussed by Perry and Chilton (6). Development of a promising computer code using basic kinetic and equilibrium data is underway. 2- and 3-CV/hr results were similar, but 1 CV/hr performance was significantly better. Near the beginning of the runs, Column 1 effluent sometimes exceeded the detection limit, presumably because of incomplete elution after the previous run. All Column 2 results were at the detection limit for up to 60 column volumes throughput, which was as far as the runs were carried.

Resin and Feed Effects

Appreciable resin batch and feed differences were found as shown by Fig. 3. The main effect was a change in resin capacity as indicated by the relative position of the 53% C/C_o point. Little change occurs in the kinetics as indicated by the relatively constant slope of the breakthrough curve on log-probability paper. Feed 2 (5.6M Na^+, 1.7M OH^-) performed appreciably better than Feed 1 (5.6M Na^+, 1.0 OH^-) as would be expected from K_D data on the effect of increasing hydroxyl ion concentration. No difference could be found between Feeds 2 and 3, the latter containing added sodium carbonate.

The 1973 resin, which Wiley (1) had used, showed less than half the capacity of the 1978 resin in the process tests. Bench scale tests with radioactive tracer confirmed that the 1973 resin had deteriorated between 1975, the time of Wiley's tests, and 1978, the time of the process tests. The cause of the deterioration will be studied further. The 1977 resin showed about the same capacity as the 1973 resin but poorer kinetics due to coarser particle size. Its initial low cesium capacity relative to the 1978 resin is still not explained.

Increased feed cesium concentration reduced relative resin capacity. With the 1973 resin at 53% breakthrough, 19.5 CV's were achieved at $C_o = 2 \times 10^{-4}$ molar whereas 16 CV's were obtained at $C_o = 7.7 \times 10^{-4}$ molar. The absolute quantity of cesium absorbed was higher in the higher C_o run, namely, 0.0566 mole vs. 0.0179 mole.

Elution, Regeneration and Backwashing

The 2M ammonium carbonate-2M ammonium hydroxide elutriant was adopted based on earlier studies (2). In one typical experiment with 60 column volumes fed during the load step, the eluted cesium band appeared after 1 column volume of the upflow elutriant had been fed.

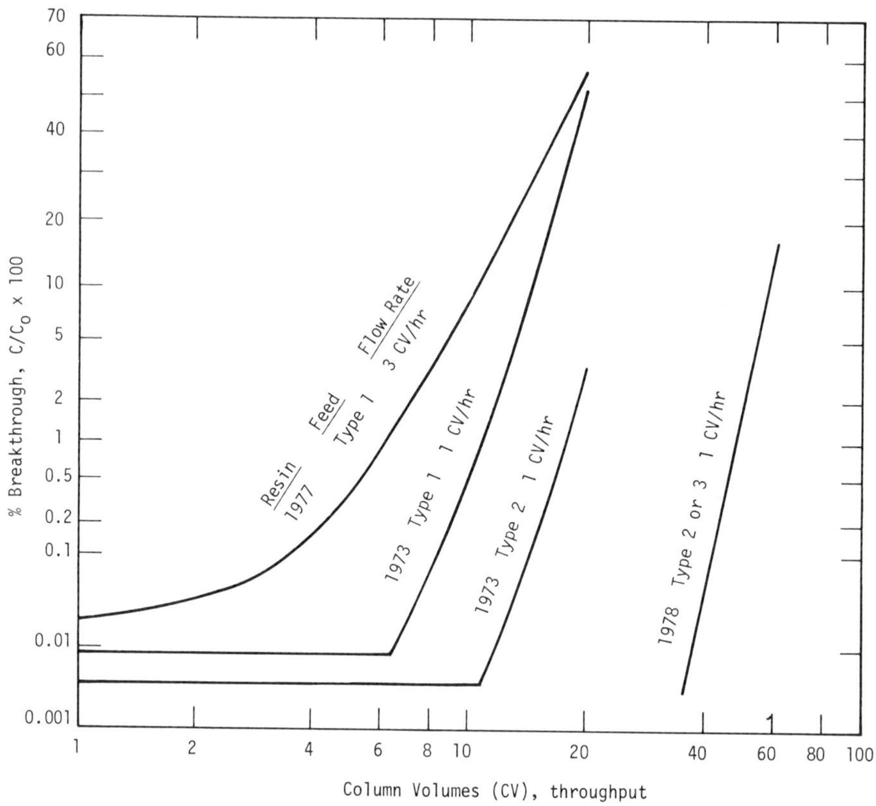

Fig. 3. Process Tests -- Effect of Resin Batch and Feed Type. Column 1 Performance: $C_o = 2 \times 10^{-4}$M.

Feed 1. 5.6M Na^+; 1.0M OH^-
Feed 2. 5.6M Na^+; 1.7M OH^-
Feed 3. 5.6M Na^+; 1.6M OH^-; added Na_2CO_3

The band peaked at 18.5 times the feed concentration (C_o) at 4.5 CV's and was down to 3.5% of C_o at 10 column volumes.

In tests to determine whether elutriant could be recycled and how pure it would have to be, cesium ion was intentionally added to elutriant fed to the columns. Elution proceeded normally, meaning that a relatively simple evaporator/stripper could adequately separate NH_3, CO_2 and H_2O from cesium concentrate. These separated components would be combined with make-up quantities to form the recycle elutriant.

Complete regeneration with NaOH is important, otherwise aluminum hydroxide will precipitate. This is due to sodium aluminate in the feed contacting ammonia residues on or surrounding the resin particles. An initially contemplated 3-CV regeneration step was insufficient to convert all the resin from the ammonium form to the sodium form, and column pluggage resulted. The length of the regeneration step was therefore increased to 5 CV's. The extent of regeneration was followed by measuring the pH, electrical conductivity and the concentrations of sodium and ammonium ion in the regenerant effluent. Electrical conductivity rose tenfold at the end of the regeneration step, at which point the ammonium ions in the effluent had been reduced to 1% of the elutriant ammonium concentration. The pH rose correspondingly, but electrical conductivity monitoring is preferred for plant use as it is more reliable than pH. Backwashing with water eliminates airpockets, reduces channeling, gets rid of accumulated fines and size-classifies the resin. A backwashing test after seven repeat ion exchange cycles produced no significant performance change in the next cycle. A separate backwash column may not be feasible for plant use, however, and alternatives are being looked at.

PROPOSED SCALE-UP TESTS

An intermediate-scale (~1/13) semiworks is now being designed to demonstrate the cesium and strontium ion exchange flowsheet and performance on an expanded scale. The particular size was selected on the basis of ability to demonstrate distributor design, valveless operation and resin handling.

CONCLUSIONS

Resin characteristics, i.e., sodium and cesium equilibria and diffusion rates, were determined and can now guide resin selection and aid in interpreting column performance. The K_D-cesium ion concentration relation gives evidence of three or more different types of ion exchange sites.

The countercurrent load/elution/regeneration cycle for the removal of cesium by ion exchange repeatedly reached the goal decontamination factor (DF) of 10,000 at throughputs up to 60 column volumes.

The proposed ammonium elutriant is satisfactory. Regeneration end-point can be monitored by measuring electrical conductivity of the effluent.

REFERENCES

1. J. R. Wiley, "Decontamination of Alkaline Radioactive Wastes by Ion Exchange," Ind. Eng. Chem. Process Design & Devel. 17, 67-71 (1973).

2. K. C. Knoll, Removal of Cesium from Redox Alkaline Supernatant Wastes by Ion Exchange, USAEC Report HW-84105, General Electric Co., Hanford Atomic Products Operation, Richland, WA (1963).
3. I. M. Abrams, "Countercurrent Ion Exchange with Fixed Beds," Indust. Water Engineering 10, 18-26 (1973).
4. J. R. Wiley and R. M. Wallace, Removal of Cesium from Savannah River Plant Waste Supernate, ERDA Report DP-1388, E. I. du Pont de Nemours and Co., Savannah River Laboratory, Aiken, SC (1975).
5. J. Crank, The Mathematics of Diffusion, 91 (Clarendon Press, London, 1975).
6. R. H. Perry and C. H. Chilton, Chemical Engineers Handbook, 5th Ed., 16-25, 16-31, 16-33 (McGraw-Hill, New York, 1973).

ACTINIDE SEPARATION FROM HLLW

J. O. Liljenzin, G. Persson, I. Hagström and I. Svantesson

Department of Nuclear Chemistry
Chalmers University of Technology
S-412 96 Goteborg, Sweden

INTRODUCTION

Spent nuclear fuel can either be incapsulated and stored permanently in a deep underground repository or reprocessed to recover valuable U and Pu. The high level liquid waste (HLLW) solution resulting from PUREX reprocessing contains the fission products, small fractions of the U and Pu, and most of the Np, Am and higher actinides originally present in the spent fuel. The α-activity in the HLLW amounts to more than 99% of all such activity in wastes from the nuclear fuel cycle (1). This solution can be vitrified and stored as high active glass in deep underground repositories in a similar way as the unprocessed fuel. The Swedish KBS study (2, 3) shows that such repositories can be placed in granite bedrock at 500 m depth. Careful analysis of the possible releases from a repository of this kind shows that U and Np isotopes, with daughters, are responsible for the main radiation doses generated by the repository (2, 4). Because of the enormous time scale, a remedy seems to be a reduction of actinide content by about a factor 100 before solidification of the waste. In anticipation that a separation process might eventually be required or at least had to be considered as an option, we began development work on an actinide separation process in 1974. The general philosophy and a broad outline of our flowsheet was presented in 1976 at the Symposium on Waste Management in Tucson (5).

METHOD SELECTION

In general, it was felt that a separation method for the actinides present in PUREX HLLW solution should: 1) be continuous,

2) avoid handling of solids as far as possible, 3) be remotely operable, 4) not contain steps, like ion-exhange columns, where large amounts of the separated actinides accumulated, 5) not involve processes leading to an increase in salt content of the resulting actinide free fission product solution, and 6) generate the smallest amount possible of secondary wastes. Only solvent extraction processes seemed able to meet all these requirements. Among the many extracting agents possible, the selected reagent should have well known and sufficient radiation resistance and also be available commercially in sufficient quantities. This led us to investigate the suitability of di(2-ethylhexyl)phosphoric acid (HDEHP) as extracting agent for actinides from HLLW solution. Experiments on synthetic HLLW solutions (6) showed that 1M HDEHP in a kerosene diluent (Nysolvim 75A) extracts Pa, U, Np and Pu from a HLLW solution 6M in HNO_3 with distribution ratios (D-values) of 70 or more. Am and the higher actinides, on the other hand, have small D-values at 6M HNO_3 (0.01) like the lanthanides but reasonably high D-values (>50) at 0.1M HNO_3, where also the lanthanides are well extracted, see Fig. 1. This behavior can be used to separate the lower actinides from Am and the other higher actinides, provided a good method is used for reduction of the HNO_3 content.

After a set of preliminary experiments, destruction of HNO_3 by the addition of formic acid was ruled out due to the slowness of the reaction at low HNO_3 concentration and sludge formation (5). It was then decided to utilize the HNO_3 extraction capacility of tri-n-butyl phosphate (TBP) which has been used industrially for the recovery of

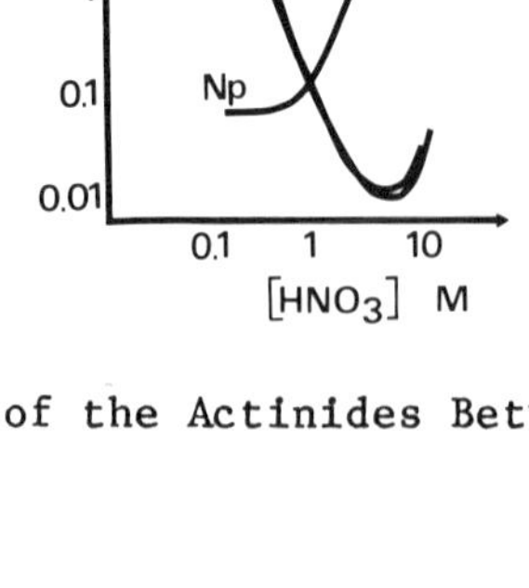

Fig. 1. Distribution of the Actinides Between 1M HDEHP and HNO_3

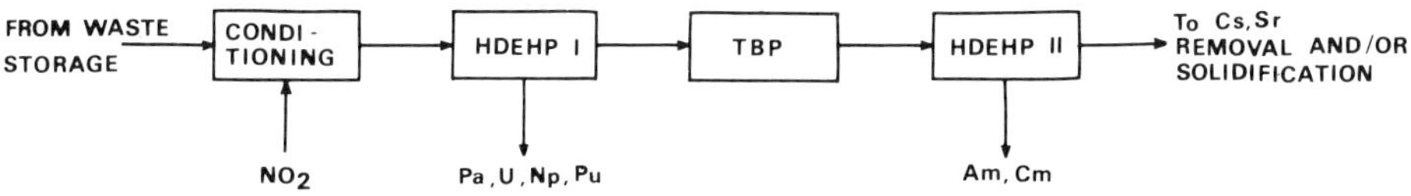

Fig. 2. General Flowsheet

HNO_3 and HF from spent pickling baths (7, 8). A large series of experiments on synthetic HLLW solution (9) showed that it was possible to reduce the HNO_3 concentration to below 0.1M in counter current operation using 50% TBP in kerosene (Nysolvin 75A) (5), without coextraction of elements other than Tc, Ru and Pd. Fig. 2. shows the resulting flowsheet where all details have been omitted for clarity. The process can be divided in three main cycles: 1) extraction of light actinides by 1M HDEHP, clean-up and recovery of separated elements; 2) extraction and recovery of HNO_3, Tc, Ru and Pd by 50% TBP; and 3) extraction of heavy actinides and the lanthanides by 1M HDEHP, separation of actinides from lanthanides using diethylenetriaminepentaacetic acid (DTPA) and recovery of the separated elements. The final effluent is then optionally passed through zeolite columns or vitrified. The process was tested in mixer-settler batteries after computer optimization using batch experimental data for the HDEHP and TBP systems (6, 9, 10).

FIRST HDEHP CYCLE

To establish a fixed valence state distribution for the light actinides (Pa, U, Np, Pu), nitrosyl complexes of Ru, and a proper HNO_3 concentration, the HLLW feed solution first enters a feed adjustment stage where HNO_3 is adjusted to 6M concentration and NO_x are bubbled through the heated solution. The NO_x addition may be omitted if sufficient NO_x is produced by radiolysis. The solution is then cooled to 20°C and fed by a metering pump to the first extraction battery; see Fig. 3. In the first mixer-settler battery, the light actinides are extracted together with Fe, Zr, Nb, Mo, In and some Y and Sn. Due to the slow kinetics of iron extraction, two mixers are used in series at the feed point to provide sufficient holdup time. Most of the extracted Y and part of the extracted Sn, Nb and In, together with all of the minute amounts of other elements extracted, are then stripped with 6M HNO_3 in the second battery. The aqueous waste from this operation is fed to the lower part of the TBP-cycle; see Fig. 4. To prevent accumulation of Zr and Nb in the organic phase, these elements, together with Pa and some Mo and Sn, are stripped from the organic phase in the third battery using a dilute mixture of HNO_3 and HF. The effluent from this battery is treated with $Al(NO_3)_3$ to complex free HF and joined with the final waste stream. However, a small amount of HF is extracted.

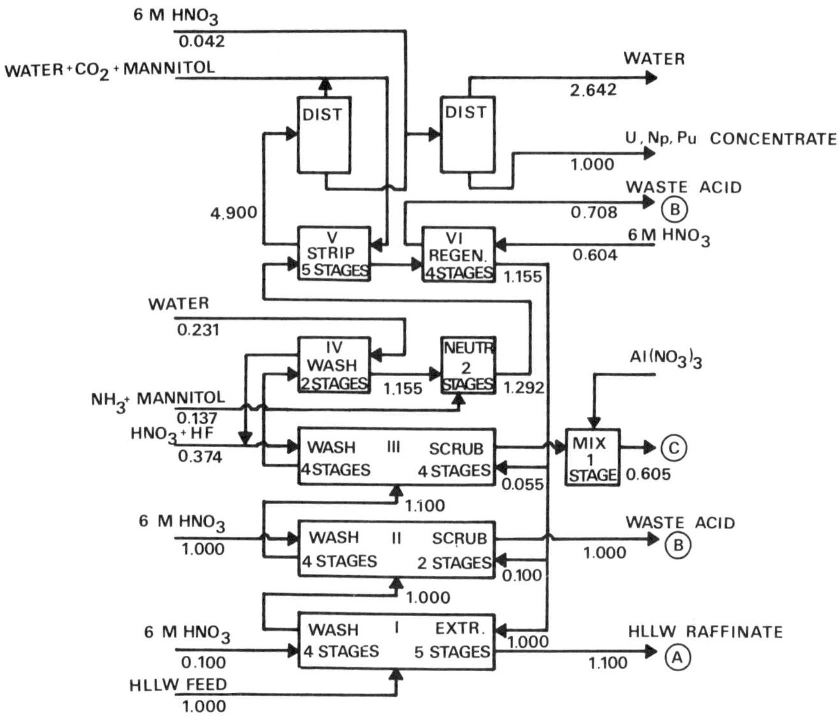

Fig. 3. Flowsheet for the First HDEHP Cycle

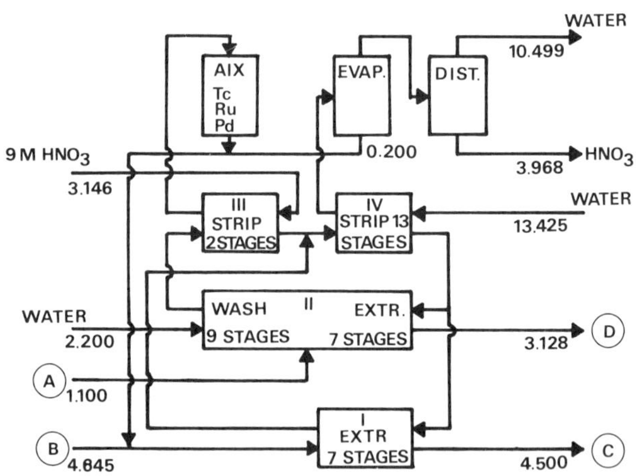

Fig. 4. Flowsheet for the TBP Cycle

This is removed by a slightly acidic aqueous wash in the 4th battery and recycled to the previous battery.

The organic phase is then neutralized with an aqueous solution containing about 8.7M ammonia and 0.3M mannitol using two mixers in series. The effluent is monitored using a combined glass-calomel electrode calibrated in standard buffers. The addition rate is regulated to keep a "pH" reading of 11 ± 0.2. Mannitol is necessary in this stage to prevent sludge formation due to precipitation of Fe (III)-hydroxide, and the actual concentration needed depends on the amount of iron in the HLLW feed. In the 5th battery, all metals are stripped from the neutralized organic phase using 0.5M $(NH_4)_2CO_3$ containing 0.1M mannitol; again the mannitol concentration needed depends on the iron concentration, which for the value given was 1.5g/l in the HLLW feed. To recover most of the $(NH_4)_2CO_3$ used in the strip operation, the effluent is fed to an evaporator where water and $(NH_4)_2CO_3$ are distilled off to a volume reduction of about 25%. The condensate contains an excess of NH_3 which is neutralized using gaseous CO_2, mannitol and water are added, and the solution is recirculated to the strip circuit. To obtain a concentrated U, Np, Pu solution, HNO_3 is added to the practically carbonate free concentrate and the volume is then further decreased by evaporation of water, which is reused in the process. The product obtained is a concentrated metal nitrate solution which can: 1) be acidified, extracted with TBP from the PUREX process and the extract containing U and Pu returned to the PUREX plant, or 2) be calcined and used for target fabrication. Finally, the clean organic phase is regenerated in the 6th battery using 6M HNO_3 and recirculated. To recover HNO_3, the aqueous waste is fed to the lower part of the TBP cycle. Experimentally determined removal efficiencies show that more than 99.99% U, 98.9% Np and 99.99% Pu are recovered from the HLLW feed. This has been verified using a small batch (30 ml) of old HLLW from low burnup fuel (400 MWd/ton).

TBP CYCLE

In the TBP cycle, HNO_3 is recovered both from the HLLW stream and from secondary acid waste streams using 50% TBP in kerosene. Separate extraction batteries are used for the HLLW stream and the waste acid stream. This is advantageous as the waste acid stream contains only minute amounts of lanthanides, Ru, Pd, and heavy actinides, and hence the aqueous wash stages can be omitted resulting in a smaller aqueous effluent volume than if it was joined with the HLLW stream. Both aqueous effluents from the extraction batteries are below 0.1M in H^+. The use of water as feed to the wash stages rather than HNO_3 is necessary for several reasons, because otherwise: 1) Ru and Pd circulate internally in the battery until their concentrations have built up to a value where their salts precipitate, and 2) part of the heavier actinides and lanthanides would be present in the organic effluent.

The organic extract stream is fed to the Tc, Ru, Pd strip battery where these elements are stripped using 9M HNO_3. The aqueous product is about 3M HNO_3 and the noble elements are recovered from this stream, e.g., using anion exchangers. The effluent from the ion exchangers is joined with the waste acid stream. HNO_3 is finally stripped with water, yielding a 3M HNO_3 and a clean TBP stream which is recycled to the extraction batteries. The HNO_3 product is purified by evaporation giving a strongly acidic residue which is joined with the anion exchange effluent stream and a condensate from which HNO_3 and water are recovered by distillation. Part of the HNO_3 is reused in the process. As the flow rates were optimized for minimum aqueous waste volume and the 400 MWd/ton HLLW solution, Tc, Ru and Pd recoveries are 67%, 74% and 98%, respectively, which is not as high as possible permitting a larger aqueous waste volume or given a different feed. Mixer-settler batteries have been used in the testing of this cycle, but pulsed columns could be used with advantage.

SECOND HDEHP CYCLE

Lanthanides and heavy actinides are coextracted in the first battery; see Fig. 5. The actinides are selectively stripped in the second battery using a solution of 0.05M ammonium-DTPA and 1.5M lactic acid. This solution is then fed to the 4th battery where the actinides are extracted again with 1M HDEHP in kerosene. The DTPA-lactic acid solution is recirculated after pH adjustment with NH_3, thus avoiding a large waste source (10). To be able to strip into and extract from the same solution, the conditions are selected so

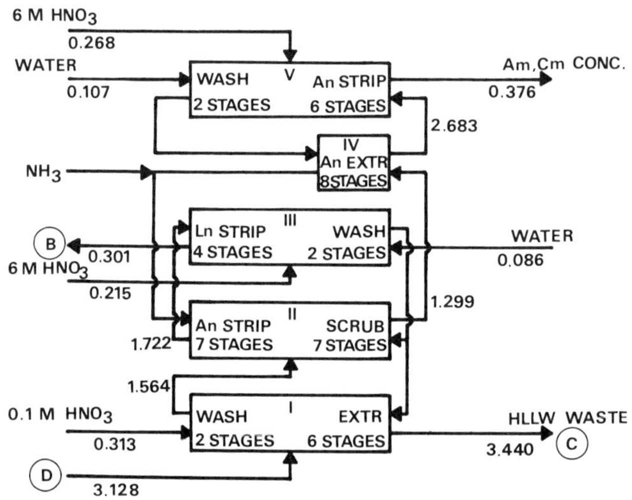

Fig. 5. Flowsheet for the Second HDEHP Cycle

that the D-values for the actinides are around one. Then stripping or extracting action can be selected by choice of a low or a high organic to aqueous flow ratio (Θ) because the extraction in a counter current battery is governed by the product ΘD. The extraction is further facilitated by a decrease in pH in the stripping and extraction batteries caused by the extraction of ammonia. The extracted heavy actinides are stripped with a small flow of 6M HNO_3 in the fifth battery. This stream can be evaporated and calcined to produce, e.g., feed material for target production. After the actinides have been stripped in the second battery, remaining lanthanides are stripped with 6M HNO_3 in the third battery and the organic phase recirculated. The lanthanide waste stream is fed back to the waste acid stream. Careful pH control is needed in the selective strip battery to provide a good actinide/lanthanide separation. Separation factors between Am and Nd, which is the critical pair, of about 8000 are achievable. The Am and Cm recoveries are 99.8% and more than 99.9%, respectively.

DISCUSSION

An actinide separation process is of no use if the separated elements cannot be disposed of in a better way than the bulk of the HLLW solution. One such possibility is offered by nuclear transmutation in light water reactors. It has been suggested (11) that the separated Np could advantageously function as burnable poison in PWRs, replacing the boron pins currently used. This would reduce tritium poisoning effects, avoid positive moderator temperature coefficients and permit better control of reactivity and power shapes, especially during expected, prolonged, future fuel cycles. A similar effect is obtained if the separated Am-Cm mixture is incorporated in the proper amount within the normal fuel pins. Finally, both applications can be used together to optimize fuel burnup and core temperature profiles. If this suggested application for the minor actinides is used by the nuclear industry, it may turn out that available HLLW resources will not be sufficient, and we will have a supply rather than a disposal problem. The separation process described is by no means the only one possible. Certainly several other alternatives exist. However, with a limited amount of manpower and funding, we felt that this process could be made to work in the desired way, and this has been found to be the case.

ACKNOWLEDGEMENT

We are grateful to Professor J. Rydberg for initiating this work and many discussions and to B. Hjorth, M. Svensson and E. Jomar for skillful laboratory and drawing assistance. This work was supported by the Swedish Council for Radioactive Waste (PRAV) and the Swedish Natural Research Council (NFR).

REFERENCES

1. J. O. Liljenzin, J. Rydberg and G. Skarnemark, "Reducing the Long Term Hazard of Reactor Waste Through Actinide Removal and Destruction in LWR's," paper presented at the Symposium on Separation Science and Technology, Gatlinburg, Oct. 30-Nov. 2 (1979).
2. Handling of Spent Nuclear Fuel and Final Storage of Vitrified High Level Reprocessing Waste," Karnbranslesakerhet (1978).
3. "Handling and Final Storage of Unprocessed Spent Nuclear Fuel," Karnbranslesakerhet, Stockholm (1978).
4. J. O. Liljenzin, "The Release of a Nuclide Into a Recipient From Buried Canned High Level Glass," Disposal of High Active Nuclear Fuel Waste. A Critical Review of the Nuclear Fuel Safety (KBS) Project on Final Disposal of Vitrified High Active Nuclear Fuel Waste, Industridepartementet, DsI 1978:17, Stockholm (1978).
5. J. O. Liljenzin, I. Svantesson and I. Hagstrom, "A Possible Solution to the Long Time Storage Problem for High Level Waste," Proceedings of the Symposium on Waste Management (R. G. Post, ed.), ERDA CONF-761020.
6. I. Svantesson, I. Hagstrom, G. Persson and J. O. Liljenzin, "Distribution Ratios and Empirical Equations for the Extraction of Elements in PUREX High Level Waste Solution. II: HDEHP," submitted to J. Inorg. Nucl. Chem., (1979).
7. J. O. Liljenzin, H. Reinhardt and J. Rydberg, "Satt Att Avlagsna Envarda Oorganiska Syror ur Vattenlosningar (A Method for Removing Monovalent Inorganic Acids From an Aqueous Solution Containing Metal Salts of the Acids)," SW. Pat. Appl. 760/62 1975, U.S. Pat. 3,870,060, Canadian Pat. 998841, etc.
8. J. H. Dempster and P. Bjorklund, "Operating Experience in Recovery and Recycling of Nitric and Hydrofluoric Acids From Waste Liquors," Proc. 4th Annual Meeting Hydrometallurgy Section, C.I.M., Toronto (1974).
9. I. Svantesson, I. Hagstrom, G. Persson and J. O. Liljenzin, "Distribution Ratios and Empirical Equations for the Extraction of Elements in PUREX High Level Waste Solution. I: TBP," J. Inorg. Nucl. Chem. 41, 383 (1979).
10. I. Svantesson, I. Hagstrom, G. Persson and J. O. Liljenzin, "Separation of Am and Nd by Selective Stripping and Subsequent Extraction with HDEHP Using DTPA-Lactic Acid Solution in a Closed Loop," Radiochem. Radioanal. Letters 37, 215 (1979).
11. G. Andersson, Preliminary Studies of 'Minor' Actinide Recycling in LWR, Studsvik Technical Report K2-78/6 (1978).
12. G. Andersson, On Waste Utilization in PWR, Studsvik Technical Report K2-79/91 (1979).

METHODS FOR SEPARATING ACTINIDES FROM REPROCESSING AND REFABRICATION PLANT WASTES*

D. W. Tedder,** B. C. Finney, J. O. Blomeke

Oak Ridge National Laboratory
Oak Ridge, TN 37830

ABSTRACT

Chemical processing flowsheets have been developed to partition actinides from all actinide-bearing LWR fuel reprocessing and refabrication plant wastes. These wastes include high-activity-level liquids, scrap recovery liquors, HEPA filters and incinerator ashes, and chemical salt wastes such as sodium carbonate scrub solutions, detergent cleanup streams, and alkaline off-gas scrubber liquors. The separation processes that were adopted for this study are based on solvent extraction, cation exchange chromatography, and leaching with Ce^{4+}-HNO_3 solution.

INTRODUCTION

A research and development program(1,2) coordinated by Oak Ridge National Laboratory has been underway since 1977 to evaluate, for waste management purposes, the feasibility and incentives for partitioning (separating) the actinides from fuel reprocessing and refabrication plant waste streams and recycling them to reactors for transmutation to short lived and stable fission products. This paper describes the processes that were chosen to establish the technical feasibility of carrying out such separations (3).

---

*Research sponsored by the Office of Nuclear Waste Management, U. S. Department of Energy, under contract W-7405-eng-26 with the Union Carbide Corporation.

**Present Address: School of Chemical Engineering, Georgia Institute of Technology, Atlanta, GA 30332

A four-point strategy was adopted in order to develop actinide partitioning flowsheets that are defensible from the standpoint of chemical feasibility:

1. utilize only demonstrated technology that is readily applicable to commercial recycle operations,
2. develop generic capabilities to deal with all types of wastes,
3. choose processing systems that complement each other and provide actinide recovery opportunities in depth, and
4. experimentally evaluate the proposed methods.

Although the experimental work was limited to laboratory and hot-cell investigations of the various treatment subsystems, we believe that the selected processes and flowsheets are sufficiently advanced to permit recoveries of greater than 99% of the actinides.

WASTE TREATMENT SUBSYSTEMS

It is envisioned that the separation processes would be incorporated into separate waste treatment facilities (WTFs) that are located adjacent to the fuel reprocessing and refabrication plants.

A schematic diagram depicting fuel reprocessing with actinide partitioning is shown in Fig. 1. The various wastes are transferred to the WTF for actinide partitioning and afterward are returned to the reprocessing plant for immobilization and packaging. The partitioned actinides are returned as a concentrated nitrate solution to the reprocessing plant for conversion to the oxides.

The subsystems incorporated in the WTF are: 1) High level Solid Waste Treatment, 2) High Level Liquid Waste Treatment, 3) Solid Alpha Waste Treatment, 4) Cation Exchange Chromatography, 5) Salt Waste Treatment, 6) Actinide Recovery, 7) Solvent Cleanup and Recycle, 8) Off-Gas Treatment, 9) Actinide Product Concentration, and 10) Acid and Water Recycle. The WTF supporting a fuel refabrication facility, although similar, does not contain subsystems 1) and 2).

High Level Solid Waste Treatment

Cladding hulls and dissolver solids are generated as wastes during the reprocessing of LWR fuels. Although the alpha activity in these wastes is generally low, the WTF has the provisions for giving them an additional leach with HNO_3-HF solution.

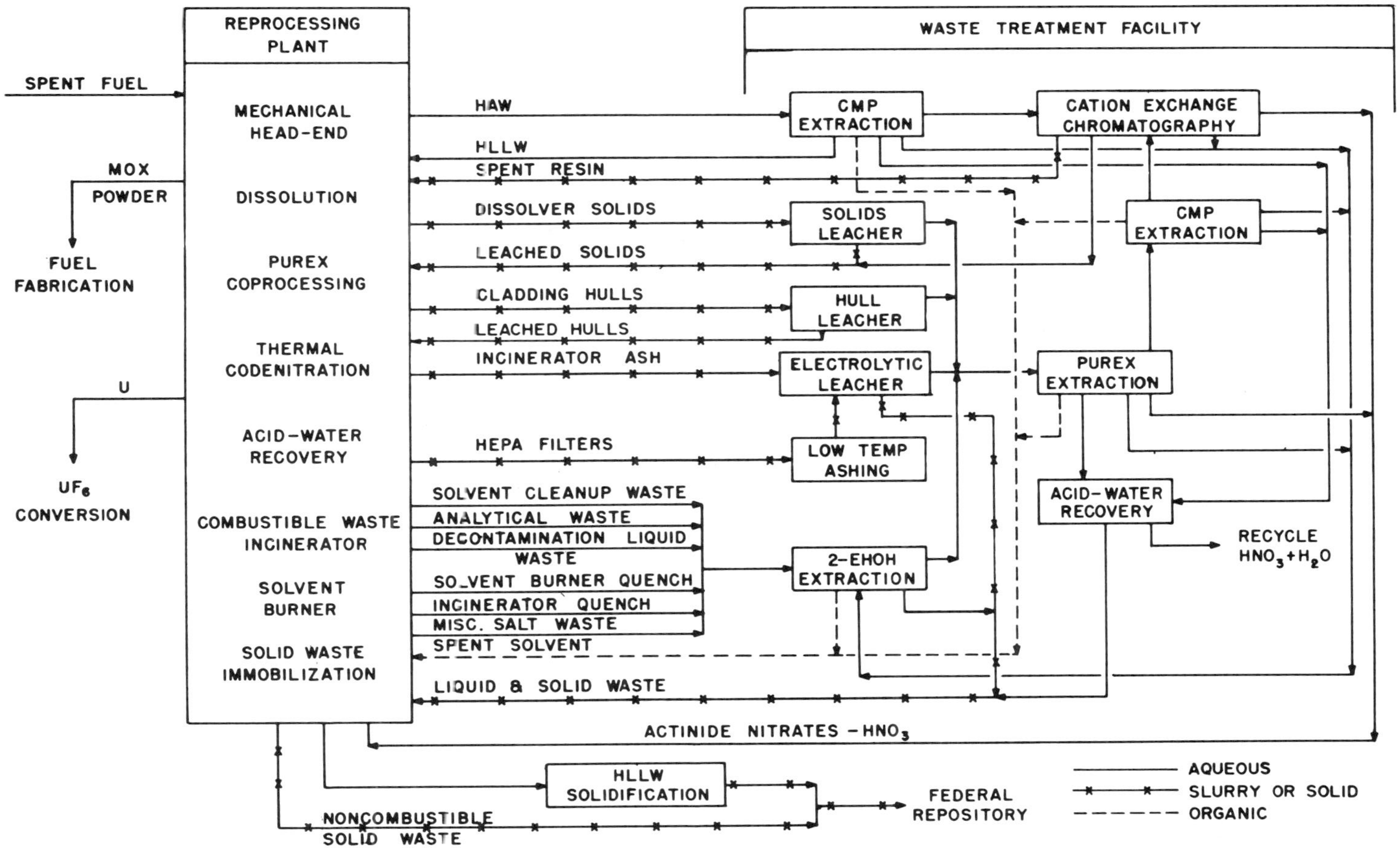

Fig. 1. Schematic Diagram of LWR Fuel Reprocessing With Actinide Partitioning

High Level Liquid Waste (HLLW) Treatment

The HLLW, after generation, is transferred to the WTF where it is contacted with a bidentate extractant composed of 30% dihexyl-N, N-diethylcarbamylmethylene phosphonate (CMP) in diisopropylbenzene. All the actinides and lanthanides are removed from the HLLW to some degree under high acid conditions (4) and all other fission products that tend to be extracted can be controlled by scrubbing or by adjusting the extraction conditions to reject those that exist as anions.

The coextracted actinides and lanthanides are stripped from the solvent by sequential treatment in two columns: a dilute nitric acid strip column containing a reductant, followed by a dilute oxalic acid strip column. After the oxalic acid has been destroyed, the two strip solutions are combined with other recycle streams, partially denitrated, concentrated, and finally sent to Cation Exchange Chromatography for lanthanide removal.

Experimental results (4) using actual HLLW suggest that greater than 99.99% of the tetravalent actinides and 99.9% of the trivalent actinides may be removed using CMP extraction.

Solid Alpha Waste Treatment

Solid Alpha Waste Treatment consists of the processing steps that have been developed to recover actinides from HEPA filters and incinerator ashes. Such wastes are generated primarily in the nitrate-to-oxide conversion step during reprocessing, the oxide powder handling and pellet grinding operations in fuel fabrication, and the incineration of alpha-contaminated combustible wastes in both reprocessing and fabrication.

For the purposes of this study, it is assumed that the HEPA filters are dismantled and the filter media are then shredded and packaged in 55-gal drums at the incinerator area. The incinerator ashes are also packaged in drums and then transferred from the incinerator area to the WTF.

At the WTF, the shredded HEPA media are first metered through a low-temperature ($\sim$ 300°C) asher to oxidize any high-molecular-weight glues or binders that may be present. The asher product is collected in a hopper and loaded into the leaching equipment, which consists of a vertical, cylindrical column and a stirred electrolytic oxidizer tank. A leaching cycle consists of successively loading the vertical column with a batch of waste and circulating a mixture of nitric acid, cerous nitrate, and gadolinium nitrate by thermosiphon through the fixed bed and back to the oxidizer tank. As the leachant circulates and approaches reflux temperatures, an electrical potential is applied across the electrodes in the oxidizer tank to convert the cerous nitrate to ceric nitrate.

On completion of the leaching, oxalic acid is added to the leachant to reduce the corrosive ceric ion. The leachant is then transferred to a holding tank for subsequent centrifugation and actinide removal. The leached HEPA material is washed with nitric acid and water, compacted to near its original volume with air pressure, and discharged into drums. The drums are sent to the on-site immobilization and storage area for ultimate disposal.

Incinerator ashes are decontaminated simply by loading them into a separate, stirred electrolytic oxidizer tank and subjecting them to essentially the same leaching operations as those described for the HEPA waste.

The leaching of HEPA filter waste and incinerator ashes with Ce^{4+}-HNO_3 is complex, but results of experimental studies (5,6) suggest that this method is probably technically feasible and, in some respects, preferable to leaching with fluoride. The Ce^{4+}-HNO_3 is about as effective as fluoride in accelerating the dissolution of all transuranic actinides but does not dissolve or greatly soften the HEPA filter waste. Fluoride converts the fiberglass into a viscous, sticky mass and totally dissolves the incinerator ashes, whereas the ceric acid does not affect the fiberglass and only partially dissolves the incinerator ashes. In terms of decontamination, the ceric acid has achieved 99.99% dissolution of all actinides in the filter media; however, as much as 5% of the alpha activity initially in the incinerator ashes does not dissolve.

Cation Exchange Chromatogrphy

Separation of the actinides and, in particular, the trivalent actinides from the lanthanides is accomplished by cation exchange chromatography (CEC) (7). A number of small batches must be run in parallel in order to deal with the large quantities of curium and associated heat that build up due to the recycle of the transuranics. Conceptually, this problem is dealt with by a modular approach in the design of cation exchange racks. Each rack (or module) consists of a single sorption column and two elution columns mounted on a balanced frame that has overall dimensions about the size of an office file cabinet. In this manner, the number of watts per batch of radioactive material loaded onto the module is kept acceptably small.

The experimental work relating to this processing area has focused on identifying any interactions between the CEC area of the WTF and the HLLW Treatment area that produces the CEC feed, and on determining the elution sequence for the tetravalent actinides (8). The expected impurities in the feed to the CEC will not significantly affect system performance, and there is no evidence that CEC feed adjustment beyond the control of acidity level is necessary. Also, results of tests have confirmed that the tetravalent actindes will elute ahead of the barrier ion in CEC; therefore, these elements will report to the Salt Waste Treatment system as desired.

Although CEC is capable of a very high degree of actinide partitioning, its principal disadvantage lies in the high rate of ion exchange resin replacement caused by radiation degradation.

Salt Waste Treatment

Both alkaline and acidic salt wastes are produced during fuel reprocessing and refabrication. The most important of these, with respect to actinide losses, is the alkaline sodium carbonate scrub solution that is generated during the cleanup of tributyl phosphate (TBP) for recycle. This waste may contain up to 0.5% of the actinides in the feed. Difficulties have also been experienced with the alkaline detergent wastes containing surfactants that extract into TBP and the sodium carbonate scrub wastes resulting from CMP recycle and cleanup.

A process was developed to recover actinides from these wastes that consists of first acidifying them and then contacting them with 2-ethylhexanol (9). This alcohol extracts the acidic degradation products such as dibutyl and monobutyl phosphoric acids but leaves the actinides in the raffinate. Subsequently, the raffinate may be contacted with TBP and CMP for actinide recovery since it no longer contains those acidic degradation products that would interfere with the separation. The alcohol extractant, containing the degradation products and detergents, is recycled after the acidic species have been removed by washing with sodium carbonate.

Actinide Recovery

Both WTFs require an Actinide Recovery area where actinides are recovered from the liquors being produced in the above-mentioned separation processes. The WTF supporting the fuel reprocessing plant requires both TBP and CMP extraction cycles, but the WTF supporting the fuel refabrication plant can be operated with a CMP extraction cycle alone by utilizing the existing, on-site TBP scrap recovery system.

In each case, the various actinide-bearing liquors are concentrated, acidified, and clarified by centrifugation as needed. Subsequently, the concentrate is contacted with TBP and then with CMP. The raffinate, now essentially actinide-free, is further concentrated and denitrated with formic acid; the resulting slurry is then converted to a concrete and packaged in drums. The TBP is stripped in the usual manner, and the recovered actinides are sent back to the main processing facility. The CMP is stripped in two columns, similar to the operation of the HLLW treatment strip columns, to recover the stable cerium and gadolinium for recycle to the Solid Alpha Waste Treatment area. Low concentrations of trivalent actinides are maintained in the cerium recycle stream by bleeding a fraction of this stream to the CEC system. Thus, the major differences between the Actinide Recovery area and the HLLW Treatment area are the somewhat higher levels of radioactivity in the latter system.

Miscellaneous Support Systems

Both TBP and CMP solvents are cleaned with water and sodium carbonate washes and then equilibrated with acid before they are recycled. This treatment aids in the operation of the extraction columns and also serves to strip the ruthenate and pertechnetate anions from the CMP solvent. Activity levels are controlled in the Actinide Recovery Area by bleeding a fraction of the CMP to the HLLW Treatment Area; a small fraction of the CMP from the HLLW Treatment Area is sent to the on-site solvent burner.

Both WTFs require extensive off-gas treatment. Acid denitration requires the recovery of nitrogen oxides. Ruthenium absorption beds are provided in the event of ruthenium volatilization in the Solid Alpha Waste Treatment areas. The WTF associated with a fuel reprocessing plant will also require iodine retention systems. Both WTFs utilize HEPA filtration and catalytic destruction of NO_x.

Extensive acid and water recycle is carried out in both WTFs. The process streams generated internally are voluminous, have low levels of alpha activity, and contain large quantities of dissolved solids. Thus, large quantities of dilute acid and water need to be recycled as evaporator overheads.

REFERENCES

1. J. O. Blomeke, "A Program to Establish the Technical Feasibility and Incentives for Partitioning" Management of Radioactive Waste: Waste Partitioning as an Alternative NR-CONF-001 (1976).
2. A. G. Croff, D. W. Tedder, J. P. Drago, J. O. Blomeke, and J. J. Perona , A Preliminary Assessment of Partitioning and Transmutation as a Radioactive Waste Management Concept, ORNL/TM-5808, Oak Ridge National Laboratory, Oak Ridge, TN (1977).

3. D. W. Tedder, B. C. Finney, and J. O. Blomeke, Actinide Partitioning-Transmutation Program Final Report. II. Partitioning Processes for LWR Fuel Reprocessing & Refabrication Plant Wastes, ORNL/TM-6982, Oak Ridge National Laboratory, Oak Ridge, TN (In Press).
4. L. D. McIsaac, J. D. Baker, J. F. Krupa, R. E. LaPointe, D. H. Meikrantz, and N. C. Schroeder, Study of Bidentate Compounds for Separation of Actinides from Commercial LWR Reprocessing Wastes, ICP-1180, Idaho National Engineering Laboratory, Idaho Falls, Idaho (1979).
5. F. M. Scheitlin and W. D. Bond, Recovery of Plutonium from HEPA Filter Media by Continuously Leaching in a Packed Column with Electrolytically Produced Ce(IV) Nitrate and Nitric Acid, ORNL/TM-6802, Oak Ridge National Laboratory, Oak Ridge, TN (In Press).
6. G. H. Thompson, Final Report - Program for Waste Management Analysis of the Nuclear Fuel Cycle - ORNL Actinide Recovery from Combustible Waste: The Ce(IV)-HNO_3 System, RFP-2907, Rocky Flats Plant, Golden, CO (In Press).
7. E. J. Wheelwright, L. A. Bray, H. H. Van Tuyl, and H. T. Fullam, Flowsheet for Recovery of Curium from Power Reactor Fuel Reprocessing Plant Waste, BNWL-1831, Battelle Pacific Northwest Laboratory, Richland, WA (1974).
8. C. W. Forsberg, The Elution Sequence of Neptunium and Plutonium from a Cation Exchange Chromatography Column Eluted with Diethylenetriaminepentaacetic Acid (DTPA), ORNL/TM-6737, Oak Ridge National Laboratory, Oak Ridge, TN (1979).
9. E. P. Horwitz, C. A. A. Bloomquist, G. W. Mason, R. A. Leonard, and A. A. Ziegler, The Recovery of Actinides from TBP-Na_2CO_3 Scrub Waste Solutions: The Aralex Process, ANL-79-74, Argonne National Laboratory, Argonne, IL (1979).

RUTHENIUM VOLATILITY BEHAVIOUR DURING HLLW-VITRIFICATION IN A LIQUID-FED CERAMIC WASTE MELTER

S. Weisenburger and K. Weiss

Karlsruhe Nuclear Research Centre
P.O. Box 3640
7500 Karlsruhe, Fed. Rep. of Germany

INTRODUCTION

During vitrification of high-level liquid waste (HLLW) some elements contained in the waste can transform to volatile or semivolatile compounds and can enter into the process off-gas as gaseous effluents. One of the most troublesome candidates concerning volatilization is ruthenium. HLLW can contain significant concentrations of this element. Typically, a 3 year old waste, generated from spent LWR fuel with 28 700 MWD/to burn-up and concentrated to 379 l/to, will contain up to 5.3 g/l of nonradioactive ruthenium and 0.07 g/l of radioactive Ru-106. This corresponds to a Ru-106 activity of 240 Ci/l in the HLLW (1).

Volatile ruthenium compounds are predominantly formed during the vitrification process by complex chemical reactions between ruthenium and oxygen. Volatilized ruthenium compounds can remain as a stable gaseous form even in the cooler regions of the off-gas train, where vaporized materials are quickly condensed and removed, and be released as a gas. Under appropriate conditions however, the volatile ruthenium does not remain gaseous. When it exits the hot zone of the process, ruthenium species can be converted to solids which can either deposit in distinct parts of the off-gas pipe, or remain airborne in the off-gas as extremely fine particulates. In the case of extended deposition, a potential exists for plugging the off-gas pipe.

The complex and unpredictable behavior of volatilized ruthenium along with the possible complications in the off-gas pipe, and the potential difficulties for controlling release of gaseous radioactive Ru-106 makes it desirable to suppress the ruthenium volatility as much as reasonably possible. This is additionally advantageous because any ruthenium collected from the off-gas, either in liquid scrub solutions or in solid adsorber beds, can create additional highly active waste must must be treated or recirculated into the process.

RUTHENIUM BEHAVIOR IN SOME TECHNICAL PROCESSES

The behavior of ruthenium is specific for each type of vitrification process and depends on a variety of conditions including waste composition and actual process parameters. Ruthenium losses into the off-gas have been reported to be between 0.002% and virtually 100% (2). Conditions during calcination are obviously of primary importance. Spray calciners seem to have low potential for generation of volatile ruthenium. Volatilized ruthenium passing into calciner process off-gas has been determined during processing of a simulated commercial LWR-waste to lie below 1 wt % of total ruthenium processed (3). Pot processes are known for generation of ruthenium volatilities above 15 wt %, and for the rotary kiln calcining unit of AVM ruthenium losses into the off-gas of about 38 wt % have been reported (4). In a fluidized bed calciner volatile ruthenium leaving the calcining unit varied between 0.002 wt % and 84 wt % indicating a strong influence of process parameters such as type of heating and bed temperature (1). Ruthenium behavior in the liquid fed ceramic-lined waste melter process is not known. HLLW-vitrification in such a unit is a relatively new technique, and the extent of potential ruthenium volatilization must be clarified.

The current state of knowledge of the ruthenium chemistry involved during drying, calcining and melting of HLLW has not yet defined all the volatile ruthenium species. It is commonly observed that oxidizing conditions during processing of HLLW can appreciably enhance formation of volatile ruthenium species. The extent of the attack of oxygen on ruthenium or ruthenium compounds will thus depend primarily on conditions influencing the oxygen concentration in the calcining and melting unit. These include nitrate concentration of HLLW, and also the amount of air introduced. Decomposition of nitrates during calcination produces NO_x and oxygen. A major part of volatilized ruthenium species produced by the attack of oxygen may be the rutheniumtetraoxide (RuO_4) which is gaseous above 110°C and metastable up to 600°C. Other volatile ruthenium species may be produced due to decomposition of ruthenium compounds present in the waste. The compounds are primarily nitrato complexes of nitrosyl ruthenium, RuNO(III), having one, two or three nitrato groups, NO_3, associated with each ruthenium nitrosyl group. The question of primary concern here is how much ruthenium will be volatilized under the process conditions of a liquid fed ceramic waste melter where the HLLW is fed onto a large molten glass pool. At the glass pool surface the HLLW is continuously dried, calcined and melted under a layer of boiling liquid.

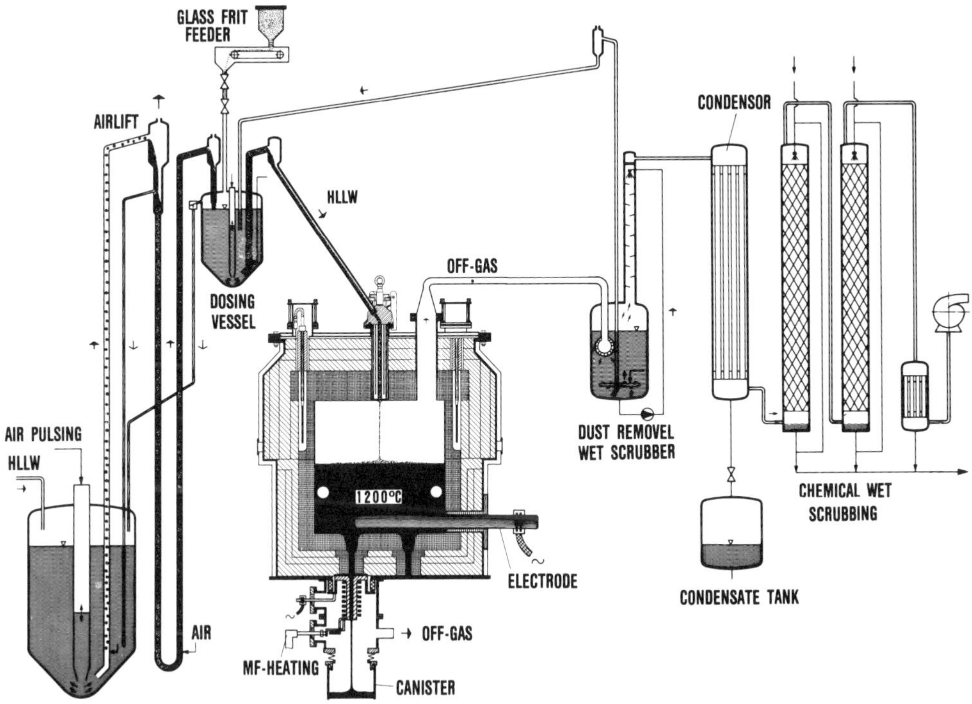

Fig. 1. Schematically illustrated vitrification unit with the liquid-fed ceramic melter and the wet off-gas system.

EXPERIMENTAL

Ruthenium behavior has been investigated in a pilot-scaled vitrification plant. The simulated HLLW contained 1 g/l of non-radioactive ruthenium introduced to the waste as a 20% solution of nitrato complexes of nitrosyl ruthenium. The waste was liquid-fed into a Joule-heated ceramic melter having a quadratic glass pool surface of 0.36 m^2 and operated at 1200°C. The power was provided by passing electric current through the glass pool between two pairs of immersed electrodes made of molybdenum. The process off-gas was directed first to a dust-removal wet scrubber followed by a condenser and chemical scrubbers. The facilities are shown schematically in Fig. 1 and described in more detail in Ref. 5.

Several continuous long-term runs of approximately 100 hours have been performed and the ruthenium collected in the scrub solution of the dust removal wet scrubber, in the condensate and in the scrub solution of the chemical wet scrubber was determined by special

Table 1. Flow Path of Ruthenium in the Off-Gas System (Glass Pool Flooded to 70-90%, Waste Solution 3.4 m Nitrates Per Liter in Total).

Operation Time t (hour)	Ruthenium (wt.% of total Ru processed up to time t)				
	Off-gas pipe	Dust scrubber		Condensate	Chemical scrubbers
		within dust	dissolved or <3μm		
12	-	5.9	4.8	0.1	0.04
20	-	2.6	5.9	0.3	0.04
28	-	2.4	5.9	0.3	0.04
36	-	3.3	6.8	0.4	0.03
44	-	2.8	13.0	0.5	0.03
52	-	1.2	11.0	0.9	0.03
60	-	4.2	10.3	0.9	0.03
68	-	2.5	10.0	0.8	0.03
76	-	2.5	10.2	0.8	0.03
84	-	2.8	10.0	0.8	0.03
92	-	2.4	12.4	1.0	0.03
100	4.3	2.3	10.4	1.0	0.03
Average	4.3	2.9	9.2	0.8	0.03

techniques described in Ref. 6. Liquid samples were taken and prepared for analysis either by X-ray fluorescense or by atomic absorption, depending on ruthenium concentration levels in the samples. After termination of each run solid deposits were removed from the off-gas pipes and the ruthenium concentrations in this material was determined. A ruthenium absorber was not used but will be applied in future runs with an advanced melter system presently under construction. A silica gel adsorber bed will be installed after chemical wet scrubbing. The feed solution contained 2.5 moles of free nitric acid and 0.9 moles of nitrate per liter. For demonstration of the influence of nitric acid concentration on ruthenium behavior, runs were performed with the original solution as well as with solutions denitrated by formic acid at 100°C in a pilot scale denitration plant. The residual total nitrate concentration of the denitrated solution was below 0.7 moles per liter.

RESULTS

Typical results of ruthenium losses from the ceramic melter into the off-gas train are given in Fig. 2. The ruthenium loss was found to be lowest with a relatively high feed rate above about 25 l/h which led to a complete flooding of the molten glass pool with liquid waste. Results for the loss of ruthenium during operation of the melter, expressed as wt % of the total ruthenium processed up the actual operation time are shown in Fig. 2. A second long-term run was conducted at a feed rate of 20 l/h. The molten glass pool was covered to an extent of 70-90% with liquid or layers of calcining solid and 10-30% remained uncovered as molten glass at temperatures between 900-1000°C. The ruthenium loss increased and followed curve II in Fig. 2. Curve III, indicating even more ruthenium loss,

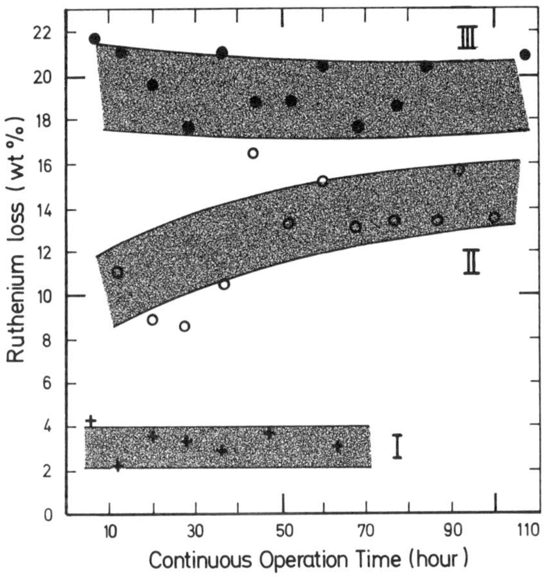

Fig. 2. Ruthenium passing into the off-gas train (without Ru deposited in the off-gas pipe; Curve I: Molten glass pool completely flooded with waste; Curve II: Flooded between 70-90%; Curve III: Flooded below 50%

represents results when only 8-10 l/h was processed. In this case only 30-50% of molten glass pool was covered with liquid or calcining solids.

Each value of ruthenium loss shown In Fig. 2 represents the ruthenium collected during operation in the wet scrubber, in the condenser and in the chemical scrubbers. Ruthenium deposited in the off-gas pipe is not included, but was determined separately after the end of each run by analysis of solid material collected from the pipe. In the case of complete flooding of the molten glass pool (curve I of Fig. 2), 1.28 wt % of the total ruthenium processed was deposited in the off-gas pipe. At lower feed rates the ruthenium amounts in the deposits increased; when flooding of the glass pool was below 50% 4.3 wt % of total ruthenium processed was found in the off-gas pipes after a 100 hour run. Most of the deposited material was in the 2.5 m long 0.1 m in diameter off-gas pipe which connects the ceramic melter with the dust-removal wet scrubber. The color of the deposited material was dark-brown to black and slightly sintered in the inner surface of the pipes.

Based on the composition of the simulated feed solution the ruthenium concentration in the calcined material would be expected to be 1.6 wt %. In all cases the deposited material in the off-gas pipe exceeded this concentration considerably. All materials deposited in the pipes showed in all runs a higher ruthenium concentration of 4.5 wt % to 6.5 wt % which is obviously higher than the 1.6 wt % in the calcine from this feed and may be due to those volatilized ruthenium species which are converted to the solid state when passing through the pipe between the ceramic melter and the dust-removal wet scrubber.

From the results given in Fig. 2 it is evident that an easy way to suppress ruthenium loss from a liquid-fed ceramic waste melter system would be the complete flooding of the molten glass pool with liquid waste sölution. However, practical experience showed that this is not the preferred operation mode for the melter (5). Optimum conditions for both reliable melter control and suppression of ruthenium volatility may be achievable at a feed rate which achieves nearly complete flooding (90-95%) in the steady-state condition on the molten glass pool.

RUTHENIUM FLOW PATH IN THE OFF-GAS TRAIN

Most the the ruthenium which leaves the melter and is not deposited in the off-gas pipe was removed in the hot scrub solution of the dust removal wet scrubber (about 12.1 wt % of total ruthenium processed). An appreciable part of the ruthenium was deposited in the off-gas pipe (4.3 wt %) while minor amounts were found in the condensate (0.8 wt %), and in the scrub solution of the chemical wet scrubbers (0.03 wt %). Results are given in detail in Table 1 along with process conditions maintained during the run.

It is interesting to note the two different forms in which the ruthenium was collected in the scrub solution of the dust removal wet scrubber. Most of the ruthenium was in solution or in very fine particles below 3 μm in diameter (9.2 wt % of total ruthenium processed) while a smaller portion (2.9 wt %) was contained in the dust particles above about 15 μm in diameter. The dissolved ruthenium in the scrub solution may represent volatile ruthenium compounds which came into contact with the scrub solution. However, the ruthenium contained in the dust particles apparently were carried-over in a solid form as an integral part of particulates entrained in the off-gas stream and is unavoidable. It could, of course, only be decreased by selecting specific process conditions in the melter, for example, by complete flooding of the molten glass pool. The results in Table 1 represent a run in which the glass pool was 70-90% covered with liquid. Process conditions directed at suppression of ruthenium volatilization, e.g., denitration, do not effect the carry-over of solid ruthenium by particulate materials as will be shown below.

The ruthemium collected in the condensate and chemical scrubbers is almost completely dissolved. The ruthenium which passes through the chemical wet scrubbers has not yet been totally controlled. It appears that the bulk of ruthenium is removed from the off-gas stream. An adsorber bed will be required for complete clean-up and will be installed and used in future runs. Based on the average values given in Table 1 and assuming a decontamination factor of 1000 for the ruthenium adsorber bed an overall decontamination factor for ruthenium for the system should be in the range of 5.10^5.

EFFECT OF DENITRATION ON RUTHENIUM BEHAVIOR

As has been shown in Fig. 2 ruthenium behavior is closely related to process conditions at the glass-pool surface. Ruthenium in the off-gas decreased strongly when the waste solution covered the entire surface of the molten glass pool. Another way of suppressing ruthenium loss, but without flooding the entire glass pool with the liquid waste, may be the denitration of the waste solution prior to processing it in the ceramic melter. Denitration prevents oxidation of ruthenium in the melter by nitrates during calcination and is essential in suppressing formation of volatile ruthenium compounds.

For several vitrification runs a waste solution with a total nitrate concentration of 3.4 m per liter was denitrated. The amount of ruthenium collected in the off-gas system during processing of this waste is shown in Fig. 3. The values are compared with the

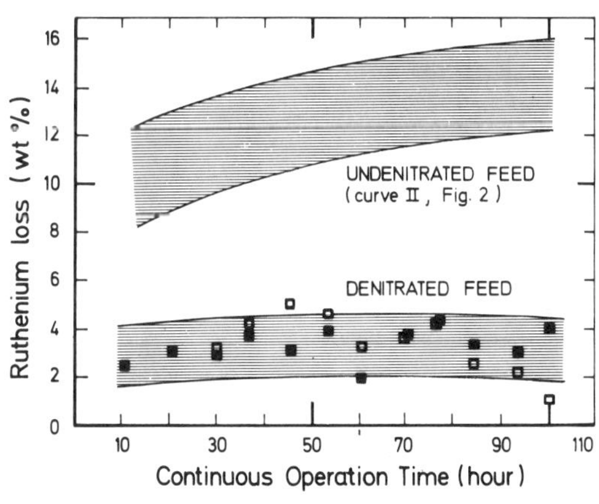

Fig. 3. Ruthenium passing into the off-gas stream for denitrated waste solution (0.7 mm nitrates per liter) compared with the loss for undenitrated waste solution (3.4 m nitrates per liter). Glass pool flooded to 70-90%; ■ □ represents two different runs under the same process conditions.

corresponding results obtained with waste solutions which was not denitrated, under the same process conditions (feed rate 20 l/h, 70-90% of the molten glass pool covered with liquid). It can be seen that the ruthenium loss in a liquid-fed ceramic melter is strongly decreased by denitration. The differences in ruthenium behavior between processing of denitrated and of undenitrated waste respectively can be seen in more detail in Table 2. First the ruthenium deposited in the off-gas pipe is reduced from 4.3 wt % (of total ruthenium processed) to 3.1 wt %. The main contrast, however, is in the scrub solution of the first wet scrubber where the ruthenium is nearly completely contained in the collected dust particles and only a negligible portion of the ruthenium is present in the dissolved form. It is evident that the formation of volatile ruthenium has been suppressed very effectively by denitration.

From comparison of the data in Table 2 it may be noted also that only small differences exist in the amount of ruthenium reaching and being removed in the chemical wet scrubbers. Neither process conditions in the liquid-fed ceramic melter nor denitration of the waste solution seems to have a marked effect on the ruthenium concentration in the scrub solution.

From an overall point of view, however, the results show that denitration can have a pronounced effect in suppressing carryover of volatilized ruthenium into the off-gas train. The effect, of course, is not independent of other parameters as for instance the conditions at the glass pool surface, or rate of leakage of air into the system. If the amount of oxygen introduced by leakage were

Table 2. Comparison of Ruthenium Flow Path in the Off-Gas Train for Denitrated and Undenitrated Waste Solution.

Collected in	Ruthenium (wt % of total Ru processed in the run)		
	Denitrated 1) solution	Undenitrated 1) solution	Undenitrated 2) solution
Off-gas pipe	3.1	4.3	1.3
Dust scrubber			
as dissolved	0.4	9.2	2.0
within dust >15μm	2.6	2.9	1.3
Condensate	0.3	0.8	0.6
Chemical scrubbers	0.08	0.03	0.05

1) 70-90% of glass pool flooded with waste solution.
2) Entire glass pool surface flooded with waste solution.

high in the melter compared to the oxygen from nitrates then a significant effect of denitration in suppression of ruthenium volatilization would not be observed. The melter was operated at a pressure of 20-30 mm WS below atmospheric pressure and the air leakage rate was 5-8 m^3/h. Although denitration could be effective in suppressing ruthenium volatility, its application in an active plant is an additional process step in the course of vitrification. Beyond this the overall decontamination factor is not strongly influenced when a liquid-fed ceramic melter and a wet off-gas clean-up device is used. However, denitration would be helpful in diminishing circulation of ruthenium within the process by maintaining the ruthenium concentration in the scrub solution of the first scrubber at a markedly lower level (this solution is discontinuously recycled into the process stream).

Premature plugging of the off-gas pipe due to deposits of ruthenium compounds seems unlikely with a liquid-fed ceramic melter system. The results have shown that independent of denitration the deposited material in the pipes is enriched in ruthenium compounds (the deposits contain between 4.5-6.5 wt % ruthenium) but the bulk of deposited material (95 wt %) is made up of elements other than ruthenium. Thus, the potential for premature plugging of the off-gas pipe, especially between the ceramic melter and the first scrubber, must be reduced by suppressing particulate entrainment rather than by reducing ruthenium volatilization.

SUMMARY

There are two effective and independent ways to suppress generation of volatilized ruthenium in a liquid-fed ceramic lined electric waste melter. The first one is to flood the molten glass pool completely and continuously with liquid waste. The second one is denitration of the waste solution prior to feeding into the melter. In both cases carry-over of ruthenium into the off-gas can be maintained below approximately 4 wt % of the total ruthenium processed. Denitration could be helpful in diminishing the ruthenium circulating within the process with the scrub solution from the dust-removal wet scrubber. Premature plugging of the off-gas pipe, connecting the ceramic melter with the first scrubber, by ruthenium compounds alone is not probable because the bulk of deposited material (95 wt %) is composed of elements other than ruthenium.

REFERENCES

1. J. D. Christian and D. T. Pence, Critical Assessment of Methods for Treating Airborne Effluents From HLLW-Solidification Processes, PNL-2486, 69 (1977).

2. J. D. Christian and D. W. Rhodes, Ruthenium Containment During Fluid-Bed Calcination of HLLW From Commercial Nuclear Fuel Reprocessing Plants, ICP-1091, 1 (1977).
3. M. A. Hanson and J. J. McElroy, Semivolatile Fission Product Behavior With HLLW-Solidification, BNWl-SA-6461, 16 (1977).
4. R. Bonniaud, "Large Scale Waste Glass Production", Proc. of the Conf. of High-Level Solid Waste Forms, Denver, Colorado NUREG/CP-0005, 170 (1978).
5. S. Weisenburger, "Vitrification of HLLW in a Continuous Liquid-Fed Ceramic Melter", Proc. of the Cincinnati Conf. Ceramics in Nuclear Waste Management, CONF-790420, 86-92, (1979).
6. D. Hentschel, S. Weisenburger, Einsatz Analytischer Methoden zur Prozeßüberwachung einer inaktiven HAW-Verglasungsanlage, KfK-Nachrichten 3/79, 57, (1979).

EFFECT OF DENITRATION ON RUTHENIUM VOLATILIZATION

R. Odoj, E. Merz, and R. Wolters

Institute for Chemical Technology
Kernforschungsanlage Jülich
Fed. Rep. of Germany

ABSTRACT

The volatilization of Ru depends on the compound in which the element is fixed. The ruthenium that dissolves in nitric acid solutions of fuel reprocessing plants exists principally as complexes of nitrosyl ruthenium with the general formula, $RuNO(NO_3)_xOH_{3-x} \cdot yH_2O$. As the molarity of the HNO_3 is increased, the concentration of trinitrato-complexes increases. After the denitration of $RuNO(NO_3)_3$ with HCHO in 8 M HNO_3 solution, heating in the temperature range of 130-200°C caused all denitrated ruthenium compounds to decompose to stable ruthenium dioxide. The volatility of RuO_2 in air is 1-2% up to 1050°C in contrast to elemental Ru.

INTRODUCTION

The first step in the German concept for high level liquid waste (HLLW) solidification is the chemical denitration of HLLW with formaldehyde (HCHO). This technique is in contrast to the present French vitrification process (AVM) which is the reference system for the German fuel cycle center under design. The process of denitration should fulfill two conditions:

1. Destroy the nitric acid and the nitrates of the fission products;
2. Reduce the oxidation state of Ru(VIII) to obtain a stable and nonvolatile compound for the solidification process.

During waste treatment, the volatility of Ru is of concern for both the off-gas treatment and the safe re-use of the distilled nitric acid. The concern with regard to the off-gas treatment arises because the annual Ru inventory produced for the German "Entsorgungszentrum," capable of supporting a 52000 MWe reactor economy, is approximately 3300 kg of non-radioactive Ru and 8 kg of ^{103}Ru and ^{106}Ru (total of $2.8x10^7$ Ci), which, if volatilized, could condense and plug the off-gas pipes.

RUTHENIUM SPECIES

Collodial or near molecular forms of Ru and RuO_2 which are formed in irradiated fuel elements (1) are not readily soluble in nitric acid solutions. Studies indicate that more than 50% of the ruthenium present in spent oxide fuel may remain undissolved in nitric acid (2,3). The ruthenium that dissolves exists principally as complexes of tervalent nitrosyl ruthenium cation (RuNO), in which the ligands are nitrato, nitro, hydroxo and aquo groups, which can be represented (4) by the general formula $RuNO(NO_3)_x(OH)_{3-x}(H_2O)_2$. The fraction of the different RuNO nitrato complexes at equilibrium in 5M and 8M HNO_3 can be calculated from the data of Fletcher, et al. (4) as shown in Table 1. Careful note should be made that some results published in the literature about the volatility of ruthenium have been obtained using simulated waste containing insoluble RuO_2 as a component. Its thermal behavior in solution differs very much from that of the actual ruthenium nitrosyl compounds in HLLW.

DENITRATION

The reduction of HNO_3 by HCHO (7) is illustrated by Equation 1)

$$4HNO_3 + HCHO \rightarrow 4NO_2 + CO_2 + 3\ H_2O \qquad 1)$$

$$4HNO_3 + 3\ HCHO \rightarrow 4NO + 3CO_2 + 5\ H_2O \qquad 2)$$

for the reaction at high acidity (8-16 M HNO_3) and Equation 2) at low acidity (2-8M HNO_3). They indicate that the addition of equimolar

Table 1. Nitrato Complexes in Nitric Acid Solutions

Complex	HNO_3 Concentration 5M	8M
trinitrato	20%	50%
dinitrato	20%	20%
mononitrato	40%	30%
non-nitrato	20%	0%

amounts of HCHO ensures the destruction of all HNO_3 independent of the acidity of the solution.

Fig. 1 is a schematic representation of the denitration arrangement used in the laboratory. The apparatus could be used with either a reflux condensor or a column with Raschig rings. The latter modification provided a partial distillation of the nitric acid in addition to denitration which helped to reduce the nitrate ion content in the denitration vessel. In any case about 99% of the nitrate ions in solution were destroyed by the denitration.

Depicted in Fig. 2. are three weight loss curves determined by thermogravimetry (TG). The two upper curves are obtained by heating the products after denitration of 8M nitric acid solutions containing ruthenium, denitrated in the apparatus shown in Fig. 1, using either the Raschig rings or reflux condenser. Chemical analysis, infra-red investigations and further thermogravimetric studies indicated the compounds remaining were $(RuNO)_2NO_3(OH)_5$ and $RuNO(NO_3)OH_2$ respective-

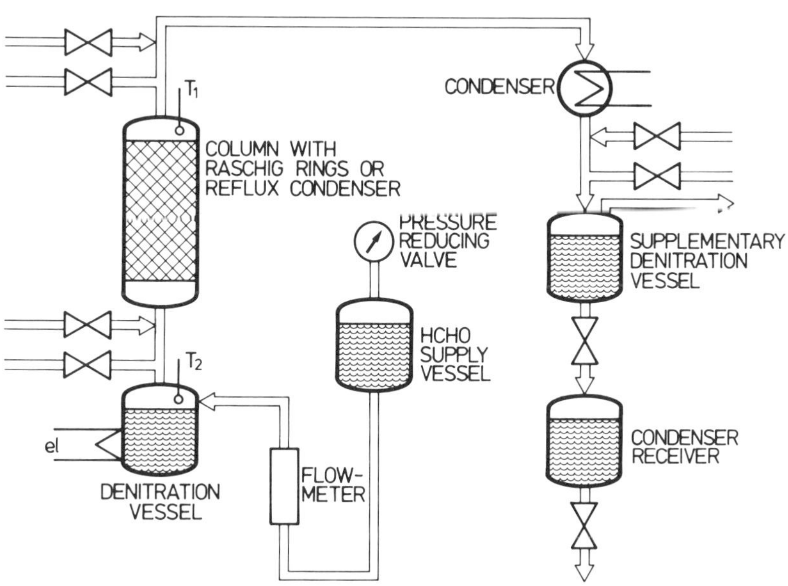

Fig. 1. Schematic Denitration Installation

ly. The third curve was obtained from the compound, $RuNO(NO_3)_3$. Weight losses of the nitrosyl compounds always corresponded to their decomposition to RuO_2. However, the weight loss of $Ru(NO_3)_3$ in TG analysis was observed to be 10% greater than expected. This suggests possible oxidation of ruthenium to RuO_4 by excess nitrate. If denitration was carried out in 5M HNO_3 instead of 8M, the TG weight loss of the product was somewhat smaller. In 3M nitric acid no denitration took place. Optimization studies of the denitration process using HCHO are being completed and will be published in the near future (6).

VOLATILITY

Ruthenium can leave the denitration vessel through entrainment and volatilization. Entrainment is the carry-over of spray into the off-gas or distillate. The entrainment factor for all soluble elements or compounds during distillation is about the same order of magnitude, whereas the volatilization factor depends on the vapor pressure of the compound in which the element is fixed. From the literature, entrainment factors vary from about 10^{-4} to 10^{-7} (5). The ruthenium distillation factor usually is some orders of magnitude greater, varying from 10^{-5} to 2%, depending on the nitric acid concentrations. Therefore, the entrainment factor has not been considered further.

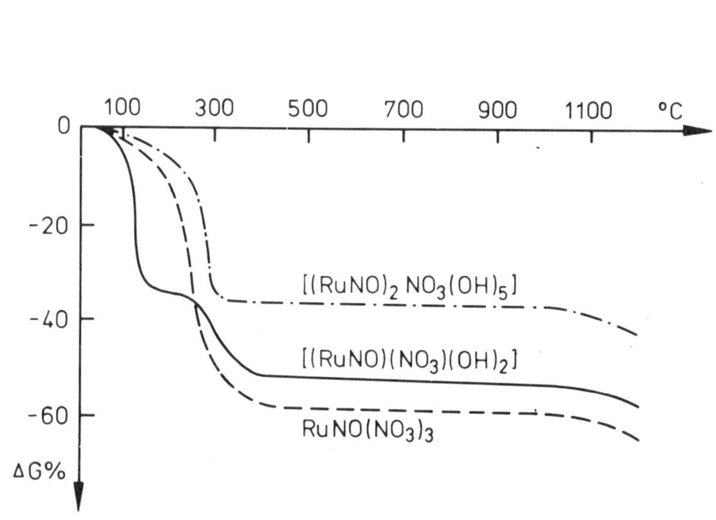

Fig. 2. TG-curves for Ruthenium Nitrosyl Nitrates

Ruthenium tetroxide (RuO_4) has been generally considered to be the volatile species responsible for the partial pressure of ruthenium over the solutions. However, the large amount of NO-gas produced during denitration should reduce RuO_4 to RuO_2 as shown in Equation 3).

$$RuO_4(g) + 2NO \rightarrow RuO_2 + 2NO_2 \qquad 3)$$

$$2NO_2 + H_2O \rightarrow HNO_3 + HNO_2 \qquad 4)$$

$$3HNO_2 \rightarrow 2NO + H_2O + HNO_3 \qquad 5)$$

Nitrogen dioxide reacts with water to produce nitrous acid (Eq. 4)) which is in equilibrium with HNO_3 and NO as shown in Eq. 5). Nitrous acid is a reducing agent in contrast to nitric acid. Therefore, considering the reducing properties of HCHO, and that both aqueous and gaseous reductants are produced in the denitration process (Eqs. 1) and 5)), it is reasonable to assume that the volatility of ruthenium during denitration and distillation is negligible if the nitric acid concentration is about 5M and air is excluded. Investigations with radioactive ruthenium are presently in progress in our laboratories to verify this supposition.

All denitrated products had to be heated to about 120-130°C for about 30 minutes, then the temperature could be raised further.

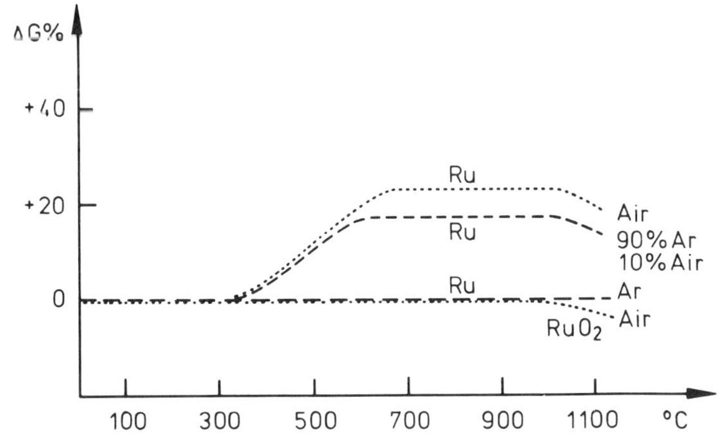

Fig. 3. Behavior of Ru and RuO_2 During TG Studies

However, if the temperature was elevated immediately to >130°C, an exothermic decomposition of the denitrated samples and the generation of gaseous species (NO, NO_2 and H_2O) was so violent that deflagration occured and parts of the sample were ejected from the crucible. In any case, thermogravimetric investigations and X-ray analysis pointed out that the resulting product of heating the denitrated samples was always RuO_2, provided the maximum temperature used was less than 1000°C. Above this temperature, oxidation of RuO_2 to the volatile RuO_3 occurs in the atmosphere (8). As shown in Fig. 3. RuO_2 is itself volatilized in the temperture range above 1100°C following oxidation of Ru to RuO_2 at 300°C.

SUMMARY

Ruthenium volatility can be as high as 50-100% during calcining of reactor waste in the absence of reducing additives. High volatility would also be expected for the rotary kiln (AVM) vitrification process. Because of the radioactivity and the plugging danger of condensed ruthenium compounds in the off-gas system, increased attention should be given to ruthenium control in the treatment of HLLW solutions. There are three possibilities for diminishing the danger of uncontrolled volatilization:

1. Ruthenium can be removed by intentional volatilization by oxidation to RuO_4 in the presence of a catalyst,
2. An effective ruthenium trap can be developed and installed in the off-gas system,
3. Reduce the volatility of ruthenium by denitration of the HLLW solution.

The third alternative is the German concept whereby ruthenium volatility is controlled by conversion of trinitrato-ruthenium nitrosyl complexes to lower nitrato and nitroso complexes. During the denitration, most of the nitrates of the fission products and nitric acid can be destroyed. This would minimize the possibility of the oxidation of RuO_2 to RuO_4 by NO_2 which seems likely to be the most important process during the calcination or vitrification when using an inert purge gas.

Since there is an exothermic decomposition with large amounts of gaseous species released when the denitrated products are heated, it seems best to use the equipment of the Jülich FIPS-process (9). The drum drier temperature is in the 120-150°C range, thereby ensuring a mild and controlled thermal decomposition of the denitrated ruthenium compounds. Consequently, there will be no, or minimal, nitrates in the melter or in the inert gas atmosphere over the melter.

REFERENCES

1. J. H. Goode, C. L. Fitzgerald and V. C. A. Vaughen, The Dissolution of Unirradiated and Irradiated $(U,Pu)O_2$ in Nitric Acid, ORNL-5015 (Feb. 1975).
2. KFK, priv. communication, May 1979.
3. J. D. Christian, "Process Behavior and Control of Ruthenium and Cerium," Proc. of ANS-AIChE Meeting, Sun Valley, ID Aug. 5-6 (1976).
4. J. M. Fletcher, P. L. Jenkins, F. S. Martin, A. R. Powell and R. Todd, "Nitrato and Nitro Complexes of Nitrosylruthenium," J. Inorg. Chem. 1, 378 (1955).
5. A. S. Wilson, "Ruthenium Volatilization in the Distillation of Nitric Acid," J. Chem. Eng. Data 5(4), 521-524 (1960).
6. R. Odoj and H. Gölitzer, "Prozeßverlauf und Reaktionsmodell der Denitrierung von HLLW mit HCHO," Jül-Bericht (1980).
7. T. V. Healy, "Reaction of Nitric Acid with Formaldehyde," J. Appl. Chem. 8, 553-561 (1958).
8. C. B. Alcock and G. W. Hooper, "Thermodynamics of the Gaseous Oxides of the Platinum Group Metals," Proc. Roy. Soc. London Ser. A 254, 551 (1960).
9. S. Halaszovich, E. Merz and R. Odoj, "Progress in Fission Product Solidification and Characterization Using the Jülich FIPS-Process," Proc. of Meeting of the Am. Ceram. Soc., Cincinnati, Ohio (1979).

ADVANCES IN THE FLUOROCARBON PROCESS FOR DECONTAMINATION OF NUCLEAR FACILITY OFF-GASES*

B. E. Kanak

Union Carbide - Nuclear Division
Oak Ridge, TN

INTRODUCTION

The Nuclear Division of Union Carbide is responsible for the development of the fluorocarbon-based selective absorption process for removing noble gas fission products, carbon-14 and other radioactive contaminants from the gaseous wastes of nuclear facilities. In order to accomplish the dual objectives of component removal and subsequent concentration, the process utilizes three operations: absorption, fractional stripping and final stripping. Each of these performs a separation by exploiting the solubility differences between the various off-gas components in the process solvent, dichlorodifluoromethane (R-12). The performance and reliability of the process have been demonstrated on an engineering scale with 10 years of pilot plant operation at the Oak Ridge Gaseous Diffusion Plant (ORGDP) which included extended testing with ^{85}Kr, ^{133}Xe and ^{131}I. Removal efficiencies of greater than 99.99% for Xe, CO_2, I_2, and CH_3I, greater than 99.9% for Kr and H_2O, and greater than 99% for NO_2 and N_2O have been experimentally obtained. With a feed stream containing 150 ppm of noble gas, a 97% noble gas product has been achieved. In addition to demonstrating the required removal efficiencies, the process has shown a remarkable tolerance for many of the normally troublesome off-gas components and does not rely upon elaborate feed pretreatment steps. Recent work has led to a major flow scheme simplification, reducing the three-column process to a single patented column. A third generation pilot plant facility employing this new concept has successfully undergone initial testing this last year at ORGDP.

---

*This document is based on work performed at the Oak Ridge Gaseous Diffusion Plant operated by Union Carbide Corporation under Contract W-7405 Eng 26 with the U. S. Department of Energy.

BACKGROUND

Throughout its history, the fluorocarbon development effort has addressed off-gas applications in almost every aspect of the nuclear fuel cycle. Steinberg (1) originally suggested using R-12 as a solvent in a three-column process to selectively absorb Kr and Xe from the gaseous wastes of a reprocessing plant in 1959. In 1966, a pilot plant was constructed and a development effort was initiated at ORGDP to establish the general feasibility of the fluorocarbon process. At that time, the work was aimed at the development of a mobile processing system which could be transported to the site of a hypothetical reactor within 24 hours of a fuel failure accident and recover the noble gases and other volatile radionuclides released to the containment vessel. As envisioned, all of the recovery equipment was to be situated in a series of trailer trucks (2). Unfortunately, with respect to the Three Mile Island incident, this work did not proceed past the conceptual design stage.

In 1970, the program emphasis shifted to the routine cleanup of LWR off-gases (3), and the fluorocarbon process was subsequently offered commercially for LWR application (4). Based on the demonstrated operability and performance of the ORGDP pilot plant, process safety considerations and, in particular, the lack of any required feed pretreatment, the fluorocarbon process was adapted for reprocessing plant applications in 1971. A second generation pilot plant facility was built at ORGDP in 1972 which offered greater flexibility and the more sophisticated analytical equipment necessary for detailed component analysis (5). This facility demonstrated that in addition to providing Kr and Xe removal, the fluorocarbon process could effectively be utilized to contain carbon-14 as CO_2, various nitrogen oxides, elemental and organic iodine and water. Rigorous process models and scale-up studies (6, 7, 8) were completed to allow the confident design and optimization of a full scale facility. These works laid the groundwork for the construction of the single-column process in 1978.

The concept of the fluorocarbon process as a mobile emergency reactor off-gas decontamination system has recently been revived in the wake of the Three Mile Island incident and is undergoing review by responsible persons in the industry. It would be possible to have a mobile unit ready for deployment within three to four years.

PROCESS BASIS

Fluorocarbon solvents have been identified as a valuable and somewhat unique group of solvents unusually suited for separating a number of industrially important components from various gas mixtures (9, 10). Steinberg selected R-12 for noble gas removal primarily because of its capacity, separation factor, and thermal and radiation stability as well as process safety and economic features (1). The physical properties of R-12 are well known. The basic thermodynamic properties are detailed by Mr. Harness, et al. (11).

Distribution coefficients for several of the off-gas components of interest are shown in Fig. 1. A substantial amount of equilibrium data now exists for krypton and xenon in R-12 solution. The initial work was performed by Steinberg at Brookhaven National Laboratory. Later work was reported from the University of Tokyo by Yamamoto and Takeda (12). The most recent data are given by Shaffer at ORNL (13). Merriman utilized several techniques based on Hildebrand's regular solution theory to estimate krypton and xenon equilibrium coefficients (14). All investigators show good agreement.

Toth, et al., (15) recently completed a laboratory study to define the behavior of other nuclear fuel reprocessing plant off-gas components such as CO_2, NO_2, I_2 and CH_3I. While I_2 was found to be fairly soluble in R-12, CH_3I, NO_2, and CO_2 were found to be totally miscible in the temperature range of interest. Distribution coefficients were also determined for these species. Beyond this work, some estimated solubility relationships have also been given (14).

PROCESS DESCRIPTION

Fig. 2. is a schematic of the selective absorption process as it was originally conceived. The main separation of radioactive contaminants from the bulk gas, either air or nitrogen, is effected in the absorber. The intermediate or fractional stripper serves to remove the coabsorbed carrier gas from the solvent, thereby enriching the remaining dissolved gas in the more soluble components. The final stripper removes all the remaining gas from the process solvent for collection and regenerates the solvent for recycle back to the absorber. The absorption step is carried out in a packed column only, while both stripping operations are performed in packed columns with reboilers and overhead condensers. In addition, the intermediate stripper includes a flash drum. Typically, the absorber is operated at 100 psig, the intermediate stripper at 50 psig, and the final stripper at about 18 psig. Support equipment items for the basic process include a feed gas heat exchanger, process gas compressor, solvent cooler, storage tanks and several refrigeration compressors. If the feed gas contains significant quantities of high boiling components such as H_2O, I_2, CH_3I, or NO_2, a solvent purification still is available as an in-line option to prevent these materials from accumulating in the recirculating solvent. A solvent recovery system is necessary to remove solvent vapor from the process off-gas. Fig. 3. is a photograph of the second generation ORGDP pilot plant which will process a nominal 15 SCFM feed stream. Detailed engineering drawings and descriptions of this facility are available elsewhere (16).

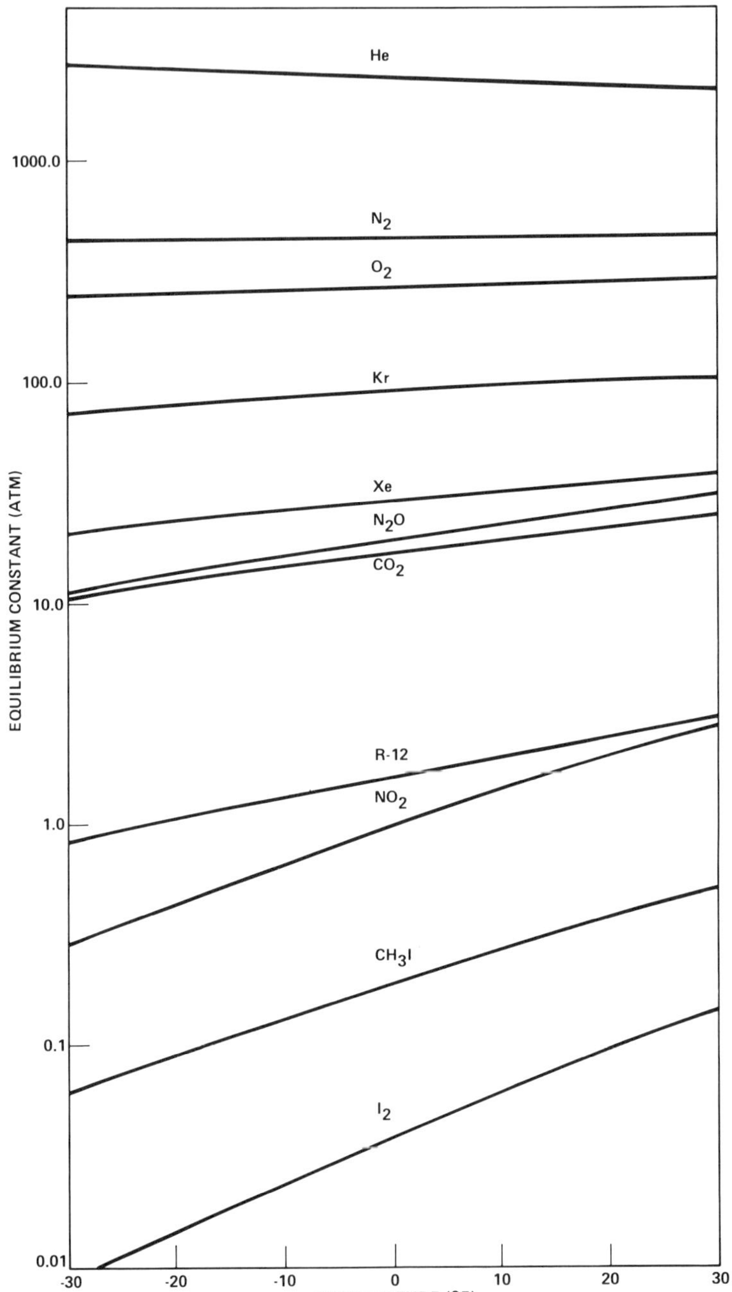

Fig. 1. Equilibrium Distribution Coefficients for Various Feed Gas Components in R-12

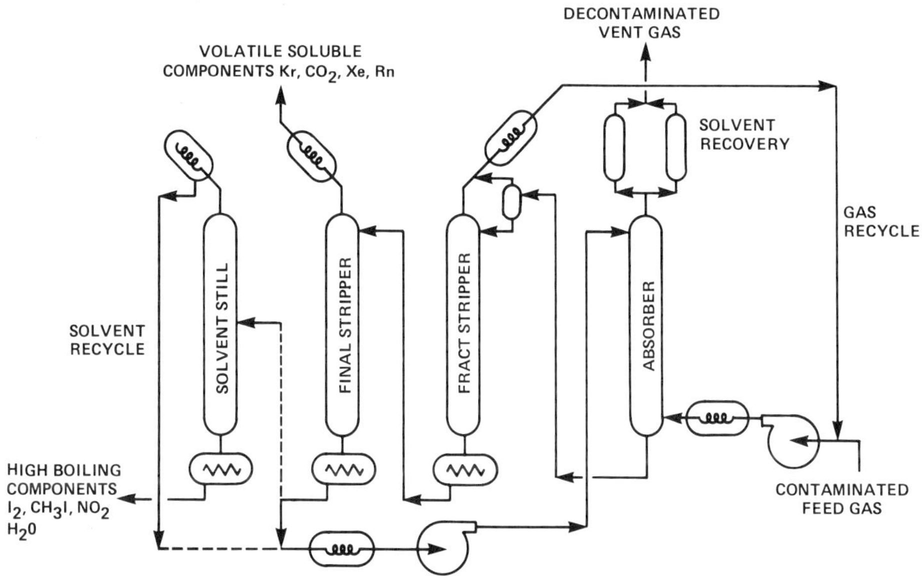

Fig. 2. Schematic of Three-Column Selective Absorption Process

Fig. 3. Overview of ORGDP Three-Column Pilot Plant

In the course of operation and analysis of this facility, a soluble gas peak was observed in the fractional stripper. It was found that this phenomenon, caused by gross internal condensation of the upflowing stripping vapor (6), could be controlled and the magnitude of the peak could be increased dramatically. It became apparent that if sufficient stripping stages were provided below the condensation zone, the final stripper could be eliminated with the product being collected as a side-stream. Furthermore, it also seemed feasible to place the intermediate stripper directly below the absorber and operate the entire assembly at a common pressure. Subsequently, a column was designed that combined the three functional steps of absorption, fractional stripping and final stripping into one continuous contactor (17). Fig. 4. is a schematic of this piece of equipment. Decontaminated off-gas flows from the top of the combination column and regenerated solvent is pulled from the bottom. The fission product gases are collected as a side-stream. The combination column re-

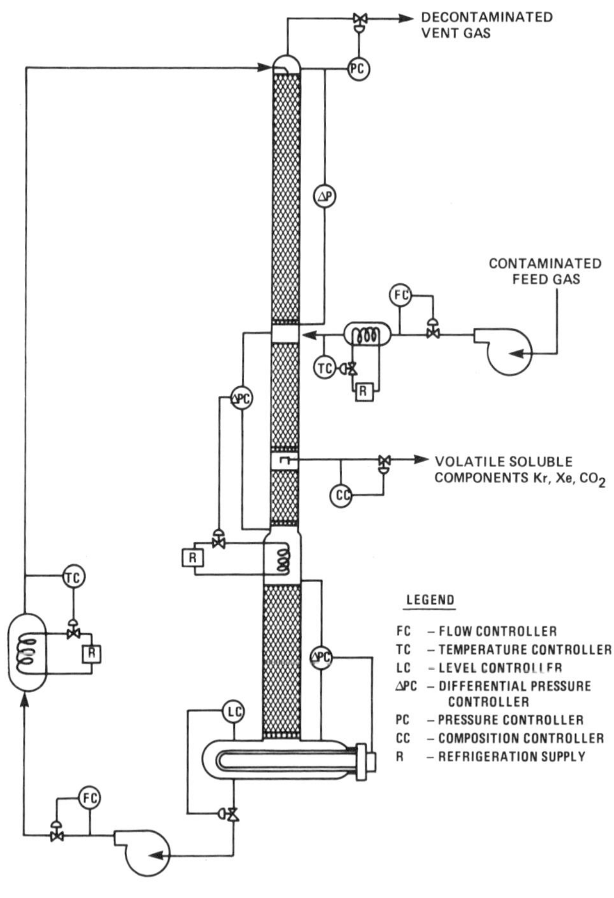

Fig. 4. Schematic of Single-Column Selective Absorption Process

Fig. 5. ORGDP Single-Column Pilot Facility

quires substantially less equipment and control instrumentation than the conventional flow sheet; and because of its simplicity, it offers numerous operational, reliability, and economic advantages. Due to the great potential and design uncertainties, a combination column was built and installed at ORGDP for evaluation. Fig. 5. is a photograph of the column taken during construction. The column is approximately 24 feet tall and has the same flow capacity as the three-column development facility. The absorption and intermediate sections are 3 inches in diameter, while the final stripper section has a diameter of 6 inches. High efficiency wire mesh packing is used throughout. Pilot plant tests have shown that scale-up on the column area is direct with this packing if well designed gas and liquid distributors are employed (7). Hence, larger equipment utilizing the same packing material can be designed with confidence.

The combination column has been undergoing performance evaluation for almost 1 year. These tests not only established the overall feasibility of the concept but also showed conclusively that the combination column can perform essentially as well as the three-column process. On the basis of a one-to-one comparison of the two options, the combination column has recently been selected as the preferred version for future applications.

REFERENCES

1. M. Steinberg, USDOE Report BNL-542, Brookhaven National Laboratory, Upton, NY (1959).
2. J. R. Merriman, A Mobile Processing Unit for Krypton-Xenon Absorption, USDOE Report K-1725, Oak Ridge, TN (1967).
3. M. J. Stephenson, J. R. Merriman and D. I. Dunthorn, Application of the Selective Absorption Process to the Removal of Kr and Xe From Reactor Off-Gas, USDOE K-L-6288, Oak Ridge, TN (1972).
4. G. Griffith, "99% Cleanup of Nuclear Gaseous Wastes," Power Engineering March, 62 (1973).
5. M. J. Stephenson and R. S. Eby, Development of the FASTER Process for Removing Krypton-85, Carbon-14 and Other Contaminants From the Off-Gas of Fuel Reprocessing Plants, USDOE Report K-GD-1398, Oak Ridge, TN (1976).
6. M. J. Stephenson, Analysis of a Fractional Gas Stripper, USDOE Report K-1895, Oak Ridge, TN (1978).
7. B. E. Kanak, Analysis of a Gas Absorption Column with Soluble Carrier Gas and Volatile Solvent, USDOE Report K-2007, Oak Ridge, TN (1979).
8. D. E. Wood, USDOE Report ORNL-TN-5797, Oak Ridge, TN (1977).
9. J. R. Merriman, J. H. Pashley, M. J. Stephenson and D. I. Dunthorn, "Process for the Separation of Components from Gas Mixtures," U. S. Patent 3,762,133 (1973).
10. J. R. Merriman, J. H. Pashley, M. J. Stephenson and D. I. Dunthorn, "Removal of Purified Helium or Hydrogen from Gas Mixtures," U. S. Patent 3,785,120 (1974).
11. R. C. McHarness, B. J. Eiseman and J. J. Martin, "Freon-12," Refrigeration Engineering, 32, Sept. (1955).
12. Y. Yamamoto and H. Takeda, "Solubility of ^{85}Kr in Some Organic Solvents," J. Fac. Eng. U. Tokyo, A-7, 44 (1970).
13. J. H. Shaffer, W. E. Shockley and C. E. Green, The Solubility of Krypton and Xenon at Infinite Dilution in Dichlorodifluoromethane, USDOE Report ORNL-TM-6652, Oak Ridge, TN (1978).
14. J. R. Merriman, USDOE Report KY-G-400, Paducah, KY (1977).
15. L. M. Toth, J. T. Bell and D. W. Fuller, USDOE Report ORNL-TM-6484, Oak Ridge, TN (1978).
16. M. J. Stephenson, R. S. Eby and V. C. Huffstetler, USDOE Report K-1876, Oak Ridge, TN (1977).
17. M. J. Stephenson and R. S. Eby, "Gas Absorption Process," U. S. Patent 4,129,425 (1978).

AUTHOR INDEX

A

B

C

SUBJECT INDEX

G (Continued)

H

I

I (Continued)

J

K

L

R (Continued)

S

S (Continued)

T

U

V

W

W (Continued)

X

Z